Phylogeny and Evolution of Angiosperms

PHYLOGENY
and
EVOLUTION
of
Angiosperms

DOUGLAS E. SOLTIS
University of Florida, Gainesville

PAMELA S. SOLTIS
University of Florida, Gainesville

PETER K. ENDRESS
University of Zurich

MARK W. CHASE
Royal Botanic Gardens, Kew

SINAUER ASSOCIATES, INC. PUBLISHERS
Sunderland, Massachusetts

Sinauer Associates, Inc.
23 Plumtree Road
Sunderland, MA 01375-0407   USA

FAX: 413-549-1118
EMAIL: orders@sinauer.com, publish@sinauer.com
www.sinauer.com

Library of Congress Cataloging-in-Publication Data

Phylogeny and evolution of angiosperms / Douglas E. Soltis ... [et al.].
    p. cm.
  Reprint. Originally published: Washington : Smithsonian Books, 2005.
  Includes bibliographical references.
  ISBN 0-87893-817-6 (pbk.)
  1. Angiosperms--Phylogeny. 2. Angiosperms--Evolution. I. Soltis, Douglas
E.

QK989.P49 2005b
581.3'8--dc22

2005011564

Printed in the U.S.A.

5 4 3 2 1

*We dedicate this book to those who have supported
and shared our interest in botany:
Parents, family members, spouses, students, friends, and colleagues.*

Contents

Preface

The primary motivating force for preparing this book was the dramatic change in our understanding of angiosperm phylogeny during the past 10 years. Many long-standing views of deep-level relationships were altered suddenly and substantively as a direct result of molecular analyses. Many of the major angiosperm clades do not correspond to the classes, subclasses, and orders of modern classifications. For example, monocots (class Liliopsida) were derived from within a basal grade of families traditionally considered to be dicots (class Magnoliopsida). Hence, the long recognized monocot–dicot split does not accurately reflect phylogenetic relationships revealed by analyses of DNA sequence data. Furthermore, long-recognized subclasses of both monocots and "dicots" required revision or abandonment. Although significant differences exist among modern classifications at the ordinal level, clades identified by molecular analyses often fail to match the orders of any recent classifications. The circumscriptions of many families have also required modification for the same reasons.

As a result of the major modification in our understanding of angiosperm relationships, the usefulness of the classifications of Cronquist (1981) and Takhtajan (1997) has therefore declined, although these references are still widely consulted and are still a rich source of information. The Angiosperm Phylogeny Group's (APG 1998; APG II 2003) reclassification of angiosperms, based on recent inferences of phylogeny, offers an alternative to the classification systems of previous decades. We follow the APG classification in this book. In addition, a wealth of recent data, coupled with our current understanding of phylogenetic relationships, permits reevaluation of many evolutionary hypotheses posed by Cronquist (1981, 1988), Takhtajan (1997), and Stebbins (1974).

Our major goals in this book are threefold. First, we provide a comprehensive summary of current concepts of angiosperm phylogeny. We then illustrate the profound effect that this phylogenetic framework has on interpretations of character evolution. It has not been possible to summarize all aspects of angiosperm evolution, of course, so we selected examples involving a diverse array of characters, including chromosomes, chemistry, and a sample of morphological features, especially floral and fruit characters. The coverage provided is meant to illustrate several themes. For example, some chemical characters or pathways previously considered to have evolved many times have instead been demonstrated to have originated rarely in the course of angiosperm evolution. Conversely, other characters long thought to be of taxonomic utility are highly dynamic in their evolution. Finally, we point to inadequacies in current understanding of both phylogeny and morphology and to the need for additional study. We hope that our efforts will stimulate future research to identify morphological synapomorphies for DNA-based clades and to refine the hypotheses of angiosperm phylogeny and evolution presented here.

Doug Soltis
Pam Soltis
Peter Endress
Mark Chase

Acknowledgments

Many people have helped with this book. Jim Thompson prepared several figures and Ashley Morris assisted with others. Pam Williams prepared several of the tables and figure legends. Katie Soltis assembled most of the literature cited, with contributions from Pam Williams, Jim Thompson, and Sandy Morrison. A special thanks to those individuals who read chapters or portions of the text—Victor Albert, Todd Barkman, Clemens Bayer, Peter Crane, Jonathan Davies, Jerrold Davis, Claude de-Pamphilis, David Dilcher, Jim Doyle, Jeff Doyle, Matt Gitzendanner, Sara Hoot, Walter Judd, Jesper Kårehed, Robert Kuzoff, Ilia Leitch, Richard Olmstead, Paul Manos, Greg Plunkett, Yin-Long Qiu, Jennifer Tate, Jenny Xiang, and Jonathan Wendel. Other colleagues shared data, unpublished manuscripts, provided helpful information, or assisted by providing figures for use in the book—Victor Albert, Randy Bayer, Alex Bernard, David Baum, Birgitta Bremer, Käre Bremer, Matyas Buzgo, Sherwin Carlquist, Peter Crane, Claude dePamphilis, Jonathan Davies, David Dilcher, Andrew Douglas, Andrew Doust, Taylor Feild, Mark Fishbein, Else Marie Friis, Amy Litt, Steve Johnson, Elizabeth Kellogg, Sangtae Kim, Johannes Lundberg, Steven Manchester, Paul Manos, Regis Miller, Mark Mort, Dan Nickrent, Chris Pires, Vincent Savolainen, Ed Schneider, Andrea Schwarzbach, Mark Simmons, Victoria Sosa, Jim Smith, Marshall Sundberg, Ken Sytsma, Susan Swensen, Shirley Tucker, and Maria von Balthazar. A special thanks to Carroll Wood for granting permission to use many of his beautiful line drawings of flowering plants. Sheila Conner and Amy McPherson helped obtain the permissions required from the Journal of the Arnold Arboretum and Annals of the Missouri Botanical Garden, respectively. A very special thanks to Fran Aitkens for her careful and tireless editing of the manuscript and figures. We are also very grateful to Andy Sinauer, Chelsea Holabird, Christopher Small, and Joanne Delphia, who did such a wonderful job in bringing this book to completion.

This work was made possible in part by a Fulbright Award to Doug and Pam Soltis and by a sabbatical leave provided by the University of Florida. Discussions and interactions fostered by the Floral Genome Project and Deep Time Research Coordination Network were also instrumental in preparation of parts of the text. Doug and Pam Soltis also thank the Royal Botanic Gardens, Kew, where part of the text was written.

1

Relationships of the Angiosperms to Other Seed Plants

Introduction

Five major lineages of seed plants are extant: angiosperms, cycads, conifers, gnetophytes or Gnetales, and *Ginkgo*. These groups have typically been treated as the distinct phyla Magnoliophyta or Anthophyta, Cycadophyta, Coniferophyta, Gnetophyta, and Ginkgophyta, respectively. Of the extant groups of seed plants, the angiosperms are by far the most species-rich, with more than 13,000 genera and 250,000 species. The conifers are the second largest group of seed plants, with approximately 70 genera and nearly 600 species. The cycads comprise approximately 9 genera and 120 species. Gnetales consist of three morphologically disparate genera, *Gnetum, Ephedra,* and *Welwitschia,* and about 90 species. There is a single species of *Ginkgo, G. biloba.* Several extinct lineages of seed plants, including Caytoniales, Bennettitales, *Pentoxylon,* corystosperms, and glossopterids, have been proposed as putative close relatives and possible progenitors of the angiosperms.

The seed plants represent, by botanical standards, an ancient radiation, with the first seeds appearing near the end of the Devonian (~370 million years ago; Mya); by the Early to Middle Carboniferous, a diversity of seed plant lineages already existed (Kenrick and Crane 1997). The fossil record of conifers dates to the Middle Pennsylvanian and that of true cycads to the Early Permian. Available data indicate that by the Late Carboniferous (~290–320 Mya), at least three (cycads, conifers, *Ginkgo*) of the five extant lineages of seed plants had probably diverged (Kenrick and Crane 1997; Donoghue and Doyle 2000). In contrast, the angiosperms are relatively young—their earliest unambiguous fossil evidence is from the Early Cretaceous (~130 Mya; see Chapter 3).

Relationships among these lineages of extant seed plants, as well as the relationships of extant groups to fossil lineages, have been issues of longstanding interest. A topic of particular intrigue has been the closest relatives of the angiosperms. At some point, nearly every living and fossil group of gymnosperms has been proposed as a possible ancestor of the angiosperms (e.g., Wieland 1918; Thomas 1934, 1936; Melville 1962, 1969; Stebbins 1974; Meeuse 1975; Long 1977; Doyle 1978, 1998a, 1998b; Retallack and Dilcher 1981; Crane 1985; Cronquist 1988; Crane et al. 1995). Among extant seed plants, the relationship between angiosperms and Gnetales has received exceptional attention. Determining the closest relatives of the angiosperms is not only of great systematic importance but also critical for assessing character evolution. For example, the outcome of investigations of character evolution among basal angiosperms, including studies focused on the origin and diversification of crucial angiosperm structures, may be influenced by those taxa considered their closest relatives (see Chapter 3). The effect of outgroup choice on the reconstruction of character evolution within angiosperms is readily seen with Gnetales. Until recently, Gnetales were considered by many to represent the closest living relatives of the angiosperms; the use of Gnetales as outgroup can profoundly influence character-state reconstruction within the angiosperms (see "The Anthophyte Hypothesis" section).

Clarifying relationships among seed plants, both extant and fossils, has proved to be difficult. Although progress has been made recently in elucidating relationships among extant seed plants using DNA sequence data, a complete understanding of seed plant phylogeny is not possible without the integration of fossils. Factors that have contributed to the difficulties in phylogeny reconstruction of seed plants include the great age of these groups and the considerable morphological divergence among them, as well as the extinction of many lineages. The tremendous morphological gap among extant and fossil seed plant lineages has complicated and ultimately compromised efforts to reconstruct relationships with morphology because of homoplasy and uncertainty about the homology of structures (e.g., Doyle 1998b; Donoghue and Doyle 2000). Resolution of relationships among extant seed plants by using DNA characters has also been difficult because some lineages have relatively short branches, whereas other groups (e.g., Gnetales) are characterized by long branches. This problem is compounded by the presence of long branches to the sister group of seed plants.

Seed plant relationships and the closest relatives of the angiosperms have been the focus of several reviews (e.g., Crane 1985; Doyle and Donoghue 1986; Doyle 1996, 1998a, 1998b, 2001; Frohlich 1999; Donoghue and Doyle 2000), on which we rely in this overview; we also

draw on recent developments. Although we consider seed plant relationships in general, the major focus of this chapter is the closest relative(s) of the angiosperms.

Phylogenetic Studies: Extant Taxa

Since the 1980s, considerable effort has been focused on resolving relationships among seed plants. Both morphological and molecular cladistic studies have played a major role in the study of extant seed plant phylogeny, although molecules and morphology have so far yielded different conclusions about the relationships of Gnetales and angiosperms. Based on the morphological data now available, one could conclude that seed plants represent an example in which cladistic analyses of morphology alone failed on some major questions. In this section, we review efforts to reconstruct the phylogeny of seed plants, for simplicity, focusing first on extant taxa; we also provide a brief history of the placement of Gnetales relative to the angiosperms. In the next section, we focus on cladistic analyses that include fossil seed plants and provide our own analyses.

Early views of the relationships of Gnetales

Wettstein (1907) and Arber and Parkin (1908) first proposed that Gnetales were closely related to angiosperms on the basis of several shared features: vessels, net-veined leaves (present in *Gnetum*), and "flower-like" reproductive organs (see also Doyle 1996; Frohlich 1999). However, their hypotheses to explain the close relationship of Gnetales and angiosperms differed dramatically. Wettstein (1907) based his proposal that Gnetales were ancestral to the angiosperms on his view that the former Amentiferae, a group that included wind-pollinated taxa such as Juglandaceae, Betulaceae, and Casuarinaceae, are the most "primitive" living angiosperms. Wettstein maintained that the distinctive inflorescences (termed *aments*) of Amentiferae, consisting of simple unisexual flowers, are homologous with the unisexual strobili of Gnetales. Arber and Parkin (1908) also proposed a close relationship of angiosperms and Gnetales but argued that the reproductive structures of Gnetales are not primitively simple, but reduced, derived from ancestral lineages having more parts.

By the mid-1900s, most authors did not consider Gnetales and angiosperms as closest relatives. Bailey (1944a, 1953) noted that the vessels in the two groups are derived from different kinds of tracheids. In addition, Gnetales bear ovules directly on a stem tip, whereas in angiosperms, the ovules are produced within the carpel, a structure that probably represents a modified leaf. Views on the earliest angiosperms also changed, with Magnoliaceae and other angiosperms with large strobiloid flowers considered most ancient, whereas the simple flowers found in Amentiferae were considered secondarily reduced rather than primitively simple (Arber and Parkin 1907; Cronquist 1968; Takhtajan 1969). This disrupted the link between Gnetales and angiosperms (via a basal Amentiferae) envisioned by Wettstein. Issues of relationship also became more complex when Eames (1952) proposed that Gnetales were not monophyletic. He considered *Ephedra* to be related to the fossil group Cordaites and conifers; *Gnetum* and *Welwitschia* were thought to be closer to another extinct lineage of seed plants, Bennettitales. Concomitantly, paleobotanists focused attention on fossil taxa such as *Caytonia* and glossopterids as the closest relatives of angiosperms (reviewed in Doyle 1996; Frohlich 1999; see "The Fossils: In Search of the Sister Group of Angiosperms," below), shifting attention away from Gnetales as possible close relatives of the angiosperms. Hence, by the mid-1900s, Gnetales were in disfavor as possible close relatives of the angiosperms. Gnetales reemerged, however, as putative close relatives of angiosperms when cladistic approaches were first used to investigate seed plant relationships.

The anthophyte hypothesis

Seed plant relationships were first assessed using cladistic analyses of morphological characters in the 1980s. Several of these early studies of extant, as well as fossil, taxa (e.g., Parenti 1980; Hill and Crane 1982; Crane 1985; Doyle and Donoghue 1986) not only revealed that Gnetales are monophyletic but also implicated Gnetales as the closest living relatives of angiosperms. Subsequent analyses of morphological characters (e.g., Loconte and Stevenson 1990; Doyle and Donoghue 1992; J.A. Doyle 1994, 1996), some of which also included fossils, often differed in the relationships suggested among extant seed plants, although some results are consistent with the earlier studies (Figure 1.1).

In morphological cladistic analyses, gymnosperms are not monophyletic to the exclusion of the angiosperms, and the positions of some lineages also seem unstable. For example, Crane (1985) found that cycads are sister to other extant seed plants; conifers and *Ginkgo* form a clade that is sister to angiosperms + Gnetales (Figure 1.1A). In contrast, the shortest trees of Doyle and Donoghue (1986) indicated that conifers + *Ginkgo* are sister to a clade in which cycads are the sister to angiosperms + Gnetales (Figure 1.1B). Loconte and Stevenson (1990) found cycads, followed by *Ginkgo*, then conifers, to be subsequent sisters to Gnetales + angiosperms (Figure 1.1C). Despite the diversity of topologies, most cladistic analyses of morphology agreed that Gnetales are sister to the angiosperms (e.g., Crane 1985; Doyle and Donoghue 1986, 1992; Loconte and Stevenson 1990; Doyle 1996). Thus, although these cladistic studies often differed in the interpretation of

characters, attention was again focused on Gnetales as sister to the angiosperms.

The cladistic analysis of morphology by Nixon et al. (1994) found that Gnetales are not monophyletic. In Nixon et al.'s shortest trees, the angiosperms are derived from within a paraphyletic Gnetales; *Ephedra* is followed by *Welwitschia* + *Gnetum*, with the latter clade sister to the angiosperms. Doyle (1996) reconsidered the five key morphological characters that linked angiosperms with *Gnetum* and *Welwitschia* but not *Ephedra*—cellularization of the female gametophyte, presence or absence of archegonia, free nuclear versus cellular embryogeny, number of nuclei in the male gametophyte, and presence or absence of a stalk cell in the male gametophyte. Subsequent reanalyses by Doyle (1996) found an angiosperm clade and a mono-

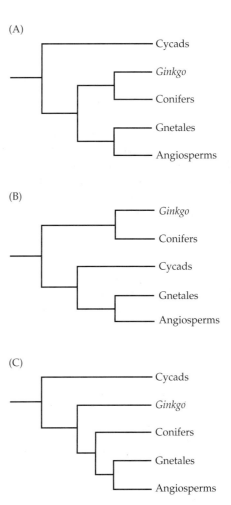

FIGURE 1.1 Simplified topologies depicting relationships among extant seed plants based on phylogenetic analyses of morphological data. Fossil taxa have been removed from these topologies. (A) Parenti (1980); Crane (1985); Doyle and Donoghue (1986, 1992); Doyle (1996). (B) Doyle and Donoghue (1986, 1992); Doyle (1996). (C) Loconte and Stevenson (1990).

phyletic Gnetales (Figure 1.1B), the latter with low bootstrap support (75%). This example illustrates the general problems involved in interpreting and coding morphological characters in phylogenetic studies of seed plants. The monophyly of Gnetales is now well supported by both morphology and molecules (see "DNA Sequence Data: Demise of the Anthophyte Hypothesis," section).

Crane (1985) conducted two cladistic analyses of extant and fossil seed plants and in one analysis recovered a clade of Bennettitales, *Pentoxylon,* and Gnetales + angiosperms (Figures 1.1A and 1.2). Doyle and Donoghue (1986, 1992) similarly found shortest trees in which Gnetales and angiosperms appeared in a clade with the fossil taxa Bennettitales and *Pentoxylon;* however, Gnetales and angiosperms were not exclusive sisters (Figure 1.3). Doyle and Donoghue (1986) named this clade of angiosperms, Gnetales, Bennettitales, and *Pentoxylon* the "anthophytes" in reference to the flower-like reproductive structures shared by all members. Rothwell and Serbet (1994) later recovered the same anthophyte clade. *Caytonia* was later found to be sister to the anthophyte clade (Doyle 1996). The anthophyte clade remained a focal point of study and controversy for about 15 years.

Importantly, the close relationship inferred between Gnetales and angiosperms was not robust in any morphological cladistic analysis. Doyle and Donoghue (1986, 1992), for example, found topological differences in trees that were only one or two steps longer than the shortest trees they obtained. In some trees only one step longer than the shortest trees, angiosperms appeared as

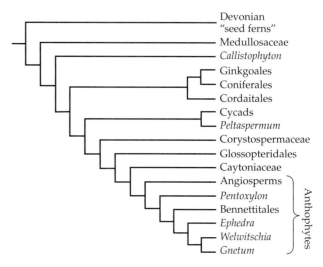

FIGURE 1.3 Shortest tree recovered by Doyle and Donoghue (1992). (Redrawn from Doyle and Donoghue 1992.)

sister to *Caytonia* and glossopterids, rather than with Gnetales, Bennettitales, and *Pentoxylon.* In other one-step-longer trees, the anthophyte clade was retained, but relationships among anthophyte taxa varied (Doyle and Donoghue 1992; reviewed in Doyle 1996). In some studies, Gnetales appeared sister to the angiosperms even when data for fossils were included (e.g., Crane 1985), whereas in others (e.g., Doyle and Donoghue 1986) the sister relationship between Gnetales and angiosperms emerged only when the fossils were removed.

One limitation of early cladistic studies of morphology is that investigators often treated the angiosperms as a single terminal. This approach required assumptions about the ancestral states of the angiosperms; criticisms prompted additional analyses in which several different, putatively basal angiosperm lineages were represented (e.g., Doyle et al. 1994; Nixon et al. 1994; Doyle 1996, 1998b). Although the sister relationship of Gnetales and angiosperms was again recovered in these analyses, strong bootstrap support for this relationship was lacking, and suboptimal trees again yielded diverse topologies.

Despite the lack of internal support and other concerns, the "anthophyte hypothesis" quickly became widely accepted. Concomitantly, acceptance of the anthophyte hypothesis had a profound impact, stimulating the reinterpretation of character evolution (reviewed in Frohlich 1999; Donoghue and Doyle 2000), including the origin of the carpel, the angiosperm leaf (J.A. Doyle 1994, 1998b), and double fertilization (Friedman 1990, 1992, 1994). For example, the "double fertilization" process in Gnetales was considered a possible precursor to the double fertilization of angiosperms (Friedman 1990, 1992, 1994) and ultimately a putative synapomorphy for Gnetales + angiosperms (Doyle 1996).

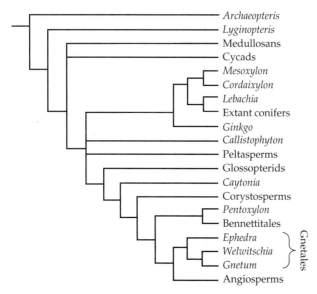

FIGURE 1.2 One of two shortest trees recovered by Crane (1985) in a cladistic analysis of extant and fossil seed plants (this represents the "preferred topology" of Crane 1985). (Redrawn from Crane 1985.)

DNA sequence data: demise of the anthophyte hypothesis

In this section, we summarize the major features of the many molecular phylogenetic analyses of seed plants, provide summary topologies of the relationships hypothesized among extant seed plants (Figure 1.4), and discuss our best current estimate of phylogeny among extant seed plants.

Molecular phylogenetic studies have provided strong support for the monophyly of Gnetales (e.g., Hamby and Zimmer 1992; Hasebe et al. 1992; Chase et al. 1993; Albert et al. 1994; Goremykin et al. 1996; Chaw et al. 1997; Stefanovic et al. 1998; Qiu et al. 1999; P. Soltis et al. 1999a; Bowe et al. 2000; D. Soltis et al. 2000). However, analyses of single genes have provided conflicting results regarding the relationship of Gnetales and angiosperms (Figure 1.4; see also Doyle 1998a, 1998b). Some analyses of *rbcL* alone and some analyses of partial 18S and 26S rRNA sequences placed Gnetales as sister to all other seed plants and angiosperms as sister to a clade of cycads, *Ginkgo,* and conifers (Figure 1.4A; e.g., Hamby and Zimmer 1992; Albert et al. 1994). Other parsimony analyses of *rbcL* placed angiosperms as sister to a clade of gymnosperms; within the latter clade, Gnetales were sister to cycads, which in turn, were sister to *Ginkgo* + conifers (Figure 1.4B; Hasebe et al. 1992). A maximum likelihood analysis of *rbcL* also placed angiosperms as sister to the monophyletic gymnosperms, but relationships among gymnosperms were different than in the parsimony topology (compare Figures 1.4C and 1.4B; Hasebe et al. 1992). Using partial 26S rDNA data, however, Stefanovic et al. (1998) recovered an "anthophyte" topology with angiosperms sister to Gnetales.

As single-gene studies accumulated (Figure 1.4, A to F), the evidence generally indicated that Gnetales were not sister to the angiosperms (but see Stefanovic et al. 1998). Single-gene investigations of plastid (ITS, *rpoC1*), nuclear (18S rDNA), and mitochondrial (*cox1*) sequences, for example, revealed a sister-group relationship between Gnetales and conifers (Figure 1.4, D to F; Goremykin et al. 1996; Malek et al. 1996; Chaw et al. 1997; P. Soltis et al. 1999a). Few of these studies provided any measure of internal support for relationships. However, analyses of 18S rDNA sequences provided moderate or low bootstrap support (84% and 64%, respectively) for a Gnetales + conifers sister-group relationship, depending on the size of the dataset (Figure 1.4, E and F; Chaw et al. 1997; P. Soltis et al. 1999a). Some of these single-gene analyses also indicated that gymnosperms were sister to the angiosperms (e.g., Goremykin et al. 1996; Chaw et al. 1997, 2000; P. Soltis et al. 1999a).

Other DNA sequence analyses provided additional evidence for the monophyly of the gymnosperms and for a close relationship of Gnetales and conifers, although the taxon sampling in these studies was sparse (e.g., Hansen et al. 1999; Winter et al. 1999; Frohlich and Park-

er 2000). For example, Hansen et al. (1999) obtained sequence data for a 9.5-kb portion of the plastid genome, but included only *Pinus, Gnetum,* and three angiosperms and used *Marchantia* as the outgroup. Winter et al. (1999) analyzed MADS box genes, but again only *Gnetum* was used to represent Gnetales, and cycads and *Ginkgo* were not included. The results of Winter et al. (1999) are significant, nonetheless, in that similar results were obtained for five homeotic genes; in each case the sequences of MADS box genes from *Gnetum* were found to be more closely related to those of conifers than to angiosperms. Analyses of *Floricaula/ LEAFY (FLO/LFY)* sequences also indicated that gymnosperms were monophyletic; within the gymnosperm clade, Gnetales appeared as sister to Pinaceae (Frohlich and Parker 2000).

Because sample sizes were often small and internal support low, these single-gene analyses were considered equivocal for seed plant relationships in general and the relationship between Gnetales and angiosperms in particular (Doyle 1998b; Donoghue and Doyle 2000). Nonetheless, the results of these single-gene analyses posed a serious challenge to the widespread acceptance of the anthophyte hypothesis.

Recent analyses in which sequences from multiple genes were combined (Figure 1.4, G to I) have indicated, with strong support, that Gnetales are not closely related to angiosperms (e.g., Qiu et a. 1999; P. Soltis et al. 1999b; Bowe et al. 2000; Chaw et al. 2000; Graham and Olmstead 2000; Pryer et al. 2001; Magallón and Sanderson 2002; D. Soltis et al. 2002a; Burleigh and Mathews 2004). In some of these analyses, the number of gymnosperms used was small because the investigations were aimed at relationships among the angiosperms (e.g., Qiu et al. 1999; P. Soltis et al. 1999b; Graham and Olmstead 2000) or vascular plants (Pryer et al. 2001). Several of these analyses were focused on seed plant relationships, however, and indicated that Gnetales were derived from within conifers and sister to Pinaceae (Figure 1.4H; Bowe et al. 2000; Chaw et al. 2000). Bowe et al. (2000) analyzed a four-gene dataset (*rbcL,* 18S rDNA, and mitochondrial *atpA* and *cox1*) and found strong support for the sister-group relationship of Pinaceae and Gnetales. Analysis of combined sequences of mitochondrial small subunit (SSU) rDNA, 18S rDNA, and *rbcL* similarly provided strong internal support for a Gnetales + Pinaceae sister-group relationship, referred to as the "gne-pine" hypothesis (Chaw et al. 2000; Figure 1.4H). An eight-gene analysis (D. Soltis et al. 2002a), which assembled a dataset of four plastid DNA genes (*rbcL, atpB, psaA,* and *psbB*), three mtDNA genes (*mtSSU, cox1,* and *atpA*), and one nuclear gene (18S rDNA; 15,772 bp per taxon), also provided strong support for a Gnetales + Pinaceae sister-group relationship. Burleigh and Mathews (2004) compiled a 13-gene data set and similarly found strong support for Gnetales + Pinaceae.

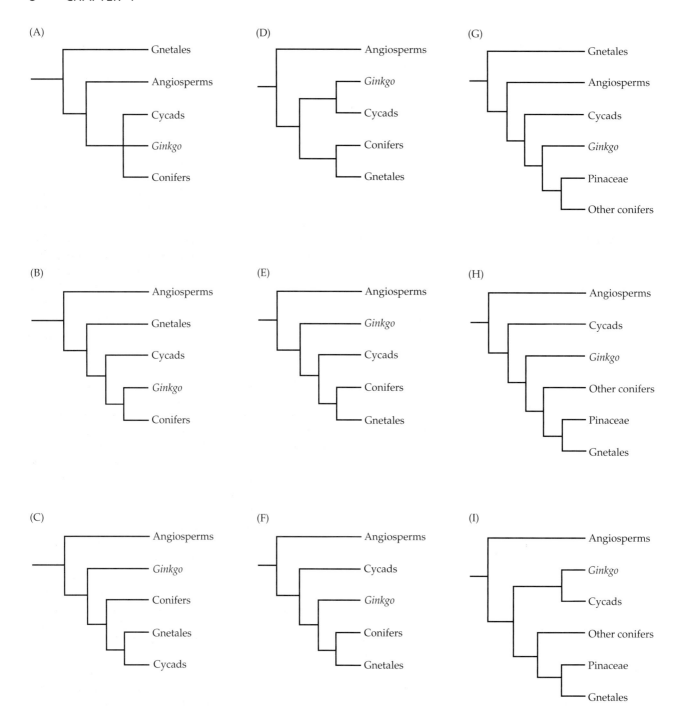

FIGURE 1.4 Simplified topologies depicting relationships among extant seed plants based on phylogenetic analysis of gene sequence data. A–F, single-gene datasets; G–I, combined, multigene datasets. (A) rRNA sequence data, Hamby and Zimmer (1992); *rbcL*, Albert et al. (1994). (B) *rbcL* with parsimony, Hasebe et al. (1992). (C) *rbcL* with maximum likelihood, Hasebe et al. (1992). (D) Plastid ITS (=cpITS), Goremykin et al. (1996). (E) 18S rDNA, Chaw et al. (1997). (F) 18S rDNA, P. Soltis et al. (1999a). (G) plastid genes with parsimony, D. Soltis et al. (2002a); *rbcL*, all positions with parsimony, Chaw et al. (2000); combined genes, representing all three genomes, with parsimony, Burleigh and Mathews (2004); *psaA* + *psbB*, all positions with parsimony and with maximum likelihood, Magallón and Sanderson (2002); *psaA* + *psbB*, third positions with parsimony, Magallón and Sanderson (2002). (H) Combined genes representing all three genomes, Bowe et al. (2000) and Chaw et al. (2000); combined genes, representing all three genomes, with maximum likelihood and Bayesian inference, D. Soltis et al. (2002a); plastid genes with maximum likelihood, D. Soltis et al. (2002a); *psaA* + *psbB*, first + second positions with parsimony and also with maximum likelihood, Magallón and Sanderson (2002); combined genes, representing all three genomes, fast-evolving sites excluded, Burleigh and Mathews (2004). (I) All genes, representing all three genomes, with parsimony, D. Soltis et al. (2002a).

The possible derivation of Gnetales from within conifers (Bowe et al. 2000; Chaw et al. 2000) was initially supported largely by mitochondrial DNA sequence data. Nuclear 18S rDNA sequences, in contrast, placed Gnetales as sister to all conifers with bootstrap support above 50% (Figure 1.4, E and F; Chaw et al. 1997; P. Soltis et al. 1999a). However, analyses of plastid DNA gene sequences provided varying results for the placement of Gnetales (Figure 1.4, G and H; reviewed in Rydin et al. 2002; D. Soltis et al. 2002a). Using all codon positions for *rbcL,* Chaw et al. (2000) found Gnetales to be sister to other seed plants with both maximum parsimony and maximum likelihood methods (Figure 1.4G); with third position transitions removed, Gnetales were sister to Pinaceae. Gnetales were sister to other seed plants in several multigene analyses using parsimony (e.g., Rydin et al. 2002; D. Soltis et al. 2002a).

The difficulties involved in resolving seed plant relationships were revealed by a detailed analysis of two plastid photosystem genes, *psaA* and *psbB* (Sanderson et al. 2000; Magallón and Sanderson 2002). When first and second positions were considered, gymnosperms were found to be monophyletic and Gnetales derived from within conifers, allied with *Pinus,* as in the analyses that included mtDNA sequences. However, when third positions were considered, Gnetales appeared instead as the sister to all other seed plants (Figure 1.4G). This conflict between first and second versus third codon positions in plastid genes was also reported in subsequent investigations of seed plant relationships (Rydin et al. 2002; D. Soltis et al. 2002a).

The signal in plastid DNA datasets for seed plants is complex. D. Soltis et al. (2002a) analyzed an eight-gene dataset with maximum parsimony, maximum likelihood, and Bayesian approaches and retrieved a gymnosperm clade sister to angiosperms (Figure 1.4, H and I). Within gymnosperms, the conifer clade included Gnetales as sister to Pinaceae. Analyses of the mtDNA partition using maximum parsimony, maximum likelihood, and Bayesian inference also yielded the "gne-pine" topology. However, maximum parsimony analyses of the combined plastid DNA genes placed Gnetales as sister to all other seed plants with strong bootstrap support (Figure 1.4G), whereas maximum likelihood and Bayesian analyses of the plastid DNA dataset placed Gnetales as sister to Pinaceae (Figure 1.4H). Maximum parsimony and maximum likelihood analyses of first and second codon positions of the plastid DNA partition also placed Gnetales as sister to Pinaceae. In contrast, maximum parsimony analyses of third codon positions placed Gnetales as sister to other seed plants; maximum likelihood analyses of third codon positions placed Gnetales with Pinaceae.

The analyses of Sanderson et al. (2000), D. Soltis et al. (2002a), Rydin et al. (2002), and Magallón and Sander-son (2002) indicated that most of the discrepancies in seed plant topologies involve third codon positions of plastid DNA genes. The nature of this conflict appears to be complex (see also Burleigh and Mathews 2004). Sanderson et al. (2000) and Magallón and Sanderson (2002) reported conflict between first and second versus third codon positions in the plastid genes *psaA* and *psbB.* Rydin et al. (2002) analyzed the plastid genes *rbcL* and *atpB* across seed plants and found that the conflict in codon position is more pronounced in *rbcL* than in *atpB,* but the conflict in both is less than that observed for *psaA* and *psbB.* Although third codon positions of plastid genes generally have most of the phylogenetic signal (e.g., Källersjö et al. 1998; Olmstead et al. 1998), the third positions may be saturated in some instances (Rydin et al. 2002), depending on taxon sampling. These results may also reflect short branches within the seed plant radiation, as well as high rates of molecular evolution in Gnetales and the outgroups (reviewed in Palmer et al. 2004). Adding to the complexity of the conflict between first plus second versus third positions is the fact that transitions within each codon position conflict with transversions (Chaw et al. 2000; Rydin et al. 2002). Burleigh and Mathews (2004) further examined the complexity of molecular datasets for seed plants. Conflicting signal in the datasets could be explained by differences in trees obtained with rapidly versus slowly evolving sites. Trees in which Gnetales are sister to all other seed plants appear to be the result of signal in the fastest evolving sites, whereas when these sites are excluded, gne-pine trees are obtained.

Importantly, some of the same molecular analyses noted above also provided moderate to strong support for the monophyly of extant gymnosperms. Analyses of 18S rDNA alone indicated that the gymnosperms constitute a clade (Figure 1.4, E and F; Chaw et al. 1997; P. Soltis et al. 1999a). In most of Bowe et al.'s (2000) analyses, bootstrap support for the monophyly of the gymnosperms was above 80% (Figure 1.4H). Chaw et al. (2000) obtained bootstrap values above 90% for the monophyly of the gymnosperms in their analyses of a combined three-gene dataset (Figure 1.4H). D. Soltis et al. (2002a) and Burleigh and Mathews (2004) also found strong support for the monophyly of extant gymnosperms (Figures 1.4H and 1.5). In addition, both Bowe et al. (2000) and Chaw et al. (2000) obtained strong bootstrap support (>90%) for cycads followed by *Ginkgo* as successive sisters to a conifer clade that includes Gnetales (Figure 1.4H). Using eight genes, D. Soltis et al. (2002a) also obtained strong support for cycads followed by *Ginkgo* as sister to the remaining gymnosperms with Bayesian inference and maximum likelihood analyses (Figures 1.4H and 1.5); however, with parsimony they found a weakly supported clade of cycads + *Ginkgo* (Figure 1.4I). Burleigh and Mathews (2004) also provided strong support (98%) for cycads followed by *Ginkgo* as sister to an expanded conifer clade in which Gnetales are sister to Pinaceae (Figure 1.5).

Perhaps one of the strongest inferences obtained from DNA data is the placement of Gnetales with conifers (Figure 1.5; see Palmer et al. 2004), but the precise relationship of Gnetales to conifers is unclear. Some analyses of both plastid DNA and mitochondrial DNA genes support a placement of Gnetales within conifers as sister to Pinaceae (the gne-pine hypothesis; Bowe et al. 2000; Chaw et al. 2000). However, in most molecular studies of seed plants, the number of conifers included was small. Magallón and Sanderson (2002) attempted to remedy this problem by constructing a taxonomically comprehensive dataset for *psaA* and *psbB*, but Gnetales remained sister to Pinaceae in some analyses. With a dataset that included 15 conifers and 13 loci, Burleigh and Mathews (2004) found strong support for Gnetales + Pinaceae. However, Rydin et al. (2002), using 30 conifers in an analysis of a four-gene dataset, found strong support for a monophyletic conifer clade that excluded Gnetales, suggesting that the gne-pine hypothesis was an artifact of inadequate taxon sampling in some analyses. Furthermore, use of the parametric bootstrap in analyses of a dataset involving many base pairs but few taxa revealed that the alternative placement of Gnetales as sister to conifers (rather than as sister to Pinaceae) could not be rejected (D. Soltis et al. 2002a). With 18S rDNA sequences (Chaw et al. 1997; P. Soltis et al. 1999a) and with multiple genes that included a broad representation of conifers (Rydin et al. 2002), Gnetales are sister to conifers. Lastly, a structural mutation in the plastid genome (Raubeson and Jansen 1992) also indicates that a Gnetales-conifer sister-group relationship may be a more parsimonious explanation of the data. Most land plants, including Gnetales, have two copies of the ribosomal genes in the plastid genome (the inverted repeat region), but conifers have only a single ribosomal coding region. Placement of Gnetales within conifers would necessitate that the ribosomal genes were lost in the conifers and then subsequently regained in Gnetales. Recent fossil evidence also supports a relationship of Gnetales and conifers (Wang 2004).

Combined datasets of multiple gene sequences have begun to provide a more consistent picture of relationships among extant seed plants than emerged from studies that used single genes. Perhaps the current "best estimate" of relationships for extant seed plants would have gymnosperms monophyletic with cycads and then *Ginkgo* as successive sisters to Gnetales + conifers (Figure 1.5). The relationship between cycads and *Ginkgo* remains somewhat unclear, however. Most recent analyses indicate that cycads and *Ginkgo* are successive sisters to other gymnosperms; some analyses indicate, albeit with weak support, that they form a clade that is sister

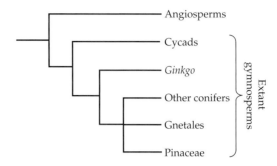

FIGURE 1.5 Current, best estimate of relationships among extant seed plants; Gnetales are either sister to conifers, or embedded within conifers and sister to Pinaceae.

to other gymnosperms. The relationship of Gnetales is unclear; they are either sister to all conifers or sister to just Pinaceae (Figure 1.5). It is unwarranted, however, to conclude that <u>all</u> gymnosperms are monophyletic just from the analysis of extant taxa. There are many fossil lineages of gymnosperms, and it is crucial that these be integrated into more comprehensive cladistic analyses of seed plants. Recent analyses that include fossils (see "Phylogenetic Studies: Integrating Fossils," below) continue to place extinct gymnosperms, including Bennettitales and Caytoniales, in a clade with angiosperms rather than with extant gymnosperms.

Morphology revisited

With a single exception (Nixon et al. 1994), DNA sequence data and morphology strongly agree on the monophyly of the Gnetales, a noteworthy finding given the great morphological divergence among these three genera. Of more interest, however, is the apparent disagreement between molecules and morphology regarding the affinities of Gnetales. Whereas cladistic analyses of morphological data have indicated that angiosperms and Gnetales are each other's closest living relatives (albeit without strong support), most DNA sequence data placed Gnetales with conifers. Doyle (1998b) initially attempted to reconcile this conflict, but because additional molecular studies argued against a close relationship between Gnetales and angiosperms, such efforts diminished (Donoghue and Doyle 2000).

Because morphology did not strongly support the anthophyte hypothesis, it is instructive to reconsider those few characters that united angiosperms and Gnetales in morphological cladistic analyses (Crane 1985; Doyle and Donoghue 1987, 1992; Loconte and Stevenson 1990; D.W. Taylor and L.J. Hickey 1992; Nixon et al. 1994; Doyle 1996). Some of these features are actually shared by all "anthophytes" (i.e., angiosperms, Gnetales, Bennettitales, *Pentoxylon*), as well as by Caytoniales. (Caytoniales were found to be sister to the anthophyte clade; however, Doyle 1996 discarded the name "anthophyte" for this larger clade because adding *Caytonia* violated his original concept of the clade.) Furthermore, different unifying features of Gnetales and angiosperms have been suggested by different investigators. Careful scrutiny leads to the conclusion that the homology of many of these shared characters is, in fact, dubious (see also Donoghue and Doyle 2000; Doyle 2001).

On the basis of Crane's (1985) analysis, two key features unite angiosperms, Gnetales, and other anthophytes (i.e., Bennettitales and *Pentoxylon*): a thin rather than thick megaspore membrane as in other seed plants, and microsporophylls aggregated in a whorl, or pseudo-whorl, a structure distinct from the pollen cones of conifers. Doyle and Donoghue evaluated the relationship between Gnetales and angiosperms in several papers (e.g., J.A. Doyle 1978, 1994, 1996, 1998a, 1998b; Doyle and Donoghue 1986, 1992); we summarize here the non-DNA characters that support a sister group of Gnetales + angiosperms in Doyle's (1996) most recent analysis. Whereas a tunica layer in the apical meristem appears to be lacking in other seed plants (except Araucariaceae), it is present in angiosperms and Gnetales. Angiosperms and Gnetales also share similar lignin chemistry (presence of a Mäule reaction, which is absent from other seed plants; McLean and Evans 1934; Gibbs 1957), double fertilization, microsporangia fused at least basally, an embryo derived from a single uninucleate cell via cellular divisions, a thin megaspore wall (as in Crane 1985 and Loconte and Stevenson 1990), siphonogamy, and a granular exine structure. Doyle (1996) also scored vessels in angiosperms and Gnetales as homologous. Loconte and Stevenson (1990) analyzed only extant taxa and provided three synapomorphies of Gnetales and angiosperms—thin megaspore wall (following Crane 1985), short cambial initials, and lignin syringial groups (equivalent to the Mäule reaction of Doyle 1996).

Some of the putative synapomorphies for angiosperms and Gnetales from Crane (1985), Doyle (1996), and Loconte and Stevenson (1990) are more complex than was initially suggested in these analyses and may in fact not be homologous (see also Donoghue and Doyle 2000). For example, although the angiosperms and Gnetales were coded the same for the presence of a tunica layer in the vegetative shoot apex, the tunica is two cells thick in many angiosperms and only one cell thick in Gnetales. Similarly, although angiosperms and Gnetales were coded as having the same state for the thickness of the megaspore wall, the megaspore wall is thin in Gnetales and absent in angiosperms. The pollen exine character used in some studies is now known to be inappropriate because a granular exine is not ancestral in angiosperms, as once was maintained (e.g., Doyle and Endress 2000; Doyle 2001; see Chapter 3 on basal angiosperms). Furthermore, the homology of vessels in angiosperms and Gnetales has long been doubted (Bailey 1944a, 1953), and Carlquist's (1996) re-

cent analysis concluded that they are not homologous. Angiosperms and Gnetales should therefore not be scored identically for these features of the tunica, megaspore, and vessel elements.

The homology of double fertilization in angiosperms and Gnetales has also been questioned (see Friedman 1994, 1996; Doyle 1996, 2000), but this issue is complex. In most angiosperms, a second sperm nucleus fuses with two nuclei of the megagametophyte (producing triploid endosperm), whereas in Gnetales a second sperm fuses with only one nucleus of the megagametophyte, yielding a diploid nucleus. However, double fertilization in some basal angiosperms is like that of Gnetales. In families such as Nymphaeaceae (but not Amborellaceae), a second sperm nucleus fuses with a single megagametophyte nucleus (Williams and Friedman 2002; see Chapter 3). Furthermore, double fertilization events that seem similar to those documented for *Ephedra* (Gnetales) have been reported for conifers, including *Thuja* and *Abies*. In addition, developmental events in cycads and *Ginkgo* are consistent with double fertilization (reviewed in Friedman and Floyd 2001). Thus, double fertilization may be a synapomorphy for all extant seed plants (Friedman and Floyd 2001).

Divergence times and the age of the angiosperms

The monophyly of extant gymnosperms, based on molecular evidence, and the sister-group relationship of angiosperms and extant gymnosperms suggest that the angiosperms may be much older than their fossil record indicates. Given that by the Late Carboniferous (~290–320 Mya) or Permian, cycads, conifers, and Ginkgoales were already represented in the fossil record, some have argued that their angiosperm sister group must be that old as well (see Axsmith et al. 1998; Doyle 1998a). However, the fact that extant gymnosperms are monophyletic does not necessarily imply that the angiosperm clade similarly dates to the Carboniferous. There could be several now-extinct lineages of seed plants along the branch to the angiosperms; in fact, our reanalyses (see "Demise of the Anthophyte Clade: In Search of Alternatives," below) indicate that this is the case, with lineages such as Caytoniales, Bennettitales, and *Pentoxylon* appearing on the branch to the angiosperms. Therefore, the argument for a Carboniferous origin of the angiosperm stem lineage could be an artifact of misinterpreting the molecular-based evidence for the monophyly of extant gymnosperms as evidence for the monophyly of all gymnosperms. Further study of the phylogeny of all seed plants, modern and fossil, is needed.

When did the clade containing extant gymnosperms and the branch leading to angiosperms diverge? Using the tracheophyte topology and multigene dataset of Pryer et al. (2001) and Sanderson's (1997) method of nonparametric rate smoothing, P. Soltis et al. (2002) estimated divergence times for major groups of tracheophytes. Rather than assuming a molecular clock with constant rates of evolution among lineages, the nonparametric rate smoothing method allows rates to vary, but assumes that such variation is autocorrelated—that is, rates are inherited from an ancestral lineage by the immediate descendants of that lineage (Sanderson 1997). Estimates of the divergence time between angiosperms and extant gymnosperms have varied dramatically, 346 to more than 900 Mya (P. Soltis et al. 2002). The latter age is clearly much too old, given the geological record of life on Earth. However, most values are between 346 and 367 Mya, a period that corresponds to the Late Devonian and Early Carboniferous and agrees with the earliest records of gymnosperm seeds. However, the seeds and pollen of these earliest gymnosperms do not correspond to those of crown group seed plants, a difference that reinforces the view that extant seed plants are not this old. Although the split between the clade containing extant gymnosperms and the clade containing angiosperms is likely an ancient one, angiosperms themselves need not be this old. Some extinct seed plant lineages may be more closely related to angiosperms than are any extant groups (see "Caytoniales" and "Bennettitales," below). Further studies of divergence times using additional methods that relax assumptions of constant evolutionary rates among lineages and allow multiple calibration points—penalized likelihood (Sanderson 2002) and Bayesian methods (Thorne and Kishino 2002)—may clarify the timing of the split between extant angiosperm and gymnosperm lineages.

Phylogenetic Studies: Integrating Fossils

To this point, we have focused largely on extant seed plant lineages. We now consider those possible close relatives of the angiosperms that are known from the fossil record. The importance of integrating fossils, and thus morphology, into datasets to understand the phylogeny of seed plants has long been emphasized (e.g., Doyle and Donoghue 1987; Kenrick and Crane 1997; Donoghue and Doyle 2000; Rydin et al. 2002; D. Soltis et al. 2002a; Crane et al. 2004). Even if we assume that molecular data have largely resolved relationships among living seed plant groups (a point that can still be debated), a complete understanding of seed plant relationships and the origins of angiosperm structures such as floral organs still requires the integration of fossil taxa. Furthermore, integration of fossil taxa into datasets can affect the phylogenetic placement of extant taxa. For example, relationships among lineages of anthophytes have varied among studies depending on

whether fossils were included. In some cases, Gnetales were sister to the angiosperms, even when fossils were included (e.g., Crane 1985); in others (e.g., Doyle and Donoghue 1986), the sister relationship between Gnetales and angiosperms appeared only when fossils were removed from the matrix.

Morphological cladistic analyses

Although the results of various morphological cladistic analyses differ, several extinct lineages of seed plants consistently appear—albeit in different positions—as possible close relatives of the angiosperms (Figures 1.1 to 1.3). Crane (1985) conducted two separate analyses that involved different codings of cupules. In the first analysis, the "cupules" of Bennettitales and *Pentoxylon* were coded as not homologous to the cupules of glossopterids, *Caytonia*, corystosperms, and angiosperms. In the second analysis, the cupules were coded as homologous. In the first analysis, Crane found a clade of Bennettitales + *Pentoxylon* as sister to angiosperms + Gnetales; glossopterids, *Caytonia*, and corystosperms were part of a separate clade. In the second analysis, Bennettitales + *Pentoxylon* were sister to angiosperms + Gnetales; corystosperms, *Caytonia*, and glossopterids were successive sisters to this clade (Figure 1.2). Crane (1985) considered the latter his preferred topology.

Doyle and Donoghue (1986, 1992) found a clade of angiosperms, Gnetales, Bennettitales, and *Pentoxylon*—their "anthophyte" clade (Figure 1.3). In their analyses, *Pentoxylon* and Bennettitales appeared between Gnetales and angiosperms (Doyle and Donoghue 1986, 1987, 1992), and *Caytonia* was sister to the anthophyte clade (Doyle and Donoghue 1986, 1992) (Figure 1.3). Rothwell and Serbet (1994) added new taxa, as well as new characters not used in previous cladistic studies, and recovered an anthophyte clade comprising angiosperms, Gnetales, Bennettitales, and *Pentoxylon*. However, in their anthophyte clade, *Pentoxylon*, followed by Bennettitales, appeared as successive sisters to a clade of angiosperms + Gnetales. Extant conifers were sister to their anthophyte clade, and *Caytonia* and *Glossopteris* were phylogenetically well removed from their anthophyte clade. In a more recent analysis (Doyle 1996), a "glossophyte clade" consisted of *Caytonia* + angiosperms, with glossopterids, *Pentoxylon*, Gnetales, and Bennettitales all part of a polytomy.

Demise of the anthophyte clade: in search of alternatives

With the demise of the anthophyte hypothesis, other concepts of angiosperm origin and relationship need to be evaluated and new analyses conducted. Perhaps the most critical need at this point is the rigorous reassessment of the homology of non-DNA characters, as well as a search for new characters. These approaches are beyond the scope of this book; we have, however, evaluated alternatives to the anthophyte hypothesis in general and a sister-group relationship between Gnetales and the angiosperms in particular. We used several approaches to reassess those groups (fossil and extant) that are most closely related to angiosperms. Specifically, we conducted cladistic analyses using existing morphological matrices and (1) removing "contentious" non-DNA characters; (2) adding molecular data to both original and modified non-DNA matrices; and (3) constraining Gnetales to be sister to Pinaceae or conifers.

In our first analysis, we recoded the non-DNA matrices of Crane (1985), Rothwell and Serbet (1994), and Doyle (1996). Only some of the analyses and topologies are described here. We focused on the more contentious characters—presence of a tunica, double fertilization, vessels, megaspore wall thin or absent. To a lesser extent we also reconsidered two other characters—microsporangia fused at least basally and an embryo derived from a single uninucleate cell via cellular divisions.

Beginning with the matrix of Doyle (1996), we reconsidered five of the characters linking Gnetales with angiosperms—presence of a tunica, lignin chemistry (presence of a Mäule reaction), double fertilization, vessels, megaspore wall thin or absent. We either omitted them from the matrix (e.g., Mäule reaction) or recoded them as nonhomologous in angiosperms and Gnetales. With either approach, a clade of anthophyte taxa remained fairly stable. Even with the removal of all five of these characters, plus removal of additional characters from the matrix of Doyle (1996) (e.g., characters 11, stomata anomocytic versus some or all paracytic; 41, microsporangia free versus fused at least basally; 86, embryo derived from several free nuclei versus from a single uninucleate cell by cellular divisions; 87, proembryo not tiered versus tiered), a clade of *Pentoxylon, Caytonia,* Bennettitales, angiosperms, and Gnetales (+ *Piroconites*) remained intact. We also conducted these analyses without *Piroconites* because recent palynological data (Osborn 2000) have challenged its close relationship with Gnetales. With or without *Piroconites,* these same seed plant lineages often form a polytomy, or in some cases *Caytonia* alone is sister to the angiosperms. Thus, even as controversial characters are removed, a clade of *Pentoxylon, Caytonia,* Bennettitales, angiosperms, and Gnetales remains.

Adding molecular data to these modified morphological matrices quickly removed Gnetales from the anthophyte clade. For example, adding only the sequences for the mitochondrial gene *atpA* (= *atp1*) to a modified matrix of Doyle (1996; five contentious characters removed; we also removed *Piroconites* from the matrix) resulted in the placement of Gnetales with Pinaceae, as in other recent analyses of molecules alone

(e.g., Bowe et al. 2000; Chaw et al. 2000; Magallón and Sanderson 2002; D. Soltis et al. 2002a; Burleigh and Mathews 2004). *Caytonia,* followed subsequently by Bennettitales, *Pentoxylon,* and then glossopterids, then appeared as the closest relatives of the angiosperms (Figure 1.6).

We also constructed a molecular scaffold (Springer et al. 2001) on the basis of recent molecular phylogenetic analyses, constraining Gnetales to be sister to the conifers. We again removed *Piroconites* from the matrix and then analyzed Doyle's (1996) original morphological matrix. These analyses found a revised "anthophyte" clade of angiosperms + *Caytonia,* Bennettitales, and *Pentoxylon,* with glossopterids sister to the remaining members of the clade (similar to Figure 1.6). Doyle (2001) independently conducted a similar constraint analysis and obtained results identical to ours.

The fossils: in search of the sister group of angiosperms

In this section, we provide brief coverage of five fossil lineages that may be the closest relatives of angiosperms; the possible close relationship to angiosperms is based on cladistic analyses and/or noteworthy morphological similarities: Caytoniales, Corystospermatales, Bennettitales, *Pentoxylon,* and glossopterids. We discuss these lineages with the caveat that our understanding of the relationships of fossil seed plants is now in a state of flux.

The search for fossil groups as possible angiosperm ancestors has focused on features of the ovule and structures that may be homologous to the angiosperm carpel. The ovules of angiosperms and gymnosperms differ in several important respects. Whereas angiosperm ovules generally have two protective layers or integuments, all gymnosperms have a single integument (Figure 1.7). (Some angiosperms have only a single integument due either to loss of an integument or to fusion of the two integuments; see Chapter 9 on asterids, for example.) Furthermore, the gymnosperm micropyle is located opposite the stalk bearing the ovule, whereas in many angiosperms the ovule is curved back on itself with the micropyle close to the stalk (i.e., an anatropous ovule; Figure 1.7).

Several lines of evidence have indicated that the outer integument of angiosperms has a different origin from the inner integument. A mismatch of the integuments produces a "zig-zag" micropyle in many basal angiosperms (Stebbins 1974). This feature is actually widely distributed in the angiosperms, occurring in diverse eudicots, including Bixaceae, various Brassicales, Fabaceae, Malvaceae, Hamamelidaceae, and Dilleniaceae. In addition, stomata are occasionally produced on the outer but not the inner integument (see Stebbins 1974; Corner 1976).

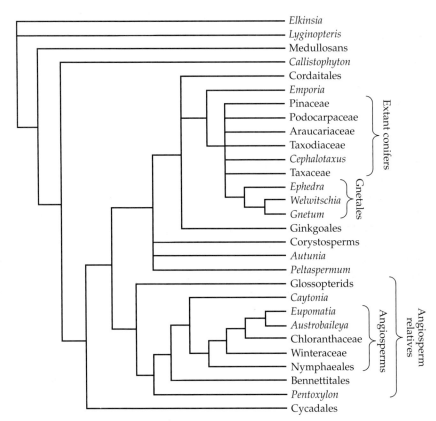

FIGURE 1.6 Strict consensus tree of 48 trees (length, 704 steps; CI, 0.714; RI, 0.734) resulting from phylogenetic analysis of the morphological matrix of Doyle (1996) with the addition of gene sequence data for *atpA* for extant taxa.

Developmental studies have shown that in many basal angiosperms, the outer integument is hood-shaped, indicating an origin from a leaf (Yamada et al. 2001a, 2001b). The gene *INNER NO OUTER (INO)* participates in the regulation of dorsoventrality of lateral organs and is expressed in the abaxial side of leaves, as well as in the outer epidermis of the outer integument in *Arabidopsis* (e.g., Bowman 2000). This same expression pattern has been observed in ovules of *Nymphaea*, supporting the hypothesis that the outer integument is homologous with a leaf and that ovules are located on the adaxial surface, away from the zone of expression (Yamada et al. 2003). This differential pattern of *INO* gene expression in inner and outer integuments is consistent with separate origins of the two integuments of angiosperms.

Fossil groups such as Caytoniales, glossopterids, *Pentoxylon*, corystosperms, and Bennettitales have been proposed as possible close relatives of angiosperms because the ovule is surrounded by a cupule, a structure that may be homologous to the outer integument of angiosperms. In Caytoniales and corystosperms, the cupule plus ovule

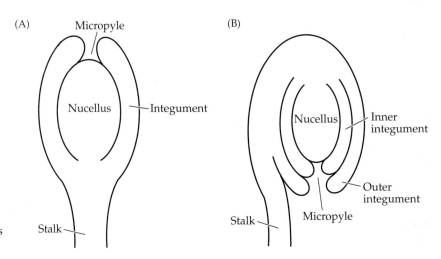

FIGURE 1.7 Ovules of gymnosperms and angiosperms. Gymnosperms (A) have a single integument; angiosperms (B) have inner and outer integuments.

structure is curved, which has been considered a possible antecedent of the anatropous ovule of angiosperms. Corystosperms produced ovules on the abaxial surface of the megasporophyll and therefore probably were not angiosperm ancestors. However, *Caytonia*, *Petriellaea* (a Mesozoic seed fern; Taylor et al. 1994), and *Glossopteris* (Taylor and Taylor 1992) produced ovules on the adaxial surface of a megasporophyll and, on this basis, could be angiosperm relatives. The seed-enclosing structures of *Petriellaea* and *Glossopteris* seem to have evolved via different structural modifications: in *Petriellaea* by transverse folding of the leaf and in some glossopterids by longitudinal enrolling of the leaf margin (Taylor and Taylor 1992). The folding of the megasporophyll in both fossils differs from the presumed origin of the angiosperm carpel via longitudinal folding of a megasporophyll. This leaves *Caytonia* as a possible ancestor or close relative of angiosperms.

CAYTONIALES   Before morphological cladistic analyses placed Gnetales as sister to the angiosperms (Crane 1985; Doyle and Donoghue 1986), Caytoniales had received considerable attention as a possible angiosperm ancestor (e.g., Gaussen 1946; Stebbins 1974; Doyle 1978, 1996). Although not initially considered an "anthophyte" in early cladistic analyses (Crane 1985; Doyle and Donoghue 1986), Caytoniales emerged as part of an expanded anthophyte clade in later analyses (Doyle and Donoghue 1992), appearing in some analyses as sister to the angiosperms (Doyle 1996).

In Caytoniales, male and female reproductive structures do not appear to have been produced together; in fact, neither has been found attached to stems (Figure 1.8). Caytoniales were considered a possible ancestor of the angiosperms because of their ovules. The morphology of Caytoniales seemed to explain the origin of both the two integuments and the anatropous ovule characteristic of most angiosperms. In Caytoniales, each cupule contained several ovules, and the cupule almost completely enclosed these ovules, leaving only a small opening between the ovule and the stalk of the ovary. Within the cupule, numerous unitegmic ovules were present, each with a micropyle oriented toward the opening (or mouth) of the cupule (Figure 1.8). It was argued that if the *Caytonia* cupule contained only a single ovule rather than multiple ovules, then the resultant structure would resemble a typical angiosperm ovule in being anatropous and possessing two integuments (Gaussen 1946; reviewed in Stebbins 1974; Frohlich and Parker 2000). This hypothesis was accepted by Stebbins (1974) and Doyle (1978).

The Caytoniales model for the origin of the angiosperms has several problems. One criticism is that no Caytoniales fossils have been reported that possess just one ovule per cupule. The counter argument is that only a simple reduction in ovule number is required.

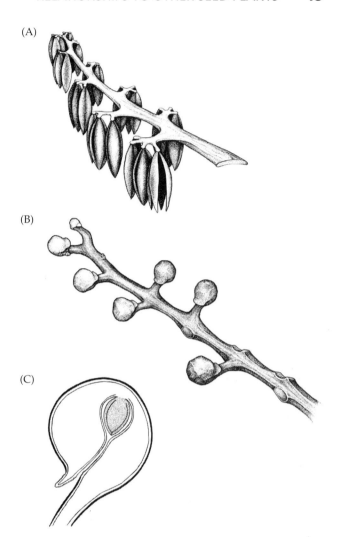

(A)

(B)

(C)

FIGURE 1.8 Reconstructions of Caytoniales (from Crane 1985). (A) Male structures of *Caytonanthus arberi*, based on Harris (1941). (B) *Caytonia nathorstii* megasporopyhlls, based on Harris (1964). (C) *Caytonia* cupule containing seeds, based on Reymanówna (1973).

A second problem is that the origin of the carpel from Caytoniales cannot easily be explained because the ovules of these plants were located on opposite sides of a narrow stalk, a structure that is difficult to envision as forming a carpel (Figure 1.8B). It was proposed that this stalk became wide and flat and eventually enclosed the ovules, forming the angiosperm carpel (Gaussen 1946; Doyle 1978; Crane 1985). Another difficulty with Caytoniales as a possible angiosperm ancestor is that the microsporophylls in Caytoniales were highly divided, differing greatly in morphology from angiosperm stamens (Figure 1.8A).

Because of the large morphological gap between Caytoniales and angiosperms, the hypothesis that Caytoniales are closely related to the angiosperms has been criticized (e.g., Nixon et al. 1994; Frohlich and Parker 2000).

CORYSTOSPERMATALES The possibility of a close relationship of corystosperms to angiosperms has been debated. The first analysis to place corystosperms with angiosperms and other anthophytes was that of Crane (1985). In one of his two analyses, corystosperms appeared as sister to the clade of Bennettitales + *Pentoxylon* and Gnetales + angiosperms. In Doyle and Donoghue's (1992) tree, corystosperms, followed by glossopterids, appeared as the sister taxa of the anthophytes. Corystosperms did not appear close to any anthophytes in the analyses of Doyle (1996, 1998), appearing, instead, with other Permian and Triassic seed ferns as sister to a clade of conifers. Nor were corystosperms part of the anthophyte clade in the analysis of Rothwell and Serbet (1994). Instead, they appeared in a clade with glossopterids, Caytoniales, *Callistophyton*, and peltasperms, well removed from the anthophyte clade of *Pentoxylon*, Bennettitales, angiosperms, and Gnetales.

Corystosperms have been discussed as possible angiosperm relatives because, like Caytoniales, they possessed cupules that appear to be reasonable antecedents of the angiosperm bitegmic ovule (Stebbins 1974; Doyle 1978; Frohlich and Parker 2000). Like Caytoniales, corystosperms had recurved cupules and have therefore been used to explain the origin of the anatropous ovule. Unlike Caytoniales, corystosperms possessed one ovule per cupule. An evolutionary series or progression in megasporophyll diversification has been proposed from Caytoniales through corystosperms to angiosperms (Thomas 1955; see Crane 1985) (Figure 1.9). However, in corystosperms the cupules were borne on slender stalks rather than on broader structures; these slender structures were viewed as unlikely precursors of the angiosperm carpel.

Recent work has indicated, however, that corystosperms were not a likely ancestor of angiosperms, despite renewed interest in the group by some workers (see "The Mostly Male Hypothesis," below), because the ovule curvature in corystosperms was in the opposite direction from that of angiosperms. That is, in corystosperms the ovules were located on the abaxial side of the leaf or cupule (Klavins et al. 2002). However, an increasing body of evidence, including gene expression data (Yamada et al. 2001a, 2001b; reviewed above) has indicated that the outer integument of angiosperms is derived from part of a leaf that bore ovules on its adaxial side (see Stebbins 1974; Kato 1990; Stewart and Rothwell 1993).

THE MOSTLY MALE HYPOTHESIS Corystosperms received renewed attention as a putative angiosperm precursor through the work of Frohlich and Parker (2000). Frohlich and Parker's mostly male hypothesis for the origin of the flower (Figure 1.10) was based on their work with the homeotic gene *Floricaula/LEAFY (FLO/LFY)*. Although,

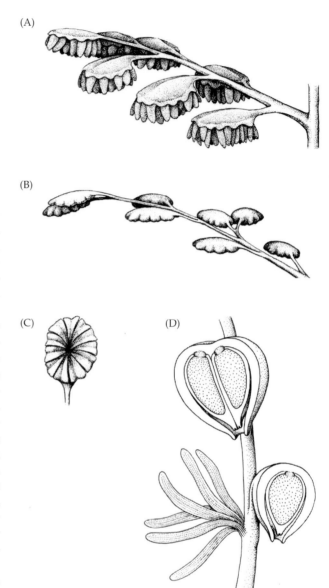

FIGURE 1.9 Reconstructions of corystosperms (from Crane 1985; Frohlich and Parker 2000). (A) Male structures of *Pteruchus africanus*, redrawn from Townrow (1962). (B, C) Synangia of *Pteroma thomasii*, redrawn from Harris (1964). (D) Cupules of *Ktalenia*, from Taylor and Archangelsky (1985).

as noted, corystosperms now appear to be unlikely close relatives of angiosperms, the mostly male hypothesis still merits attention. Gymnosperms generally have two copies of the *LFY* gene (although *Gnetum* appears to have only a single copy; M.W. Frohlich, pers. comm.; see Figure 1.10A), referred to as the "needle" and "leaf" families. The function of the *LFY* genes of pine has been studied in detail. The "needle" (*NEEDLY*) paralogue is expressed only in early-developing female reproductive structures (hereafter referred to as LFY_f) (Mouradov et al. 1998). The "leaf" paralogue is expressed in male re-

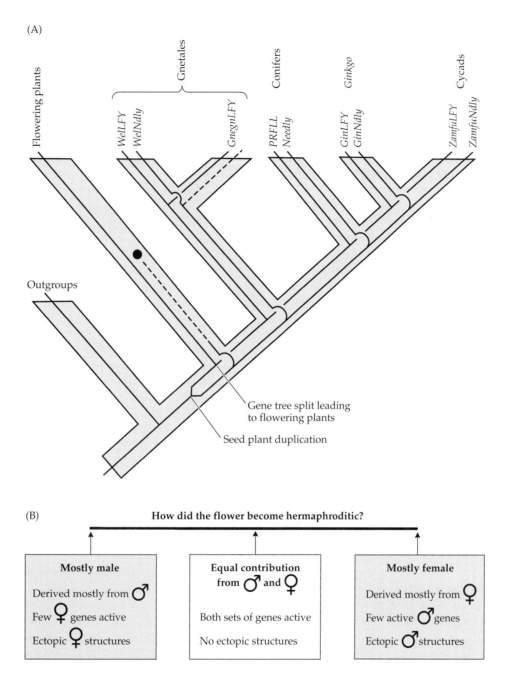

FIGURE 1.10 (A) Proposed evolution of the homeotic gene *Floricaula/LEAFY (FLO/LFY)* in seed plants (Frohlich and Parker 2000). A gene duplication occurred in seed plants followed by a loss of one copy (*LFY$_f$*) in the ancestor of angiosperms (filled dot). *Gnetum* seems to have only *LFY$_m$*, suggesting an independent loss of one copy of *LFY* in this gymnosperm (dashed line). (B) Mostly male hypothesis (Frohlich and Parker 2000) for the origin of the bisexual flower; an alternative hypothesis is provided by Albert et al. (2002; see text).

productive structures of gymnosperms (hereafter *LFY$_m$*). Frohlich and Meyerowitz (1997) speculated that these two gene families show such specialization because gymnosperms have had separate male and female reproductive structures since the Devonian. The *LFY* duplication in gymnosperms may have accompanied or perhaps facilitated the specialization of separate male and female reproductive structures (Figure 1.10B).

Angiosperms have only one copy of *LFY*, which is most closely related to *LFY$_m$*. Angiosperms have apparently lost the female-specifying "needle" paralogue (*LFY$_f$*) and retained the male-related "leaf" paralogue (*LFY$_m$*). These data prompted Frohlich and Parker (2000) to propose the mostly male theory of floral origins (Figure 1.10). Their hypothesis offers one developmental genetic mechanism by which a plant with separate male and female reproductive structures, a feature of all gymnosperms, could produce a bisexual structure such as a flower. Frohlich and Parker (2000) suggested that developmental control of floral organization derives more from systems operating in the male reproductive structures of the gymnosperm ancestor of angiosperms than from the female reproductive structures; hence, "mostly male." In the mostly male hypothesis, ovules are considered ectopic in origin on male reproductive structures (i.e., stamens) in early flowers (Figure 1.10).

Frohlich and Parker proposed a suite of morphological features that might have been present in the gymnosperm ancestor of the angiosperms on the basis of the requirements of the mostly male theory. In their view, Caytoniales are a poor fit because the microsporophylls were deeply divided into numerous narrow segments (Figure 1.8) and thus differed greatly from angiosperm stamens (see above). However, *Caytonia* is a better fit in other features, including leaf venation, stomata, and cuticle. Frohlich and Parker suggested that corystosperms possessed male reproductive structures appropriate for the mostly male theory. In the corystosperm *Pteroma,* for example, the male organs were flat, broad, and unlobed (Figure 1.9). There was a single ovule per cupule, and the corystosperm cupule plus ovule was recurved, providing a potential antecedent of the bitegmic, anatropous ovule characteristic of angiosperms (Figure 1.9). Although the corystosperm cupules were produced on slender stalks and have therefore been considered problematic in terms of a precursor to the angiosperm carpel, this is not an issue in the mostly male origin of the carpel because the carpel was produced ectopically on the microsporophyll. As noted, however, corystosperms produced ovules on the abaxial surface of the megasporophyll (rather than adaxial as in angiosperms) and therefore may not have been close relatives of angiosperms.

Researchers have challenged the mostly male hypothesis on the basis of additional studies of *FLO/LFY.* Shindo et al. (2001) demonstrated that, although the *FLO/LFY* gene *GpLFY* from *Gnetum* is in the "leaf" clade (LFY_m) of Frohlich and Parker (2000), *GpLFY* is expressed in female strobili of *Gnetum.* This does not agree with the Frohlich and Parker assumption that only gene members of the "needle" clade (i.e., LFY_f) were expressed in gymnosperm female structures. These observations prompted Shindo et al. to conclude that the mostly male theory is not plausible.

Albert et al. (2002) provided an alternative to the mostly male hypothesis. Their model assumes that pleiotropic interactions between LFY_m and LFY_f were critical for stabilizing the retention of these two genes in gymnosperms and suggests that disruption of this delicate balance between the two *LFY* genes occurred in an ancestor of modern angiosperms. This ancestral

taxon might have had unisexual flowers together on the same plant, or might have been loosely bisexual. LFY_f would then have been lost through selection for an integrated bisexual reproductive axis. Albert et al. (2002) further suggested that their hypothesis would help explain early fossil angiosperms such as *Archaefructus,* which produced stamens and carpels on loosely integrated terminal axes. However, the interpretation of the reproductive axes in *Archaefructus* varies (see Chapter 3). According to the Albert et al. model, one could hypothesize that LFY_f had not yet been lost in *Archaefructus.* This might also indicate that a bisexual flower came later in angiosperm evolution.

Albert et al. (2002) provided genetic criteria for evaluating their model as well as the mostly male hypothesis of Frohlich and Parker (2000). The mostly male hypothesis predicts that early stages of flower development will be controlled primarily by genes orthologous to those expressed in male structures of gymnosperms. However, the Albert et al. hypothesis does not have this requirement. They proposed instead that loss of LFY_f would result in male and female coexpression in the same apical meristem and not ectopic female expression on a male structure. Hence, in their model, gene copies that were previously expressed as male or female would be eliminated with equal probability. It should be possible to distinguish between these alternatives by analyzing sequences for many genes expressed in early stages of flower development in diverse lineages of angiosperms (see D. Soltis et al. 2002b).

BENNETTITALES Bennettitales have long been considered close relatives of angiosperms. Arber and Parkin (1908) proposed a close relationship between angiosperms, Gnetales, and Bennettitales. In Crane (1985), Bennettitales and *Pentoxylon* appeared as sister to Gnetales + angiosperms. The analyses of Doyle and Donoghue (1992) and Rothwell and Sorbet (1994) similarly placed angiosperms, *Pentoxylon,* Bennettitales, and Gnetales together in one clade.

Bennettitales had strobili that were either unisexual or bisexual. In bisexual (hermaphroditic) forms, the microsporangia and ovules were associated on lateral branches that arose among the leaf bases (Figure 1.11). This strobiloid reproductive structure was important in the formulation of the "anthostrobilus" theory for the origin of the angiosperm flower (Arber and Parkin 1907). The strobiloid reproductive structure of Bennettitales was considered similar to the strobiloid flowers of Magnoliaceae, and this similarity was the rationale for considering members of "Ranales" (Magnoliaceae, Ranunculaceae, and their relatives) as the most primitive extant angiosperms (e.g., Bessey 1915). Many Bennettitales had flat microsporophylls with adaxial sporangia, which could be considered suitable precursors to the angiosperm stamen. However, in other regards,

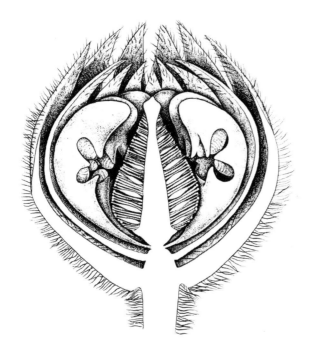

FIGURE 1.11 Reconstruction of Bennettitales (from Crane 1985). *Williamsoniella coronata*, longitudinal section through "flower," based on Harris (1964).

Bennettitales are problematic as close relatives of the angiosperms. The cupules of Bennettitales had an orthotropous orientation rather than potentially anatropous as in Caytoniales, corystosperms, and most angiosperms (see Crane 1985). Discounting Bennettitales on the basis of ovule orientation assumes, however, that the anatropous ovule is ancestral for angiosperms, an issue discussed in Chapter 3. Furthermore, ovule ori-

entation is problematic when carpels are not present. Finally, some Bennettitales did not possess cupules (Rothwell and Stockey 2002). Bennettitales have sometimes been considered part of a separate lineage, perhaps sharing a common glossopterid ancestor with the angiosperms (see "Glossopterids," below), but not as close relatives of the angiosperms. Importantly, cladistic analyses consistently place Bennettitales close to the angiosperms.

PENTOXYLON Before Crane's (1985) analysis, *Pentoxylon* was considered an isolated gymnosperm of uncertain affinity. *Pentoxylon* was first associated with anthophytes in the cladistic analyses of Crane (1985) because of its flower-like arrangement of microsporophylls, aggregation of ovules into a head (as in Bennettitales), and similarity of ovules to those of Bennettitales (Figure 1.12). *Pentoxylon* was also placed in the anthophyte clade of Doyle and Donoghue (1992) and Rothwell and Serbet (1994). However, other investigations provided new or differing interpretations of features of *Pentoxylon*. Bose et al. (1985) challenged some of the anthophyte-like anatomical features of *Pentoxylon*. However, when Doyle (1996) rescored *Pentoxylon* with the data of Bose et al. (1985) and Rothwell and Serbet (1994), the genus continued to show relationships to Gnetales, angiosperms, Bennettitales, and *Caytonia*.

In *Pentoxylon*, the ovules were sessile and putatively helically arranged into compact heads. The ovulate heads were borne terminally on short shoots. There are several problems with *Pentoxylon* as a close relative of the angiosperms. For example, there is no clear carpel prototype in *Pentoxylon*. Furthermore, like Bennettitales, the ovules of *Pentoxylon* had an orthotropous,

(A)

(B)

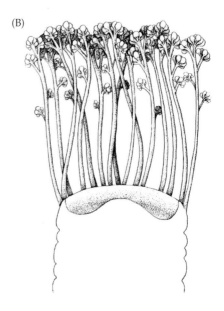

FIGURE 1.12 Reconstructions of *Pentoxylon* plants (from Crane 1985). (A) Ovulate heads of *Carnoconites cranwelliae*, based on Harris (1964). (B) Microsporangiate structures of *Sahnia*, based on Vishnu-Mittre (1953).

rather than anatropous, orientation. Researchers also debate the presence of a cupule in *Pentoxylon*, which would ultimately form the outer integument. Crane (1985) suggested that the ovules were surrounded by a cupule, whereas Nixon et al. (1994) and Rothwell and Serbet (1994) maintained that *Pentoxylon* did not possess a cupule. The megasporophylls of *Pentoxylon* were not leaf-like in appearance; rather, each ovulate head bore 10 to 20 spirally arranged, stalked, unilocular sporangia (Figure 1.12; Crane 1985). As a result of these concerns, *Pentoxylon*, like Bennettitales, has been envisioned as part of a separate lineage, perhaps sharing a common glossopterid ancestor with the angiosperms (Figure 1.13; see next section), but not as a close relative or direct ancestor of the angiosperms.

GLOSSOPTERIDS Plumstead (1956) and Melville (1962) suggested a close relationship of angiosperms and glossopterids on the basis of reticulate leaf venation. This close relationship was later questioned because the initial interpretations of glossopterid fructifications were found to be incorrect (see Doyle and Hickey 1976; Schopf 1976; Hickey and Doyle 1977). Retallack and Dilcher (1981) revived interest in a glossopterid ancestry of angiosperms, suggesting that the ovule-bearing organs of glossopterids have structures that might be homologous with both the outer integument and the carpel of angiosperms (Figure 1.13). They described the fossil glossopterid *Denkania* using terminology applied to angiosperms, stating that this fossil taxon had bitegmic, orthotropous ovules with the cupule homologous with the outer integument of angiosperms. These "bitegmic" ovules were arranged on a leaf surface but were not enclosed. The leaf surface was considered homologous with the angiosperm carpel (Figure 1.13). However, it may be inappropriate to characterize the

ovules of glossopterids as either orthotropous or anatropous given that the ovules are borne on essentially flat, leaf-like structures rather than within a carpel (Taylor and Taylor 1992).

Doyle (1996) also suggested a close relationship of glossopterids and angiosperms. He proposed not only that glossopterids might have been the ancestral group from which Gnetales arose via intermediate fossil forms such as *Piroconites* and *Dechellyia* but also that the common ancestor of *Caytonia* and the angiosperms might have had "glossopterid-like bract-sporophyll complexes." He suggested that these complexes would have had several cupules per bract (leaf) that would represent the future "anatropous cupules" of *Caytonia*, which have often been considered homologous to the bitegmic anatropous ovules of angiosperms (Figure 1.8). The underlying bract would then only have to be folded lengthwise to produce a carpel or reduced to produce a *Caytonia* sporophyll. This scenario corresponds with the views of Stebbins (1974) and Retallack and Dilcher (1981). Doyle (1996) maintained that this scenario is perhaps more plausible than the origin of a carpel through the expansion and folding of the *Caytonia* rachis (Stebbins 1974; Doyle 1978).

Cladistic analyses have often indicated a more distant relationship between glossopterids and angiosperms than between Bennettitales, *Pentoxylon*, Caytoniales, and angiosperms (Crane 1985; Rothwell and Serbet 1994; Doyle 1996, 1998a, 1998b; see also our analyses above). Glossopterids were not, for example, considered part of the "anthophyte clade." However, in Doyle and Donoghue (1992), Caytoniales followed by glossopterids were the sister taxa to the anthophyte clade. In a later analysis, Doyle (1996) recovered a "glossophytes" clade in which glossopterids were sister to the anthophyte lineages—angiosperms, Gnetales, Ben-

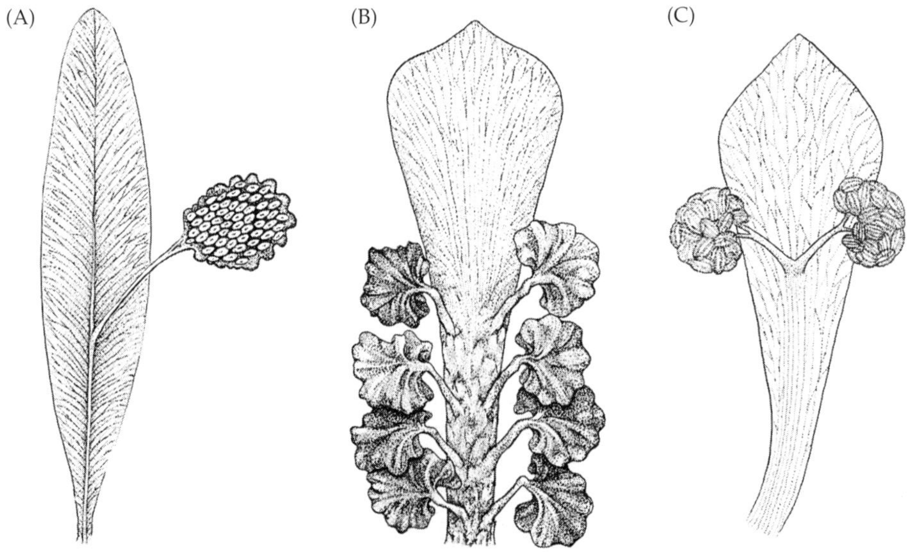

FIGURE 1.13 Reconstructions of glossopterids (from Crane 1985). (A) Megasporophyll of *Ottokaria* and associated leaf, redrawn from Pant (1977). (B) Megasporophyll of *Lidgettonia africana*, based on Thomas (1958). (C) Microsporophyll of *Eretmonia*, redrawn from Surange and Chandra (1975).

nettitales, and *Caytonia*. This topology is consistent with the hypothesis that both angiosperms (Retallack and Dilcher 1981) and Gnetales (Schopf 1976) were derived from glossopterids. These hypotheses are now more difficult to evaluate with the demise of the anthophyte hypothesis and placement of Gnetales with conifers. However, it is possible that glossopterids are the immediate sister to a clade that includes angiosperms and their closest relatives and may, therefore, have played a crucial role in angiosperm origins.

Future Studies

It has been difficult to resolve relationships among extant seed plants with molecular data. In most analyses, cycads and *Ginkgo* are reconstructed as successive sisters to other gymnosperms, although in some cases they form a clade that is sister to other gymnosperms. Additional research is needed to establish the relationship between cycads and *Ginkgo*. One of the strongest inferences of DNA data is the placement of Gnetales with conifers, but the precise relationship of Gnetales to conifers is unclear. Most analyses of both plastid and mitochondrial genes support a placement of Gnetales within conifers as sister to Pinaceae (the gne-pine hypothesis). In other analyses, Gnetales are sister to conifers; the gne-pine hypothesis may be an artifact of inadequate taxon sampling and spurious attraction. Phylogenetic studies that include more taxa and more genes are needed to resolve the position of Gnetales relative to conifers. Although approximately 85 genera of nonangiospermous seed plants are extant, most studies have included only small numbers of taxa (usually fewer than 10). Sanderson et al. (2000) used only three conifers, for example, in their analyses of seed plants. In the D. Soltis et al. (2002a) analyses of a combined matrix of 15,772 bp per taxon, only five conifers were used, only one of which represented Pinaceae. Similarly, Bowe et al. (2000) included only *Pseudotsuga* and *Abies* as representatives of Pinaceae. However, Burleigh and Mathews (2004) included 15 conifers in their multigene analyses and typically found strong support for the gne-pine hypothesis. Rydin et al. (2002) included the broadest representation of conifers in their four-gene analysis of seed plants. Additional outgroup genera could also be used; only two outgroup taxa have been consistently used (*Lycopodium* and *Asplenium*) for most of the genes sequenced to date.

Recent studies have also revealed the complexity of the molecular data used to investigate seed plant relationships (Sanderson et al. 2000; Rydin et al. 2002; D. Soltis et al. 2002a; Burleigh and Mathews 2004). Conflict exists among data partitions, and we are only beginning to understand the nature of this conflict. More investigations are needed to tease apart the incongruence among partitions of the molecular data themselves.

Recent molecular analyses have also provided strong support for the monophyly of extant gymnosperms. Despite this strong molecular inference, one should not conclude that all gymnosperms (fossil and extant) are monophyletic. In fact, cladistic analyses of morphological data involving extant and fossil taxa have indicated that gymnosperms are not monophyletic. A monophyletic gymnosperm clade would be at odds with some non-DNA characters, such as the presence of siphon-ogamy (nonmotile sperm) rather than zoidiogamy (motile sperm). Motile sperm are present in cycads and *Ginkgo,* but other seed plants (extant and fossil) possessed nonmotile sperm. A monophyletic gymnosperm clade would necessitate nonmotile sperm evolving at least twice, once in the remaining gymnosperms and once in angiosperms.

Although additional molecular work should be pursued, perhaps the most pressing need is a comprehensive analysis of extant and fossil seed plant lineages (e.g., D. Soltis et al. 2002a; Burleigh and Mathews 2004; Crane et al. 2004). Many extinct taxa still need to be incorporated into seed plant analyses. Furthermore, although Gnetales are no longer considered the sister group of angiosperms and the anthophyte hypothesis has been dismissed, our understanding of the closest fossil relatives of the angiosperms is still based on morphological cladistic analyses that find an anthophyte clade of Bennettitales, *Pentoxylon*, Gnetales, and sometimes Caytoniales. A new paradigm is needed to promote further progress on seed plant relationships in general and angiosperm origins in particular. In achieving this new synthesis, several areas of research should be pursued.

One crucial research need involves the careful reexamination of those morphological characters previously used in cladistic analyses of seed plants (e.g., Doyle 1998b; Donoghue and Doyle 2000; Crane et al. 2004). The homologies of the non-DNA characters that have been proposed to link Gnetales and angiosperms need to be reassessed; these and other non-DNA characters should be reevaluated throughout the seed plants. Although an effort has been made to use as many non-DNA characters as possible in cladistic analyses (a more-is-better approach that parallels the philosophy adopted in molecular analyses), it may be more appropriate to use fewer "high quality" characters in such analyses. Another important research need is the extensive integration of fossil and molecular datasets for seed plants. Resolving relationships among seed plants (fossil and extant), determining the sister group (and next outgroups) of the angiosperms, and elucidating the evolution of angiosperm features such as the flower will by necessity require the integration of fossils and the use of morphological data.

An important issue to consider is the best approach for combining morphological and molecular datasets.

Currently, three general approaches exist for placing fossils in the correct phylogenetic position: (1) Constrain the taxa in the morphological matrix to conform to the DNA-based topology already available and conduct a phylogenetic analysis of the morphological matrix with fossils included. One attractive approach is to use groups that are well supported in molecular analyses to construct a molecular scaffold. Morphological data would then be used for extant and fossil data to integrate the fossil taxa into the molecular scaffold. (2) Alternatively, analyze the morphological matrix, with and without fossils. This approach does not take advantage of the wealth of information provided by molecular analyses, but it allows relationships among extant taxa to change with the addition of fossils. (3) Use all characters together, morphological and DNA, to construct cladograms; this can be done both with and without fossils to evaluate the effect of fossil taxa on the topology. The consideration of new methods for integrating fossils into datasets for extant taxa is crucial. Such analyses now seem particularly important given that no living group of seed plants appears to represent a close relative of the angiosperms.

2

Phylogeny of Angiosperms: An Overview

Introduction

Enormous interest and effort have been invested in reconstructing phylogenetic relationships of the angiosperms since the 1990s. The importance of using a phylogenetic approach to evaluate evolutionary hypotheses has long been recognized and many comparisons are best made in light of a firm understanding of angiosperm relationships. A robust phylogenetic underpinning for flowering plants is of value not only to systematists but also to scientists in diverse disciplines, including evolutionary biologists, physiologists, ecologists, molecular biologists, and genomicists (see "Implications for the Study of Model Organisms," below).

In this chapter, we review efforts to reconstruct angiosperm phylogeny using morphology, RNA and DNA sequence data, and combined morphology and DNA (or RNA) datasets (see also P. Soltis and Soltis 2004). Because some of the largest phylogenetic analyses conducted to date for any group of organisms have involved the angiosperms (e.g., Chase et al. 1993; Källersjö et al. 1998; D. Soltis et al. 2000), valuable lessons in data analysis have also been learned, and these have general implications that we also briefly review here. In addition, we review the best estimate of phylogeny now available for the angiosperms and compare these relationships with recent classifications. The broad topologies for angiosperms also provide evidence for several episodes of phyletic radiations in the flowering plants and facilitate estimation of the ages of major diversification events.

History of Efforts to Reconstruct Angiosperm Phylogeny

Morphology

Although many intuitive reconstructions of angiosperm evolution (e.g., Bessey 1915; Cronquist 1968, 1981; Takhtajan 1969, 1980, 1987, 1997; Stebbins 1974; Thorne 1976) have appeared during the past century, it was not until the late 1980s that explicit analyses of angiosperm phylogeny were initiated. These pioneering initial investigations of angiosperm phylogeny relied solely on evidence from morphology (e.g., Donoghue and Doyle 1989a, 1989b; Loconte and Stevenson 1991) and set the stage for large phylogenetic analyses that have been based on both DNA and morphology. These early phylogenetic studies provided important initial insights into relationships, indicating, for example, that dicotyledonous angiosperms were not monophyletic. Instead, the monocots were nested among clades of "primitive dicots" (see Figure 3.1 in Chapter 3). In contrast, most dicotyledonous taxa formed a well-marked clade referred to as eudicots (Doyle and Hotton 1991) or tricolpates (Donoghue and Doyle 1989a, 1989b; Judd and Olmstead 2004).

Initial phylogenetic studies using morphology provided conflicting results regarding the root of the angiosperm tree, focusing attention on several possible groups as basal angiosperms. The analysis of Donoghue and Doyle (1989a), which used 62 binary characters for 20 placeholders, placed a Magnoliales clade as sister to all other angiosperms, followed by other woody magnoliids such as Laurales and a winteroid group. However, the composition of these clades did not agree with DNA-based circumscriptions (see Chapter 3). Donoghue and Doyle (1989a) also focused attention on a group they referred to as paleoherbs that included Piperaceae, Saururaceae, Aristolochiaceae, Nymphaeales (= Nymphaeaceae *sensu* APG II 2003), and monocots. In contrast, the phylogenetic analysis of morphological data conducted by Loconte and Stevenson (1991) placed Calycanthaceae as sister to all other angiosperms.

Analysis of a larger morphological dataset for basal angiosperms (108 morphological characters for 52 taxa; Doyle and Endress 2000) found trees that were more similar to those obtained with DNA sequence data (see Figure 3.2 in Chapter 3). Analysis of a large non-DNA dataset, consisting of morphological, anatomical, and chemical data for 151 angiosperms (Nandi et al. 1998) recovered a topology that was similar to trees obtained with molecular datasets. For example, the basal angiosperms formed a grade that included the monocots, and the eudicots formed a clade. Thus, there appears to be considerable phylogenetic signal in these non-DNA datasets that in many instances agrees with topologies derived from sequence data (Chase et al. 2000; also discussed below in "Single-Gene Studies" and "Combining DNA Datasets").

Single-gene studies

Initial efforts to reconstruct angiosperm phylogeny on a broad scale with sequence data began in the early 1990s. The first attempt was that of Hamby and Zimmer (1992) who constructed a dataset that was based on partial 18S and 26S rRNA sequences for 60 taxa. Although small by present standards, this analysis was a huge undertaking at that time. Hamby and Zimmer (1992) identified Nymphaeales (= Nymphaeaceae *sensu* APG II 2003) as sister to other angiosperms, showed monocots to be nested within a grade of early-branching dicots, and recovered a eudicot clade.

The cladistic analysis of *rbcL* sequence data for 499 species of angiosperms (Chase et al. 1993), which is still among the largest phylogenetic analyses ever conducted, has emerged as a landmark study for several reasons. First, Chase et al. ushered in a new era of investigation in which numerous representatives were used to reconstruct phylogeny for large, enigmatic groups. From a historical standpoint this is significant in that the prevailing view was that large datasets could not be analyzed phylogenetically (e.g., Graur et al. 1996). Instead, it was recommended that large datasets be broken into smaller datasets or pruned to include only a few exemplars. However, it was evident to many angiosperm systematists that smaller matrices did not produce the same topologies obtained with more exhaustive sampling (reviewed in Chase and Albert 1998; D. Soltis et al. 1998).

A second important aspect of the large *rbcL* analysis is that it stretched the limits of analytical capabilities available at that time (reviewed in Chase and Albert 1998). Initial efforts to analyze the large *rbcL* data matrix actually failed because versions of PAUP then available could not handle the number of terminals (500 sequences). Thus, coincident with the compilation of this large matrix was the development of software and search strategies that were appropriate for such large datasets. As reviewed in "Lessons Learned" below, advances in data analysis have continued, spurred, in part, by the large datasets assembled by angiosperm systematists.

Finally, the Chase et al. (1993) paper represented a massive collaboration of 42 investigators, most of whom contributed unpublished sequences to the effort because they appreciated the central importance of such an endeavor to angiosperm systematics. This spirit of collaboration among botanists has generally continued and has enabled plant systematists to achieve a tremendous understanding of angiosperm phylogeny in a short period. This general approach has subsequently emerged as a model for other groups of organisms, including the Deep Green (http://ucjeps.herb.berkeley.edu/bryolab/greenplantpage.html), Deep Time, and Deep Gene (http://ucjeps.herb.berkeley.edu/bryolab/deepgene/index.html) initiatives, as well as the Grass Phylogeny Working Group (GFWG 2000), Ericaceae Working Group (Kron et al. 2002), and the Compositae Alliance (Deep Achene).

In the trees of Chase et al. (1993), *Ceratophyllum* was sister to all other angiosperms (Figure 2.1A). The remaining basal (i.e., noneudicot) angiosperms formed a

clade sister to the eudicot clade. The basal angiosperms included several clades of traditional dicots, as well as the monocots. Within the eudicots, several large clades were retrieved, including the asterid, rosid, and caryophyllid clades. Large subclades were also suggested within some of the major clades of angiosperms (e.g., rosid I, rosid II, asterid I, asterid II). Much of the basic topology from this initial broad analysis of *rbcL* sequences has been further supported and refined by additional analyses (reviewed in "Combining DNA Datasets" and "Lessons Learned" sections).

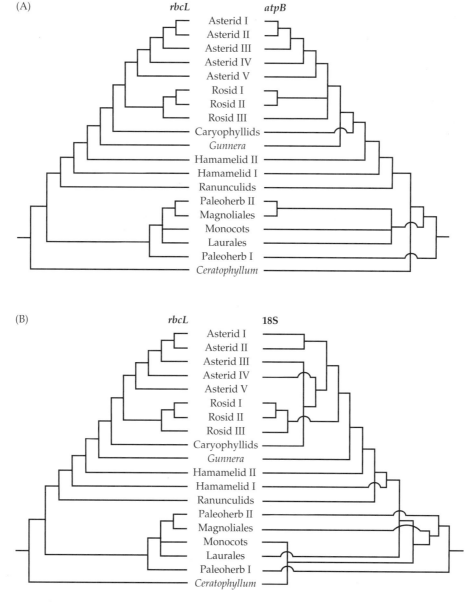

FIGURE 2.1 Comparison of simplified, general topologies for angiosperms based on three analyses of single-gene datasets: *rbcL* (the 500-terminal search of Chase et al. 1993); 18S rDNA (D. Soltis et al. 1997a); *atpB* (Savolainen et al. 2000a). (Modified from Chase and Albert 1998.) (A) Comparison of *rbcL* and *atpB*. (B) Comparison of *rbcL* and 18S rDNA.

The success of the Chase et al. (1993) collaboration prompted several questions, such as, which topologies would be found by other genes, particularly nuclear genes? Although partial sequences of 18S rRNA had previously been used to infer angiosperm relationships (Hamby and Zimmer 1992), doubts persisted about the usefulness of 18S rDNA for phylogeny reconstruction, primarily because some investigators maintained that alignment of sequences would be problematic (reviewed in Nickrent and Soltis 1995; D. Soltis et al. 1997a, 1997b). A pilot study comparing the performance of entire 18S rDNA and *rbcL* sequences confirmed the phylogenetic utility of 18S rDNA across the angiosperms (Nickrent and Soltis 1995). A subsequent analysis of 232 entire 18S rDNA sequences (D. Soltis et al. 1997b) revealed that although 18S rDNA provided lower resolution than *rbcL*, the 18S rDNA and *rbcL* topologies were highly concordant (Figure 2.1). Although some features of the 18S rDNA and *rbcL* gene trees differed, none of these differences received internal support (as measured by bootstrap values) above 50%. Two of the four 18S rDNA analyses placed *Amborella*, a clade of Austrobaileyaceae + Illiciales, and Nymphaeaceae as successive sisters to all other angiosperms (referred to as "paleoherb I" in Figure 2.1B); the positions of *Amborella* and the clade of Austrobaileyaceae + Illiciales were simply reversed in the other two 18S rDNA analyses. The basal angiosperms, including the monocots, formed a grade followed by a eudicot clade. The same major groups of eudicots revealed by *rbcL* were again recovered (e.g., asterid, caryophyllid, Santalales, Saxifragales, and rosid clades).

At the same time that a large 18S rDNA dataset was being constructed, sequences for *atpB*, a second plastid gene, were also being amassed (Savolainen et al. 2000a). Initial analyses indicated that, in length and rate of evolution, *atpB* was a close match of *rbcL* and would be ideal for phylogeny reconstruction across the angiosperms (Ritland and Clegg 1987; Hoot et al. 1995a). Analyses of *atpB* sequences alone (Savolainen et al. 2000a) also placed *Amborella*, Nymphaeaceae, and a clade of Austrobaileyaceae/Illiciales as sister to all other angiosperms (referred to as "paleoherb I" in Figure 2.1A). The basal angiosperms formed a grade, and the same major groups of eudicots revealed by *rbcL* and 18S rDNA were again recovered (asterid, caryophyllid, Santalales, Saxifragales, and rosid clades), with the same major subgroups present within the rosid and asterid clades (e.g., rosid I, II; asterid I, II).

Comparisons of the trees from analyses of single genes (*atpB*, *rbcL*, and 18S rDNA) again revealed striking overall similarities; differences among the topologies did not receive support above 50% (D. Soltis et al. 1997a, 1997b, 1998; Chase and Albert 1998; Chase and Cox 1998). Summary trees for *rbcL*, *atpB*, and 18S rDNA

are compared in Figure 2.1, illustrating this high degree of congruence. Furthermore, use of the partition homogeneity test revealed that the greatest amount of incongruence among the datasets from these three genes actually involved the two plastid datasets (*atpB* and *rbcL*), datasets that many would argue could be combined *a priori* (see D. Soltis et al. 1998).

Initial criticisms of the Chase et al. (1993) *rbcL* analysis had less to do with the topology itself than with issues of data analysis. Some maintained that the analysis was flawed in that the shortest trees had not been obtained. In fact, Chase et al. (1993) knew that shorter trees could be obtained but did not feel that lengthy searches for shorter trees would significantly alter the outcome (Chase and Albert 1998). Rice et al. (1997) conducted a lengthy reanalysis of the Chase et al. (1993) *rbcL* data matrix, investing nearly one year of computer time. Although they did find shorter trees than those reported in 1993, these did not significantly alter the big picture of angiosperm phylogeny. In fact, the Rice et al. (1997) reanalysis essentially reinforced the general phylogenetic conclusions of Chase et al. Subsequent studies have indicated that such lengthy searches of matrices with many taxa and an insufficient number of characters are ineffective and inefficient (e.g., D. Soltis et al. 1997a, 1997b, 1998; Chase and Cox 1998; Nei et al. 1998; Savolainen et al. 2000a; Takahashi and Nei 2000; see "Lessons Learned," below).

Combining DNA datasets

Although the single-gene trees for angiosperms were highly similar (Figure 2.1), internal support for many major clades and the spine of the tree was lacking (e.g., D. Soltis et al. 1997a, 1997b, 1998; Chase and Albert 1998). Thus, an immediate interest in combining datasets emerged. Because of the cooperation and collaboration that existed among many angiosperm systematists, the sequences that had been amassed for *rbcL*, *atpB*, and 18S rDNA were often for the same species or genus; in many cases the same DNA extraction had been used. This facilitated the next step in the reconstruction of angiosperm phylogeny—that of directly combining multiple molecular datasets.

Initial combinations of large DNA datasets were conducted primarily as experiments to assess the effect of additional data on the phylogenetic analysis of large datasets (D. Soltis et al. 1997b, 1998; Chase and Cox 1998). These initial efforts in combining large datasets were highly successful and provided the impetus for two projects that combined DNA datasets on a large scale across the angiosperms. One combined *rbcL* and *atpB* sequences for 357 taxa (Savolainen et al. 2000a); the second combined *rbcL*, *atpB*, and 18S rDNA sequences for 567 taxa (P. Soltis et al. 1999b; D. Soltis et al. 2000). These analyses provided topologies for the flowering plants with higher resolution and support than obtained in the single-gene trees. The major features of the three-gene topology are discussed in "Angiosperm Topology," below. Other studies have combined multiple molecular datasets for basal angiosperms, monocots, eudicots, and asterids, and these efforts are discussed in later chapters. Molecular and morphological datasets for angiosperms have also been combined (e.g., morphology and partial 18S and 26S rRNA sequences, Doyle et al. 1994; non-DNA characters and *rbcL*, Nandi et al. 1998; morphology and *rbcL, atpB,* and 18S rDNA sequences, Doyle and Endress 2000).

Lessons learned: the analysis of large datasets

The analyses of large datasets conducted on angiosperms serve as an important model, providing valuable insights into how such analyses can be conducted in other groups. However, only a decade ago some maintained that the analyses now completed for angiosperms were impossible to conduct successfully (e.g., Patterson et al. 1993; Hillis et al. 1994; Hillis 1995). The simulation studies of Hillis et al. (1994) indicated that in some instances correct phylogeny reconstruction for only four taxa would require more than 10,000 bp of DNA sequence data, implying much more difficulty with large datasets. This complexity stimulated some to suggest that phylogenetic problems be broken into a series of smaller problems, one extreme being a large number of four-taxon questions (Graur et al. 1996). Large datasets also pose problems for phylogenetic analysis because of the large number of possible trees that arise with increasing numbers of taxa and a consequent lack of confidence that an analysis has found the optimal tree or trees. The number of potential trees increases logarithmically as taxa are added (Felsenstein 1978a); for 228 taxa (the number of species analyzed by D. Soltis et al. 1997b), the number of possible trees is 1.2×10^{502}, which is more than the number of atoms in the universe (D.M. Hillis, pers. comm.). Despite these dire predictions, angiosperm systematists continued to conduct analyses involving hundreds of species, with promising results.

MORE IS BETTER Initial combinations of large DNA datasets for two and three genes revealed the importance of a "total evidence" approach (D. Soltis et al. 1997b, 1998; Chase and Cox 1998). One solution to the computational problems posed by large datasets was simply to add both taxa and characters. This approach resulted in much faster run times, which have the advantage that more thorough analyses of tree space can be conducted in a given time. In addition, internal support for clades is much higher in trees from combined datasets than in trees from separate datasets (Figure 2.2;

D. Soltis et al. 1998). In trees from the combined datasets, some clades achieved internal support above 50%, whereas these same clades did not obtain support above 50% in trees from the separate datasets. Thus, through combining large datasets for two or three genes, strong jackknife or bootstrap support was evident for the first time for most of the major clades of flowering plants. One reason for the success of these combined analyses was the greater signal than that in the individual datasets (see Chase and Cox 1998; D. Soltis et al. 1998; Bremer et al. 1999). With increased signal, starting trees are much closer to the shortest trees ultimately obtained, thereby speeding up the analyses. Furthermore, the combined signal from multiple genes clarifies historical patterns relative to noise in the dataset. These empirical results for large datasets complement the simulation studies of Hillis (1996) and Graybeal (1998), which similarly revealed the importance of adding taxa and characters in phylogenetic analyses. Although debate has continued regarding the importance of adding taxa, it is now widely agreed that increased taxon sampling is valuable in phylogeny reconstruction (e.g., Pollock et al. 2002; Zwickl and Hillis 2002; Hillis et al. 2003; D. Soltis et al., 2004; but see

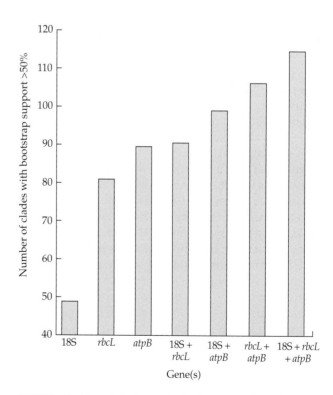

FIGURE 2.2 More is better—the importance of adding characters. The number of angiosperm clades (*y*-axis) having bootstrap support above 50% in phylogenetic analyses of separate and combined DNA sequence datasets increases as genes are added. (Modified from D. Soltis et al. 1998.)

Rosenberg and Kumar 2001). However, recent analyses of whole organellar genomes (e.g., Goremykin et al. 2003, 2004; Rokas et al. 2003) have relied on a large number of characters for a small number of taxa. The results for angiosperms (Goremykin et al. 2003, 2004) are clearly at odds with all large-scale phylogenetic trees and attributable to limited taxon sampling (D. Soltis and Soltis 2004; D. Soltis et al. 2004).

DEVELOPMENTS IN PARSIMONY ANALYSIS Programs for finding shortest trees, such as Hennig86 (Farris 1988), NONA (Goloboff 1993), and PAUP (Swofford 1993), were developed for use on what would now be considered small datasets. The numerous improvements available in PAUP* 4.0, such as the faster speed at which heuristic parsimony searches are conducted, have been a great asset to those interested in analyzing large datasets. An important development in the analysis of large datasets is a computer program called the RATCHET that greatly facilitates the ability to find shorter trees using parsimony (Nixon 1999; implemented in Winclada, Nixon 2000; and with NONA, Goloboff 1993; www.cladistics.com). When applied to the 567-taxon matrix of three genes for angiosperms, the RATCHET quickly found trees shorter than those found using PAUP* 4.0 (Swofford 1999; see D. Soltis et al. 2000).

QUICK SEARCHES FOR WELL-SUPPORTED CLADES A significant problem with the analysis of large datasets is the difficulty in assessing internal support for clades. Large datasets are not amenable to standard bootstrapping (Felsenstein 1985) and Bremer support (decay) analysis (Bremer 1988). The large angiosperm datasets were therefore also a stimulus for the adoption of computational approaches, such as "fast" or "quick" search techniques using bootstrapping (Swofford 1999) or jackknifing (Farris et al. 1996). (A fast jackknife program is also available in PAUP* 4.0; the two fast jackknife methods yield similar values; Mort et al. 2000.)

Recent evidence has indicated that these fast methods may actually be preferable to longer search strategies. The rationale for quick searches is that they sample trees from many starting points at the expense of extensive swapping on trees from a single starting point. Only those clades above some minimal threshold of internal support (e.g., a bootstrap or jackknife value of 50%) appear in the consensus of trees from these replicate searches. The well-supported clades appear relatively quickly in the analysis of big datasets—continued branch-swapping involving poorly supported branches can never result in strongly supported clades. Simulation studies (Nei et al. 1998; Takahashi and Nei 2000), as well as recent analyses of the large angiosperm datasets (e.g., D. Soltis et al. 1998; 2000; Savolainen et al. 2000a), provided support for the prem-

ise that lengthy parsimony analyses of large datasets may be unnecessary.

The parsimony jackknife method (Farris 1996; Farris et al. 1996) is well suited for the quick analysis of large datasets. It was applied, for example, to a dataset of 2538 *rbcL* sequences (Källersjö et al. 1998). A second version of the *Jac* program (Källersjö et al. 1998) has a faster tree-building algorithm and also allows branch-swapping to be conducted at each replicate. The new *Jac* program was applied to the 567-taxon dataset for angiosperms; 1000 replicates were completed in 60.63 hours (D. Soltis et al. 2000).

The correspondence between values obtained using the fast bootstrap or fast jackknife and those obtained with the standard bootstrap (with extensive branch swapping) was not clear when these fast methods were first applied. Mort et al. (2000) found that the fast and standard methods are highly comparable, particularly at higher levels of support (85% and higher). However, as support for clades decreases, percentages from the fast methods tend to be lower than values obtained from the standard bootstrap. For a further comparison of fast and standard methods on empirical and simulated datasets, see Mort et al. (2000), Debry and Olmstead (2000), and Salamin et al. (2003).

BAYESIAN INFERENCE Bayesian approaches to phylogeny reconstruction have recently been implemented (Huelsenbeck et al. 2001, 2002). This approach affords a new opportunity to analyze large datasets with support, although the MrBayes program has enormous computer memory requirements. Using MrBayes and a large *atpB* dataset for more than 300 angiosperms (from Savolainen et al. 2000a), Huelsenbeck et al. (2001, 2002) obtained a topology with *atpB* alone that was well supported and highly similar to that obtained using three genes (D. Soltis et al. 2000). However, the frequencies for clades (posterior probabilities) obtained using MrBayes are generally higher than corresponding bootstrap or jackknife percentages (e.g., Huelsenbeck et al. 2001; Miller et al. 2002; Wilcox et al. 2002), perhaps providing misleading estimates of clade support (Suzuki et al. 2002).

MOLECULAR EVOLUTION The large datasets of DNA sequences that have been assembled for *rbcL*, *atpB*, 18S rDNA and other genes such as *matK*, ITS, and *ndhF* have provided a wealth of data for studies of molecular evolution. We will not review these studies here, but instead point readers to appropriate references (see D. Soltis and Soltis 1998 for general overview; see also Albert et al. 1994; Hoot et al. 1995a; Kellogg and Juliano 1997; Chase and Albert 1998; Chase and Cox 1998; Olmstead et al. 1998; P. Soltis and Soltis 1998; Savolainen et al. 2000a, 2000b).

Fundamental differences in molecular evolution, perhaps related to functional constraints, may exist between the plastid genome of angiosperms and the mitochondrial genome of some animals, such as vertebrates (Savolainen et al. 2002). The large datasets of plastid gene sequences available for angiosperms have made it possible to conduct rigorous comparisons of plant and animal organellar gene evolution. Whereas some have suggested that functional requirements, including chemical properties, charge, and hydrophobicity, in the mtDNA genome of animals have seriously compromised estimation of relationships (Naylor and Brown 1997), such is not the case with the plastid genome (Savolainen et al. 2002). Thus, much of the success in phylogeny reconstruction at deep levels in flowering plants may be due to properties of the plastid genome itself.

Large-scale analyses of angiosperm datasets have also prompted a reassessment of the utility of third codon positions for phylogeny reconstruction. Rapidly evolving nucleotide sites have often been considered less informative than those that evolve more slowly, particularly in the analysis of deep branches. However, several studies have shown that third codon positions of plastid genes convey the most information in angiosperms (Källersjö et al. 1998; Olmstead et al. 1998), whereas in the animal mtDNA genes, these positions performed most poorly (Savolainen et al. 2002). Källersjö et al. (1998, 1999) demonstrated this on an extremely broad scale in their analysis of 2538 *rbcL* sequences across photosynthetic life. Although rapidly evolving and highly homoplasious, third positions contained most of the phylogenetically informative sites and performed well. With broad sampling, homoplasy is dispersed across the tree, allowing the historical signal carried in third positions to emerge.

Angiosperm Topology: Overview and Comparison with Modern Classifications

In this section, we provide a brief overview of angiosperm relationships that relies primarily on the 567-taxon, three-gene (*rbcL, atpB,* 18S rDNA) topology (P. Soltis et al. 1999b; D. Soltis et al. 2000). Although additional genes (e.g., *matK, matR,* 26S rDNA) will be added to create an even larger matrix for approximately 600 angiosperms, the large three-gene study remains the most comprehensive published analysis of the angiosperms as a whole. We have updated the summary tree (Figure 2.3) to reflect the results of several more-recent investigations. The basal angiosperms have been the focus of analyses that include up to 11 genes and more than 15,000 bp per taxon (e.g., Qiu et al. 1999; Zanis et al. 2002; see Chapter 3). A dataset of entire 26S

rDNA sequences has also been added to the three-gene dataset for eudicots (D. Soltis et al. 2003a). As a result of these and other investigations, the placements illustrated for some taxa, as well as percentages of internal support, have been revised from those depicted in the original three-gene topology. For example, (1) Ceratophyllaceae appear in our summary tree as sister to the monocots with support above 50%; (2) a magnoliid clade is added, comprising Laurales + Magnoliales and Piperales + Canellales (referred to as Winterales in some recent papers); and (3) among core eudicots, Gunneraceae + Myrothamnaceae (Gunnerales) appear as sister to all other core eudicots.

In this overview, we compare the current best estimate of phylogeny with some of the major features of the most prominent modern morphology-based classification systems—Dahlgren (1980), Cronquist (1981), Takhtajan (1987, 1997), and Thorne (1992a, 1992b, 2001) (see also Dahlgren et al. 1985 and G. Dahlgren 1989). In many ways, these four systems were highly similar; each recognized a basic split between monocotyledons and dicotyledons (recognized as separate classes; Magnoliopsida and Liliopsida of Cronquist 1981). The monocotyledons and dicotyledons were, in turn, divided into a series of subclasses or superorders. Relationships among these subclasses were often depicted in terms of phylogenetic shrubs or "bubble" diagrams that represented each author's intuition of evolutionary relationships (see also Stebbins 1974). The six subclasses of dicots recognized by Cronquist and their interrelationship are depicted in Figure 2.4A; ordinal interrelationships were expressed in a similar fashion (Figure 2.4, A and B). Noteworthy differences exist, however, among these four systems. Dahlgren, for example, was the only modern author to emphasize chemical characters, as well as the only one to incorporate cladistic principles into his classification. However, numerical analyses of the content of orders and the overall structure of these four classifications have indicated that they are not significantly different (Cuerrier et al. 1997).

There are both striking similarities and differences between recent morphology-based classifications (e.g., Cronquist 1981; Takhtajan 1987; Thorne 1992a, 1992b) and DNA-based phylogenetic hypotheses. Our comparisons illustrate the importance of consulting these DNA-based phylogenetic trees; it is no longer appropriate to use recent morphology-based classification systems to infer higher-level relationships in the angiosperms. The reclassification of angiosperms is discussed in more detail below in "A Reclassification Based on Sequence Data." A useful source of phylogenetic trees and other information for angiosperms is provided by Stevens (2004; www.mobot.org/MOBOT/research/APweb/).

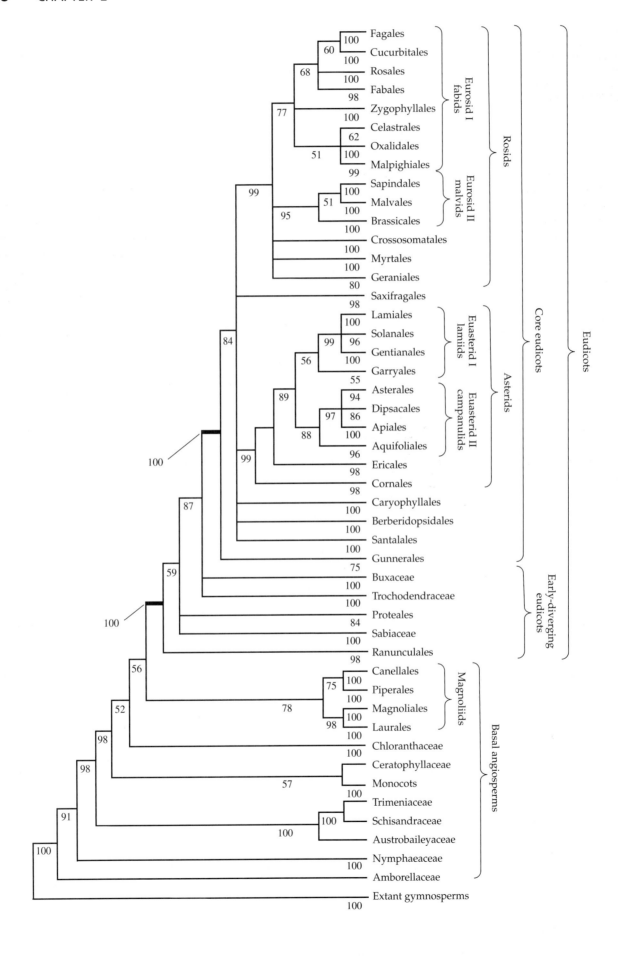

◀ **FIGURE 2.3** Summary tree for angiosperms. This tree is modified from the three-gene topology for 560 angiosperms (D. Soltis et al. 2000) to reflect current understanding of relationships among basal angiosperms (from Zanis et al. 2002) and the relationship of Gunnerales (D. Soltis et al. 2003a). Only nodes receiving bootstrap or jackknife support above 50% are depicted. The jackknife values (below branches) provided for basal angiosperms are from Zanis et al. (2002); values for eudicots are from D. Soltis et al. (2000), with the placement of Gunnerales from D. Soltis et al. (2003a). Bold lines are branches to eudicots and core eudicots. Several eudicot lineages are not depicted here (e.g., Dilleniaceae, Vitaceae). Clade names follow APG II.

In the discussions that follow, we generally refrain from using phrases such as "basal group" or "early-diverging group" because a node marks the divergence of two lineages, neither of which is basal to the other or diverges earlier than the other. It is more appropriate to refer to clades as sister to other clades or groups of clades. However, some lineages clearly diverged before others, such as the origins of Nymphaeaceae and Austrobaileyales before the origins of any of the eudicot lineages. Thus, nodes or branches can be "basal" or "early-diverging" relative to other clades, even though the extant groups, which are terminals in the tree, are not. As reviewed below (and in Chapters 3 and 5 of this book), a graded series of families is sister to a clade comprising the vast majority of all other angiosperms (the eudicots), and for convenience we will refer to this grade as the "basal angiosperms." Similarly, within the eudicot clade, a graded series of families appears as successive sisters to the core eudicots. For convenience, we will refer to this grade as the "early-diverging" (or "basal") eudicots.

Basal angiosperms

The basal angiosperms (all angiosperms other than eudicots) form a grade in the three-gene trees (P. Soltis et al. 1999b; D. Soltis et al. 2000), as well as in analyses focused specifically on these taxa (e.g., Mathews and Donoghue 1999; Parkinson et al. 1999; Qiu et al. 1999, 2000; Barkman et al. 2000; Doyle and Endress 2000; Graham and Olmstead 2000; Zanis et al. 2002, 2003). One of the most conspicuous features of both DNA- and morphology-based topologies is the absence of a major split between monocots and dicots. Instead, all of these trees indicated that, whereas the monocots are monophyletic, the dicots are not. Instead, the monocots are derived from within a basal grade of taxa that are all traditional dicots corresponding largely to Magnoliidae of Cronquist (1981).

The remaining dicots, the eudicots, which include about 75% of all species of flowering plants (Drinnan et al. 1994), form a well-supported clade (Figure 2.3). Most of the basal angiosperms are characterized by uniaperturate pollen, although there is considerable variation in pollen morphology among these taxa (reviewed in Sampson 2000). In contrast, the eudicots are characterized by triaperturate or triaperturate-derived pollen (Doyle and Hotton 1991), which some authors have suggested might have been a crucial factor for eudicot success (e.g., Furness and Rudall 2004). Other morphological features, in addition to triaperturate

(A)

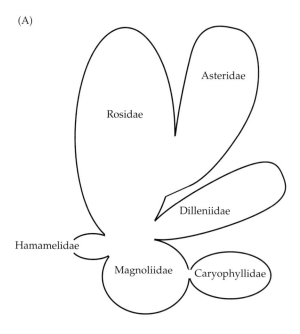

(B)

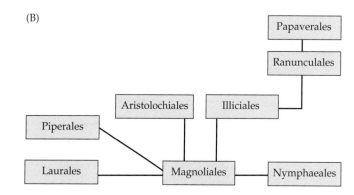

FIGURE 2.4 Relationships among major groups of angiosperms as proposed by Cronquist (1981). (A) Bubble diagram illustrating putative evolutionary relationships among subclasses of dicots. The size of the balloons is proportional to the number of species in each group (modified from Cronquist 1981). (B) Proposed relationships among orders of Cronquist's Magnoliidae. (Modified from Cronquist 1981.)

pollen, are shared by many of the core eudicots (reviewed in Nandi et al. 1998 and Savolainen et al. 2000a), including ellagic acid (apparently replaced in the euasterids by iridoids and other compounds), nonlaminar/marginal placentation, and differentiation of the calyx and corolla (Albert et al. 1998; Nandi et al. 1998).

In most recent analyses, Amborellaceae are sister to all other angiosperms (see Chapter 3), followed in succession by Nymphaeaceae (*sensu* APG II 2003), and then a small clade of Austrobaileyaceae, Trimeniaceae, and Schisandraceae (*sensu* APG II 2003) (Mathews and Donoghue 1999; Qiu et al. 1999; Renner 1999; P. Soltis et al. 1999b; D. Soltis et al. 2000; Zanis et al. 2002), a clade now referred to as Austrobaileyales. The grade itself was referred to as the "ANITA" grade (Qiu et al. 1999; Figure 2.3), but for reasons given in Chapter 3, we will avoid use of that term in this book. In contrast to previous analyses involving single genes, analyses of combined datasets (Qiu et al. 1999; Barkman et al. 2000; D. Soltis et al. 2000; Zanis et al. 2002, 2003) usually recovered these basal nodes with internal support above 50% (Figure 2.3).

The remaining basal angiosperms consist of six well-supported subclades (Figure 2.3): Chloranthaceae, monocots, Magnoliales, Laurales, Canellales, and Piperales. In the three-gene analysis, relationships among these lineages are essentially unsupported, and Ceratophyllaceae appear as the weakly supported (53%) sister to the eudicots. The addition of more gene sequences indicated, however, that Ceratophyllaceae are the sister group of the monocots and also clarified relationships among many remaining noneudicot lineages (Qiu et al. 1999; Zanis et al. 2002). There is strong evidence for a magnoliid clade of Magnoliales, Laurales, Canellales (Winterales of some recent publications), and Piperales. Within this magnoliid clade, Canellales and Piperales are sister taxa, as are Laurales and Magnoliales. Elevated support for the spine of the tree is also apparent in the Zanis et al. (2002) analysis (see Chapter 3).

Phylogenetic analyses of monocots (Chase et al. 2000, 2005; D. Soltis et al. 2000a; Chase 2004) have indicated some relationships and circumscriptions that differ substantially from many modern classification systems. The classification of Dahlgren et al. (1985) most closely resembles the topologies obtained. One conspicuous difference between DNA topologies and classifications involves the enigmatic genus *Acorus*. Typically placed in Araceae, *Acorus* appears with strong support as sister to all other monocots (see Chapters 3 and 4). Following *Acorus*, an Alismatales clade is sister to all other monocots; Alismatales correspond largely to Aranae and Alismatanae of Dahlgren et al. (1985). Following Alismatales, there are several well-supported clades of monocots (commelinids, Asparagales, Lil-

iales, Dioscoreales, Pandanales), the interrelationships of which were unresolved in the three-gene analysis. The seven-gene analysis of Chase et al. (2004) has resolved most higher-level monocot relationships, nearly all with high bootstrap percentages (see Chapter 4).

Although Lilianae of Dahlgren et al. (1985) are paraphyletic in phylogenetic analyses, they include groups that correspond well to the clades retrieved in molecular phylogenetic analyses (Dioscoreales, Liliales, and Asparagales). The phylogenetic placements of only a few families and genera do not agree with the circumscriptions of these orders by Dahlgren et al. (1985). For example, Iridaceae and Orchidaceae were placed in Liliales by Dahlgren et al. but appear in Asparagales in DNA-based analyses. Although Dahlgren et al. also recognized the presence of a commelinid group, some ordinal circumscriptions differ from those inferred from molecular analyses. Within the commelinid clade of molecular-based trees are several well-supported subclades: Arecales, Zingiberales, Poales, and Commelinales. Arecales and Zingiberales correspond well to the Arecanae and Zingiberanae of Dahlgren et al. (1985), whereas their Bromelianae and Commelinanae are both polyphyletic. Chapter 4 covers the monocots in more detail.

Eudicots

Three-gene analysis established the monophyly of the eudicots with a high degree of confidence (100%; Figure 2.3, indicated by a bold line; P. Soltis et al. 1999b; D. Soltis et al. 2000; see also D. Soltis et al. 1998 and Hoot et al. 1999). In contrast, whereas previous studies based on single genes all found a eudicot clade (e.g., Chase et al. 1993; D. Soltis et al. 1997a; Savolainen et al. 2000a, 2000b), only *rbcL* provided internal support of 50% or higher for this large clade (Chase and Albert 1998; Savolainen et al. 2000a, 2000b). Combining *rbcL* and *atpB* (D. Soltis et al. 1998; Savolainen et al. 2000a) still provided only moderate support for the monophyly of the eudicots (73%).

Within the eudicot clade is a basal grade of putatively ancient lineages (Figure 2.3): Ranunculales, Proteales, Sabiaceae, Buxaceae (including *Didymeles*), and Trochodendraceae (including *Tetracentron*). With the exception of Ranunculales and Proteales, most of these clades consist of relatively few genera and species. Ranunculales are well supported and sister to the remaining eudicots. The relationships among the remaining basal eudicots are unclear (Chapter 5). Proteales and Sabiaceae are each well supported and follow Ranunculales, either as subsequent sisters to all other angiosperms, or as a clade that is sister to all other angiosperms. The monophyly of all remaining eudicots receives strong support (Figure 2.3; indicated by a bold line); two strongly supported small clades, Buxaceae and Trochodendraceae, then appear either as successive sister groups to the core eudicots, or as a clade that is the immediate sister to core eudicots (Chapter 5). This grade of eudicots (sometimes referred to as early-diverging or basal eudicots) agrees with previous analyses (Chase et al. 1993; D. Soltis et al. 1997a; Qiu et al. 1998; Savolainen et al. 2000a, 2000b), although the analysis of Hoot et al. (1999) and the three-gene analyses (P. Soltis et al. 1999b; D. Soltis et al. 2000) are the first to provide internal support of 50% or higher for many of these relationships, particularly along the spine of the tree (Figure 2.3).

The early-diverging eudicots are followed by a strongly supported clade (100%) of core eudicots (Figure 2.3, indicated by bold line). The core eudicot clade comprises seven subclades, each of which received moderate to strong jackknife support in the three-gene analyses (D. Soltis et al. 2000): (1) Gunnerales (Myrothamnaceae and Gunneraceae); (2) Berberidopsidales (Berberidopsidaceae and Aextoxicaceae; ordinal designation was not given to this clade in APG II 2003); (3) Saxifragales; (4) Santalales; (5) Caryophyllales; (6) rosids; and (7) asterids (Figure 2.3). In the three-gene analyses (D. Soltis et al. 2000), Saxifragales appear with weak support as sister to the large rosid clade. However, due to the weak support for this relationship, and also because the use of additional genes indicates other placements of Saxifragales (e.g., D. Soltis et al. 2003a), the most conservative approach is to recognize seven clades of core eudicots. Although the monophyly of each of these subclades receives moderate to high support, relationships among them are not clear in the three-gene analysis.

More-recent analyses have progressed in elucidating relationships among core eudicots, and our summary tree reflects these results. The shortest three-gene trees found that Gunnerales are sister to the remaining core eudicots; recent analyses of a four-gene dataset (including 26S rDNA) provided more support for this relationship (Figure 2.3; D. Soltis et al. 2003a). Several earlier analyses, as well as the three-gene analysis, indicated that Santalales are sister to Dilleniaceae + Caryophyllales. With the addition of entire 26S rDNA sequences, weak jackknife support was found for a sister-group relationship of Santalales and asterids. Resolving relationships among these subclades of core eudicots now represents one of the major remaining problems of angiosperm phylogeny reconstruction.

Within the core eudicots, the rosid, asterid, and Caryophyllales clades identified with DNA data correspond roughly to Rosidae, Asteridae, and Caryophyllidae, respectively, of modern morphology-based classifications; however, the DNA-based versions are all significantly larger than their counterparts in the modern classification schemes. For example, in addition to Rosidae of modern morphology-based classi-

fications, the rosid clade also contains many families of Dilleniidae and Hamamelidae. A few families typically placed in Rosidae in modern classifications actually occur outside of the rosid clade, such as some members of Saxifragales (e.g., Crassulaceae, Haloragaceae, and Saxifragaceae), as well as the small clades of Gunnerales and Berberidopsidales. Santalales, placed in Rosidae by Cronquist and others, form a distinct clade. The asterid clade (referred to as Asteridae s.l. [*sensu lato*] by Olmstead et al. 1992, 1993) is also larger than Asteridae of modern classifications due to the inclusion of several groups of Dilleniidae, such as Ericales, and also some Rosidae, such as Cornales and Apiales (Figure 2.3). Recent studies have provided more resolution of relationships within the asterid clade than is shown in Figure 2.3 (e.g., Albach et al. 2001a; Bremer et al. 2002; see Chapter 9). Caryophyllales are also an expanded version of Caryophyllidae of modern classifications. In addition to core families of Caryophyllidae (e.g., Caryophyllaceae, Aizoaceae, Nyctaginaceae, Cactaceae, and close relatives, plus Plumbaginaceae and Polygonaceae), the DNA-based Caryophyllales clade also includes the carnivorous plant families Nepenthaceae and Droseraceae, as well as several other former families of Dilleniidae such as Tamaricaceae and Frankeniaceae.

Subclasses Hamamelidae and Dilleniidae of modern classifications are both grossly polyphyletic. Families of Hamamelidae are scattered across the three-gene topology, with some occurring as early-diverging eudicots (e.g., Platanaceae and Trochodendraceae; see Chapter 5), another (Eucommiaceae) as part of the asterid clade (see Chapter 9), and several families as part of Saxifragales (e.g., Cercidiphyllaceae, Hamamelidaceae, and Daphniphyllaceae; see Chapter 6). However, most families assigned previously to Hamamelidae (particularly the "higher hamamelids" of some authors) are found in the rosid clade (e.g., Betulaceae, Casuarinaceae, Fagaceae, Juglandaceae, Ulmaceae, and Urticaceae; see Chapter 8), but even these families of Hamamelidae do not form a monophyletic group. Betulaceae, Casuarinaceae, Fagaceae, and Juglandaceae are in Fagales, whereas Ulmaceae and Urticaceae are in Rosales. Families of Dilleniidae also appear in several different places in the eudicot trees, occurring in the asterid, rosid, or Caryophyllales clades. Paeoniaceae, treated as a dilleniid by Cronquist, are placed in Saxifragales.

Circumscriptions of families and orders

DNA sequence data provide strong support for nearly all flowering plant families recognized in modern classifications. However, some long-recognized families are clearly not monophyletic based on broad molecular phylogenetic analyses. More focused molecular and morphological cladistic analyses, using higher taxon density, often provide additional crucial information regarding familial composition. In a few cases, well-known families have been shown to be highly polyphyletic. Both Scrophulariaceae and Liliaceae as traditionally recognized are polyphyletic, consisting of several distantly related lineages. In other instances, phylogenetic analyses have indicated broader circumscriptions of traditional families (e.g., Apocynaceae, Lamiaceae, and Malvaceae; Judd et al. 1994; Judd and Manchester 1997; Bayer et al. 1999). Asclepiadaceae are nested within Apocynaceae (Judd et al. 1994). Lamiaceae are monophyletic only with the inclusion of some genera of Verbenaceae, such as *Callicarpa* (Wagstaff and Olmstead 1997). The distinction of Malvaceae from Bombacaceae, Tiliaceae, and Sterculiaceae has not been supported by recent phylogenetic analyses (e.g., Judd and Manchester 1997; Bayer et al. 1999), which indicated instead that a single broadly defined family, Malvaceae, was more appropriately recognized. Other angiosperm families also now known to be nonmonophyletic include Sapindaceae, Flacourtiaceae, Euphorbiaceae, Chenopodiaceae, Ericaceae, and Saxifragaceae (discussed in Chase et al. 2000; D. Soltis et al. 2000; Savolainen et al. 2000a, 2000b; see also APG 1998 and APG II 2003).

Although many families of angiosperms have received strong support in DNA-based analyses, most orders have not. Well-supported clades of families often show little agreement with orders recognized in modern classification systems (e.g., Dahlgren 1980; Cronquist 1981; Takhtajan 1987, 1997; Thorne 1992a, 1992b, 2001). For example, Saxifragales as delineated by DNA sequence analyses are a highly eclectic assemblage of taxa previously placed in three subclasses (*sensu* Cronquist 1981). The Ericales clade recovered in phylogenetic analyses contains families that were placed in several orders of subclasses Dilleniidae and Rosidae. Rosales of molecular analyses bear only slight similarity to Rosales of modern classifications. In addition to the presence of the name-bearing family, Rosaceae, the Rosales clade also contains several families of Hamamelidae, as well as additional families of Rosidae previously placed in other orders. Some noteworthy exceptions of instances in which ordinal circumscriptions in modern classifications (largely) agree with those in DNA-based topologies include Myrtales, Sapindales, and Zingiberales.

A reclassification based on sequence data

The high degree of resolution and internal support obtained for the angiosperms through molecular investigations indicated the need for a revised higher-level classification. Although many families of angiosperms received strong support in DNA-based analyses, the monophyly of other families was challenged, and higher levels in the taxonomic hierarchy were not support-

ed. As molecular phylogenetic studies continued at a rapid pace, different names were used by different laboratories for the same groups of angiosperms. The rapidly growing need for a standardized system of names that corresponded to the DNA-based clades prompted the Angiosperm Phylogeny Group (APG), a group of angiosperm systematists, to provide a revised higher-level (family and above) system of classification for flowering plants (APG 1998). The enormous progress that angiosperm systematists have continued to make since the APG (1998) classification has, as anticipated, prompted an updating of that classification (APG II 2003), which is discussed in Chapter 10.

The angiosperms are the first major group of organisms to be reclassified on the basis of DNA sequence data. As with most of the large molecular phylogenetic analyses, this reclassification was conducted collaboratively by many systematists, some of whom were molecular systematists and others who were classically trained. This collaborative approach is a major departure from traditional and modern classifications in which one or a few "experts" provided systematic treatments of large groups such as the angiosperms. In contrast, higher-level classification of the angiosperms now involves large collaborative networks of systematists, and this trend (at least at higher levels) will continue. This approach may, in fact, serve as a useful model for investigators of other large problematic groups of organisms. For example, a revised classification of the ferns is being developed in a manner similar to the APG format used for angiosperms (A. Smith et al., in prep.).

Implications for the study of model organisms

Examples of the importance of a phylogenetic underpinning to diverse areas of research abound. One obvious example is the value of placing model organisms in the appropriate phylogenetic context to obtain a better understanding of both patterns and processes of evolution. The fact that tomato, *Lycopersicon esculentum*, is actually embedded within a well-marked subclade within *Solanum* (and therefore is more appropriately referred to as a species of *Solanum, S. lycopersicon*) is a powerful statement (Spooner et al. 1993). This phylogenetic result is important to geneticists, molecular biologists, and plant breeders in that it points to a few close relatives of *Solanum lycopersicon* (out of a genus of several hundred species) as focal points for comparative research.

A phylogenetic perspective also provides unique opportunities for comparative genomics (e.g., D. Soltis and Soltis 2000, 2003; Walbot 2000; Kellogg 2001; A. Hall et al. 2002; Mitchell-Olds and Clauss 2002; Pryer et al. 2002; Doyle and Luckow 2003). Excellent opportunities for comparative genomics are afforded by families that are home to model organisms, such as Brassicaceae (*Arabidopsis*; A. Hall et al. 2002), Plantaginaceae (*An-*

tirrhinum; Reeves and Olmstead 1998), Fabaceae (Doyle and Luckow 2003), and Poaceae (*Zea, Oryza*; Kellogg 2001). Importantly, however, the most fruitful phylogenomic comparisons will be those made in light of the best tree, and our phylogenetic understanding of some groups containing model organisms is much better than others.

Some of the new familial circumscriptions indicated by molecular phylogenetic analyses have important implications for the study of model organisms. Brassicaceae (containing *Arabidopsis*) are only monophyletic with the inclusion of Capparaceae (Judd et al. 1994; Rodman et al. 1998). Thus, *Cleome, Capparis,* and most former Capparaceae, together with traditional Brassicaceae (e.g., *Brassica, Arabidopsis*), form an expanded family Brassicaceae (APG II 2003). One alternative treatment is to recognize three separate families, Capparaceae, Cleomaceae, and Brassicaceae (J. Hall et al. 2002; see Chapter 8). This expanded family Brassicaceae is part of a well-supported clade, Brassicales (also referred to as the glucosinolate clade), that also contains Akaniaceae, Batidaceae, Caricaceae, Limnanthaceae, Tropaeolaceae, and several other families. This knowledge of phylogenetic relationships should be used in efforts to extend the knowledge garnered from detailed genomic and developmental analyses of *Arabidopsis* to close relatives in Brassicaceae, as well as to other Brassicales (cf. A. Hall et al. 2002; Mitchell-Olds and Clauss 2002).

The snapdragon, *Antirrhinum majus,* represents one of the best model systems for the study of floral developmental genetics. Although *Antirrhinum* has long been placed in Scrophulariaceae, the family is now known to be polyphyletic, comprising perhaps four distinct lineages (Olmstead and Reeves 1995; Young et al. 1999; Olmstead et al. 2001; Chapter 9). One lineage (now Orobanchaceae s.l.; APG II 2003) contains the parasitic members of the family (*Pedicularis, Castilleja*) and Orobanchaceae s.s., which are also parasitic. A second lineage (now Schrophulariaceae s.s.) contains *Scrophularia* and *Verbascum*. A third lineage (now Plantaginaceae) contains many well-known "scrophs," such as *Antirrhinum, Digitalis, Veronica,* and genera usually placed in their own families (*Plantago, Callitriche,* and *Hippuris*). *Mimulus,* another former "scroph," may be best placed in Phrymaceae, although bootstrap support for this relationship is low (Beardsley et al. 2001; Beardsley and Olmstead 2002). One of the next challenges in evaluating current models of floral development is the intensive study of a clade in which members exhibit extensive floral variation. The clade to which *Antirrhinum* belongs represents such a group. Donoghue et al. (1998) and Reeves and Olmstead (1998) have extrapolated from what is known about the genetics of floral development in *Antirrhinum* to other asterids, doing so in a phylogenetic context (see Chapter 9).

Both broad and focused phylogenetic analyses also have implications for the study of close relatives of crops that have become the focus of considerable genetic research, such as cotton (*Gossypium*), now known to be part of an expanded Malvaceae (see Chapter 8). The closest relatives of Poaceae, the grass family, have been identified (Flagellariaceae, Joinvilleaceae, and Ecdeiocoleaceae), and phylogenetic relationships within Poaceae are now generally well resolved (GPWG 2000; see Chapter 4). Numerous legumes are of economic importance, and genome structure and evolution have been studied intensively in several species. The closest relatives of Fabaceae have long been debated; however, broad phylogenetic analyses point to a clade of Quillajaceae, Polygalaceae, and Surianaceae as sister to Fabaceae (see Chapter 8). Additional examples of phylogenetic studies that encompass model organisms in Solanaceae and Asteraceae are discussed in D. Soltis and Soltis (2000, 2003).

Sequence Variation: General Patterns and Rapid Radiations

One of the most striking aspects of the three-gene tree (D. Soltis et al. 2000) and other gene trees for the angiosperms (e.g., Chase et al. 1993; D. Soltis et al. 1997a, 1997b; Savolainen et al. 2000a, 2000b) is the highly uneven pattern of branch lengths (reviewed in Chase et al. 2000b; D. Soltis et al. 2000). Groups that are highly divergent for plastid genes are also typically highly divergent for nuclear genes. This pattern not only occurs with slowly evolving plastid and nuclear genes such as *atpB, rbcL, ndhF,* and 18S rDNA but also with noncoding regions such as plastid *trnL-F* (Sheahan and Chase 2000) and nuclear ITS (Whitten et al. 2000). Groups with unusual or atypical life-history strategies (e.g., parasites, mycotrophs, xerophytes, aquatics) appear to have higher rates of sequence divergence than close relatives with more typical or "standard" life histories (e.g., Nickrent et al. 1998). Examples from the three-gene trees for angiosperms (see D. Soltis et al. 2000) include *Ceratophyllum* relative to all other angiosperms; *Burmannia* (Burmanniaceae) relative to *Tacca* and *Dioscorea* (Dioscoreaceae); *Stackhousia* and *Siphonodon* (Celastraceae); *Marathrum* (Podostemaceae) and Clusiaceae; *Hydrostachys* and Hydrangeaceae; *Callitriche* and other members of its clade; and *Nymphoides* and *Menyanthes* (Menyanthaceae). This same pattern is also seen throughout Caryophyllales, in which *rbcL* can be used at the species level within Droseraceae and Plumbaginaceae (Williams et al. 1994; Lledó et al. 1998).

It appears, therefore, that rates of molecular evolution are not consistent across all lineages of the angiosperms, but may be much faster in some taxa, particularly in those that have undergone specialization or modification of vegetative organs. These unexpected findings provide the impression that both morphological and molecular evolution sometimes experience parallel alterations of evolutionary rates.

In contrast to clades exhibiting noticeable bursts of molecular evolution, other taxa appear to display a marked slowdown in rates of molecular evolution. Such taxa have been termed "molecular living fossils," and examples have been provided in vascular plants (e.g., tree ferns; P. Soltis et al. 2002), basal angiosperms including Winteraceae (Suh et al. 1993), and the early-diverging eudicots *Platanus* and *Nelumbo* (Sanderson and Doyle 2001). In addition to rate accelerations and decelerations, there is also evidence that change at some loci, in some taxa, is nearly clock-like or "quasi-ultra-metric" (Albert et al. 1994). The mixture of long and short branches that marks the major groups of angiosperms appears to provide evidence for several episodes of rapid radiations.

Two deep-level radiations are suggested for the angiosperms. The root of the angiosperms is now clear, with the initial lineages well supported: Amborellaceae, Nymphaeaceae, and Austrobaileyales (Figure 2.3). In contrast, the remaining clades of basal angiosperms are each well supported (e.g., Chloranthaceae, Ceratophyllaceae, monocots, Magnoliales, Laurales, Canellales, and Piperales), but relationships among these lineages are much more difficult to tease apart—in fact with three genes and nearly 5000 bp of sequence data, these lineages essentially form a polytomy. Even with more than 15,000 bp of sequence data (Zanis et al. 2002), relationships among these lineages of basal angiosperms do not always receive strong internal support (see Chapter 3). Phylogenetic trees indicate an early radiation of basal angiosperms following the origin of the angiosperms (Figure 2.3). In addition, the fact that the first branches of the topology are well supported and not species-rich and are followed by several species-rich clades indicates that the initial explosive radiation of the angiosperms did not coincide directly with the origin of flowering plants, but likely occurred later (cf. Mathews and Donoghue 1999; P. Soltis et al. 2000). That is, after the origin of the angiosperms, several basal lineages appeared, and Amborellaceae, Nymphaeaceae, and Austrobaileyales are the extant remnants of these early lineages. A radiation of basal angiosperm lineages subsequently occurred, corresponding to the origin of the remaining noneudicot clades (e.g., monocots, Chloranthaceae, Ceratophyllaceae, and a magnoliid clade of Laurales, Magnoliales, Canellales, and Piperales) and perhaps even the eudicots themselves. The monocots themselves also exemplify this mixture of long and short branches and appear to represent a rapid radiation (see Chapter 4).

This same basic pattern for basal angiosperms is repeated in the eudicots, indicating a second major deep-level radiation of flowering plants (Figure 2.3). A grade of early-diverging (or basal) eudicots is identified, with the relationships among at least some of these lineages well resolved and strongly supported (Figure 2.3). Following this grade are the core eudicots, which include most of the angiosperms. The core eudicot clade is well supported, and it, in turn, consists of several well-supported major lineages, including the rosids, asterids, Santalales, Caryophyllales, and Saxifragales. Although the monophyly of each subclade of core eudicots is strongly supported, relationships among these core eudicot lineages are poorly resolved and weakly supported. In the three-gene analysis, relationships among lineages of core eudicots are essentially a polytomy (Figure 2.3), indicating a second major radiation within the angiosperms. This hypothesized radiation is in general agreement with the fossil record. The oldest eudicot fossils represent basal lineages (Platanaceae and Buxaceae). The fossil record also suggests an uneven distribution of species diversity across the major clades of eudicots (Davies et al. 2004), with the most species-rich groups known only from relatively young fossils. These results indicate that a large proportion of eudicot diversity is the result of recent radiations (Magallón et al. 1999).

The mixture of long and short branches noted for both basal angiosperms and eudicots continues for those clades typically recognized at the level of orders and families (Chase et al. 2000; D. Soltis et al. 2000). Simple inspection of the three-gene tree reveals that this trend is common (P. Soltis et al. 2000, 2004). For example, the branch to Malpighiales is long, and within this clade there are also several well-supported lineages (most recognized as families), but relationships among these families are largely unresolved (Figure 2.5). As a result, when only those clades within Malpighiales that receive jackknife or bootstrap support above 50% are retained, the resulting tree is essentially a huge polytomy (Figure 2.5). The same pattern can also be seen in Lamiales, Ericales, and Saxifragales and is also observed within some clades corresponding to families—Asteraceae (Jansen and Kim 1996), Orchidaceae (Cameron et al. 1999), Crassulaceae (Mort et al. 2000), and Rhamnaceae (Richardson et al. 2000).

The general phylogenetic pattern that occurs repeatedly in the flowering plants is suggestive of episodic radiations (resulting in short branches) that produce lineages that exist for long periods (creating the long branches), followed by another phyletic burst, and so on. It will be worthwhile to see if similar patterns exist for other large groups of organisms. The data for angiosperms, for example, contrast with the red algae (Freshwater et al. 1994), in which the pattern of change in *rbcL* across all branches gives the appearance of slow and clock-like lineage production without the periodic "starbursts" present in the angiosperm tree. This difference could be due simply to differences in patterns

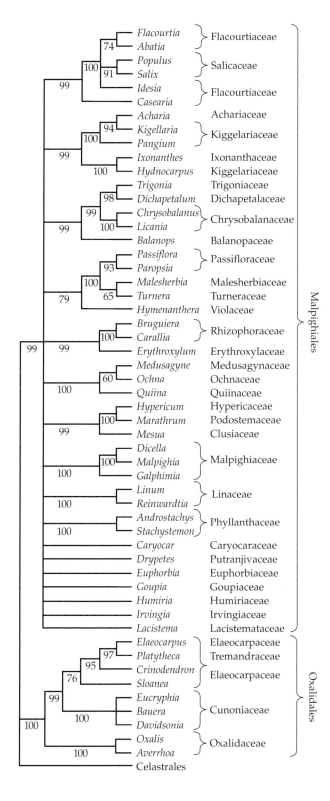

FIGURE 2.5 Possible evidence for rapid radiation. Summary tree for Malpighiales showing only those clades that received jackknife support above 50%; the resulting tree is a large polytomy. (Modified from Figure 8A in D. Soltis et al. 2000.) Family circumscriptions are pre-APG II.

an explanation (Chase et al. 2000b; D. Soltis et al. 2000). One suggestion is that most recent species-level radiations might be fundamentally different from deep-level radiations (Chase et al. 2000b). Species differences in a clade that represents a recent radiation are in fairly simple quantitative characters. In contrast, the features that unite major clades involve more fundamental characters such as chemical pathways, anatomical characters, embryology, and so on. Recent evidence for this basic pattern of the involvement of different suites of characters for species-level radiations and older radiations, resulting in the decoupling of macroevolution and microevolution, comes from the work of Kellogg (2000, 2002) on Poales.

One mechanism for this decoupling could be saltational or quantum speciation (*sensu* Grant 1981; Eldredge and Gould 1972; Stanley 1975, 1979). Bateman and DiMichele (1994) argued for the importance of saltational evolution, which is not only rapid but radical in its genotypic and phenotypic effects, and focused part of their discussion on the origin of long branches, suggesting that developmentally correlated characters have changed simultaneously during evolution. In other words, saltation might explain certain types of long branches. Topologies now available, with their uneven pattern of nucleotide change, are perhaps the "phylogenetic footprint" of these fundamental differences between species-level and deep-level radiations.

Timing of Diversification Events

The age of the angiosperms and the timing of major diversification events within this clade, such as the origin of the eudicots, monocots, and asterids, continue to generate debate. The dating of fossils is an obvious and important means of obtaining such age estimates (Magallón et al. 1999), but the identity of the fossil, its phylogenetic position, and the age of the stratum are all subject to error. Furthermore, gaps in the fossil record may have serious effects on inferences of divergence times.

As an alternative to fossils, estimates of DNA sequence divergence (i.e., molecular clocks) can also be used to estimate divergence times, given certain assumptions about evolutionary rates and a means of calibration such as fossil or geologic evidence. Several attempts have been made to use DNA sequence data and a molecular-clock assumption (Zuckerkandl and Pauling 1962) of evolution to estimate the date of major

of speciation between marine and terrestrial environments or might reflect differences in the complexity of both the organisms and their interactions with other organisms.

Whatever the reasons, an uneven pattern of nucleotide change exists for the angiosperms and requires

Era	Period	Epoch
CENOZOIC (65)	Quaternary (2)	Holocene (0.01) Pleistocene (1.8)
	Tertiary (65)	Pliocene (5.5) Miocene (24.6) Oligocene (38) Eocene (55) Paleocene (65)
MESOZOIC (248)	Cretaceous (144) Jurassic (206) Triassic (250)	

PALEOZOIC (590)	Permian (290)
	Carboniferous
	  Pennsylvanian (314)
	  Mississippian (360)
	Devonian (409)
	Silurian (439)
	Ordovician (500)
	Cambrian (540)	

PRECAMBRIAN (4560)		

Period	Epoch/Stage	Age Mya
Cretaceous	Maastrichtian	71.0 ± 2
	Campanian	84.0 ± 3
	Santonian	86.0 ± 4
	Coniacian	90.0 ± 3
	Turonian	94.0 ± 2
	Cenomanian	99.0 ± 2
	Albian	112.0 ± 2
	Aptian	121.0 ± 2
	Barremian	127.0 ± 3
	Hauterivian	132.0 ± 8
	Valanginian	137.0 ± 8
	Berriasian	144.0 ±13

FIGURE 2.6 Overview of the geologic time scale, with details for the Cretaceous Period. Numbers in parentheses indicate the age in millions of years at which the stage or period began. (Modified from Harland et al. 1989; Pan Terra Inc., 2000, www.wmnh.com. A Correlated History of Earth.)

cladogenic events in the angiosperms (e.g., Ramshaw et al. 1972; Martin et al. 1989, 1993; Wolfe et al. 1989; Brandl et al. 1992; Goremykin et al. 1997; Bremer 2000; Sanderson et al. 2000; Sanderson and Doyle 2001; Wikström et al. 2001). These molecular clock analyses typically have indicated older dates for the origin of the angiosperms, as well as for other land plants (e.g., Heckman et al. 2001), than those based on the fossil record. One of the earliest molecular analyses, for example, obtained an estimate of 350 to 420 Mya (Late Silurian to Mississippian; Ramshaw et al. 1972). Wolfe et al. (1989) and Laroche et al. (1997) estimated the age of the angiosperms as 200 Mya (Early Jurassic; Figure 2.6). Other studies have provided more recent estimates: 170 to 158 Mya (Wikström et al. 2001), 160 Mya (Goremykin et al. 1997), 140 to 190 Mya (Sanderson and Doyle 2001), and 127 and 135 Mya (P. Soltis et al. 2002). Some applications of sequence divergence and the molecular clock involved the dating of the monocot–dicot split (e.g., Wolfe et al. 1989; Martin et al. 1989, 1993; Clegg 1990), a split that is not the deepest within the angiosperms if we interpret dicots to mean eudicots.

Most of the methods used to generate this range of divergence times assume constant rates of molecular evolution across lineages and through time. When evolutionary rates are constant among lineages, the problem of estimating divergence times is straightforward,

even if the methodology to do so is not (see reviews by Sanderson 1998, Sanderson et al. 2004). If evidence is available for calibration, then absolute rates of evolution and divergence times can be estimated jointly by any of several methods. However, even under apparently constant rates, estimated divergence times might be wrong because of inaccurate calibration points, mistaken inferences of phylogeny, variance in estimates of genetic divergence, and undetected variation in evolutionary rates. Assuming rate constancy when evolutionary rates truly vary among lineages would obviously lead to inaccurate estimates of divergence times, even when phylogenetic branching patterns are correctly inferred and calibrating dates have been applied correctly. Unfortunately, rate constancy is often assumed, even when it has not been evaluated, and tests of rate constancy are generally not performed before estimating divergence times, despite the availability of many such methods, ranging from pair-wise (e.g., Sarich and Wilson 1967; Wu and Li 1985) to global tree-based (e.g., Langley and Fitch 1974; Felsenstein 1988) methods. Of course, tests of rate constancy have potential problems of their own, but we will not discuss them here; see review by Sanderson (1998).

Most phylogenetic trees for angiosperms at any hierarchical level portray unequal branch lengths, implying unequal rates of evolution among lineages (Chase et al. 2000b; D. Soltis et al. 2000; Sanderson and Doyle 2001). Thus, the assumption of rate constancy is likely violated in most estimates of divergence times in angiosperms. Violation of this assumption may be one of the major sources of error in many estimates of divergence times in angiosperms, such as the age of the angiosperms themselves or the age of the monocots. Therefore, methods that do not require constant rates of evolution across lineages will likely be more valuable for estimating the timing of major divergences in the angiosperms. Several such methods have recently been described—nonparametric rate smoothing (Sanderson 1997), penalized likelihood (Sanderson 2002), Bayesian analysis (e.g., Kishino et al. 2001; Thorne and Kishino 2002), and "PATH" (Britton et al. 2002). Because these methods have been developed only recently, most have been applied to only a small number of cases, and their relative effectiveness has not yet been thoroughly assessed.

Sanderson's (1997) nonparametric rate smoothing method has been applied in several cases in which evolutionary rates appeared—or were demonstrated—to vary among lineages (e.g., Sanderson and Doyle 2001; Wikström and Kenrick 2001; Wikström et al. 2001, 2003; P. Soltis et al. 2002). Rather than assuming a molecular clock and rate constancy, the nonparametric rate smoothing method allows the rate to change but assumes that such changes are autocorrelated—that rate change is inherited from an ancestral lineage by the immediate descendants of that lineage (Sanderson 1997). There is some evidence for this pattern of evolution from phylogenetic trees (e.g., Chase et al. 2000b; D. Soltis et al. 2000); some lineages, such as Santalales and Droseraceae, appear to have higher rates of molecular evolution than other lineages for all genes used. Applying the nonparametric rate smoothing method to 36 *rbcL* sequences, Sanderson (1997) obtained an estimate for the age of the angiosperms of 165 Mya (Middle Jurassic).

Sanderson and Doyle (2001) analyzed the sources of error in calculating divergence times and estimated confidence intervals for the age of the angiosperms using 18S rDNA and *rbcL* sequences. Their results clearly indicated inequality of rates. The branches to several taxa, such as *Platanus* and *Nelumbo,* are short. If these short branches are the result of a slowdown in molecular evolution, then these taxa may be "living fossils" from both a molecular and morphological standpoint. Similar results have been forthcoming for several fern lineages in analyses of vascular plants (tracheophytes); tree ferns and *Marattia/Angiopteris* also seem to be molecular and morphological living fossils (P. Soltis et al. 2002).

Wikström et al. (2001, 2003) used the nonparametric rate smoothing approach to estimate divergence times in the angiosperms with the 567-taxon, three-gene dataset (P. Soltis et al. 1999b; D. Soltis et al. 2000). Wikström et al. calibrated the tree by fixing the split between Fagales and Cucurbitales at 84 Mya by the occurrence

of the fossils *Protofagacea* and *Antiquacupula;* their chronogram showing age estimates for major divergences within the angiosperms is shown in Figure 2.7. From their analysis, angiosperms are estimated to have originated in the Middle to Late Jurassic (179–158 Mya), with the eudicots appearing in the Late Jurassic to Early Cretaceous (147–131 Mya) (Figure 2.7). These estimates are similar to other DNA sequence–based estimates (Brandl et al. 1992; Goremykin et al. 1997; Sanderson and Doyle 2001; P. Soltis et al. 2002). However, despite the potential of nonparametric rate smoothing for estimating divergence times when evolutionary rates vary among lineages, simulation studies have shown that it is difficult to estimate divergence times accurately when rates vary (Sanderson 1997). Furthermore, variation in rate within a lineage is difficult to evaluate. Thus, even the estimates derived using this method should be viewed as rough approximations. Nonetheless, it is noteworthy that most of the recent attempts to estimate the age of the angiosperms place the crown node at 145 to 208 Mya (Sanderson et al. 2004). Recent studies have evaluated the performance of nonparametric rate smoothing relative to penalized likelihood and Bayesian methods (both allow rates to change across the tree and permit multiple calibration points); these methods provide similar estimates (Bell et al., in press).

Although there are fossil-based claims of a pre-Cretaceous origin and diversification of the angiosperms (e.g., Cornet 1989; Sun et al. 1998), molecular-based estimates for both the early diversification of the angiosperms and the origin of the eudicots are older—and sometimes substantially older—than current fossil-based estimates. The earliest fossils generally accepted as angiosperms are from the earliest Cretaceous and are of Valanginian–Hauterivian age, 137 to 132 Mya (see Figure 3.2 in Chapter 3; Trevisan 1988; Hughes 1994; Crane et al. 1995; Brenner 1996; Friis et al. 2000). These early remains are primarily pollen with occasional leaf fossils (Crane et al. 1995). The eudicots date from the Barremian at approximately 125 Mya (Friis et al. 1999). No unambiguous fossil evidence supports a pre-Cretaceous origin of flowering plants (Crane et al. 1995; Friis et al. 2000). Although the angiosperm fossil *Archaefructus liaoningensis* was proposed to be from the Late Jurassic (Sun et al. 1998), recent evidence has placed it at a much younger period, likely of Barremian–Aptian age (Swisher et al. 1999).

The fossil evidence appears to support the hypothesis that the time intervals separating basal branches are short, meaning that the major angiosperm lineages diverged in a short time span (Crane et al. 1995; Friis et al. 2000). Molecular data, in contrast, placed substantial nucleotide change on these branches (Wikström et al. 2001; Figure 2.7), and both angiosperm and eudicot origins are pushed back in time. Further illustrating this point, Magallón et al. (1999) provided a comprehensive

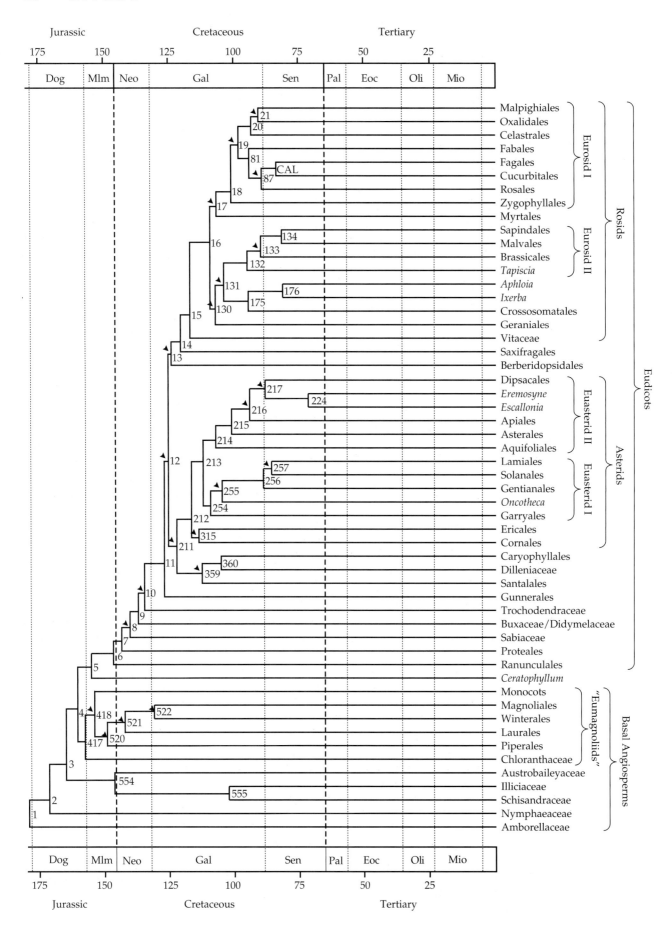

◀ **FIGURE 2.7** Estimated timing of major diversification events in angiosperms. "Eumagnoliid" is used here following D. Soltis et al (2000); the composition of the magnoliid clade is now recognized to be much narrower (see Chapters 3 and 10). (Modified from Wikström et al. 2001.) CAL = calibration point. Numbers refer to nodes (see Table 2.1). Arrowheads indicate nodes with less than 50% jackknife support. Vertical dark, dashed lines indicate interval during which most major clades arose.

summary of the early eudicot diversification times based on the fossil record. Wikström et al. (2001) compared their molecular-based estimates with these fossil-based estimates. A general pattern in these comparisons is that the molecular-based estimates are older than the fossil-based estimates (Table 2.1). Several authors have reviewed possible reasons for these discrepancies (e.g., Sanderson and Doyle 2001; Wikström et al. 2001; P. Soltis et al. 2002). Even if the branch lengths on the molecular trees are correct, the inferred rates of sequence evolution may not be constant. This would imply that the rapid morphological diversification of the angiosperms was accompanied by a rapid burst of molecular change.

It is clear that both fossil and molecular estimates of divergence times require additional study, but the investigations of Wikström et al. (2001), Sanderson and Doyle (2001), and the more recent analysis based on a supertree for angiosperms (Davies et al. 2004; see Appendix) provide useful starting points.

Future Studies

Considerable progress has been made in a short period of time in reconstructing the general picture of angiosperm phylogeny. Clades recognized as families have been circumscribed and grouped into larger, more inclusive clades (orders of APG 1998 and APG II 2003). The largest clades of angiosperms have also been discerned (rosids, asterids, eudicots, etc.), and many deep-level relationships have been elucidated. This is not to say, however, that phylogenetic work in the angiosperms is near completion—far from it. As stressed throughout this book, some crucial deep-level questions remain, such as relationships among eudicot lineages. Obviously, intergeneric relationships in most families are unclear, and interspecific relationships within many genera are uncertain. Furthermore, many currently recognized genera most likely are not monophyletic.

More characters and more taxa
One of the great lessons of angiosperm phylogenetics has been the importance of increased sampling of both taxa and characters to resolve relationships (e.g., Chase and Cox 1998; D. Soltis et al. 1998). The importance of adding characters is now obvious, but the need for ad-

equate taxon sampling is a lesson that is often forgotten—the analysis of molecular datasets that include many characters but only one or a few exemplars per major group is bound to give spurious estimates of phylogeny (e.g., Chase et al. 1993; D. Soltis et al. 1998, 2000; D. Soltis and Soltis 2004; D. Soltis et al., 2004).

Ancient, rapid radiations may be particularly difficult to resolve (e.g., Sanderson and Donoghue 1994; Fishbein et al. 2001). Clarifying some relationships and doing so with high internal support will, in some cases, require the addition of perhaps many more characters—resolution of relationships among the basal angiosperms, for example, involved analyzing more than 15,000 bp per taxon (e.g., Qiu et al. 1999; Zanis et al. 2002). Although these studies have provided tremendous insights, some questions still remain, such as the relationship of monocots and Chloranthaceae to each other and to other angiosperms. Perhaps the most significant deep-level question of phylogeny that remains in the angiosperms involves relationships among the major lineages of core eudicots. Although these clades (rosids, asterids, Caryophyllales, Santalales, Saxifragales, Gunnerales, Berberidopsidales) are each strongly supported, relationships among them are still unclear, owing, perhaps, to a rapid radiation. Similarly, relationships are still uncertain within many of the clades recognized as orders (e.g., Malpighiales, Lamiales, Zingiberales, Saxifragales); these lineages probably also reflect episodes of rapid radiation.

Collaborations among angiosperm systematists are continuing in an effort to resolve many of the remaining questions of flowering plant phylogeny. The genes *ndhF, matK,* 26S rDNA, *atp1, cox1,* and *matR* are currently being sequenced widely across the angiosperms in an effort to duplicate, or even expand, coverage used in the three-gene, 567-taxon analysis (P. Soltis et al. 1999b; D. Soltis et al. 2000) and thus create a larger, multigene matrix. Large, multigene datasets have already been constructed for large subgroups of angiosperms, such as asterids (Albach et al. 2001a; Bremer et al. 2002), basal angiosperms (Barkman et al. 2000; Zanis et al. 2002), and monocots (Chase et al. 2004).

Morphology and other non-DNA characters
Continuing efforts are needed to compile large datasets of morphological, chemical, and anatomical characters and ultimately to combine these with DNA matrices in phylogenetic analyses. The large non-DNA datasets of Doyle and Endress (2000) for basal angiosperms and of Nandi et al. (1998) for angiosperms as a whole are good examples of the types of matrices that can be compiled; other examples include recent analyses of Ericales (Kron and Judd 1997; Kron et al. 1999, 2002). These studies also illustrate one problem with these ventures—the non-DNA characters with potential to resolve phylogeny have generally not been studied in all of the taxa for

TABLE 2.1   Comparison of taxon ages estimated using nonparametric rate smoothing (Wikström et al. 2001) and ages estimated from fossils (Magallón et al. 1999)

Taxon	Estimated Age [a]	Estimated Age (Mya)[a]	Specific Fossil-Based Age[b] (Mya)	Implied Fossil-Based Age (Mya)
Ranunculales	Barremian–Tithonian (node 6)	131–147	69 (Menispermaceae)	118 (Trochodendrales)
Sabiaceae	Barremian–Valanginian (node 8)	128–140	69	
Buxaceae	Aptian (node 393)	113–124	104	
Trochodendraceae	Aptian–Hauterivian (node 10)	123–135	118	
Caryophyllales	Albian (node 360)	104–111	83 (Amaranthaceae)	
Amaranthaceae	Chattian–Bartonian (node 382)	28–40	83	
Saxifragales	Albian–Aptian (node 14)	111–121	89	
Sapindales	Campanian–Santonian (node 134)	80–84	67 (Rutaceae, Aceraceae)	69 (Malvales)
Malvales	Campanian–Santonian (node 134)	80–84	69	
Myrtales	Albian (node 17)	100–107	84 (Combretaceae)	
Urticales	Maastrichtian (node 97)	65–67	69	
Malpighiales	Coniacian–Cenomanian (node 21)	88–91	58 (Euphorbiaceae)	
Cornales	Albian–Aptian (node 315)	106–114	69 (mastixioid taxa)	
Ericales	Albian–Aptian (node 315)	106–114	69	
Apiales	Santonian–Turonian (node 225)	85–90	69 (Araliaceae)	
Dipsacales	Santonian–Turonian (node 217)	85–90	53 (Caprifoliaceae)	
Asterales	Cenomanian–Albian (node 215)	94–101	29	
Solanales	Campanian–Santonian (node 257)	82–86	53	
Lamiales	Maastrichtian (node 259)	71–74	37 (Oleaceae)	
Gentianales	Santonian–Cenomanian (node 256)	85–97	53	
Dilleniaceae	Albian (node 360)	104–111	53	
Santalales	Albian–Aptian (node 359)	111–118	53 (Olacaceae)	
Gunneraceae	Albian–Aptian (node 391)	108–118	89	

[a] Age estimate of Wikström et al. (2001). Node numbers correspond to those on the chronogram (see Figure 2.7) of Wikström et al. (2001).
[b] Age estimated from fossils (Magallón et al. 1999) and the taxon on which the minimum age is based.

which DNA sequence data have been amassed. Thus, coordinated efforts are needed to compile the large datasets required.

In addition, efforts are needed to discover new morphological characters that may be of phylogenetic value. Nonmolecular synapomorphies have typically not been recognized even for those orders that are well supported by DNA sequence data. There are exceptions, such as the glucosinolate biosynthetic pathway in Brassicales (Rodman et al. 1993, 1998) and the presence of ferulic acid in the cell walls of commelinids (Chase et al. 1995b). There are likely additional non-DNA characters of systematic significance awaiting discovery that are synapomorphies for orders or for other larger clades. These may be chemical, developmental, or micromorphological features. Only with more research will such characters be obtained. It is probably fair to say that attention to the formidable problems of morphological character analysis, as well as to other non-DNA characters, has tended to wane during the past decade in the understandable enthusiasm for molecular systematics. It is time to reverse that trend and encourage integrative training and research in the analysis of both molecular and non-DNA characters (Endress et al. 2000a; Endress 2003a, 2003b).

Integrating fossils

Molecular data have provided a robust phylogenetic assessment for extant angiosperms. Concomitantly, in recent years, paleobotanists have greatly improved our understanding of early angiosperm diversity (reviewed in Dilcher 2000; Friis et al. 2000) and great progress has been made in the following areas: (1) development of techniques for investigating early angiosperm remains, (2) increased collection and description of fossil material of early angiosperms, and (3) an intense analysis of these fossil data with special reference to floral morphological characters relative to time of occurrence (Crane et al. 1995; Friis et al. 1999, 2001; Magallón-Puebla et al. 1999; Dilcher 2000). For example, Lower Cretaceous sediments from Portugal recently yielded 105 different kinds of flowers with 13 associated pollen types through the study of mesofossils—those small floral buds, fruits, seeds, flowers, or plant parts recovered by sieving sediments (Friis et al. 1999). Mesofossils have also been studied from Cretaceous sediments in other locations, and many new taxa have been described. The macrofossil record has yielded other early angiosperm fossils, such as *Archaefructus* (Sun et al. 1998, 2002) in China.

Integrating fossils into the tree of living taxa remains essential for understanding both the origin of extant angiosperm groups and the origins of their structures (Doyle 1998a, 1998b). However, such attempts to integrate fossils and extant taxa in phylogeny reconstruction have been rare (e.g., Albert et al. 1994; Keller et al.

1996; Magallón-Puebla et al. 1996; Crepet and Nixon 1998; Eklund 1999; Springer et al. 2001; Sun et al. 2002; Friis et al. 2003; Eklund et al. 2004). Although angiosperm systematists and paleobotanists have much in common and each group has made major strides in the past decade, there has, until recently, been surprisingly little communication and integration of data between the two areas. Systematists are often unaware of the significance of fossil discoveries and of the characterizations of these fossils, and paleobotanists may not appreciate the excellent phylogenetic framework currently available for living angiosperms (but see Crane et al. 1995; Crepet and Nixon 1998; Doyle 1998a, 1998b; Doyle and Endress 2000).

The paucity of attempts to integrate fossils into a phylogenetic framework can also be attributed to a necessary reliance on morphology. Whereas a morphological matrix for living taxa into which fossils can be integrated is a necessity, attempts to formulate such matrices for angiosperms are relatively recent and, as noted, still incomplete (Nandi et al. 1998; Doyle and Endress 2000). Other factors responsible for the lack of interdisciplinary work include the difficulty in characterizing many fossils and the analytical issues that must be considered when integrating fossils into a phylogenetic framework. The timing is now appropriate to develop a new synthesis of angiosperm paleobotany and systematics/phylogenetics and a methodological framework for integrating paleontological and neontological perspectives. The Deep Time Research Coordination network (www. flmnh.ufl.edu/deeptime/) is fostering interactions between paleobotanists and phylogeneticists, with several ongoing projects aimed at integrating fossils into molecular datasets. Recent efforts to integrate fossils into molecular datasets of extant taxa by incorporating morphological characters (e.g., Sun et al. 2002; Eklund et al. 2004) should lead to new interest in both the methods and outcomes of such analyses.

Once fossils are integrated into a phylogenetic framework, they can be used to improve calibration points in the cladogram. These divergence times will open up new research possibilities, such as obtaining better estimates of the ages of particular cladogenic events and analyzing diversification rates (e.g., Sanderson and Donoghue 1994; Sanderson 1997, 1998). This information can also be used to date nodes for which fossil data are lacking. These estimates of divergence times will also facilitate the study of molecular evolution of the genes used to generate the cladograms (e.g., *rbcL, atpB,* 18S rDNA), as well as other genes that are currently under study in the angiosperms.

Megatrees: all angiosperm genera

With the sequencing technology now available, it is possible to contemplate truly massive phylogenetic efforts. An enormous number of base pairs, plus non-

DNA characters, will likely be needed to resolve some of the most vexing phylogenetic problems in the angiosperms. However, large-scale sequencing efforts that seemed daunting only a few years ago are now commonplace—the number of DNA datasets for the angiosperms and large subclades of angiosperms will only continue to grow.

We should also contemplate massive sequencing efforts at another level. It should be possible, for example, to construct comprehensive trees that involve representatives of all angiosperm genera. Given the large number of *rbcL* sequences currently available for flowering plants (>5,000), for example, it should be possible to construct a comprehensive matrix in which one species in each genus has been sequenced for this gene. Alternatively, or in addition, more rapidly evolving genes could be used for this purpose, such as *matK* or plastid DNA intergenic spacers (e.g., *trnL-F*). Given the rapidity with which the entire human genome was sequenced, coordinated efforts by systematists could produce a comprehensive generic-level angiosperm phylogeny within five years. Analyzing these large matrices would raise computational issues, but one approach would be to use the well-supported backbone of the angiosperm tree that now exists and graft onto that tree the tip clades that result from more intensive sequencing efforts to produce a megatree.

Megatrees (or grafted supertrees) should not be confused with supertrees, which have also been the subject of recent interest (e.g., Sanderson et al. 1998; Liu et al. 2001; Salamin et al. 2002). With the megatree approach, a backbone topology is used, and other topologies representing subsets of the group under investigation are simply grafted to that backbone. As an alternative to megatrees, previously inferred topologies (source trees) that share some but not necessarily all of their taxa can be combined using one of several available algorithms to create a "supertree." Although large multigene analyses are often favored and methods of supertree construction are still under debate, combinable data are not always available for all taxa, and megatrees and supertrees therefore offer possible solu-

tions. Using the grass family as a model, Salamin et al. (2002) demonstrated that supertrees offer an easy way of producing analyses with a large number of taxa. Supertrees for the grasses closely matched the combined "total evidence" phylogenetic tree provided by the Grass Phylogeny Working Group (GPWG 2000). Supertrees might therefore have a role as an exploratory tool for developing new hypotheses and indicating where future research should be focused.

A supertree of particular interest to this book is a recent attempt to assemble a comprehensive tree for angiosperms (Davies et al. 2004). This supertree for angiosperm families used 27 source trees from published and unpublished studies and represents a useful tool for future comparative studies. We have appended this supertree to the end of this book (see appendix).

Character evolution in angiosperms

Clarifying the big picture of angiosperm phylogeny has broad applications beyond systematics. The angiosperm trees and supertrees should be of value not only to systematists but also to anyone interested in character evolution. Chapters 11 to 13 are dedicated to studies of character evolution, but they represent only initial examples of topics to be explored. Evolutionary biologists can map non-DNA traits of interest (e.g., morphology, chemistry, breeding system) using these topologies, and ecologists can begin to examine interactions and community dynamics in a phylogenetic context. A phylogenetic framework is also critical for genomics comparisons. Researchers are already extending from model systems, such as *Arabidopsis* and *Oryza*, to other members of Brassicaceae (as well as Brassicales) and Poaceae, respectively. Phylogenetically based genomics initiatives will be a major research focus of the next decade, and these will be greatly enhanced by well-resolved and strongly supported topologies. Similarly, a phylogenetic framework affords the opportunity to elucidate molecular evolutionary processes, including gene evolution, gene transfer events, and structural changes in genomes.

3

Basal Angiosperms

Introduction

Basal living angiosperms have long been the focus of general intrigue. A major goal of systematists has been to determine which extant taxa occupy basal positions and have ancestral morphological features. Authors such as Takhtajan and Cronquist frequently referred to the "most primitive living angiosperms," terminology not used here because, as we note throughout this book, taxa attached to basal nodes may exhibit an array of ancestral ("primitive") and derived traits. Much of the interest surrounding these angiosperms has involved attempts to reconstruct the hypothetical ancestral angiosperm, with a particular focus on the origin and diversification of the flower. Studies of floral diversity in basal angiosperms have, in turn, influenced a broad range of disciplines, including plant systematics, ecology, pollination biology, and developmental biology.

Modern classifications (e.g., Cronquist 1981, 1988; Takhtajan 1987, 1997; Thorne 1992a, 1992b, 2001) placed most of the taxa now recognized as basal (from phylogenetic analyses of morphological data, molecular data, or combined datasets of molecules and morphology) in subclass Magnoliidae. However, Magnoliidae of these modern classifications also contained Ranunculales (e.g., Ranunculaceae, Berberidaceae, and Papaveraceae), families now recognized to be basal eudicots (see Chapter 5). The monocots (see Chapter 4) were treated as a separate class, Liliopsida, and not recognized to have arisen from an ancestral radiation of angiosperms. As reviewed in this chapter, familial composition of many orders of basal angiosperms has changed as a result of recent phylogenetic studies.

Morphological Cladistic Analyses

Several morphology-based phylogenetic analyses of basal angiosperms have been conducted, and most of these studies have focused on the relationships of angiosperms to other seed plants; some also included fossils (see Chapter 1). In early phylogenetic analyses of morphological features, angiosperms were scored as a single entity rather than as several exemplars. Doyle and Donoghue (1986, 1992) scored angiosperms as a single taxon, as a composite of Magnoliales and Winteraceae, both of which were still considered "primitive" at that time. Later studies attempted to correct this shortcoming by incorporating several "potential" basal angiosperms. Doyle et al. (1994) and Doyle (1996) included several of these angiosperms (Nymphaeales, monocots, Piperaceae, Saururaceae, Aristolochiaceae, Winteraceae, Magnoliales, Calycanthaceae, Chloranthaceae, and Laurales). Although these were improvements, the small number of families involved in early analyses illustrates one limitation of these morphological cladistic studies—not all potentially basal angiosperm families were sampled. In contrast, in several molecular phylogenetic studies (see "Molecular Analyses," below), representatives of all basal families were sampled, and, for the larger families, several genera were included. A second limitation of these early morphological analyses is that characters were assigned using composite groups, such as Laurales and Magnoliales, rather than individual species as is done with studies using DNA sequences. Furthermore, many families that we now consider critical, such as Amborellaceae and Schisandraceae (including Illiciaceae; APG II 2003; see Chapter 10), were not included in these initial morphological phylogenetic analyses, or they were lumped into orders with taxa to which they are not closely related.

Phylogenetic studies using morphology recovered an array of putatively basalmost extant angiosperms. For example, the analysis of Donoghue and Doyle (1989a) placed a composite Magnoliales clade (scored on the basis of several families including Degeneriaceae, Magnoliaceae, Myristicaceae, and Annonaceae) as sister to all other angiosperms (Figure 3.1). This result was consistent with views at that time regarding the most primitive angiosperms (e.g., Cronquist 1968, 1981; Takhtajan 1969, 1991). This Magnoliales clade was then followed by two other clades of woody magnoliids: a composite Laurales that also included Chloranthaceae and a winteroid group that included Illiciaceae/Schisandraceae (Illiciales of modern classifications such as Cronquist 1981) and Winteraceae. However, the taxonomic makeup of these "composite" clades does not agree with DNA-based circumscriptions (see "Laurales" and "Canellales," below). The remaining major

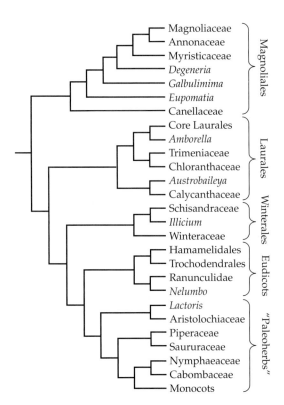

FIGURE 3.1 Representative most parsimonious tree from the phylogenetic analysis of morphological data (modified from Donoghue and Doyle 1989a). Clade names are those of Donoghue and Doyle (1989a), but the composition of these clades does not reflect our current understanding of relationships (see text; APG II 2003). As noted in the text, the term "paleoherb" has been used to define different groups by different authors; it is not of any systematic value.

clades of angiosperms identified in the shortest trees of Donoghue and Doyle (1989a, 1989b) were "dicots" with triaperturate pollen (later referred to as eudicots, and labeled as such in Figure 3.1; Doyle and Hotton 1991) and a "paleoherb" group consisting largely of herbaceous magnoliids (Aristolochiaceae, *Lactoris*, Piperales), Nymphaeaceae, and monocots. This initial use of the term "paleoherb" differs from broader applications of the term used in later molecular systematic endeavors (e.g., Chase et al. 1993). The paleoherb group is now recognized as polyphyletic, and the use of the term should be discontinued. The results of Donoghue and Doyle (1989b) were not robust; in trees only one step longer, they found that the paleoherbs were sister to all other angiosperms.

In contrast to the findings of Donoghue and Doyle (1989a, 1989b), the phylogenetic analysis of morphological data conducted by Loconte and Stevenson (1991) and Loconte (1996) placed Calycanthaceae as sister to all remaining angiosperms. Calycanthaceae were then followed by a grade of Magnoliales, Laurales, Il-liciaceae + Schisandraceae, and Lactoridales as other early-branching lineages. D.W. Taylor and Hickey (1990, 1992, 1996a, 1996b) focused attention on Chlo-ranthaceae as a possible first-branching angiosperm lineage. Not only were herbaceous taxa found early in the fossil record (Taylor 1990; D.W. Taylor and Hickey 1992) but morphological cladistic analyses (D.W. Tay-lor and Hickey 1992) also placed Chloranthaceae, fol-lowed by Piperaceae, as sister to all other angiosperms (see also Hickey and Taylor 1996).

Doyle et al. (1994) analyzed morphological data and also combined rRNA sequence data and morphology (the combined morphology plus molecular dataset is discussed in "Combined DNA and Morphology," below). Analyses of morphological data alone again in-dicated that herbaceous taxa were early-diverging, with a clade of monocots + Nymphaeaceae sister to all other flowering plants, followed by Piperaceae + Saururaceae and then Aristolochiaceae as subsequent sisters to the remaining angiosperms. However, in trees just one step longer, a different topology was again recovered, with Magnoliales sister to all other taxa, followed by Lau-rales, then Calycanthaceae; all of these groups are woody magnoliids.

Nixon et al. (1994) included 18 angiosperms in a se-ries of cladistic analyses that used extant and fossil seed plants. The unusual finding of Nixon et al. of a para-phyletic Gnetales, in which the angiosperms are em-bedded, was discussed in Chapter 1. When fossils were included, a large polytomy was obtained in the an-giosperms. When fossils were excluded, *Chloranthus* ap-peared as sister to the remaining angiosperms, followed by *Ceratophyllum*, and then *Persea*. These taxa were fol-lowed by a clade of hamamelids (*Platanus, Hamamelis, Chrysolepis, Betula,* and *Casuarina*) as sister to a clade of diverse taxa, including *Caltha, Piper, Lilium, Nymphaea, Magnolia,* and *Calycanthus*. Because of the large number of hamamelids appearing as early branches in these analyses, Doyle (1996) referred to the trees of Nixon et al. as "neo-Englerian."

The most extensive cladistic analysis of basal an-giosperms using morphology is that of Doyle and En-dress (2000). Compared to the earlier study of Donoghue and Doyle (1989a), Doyle and Endress in-creased the number of taxa from 27 to 52 and the num-ber of characters from 54 to 108. The shortest trees ob-tained by Doyle and Endress revealed what they termed as "shifts toward molecular results," due in some cases directly to the addition of new characters, such as carpel form (see "Character Evolution in Basal Angiosperms," below).These shifts in the morphology-based trees included the separation of Illiciaceae/ Schisandraceae from Winteraceae; separation of *Am-borella, Austrobaileya,* Trimeniaceae, and Chloranthaceae

from Laurales; a close relationship between Canellaceae and Winteraceae; and a close relationship of Austrobai-leyaceae with Illiciaceae and Schisandraceae. Howev-er, there are also noteworthy instances in which mor-phological and molecular trees continue to produce different results for some groups; these are discussed in "Combined DNA and Morphology," below.

Molecular Analyses

Phylogenetic analyses involving single genes played a critical role in revealing major clades of basal an-giosperms and their interrelationships. In contrast to morphology-based phylogenetic analyses, DNA-based studies were more comprehensive in their cov-erage of angiosperm families, often including all fam-ilies placed in Magnoliidae of recent classifications (e.g., Cronquist 1981), as well as multiple exemplars of many families. Analyses of two plastid genes, *rbcL* and *atpB,* and nuclear 18S rDNA (e.g., Hamby and Zimmer 1992; Chase et al. 1993; Qiu et al. 1993; D. Soltis et al. 1997a; Savolainen et al. 2000a) all agreed in recognizing the same suite of basal angiosperms as either a clade (*rbcL*) or a grade (*atpB;* 18S rDNA), which were distinct from the eudicot clade. These basal angiosperms were initially referred to as either the "uniaperturate" or "monosulcate" angiosperms (e.g., D. Soltis et al. 1997a) because many of these taxa possess pollen with a single aperture. However, the pollen morphology of these taxa is far more complex than this name implies (e.g., Sampson 2000). As a re-sult, recent molecular analyses have sometimes re-ferred to the basal angiosperms as the "noneudicots" (e.g., D. Soltis et al. 2000). Because all studies using combined datasets of multiple genes have indicated that the basal angiosperms form a grade rather than a clade, no name should actually be applied to these taxa. For convenience only, in this book, we refer to all angiosperms other than eudicots collectively as "basal angiosperms"(see Chapter 2).

The various single-gene studies identified some of the same clades of basal angiosperms, including Lau-rales and Magnoliales. However, the composition of these DNA-based clades differed from those indicat-ed in modern classifications (e.g., Cronquist 1981), al-though there was little internal support for most rela-tionships in these single-gene studies (reviewed in Chase and Albert 1998; D. Soltis et al. 2000). Hence, con-siderable uncertainty remained regarding the lineage or lineages at the basal nodes of the topology of extant flowering plants. Also unclear were the composition of major clades of basal angiosperms and the relationships among these clades. The findings of these single-gene studies are briefly reviewed in this section (see also Chapter 2).

Plastid *rbcL*

The large *rbcL* studies of Chase et al. (1993) and Qiu et al. (1993) provided the first comprehensive phylogenetic analyses of basal angiosperms. These analyses placed *Ceratophyllum* (the only genus of Ceratophyllaceae) as sister to all other extant flowering plants and also indicated that all other angiosperms regarded as "basal" formed a clade sister to the eudicots. These same analyses also provided important initial insights into relationships among basal angiosperms. Chase et al. (1993) and Qiu et al. (1993) found a clade of Canellaceae + Winteraceae (now Canellales; APG II 2003); a Laurales clade that contained Lauraceae, Monimiaceae, Idiospermaceae, Calycanthaceae, and Hernandiaceae; and a monocot clade in which *Acorus* was sister to all other monocots. The clade labeled Magnoliales in Chase et al. (1993) and Qiu et al. (1993) is a broader circumscription than that revealed by later multigene analyses (see "Combined DNA Datasets," below; APG 1998; APG II 2003; and Chapter 10), containing Canellaceae + Winteraceae as sister to a group (ultimately referred to as Magnoliales in APG 1998) of Magnoliaceae, Degeneriaceae, Himantandraceae, Eupomatiaceae, Annonaceae, and Myristicaceae.

The remaining basal angiosperms formed two clades referred to as paleoherbs because of the largely herbaceous habit of these putatively ancient lineages (Chase et al. 1993; Qiu et al. 1993) and their partial correspondence to paleoherbs *sensu* Donoghue and Doyle (1989a, 1989b). However, woody families such as Amborellaceae, Illiciaceae, and Schisandraceae were unfortunately considered paleoherbs in these studies. Hence, the application of the term paleoherb further confused the concept presented earlier (e.g., Donoghue and Doyle 1989a, 1989b; D.W. Taylor and Hickey 1992). Paleoherb I contained Aristolochiaceae, which included the monotypic *Lactoris* (Lactoridaceae) as sister to Saururaceae + Piperaceae; this paleoherb I clade would later be referred to as Piperales (APG 1998; APG II 2003). The second paleoherb clade contained Chloranthaceae as sister to an array of families that would ultimately attract attention as successive sisters to all remaining angiosperms in multigene analyses: Austrobaileyaceae, Illiciaceae, Schisandraceae (with Trimeniaceae these families now constitute Austrobaileyales; discussed below), Amborellaceae, and the water lily families (Cabombaceae and Nymphaeaceae, or Nymphaeaceae *sensu* APG 1998; APG II 2003; Chapter 10).

Internal support (as measured by bootstrap or jackknife) was not presented in Chase et al. (1993) or Qiu et al. (1993), but was later given in Chase and Albert (1998) using one of the faster methods developed after the 1993 papers were published. As in all of the single-gene analyses, internal support for the relationships revealed by analyses of *rbcL* sequences was low, with few large clades obtaining support above 50%.

Qiu et al. (1993) also conducted a constraint analysis in which Nymphaeaceae, followed by *Amborella*, *Austrobaileya*, and Illiciaceae + Schisandraceae, were made to be successive sisters to other angiosperms. This pattern was only slightly less parsimonious than the shortest trees in which *Ceratophyllum* appeared as sister to all other living angiosperms.

Nuclear 18S rDNA

Although the coverage of basal families was not as complete as in the *rbcL* studies of Chase et al. (1993) and Qiu et al. (1993), the topologies obtained with 18S rDNA (D. Soltis et al. 1997a) were highly similar to those identified using the plastid gene *rbcL*. Both 18S rDNA and *rbcL* identified the same clades of basal angiosperms (including the monocots), and both recovered a large eudicot clade. In contrast to *rbcL*, however, 18S rDNA showed that basal angiosperms formed a grade rather than a clade. Significantly, 18S rDNA was the first gene to reveal a grade of Amborellaceae, followed by Nymphaeaceae, and a small clade of Austrobaileyaceae + Illiciaceae/Schisandraceae (Trimeniaceae were not included) as successive sister taxa to all other angiosperms (D. Soltis et al. 1997a). However, internal support for relationships among basal angiosperms with 18S rDNA sequences was again low, as was observed with *rbcL*.

Plastid *atpB*

Analyses of *atpB* sequences (Savolainen et al. 2000a) provided results similar to 18S rDNA by indicating that basal angiosperms formed a grade, not a clade as with *rbcL*, and that Amborellaceae, followed by Nymphaeaceae, and Austrobaileyaceae + Illiciaceae/ Schisandraceae were successive sisters to all other angiosperms. In the *atpB* tree, *Ceratophyllum* + *Acorus* formed a clade sister to all other monocots. The remaining basal angiosperms formed a core "magnoliid" clade with Chloranthaceae sister to the remaining taxa. The subclades of this core clade are identical to those later identified in analyses of combined datasets (see "Combined DNA Datasets," below): Magnoliales and Laurales formed a clade sister to Piperales + Winteraceae/Canellaceae. Internal support for clades with *atpB* was generally low with only a few clades receiving support above 50%: Laurales (68%), Chloranthaceae (59%), and Winteraceae/ Canellaceae (94%).

Other plastid genes

Rapidly evolving plastid DNA genes have shown great potential for resolving angiosperm relationships at deep levels as well as within families. The plastid gene *matK* has a rate of evolution about three times faster than the widely sequenced *rbcL*. As a result of this faster rate of evolution, *matK* sequences were initially considered to have the greatest phylogenetic potential within families or among closely related families. However, using only 1000 bp of *matK* sequence data, Hilu et al. (2003) reconstructed a phylogenetic tree for angiosperms similar to that obtained using three genes combined. *matK* placed Amborellaceae, followed by Nymphaeaceae and Austrobaileyales as successive sisters to all other angiosperms with moderate bootstrap support. A magnoliid clade of Laurales + Magnoliales and Piperales + Canellales (the latter order is equivalent to Winterales of recent authors, including Qiu et al. 1999; Doyle and Endress 2000; D. Soltis et al. 2000; Zanis et al. 2002; see discussion of Canellales, below) was also revealed using *matK* sequences. Similarly, Borsch et al. (2003) found that phylogenetic analysis of *trnL-F*, a plastid intron and spacer region typically used for phylogenetic inference at the species and genus levels, yielded a topology for basal angiosperms similar to those obtained with three or more genes. In the shortest trees of Borsch et al. (2003), *Amborella*, followed by Nymphaeaceae and then Austrobaileyales, are once again subsequent sisters to all other angiosperms.

Combined DNA and Morphology

Three analyses have attempted to combine morphological and molecular DNA sets, either across the angiosperms (Nandi et al. 1998) or for only the basal angiosperms (Doyle et al. 1994; Doyle and Endress 2000). We discussed these studies in Chapter 2; we will therefore only briefly summarize results that are pertinent to basal angiosperm relationships.

Doyle et al. (1994) combined a morphological dataset with a dataset of partial 18S and 26S rRNA sequences from Hamby and Zimmer (1992). However, the Hamby and Zimmer dataset did not contain sequences for several important taxa (e.g., *Amborella*). Doyle et al. deleted some of the taxa for which sequence data were available, but morphological data were lacking, to produce a molecular plus morphological dataset that matched in taxon composition. As a result, some critical taxa were deleted from the molecular matrix (e.g., *Illicium*; see Hamby and Zimmer 1992). Analysis of the combined morphology and RNA dataset yielded a topology in which their Nymphaeales (= Nymphaeaceae *sensu* APG 1998; APG II 2003) appeared as sister to all other angiosperms (Doyle et al. 1994). Nymphaeaceae were followed by monocots, a clade of Piperaceae + Saururaceae, and then Aristolochiaceae as subsequent sisters to the rest of the angiosperms. Thus, this analysis again placed herbaceous taxa as sisters to all other extant angiosperms.

A recent analysis of gene sequence data plus morphology placed the herbaceous fossil *Archaefructus* (see "The Appearance of Angiosperms in the Fossil Record," below) as sister to the extant basal grade of Amborellaceae, Nymphaeaceae, and Austrobaileyales, providing additional evidence for the early appearance of the herbaceous habit (Sun et al. 2002; but see Friis et al. 2003). However, whether herbaceousness should be interpreted as an autapomorphy of *Archaefructus* or as the ancestral angiosperm state cannot be inferred.

Nandi et al. (1998) compiled a dataset of 162 extant angiosperms for which both *rbcL* sequences and nonmolecular data were available. An important contribution of this analysis is that it included a large nonmolecular dataset for basal angiosperms, as well as angiosperms in general. The nonmolecular analyses revealed some relationships similar to those recovered with DNA sequence data, such as a clade of Austrobaileyaceae and Schisandraceae (including *Illicium*), whereas other relationships are clearly spurious, such as the placement of the eudicot Eupteleaceae as an early-branching angiosperm and a magnoliid clade that also contains eudicots (e.g., Berberidopsidaceae and Nelumbonaceae). The combination of *rbcL* and nonmolecular data yielded a topology similar to that observed with *rbcL* alone.

Doyle and Endress (2000) also conducted analyses of a combined morphological and molecular (*atpB, rbcL,* and 18S rDNA sequences from P. Soltis et al. 1999b and D. Soltis et al. 2000) dataset and obtained a topology similar to that recovered using molecular data alone (Figure 3.2). However, in some instances the non-DNA characters used resulted in important differences in the placement of taxa or clades compared to DNA-based trees, including the placement in the morphology-based trees (not shown) of Chloranthaceae following *Amborella* as sister to all other angiosperms; a sister-group relationship of Nymphaeaceae and monocots; the placement of Austrobaileyaceae, Schisandraceae, and Illiciaceae after Magnoliales, Laurales, and what they (and other recent authors) referred to as Winterales (now Canellales, APG II 2003). Also of interest is the placement in the morphology-based trees of Hernandiaceae as sister to Lauraceae. In some molecular trees, in contrast, Hernandiaceae appear as sister to a clade of Lauraceae, Monimiaceae, Gomortegaceae, and Atherospermataceae. In other molecular analyses, Hernandiaceae are sister to Monimiaceae + Lauraceae (Figures 3.2 and 3.3).

Some of these conflicts between morphology-based and DNA-based trees may be a result of instances in which morphology may be too influenced by functional and ecological convergence. The close relationship of monocots and Nymphaeaceae may exemplify this phenomenon. The two groups are linked in the morphological analysis by boat-shaped pollen with a smooth sulcus membrane, an inner integument with two cell layers, and the absence of introrse anther dehiscence. The positions of monocots and Nymphaeaceae in the DNA-based trees (e.g., Qiu et al. 1999; P. Soltis et al. 1999b; Zanis et al. 2002) indicated that these are morphologi-

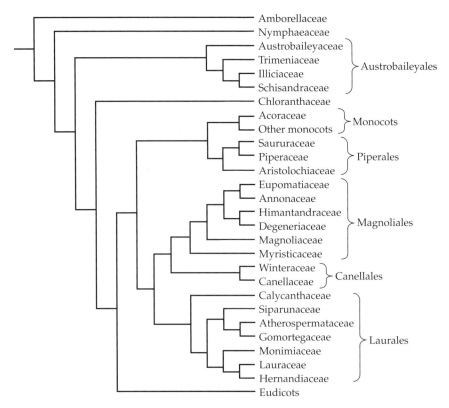

FIGURE 3.2 Single most parsimonious tree based on the phylogenetic analysis of a combined dataset of three genes (*rbcL, atpB,* 18S rDNA) plus morphology. This tree has been modified from Doyle and Endress (2000). Modifications include the use of family names as placeholders for some genera (e.g., Nymphaeaceae, Chloranthaceae), "other monocots" as a placeholder for additional monocot lineages, and the representation of eudicot exemplar families as simply "eudicots." In addition, we have used the clade name "Austrobaileyales" for their "ITA"(Illiciaceae/Schisandraceae, Trimeniaceae, Austrobaileyaceae) clade and Canellales for their Winterales clade (APG II 2003; see Chapter 10).

cal convergences. In other cases, Doyle and Endress suggested that the non-DNA characters available may be unable to tease apart some of these relationships. However, not all of the conflicts may be problems with the morphology-based analyses. In the case of Hernandiaceae and Lauraceae, the DNA-based trees may be in error. Although the two families do not appear as immediate sister taxa in DNA trees (e.g., Savolainen et al. 2000a; D. Soltis et al. 2000), there are morphological synapomorphies linking the two families. The conflict in this case may involve limited sampling of early-diverging taxa in Lauraceae in the molecular trees (Renner and Chanderbali 2000). The overall conclusion of Doyle and Endress (2000) is that morphology may be more consistent with molecular data than was previously thought; a similar conclusion was reached by Nandi et al. (1998) and Chase et al. (2000).

Combined DNA Datasets

The topologies from single-gene studies summarized above have recently been superseded by analyses of combined datasets (Mathews and Donoghue 1999, 2000; Parkinson et al. 1999; Qiu et al. 1999, 2000; P. Soltis et al. 1999b; Barkman et al. 2000; Graham and Olmstead 2000; Graham et al. 2000; Savolainen et al. 2000a; D. Soltis et al. 2000; Zanis et al. 2002, 2003). In the span of just a few years, the number of genes combined and brought to bear on the question of basal angiosperm relationships has quickly increased from two (e.g., Mathews and Donoghue 1999; Savolainen et al. 2000a) or three (P. Soltis et al. 1999b; D. Soltis et al. 2000) to as many as 11 (Zanis et al. 2002) or 17 (Graham and Olmstead 2000; Graham et al. 2000).

There are important differences among these molecular studies, not only in the genes used but also in taxon sampling and methods of phylogenetic analysis. In several studies, genes representing the nuclear, mitochondrial, and plastid genomes were combined (e.g., Parkinson et al. 1999; Qiu et al. 1999; Barkman et al. 2000; Zanis et al. 2002). The number of taxa sampled varies widely. In some cases, few taxa were used, whereas other studies sampled all families of basal angiosperms. The most complete analyses in terms of taxon sampling are those of Qiu et al. (1999, 2000), P. Soltis et al. (1999b), Savolainen et al. (2000a), D. Soltis et al. (2000), and Zanis et al. (2002, 2003). In terms of the number of base pairs used, the largest datasets so far constructed are those of Barkman et al. (2000), Graham and Olmstead (2000), and Zanis et al. (2002), involving 9 to 17 genes. Barkman et al. (2000) and Zanis et al. (2002) constructed similar datasets involving genes from all three genomes. Barkman et al.'s (2000) datasets involved 9 genes for 15 taxa (~12,000 bp per taxon) or 6 genes for 35 taxa; Zanis et al.'s (2002) datasets included 11 genes for 16 taxa (>15,000 bp per taxon) or at least 5 of the 11 genes for 104 taxa. Graham and Olmstead combined data from 17 plastid genes for 19 taxa (approximately 13.4 kb per taxon), focusing primarily on "slowly evolving genes."

Several phylogenetic approaches have also been used in the study of basal angiosperm relationships. In addition to parsimony, investigators have applied maximum likelihood in these analyses (e.g., Parkinson et al. 1999; Barkman et al. 2000; Qiu et al. 2000; P. Soltis et al. 2000; Zanis et al. 2002, 2003) and Bayesian methods (Zanis et al. 2002). Investigators have also used several other noteworthy strategies. Mathews and Donoghue (1999, 2000) used duplicate gene rooting in a parsimony analysis of phytochrome genes. Zanis et al. (2002) used a compartmentalization approach coupled with parsimony and maximum likelihood analyses in an effort to clarify relationships among basal angiosperm lineages.

Through these recent studies, many relationships among basal angiosperms have been solidified (Figure 3.3). Significantly, topologies based on sequences of nuclear, plastid, and mitochondrial genes are highly concordant. Analyses of combined datasets have indicated with a high level of internal support that basal angiosperms form a grade. Amborellaceae are sister to all other angiosperms, followed by Nymphaeaceae and a clade (now referred to as Austrobaileyales; APG II 2003) of Austrobaileyaceae, Trimeniaceae, and Schisandraceae (including Illiciaceae; Figure 3.3). These taxa were referred to as the ANITA grade (*Amborella*, **N**ymphaeaceae s.l., **I**lliciales (Illiciaceae/Schisandraceae), **T**rimeniaceae, **A**ustrobaileyaceae; Qiu et al. 1999). However, we prefer not to name grades. Furthermore, with the abandonment of the name "Illiciales" to

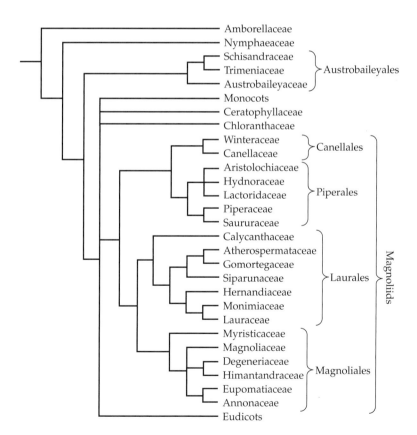

FIGURE 3.3 Conservative summary of current understanding of basal angiosperm relationships based on topologies of Zanis et al. (2002), Nickrent et al. (2002), and Qiu et al. (1999). Only nodes receiving jackknife or bootstrap support above 60% are depicted. Illiciaceae are often treated as a distinct family; however, it has been recommended that they be considered part of Schisandraceae (APG II 2003).

refer to Illiciaceae + Schisandraceae (APG II 2003), "ANITA" no longer corresponds to the names of the taxa involved and is therefore confusing.

There is some uncertainty in relationships following the basal grade of Amborellaceae, Nymphaeaceae, and Austrobaileyales (Figure 3.3). Several analyses indicated that monocots + Ceratophyllaceae are the next-branching lineage (but without strong support), followed by Chloranthaceae as sister to all other angiosperms (Qiu et al. 1999, 2000; Zanis et al. 2002, 2003). The remaining basal angiosperms then form a well-supported clade of Magnoliales + Laurales as sister to Piperales + Canellales. Our current understanding of basal angiosperm relationships is summarized in more detail in the sections that follow (see also Figure 3.3).

Amborellaceae, Nymphaeaceae, and Austrobaileyales

The sequence of nodes at the base of the tree of extant angiosperms is now known with a high degree of confidence (Figure 3.3). This is a remarkable development in that some investigators had suggested that elucidating the root of the angiosperms might not be possible because of rapid radiation and extinction of some lineages. Virtually all combined analyses involving three or more genes agree in finding the same strongly supported clades and general ordering of early-diverging angiosperms. Several groups of investigators independently and simultaneously arrived at the same conclusion regarding these nodes by using different datasets and approaches (Mathews and Donoghue 1999; Parkinson et al. 1999; Qiu et al. 1999; P. Soltis et al. 1999b; Graham and Olmstead 2000; Graham et al. 2000). Significantly, all of these analyses agree in placing *Amborella trichopoda* (Amborellaceae), a shrub or small tree endemic to New Caledonia (Figure 3.4, A and B), as sister to all other flowering plants. The placement of *Amborella* alone as sister to all other flowering plants is reinforced by recent analyses of more rapidly evolving plastid DNA regions (Borsch et al. 2003; Hilu et al. 2003), as well as by phylogenetic analyses and structural features of B-class genes (Kim et al. 2004a).

Monotypic *Amborella* is followed by the water lilies, Nymphaeaceae (including Cabombaceae *sensu* APG 1998 and APG II 2003), and a clade of Austrobaileyaceae, Schisandraceae (including Illiciaceae), and Trimeniaceae (Austrobaileyales; see APG II 2003 and Chapter 10) as successive sisters to all other angiosperms. Renner (1999) first established the position of Trimeniaceae (using *rbcL* sequence data) as a member of a clade with Austrobaileyaceae, Illiciaceae, and Schisandraceae; this position was subsequently supported with the addition of other genes.

Mathews and Donoghue (1999, 2000) identified these same basal nodes using a novel approach that in-

volved analyzing duplicated phytochrome genes by a reciprocal rooting strategy. The logic behind this method is that simultaneous analysis of sequences from a gene pair that results from a duplication event in the lineage leading to extant angiosperms should yield two identical or similar gene trees; the angiosperm topology could therefore be rooted without outgroups on the branch connecting the duplicated genes. A serious limitation, however, of Mathews and Donoghue's (1999, 2000) approach is that *Ceratophyllum* could not be included in their analyses because two copies of the gene were not detected. Hence, a crucial taxon that had been identified as sister to all other angiosperms in *rbcL* studies (e.g., Chase et al. 1993; Qiu et al. 1993) was not evaluated by Mathews and Donoghue (1999, 2000).

In the three-gene analysis, low to moderate support was obtained for most of the basal nodes. *Amborella* followed by Nymphaeaceae and then Austrobaileyales (*sensu* APG II 2003) were successive sisters to all remaining angiosperms with jackknife support of 65%, 72%, and 71%, respectively (D. Soltis et al. 2000). With the five-gene dataset of Qiu et al. (1999) and the 11-gene dataset of Zanis et al. (2002), these basal nodes were all strongly supported (above 90%; Figure 3.3). Use of both compartmentalization methods and a Bayesian approach provided additional strong support for these basal nodes (Zanis et al. 2002, 2003).

The high level of sequence divergence between angiosperms and the gymnosperm outgroups also raised the question of whether the rooting with *Amborella* sister to other extant angiosperms was affected in some way by long-branch attraction. Qiu et al. (2001) addressed this question using several types of artificial (random and nonrandom) sequences, as well as sequences of a lycopod and a bryophyte as outgroups. They concluded that the *Amborella* rooting was, in fact, based on historical signal rather than an artifact of long branches.

Amborellaceae and the root of the angiosperms

Despite considerable support for *Amborella* as sister to all other extant angiosperms (e.g., Mathews and Donoghue 1999; Parkinson et al. 1999; Qiu et al. 1999, 2000; P. Soltis et al. 1999b, 2000; Graham and Olmstead 2000; D. Soltis et al. 2000), some reservations regarding this placement have been expressed. Using the Kishino-Hasegawa test (Kishino and Hasegawa 1989), Parkinson et al. (1999) could not reject the hypothesis that Nymphaeaceae (including Cabombaceae) are sister to all other flowering plants, or that *Amborella* + Nymphaeaceae form a clade sister to all other angiosperms. In a maximum likelihood analysis of basal angiosperms using three genes, P. Soltis et al. (2000) found a clade of *Amborella* + Nymphaeaceae as sister to

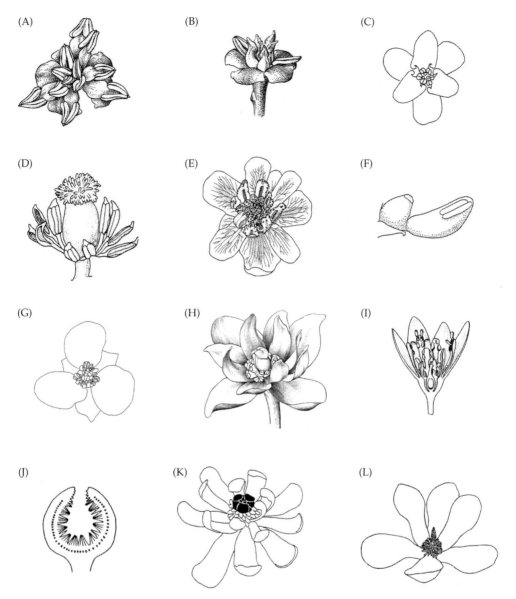

FIGURE 3.4 Flowers of basal angiosperms. (A) *Amborella trichopoda* (Amborellaceae), staminate flower (from Endress and Igersheim 2000a). (B) *Amborella trichopoda* (Amborellaceae), pistillate flower (from Endress and Igersheim 2000a). (C) *Cabomba aquatica* (Nymphaeaceae; from Endress 1994a). (D) *Trimenia papuana* (Trimeniaceae; from Endress and Sampson 1983). (E) *Austrobaileya scandens* (Austrobaileyaceae; from Endress 1980, 1992b). (F) *Sarcandra chloranthoides* (Chloranthaceae; from Endress 1987c). (G) *Saruma henryi* (Aristolochiaceae; Piperales; from Endress 1994a). (H) *Takhtajania perrieri* (Winteraceae; Canellales; from Endress et al. 2000b). (I) *Cinnamomum zeylanicum* (Lauraceae; Laurales; from Friis and Endress 1990). (J) *Tambourissa sieberi* (Monimiaceae; Laurales; from Friis and Endress 1990). (K) *Degeneria vitiensis* (Degeneriaceae; Magnoliales; from Endress 1984, 1992b). (L) *Magnolia × soulangiana* (Magnoliaceae; Magnoliales; from Endress 1987a).

all other angiosperms. Similarly, maximum likelihood favored trees with *Amborella* + Nymphaeaceae as sister to all other angiosperms in analyses of a five-gene dataset for basal angiosperms (Qiu et al. 2000). In analyses of *phyA* and *phyB* sequences, Mathews and Donoghue (2000) also could not reject alternative rootings at *Amborella* + Nymphaeaceae or at Nymphaeaceae alone using the Significantly Less Parsimonious Test (Templeton 1983). Although Graham and Olmstead

(2000) found *Amborella* as sister to all other angiosperms when both *Cabomba* and *Nymphaea* (Nymphaeaceae) were included, when *Nymphaea* was removed leaving only *Cabomba* to represent Nymphaeaceae, *Cabomba* and *Amborella* were successive sisters to all other angiosperms. The latter result could, however, be an artifact of taxon sampling; the total number of taxa used was small, and the deletion of a critical species could have a large impact.

The greatest challenge to *Amborella* alone as sister to all other angiosperms was provided by Barkman et al. (2000), who conducted analyses of two "noise-reduced" multigene datasets, a six-gene, 35-taxon dataset and a nine-gene, 15-taxon dataset. "Noise" was defined as potentially problematic characters or taxa, and an effort was made to reduce noise using the program RASA (Relative Apparent Synapomorphy Analysis; Lyons-Weiler et al. 1996). It can be argued, however, that efforts to remove noise using this approach are fundamentally flawed (Farris 2002); for example, potentially informative characters might have been removed. With this caveat in mind, Barkman et al.'s results were dependent on the method of phylogenetic analysis used. After reducing noise, weighted parsimony, neighbor-joining, and maximum likelihood analyses indicated that *Amborella* + Nymphaeaceae are sister to all other angiosperms. This placement received strong bootstrap support in some analyses (as high as 96% in the neighbor-joining analysis of the six-gene dataset). In contrast, equally weighted parsimony placed *Amborella* alone as sister to all other flowering plants (Barkman et al. 2000).

Zanis et al. (2002) used datasets similar to those used by Barkman et al. (2000). One of the datasets involved 11 genes for 16 taxa, and the second dataset included a subset of these genes for 104 taxa. Using the 16-taxon dataset, Zanis et al. explored the position of *Amborella* with both maximum parsimony and maximum likelihood analyses and nonparametric bootstrapping (standard bootstrapping). They also conducted extensive evaluations of three alternative rootings using likelihood ratio tests and the parametric bootstrap: (1) *Amborella* alone is sister to all other flowering plants; (2) *Amborella* + Nymphaeaceae are sister to all other angiosperms; and (3) Nymphaeaceae alone are sister to all other angiosperms. Zanis et al. (2002) also examined seven different data partitions using maximum likelihood analysis to locate phylogenetic signal that supports each of these three alternative phylogenetic hypotheses: nuclear genes, plastid genes, mitochondrial genes, ribosomal genes, protein-coding genes, first and second codon positions of protein-coding genes, and third codon positions of protein-coding genes. *Amborella* continued to receive strong support as sister to all other extant angiosperms. Use of a Bayesian approach to phylogenetic inference also provided strong support for *Amborella* as sister to all other flowering plants. In addition, three of four tests statistically rejected other alternative hypotheses of the angiosperm root; Nymphaeaceae alone as sister to other angiosperms was not supported in any of the analyses. Despite strong total evidence support for *Amborella* as sister to all other angiosperms, Zanis et al. could not always reject *Amborella* + Nymphaeaceae as sister to all other extant angiosperms. As noted, however, the placement of *Amborella* alone as sister to all other angiosperms has garnered considerable support from recent phylogenetic studies of additional genes.

The analyses noted above clearly illustrate some of the issues and complexities that may be involved in phylogeny reconstruction. Attention has now been focused on *Amborella*, as well as on Nymphaeaceae, as sisters to all other angiosperms. For example, floral morphology, physiological ecology, and wood anatomy of *Amborella* have recently been carefully investigated (Endress and Igersheim 2000a, 2000b; Feild et al. 2000a; Carlquist and Schneider 2001; Posluszny and Tomlinson 2003). Floral morphology and endosperm development have been compared in *Amborella* and Nymphaeaceae (Floyd and Friedman 2000; Endress 2001a; Williams and Friedman 2002; Buzgo et al. 2004); both lineages have also become the focus of analyses of floral genes (D. Soltis et al. 2002b; Stellari et al. 2004; Kim et al. 2004a).

Nymphaeaceae

The water lilies (Figures 3.4C and 3.5) have long attracted interest as a putatively ancient lineage of angiosperms. Some investigators have suggested that Nymphaeaceae might be closely allied with monocots; some even suggested that monocots might be derived from Nymphaeaceae or a now-extinct Nymphaeaceae-like lineage of ancient angiosperms (Hallier 1905; Eber 1934; Arber 1920; Cronquist 1968; Takhtajan 1969, 1991; Burger 1977). Interest in Nymphaeaceae increased after the earliest cladistic analyses based on morphology (e.g., Donoghue and Doyle 1989a, 1989b) and molecules (Hamby and Zimmer 1988) indicated that these plants might be sister to all other angiosperms. As reviewed above, recent phylogenetic studies using both non-DNA and DNA characters have strengthened support for Nymphaeaceae as one of the basalmost lineages of extant angiosperms. Friis et al. (2001) provided the earliest unequivocal fossil evidence of Nymphaeaceae, extending the history of the group into the Early Cretaceous (125–115 Mya) and into the oldest fossil assemblages that contain stamens and carpels (see "The Appearance of Angiosperms in the Fossil Record," below).

We follow the suggested treatment of APG II (2003) and use the family name Nymphaeaceae to designate a broadly defined family that also includes Cabombaceae. In contrast, other recent papers have used the ordinal name Nymphaeales to designate this clade and recognized two families, Nymphaeaceae and Cabombaceae (e.g., Les et al. 1999). Treatments of the water lilies have varied considerably (reviewed in Les et al. 1991, 1999). Cronquist (1981, 1988) recognized several families (e.g., Nymphaeaceae, Cabombaceae, and Barclayaceae), considered Ceratophyllaceae closely relat-

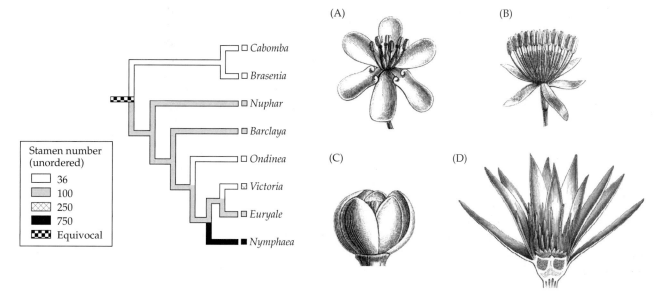

FIGURE 3.5 Stamen diversification in Nymphaeaceae. (Left) MacClade reconstruction of stamen evolution in Nymphaeaceae. Numbers are maximum number of stamens. Topology is the shortest tree of Les et al. (1999); stamen character states were compiled from data in Les et al. (1999). (Right) Line drawings of Nymphaeaceae from Engler and Prantl (1898). (A) *Cabomba aquatica*. (B) *Brasenia purpurea*. (C) *Nuphar pumilum*. (D) *Nymphaea coerulea*.

ed to these families, and also included the eudicot family Nelumbonaceae within Nymphaeales. Takhtajan (1997) recognized six families of water lilies as part of his subclass Nymphaeidae.

Nymphaeaceae (*sensu* APG 1998; APG II 2003) consist of eight genera: *Barclaya, Brasenia, Cabomba, Euryale, Nuphar, Nymphaea, Ondinea,* and *Victoria.* The most comprehensive phylogenetic analysis of the group is based on sequences from three genes plus non-DNA characters (Les et al. 1999). These investigators found strong support for a split of *Brasenia* and *Cabomba* (Cabombaceae) from the remaining genera (Nymphaeaceae in the strict sense).

Since the work of Bessey (1915), the "primitive" angiosperm flower has commonly been considered to possess numerous perianth parts, stamens, and carpels. More recent workers have elaborated on this same general theme (see Cronquist 1968, 1981, 1988; Takhtajan 1969, 1987, 1991, 1997) and envisioned a general trend of reduction from flowers with numerous parts to those with relatively few floral organs (Stebbins 1974; Cronquist 1968, 1981, 1988; Takhtajan 1969, 1991). Stebbins (1974) suggested that secondary increases could occur in some floral organs, such as stamens, but such increases were likely rare.

Nymphaeaceae are of interest in terms of floral evolution because of the large range in variation in the number of floral parts among genera. On the basis of long-standing views of angiosperm floral evolution, flowers such as *Nymphaea* and *Victoria* with numerous floral parts were considered "primitive." However, using the well-resolved tree available for Nymphaeaceae and estimates for stamen number from Les et al. (1999), our character reconstructions (Figure 3.5) indicate instead that numerous floral parts are derived, rather than ancestral, in Nymphaeaceae. The greatest number of petals, stamens, and carpels occurs in *Nymphaea* and *Victoria,* which represent derived genera within the family (we illustrate only stamen number; Figure 3.5). The results of these character reconstructions support Schneider (1979), who earlier suggested that numerous floral parts within the family were of secondary origin. The number of floral parts present in genera of Nymphaeaceae may be associated with pollination (Gottsberger 1977, 1988; Les et al. 1999; Lippok et al. 2000). Large flowers with numerous parts, such as *Nymphaea* and *Victoria,* may represent an evolutionary response to herbivory by beetles, which are major pollinators. Genera with lower numbers of floral organs have different pollinators or pollination systems. The smaller number of floral parts in *Euryale,* another derived genus (Figure 3.5), appears to represent a decrease associated with cleistogamy and self-pollination (see Williamson and Schneider 1993a; Les et al. 1999).

Austrobaileyales

In recent classifications, Illiciales have included Illiciaceae and Schisandraceae, two families that exhibit several morphological similarities and have long been considered closely related (Cronquist 1981, 1988; Takhtajan 1997). Both Illiciaceae and Schisandraceae have tricolpate pollen, which is not found elsewhere among the basal angiosperms (Cronquist 1981; Sampson 2000). Phylogenetic analyses confirmed that Illiciaceae and Schisandraceae are sister taxa (considered a single family, Schisandraceae, in APG II 2003; see Chapter 10) and also revealed that their closest relatives are two small families from Australasia, Austrobaileyaceae and Trimeniaceae (e.g., Qiu et al. 1999; Renner 1999; P. Soltis et al. 1999b; Savolainen et al. 2000a, 2000b; D. Soltis et al. 2000). Bootstrap support for this clade has been 99% to 100% in analyses of combined DNA datasets (e.g., Qiu et al. 1999; D. Soltis et al. 2000; Zanis et al. 2002, 2003). Because of the strong support for this clade, "Illiciales" has been expanded (and renamed Austrobaileyales because the latter is older and because if *Illicium* is included in Schisandraceae there would no longer be an Illiciaceae in Illiciales) to include not only Illiciaceae and Schisandraceae but also Austrobaileyaceae and Trimeniaceae (Figure 3.3; APG II 2003). However, Trimeniaceae and Austrobaileyaceae were not considered closely related to Illiciaceae and Schisandraceae in recent morphology-based classifications (e.g., Takhtajan 1997; Cronquist 1981, 1988). Trimeniaceae, for example, were usually placed in Laurales, separate from Illiciaceae and Schisandraceae (Cronquist 1981; Takhtajan 1997). A cladistic analysis of non-DNA characters (Doyle and Endress 2000) did not recover an Austrobaileyales clade. However, the analysis of non-DNA characters by Nandi et al. (1998) did reconstruct a clade of Austrobaileyaceae, Illiciaceae, and Schisandraceae. At this point, non-DNA synapomorphies uniting *Trimenia* with other members of this clade are still unclear.

Within the Austrobaileyales clade, relationships are strongly supported, with *Austrobaileya* sister to the rest of the order. *Trimenia* is then sister to a well-supported Schisandraceae. Phylogenetic relationships within *Illicium* were addressed in a recent molecular analysis (Hao et al. 2000).

Other Basal Angiosperms: Monocots, Ceratophyllaceae, and Chloranthaceae

Whereas the basalmost nodes are now well resolved and supported, relationships among some of the remaining lineages of angiosperms have been more difficult to elucidate. Analyses of multigene datasets with adequate taxon sampling all have provided strong support for the monophyly of each of the remaining clades of basal angiosperms: monocots, Ceratophyllaceae, Chloranthaceae, Laurales, Magnoliales, Canellales (Winterales of some recent papers), and Piperales. Until recently, however, the relationships among these lineages were unclear. In the three-gene analyses (D. Soltis et al. 2000), for example, these lineages and the eudicots together formed a clade with low jackknife sup-

port (71%), but the relationships among these remaining subclades of basal angiosperms were unclear. Through the analysis of numerous taxa and the use of five genes (Qiu et al. 1999, 2000), six genes (Zanis et al. 2002, 2003), and larger multigene datasets (Zanis et al. 2002), relationships among these remaining lineages have become better understood (Figure 3.3). Following the Austrobaileyales, Ceratophyllaceae + monocots form a clade that is sister to all remaining angiosperms. The Ceratophyllaceae + monocot clade is then followed by Chloranthaceae. The remaining basal angiosperms form a well-supported magnoliid clade composed of four strongly supported subclades—Magnoliales + Laurales are sister to Piperales + Canellales (Figure 3.3). The magnoliid clade is sister to the eudicots.

The support for the sister-group relationship of monocots and Ceratophyllaceae, and for the positions of this clade and Chloranthaceae, is low, even when numerous genes are combined. The sister-group relationship of Ceratophyllaceae + monocots received only 57% support, and their position as sister to other angiosperms received only 52% support in Zanis et al. (2002), based on 5 to 11 genes per taxon and a 104-taxon dataset. The sister-group relationship of Chloranthaceae to other angiosperms received only 52% support (Zanis et al. 2002).

The highest internal support for relationships among these remaining basal lineages was obtained using a compartmentalization approach (Zanis et al. 2003). Mishler (1994) proposed compartmentalization as a method to reduce a large dataset to a more manageable size, decrease the effect of "spurious homoplasy," and maximize the amount of information used in an analysis. Well-supported clades (compartments) are identified, relationships within compartments are constrained, and a hypothetical ancestor of each compartment is reconstructed. Relationships can then be inferred among compartments. This approach also facilitates the use of maximum likelihood with the compartmentalized dataset, whereas the initial dataset may be too large for this method of phylogenetic inference. There are caveats, however, to the compartmentalization approach. Because compartmentalization decreases the size of the dataset, bootstrap percentages can be artificially inflated (reviewed in Zanis et al. 2003). With this approach, all basal angiosperm relationships received support above 90%, except the node supporting Chloranthaceae as sister to the magnoliid clade plus eudicots.

Enormous progress has been made in elucidating basal angiosperm relationships. Bootstrap support is now high for many basal angiosperm relationships, even without a compartmentalized approach. Furthermore, parsimony, maximum likelihood, and Bayesian approaches are converging on highly similar topologies. Nonetheless, major problems remain. The most pressing phylogenetic questions that remain in the basal angiosperms involve the positions of Chloranthaceae, monocots, and Ceratophyllaceae, relative to the magnoliid clade and eudicots (Figure. 3.3).

Ceratophyllaceae + monocots

The sister group of the monocots has long been debated. Various morphological analyses have indicated similarities to Chloranthaceae, Aristolochiaceae, or Nymphaeaceae (e.g., Doyle et al. 1994; Doyle and Endress 2000; see section above on morphological analyses; see also Chapter 4). For example, an atactostele with numerous leaf traces diverging into the leaves and continuing in a parallel arrangement characterizes the monocots and is also found in all Nymphaeaceae and some members of Piperales. Monocots also possess PIIc-subtype sieve-element plastids defined by the presence of triangular protein crystalloids; this type of plastid is otherwise known only in some Aristolochiaceae (Behnke 1971, 1988a). Cellular endosperm is found in Piperales and some monocots such as *Acorus* (Tucker and Douglas 1996; Floyd and Friedman 2000). However, the relationships indicated by DNA sequence data indicate that these morphological characters likely developed in parallel in monocots and other lineages.

Ceratophyllaceae consist of the single genus *Ceratophyllum* and typically have been associated with the water lilies (Nymphaeaceae and segregate families) in morphology-based classifications (Cronquist 1981, 1988; Takhtajan 1997). The phylogenetic position of Ceratophyllaceae has also been of interest since initial analyses of *rbcL* sequences indicated the family might be sister to all other angiosperms (e.g., Les et al. 1991; Chase et al. 1993; Qiu et al. 1993). This position of *Ceratophyllum* has received added attention because fossils with distinctive horned fruits similar to those of *Ceratophyllum* are among early fossil angiosperms (Dilcher 1989). However, the branch leading to *Ceratophyllum* in *rbcL* topologies was exceptionally long; hence, this placement could be spurious due to long branch attraction (Felsenstein 1978b). Subsequent analyses of single genes other than *rbcL* indicated alternative placements for Ceratophyllaceae. Both *atpB* and 18S rDNA (D. Soltis et al. 1997a; Savolainen et al. 2000a) placed *Ceratophyllum* with the monocots. In three-gene analyses (D. Soltis et al. 2000), Ceratophyllaceae were sister to the eudicots with low jackknife support (53%); monocots appeared as part of a large clade that included Laurales, Magnoliales, Canellales, and Chloranthaceae, with no clear pattern of relationship within this large group.

A sister-group relationship between monocots and Ceratophyllaceae (Figure 3.3) has been indicated by several studies using nuclear, mitochondrial, and plas-

tid gene sequences, although this relationship should still be considered tentative. The five-gene analysis of Qiu et al. (1999) was the first multi-gene analysis to indicate that monocots and Ceratophyllaceae are sister taxa, albeit without support above 50%. Later studies involving additional genes ultimately provided internal support above 50% for this relationship. Using a six-gene dataset for a smaller number of taxa, P. Soltis et al. (2000) obtained low support (71%) for this sister-group relationship. Using sequences representing 17 plastid genes, Graham and Olmstead (2000) also found low support (54%) for this relationship, as did Zanis et al. (2002) using an 11-gene dataset (57%). The strongest support (>90%) for a Ceratophyllaceae + monocot clade is provided by the compartmentalization analysis of a six-gene dataset (Zanis et al. 2003). Maximum likelihood analyses also indicated that monocots and *Ceratophyllum* are sister taxa (Graham and Olmstead 2000; Qiu et al. 2000; P. Soltis et al. 2000; Zanis et al. 2002, 2003), as did Bayesian analysis (Zanis et al. 2002). Thus, available sequence data indicate that Ceratophyllaceae could be sister to the monocots. However, because internal support for this relationship is generally low, more DNA sequence data are needed to evaluate this relationship. Furthermore, non-DNA synapomorphies for this clade are unclear and also require careful evaluation. The monocots are discussed in more detail in Chapter 4.

Some of the features shared by monocots and Chloranthaceae (see the section below) may simply represent plesiomorphies subsequently lost in other lineages, based on the topologies that have emerged. Conversely, several of the morphological similarities of Nymphaeaceae and monocots could be best interpreted as parallel evolution (Doyle and Endress 2000), given the distant phylogenetic relationship between these families indicated in recent analyses (Figure 3.3).

Chloranthaceae

The position of Chloranthaceae (Figure 3.3) among basal angiosperms has long been problematic. Takhtajan (1987, 1997) placed the family in its own order and suggested a close relationship with Trimeniaceae and Piperales. Others have also suggested a close relationship of Chloranthaceae and Trimeniaceae (Endress 1986c, 1987; Thorne 2001). Cronquist (1981) placed Chloranthaceae in his Piperales. As noted, D.W. Taylor and Hickey (1992, 1996b), Doyle et al. (1994), and Hickey and Taylor (1996) focused attention on Chloranthaceae as the possible sister group to all other extant angiosperms.

Even with recent multigene analyses, the placement of Chloranthaceae remains uncertain. Multigene studies with dense taxon sampling of basal lineages indicated that Chloranthaceae follow the monocot + Cer-

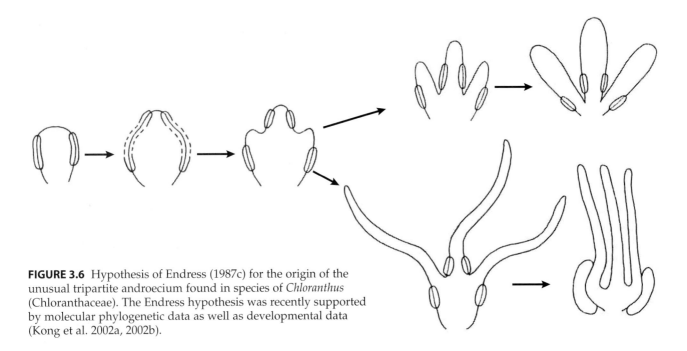

FIGURE 3.6 Hypothesis of Endress (1987c) for the origin of the unusual tripartite androecium found in species of *Chloranthus* (Chloranthaceae). The Endress hypothesis was recently supported by molecular phylogenetic data as well as developmental data (Kong et al. 2002a, 2002b).

atophyllaceae clade as sister to all remaining angiosperms (Figure 3.3; Qiu et al. 1999; Zanis et al. 2002, 2003). However, this position receives fairly strong support only in the compartmentalization analyses (86%) of Zanis et al. (2003). Phylogenetic analyses of non-DNA characters similarly indicated that Chloranthaceae occupy an isolated position as an early-diverging lineage of angiosperms (e.g., Nixon et al. 1994; Nandi et al. 1998; Doyle and Endress 2000). The phylogenetic analysis of non-DNA plus DNA characters by Doyle and Endress (2000) placed the family "above" Austrobaileyales as sister to all remaining angiosperms rather than after the monocots + Ceratophyllaceae as indicated by DNA.

The molecular placement of Chloranthaceae as sister to all other angiosperms following the branch to monocots + *Ceratophyllum* is of interest in that Burger (1977) proposed that Chloranthaceae may be closely involved in the origin of the monocots. If the topology in Figure 2.3 is correct, however, this scenario would be unlikely. In addition, a number of morphological features are shared by Chloranthaceae and Amborellaceae, Nymphaeaceae, and Austrobaileyales, including ascidiate carpels, unilacunar nodal anatomy, and chloranthoid leaf teeth, many of which likely represent symplesiomorphies. It is also significant that some of the earliest known fossil angiosperm material is similar to modern Chloranthaceae (e.g., Friis et al. 1986, 1994; Crane et al. 1995; Brenner 1996), in agreement with phylogenetic inferences that point to the antiquity of the lineage. Friis et al. (1986), using evidence from mesofossils, documented the early occurrence of the family

in the fossil record. The position of Chloranthaceae remains one of the most pressing deep-level questions of relationships within the angiosperms.

Recent analyses have clarified relationships among the four extant genera of Chloranthaceae (Zhang and Renner 2003; Eklund et al. 2004). Kong et al. (2002a) studied relationships and character evolution in *Chloranthus*. Other investigators have examined floral evolution and integrated fossils of Chloranthaceae into datasets and topologies of extant taxa (Doyle et al. 2003; Eklund et al. 2004).

Stamen morphology is highly variable within Chloranthaceae. *Sarcandra* and *Ascarina* have "normal" stamens with two pairs of microsporangia (Figure 3.4F), but most species of *Chloranthus* have stamens that are three-lobed with four thecae (Figure 3.6). It has long been controversial whether the stamens of *Chloranthus* represent a single stamen with four pairs of sporangia or three independent stamens that have become fused at the base (reviewed in Kong et al. 2002a; Doyle et al. 2003). Largely on the basis of fossil evidence, most researchers have favored the latter hypothesis (e.g., Crane et al. 1989; Herendeen et al. 1993; Eklund et al. 1997), but Endress (1987c) considered the structure to represent a single tripartite stamen. Recent phylogenetic studies of extant *Chloranthus* (Kong et al. 2002a), as well as analysis of floral organogenesis of *Chloranthus sessilifolius* (Kong et al. 2002b), a species with a tripartite androecium, have provided support for Endress's (1987c) hypothesis (Figure 3.6). Interpretations based on both extant and fossil taxa are equivocal (Doyle et al. 2003; Eklund et al. 2004).

Other Basal Angiosperms: The Magnoliids

The magnoliid clade consists of Magnoliales, Laurales, Piperales, and Canellales and may be sister to the eudicots (Figures 2.3, 3.3). In the three-gene study (D. Soltis et al. 2000), a more broadly defined magnoliid clade (referred to as the eumagnoliids) was indicated that included not only Magnoliales, Laurales, Piperales, and Canellales, but also Chloranthaceae and monocots. However, analyses of additional genes have indicated that this broader circumscription is inappropriate (e.g., Qiu et al. 1999, 2000; Zanis et al. 2002, 2003). The magnoliid clade of Magnoliales, Laurales, Piperales, and Canellales did not receive jackknife support above 50% in the three-gene analyses of P. Soltis et al. (1999b) and D. Soltis et al. (2000), but with the addition of more genes, bootstrap support for this clade increased to 67% (Qiu et al. 1999) for five genes and to 100% in a compartmentalized analysis of six genes (Zanis et al. 2003). With a 104-taxon dataset of 5 to as many as 11 genes (up to 15,000 bp of sequence data per taxon), support for the magnoliid clade was 78% (Zanis et al. 2002). Thus, the term "magnoliid clade" should be restricted to denote the clade of Magnoliales, Laurales, Canellales, and Piperales (Figure 3.3).

After the monocots, the magnoliid clade contains the largest number of basal angiosperm genera and species. Some of the families in the magnoliid clade are large, such as Annonaceae (120 genera, 2,000 species), Lauraceae (32 genera, 2,500 species), Piperaceae (5 genera, 2,000 species), and Myristicaceae (16 genera, 380 species). There are several possible non-DNA synapomorphies for the magnoliid clade (Nandi et al. 1998; Judd et al. 2002), although more survey work is needed to establish the full extent of the occurrence of these characters. Possible synapomorphies include the phenylpropane compound asarone, the lignans galbacin and veraguensin, and the neolignan licarin (Hegnauer 1962– 1994). In addition, sieve-tube plastids of the P-type are present in most of these families (Behnke 1988a), and all but Piperales share a stratified phloem with wedge-shaped phloem rays (Nandi et al. 1998).

The magnoliid clade comprises two well-supported sister groups, Magnoliales + Laurales (71% bootstrap support, Qiu et al. 1999; 100% Zanis et al. 2002, 2003) and Canellales + Piperales (83% bootstrap support, Qiu et al. 1999; 100% Zanis et al. 2002, 2003). Although a recent cladistic analysis of morphological characters (Doyle and Endress 2000) recovered the same Magnoliales, Laurales, Canellales, and Piperales clades, that study resulted in different relationships among these four clades. Doyle and Endress (2000) found that Magnoliales + Canellales were sister to Laurales, and Piperales were more distantly related, falling in a large

polytomy with monocots, Nymphaeaceae, and several clades of eudicots; none of these non-DNA relationships received bootstrap support above 50%.

These sister-group relationships within the magnoliid clade are largely unexpected from the perspective of modern morphology-based classifications. Winteraceae were considered a close relative of Magnoliaceae (e.g., Cronquist 1968, 1981, 1988); families of Piperales (Piperaceae, Aristolochiaceae, Lactoridaceae, Saururaceae) had not been considered closely related to either Winteraceae or Canellaceae in recent classifications (e.g., Cronquist 1981, 1988; Takhtajan 1987, 1997). Non-DNA synapomorphies for Magnoliales + Laurales and Canellales + Piperales are still unclear.

Relationships within each of the four orders, Magnoliales, Laurales, Piperales, and Canellales, are discussed below.

Magnoliales

Magnoliales contain Myristicaceae, Degeneriaceae, Himantandraceae, Magnoliaceae, Eupomatiaceae, and Annonaceae (Figure 3.3). The same Magnoliales clade was indicated in a cladistic analysis of non-DNA characters (Doyle and Endress 2000). Many of these families have generally been considered closely related (e.g., Cronquist 1981, 1988; Takhtajan 1987, 1997). Cronquist's (1981) Magnoliales represented a diverse assemblage, consisting of these six families, plus Winteraceae, Canellaceae, and Lactoridaceae. Winteraceae and Canellaceae are now placed in Canellales (see below), and Lactoridaceae are in Piperales; the remaining families constitute Magnoliales as currently recognized (APG 1998; APG II 2003). Takhtajan placed only Degeneriaceae, Himantandraceae, and Magnoliaceae in his Magnoliales. On the basis of morphological attributes, some families of Magnoliales (e.g., Magnoliaceae, Degeneriaceae, and Himantandraceae) had been considered to be "among the most archaic living angiosperms" (e.g., Cronquist 1968, 1981, 1988; Takhtajan 1969, 1991, 1997). However, recent topologies have revealed that these families are nested well within the angiosperm tree (Figure 3.3).

Members of Magnoliales are united by numerous non-DNA characters including reduced fiber pit borders, stratified phloem, an adaxial plate of vascular tissue in the petiole, palisade parenchyma, asterosclereids in the leaf mesophyll, continuous tectum in the pollen, and multiplicative testa in the seed (Doyle and Endress 2000).

DNA-based topologies placed Myristicaceae as sister to the remaining members of Magnoliales (Sauquet et al. 2003). This position of Myristicaceae is also supported by several non-DNA characters (Doyle and Endress 2000). Relationships among the remaining members of Magnoliales are not completely clear.

Degeneriaceae and Himantandraceae form a well-supported subclade, as do Annonaceae and Eupomatiaceae, and the monophyly of Magnoliaceae is strongly supported (e.g., Sauquet et al. 2003). However, relationships among these three subgroups are unclear, and they are depicted as a trichotomy (Figure 3.3). Magnoliaceae appeared as sister to Degeneriaceae and Himantandraceae in some analyses, but without strong support (Qiu et al. 1999; D. Soltis et al. 2000; Zanis et al. 2002). Both Magnoliaceae and Annonaceae have also been the focus of more detailed phylogenetic analyses (e.g., Doyle and Le Thomas 1997; Doyle et al. 2000; Kim et al. 2001; Sauquet et al. 2003; Mols et al. 2004).

Laurales

Laurales consist of seven families (Figure 3.3): Calycanthaceae (including Idiospermaceae), Monimiaceae, Gomortegaceae, Atherospermataceae, Lauraceae, Siparunaceae, and Hernandiaceae. Circumscription of Laurales has varied greatly among recent morphology-based classifications (e.g., Cronquist 1981, 1988; Takhtajan 1987, 1997; reviewed in Renner 2001). Cronquist (1981, 1988) placed Amborellaceae and Trimeniaceae in his Laurales together with Monimiaceae, Gomortegaceae, Calycanthaceae, Idiospermaceae, Lauraceae, and Hernandiaceae. *Amborella* and *Trimenia* were placed in Monimiaceae in earlier works (e.g., Perkins 1925). Others have placed Chloranthaceae in Laurales (e.g., Thorne 1974; Takhtajan 1987, 1997).

The best putative synapomorphy for the seven families of Laurales is a perigynous flower in which the gynoecium is frequently deeply embedded in a fleshy receptacle (Figure 3.4, I and J; Endress and Igersheim 1997; Renner 1999). This receptacle sometimes becomes woody in fruit (e.g., Calycanthaceae). Several other non-DNA characters that unite the clade are the presence of inner staminodia, ascendant ovules, tracheidal endotesta (Doyle and Endress 2000), a carpel that is intermediate between ascidiate and plicate, and possibly nodal anatomy (see below). Other morphological characters that were considered to typify Laurales as previously circumscribed (e.g., uniovulate carpels, unilacunar nodes, opposite leaves, and uniaperturate pollen) not only vary within Laurales as considered here but are also found in other basal angiosperm lineages.

Detailed phylogenetic analyses of Laurales were provided by Renner (1999) and Chanderbali et al. (2001), who used sequences from multiple DNA regions. Within Laurales, the deepest split is between Calycanthaceae (including Idiospermaceae) and the remaining six families, which in turn form two clades (Figure 3.3): (1) Siparunaceae, which are sister to Atherospermataceae + Gomortegaceae; and (2) a clade of Hernandiaceae sister to Monimiaceae + Lauraceae. Monimiaceae as traditionally recognized are clearly polyphyletic. Phylogenetic studies indicated that Atherospermataceae and Siparunaceae (then considered subfamilies within Monimiaceae, but see Schodde 1970) must be recognized as distinct families. Several morphological character-state changes are congruent with the molecular tree (Renner 1999). Calycanthaceae have disulculate tectate-columellate pollen, whereas their sister clade has inaperturate, thin-exined pollen (with the exception of Atherospermataceae, which have columellate, but meridianosulcate or disulcate pollen). Calycanthaceae also have two ovules, whereas the remaining Laurales have a solitary ovule in each carpel. Hernandiaceae and Lauraceae + Monimiaceae have apical ovules, whereas the ancestor of the clade of Siparunaceae and Atherospermataceae + Gomortegaceae is inferred to have had basal ovules, a condition lost in *Gomortega,* the only lauralean genus with a syncarpous ovary. Depending on the correct placement of Calycanthaceae-like fossil flowers, tetrasporangiate anthers with valvate dehiscence (with the valves laterally hinged) may be ancestral in Laurales and lost in modern Calycanthaceae and Monimiaceae.

Piperales

Piperales consist of Aristolochiaceae, Lactoridaceae, Piperaceae, Saururaceae, and Hydnoraceae (Figure 3.3). Several molecular phylogenetic analyses have provided strong support for the monophyly of Piperales (e.g., Qiu et al. 1999; P. Soltis et al. 1999b; Barkman et al. 2000; D. Soltis et al. 2000; Zanis et al. 2002, 2003). Molecular phylogenetic analyses also place Hydnoraceae, a family of parasitic plants (illustrated and discussed in Chapter 11), within Piperales. Although the exact placement of Hydnoraceae is uncertain, they appear close to Aristolochiaceae (Nickrent and Duff 1996; Nickrent et al. 1998, 2002). Cladistic analyses of morphological characters likewise recognized a clade of Piperaceae, Saururaceae, Aristolochiaceae, and Lactoridaceae (Doyle and Endress 2000); Hydnoraceae were not included in that analysis.

The circumscription of Piperales has varied. Saururaceae and Piperaceae have consistently been considered closely related (Cronquist 1981, 1988; Takhtajan 1987, 1997). However, the relationships of Aristolochiaceae, Lactoridaceae, and Hydnoraceae have long been problematic. Cronquist (1981, 1988) placed Lactoridaceae in Magnoliales close to Magnoliaceae (see also Lammers et al. 1986) and put Aristolochiaceae in their own order; he also added Chloranthaceae to Piperales. Some have suggested a close relationship of Lactoridaceae and Piperaceae (Weberling 1970; Carlquist 1990) and of Aristolochiaceae to Annonaceae (e.g., Dahlgren 1980; Thorne 1992a, 1992b; Takhtajan 1997). Whereas Takhtajan (1987, 1997) considered Hydnoraceae to be closely related to Aristolochiaceae, others (e.g., Cronquist 1981; Heywood 1998) placed Hydnoraceae in Rosidae.

The monophyly of Piperales is supported by distichous phyllotaxis, a single prophyll, and oil cells. Within Piperales, Piperaceae and Saururaceae form a well-supported clade (Figure 3.3). *Lactoris* is closely related to Aristolochiaceae, although the exact position of *Lactoris* within Piperales remains uncertain. In molecular phylogenetic analyses, *Lactoris* appears within Aristolochiaceae, as sister to *Aristolochia + Thottea* (Qiu et al. 1999; Zanis et al. 2003) or *Aristolochia* alone (D. Soltis et al. 2000; *Thottea* was not included), but support for a placement of *Lactoris* within Aristolochiaceae is weak, even with five genes.

The relationship of Lactoridaceae to Aristolochiaceae is often considered a "surprise" result of molecular phylogenetics, given what appear to be major gross morphological differences between the two families. However, morphological similarities between Aristolochiaceae and *Lactoris* have been noted and include strongly extrorse anthers with broad connective, almost sessile anthers, and stamens basally fused with the gynoecium (Endress 1994a). González and Rudall (2001) examined branching pattern, inflorescence morphology, and stipule development in *Lactoris* and other Piperales. They determined that whereas sympodial growth and a sheathing leaf base are present in all Piperales, the presence of stipules is confined to *Lactoris*, Saururaceae, and some Piperaceae. These characters are consistent with the placement of *Lactoris* within Piperales. A close relationship between *Lactoris* and Aristolochiaceae is indicated by the presence of similar inflorescences in both (González and Rudall 2001). In morphology-based cladistic analyses, *Lactoris* appears with Piperaceae + Saururaceae in some most-parsimonious trees and with Aristolochiaceae in others (Doyle and Endress 2000).

Given the uncertain position of *Lactoris* in both molecular and morphological trees, APG (1998) and APG II (2003) recommend that Lactoridaceae be retained until more convincing evidence of its placement is obtained. Additional taxon sampling of Aristolochiaceae will allow more precise determination of the relationship between Aristolochiaceae and *Lactoris*. Recent phylogenetic analyses of Aristolochiaceae are those of Murata et al. (2001) and González and Stevenson (2002).

Canellales

A sister-group relationship of Canellaceae + Winteraceae received bootstrap or jackknife support of 99 or 100% in all multigene analyses (e.g., Qiu et al. 1999; P. Soltis et al. 1999b; D. Soltis et al. 2000; Zanis et al. 2002, 2003). Doyle and Endress's (2000) morphological analysis also found this sister group (bootstrap support <50%). However, recent morphology-based classifications considered Winteraceae closely related to Magnoliaceae (e.g., Cronquist 1981, 1988; Heywood 1993). Canellaceae, in contrast, were often placed near Myris-

ticaceae (e.g., Wilson 1966; Cronquist 1981, 1988). Takhtajan (1997) placed Canellaceae and Winteraceae in separate orders, noting similarities of Canellaceae to both Winteraceae and Illiciaceae.

This Canellaceae + Winteraceae clade was initially referred to as Winterales (e.g., Qiu et al. 1999; Doyle and Endress 2000; D. Soltis et al. 2000; Zanis et al. 2002). However, Canellales is older than Winterales, and the former name has been used in APG II (2003; see also Chapter 10). This will undoubtedly lead to some confusion in the literature.

The phylogenetic position of Winteraceae within the magnoliid clade, well removed from the base of the angiosperms, is noteworthy in that some classifications have considered the family to be perhaps the "most archaic" living family of angiosperms (Cronquist 1981; see also Endress 1986a). Winteraceae have also been of interest because of their vesselless xylem and (in some taxa) plicate carpels. The phylogenetic position of Winteraceae indicates, however, that these features could be secondarily derived (Young 1981; Doyle 2000; see discussion of character evolution in basal angiosperms, below).

Putative synapomorphies of Canellales include well-differentiated pollen tube transmitting tissue, an outer integument with only two to four cell layers, and seeds with a palisade exotesta (Doyle and Endress 2000). Winteraceae and Canellaceae share a similar irregular, "first-rank" leaf venation (Hickey and Wolfe 1975; see also Doyle 2000). There are also anatomical similarities in stelar and nodal structure (Keating 2000). Vascularization of the seeds in Winteraceae also indicates a close relationship to Canellaceae (Deroin 2000).

Phylogenetic studies using morphology (Vink 1988, 1993) indicated that *Takhtajania,* a genus until recently thought extinct in nature, was sister to all other Winteraceae. With the rediscovery of *Takhtajania* (Schatz 2000), material became available for molecular phylogenetic and other analyses, and the placement of *Takhtajania* as sister to other Winteraceae was confirmed by DNA sequence data (Karol et al. 2000). An issue of the *Annals of the Missouri Botanical Garden* (87:3, pp. 297 to 432) was dedicated to an overview of *Takhtajania* (illustrated in Figure 3.4H), including discussion of its rediscovery and relationships. The issue contained papers discussing the morphology, anatomy, palynology, cytology, and ecology of both Canellaceae and Winteraceae (Carlquist 2000; Doust 2000; Doyle 2000; Endress et al. 2000b; Feild et al. 2000b; Karol et al. 2000; Sampson 2000; Tobe and Sampson 2000).

Character Evolution in Basal Angiosperms

The high degree of resolution and support for relationships among basal angiosperms affords the opportunity to analyze character evolution across basal lineages and throughout the angiosperms as a whole. There are, however, several important caveats in these studies. Importantly, the sister group of the angiosperms is unknown, which compromises efforts to infer ancestral angiosperm states. Furthermore, many characters of interest are confined to the angiosperms and cannot be scored in the outgroup. For example, extant gymnosperms lack a perianth. If the outgroup of the angiosperms is coded as having an indeterminate number of perianth parts, then an indeterminate number is also ancestral for the angiosperms. Alternatively, if the ancestor of the angiosperms is considered to lack a perianth, then it is equally parsimonious for the base of the angiosperms to be either trimerous or indeterminate in perianth merosity (Zanis et al. 2003). Alternative methods of character-state reconstruction that do not rely on outgroup coding, such as maximum likelihood (e.g., Pagel 1998, 1999), may be more useful for inferring patterns of character evolution in the earliest angiosperm lineages. Extant basal angiosperms such as *Amborella* and Nymphaeaceae are likely not the earliest angiosperms and may not even represent or resemble the products of the earliest diversification events. Furthermore, the likelihood of extinction of many basal angiosperm lineages (e.g., Friis et al. 2000) also compromises the accuracy of any character-state optimizations. It is also important to realize that we are attempting to reconstruct the common ancestor of all extant angiosperms, not necessarily the first flowering plants.

With these limitations in mind, in this discussion we focus primarily on characters that are among those most critical for understanding the morphology and anatomy of the earliest angiosperms and their subsequent diversification. These include habit, the presence of vessel elements, and diverse aspects of floral morphology including perianth and stamen phyllotaxis, perianth merosity, perianth differentiation, stamen form, and carpel form and sealing.

Albert et al. (1998), Doyle and Endress (2000), Zanis et al. (2003), and Ronse De Craene et al. (2003) focused on the evolution of floral features. However, Albert et al. (1998) used a topology with *Ceratophyllum* sister to all other angiosperms, reflecting the understanding of relationships available at that time from analysis of *rbcL* sequences. In their mapping studies, Doyle and Endress (2000) standardly used a combined morphology plus three-gene dataset (the latter from D. Soltis et al. 2000). However, some critical taxa (e.g., *Ceratophyllum*) were omitted; more importantly, the topology they used differs substantially from our current understanding of relationships. For example, their placement of Chloranthaceae, monocots, Piperales, Canellales, and Laurales does not agree with the placements of these taxa in re-

cent molecular analyses (e.g., D. Soltis et al. 2000; Zanis et al. 2003). In addition, Albert et al. (1998) and Doyle and Endress (2000) used families as placeholders, either exclusively or in large part, rather than individual genera, making it difficult to assess patterns of variation for characters that are polymorphic within a family. Finally, the recent clarification of basal angiosperm relationships has focused attention on Amborellaceae, Nymphaeaceae, and Austrobaileyales. As a result, a wealth of new data has been obtained for these taxa, including important information on floral development (e.g., Endress and Igersheim 2000a, 2000b; P. Endress 2001a; Posluszny and Tomlinson 2003; Buzgo et al. 2004). To reflect these new data, we have used different codings for several characters (e.g., merosity, phyllotaxis) from those used in earlier reconstructions of floral evolution (e.g., Albert et al. 1998; P. Soltis et al. 2000; Ronse De Craene et al. 2003; Zanis et al. 2003).

In our reconstructions of character evolution conducted for this book, we used the best current estimate of basal angiosperm relationships (Figure 3.3) and MacClade (Maddison and Maddison 1992). Two sets of optimizations were conducted, one for perianth phyllotaxis, perianth merosity, perianth differentiation, and stamen phyllotaxis and a second set for some of the characters analyzed by Doyle and Endress (habit and anatomical and micromorphological characters). For optimizations of perianth differentiation, merosity, and phyllotaxis and for stamen phyllotaxis, we used a 63-taxon tree. This topology represents a reduced version of a larger topology (>100 taxa) and underlying matrix assembled by Ronse De Craene et al. (2003) for the study of floral evolution. To construct the tree, one of the shortest trees of Zanis et al. (2003) was used as a backbone and additional genera were included for some families (e.g., Winteraceae and Annonaceae), using published topologies for those groups. This grafted supertree enabled us to examine patterns of variation within and among families. We typically used genera as placeholders. Because many of these genera may not be readily familiar to all readers, a list of genera used as placeholders, as well as the family and order (if placed to order) designations for each genus (*sensu* APG II 2003), is provided in Table 3.1. Floral characters and states mostly follow those given in Ronse De Craene et al. (2003), P. Endress (2001a), and the primary sources cited therein. For the second set of reconstructions, we used the tree from Zanis et al. (2003) pruned to 53 taxa. In this set of analyses, families were sometimes used as placeholders because few data were available for some characters. In addition to explicit reconstructions, we have also informally reconsidered the evolution of vessel elements and stamen form, and have summarized the reconstructions of endosperm development (Floyd and Friedman 2000).

TABLE 3.1   Taxa employed in character state reconstructions (arranged by family), providing family and ordinal designations (*sensu* APG II 2003; see also Chapter 10)

Order (if assigned)	Family	Genus	Order (if assigned)	Family	Genus
Acorales	Acoraceae	Acorus	Laurales	Lauraceae	Laurus
Alismatales	Alismataceae	Sagittaria	Magnoliales	Magnoliaceae	Magnolia
—	Amborellaceae	Amborella	Magnoliales	Magnoliaceae	Liriodendron
Magnoliales	Annonaceae	Annona	Laurales	Monimiaceae	Hortonia
Magnoliales	Annonaceae	Cananga	Laurales	Monimiaceae	Kibara
Alismatales	Aponogetonaceae	Aponogeton	Magnoliales	Myristicaceae	Knema
Alismatales	Araceae	Spathiphyllum	Magnoliales	Myristicaceae	Myristica
Piperales	Aristolochiaceae	Aristolochia	Proteales	Nelumbonaceae	Nelumbo
Piperales	Aristolochiaceae	Asarum	—	Nymphaeaceae	Brasenia
Piperales	Aristolochiaceae	Saruma	—	Nymphaeaceae	Cabomba
Piperales	Aristolochiaceae	Thottea	—	Nymphaeaceae	Nuphar
Laurales	Atherospermataceae	Daphnandra	—	Nymphaeaceae	Nymphaea
Asparagales	Asphodelaceae	Bulbine	—	Nymphaeaceae	Victoria
Austrobaileyales	Austrobaileyaceae	Austrobaileya	Ranunculales	Papaveraceae	Dicentra
Alismatales	Butomaceae	Butomus	Piperales	Piperaceae	Piper
—	Buxaceae	Buxus	Proteales	Platanaceae	Platanus
—	Buxaceae	Didymeles	Ranunculales	Ranunculaceae	Glaucidium
Laurales	Calycanthaceae	Calycanthus	Ranunculales	Ranunculaceae	Ranunculus
Canellales	Canellaceae	Canella	Piperales	Saururaceae	Houttuynia
—	Ceratophyllaceae	Ceratophyllum	Piperales	Saururaceae	Saururus
—	Chloranthaceae	Ascarina	Austrobaileyales	Schisandraceae	Illicium
—	Chloranthaceae	Chloranthus	Austrobaileyales	Schisandraceae	Schisandra
—	Chloranthaceae	Hedyosmum	Alismatales	Tofieldiaceae	Tofieldia
Magnoliales	Degeneriaceae	Degeneria	Austrobaileyales	Trimeniaceae	Trimenia
Magnoliales	Eupomatiaceae	Eupomatia	—	Trochodendraceae	Tetracentron
Ranunculales	Eupteleaceae	Euptelea	—	Trochodendraceae	Trochodendron
Laurales	Gomortegaceae	Gomortega			
Laurales	Hernandiaceae	Gyrocarpus	Canellales	Winteraceae	Zygogynum
Laurales	Hernandiaceae	Hernandia	Canellales	Winteraceae	Drimys
Magnoliales	Himantandraceae	Galbulimima	Canellales	Winteraceae	Tasmannia
Piperales	Lactoridaceae	Lactoris	Canellales	Winteraceae	Takhtajania
Laurales	Lauraceae	Cryptocarya			

Habit

The hypothesized habit (woody versus herbaceous) of early angiosperms has long been debated. Cronquist (1968, 1988), Takhtajan (1969), Stebbins (1974), Thorne (1976), Loconte (1996), and others considered the ancestral angiosperms to be woody because all gymnosperms are woody and also because this habit is prevalent in the extant angiosperms they considered to be archaic (e.g., Magnoliales). In contrast, some phylogenetic studies and fossil data focused attention on several herbaceous families (e.g., most Chloranthaceae, Piperaceae, Nymphaeaceae) as possible sister groups to all other extant angiosperms; these data indicated a

herbaceous origin of angiosperms (e.g., D.W. Taylor and Hickey 1992, 1996a, 1996b; Doyle et al. 1994).

Doyle and Endress (2000) reevaluated the evolution of habit, recognizing two states: (1) tree or shrub (woody) or (2) rhizomatous, scandent, or acaulescent (herbaceous). In their reconstructions, the ancestral habit in angiosperms is equivocal, with the rhizomatous, scandent (herbaceous) habit established as ancestral for all angiosperms except *Amborella*; the tree-shrub (woody) habit was derived independently in core magnoliids and eudicots. Doyle and Endress (2000) scored *Chloranthus* and *Ascarina* (Chloranthaceae) as herbaceous (rhizomatous, scandent, or acaulescent) and

Hedyosmum as polymorphic because the young stages of *Hedyosmum* have been considered rhizomatous and the mature plants woody (Blanc 1986).

We mapped habit using the two states used by Doyle and Endress (2000), although we coded *Lactoris* as woody rather than herbaceous; clearly more detailed anatomical studies of *Lactoris* would be useful. The coding of Chloranthaceae for habit has a large effect on character-state reconstruction. The coding of Chloranthaceae as herbaceous (per Doyle and Endress 2000) may be problematic. If *Hedyosmum* is scored as woody, rather than as polymorphic for both woody and herbaceous forms, or if all three representatives of Chloranthaceae (*Chloranthus, Ascarina, Hedyosmum*) are scored as polymorphic, then our reconstructions (Figure 3.7) indicate that the woody habit is ancestral for extant angiosperms. This is also the case if a topology is used in which Nymphaeaceae are treated as sister to *Amborella*. The sensitivity of the results to single changes in the data matrix is further illustrated by adding the fossil *Archaefructus* (illustrated in Figure 3.8) to the analysis; with *Archaefructus* added, the ancestral state for angiosperms is reconstructed as herbaceous (not shown).

Chloranthoid leaf teeth

The function of chloranthoid leaf teeth is unknown, but from evidence in ferns (Sperry 1983), they may function as release valves for root pressure (Feild et al. 2003b, 2004). Chloranthoid teeth are present in several of the basalmost angiosperm lineages, including *Amborella* (Bailey and Swamy 1948), *Kadsura* and *Schisandra* (Schisandraceae; Hickey and Wolfe 1975), and species of *Trimenia* (Trimeniaceae; Rodenburg 1971; Endress 1987c). Species of *Illicium* generally have entire leaf margins, but chloranthoid teeth have also been reported for the genus (Hickey and Wolfe 1975). We have scored both *Trimenia* and *Illicium* as polymorphic for presence/absence of chloranthoid teeth. Chloranthoid teeth are also present in Chloranthaceae, but are not found in *Austrobaileya* or Nymphaeaceae. In the latter family, which is aquatic, such structures presumably would not be needed to release root pressure. Among basal eudicots, a similar tooth type is found in Trochodendraceae, Buxaceae, and several families of Ranunculales (Berberidaceae, Papaveraceae, Ranunculaceae, and Lardizabalaceae). Doyle and Endress (2000) scored the leaf teeth of these eudicots as homologous to the chloranthoid type; we have followed that approach here.

Doyle and Endress (2000) suggested that chloranthoid leaf teeth are ancestral in the angiosperms. However, our analyses indicate that the ancestral condition of extant angiosperms is equivocal; the ancestral state for all angiosperms except *Amborella* is the absence of chloranthoid leaf teeth, with multiple origins of chloranthoid teeth (not shown). Although *Amborella* has chloranthoid teeth, the ancestor of Austrobaileyales is

reconstructed as ambiguous. This is the case regardless of whether *Illicium* and *Trimenia* are scored as polymorphic for this feature or only as possessing chloranthoid teeth. If a topology is used in which *Amborella* and Nymphaeaceae are placed as sisters (see Barkman et al. 2000), then the ancestral state for extant angiosperms is the absence of chloranthoid teeth. Regardless of topology, Chloranthaceae represent a separate origin of this tooth type. However, this is a simplistic reconstruction. As noted, if chloranthoid teeth function to release root pressure, they would not be needed in aquatics and may have been lost. Character optimizations are made especially difficult in such circumstances.

Several independent origins of chloranthoid teeth likely occurred in the basal eudicots: in the common ancestor of Papaveraceae, the ancestor of Berberidaceae and Ranunculaceae, and the ancestors of Trochodendraceae and Buxaceae (or their common ancestor if they are sister taxa). Given the apparent independent derivation of chloranthoid teeth in the eudicots, this character does not appear to be homologous to chloranthoid teeth in basal lineages. Additional studies are needed to assess the homology of chloranthoid teeth of basal angiosperms and eudicots.

Ethereal oils

Ethereal oil cells are prevalent in many basal angiosperm lineages, including all Austrobaileyales and all members of the magnoliid clade. They are also present in *Acorus*, which is sister to all other monocots (see Chapter 4). However, oil cells are not present in *Amborella* (Bailey and Swamy 1948), Nymphaeaceae, or Ceratophyllaceae. The ancestral condition for angiosperms is therefore the absence of ethereal oils. The ancestral state above the branch to Nymphaeaceae remains equivocal with the "all most parsimonious states" resolving option in MacClade. With ACCTRAN (accelerated transformation) optimization, ethereal oils are ancestral above the node to Nymphaeaceae. The absence of these oils in Nymphaeaceae and Ceratophyllaceae could easily be due to losses associated with the aquatic habit. Oil cells are reconstructed as ancestral for both Chloranthaceae and the magnoliid clade and appear to have been present early in the fossil record. Oil cells are abundant, for example, on carpels, petals, and leaves of *Archaeanthus* (Dilcher and Crane 1984). Regardless of optimization or topology, the ancestral state for eudicots is the absence of ethereal oils.

Nodal anatomy

The nodal anatomy of angiosperms has long been a topic of research interest (e.g., Canright 1955; Bailey 1956; Benzing 1967; Cronquist 1968), but the adaptive significance of changes in nodal anatomy is uncertain. Branch systems in the stele in ferns, gymnosperms, and many angiosperms usually have two traces from a sin-

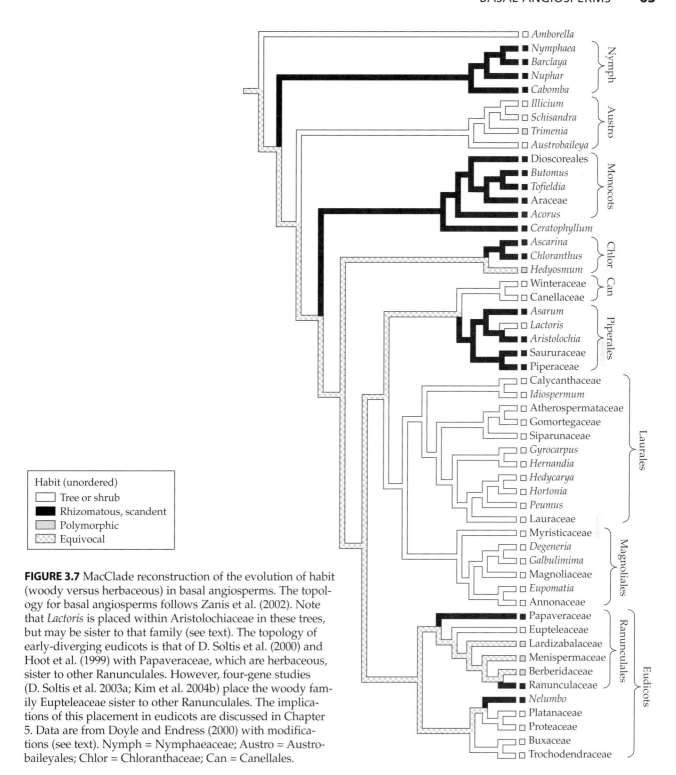

FIGURE 3.7 MacClade reconstruction of the evolution of habit (woody versus herbaceous) in basal angiosperms. The topology for basal angiosperms follows Zanis et al. (2002). Note that *Lactoris* is placed within Aristolochiaceae in these trees, but may be sister to that family (see text). The topology of early-diverging eudicots is that of D. Soltis et al. (2000) and Hoot et al. (1999) with Papaveraceae, which are herbaceous, sister to other Ranunculales. However, four-gene studies (D. Soltis et al. 2003a; Kim et al. 2004b) place the woody family Eupteleaceae sister to other Ranunculales. The implications of this placement in eudicots are discussed in Chapter 5. Data are from Doyle and Endress (2000) with modifications (see text). Nymph = Nymphaeaceae; Austro = Austrobaileyales; Chlor = Chloranthaceae; Can = Canellales.

gle gap. The most common nodal types are one-trace unilacunar, two-trace unilacunar, trilacunar, and multilacunar (Figure 3.9). Bailey (1956) proposed that unilacunar nodes with two traces are ancestral in the angiosperms. Cronquist (1968, 1988) suggested that either unilacunar nodes with two traces or perhaps trilacunar nodes were ancestral. In their reconstructions of nodal

anatomy, Doyle and Endress (2000) provided support for the former hypothesis. Our results based on analyses conducted for this book generally agree with those of Doyle and Endress (2000), but because of the different topologies used in their study and ours, there are also noteworthy differences. Furthermore, Doyle and Endress scored Amborellaceae as having unilacunar

(A)

(B)

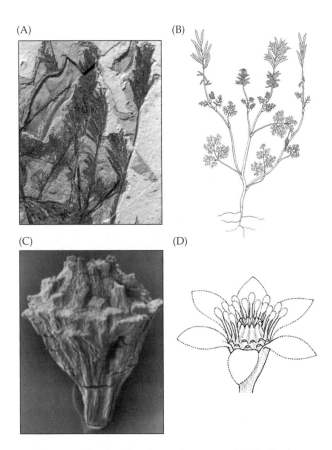

(C)

(D)

FIGURE 3.8 Fossils of early angiosperms. (A) *Archaefructus sinensis*, photograph of fossil reproductive axes (courtesy of D. Dilcher). (B) Reconstruction of *Archaefructus sinensis* (courtesy of D. Dilcher; diagram by K. Simons and D. Dilcher). (C) Lateral view of fossil water lily flower (from Friis et al. 2001). (D) Reconstruction of fossil water lily flower (from Friis et al. 2001; courtesy of P. von Knorring and E. M. Friis).

nodes with two traces, following Bailey (1956). However, a more recent anatomical study (Carlquist and Schneider 2001) indicated that *Amborella* has unilacunar nodes with one trace. It appears, therefore, that *Amborella* may have both states. We therefore used multiple scorings of *Amborella*. In one optimization, we scored *Amborella* as polymorphic, having unilacunar nodes with both one trace and with two traces. We also examined alternative reconstructions in which *Amborella* was coded as having unilacunar nodes with one trace and compared these results with those in which it was coded as having unilacunar nodes with two traces.

Because of the uncertainty surrounding the distinction between unilacunar nodes with one versus two traces, we conducted two different reconstructions. In the first, we followed Doyle and Endress (2000) and recognized four states for nodal anatomy: unilacunar with one trace, unilacunar with two traces, trilacunar, and multilacunar. Siparunaceae were scored in the manner that *Amborella* was scored (Doyle and Endress 2000). In

the second reconstruction, we combined the one- and two-trace types into a single category, unilacunar, and considered three basic states for nodal anatomy: unilacunar, trilacunar, and multilacunar.

Combining unilacunar one-trace and two-trace nodes into a single character state—unilacunar—unambiguously reconstructs the base of the angiosperms through Chloranthaceae as unilacunar (Figure 3.9). When two types of unilacunar nodes are considered, unilacunar nodes with two traces are similarly reconstructed as ancestral for the angiosperms through Chloranthaceae, if *Amborella* is coded as polymorphic, possessing both one and two traces, or as having two traces. However, if *Amborella* is coded as having only nodes that are unilacunar with one trace, then the ancestral state for the angiosperms becomes ambiguous; the ancestral state of all angiosperms except *Amborella* is then reconstructed as unilacunar with two traces. Unilacunar nodes with two traces are present in some Nymphaeaceae (other members of the family have trilacunar nodes) and in *Trimenia* (Austrobaileyales; Marsden and Bailey 1995); other Austrobaileyales (*Austrobaileya, Schisandra,* and *Illicium*) have unilacunar nodes with one trace (Dickison and Endress 1983).

The ancestral state for the magnoliids + eudicots is trilacunar (Figure 3.9C). In addition, both reconstructions indicate that nodal anatomy may provide synapomorphies for some clades. Unilacunar nodes unite Laurales when unilacunar one-trace and two-trace nodes (Figure 3.9, A and B) are combined. When one- and two-trace nodes are scored separately the ancestor of Laurales is reconstructed as having unilacunar nodes with two traces; unilacunar nodes with one trace were derived in the ancestor of Hernandiaceae. An additional, independent origin of trilacunar nodes is indicated for some Magnoliales (Myristicaceae and *Galbulimima,* Himantandraceae). These results contrast with the findings of Doyle and Endress (2000), who used a topology that did not place Canellales (referred to as Winterales by Doyle and Endress) with Piperales or Laurales with Magnoliales. Future analyses should perhaps separate nodal anatomy into two characters: one that codes the number of lacunae and one that codes the number of vascular traces.

Vessel elements

Several families of angiosperms have attracted attention because they lack vessels. Cronquist (1968, 1988), Takhtajan (1969), and others suggested that some of those families lacking vessels (e.g., Winteraceae) might be primitively vesselless. However, a morphological cladistic analysis (Young 1981) showed that it was more parsimonious to have vessels present in the common ancestor of the angiosperms and then lost many times in diverse taxa including *Amborella,* Winteraceae, Trochodendraceae, and *Sarcandra* (Chloranthaceae). Donoghue and

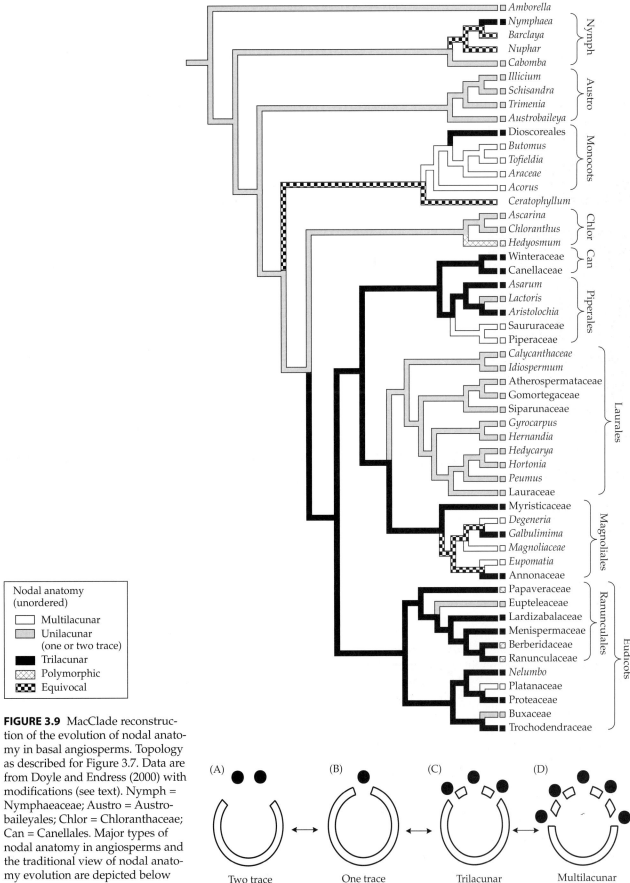

FIGURE 3.9 MacClade reconstruction of the evolution of nodal anatomy in basal angiosperms. Topology as described for Figure 3.7. Data are from Doyle and Endress (2000) with modifications (see text). Nymph = Nymphaeaceae; Austro = Austrobaileyales; Chlor = Chloranthaceae; Can = Canellales. Major types of nodal anatomy in angiosperms and the traditional view of nodal anatomy evolution are depicted below the reconstruction.

Nodal anatomy
(unordered)

☐ Multilacunar
▨ Unilacunar
(one or two trace)
■ Trilacunar
▨ Polymorphic
▨ Equivocal

(A) Two trace unilacunar
(B) One trace unilacunar
(C) Trilacunar
(D) Multilacunar

Doyle (1989b) provided additional support for Young's hypothesis of many losses of vessels.

The discovery that *Amborella,* a vesselless angiosperm, is sister to all other extant angiosperms has renewed interest and discussion about whether the angiosperms were ancestrally vesselless. Adding to the intrigue is the discovery in some angiosperms of tracheary elements that are intermediate between true vessels and tracheids (see Schneider et al. 1995; Schneider and Carlquist 1996). For example, a perforation plate (Figure 3.10) is a feature of a vessel; it is considered to have one, or a series, of modified "pits" or perforations in the end wall of a water-conducting cell. Pit membranes are lacking in the perforations of a true vessel, but some taxa, such as Nymphaeaceae, have porose pit membranes in the end walls. Adding to the complexity is the report of a combination of porose and non-

porose pit membranes in *Zygogynum* of Winteraceae (Carlquist 1983) and *Amborella* (Feild et al. 2000b; Carlquist and Schneider 2001). Recognizing this added complexity, Doyle and Endress (2000) recognized an "intermediate" condition between true vessels and tracheids. In their reconstructions, they found that the vesselless condition is ancestral in angiosperms, with both *Amborella* and Nymphaeaceae lacking vessels, and that vessels were lost in Winteraceae and Trochodendraceae.

More recently, however, Carlquist and Schneider (2002) suggested that definitions of vessels are overly simplistic. When all sources of data are considered, these authors found that changes in not one, but at least four and perhaps as many as six, characters can occur in the evolution of a vessel from a tracheid (Table 3.2). For four of these characters, delimiting the differences in character states is straightforward. Two other differences between tracheids and vessels are relative and therefore cannot be easily coded. In vessel-bearing woods that also contain tracheids, vessel elements are shorter than the tracheids in any given sample, whereas in vesselless wood, lengths of the water-conducting elements (tracheids only) in any given sample follow a unimodal distribution. In addition, tracheids of vessel-

FIGURE 3.10 Changes proposed to occur during evolution of vessel elements from tracheids in vesselless angiosperms (modified from Carlquist and Schneider 2002). (A) Vesselless angiosperm; tracheids similar to those present in *Amborella.* (B) Vessel-bearing angiosperm. The vessel (on the left) and the tracheid (on the right) are similar to those found in Chloranthaceae and *Illicium,* basal angiosperms in which the vessels exhibit "tracheid" features (see text). The six changes between tracheids and vessels are illustrated: (1) Pit membranes of end walls of tracheids typically are nonporose (not shown), but in some taxa the pit membranes become porose (C). As vessels originate, varying degrees of pit membrane remnants are visible in perforations (D). In contrast, pit membranes of the lateral walls of vessels remain intact and nonporose, just as they are on lateral walls of tracheids (E). (2) Conductive area of the end wall of the wider tracheid in A is larger than the conductive area of pits of the tracheid lateral walls because the pits are larger on the end walls. This is interpreted as an incipient form of vessel-like structure in *Amborella.* As vessels originate, the differences in size of perforations become more pronounced. (3) There are proposed changes in the morphology of perforations of end walls compared to lateral walls when vessels originate. One difference is the narrower borders on the perforations of the vessel compared to borders of lateral pit walls (shown); other differences in morphology of lateral wall pits also occur (see Table 3.2). (4) Vessel elements become wider than the tracheids that they accompany. The vessel element in B is wider than the associated tracheid. (5) Vessel elements are shorter than the tracheids that they accompany. Whereas tracheids are all the same length in wood that lacks vessels (A), vessel-bearing wood has vessel elements shorter than the tracheary elements they accompany (B). (6). Tracheids in vesselless wood (A) are longer than both the vessel elements and the tracheids of a vessel-bearing wood (B). That is, the origin of vessels results in shorter vessel elements and also shorter tracheid elements. (Legend modified from Carlquist and Schneider 2002.)

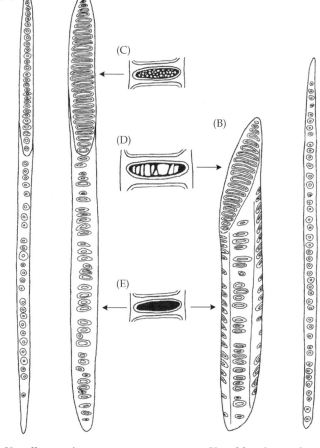

Vesselless angiosperm   **Vessel-bearing angiosperm**

TABLE 3.2 Characters that distinguish vessels from tracheids (modified from Carlquist and Schneider 2002)

1. Pit membranes of end walls
    - lacking porosities (tracheid)
    - porose or web-like (intermediate)
    - entirely lacking (true vessel)

2. Pits of end walls
    - with the same conductive area per unit wall area as pits of lateral walls
    - with more conductive area per unit wall area than the pits of lateral walls

3. Pit morphology of end walls
    - identical to pitting of lateral walls
    - different from that of end walls

4. Tracheary element diameter
    - uniform
    - bimodal, with wider elements possessing one or more other features of vessel elements

less woods are much longer than both the tracheids and vessel elements of vessel-bearing woods.

Hence, "vessel" should not be treated as a single character with two states, presence or absence. Furthermore, adding a third state, intermediate, does not adequately capture the complexity of the transition from tracheid to vessel. Unfortunately, adequate data are not currently available to map any of the characters that Carlquist and Schneider recognized as distinguishing "vessels" from "tracheids." Porosity, for example, can be measured only by SEM examination, and the number of taxa examined critically for this feature is still small.

Recent ecophysiological insights into the function of water-conducting cells in basal angiosperms also bear on the evolution of vessel elements. It is possible, perhaps probable, that the early angiosperms lacked vessels, because *Amborella* and Nymphaeaceae lack vessels and other early-diverging lineages such as *Illicium* and *Austrobaileya* have what are regarded as "primitive" vessel types with tracheid-like features (Carlquist and Schneider 2002). In *Austrobaileya*, for example, the vessels have gradually tapered end walls, numerous scalariform perforation plates, and are generally small in diameter (Bailey and Swamy 1948; Carlquist 2001a; Figure 3.10). *Illicium* and *Kadsura* have vessel elements with long, oblique perforation plates that partially retain pit membranes (Carlquist 1983; Carlquist and Schneider 2002).

Perianth phyllotaxis

To explore patterns of evolution in perianth phyllotaxis, we used both multistate and binary character codings. With multistate coding, phyllotaxis was scored as absent (no perianth), a single whorl, two whorls, multiple whorls (more than two), and spiral. We also used a second mul-

tistate coding in which we reduced the number of forms of whorled phyllotaxis (i.e., single whorl, two whorls, multiple whorls) to a single character state—whorled (see Zanis et al. 2003). In this second multistate coding, phyllotaxis was scored as absent, spiral, or whorled. For binary coding (not shown), phyllotaxis was either spiral or whorled (see Ronse De Craene et al. 2003).

Amborella has spiral phyllotaxis, as do members of Austrobaileyales (Figure 3.4; Endress and Igersheim 2000a). In some families, phyllotaxis is complex and does not fit cleanly into any of the states specified for either multistate or binary coding. In some Nymphaeaceae, phyllotaxis has often been considered to be spiral, but it now appears to be primarily whorled (Figure 3.11) or in some cases irregular (Endress 2001a). In Winteraceae, phyllotaxis is primarily whorled but occasionally spiral (Doust 2000). In *Drimys winteri,* flowers within one tree vary between spiral and whorled phyllotaxis (Doust 2001). For other taxa, phyllotaxis has not been carefully examined, and critical studies are still needed. In addition, the distinction between spiral and whorled is not always clear. Ontogenetic studies have revealed that in some cases floral organs that appear to be whorled in mature flowers actually result from spiral initiation of primordia and the rhythmic occurrence of a long plastochron (the time interval between the initiation of two consecutive organ primordia; Tucker 1960; Endress 1994c) after several short plastochrons (e.g., Huber 1980; Leins and Erbar 1985). Thus, both spiral and whorled phyllotaxis might have organs developing in a spiral sequence (Endress 1987a, 1987b). In contrast, *Illicium* has spiral phyllotaxis of all floral organs in developing buds, but in mature flowers the carpels have an apparent whorled arrangement (Figure 3.11).

Reconstruction of phyllotaxis with more than one whorled state (e.g., single whorl, two whorls, multiple whorls) results in ambiguous character-state optimizations, and evolutionary inference is difficult. Albert et al. (1998) similarly discussed the ambiguities and limitations in a multistate character coding of perianth phyllotaxis. Hence, our discussion focuses on the results of reconstructions that involve three states—spiral, whorled, and absent (Figure 3.12); for more detail on multistate reconstructions, see Albert et al. (1998), Ronse De Craene et al. (2003), and Zanis et al. (2003).

For our analysis, we scored *Ceratophyllum* as lacking a perianth (see Endress 1994b, 2001a), although the whorl of scales that precedes the reproductive organs has been variously interpreted as perianth or bracts (Aboy 1936; Les 1988, 1993; Les et al. 1991; Endress 1994b, 2001a; Albert et al. 1998). Most members of Chloranthaceae lack a perianth; however, the pistillate flowers of *Hedyosmum* have three structures variously interpreted as a perianth or as staminodes (reviewed in Endress 1987c). We scored *Hedyosmum* as having an undifferentiated perianth. As noted, the interpretation of perianth

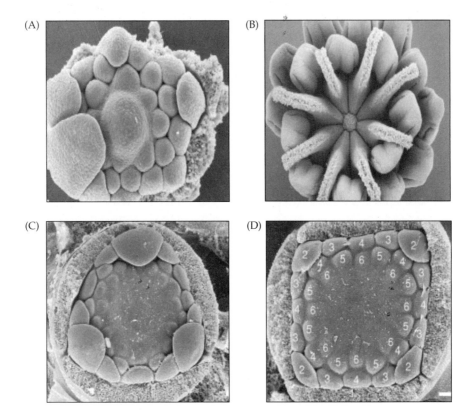

FIGURE 3.11 Phyllotaxis and merosity (merism) in floral buds of basal angiosperms. (A) Young floral bud of *Illicium anisatum* (Schisandraceae) showing spiral phyllotaxis of organs, including carpels (around central hump). (From Endress 2001a.) (B) Mature flower of *Illicium anisatum* (Schisandraceae) showing apparent "whorled" orientation of carpels. (From Endress 2001a.) (C) Young floral bud of *Victoria cruziana;* whorled phyllotaxis and trimerous. (From Endress 2001a.) (D) Young floral bud of *Victoria cruziana;* whorled phyllotaxis and tetramerous. (From Endress 2001a.)

phyllotaxis in Nymphaeaceae has been debated. Ronse De Craene et al. (2003) considered *Brasenia* and *Cabomba* to have whorled phyllotaxis; they considered the phyllotaxis of other genera (*Nymphaea, Victoria, Nuphar*) to be polymorphic, with early organs appearing in a whorled arrangement and later ones in a spiral. Zanis et al. (2003), in contrast, scored *Nymphaea, Victoria,* and *Nuphar* as spiral. In our analysis, we follow Endress (2001a), who suggested that the phyllotaxis of *Nymphaea, Victoria,* and *Nuphar* is whorled and then becomes irregular, but "the phyllotaxis of these flowers is not spiral."

Spiral phyllotaxis is present in many members of the Amborellaceae-Nymphaeaceae-Austrobaileyales grade, including *Amborella, Trimenia, Austrobaileya,* and all Schisandraceae (including *Illicium;* see Figures 3.4 and 3.11). However, because of the whorled (and irregular) phyllotaxis of Nymphaeaceae, the ancestral reconstruction for perianth phyllotaxis for the angiosperms depends on the coding of the outgroup. For example, if the outgroup (gymnosperms) is coded as lacking a perianth, then either a spiral or whorled phyllotaxis is reconstructed as equally parsimonious for the base of the angiosperms. If the outgroup is coded as having a spiral-

ly arranged perianth, then the spiral perianth is reconstructed as ancestral for the angiosperms. If the outgroup is coded as having a whorled perianth, then the whorled perianth is ancestral for the angiosperms, and a spiral perianth evolved independently several times.

Although the ancestral phyllotaxis for the angiosperms is reconstructed as ambiguous in this analysis, different results were found in earlier analyses. Albert et al. (1998) found a simple whorled perianth to be ancestral in the angiosperms because they used the *rbcL* topology (Chase et al. 1993) in which *Ceratophyllum* is sister to all other angiosperms and coded *Ceratophyllum* as having a perianth rather than as lacking a perianth (see above). In other studies (e.g., P. Soltis et al. 2000), the ancestral condition for angiosperms was reconstructed as spiral because different codings for Nymphaeaceae were used. If genera such as *Victoria, Nymphaea,* and *Nuphar* were scored as having a spiral perianth, then a spiral perianth is reconstructed as ancestral throughout the Amborellaceae–Nymphaeaceae–Austrobaileyales grade (e.g., P. Soltis et al. 2000; Zanis et al. 2003). However, if the perianth phyllotaxis of these genera is considered to be primarily whorled (see Endress 2001a, 2001b; Ronse

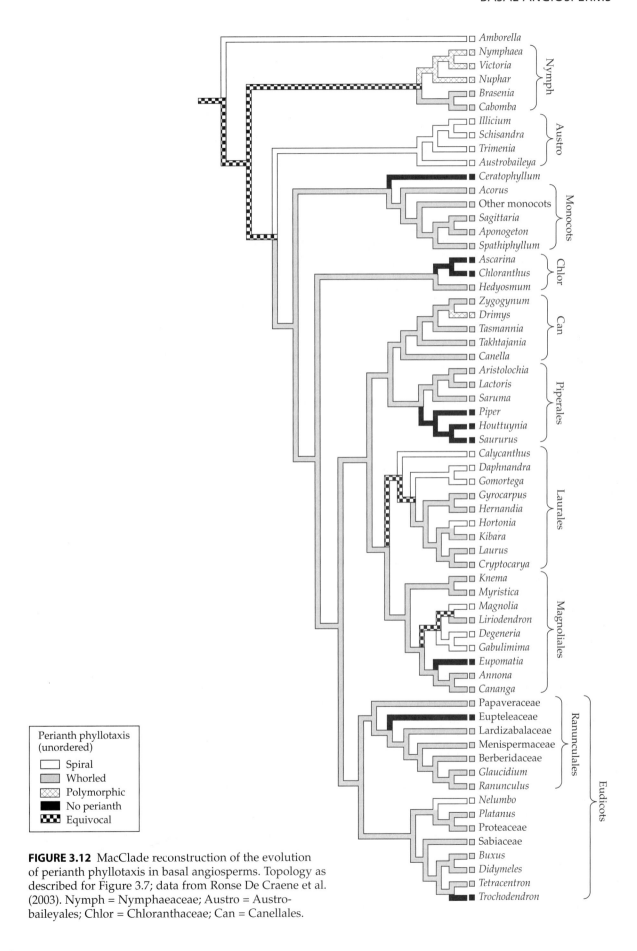

FIGURE 3.12 MacClade reconstruction of the evolution of perianth phyllotaxis in basal angiosperms. Topology as described for Figure 3.7; data from Ronse De Craene et al. (2003). Nymph = Nymphaeaceae; Austro = Austrobaileyales; Chlor = Chloranthaceae; Can = Canellales.

De Craene and Smets 1993), which developmental data indicate is the case, then a whorled perianth is present throughout the family, and the ancestral phyllotaxis for angiosperms is reconstructed as equivocal (Figure 3.12). The same ambiguous reconstruction for angiosperms is obtained if the perianth phyllotaxis of Nymphaeaceae is considered equivocal or polymorphic rather than whorled. Thus, the ancestral state for phyllotaxis in the angiosperms is equivocal.

Above the Amborellaceae–Nymphaeaceae–Austrobaileyales grade, whorled perianth phyllotaxis is reconstructed as ancestral for all remaining angiosperms with multiple shifts to a spiral perianth occurring in Calycanthaceae, some Atherospermataceae (e.g., *Daphnandra*; not all members of the family have a spiral perianth—*Dryadodaphne* has whorled phyllotaxis), Gomortegaceae, some Monimiaceae (e.g., *Hortonia*; Endress 1980a, 1980b), Degeneriaceae (*Degeneria*), and Magnoliaceae (some species of *Magnolia*; Figure 3.12). Another possible transformation from whorled to spiral phyllotaxis may have occurred in *Drimys* and *Pseudowintera* (Winteraceae), which have a complex phyllotaxis involving spirals and multiple whorls (Doust 2000, 2001; Endress et al. 2000b). Thus, perianth phyllotaxis is highly labile in basal angiosperms, as was previously suggested (Endress 1987a, 1987b, 1994a, 1994c; Albert et al. 1998; Zanis et al. 2003). Studies of floral development indicate that some basal angiosperms are not fully committed to either spiral or whorled phyllotaxis (Buzgo et al. 2004).

Perianth merosity (merism)

In our analysis, we coded perianth merosity as dimerous, trimerous, tetramerous, pentamerous, or indeterminate, the latter referring to a variable and typically large number of perianth parts. Data for merosity were taken from Endress (2001a), Ronse De Craene et al. (2003), and Zanis et al. (2003). The flowers of many lineages of basal angiosperms have multiple parts, whereas those of other clades are clearly trimerous in their basic floral plan. However, in some instances the coding of merosity is problematic. In some Nymphaeaceae, many perianth parts are present (e.g., *Victoria* and *Nymphaea*), but the basic plan for these genera is trimerous or tetramerous (Endress 2001a). Other families are particularly complex. In species of Winteraceae, the outermost floral organs are in dimerous whorls in some genera, whereas others possess tetramerous, and even pentamerous whorls (Endress et al. 2000b). Similarly, in Magnoliaceae, the perianth of some species of *Magnolia* is indeterminate, whereas that of *Liriodendron* and other species of *Magnolia* is in three trimerous whorls; Magnoliaceae may therefore represent a transition from whorled to spiral phyllotaxis (Tucker 1960; Erbar and Leins 1981, 1983). In other cases, more data are needed. Only one species of Siparunaceae has been investigated

(*Siparuna nicaraguensis*, now called *S. thecaphora*; Endress 1980). Staminate flowers have an undifferentiated perianth of four (or five) tepals and could be interpreted as dimerous. However, pistillate flowers have five or six tepals, and merism is unclear. We did not include Siparunaceae in Figure 3.13; however, Zanis et al. (2003) attempted reconstructions with Siparunaceae scored as either dimerous or uncertain.

Amborella is considered to exhibit indeterminate merosity (Figures 3.4 and 3.13). Other early-branching lineages also have indeterminate merosity with Austrobaileyales uniformly indeterminate. Within Nymphaeaceae, *Cabomba*, *Brasenia*, and *Nuphar* are trimerous; other genera (e.g., *Victoria*, *Nymphaea*) also appear trimerous or tetramerous (Figures 3.11 and 3.13; see Endress 2001a). Hence, trimery appears to be ancestral for the entire family. Importantly, these results also illustrate that trimery arose early in angiosperm evolution. As for phyllotaxis, reconstruction of the ancestral merosity of extant angiosperms depends on the coding of merosity for the outgroup. If the outgroup of the angiosperms is coded as having an indeterminate number of perianth parts, then an indeterminate number is also ancestral for the angiosperms. Alternatively, if the ancestor of the angiosperms is considered to lack a perianth, then it is equally parsimonious for the base of the angiosperms to be either trimerous or indeterminate in perianth merosity (see Zanis et al. 2003).

Regardless of the outgroup coding, above the basal grade of *Amborella*, Nymphaeaceae, and Austrobaileyales, the ancestral character state for the angiosperms is a trimerous perianth (Figure 3.13; see also Ronse De Craene et al. 2003, Zanis et al. 2003). This is the case regardless of whether *Amborella* and Nymphaeaceae are treated as a clade or as successive sisters to all other angiosperms. Thus, although the trimerous condition is typically associated with monocots, these results indicate that trimery has played a major role in the early evolution and diversification of the flower (see also Kubitzki 1987; Ronse De Craene and Smets 1994).

Following the origin of a trimerous perianth, there was a return to an indeterminate perianth in several magnoliid lineages, including Calycanthaceae (e.g., *Calycanthus*), the clade of Atherospermataceae (represented by *Daphnandra*) and Gomortegaceae (*Gomortega*), Himantandraceae (*Galbulimima*), some Monimiaceae (e.g., *Hortonia*), and some Magnoliaceae (*Magnolia*). A perianth has also been lost several times (e.g., Eupomatiaceae, Piperaceae, most Chloranthaceae, and Ceratophyllaceae; Figures 3.12 and 3.13).

These reconstructions therefore indicate that perianth merosity is also labile in basal angiosperms (see also Endress 1987a, 1987b, 1994a, 1994c; Albert et al. 1998; Zanis et al. 2003), a condition that continues through the basal nodes of the eudicots (see Chapter 5) and contrasts with that of core eudicots and monocots.

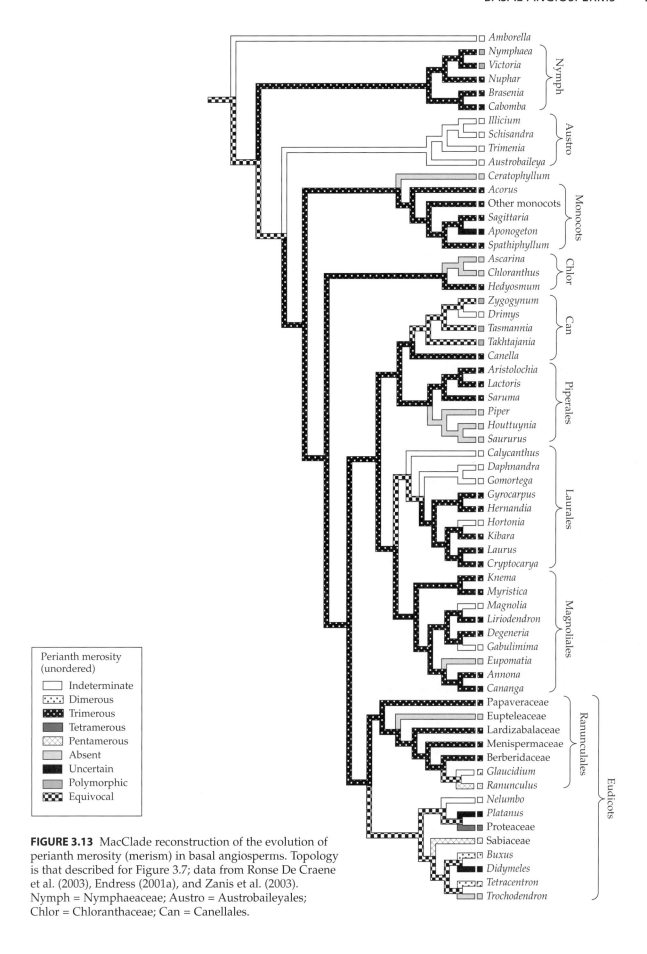

FIGURE 3.13 MacClade reconstruction of the evolution of perianth merosity (merism) in basal angiosperms. Topology is that described for Figure 3.7; data from Ronse De Craene et al. (2003), Endress (2001a), and Zanis et al. (2003). Nymph = Nymphaeaceae; Austro = Austrobaileyales; Chlor = Chloranthaceae; Can = Canellales.

Perianth differentiation

By definition, a differentiated or bipartite perianth has an outer whorl of sepals clearly differentiated from the inner whorl(s) of petals. In contrast, an undifferentiated perianth lacks clear distinction between the outer and inner whorls, or the perianth may consist of undifferentiated spirally arranged parts; these undifferentiated perianth organs have traditionally been referred to as tepals. The term "tepal" was coined by de Candolle (1827) to describe a calyx and corolla that are not differentiated; thus, the entire perianth may be petaloid. Takhtajan (1969, 1987, 1997), in contrast, used the term tepal in a phylogenetic sense such that all monocots have tepals, as derived from magnoliid tepals or bracts (Endress 1994a). Takhtajan's definition of tepal limits the application of this term to specific groups of angiosperms and requires different terms for an undifferentiated perianth in other groups. In our discussion, we use the term tepal as defined by de Candolle—an undifferentiated calyx and corolla.

Distinguishing sepals from petals is not always straightforward (see Endress 1994a; Albert et al. 1998). Whereas sepals and petals are readily distinguished in most eudicots, this is not usually the case in basal angiosperms, many of which have numerous spirally arranged, undifferentiated perianth parts, a condition long considered ancestral (e.g., Bessey 1915; Cronquist 1968, 1988; Takhtajan 1969). Perianth differentiation has also been of interest with regard to the origin of the sepal and petal (e.g., Eames 1931; Hiepko 1965a; Kosuge 1994; Albert et al. 1998; Kramer and Irish 1999, 2000; Zanis et al. 2003). It has been suggested that petals evolved first and that sepals evolved later (e.g., Albert et al. 1998) and that petals have evolved multiple times from different floral organs in different groups (e.g., Eames 1961; Takhtajan 1969; Kosuge 1994; Albert et al. 1998; Zanis et al. 2003). The petals present in some families, such as Magnoliales and Laurales, are proposed to have evolved from an undifferentiated perianth, whereas in other groups, such as Nymphaeaceae, Ranunculales, and other eudicots, petals are similar to stamens in morphology and ontogeny and may, in fact, be derived from stamens. Along these lines, Takhtajan (1969, 1987) suggested two separate origins of petals, one from stamens (e.g., Ranunculales) and one from bracts (e.g., Magnoliales).

Support for independent origins of petals has come from morphological studies showing that "petals" of various angiosperms exhibit major differences (reviewed in Endress 1994a; Kramer and Irish 2000) and can be grouped into two basic classes. In one group are petals that resemble stamens. The characters shared by these petals and stamens include the following: the petals are developmentally delayed relative to the stamens; petals are similar in appearance to stamen primordia at inception; petals are supplied by a single vascular trace; and,

in some cases, petals possess nectaries (Endress 1994a). These petals have sometimes been termed "andropetals." The second type of petaloid organ is found in undifferentiated perianths (standardly termed tepals, Cronquist 1981) and is more leaf-like in general characteristics. These structures initiate and mature much earlier than the stamens, often have three vascular traces, and are generally more leaf-like in appearance than other petals (Smith 1928; Tucker 1960; Takhtajan 1969).

According to Albert et al. (1998), two or more whorls of perianth parts must be present for an unambiguous interpretation of sepals and petals. If only a single perianth whorl is present, it may be difficult to score as "sepals" or "petals" (see also Endress 1994a, 1994c, and Chapter 11). Is a single whorl an undifferentiated perianth of neither sepals nor petals? Or, is the single whorl either sepals or petals with the other perianth whorl absent? A single-whorled perianth has traditionally been referred to as being composed of "sepals" as a matter of convention (e.g., Cronquist 1968). Families of basal angiosperms that contain taxa with a single-whorled perianth include nearly all Aristolochiaceae (except *Saruma*), all Myristicaceae, and Chloranthaceae (*Hedyosmum*). In some cases, however, the nature of a single-whorled perianth can be determined through comparison with the perianths of closely related taxa. In Aristolochiaceae, most taxa have a single-whorled perianth that is considered a calyx (Cronquist 1968, 1981; Takhtajan 1991; Tucker and Douglas 1996). This determination is supported by observations for another genus of the family. *Saruma* has two perianth whorls that are differentiated into sepals and petals. Furthermore, in some species of *Asarum*, petals apparently begin to develop, but the only traces are small thread-like structures (see Leins and Erbar 1985). A well-resolved and supported topology for Aristolochiaceae is not available. However, current topologies place *Asarum* and *Saruma* as sister to the rest of the family; hence, the ancestral condition for the family could be a differentiated perianth with the single-whorled perianth derived through the loss of petals.

In our analysis, we scored the single whorls of Myristicaceae and *Hedyosmum* (Chloranthaceae) as representing an undifferentiated perianth, although others have considered these to be sepals for the reasons discussed above (e.g., Cronquist 1981; Zanis et al. 2003). The perianths of some taxa are difficult to characterize. In Himantandraceae (composed of the single genus *Galbulimima*), Cronquist (1981) described the perianth as consisting of both sepals and petals, whereas others have considered the outer structures bracts, rather than sepals, meaning that the remaining petaloid structures represent an undifferentiated perianth or staminodes (e.g., Endress 1977). We have tentatively interpreted these structures as an undifferentiated perianth (Figure 3.14). Adding to

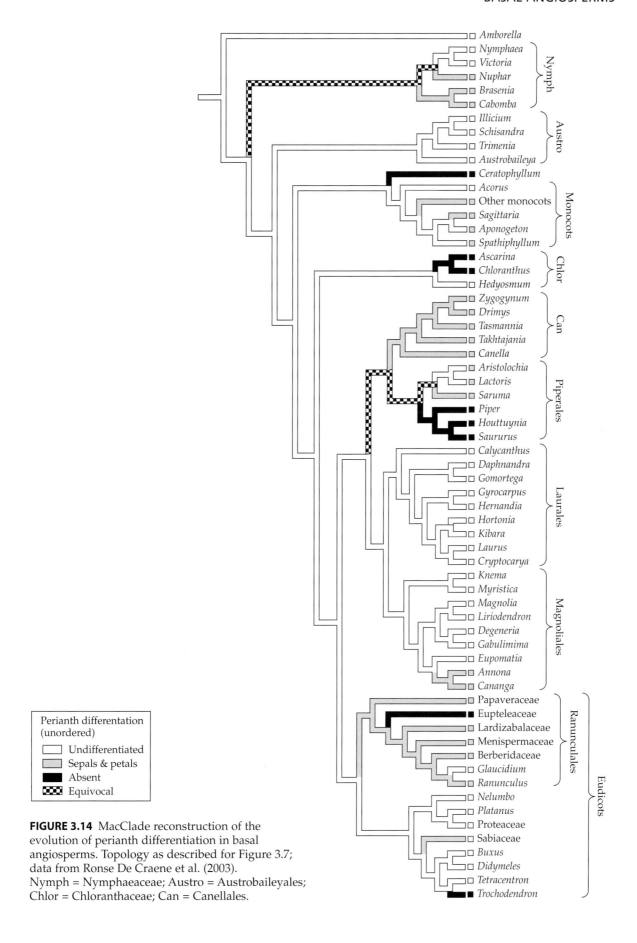

FIGURE 3.14 MacClade reconstruction of the evolution of perianth differentiation in basal angiosperms. Topology as described for Figure 3.7; data from Ronse De Craene et al. (2003). Nymph = Nymphaeaceae; Austro = Austrobaileyales; Chlor = Chloranthaceae; Can = Canellales.

the complexity of cleanly characterizing the perianths of basal angiosperms is the recent observation that even in the "undifferentiated" perianth of *Amborella* there is some morphological differentiation between inner and outer members of the perianth (Endress and Igersheim 2000a; Endress 2001a; Buzgo et al. 2004).

In our reconstructions (Figure 3.14), the ancestral state for the angiosperms is an undifferentiated perianth; this is the case regardless of whether *Amborella* and Nymphaeaceae are treated as sister taxa or as successive sister groups to all remaining angiosperms. *Amborella* and Austrobaileyales have an undifferentiated perianth. In contrast, the ancestral state for Nymphaeaceae is reconstructed as a differentiated perianth because the early-branching Nymphaeaceae (e.g., *Cabomba, Brasenia, Nuphar*) have a differentiated perianth, whereas more derived taxa (*Victoria, Nymphaea*) have an undifferentiated perianth. Above Austrobaileyales, the undifferentiated perianth continues to be ancestral for the remaining angiosperms (Figure 3.14). All reconstructions indicate that a differentiated perianth evolved many times. Separate origins include the monocots, some Magnoliaceae, Annonaceae, Canellales, some Aristolochiaceae, and Siparunaceae, with additional origins in early-diverging eudicots and core eudicots (see Chapter 5). The data indicate that "sepals" and "petals" in these various lineages are not homologous.

Our coding of a single-whorled, trimerous perianth as undifferentiated in *Hedyosmum* (Chloranthaceae) and Myristicaceae is a conservative approach. Zanis et al. (2003) noted that many groups of basal angiosperms such as Canellaceae, Magnoliaceae, and Nymphaeaceae have a trimerous outer whorl and an inner whorl or whorls that are variable in merosity. Thus, the trimerous outer whorl in the basal angiosperms may be homologous to the trimerous perianth found in taxa having a single-whorled perianth. If the outer whorl is differentiated from the inner whorl (petals), then the single trimerous whorl of Chloranthaceae, Myristicaceae, and Aristolochiaceae would be considered a single whorl of sepals, in agreement with the traditional interpretation. If we assume that the single-whorled perianth is in all cases sepaloid, then the ancestral reconstruction above Austrobaileyales changes to a differentiated perianth (not shown; see Zanis et al. 2003).

Stamen phyllotaxis

Reconstructions of the evolution of stamen phyllotaxis (not shown) closely mirror the differentiation of perianth phyllotaxis discussed above (see also Ronse De Craene et al. 2003). A comparison of stamen phyllotaxis and perianth phyllotaxis (as binary characters) using the concentrated change test (Maddison 1990) indicates that stamen and perianth phyllotaxis are closely associated.

Stamen form

Relative to eudicots, the stamens of many basal angiosperms are distinctive because of their large size, extensive quantity of sterile tissue surrounding the pollen sacs, and the lack of differentiation between filament and anther (Endress and Hufford 1989). The leaf-like, or laminar, stamens present in many basal taxa have long been of interest (Figure 3.15). Authors have often proposed that the earliest angiosperms possessed leaf-like (laminar) stamens (e.g., Canright 1952; Eames 1961; Cronquist 1968; Takhtajan 1969). From this putatively ancestral type of stamen, a gradual modification in size and shape was proposed, resulting in the typical stamen that possesses a well-defined anther and filament region (Figure 3.15). Hufford (1996a) attempted to reconstruct stamen evolution across basal angiosperms by using topologies from Donoghue and Doyle (1989a, 1989b) with either Magnoliales or Nymphaeaceae and monocots as sister to other angiosperms; two basic states were used, laminar and linear. With the first topology, Hufford reconstructed laminar stamens as ancestral; with the second topology, as not ancestral. We have elected not to reconstruct the evolution of stamen form in basal angiosperms because of the numerous intermediates between laminar stamens and those with well-defined anther and filament regions (Figure 3.15; Endress and Hufford 1989; Endress 1994a). Appropriate assessments of characters and character states require additional developmental studies. Until those data are available, character-state reconstructions are premature. In the following paragraphs we review the diversity of stamen form in basal angiosperms.

Among basal angiosperms, *Amborella* and *Austrobaileya* have distinctly laminar stamens as do some Nymphaeaceae (Figures 3.4 and 3.15). Stamen form in the latter family is diverse, with laminar stamens in *Nuphar* and *Victoria* (Endress and Hufford 1989; Hufford 1996a) and stamens with well-defined anther and filament regions present in *Cabomba* and *Brasenia* (Cronquist 1988); laminar stamens and stamens intermediate between laminar and linear occur in *Nymphaea* (Figure 3.15; Hufford 1996a). In Trimeniaceae, the stamens have well-defined anther and filament regions. The stamens of Schisandraceae (including *Illicium*) have been considered "relatively undifferentiated," with the filament region often broader than the "anther" region (Endress and Hufford 1989); although not truly laminar, they also do not exhibit the well-defined anther and filament regions typical of eudicots (Figure 3.15). Hence, within Amborellaceae, Nymphaeaceae, and Austrobaileyales, laminar, linear, and intermediate stamen forms are found.

Laminar stamens are present in Magnoliaceae, Degeneriaceae, Himantandraceae, and Eupomatiaceae (all members of Magnoliales). In Winteraceae, the stamens are only slightly differentiated into anther and filament

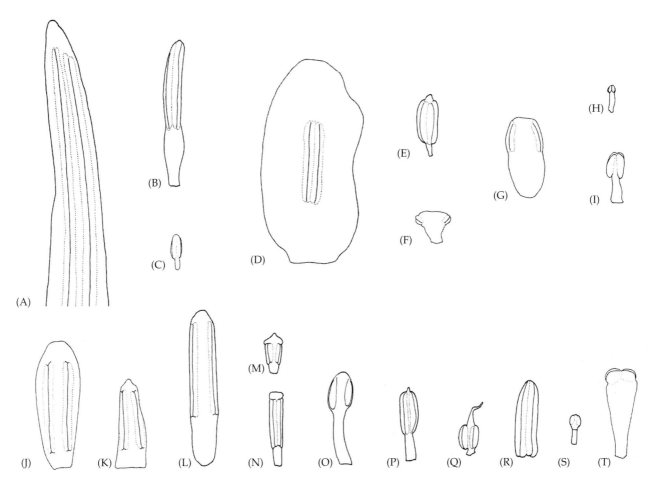

FIGURE 3.15 Stamen form in basal angiosperms, illustrating the complete range of forms from laminar to well-defined anther and filament region, as well as intermediate morphologies. Stamens are drawn to comparable scale and are also arranged in a generally phylogenetic manner. A–C, Nymphaeaceae; D–F, Austrobaileyales; G, Chloranthaceae; H–I, monocots; J–N, Magnoliales; O–P, Laurales; Q–S, Piperales; T, Canellales. (From Endress 1994a, 1996a). (A) *Nymphaea* (Nymphaeaceae). (B) *Nymphaea* (Nymphaeaceae). (C) *Cabomba* (Nymphaeaceae). (D) *Austrobaileya* (Austrobaileyaceae). (E) *Trimenia* (Trimeniaceae). (F) *Kadsura* (Schisandraceae). (G) *Sarcandra* (Chloranthaceae). (H) *Acorus* (Acoraceae). (I) *Sagittaria* (Alismataceae). (J) *Degeneria* (Degeneriaceae). (K) *Eupomatia* (Eupomatiaceae). (L) *Magnolia* (Magnoliaceae). (M) *Artabotrys* (Annonaceae). (N) *Annona* (Annonaceae). (O) *Laurus* (Lauraceae). (P) *Chimonanthus* (Calycanthaceae). (Q) *Asarum* (Aristolochiaceae). (R) *Aristolochia* (Aristolochiaceae). (S) *Piper* (Piperaceae). (T) *Zygogynum* (Winteraceae).

(Figure 3.15). Winteraceae and Magnoliales appear to represent an independent origin (or origins) of the laminar stamen type that is separate from basal families such as Amborellaceae and Austrobaileyaceae. Many remaining families of basal angiosperms have stamens with well-differentiated filament and anther: Monimiaceae, Chloranthaceae, Myristicaceae, Lauraceae, Piperaceae, Saururaceae, Aristolochiaceae, and Lactoridaceae (Figure 3.15). The stamens of *Ceratophyllum* lack a filament region, but basal monocots and Chloranthaceae possess stamens with well-differentiated anther and filament regions (Figure 3.15).

Not all laminar stamens appear to be homologous (Endress and Hufford 1989; Hufford 1996a). For exam-ple, the laminar stamens of *Galbulimima* and *Degeneria* differ substantially from those in *Austrobaileya* (Figure 3.15) in radial thickness, sporangial dehiscence, location of pollen sacs, and the manner in which the sporangia are embedded within the microsporophyll (Endress 1994a). The possible multiple origins of laminar stamens was suggested earlier (e.g., Endress and Hufford 1989; Hufford 1996a).

The prevalence of laminar and intermediate stamen morphologies in the basalmost angiosperms agrees with the fossil record. Angiosperm stamens in Early Cretaceous fossil floras typically exhibit little differentiation between anther and filament (e.g., Friis et al. 2000). It is therefore tempting to speculate that laminar stamens are ancestral in the angiosperms. However, there is room for caution, given the diversity of forms present within just Amborellaceae, Nymphaeaceae, and Austrobaileyales. *Cabomba* and *Brasenia,* the sister group to other Nymphaeaceae, both have stamens with well-differentiated anther and filament regions, and it has been suggested, in fact, that laminar stamens are derived within Nymphaeaceae in association with beetle pollination (Gottsberger 1988).

Pollen

Cronquist (1988) considered pollen with granular exine to be the ancestral condition in the angiosperms because of its prevalence in his Magnoliales. Furthermore, granular pollen was one of the putative synapomorphies of Gnetales and angiosperms (Crane 1985; Doyle and Donoghue 1986; Doyle 1996) and a feature that placed Magnoliales at the base of the angiosperms in earlier investigations (Donoghue and Doyle 1989a, 1989b).

Doyle and Endress (2000) examined the evolution of infratectal pollen wall structure. In addition to the granular and columellar states recognized in an earlier study (Donoghue and Doyle 1989b), they recognized an intermediate condition characterized by irregular radial (columellar) elements mixed with apparent granules. It is not entirely clear whether this "intermediate" condition is an appropriate independent state; additional developmental studies are needed to evaluate this state more critically.

Our reconstructions agree well with those of Doyle and Endress (2000). The intermediate condition is found in *Amborella* and some Nymphaeaceae, indicating that this could be the ancestral state for angiosperms. However, above the node to Nymphaeaceae, the ancestral state for the remaining angiosperms is columellar. Most of the transitions to granular and intermediate forms occur within Laurales + Magnoliales; Piperales + Canellales share columellar pollen. These results were not evident in Doyle and Endress (2000) because they used a tree in which neither Laurales + Magnoliales nor Canellales + Piperales appeared as sister groups.

Our reconstructions indicate that a granular infratectum is not ancestral in the angiosperms but likely arose independently within some Nymphaeaceae (e.g., *Nuphar* and *Nymphaea*), Laurales (e.g., Lauraceae and Hernandiaceae), and Magnoliales (Annonaceae, Eupomatiaceae, Degeneriaceae, Himantandraceae). Doyle and Endress (2000) indicated that granular pollen is derived from the intermediate condition within both Laurales and Magnoliales. However, our reconstructions of the origins of granular pollen are equivocal; the ancestral state for the Laurales + Magnoliales clade is uncertain, as are the ancestral states for the individual orders Laurales and Magnoliales.

Carpel form

There are two extremes of carpel form (Doyle and Endress 2000), plicate and ascidiate (Figure 3.16). The plicate form can be compared to a leaf folded down the middle with ovules on the folded (plicate) zone or area. The plicate condition was considered ancestral for angiosperms by Bailey and Swamy (1951) and Eames (1961). In the ascidiate form (Figure 3.16), in contrast, the carpel grows like a tube; this form was considered ancestral by Leinfellner (1969) and van Heel (1981, 1983). Doyle and Endress (2000) noted that intermediate carpel conditions also exist between the extremes of plicate and ascidiate. Such intermediate types are found in several basal angiosperms, including *Barclaya* (Nymphaeaceae), *Illicium* (Schisandraceae, *sensu* APG II 2003), and *Acorus* (Acoraceae). Although we follow Doyle and Endress (2000) in recognizing an intermediate state, it is not clear whether this is an appropriate, independent state; additional developmental studies are encouraged.

Reconstructions indicate that ascidiate carpels are ancestral in the angiosperms (Figure 3.17; Doyle and Endress 2000). Ascidiate carpels characterize *Amborella*, most Nymphaeaceae, Austrobaileyaceae, most Schisandraceae, and Trimeniaceae (Figure 3.16; Endress and Igersheim 2000b). Intermediate carpel forms are present in *Barclaya* (Nymphaeaceae) and also in *Illicium* (Schisandraceae). Above Amborellaceae, Nymphaeaceae, and Austrobaileyales, *Ceratophyllum* and Chloranthaceae also have ascidiate carpels. The first derivation of plicate carpels occurs in the monocots. This character reconstruction also agrees with the prevalence of putative ascidiate carpels in fossils from the Early Cretaceous (Crane et al. 1995).

Above the branch to Chloranthaceae, the plicate form is reconstructed as ancestral in the angiosperms. Several reversals to the ascidiate carpel form are also apparent in basal angiosperms and early-diverging eudicots. A reversal occurs in Monimiaceae; additional reversals may have occurred in some Araceae. Reversals to an ascidiate carpel also occurred in several basal eudicot lineages (Berberidaceae and *Nelumbo*). An intermediate carpel type characterizes Laurales. If this intermediate condition is truly an independent character state, it represents a possible morphological synapomorphy of that clade. Intermediate carpel form has also evolved independently in *Acorus*, Myristicaceae, and the eudicot *Euptelea* (Ranunculales). It would be of interest to characterize these intermediate types in more detail.

Carpel sealing

In most taxa having ascidiate carpels, the carpels are fused by secretion rather than by postgenital fusion (Figure 3.16; Endress and Igersheim 2000b). Carpel form and

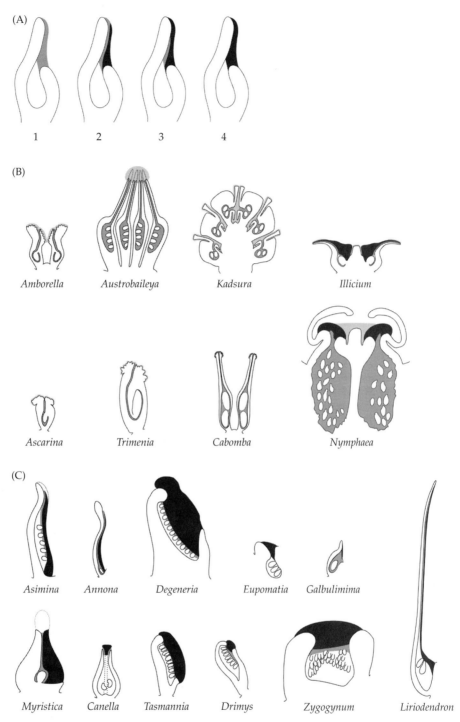

FIGURE 3.16 Carpel form and sealing in basal angiosperms. (From Igersheim and Endress 2000a). (A) Types of angiospermy. Carpels in median longitudinal section. Gray areas: ventral slit closed by secretion; dark areas: ventral slit closed by postgenital fusion. Type 1: angiospermy by secretion; Type 2: angio-spermy by a continuous secretory canal and partial postgenital fusion at the periphery; Type 3; angiospermy by a partial secretory canal and complete postgenital fusion at the periphery; Type 4: angiospermy by complete postgenital fusion. (B) Illustrations of carpel types in Amborellaceae, Nymphaeaceae, Austrobaileyales, and Chloranthaceae. Median longitudinal sections of gynoecia showing carpel structure and gynoecium structure. Gray areas designate secretion inside of carpels; light gray areas: secretion at the outer surface of carpels; dark areas: postgenital fusion; not drawn to scale. (C) Illustrations of carpel types in magnoliid clade. Median longitudinal sections of carpels, ventral side at right (the figure of *Canella* represents an entire gynoecium because the carpels are completely united). Gray areas: secretion inside of carpels; dark areas: postgenital fusion; secretion at the outer carpel surface is not indicated. In microtome sections of *Eupomatia* and *Lirioden-dron* secretion is not visible inside the carpels, although it is likely to be present. (From Endress and Igersheim 2000b.)

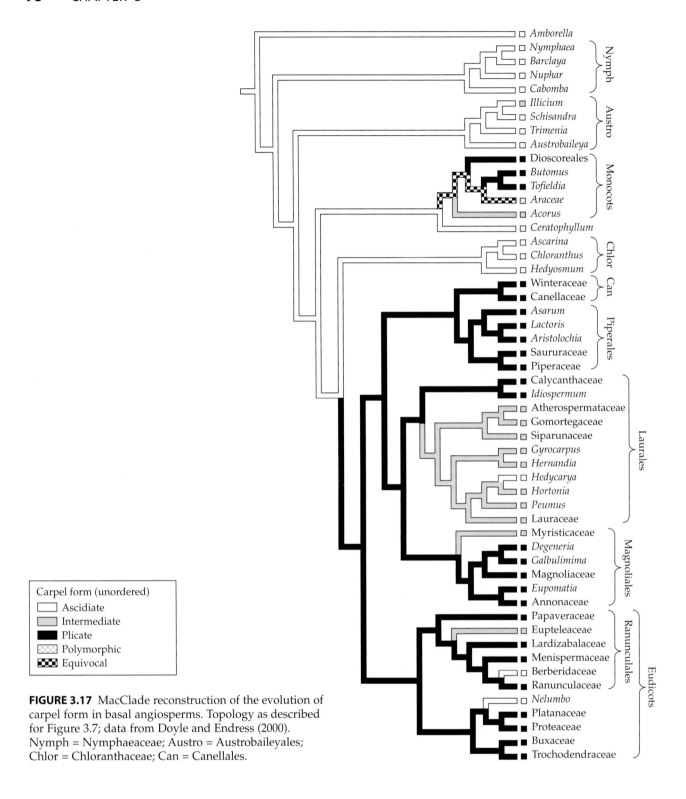

FIGURE 3.17 MacClade reconstruction of the evolution of carpel form in basal angiosperms. Topology as described for Figure 3.7; data from Doyle and Endress (2000). Nymph = Nymphaeaceae; Austro = Austrobaileyales; Chlor = Chloranthaceae; Can = Canellales.

carpel sealing are therefore closely associated. Endress and Igersheim (2000b) recognized four types of carpel sealing: (1) a complete lack of postgenital fusion but occlusion of the inner space by secretion; (2) a combination of postgenital fusion at the periphery and no fusion but occlusion by secretion in the center with a complete un-

fused canal up to the stigma; (3) a combination of postgenital fusion encompassing the entire periphery and an unfused secretory canal that is not complete but ends below the stigma; and (4) complete postgenital fusion of the inner space between the ovary and the stigma. *Amborella* exhibits the first of these types (Figure 3.16), as do most Austrobaileyales (however, *Illicium* has type 2) and *Cabomba* (Nymphaeaceae). In other Nymphaeaceae such as *Barclaya* and *Nymphaea,* type 3 is present. Most eudicots are characterized by type 4.

The fact that Amborellaceae, some Nymphaeaceae, and most Austrobaileyales have carpels that are not postgenitally fused but are sealed by secretion has important evolutionary implications. Reconstructions indicate that carpels sealed by secretion might represent the ancestral condition for angiosperms. These data therefore imply that angiospermy might have evolved first in early flowering plants and postgenital fusion of the carpel evolved later (Endress and Igersheim 2000b).

Apocarpy versus syncarpy

It has long been suggested that the hypothetical primitive angiosperms had numerous free carpels (apocarpy; Cronquist 1968; Takhtajan 1969; Stebbins 1974). Our reconstructions unambiguously reconstruct the ancestral state of the angiosperms, as well as that of early-diverging eudicots, as apocarpous. *Amborella, Cabomba,* and *Brasenia* (Nymphaeaceae; although other Nymphaeaceae have united carpels), Austrobaileyales, *Ceratophyllum,* Laurales, Magnoliales, and most Ranunculales are all characterized by apocarpy. The reconstructions indicate that syncarpy is derived within Nymphaeaceae; the monocots may be ancestrally syncarpous (*Acorus* and Araceae are both syncarpous), with a reversal to apocarpy in *Tofieldia* and *Butomus.* Syncarpy may also be ancestral for Piperales + Canellales. These orders are characterized by parasyncarpy (cf. Doyle and Endress 2000), with a subsequent reversal to apocarpy in Winteraceae. "Parasyncarpous" describes a gynoecium in which the carpels are united and a unilocular ovary is present (no fusion in the center, no septa or only partial septa present). "Eusyncarpous" describes a gynoecium with a plurilocular ovary in which the septa meet in the center and are usually fused.

Ovule number

Doyle and Endress (2000) considered the evolution of ovule number and scored character states as one, mostly two, and more than two, using their DNA plus morphology tree described in "Combining DNA and Morphology," above. With these states and their topology, the ancestral state for the angiosperms—as well as for most basal nodes—is equivocal. This equivocality appears to be due more to variability in this character and the character-state coding used than to the topology. Many basal angiosperms have a small number of ovules per carpel. *Amborella* and Trimeniaceae have a single ovule per carpel; Schisandraceae (including *Illicium*) typically have one or two. Chloranthaceae and *Ceratophyllum* also have one ovule per carpel. However, *Austrobaileya* has 8 to 15 ovules per carpel; the ovule

number per carpel in *Acorus* is 2 to 6; the number in Nymphaeaceae ranges from 1 to 300, but in *Cabomba* and *Brasenia,* the number is low, typically 1 to 3. Scoring these critical taxa as "more than two" ovules per carpel hinders attempts to reconstruct the number of ovules per carpel in basal angiosperms by obscuring diversity by creating a single large category (more than two) that includes a low to moderate number of ovules per carpel as well as a very high number of ovules per carpel. The fact that some basal angiosperms have a relatively low number of ovules (5–20) is lost by lumping all taxa together that have more than two ovules per carpel. In addition, some families have substantial variation in ovule number (e.g., Nymphaeaceae, many families of Canellales and Piperales, and the basal eudicot family Ranunculaceae).

In our reconstruction for this book, we added additional states: 1, mostly 2 (but including 3 or 4), 5 to 20, and more than 20 ovules per carpel. We also coded taxa of Nymphaeaceae following the information in Endress (2001a). Importantly, both *Cabomba* and *Brasenia* have low numbers of ovules (typically 1 to 3). We also coded the polymorphism evident in some families; this was not captured using the coding of Doyle and Endress (2000).

In our reconstruction (not shown), a single ovule is ancestral for the angiosperms and is ancestral through the branch leading to Chloranthaceae. The ancestral state for the magnoliid clade is ambiguous. Above the magnoliid clade, the ancestral state may also be low, but depends on the placement of Papaveraceae. If Papaveraceae are sister to other Ranunculales, then the ancestral state for the basal node of the eudicots is two ovules per carpel; if Eupteleaceae are sister to other Ranunculales, then the ancestral state for this node is equivocal, apparently because of the great variability in ovule number in some of these basal eudicot lineages.

Ovule curvature

Most angiosperms have anatropous ovules (see Figure 1.8 in Chapter 1); however, some angiosperms, including several early-branching lineages, have orthotropous ovules. Gymnosperms (living and extinct) also have orthotropous ovules. Because of the prevalence of anatropous ovules in the angiosperms, this state typically has been considered to represent the ancestral condition in flowering plants (Cronquist 1968; Takhtajan 1969; Stebbins 1974). However, *Amborella* and several other basal lineages, including Chloranthaceae, some Nymphaeaceae (i.e., *Barclaya*), *Ceratophyllum,* and *Acorus*, possess orthotropous ovules.

In our mapping studies, we scored ovules as anatropous or orthotropous (including hemitropous). The ancestral condition for the angiosperms (not shown) is ambiguous. With the MacClade option "all most parsimonious states," the ancestral condition remains ambiguous for the early-branching lineages; the anatropous ovule is not reconstructed as ancestral until the node leading to the magnoliid clade and eudicots (i.e., above the branch to Chloranthaceae). With ACCTRAN optimization, the ancestral state for angiosperms is ambiguous, with the branches to Nymphaeaceae and Austrobaileyales reconstructed as anatropous; this is followed by a switch to an orthotropous ancestral condition for monocots + Ceratophyllaceae and Chloranthaceae. Above Chloranthaceae, anatropous ovules are ancestral.

However, developmental data have indicated that the interpretation of ovule evolution is more complex than the reconstructions noted above. In many basal angiosperms, the outer integument is not cup-shaped, but rather markedly asymmetric and hood-shaped. Yamada et al. (2001a) found that the outer integument of both *Amborella* and *Chloranthus* is asymmetrical. They concluded that the developmental pattern of the outer integument and the ovule developmental pattern seen in both *Amborella* and *Chloranthus* are not equivalent to the orthotropous ovules of eudicots. That is, developmental data indicate that the orthotropous ovules found in ascidiate carpels (as in both *Amborella* and *Chloranthus*, see above) are derived (see Endress 1986a, 1994; Yamada et al. 2001a). Further support for the antiquity of ascidiate carpels with anatropous ovules is provided by the fossil record, including the fossil *Couperites* (Chloranthaceae) with an anatropous ovule (Pedersen et al. 1994) and fruits with anatropous ovules from the Barremian of Portugal (one of the oldest strata bearing angiosperm fossils; Friis et al. 1999, 2000).

Endosperm

Until recently, any review of angiosperm evolution would have indicated that the ancestral state for angiosperms was double fertilization with the second fertilization event forming triploid endosperm (the latter formed via fusion of a sperm and two polar nuclei from the female gametophyte). However, Williams and Friedman (2002) demonstrated that diploid endosperm is formed in *Nuphar* (Nymphaeaceae). Rather than possessing a female gametophyte of eight nuclei and seven cells, which is typical of angiosperms, *Nuphar* has four nuclei in four cells (Figure 3.18). The second fertilization event involves a sperm nucleus and just one nucleus of the female gametophyte, resulting in diploid rather than triploid endosperm. Previous studies provide indirect developmental data consistent with a four-celled female gametophyte in four other basal angiosperms: *Nymphaea, Cabomba, Schisandra,* and *Illicium* (reviewed in Williams and Friedman 2002). However, *Amborella* has a typical female gametophyte of eight nuclei and seven cells (Tobe et al. 2000). This should yield triploid endosperm; however, the fertilization process in *Amborella* has not yet been investigated (Williams and Friedman 2002). Because some basal angiosperms have four-celled female gametophytes, character reconstructions using available data indicate that the ancestral state for the female gametophyte (four-celled versus seven-celled) and endosperm (diploid versus triploid) in the angiosperms is equivocal (Figure 3.19A). Above Amborellaceae, Nymphaeaceae, and Austrobaileyales, however, the ancestral state for the remaining angiosperms is triploid endosperm. It is important to obtain data on female gametophyte form (seven-celled versus four-celled) for *Austrobaileya, Trimenia,* and Chloranthaceae. If *Austrobaileya* and *Trimenia* possess the seven-celled female gametophytes typical of most angiosperms, then MacClade reconstructions suggest that the four-celled type evolved twice, with seven-celled female gametophyte and triploid endosperm formation

(A) Typical double fertilization

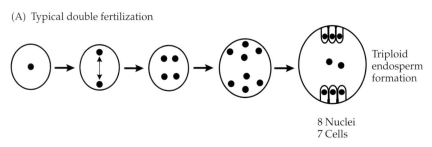

Triploid endosperm formation

8 Nuclei
7 Cells

FIGURE 3.18 Embryo sac (female gametophyte) formation in *Nuphar* compared with other angiosperms. (A) Seven-celled (eight-nucleate) female gametophyte and triploid endosperm formation typical of most angiosperms. (B) Four-celled (four-nucleate) female gametophyte and diploid endosperm formation (as reported in *Nuphar*) (see text and Williams and Friedman 2002).

(B) Double fertilization in Nymphaeaceae

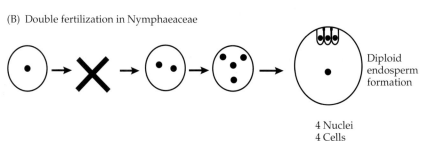

Diploid endosperm formation

4 Nuclei
4 Cells

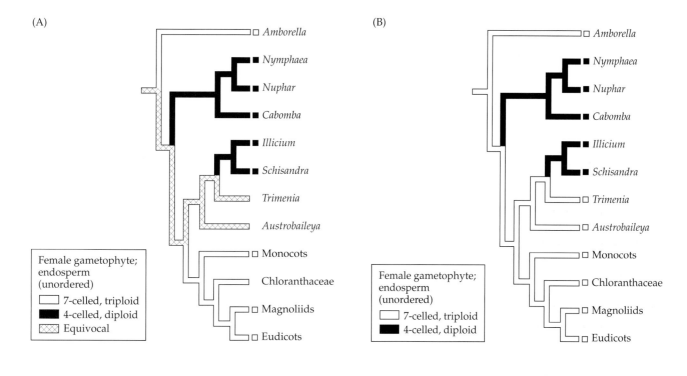

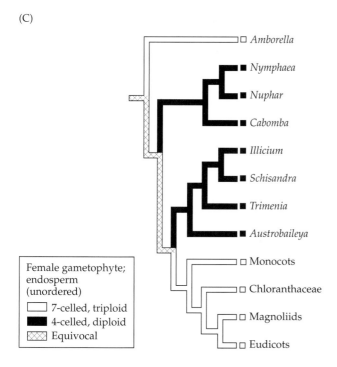

FIGURE 3.19 MacClade reconstructions of the diversification of female gametophyte (seven-celled versus four-celled) and endosperm (triploid versus diploid) in angiosperms. (A) Reconstruction based on existing data. The number of cells in the female gametophyte of several critical basal lineages is unknown (e.g., Chloranthaceae, Trimeniaceae, Austrobaileyaceae). (B) Reconstruction with the assumption that Chloranthaceae, Trimeniaceae, Austrobaileyaceae have seven-celled female gametophytes and triploid endosperm. (C) Reconstruction with the assumption that Chloranthaceae have seven-celled female gametophytes and triploid endosperm; Trimeniaceae, Austrobaileyaceae have four-celled female gametophytes and diploid endosperm.

ancestral for the angiosperms (Figure 3.19B). However, if *Austrobaileya* and *Trimenia* possess four-celled female gametophytes, then the ancestral state for angiosperms remains equivocal (Figure 3.19C), regardless of the condition ultimately found in Chloranthaceae.

In addition to the remarkable discovery of diploid endosperm, subsequent patterns of early endosperm development are more complex than described in the traditionally used categories of cellular, free nuclear,

and helobial (Floyd and Friedman 2000). Through character reconstructions, Floyd and Friedman (2000) suggested that basal angiosperms have retained plesiomorphic features. All basal angiosperms exhibit extensive endosperm development as well as differential developmental fates of the chalazal and micropylar regions of the endosperm (bipolar endosperm development). In most basal angiosperms, endosperm development in both domains is cellular (cell walls are formed), an ontogeny referred to as "bipolar cellular." Furthermore, endosperm development appears to be highly conserved throughout the radiation of basal angiosperms, with the exception of a few lineages (e.g., *Cabomba, Saururus,* Lauraceae). The diversification of endosperm development appears to coincide with the radiations of monocots and eudicots (Floyd and Friedman 2000).

Genetics of Floral Development: Implications for the Early Diversification of the Flower

Floral initiation and development in the model organisms *Arabidopsis* and *Antirrhinum* are controlled by many genes in several gene families that exhibit a diversity of functions (e.g., Bowman et al. 1989, 1991; Schwarz-Sommer et al. 1990; Coen and Meyerowitz 1991; Lee et al. 1997; Riechmann and Meyerowitz 1997; Ma 1998; Theissen and Saedler 1999; Ma and dePamphilis 2000; Theissen et al. 2000; Zhao et al. 2001; D. Soltis et al. 2002b). Among those genes associated with the genetic control of reproductive meristems, several MADS-box genes play key roles in specifying floral organ identity. During flower development in *Arabidopsis,* the identity of the floral organs is specified by at least three classes of homeotic genes, the best known of which are the A-, B-, and C-function genes (Figure 3.20; e.g., Bowman et al. 1989, 1991; Schwarz-Sommer et al. 1990; Coen and Meyerowitz 1991; Bradley et al. 1993). In addition to the ABC-function genes, the *Arabidopsis SEPALLATA-1, -2,* and *-3* genes are also involved in the specification of petals, stamens, and carpels (Pelaz et al. 2000; Honma and Goto 2001; Theissen and Saedler 2001). These have been referred to as E-function genes (Figure 3.20), but it is not yet clear how widely conserved their functions are among angiosperms (e.g., Kotilainen et al. 2000; Theissen and Saedler 2001).

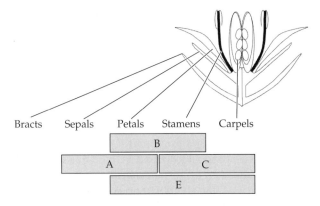

Bracts Sepals Petals Stamens Carpels

FIGURE 3.20 The well-known ABC model (Bowman et al. 1989; Schwarz-Sommer et al. 1990; Coen and Meyerowitz 1991) of floral development, updated to show a suggested role of *SEPALLATA-1, 2,* and *3* genes (E-function genes; Honma and Goto 2001; Theissen and Saedler 2001).

The model organisms *Arabidopsis* and *Antirrhinum* are highly derived eudicots in the rosid and asterid clades, respectively. Although the eudicots represent approximately 75% of angiosperm species (Drinnan et al. 1994), most of the diversity in arrangement and number of floral parts actually occurs among basal angiosperm lineages. Floral organization and development are considered "open" and highly labile in basal angiosperms (Endress 1987a, 1987b, 1990a). In contrast, in most eudicots, numbers of floral parts are low (i.e., 4 or 5) and fixed, and floral organs are arranged in whorls, indicating that the basic floral bauplan became highly canalized during the early diversification of the eudicots (Endress 1987a, 1987b, 1990a, 1994c; Albert et al. 1998). Thus, critical components of the floral genetic program may have evolved among the most basal lineages of angiosperms.

Until recently, little was known about the underlying genetic architecture of floral development in basal angiosperms (Albert et al. 1998; Baum 1998; Kramer and Irish 2000; Ma and dePamphilis 2000; D. Soltis et al. 2002b). The morphological distinction between sepal and petal is not straightforward in many basal angiosperms, as well as in some early-diverging eudicots (see Chapter 5). Whereas eudicots typically have a well-differentiated perianth of sepals and petals, many basal angiosperms bear numerous petaloid structures. It seems likely that "petals" have evolved several times and are therefore not homologous across all angiosperms (Albert et al. 1998; Figure 3.14; see "Perianth Differentiation," above). In other basal angiosperm families, it is not clear whether a perianth is even present. In Eupomatiaceae, the flower is interpreted as lacking a perianth, but it is initially enclosed when young in a deciduous "calyptra," leaving a flower with numerous stamens and carpels. This calyptra has been interpreted as a bract (Endress 1977, 2003c), but could this calyptra be a modified perianth (Cronquist 1981; reviewed in Endress 2003c)?

From our current understanding of the genetics of floral development in model eudicots (Figure 3.20), expression of B-function genes provides the critical difference between sepals and petals and stamens and carpels in eudicot models. Albert et al. (1998) suggested that, for consistency, an organ terminology based on the underlying developmental genetic program could therefore be adopted. Sepals (A), petals (AB), and stamens (BC) would be readily distinguished from each other with this approach. Transitions between undifferentiated and differentiated perianths result from changes in the expression domain of the B-group genes (Bowman 1997; Albert et al. 1998).

Albert et al. (1998) suggested that the ancestral condition for angiosperms was a single petaloid whorl expressing both A- and B-functions; the sepaloid outer whorl was then added externally (Figure 3.21, a and b).

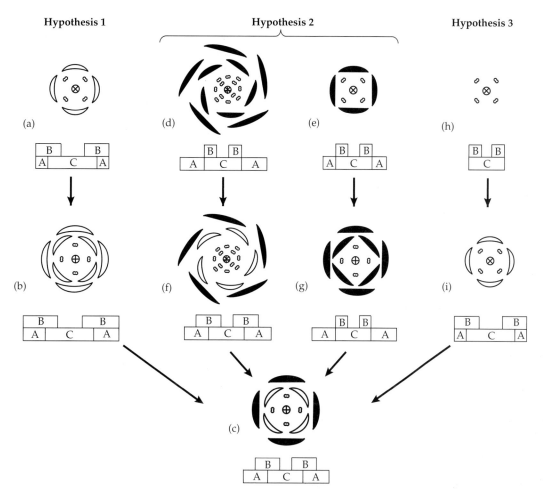

FIGURE 3.21 Hypotheses for the underlying developmental genetics of floral evolution in early angiosperms. Hypothesis 1: The model of Albert et al. (1998) suggests that: (a) The ancestral condition for angiosperms was a single petaloid whorl expressing both A- and B-functions. (b) An outer whorl was then added externally. (c) B-function gene products would then be excluded from this outer whorl. A possible modification of this hypothesis (not shown) assumes that the ancestral angiosperm condition was an undifferentiated perianth of spirally arranged tepals. In this model, these perianth parts express both A- and B-functions and are petaloid. A sepaloid outer whorl (which possesses only A-function expression) was then added externally. B-function gene products would then be excluded from this outer whorl. Hypothesis 2: Baum (1998) assumed (d, e) an ancestral angiosperm with a sepaloid perianth. B-function expression is limited in these ancestral angiosperms to stamens. The original sepaloid organs in this model have only A-function products. (f, g) The distinction between sepals and petals evolved as a result of the outward expansion of B-function gene products into the inner of two sepaloid perianth whorls to produce petals. In a multipartite ancestor (d), the first event is expansion of B-class gene expression to the second whorl to produce petals (f). In a unipartite ancestor (e), an outer perianth is first added (g) with subsequent expansion of B-function gene products to the second whorl to produce petals (c) (modified from Baum 1998). Hypothesis 3: (h) Early angiosperms lacked a perianth with only B and C genes present and expressed in stamens and carpels, respectively. (i) A perianth was added later with the expansion of B-class gene expression into the inner perianth whorl (petals). A-class expression was added to the outer perianth whorl (sepals).

B-function gene products were then excluded from this outer whorl, leaving a whorl with only A-function. The model of Albert et al. implies that all living angiosperms are descended from a common ancestor possessing a petaloid perianth and the developmental control of these structures was determined by A- and B-function genes in a fashion similar to that observed in eudicots (Baum 1998; Baum and Whitlock 1999).

Baum (1998) proposed an alternative hypothesis that assumes, instead, an ancestral angiosperm with a sepaloid perianth with B-function limited to stamens (Figure 3.21). The original sepaloid organs in this model have only A-function products. In one variant of this model, early angiosperms possessed a bipartite perianth (Figure 3.21d), and the distinction between sepals and petals evolved as a result of the expansion of B-function gene products from stamens into the inner of two sepaloid perianth whorls to produce petals. If the ancestral angiosperm had a unipartite perianth, then the initial step would have been the addition of another perianth whorl (Figure 3.21, e and g).

Genetic and molecular evidence for the conservation of portions of the ABC model have indicated that this regulatory network may be applicable to most angiosperms (Ma and dePamphilis 2000). Importantly, however, modifications of components of the ABC model appear to have occurred in different lineages of angiosperms (Ma and dePamphilis 2000; Johansen et al. 2002; D. Soltis et al. 2002b). Research on representatives of diverse families including Magnoliaceae, Liliaceae, Fabaceae, Asteraceae, and Papaveraceae indicates that the function of *APETALA3* (*AP3*) and *PISTILLATA* (*PI*) as B-class organ identity genes is not rigidly conserved across all flowering plants (e.g., Heard and Dunn 1995; Kramer and Irish 1999, 2000; Yu et al. 1999; Kramer et al. 2003; Kanno et al. 2003). Furthermore, although the B and C functions of this model also extend to maize and rice (Ambrose et al. 2000; Ma and dePamphilis 2000; Nagasawa et al. 2003), conservation of the A function remains to be demonstrated, and only B and C functions are present in gymnosperms.

Floral gene data and expression patterns have emerged for basal angiosperms that permit a reevaluation of models for the origin of the perianth (e.g., the hypotheses of Albert et al. and Baum; Figure 3.21). Kramer and Irish (1999, 2000) found that expression of *AP3* and *PI* orthologues (B-class genes) in stamens is conserved in the basal angiosperms *Calycanthus* (Calycanthaceae), *Liriodendron* and *Michelia* (Magnoliaceae), *Asarum*, (Aristolochiaceae), *Sagittaria* (Alismataceae), and in the early-diverging eudicots *Ranunculus* (Ranunculaceae), *Papaver*, and *Dicentra* (Papaveraceae), but the expression patterns in petaloid organs differ from those in core eudicots such as *Arabidopsis*. Whereas the petaloid organs of core eudicots exhibit moderate to strong expression of *AP3* and *PI* orthologues at early stages of development, this is not the case for the petaloid organs of *Calycanthus* and *Liriodendron* (Kramer and Irish 2000). In the monocot *Sagittaria* (Alismataceae), as well as in the eudicot *Sanguinaria* (Papaveraceae), *AP3* and *PI* orthologues also are expressed in developing sepals (Kramer and Irish 2000). In the monocot *Tulipa* (Liliaceae), which has two whorls of tepals, B-class genes are expressed in both whorls, as well as in the stamens (Kanno et al. 2003). Similarly, B-class gene expression (both *AP3* and *PI* orthologues) is also apparent in the spirally arranged tepals of the basal angiosperm *Amborella* (Amborellaceae; D. Soltis et al. 2004; S. Kim, pers. comm.). The outer perianth whorl of *Nuphar* has sometimes been referred to as sepals, and the inner whorl as petals (sometimes referred to collectively as tepals); B-class gene expression is apparent in both whorls (D. Soltis et al. 2004;

S. Kim, pers. comm.). If gene expression is used as a criterion for organ identity following Albert et al. (1998), then the perianth of these basal angiosperms and Ranunculales in the eudicots is best considered as petaloid not sepaloid. Thus, the available data clearly favor a modified ABC model (Figure 3.21). Furthermore, the study of floral development in *Amborella* reveals a gradual change of morphological characters between floral organs, posing the hypothesis that genes correlated with stamen identity (e.g., B-class genes) may show a gradual change in expression pattern throughout the floral parts of *Amborella* and perhaps other basal angiosperms (Buzgo et al. 2004; see also Kramer et al. 2003).

Data for A-class genes and gene expression in basal angiosperms remain fragmentary. Homologues of the A-class gene *APETALA1* (*AP1*) have been isolated and sequenced from magnoliids (e.g., *Magnolia*) and early-diverging eudicots (Litt and Irish 2003). *AP1* homologues are also present in Nymphaeaceae (D. Soltis et al. 2004; S. Kim, pers. comm.). It may also be that the function of A-class genes is heterogeneous among angiosperms, with different loci serving the same purpose of spatially limiting B-function gene expression. This appears to be the case for rosids versus asterids, for example, in that the asterids *Petunia* and *Antirrhinum* do not possess orthologues of *AP1* or *AP2* that function as A-class genes as in the rosid *Arabidopsis*. Alternatively, there may have been no true A-function in early angiosperms. Litt and Irish (2003) observed that *AP1* sequences from magnoliids are highly similar to those of *SEPALLATA* (*SEP*) genes (E-function). Thus, the perianth of the earliest angiosperms may have exhibited expression of a gene ancestral to A and E—an AE model (Figure 3.21; Hypothesis 3; V.A. Albert, pers. comm.). *SEPALLATA* genes (E-genes) appear to have been present before the gymnosperm–angiosperm divergence. The angiosperm *SEPALLATA* clade is sister to the *AP1* clade, so the ancestral function of the *SEP/AP1* clade may have been crucial for the origin of the flower.

Alternative hypotheses to those of Albert et al. (1998) and Baum (1998) could be considered. For example, early angiosperms may have completely lacked a perianth (Figure 3.21, hypothesis 3), as in *Archaefructus* from the Early Cretaceous (Figure 3.7; Sun et al. 2002). This hypothesis does not conflict with that of Albert et al. but suggests that a perianth was added later to the flower. In this scenario, B- and C-class genes were expressed in the stamens and C-class genes in the carpels (Figure 3.21h). B-class gene expression may have been "leaky," with some B-class gene expression in carpels. This possibility is indicated by the demonstration of B-class gene expression in carpels of several basal angiosperms (Kramer and Irish 2000; D. Soltis et al. 2004; S. Kim, pers. comm.). The addition of the first perianth parts, either as a whorl or spiral, would have been accompanied by expansion of B-class gene expression into these organs (Figure 3.21i) and the addition of A (or AE; see above).

As this overview illustrates, the processes responsible for the origin and subsequent evolution of the flower remain fundamental problems in plant biology. Major questions include (D. Soltis et al. 2002b): How did flowers become bisexual, given that male and female reproductive organs are in separate structures in all extant gymnosperms? How did the major organs (sepals, petals, stamens, and carpels) originate? What genes control the number, arrangement, and fusion of floral organs? What effector genes generate the characteristic features of these major organs? What genes control floral initiation and development throughout the angiosperms, especially in the most basal lineages, and how do they compare with those in model organisms? How much of the developmental machinery is common to most lineages, and how much is peculiar to restricted groups of angiosperms? It is also important to distinguish expression of a gene from the function of a gene product in organ determination. B-class genes, for example, are expressed in diverse organs, including ovules, but their function in organs other than petals and stamens has yet to be clearly demonstrated (e.g., Yu et al. 1999; Tzeng and Yang 2001; Skipper 2002; Sundstrom and Engström 2002).

Several authors have stressed that to understand the genetic architecture of floral development, including the origin and subsequent diversification of the flower, data are needed not just for a few model organisms but for gymnosperms, basal angiosperm lineages, and early-diverging eudicots (Albert et al. 1998; Baum 1998; Kramer and Irish 1999; D. Soltis et al. 2002b). Importantly, the list of "new" model organisms has been expanding rapidly to include a diverse array of species that better represent the phylogenetic diversity of angiosperms. As a result of these ongoing studies, a more comprehensive paradigm of floral developmental genetics will emerge within the next several years.

The Appearance of Angiosperms in the Fossil Record

Many molecular clock estimates indicate that angiosperms arose before the Cretaceous (see Chapter 2). However, no unambiguous fossil evidence supports a pre-Cretaceous origin of flowering plants (Crane et al. 1995; Friis et al. 2000). Although the angiosperm fossil *Archaefructus liaoningensis* was originally proposed to be from the Late Jurassic (Sun et al. 1998), more recent evidence indicates it is younger and likely of Barremian–Aptian age (Swisher et al. 1999). Nonetheless, *Archaefructus* (Figure 3.7) represents perhaps the most complete early fossil angiosperm (Sun et al. 2002). The oldest angiosperm fossils are from the earliest Cretaceous and are of Valanginian–Hauterivian age (about

140–130 Mya; Trevisan 1988; Hughes 1994; Brenner 1996; for geologic time scale, see Figure 2.6). These early remains are primarily pollen with occasional leaf fossils (Crane et al. 1995). The oldest reproductive structures of angiosperms are flowers, fruits, and seeds from the Barremian or Aptian (130–115 Mya).

Members of Amborellaceae, Nymphaeaceae, and Austrobaileyales were present in the Early Cretaceous. There is, in fact, evidence for more diversity in these groups in the fossil record than is exhibited by extant members (Friis et al. 2000, 2001). Fossil leaves identified as Illiciaceae (now part of Schisandraceae) and Austrobaileyaceae have been reported from the Early Cretaceous (latest Barremian or earliest Aptian; Upchurch 1984). Stamens similar to those of *Amborella* and *Illicium* have also been recovered from the early Cretaceous (Friis et al. 2000). Pollen of Hauterivian age has been identified that is similar to the pollen of *Amborella* (Doyle 2001). Another fossil consisting of densely aggregated, spirally arranged carpels shows similarity to Schisandraceae (Friis et al. 2000). The discovery of a fossil flower of Nymphaeaceae (Figure 3.8) from the Early Cretaceous (125–115 Mya; Friis et al. 2001) adds to the growing congruence between molecular phylogenetic results and the paleobotanical record. Early Cretaceous floras also included fossils that are attributable to Chloranthaceae, including flowers that are similar to *Hedyosmum* and stamens similar to those of *Ascarina* (Friis et al. 1986, 1997, 1999). The early appearance of Chloranthaceae in the fossil record is congruent with molecular phylogenetic results in that Chloranthaceae diverge just one or two nodes above Amborellaceae, Nymphaeaceae, and Austrobaileyales in DNA-based topologies (Qiu et al. 1999; Zanis et al. 2002). Monocots appear to be a possible exception to the rough correspondence between fossils and DNA-based topologies for extant taxa. The earliest mesofossils of monocots are Turonian in age (Gandolfo et al. 1998), later than might be expected from the early divergence of monocots in DNA-based trees (Crepet 2000) and the estimated ages of monocots based on molecular dating methods (e.g., K. Bremer 2000; Chapters 2 and 4).

Although it is tempting to conclude from these fossil data that the earliest angiosperms were morphologically similar to extant members of Amborellaceae, Nymphaeaceae, and Austrobaileyales, caution must be exercised because many early angiosperm fossils cannot be associated closely with any extant taxa. These difficulties are sometimes the result of a poor understanding of the fossil, but in many cases these difficulties arise because the character combinations observed in the fossils differ from those of extant flowering plants. Some early angiosperm fossils show a mosaic of features. Although some fossils show similarities to extant Amborellaceae, Nymphaeaceae, and *Illicium*, few of these fossils are clearly assignable to any of these extant groups. Some fossils, such as *Archaefructus*, placed in its own family, may represent distinct lineages (Sun et al. 2002; but see Friis et al. 2003), whereas the fossil described by Friis et al. (2001) may be a member of Nymphaeaceae. There is also evidence for significant extinction. All of these observations point to the difficulty of using extant basal lineages to infer characteristics of the earliest angiosperms.

Examination of fossil flowers has recently provided some generalizations about the structural organization and diversity of early angiosperm flowers (Friis et al. 2000), but generalizing about other features is more difficult. In addition, complete reconstructions of floral organization are possible for fewer than 10% of the angiosperm taxa present in some Early Cretaceous floras, further prompting caution and suggesting that for some characters it is premature to offer generalizations. It is clear, however, that many angiosperm reproductive structures from the Early Cretaceous were small. For example, none of the fossil seeds or fruits in the Famalicão assemblage exceeds 3.9 mm in length (Eriksson et al. 2000). Even allowing for shrinkage during fossilization, these flowers are small. The small size of Early Cretaceous flowers generally agrees with the floral characteristics of Amborellaceae, Austrobaileyales, and some Nymphaeaceae, most of which have small or moderate-sized flowers. The fossil record seems to argue against large, magnoliid flowers as the ancestral flower type as suggested by Cronquist (1968) and Takhtajan (1969).

The fossil record is more ambiguous on the organization (spiral versus whorled) of floral parts in Early Cretaceous flowers because this feature can be difficult to reconstruct even in well-preserved fossil materials (Friis et al. 2000). The arrangement of the carpels of *Archaefructus* is uncertain (Sun et al. 2002). However, several characteristics of Early Cretaceous flowers are clear. Both unisexual and bisexual flowers are present in Early Cretaceous floras. Flowers with few floral parts predominate, and flowers with many parts are uncommon. Differentiation into sepals and petals has not been observed in Early Cretaceous flowers. Flowers of the fossil *Archaefructus* lacked a perianth, but it not clear whether flowers of *Archaefructus* represent an ancestral state for angiosperms or are modifications associated with a submerged aquatic habit. The latter must be considered a strong possibility because extant submerged aquatic families (e.g., Podostemaceae; Hydrostachyaceae) have highly modified flowers. The relationships of these families could be ascertained only by molecular analysis, and even then only with some difficulty (see Chapters 8 and 9).

Stamens in Early Cretaceous fossils typically exhibit poor differentiation between filament and anther. Fossils representing early angiosperm lineages also appear to lack a well-differentiated style. Often it is not possible to determine whether fossil carpels were asciiate or plicate because ontogenetic data are usually needed to make this determination. However, it appears that both types were present in Early Cretaceous floras. Most fruits examined have a single seed, but a few have two to as many as eight seeds. These data suggest that the early carpel possessed a small number of ovules. Friis et al. (2000) inferred that both insect-pollinated and wind-pollinated taxa were present in Early Cretaceous floras and that pollen was probably the major reward for visiting insects. Ethereal oil cells have also been documented in many fossil flowers.

Crepet (1996) concluded that there was a "dramatic modernization" of the angiosperms by the Turonian (approximately 90 Mya), which appears to coincide with a much improved fossil record of insects (Grimaldi 1999). Our understanding of the timing of angiosperm diversification versus insect diversification (and the timing of angiosperm–pollinator relationships) has improved; the pattern and timing of early angiosperm diversification is consistent with a similar pattern of radiation in anthophilous insects (Crepet 2000). Furthermore, there is "a compelling similarity between the rate of floral innovation per million years and the rate of angiosperm diversification during the Cenomanian–Turonian interval, which coincided with the first occurrences of many derived insect pollinators" (Crepet 2000).

Reconstructing the Earliest Angiosperms

Botanists and evolutionary biologists have frequently attempted to reconstruct the floral morphology, habit, ecology, and other characteristics of the earliest angiosperms (e.g., Arber and Parkin 1907; Axelrod 1952; Cronquist 1968, 1981, 1988; Takhtajan 1969; Stebbins 1974; Doyle and Hickey 1976; Doyle 1978; Retallack and Dilcher 1981; D.W. Taylor and Hickey 1996b). Cronquist (1968, 1988), Takhtajan (1969), and Stebbins (1974) attempted to reconstruct a hypothetical ancestral angiosperm in great detail. Cronquist and Takhtajan took the view that extant members of Magnoliidae, Magnoliales in particular, are the most archaic living angiosperms, with each species retaining some presumably ancestral features. They suggested that the earliest angiosperms had simple, entire leaves, were evergreen, woody plants, but "not necessarily trees," of moist tropical habitats with an equable climate. Cronquist maintained that the nodes were unilacunar with two leaf traces or perhaps trilacunar with three traces. The wood of these plants had no vessels, but possessed long, slender tracheids, with long, tapering ends, and numerous scalariform-bordered pits. The flowers were large, bisexual, and strobiloid, with numerous spirally arranged parts (e.g., *Magnolia*-like). The perianth was not differentiated into sepals and petals, but consisted of spirally arranged tepals. The stamens were relatively large and laminar with no differentiation into filament and anther. The pollen had a single long sulcus; the exine had a raised-reticulate surface ornamentation and internally was either solid or, more probably, tectate-granular. No nectar or nectaries were present. The flowers contained several to many carpels, each of which was folded along the midrib (plicate) and unsealed, with the margins loosely appressed rather than "anatomically joined," effectively closed by "a tangle of hairs." The embryo was small; double fertilization occurred, and the endosperm was triploid. The endosperm was probably copious and of the nuclear type (i.e., it had a free-nuclear stage early in ontogeny).

Stebbins (1974) had a somewhat different view of the ancestral angiosperm, reflecting his belief that novelty evolved in marginal habitats. Nonetheless, he also maintained that these first angiosperms were similar in many respects to modern Magnoliales. He suggested that the first angiosperms were shrubs that grew in semi-arid or seasonally dry tropical or subtropical habitats, with leaves strongly reduced in size in association with this dry habitat. A seasonal climate favored compression of the reproductive cycle into a relatively short period. The distinctive characteristics of angiosperms arose in response to seasonal drought (Axelrod 1970, 1972; see also Hickey and Doyle 1977), including closure of the carpel, reduction in the size of gametophytes, evolution of double fertilization, and a shift from wind- to insect-pollination.

Another hypothesis was offered by Retallack and Dilcher (1981) who maintained that the "xeromorphic bottleneck" (Hickey and Doyle 1977) in angiosperm origins and early evolution was induced by near-marine environments rather than by arid environments. Building on their "herbaceous origin" hypothesis, D.W. Taylor and Hickey (1996b) suggested that angiosperms evolved in a fluvial regime, in sites characterized by high disturbance and moderate amounts of alluviation. These areas would have had high nutrient levels and frequent loss of plant cover due to periodic disturbance.

Whereas Cronquist, Takhtajan, and Stebbins focused on Magnoliales, and D.W Taylor and Hickey (1996b) focused on Piperales for inferring ancestral characteristics of the angiosperms, recent phylogenetic studies convincingly point to the Amborellaceae–Nymphaeaceae–Austrobaileyales grade as the basalmost extant angiosperms. Endress (1986a) was perhaps the only investigator to suggest, from morphology, that some of these groups were among the earliest-diverging extant angiosperms. He indicated that Amborellaceae, Austrobaileyaceae, Trimeniaceae, and Winteraceae (which is not one of the basalmost angiosperms in recent phylogenetic analyses) were potential candidates as some of the earliest angiosperm families.

With improved understanding of angiosperm phylogeny, it is now possible to make more informed re-

constructions of the morphology and ecology of the ancestors of extant early-diverging angiosperms, as well as subsequent evolutionary patterns. Of course, extant basal angiosperms may not be representative of the earliest, now extinct, lineages and may not even represent the earliest diversification of the angiosperms. Furthermore, some Early Cretaceous fossils cannot be clearly assigned to any extant families (Friis et al. 2000); *Archaefructus* is just one example (Sun et al. 2002; Friis et al. 2003).

Thus, although we have made great advances in our understanding of many aspects of angiosperm phylogeny, morphology, floral genetics, and paleobiology, in many respects the origin of the angiosperms remains an "abominable mystery." We and others have gone to great lengths to reconstruct various features of early angiosperms, yet it is not at all clear that these reconstructions are representative of the earliest angiosperms. Because gymnosperms lack many of the characters of interest in early angiosperms, in many cases, our reconstructions are ambiguous regarding the ancestral state of a character for angiosperms. Furthermore, reconstructions based on extant taxa do not take into account the diversity of fossils. The fact that *Archaefructus* may be an early-branching angiosperm further reveals the enormous diversity in morphology and form present in early angiosperms. For example, whereas extant early-diverging angiosperms all exhibit a perianth, *Archaefructus* does not. Interpretation of the flowers in this fossil is unclear. *Archaefructus* may have possessed a loose aggregation of stamens and carpels, or, alternatively, the flowers were unisexual.

With these important caveats in mind, we offer a reconstruction of early angiosperms that is based on the features of extant species with input from the fossil record. Perhaps the best we can hope to do is to reconstruct early motifs among basal angiosperms.

It remains unclear whether the early angiosperms were woody or herbaceous. As noted, several prominent authors have favored a terrestrial origin (Cronquist 1968; Takhtajan 1969; Stebbins 1974), with the first angiosperms as shrubs or small trees, which agrees with early-diverging extant taxa such as *Amborella* and members of Austrobaileyales. However, other authors proposed that angiosperms were ancestrally herbaceous (e.g., D.W. Taylor and Hickey 1992, 1996a, 1996b; Doyle et al. 1994). Nymphaeaceae, which follow *Amborella* as sister to all remaining angiosperms, are aquatic herbs. The early angiosperm *Archaefructus* was also herbaceous and aquatic. The aquatic habit clearly appeared very early in the diversification of flowering plants. Some of our reconstructions based on extant taxa suggest that angiosperms were ancestrally woody; only if Chloranthaceae are coded as herbaceous is the ancestral state for angiosperms ambiguous. *Amborella* and Austrobaileyales are small trees, shrubs, or woody vines; all are shade tolerant and exhibit morphological and physiological traits that enhance the capture of light in a forest understory (Feild et al. 2000a, 2003a, 2003b). Based on characteristics of Amborellaceae and Austrobaileyales, Feild et al. (2004) proposed that angiosperms arose in "dark and disturbed" understory habitats.

Our analyses indicate that early angiosperms also possessed unilacunar (one- or two-trace) nodal anatomy, although the evolutionary significance of this anatomical characteristic is unknown. It is still unclear whether the ancestor of extant angiosperms possessed leaves with chloranthoid teeth. Another important issue is whether the earliest angiosperms possessed vessels. Vessels are considered more efficient in conducting water than tracheids and provide water to leaves at an increased rate (Carlquist 1975; Takhtajan 1969). The xylem features of *Amborella, Austrobaileya, Illicium,* and *Kadsura* suggest that early angiosperms likely lacked vessels or possessed "primitive" vessel elements with tracheid features (Figure 3.10). However, because many early angiosperms may have lived in wet understory conditions, a low water-transport capacity of the stems would not have been problematic (Feild et al. 2000a, 2003a, 2003b, 2004). For example, *Amborella* lacks vessels, and thus the water-conducting capacity of the stems is low, but the plants grow in areas of low evaporative demand and high rainfall. Vessels are present in *Austrobaileya,* but these vessels have tracheid-like features such as gradually tapered end walls, frequent scalariform perforation plates, and small diameters (Bailey and Swamy 1948; Carlquist 1996; Carlquist and Schneider 2002). Although the hydraulic capacity of the xylem in *Austrobaileya* is higher than that of *Amborella,* the values are still lower than in other climbing angiosperms (Tyree and Ewers 1991, 1996). The stem xylem water-transport capacities of *Illicium* and *Kadsura* are also much lower than values for most vessel-bearing shrubs and deciduous angiosperm trees and lianas, respectively (Tyree and Ewers 1991, 1996; Sperry et al. 1994; see Feild et al. 2000a, 2003a, 2003b, 2004). The low water-conducting values in *Illicium* and *Kadsura* are probably the result of the unspecialized vessels, which are short, with long, oblique scalariform perforation plates that partially retain pit membranes (Carlquist 1982).

Early terrestrial angiosperms may have been successful because they exploited wet, understory forest environments more effectively than ferns and other seed plants, such as Bennettitales, Caytoniales, cycads, and Gnetales, present in these understory environments at that time. Furthermore, these early angiosperms may have been successful because they had more developmental flexibility than these other plant groups. Production of different types of leaves in varying light environments may have enabled early angiosperms to compete more effectively across a wider range of light regimes than other groups with more limited morphological plasticity (Feild et al. 2004).

In contrast to *Amborella* and Austrobaileyales, Nymphaeaceae possess ecological and physiological traits associated with sunny to shady aquatic habitats, but some authors have proposed that their ecology and morphology may not be representative of the earliest angiosperms (e.g., Schneider and Williamson 1993). Nymphaeaceae are often believed to represent a separate and distinct evolutionary experiment that does not represent the ecology associated with the origin of the angiosperms (Feild et al. 2004). However, as we have stressed, *Amborella* and Austrobaileyales may also not be representative of the earliest angiosperms. Fossils such as *Archaefructus* (Sun et al. 2002) and those attributable to Nymphaeaceae (Friis et al. 2001; Gandolfo et al. 2004) indicate that the aquatic habit appeared early in angiosperm evolution.

The floral features of *Amborella,* Nymphaeaceae, and Austrobaileyales suggest that the flowers of early angiosperms were not the large strobiloid structures envisioned by Cronquist (1968, 1981) and Takhtajan (1969), but more likely were small to moderate in size. Further support for this view comes from the fossil record. The angiosperm reproductive structures (flowers, fruits, seeds) from the Early Cretaceous are small (Friis et al. 2000; see also Chapter 1). The fossil *Archaefructus* only adds to the puzzle—it is not clear what constitutes a flower in these plants.

Our reconstructions of basal angiosperm phyllotaxis were equivocal about the ancestral condition; *Amborella* and Austrobaileyales have spiral phyllotaxis, Nymphaeaceae have whorled phyllotaxis, but the distinction between spiral and whorled is not always clear. Different flowers on the same plant of *Drimys winteri* may have spiral or whorled phyllotaxis (Doust 2001). Ontogenetic studies have revealed that in some cases, floral organs that appear whorled in mature flowers actually result from spiral arrangement and initiation of primordia; both spiral and whorled phyllotaxis may have the organs developing in a spiral sequence (see above and Chapter 12). The fossil record is equivocal about the phyllotaxis of many early angiosperm flowers (Friis et al. 2000).

Although Cronquist (1968), Takhtajan (1969), and Stebbins (1974) envisioned early angiosperms with numerous (indeterminate) perianth parts, our reconstructions of the ancestral state for angiosperms are equivocal. Although *Amborella* and Austrobaileyales have indeterminate perianths, our reconstructions point to an early and significant role of trimery in basal angiosperms. The ancestral condition for Nymphaeaceae is a trimerous perianth, as is the ancestral state for all angiosperms above the node to Austrobaileyales.

The perianth of the early angiosperms was likely undifferentiated. An alternative scenario is that early angiosperms did not possess a perianth—this was added later. The stamens probably were not differentiated into anther and filament regions, but were leaf-like (laminar). The infratectal structure of the pollen was not granular, as previously suggested (e.g., Cronquist 1988), but exhibited an "intermediate" morphology (*sensu* Doyle and Endress 2000) characterized by irregular radial (columellar) elements mixed with granules. The early carpel was ascidiate (not plicate as suggested by Eames 1961; Cronquist 1968, 1988) and not postgenitally fused, but rather closed by secretion (Endress and Igersheim 2000b). The ploidy of the endosperm of early angiosperms is now unclear—triploid and diploid endosperm are equally parsimonious alternatives (Williams and Friedman 2002). Endosperm formation was likely "bipolar cellular" (Floyd and Friedman 2000); seeds were likely few in number and small, with copious endosperm produced.

Future Research

Darwin (1859) considered the origin and rapid radiation of the angiosperms an "abominable mystery." The 1990s and early 2000s saw enormous progress in several areas, including (1) elucidating important patterns of morphological development, (2) recovering and placing ancient angiosperm fossils (see Crepet 2000; Crane et al. 2004), (3) identifying the most ancient clades of extant angiosperms and reconstructing the likely features of the earliest flowering plants, and (4) elucidating some of the key organ identity genes involved in floral development. Nonetheless, the origin of the angiosperms remains a monumental problem; the abominable mystery is far from solved.

The basal angiosperms provide an excellent example of progress on what was recently considered a potentially intractable phylogenetic problem. Many major questions of relationship are well resolved and supported. *Amborella,* Nymphaeaceae, and Austrobaileyales represent the earliest lineages of extant angiosperms. Most analyses point strongly to *Amborella* as sister to all other angiosperms, although the possibility of *Amborella* + Nymphaeaceae as sister to all angiosperms cannot be completely ruled out. The relationship of *Ceratophyllum* and monocots, as well as the position of Chloranthaceae, remain among the most significant deep-level phylogenetic problems in the basal angiosperms. Additional sequence data may help to resolve such difficult phylogenetic problems, although analysis of more than 15,000 bp per taxon has not convincingly resolved these relationships. The origin of monocots, Chloranthaceae, magnoliids, and eudicots appears to represent a rapid radiation after the initial origin and diversification of the angiosperms. Improved inferences of basal angiosperm relationships, as well as character-state reconstructions, will require additional data from fossils—both the discovery of new fossils, which by nature is a stochastic process (Crepet 2000), and reanalysis of existing fossil data of both angiosperms and lineages that may be the sister group to the angiosperms. As several investigators have stressed (Donoghue and Doyle 2000; Zanis et al. 2002, 2003), future analyses must focus on the integration of molecular and fossil datasets, as well as on further morphological and functional characterization of extant angiosperms. The placement of the fossil *Archaefructus* as sister to all extant angiosperms represents an example of the integration of fossils into largely molecular datasets of extant taxa. Additional fossil taxa need to be examined, and various methods of integrating fossils into phylogenetic trees of extant taxa need to be used and compared (Crane et al. 2004; Crepet et al. 2004).

Another interesting challenge offered by the basal angiosperms is the elucidation of the developmental genetics of floral diversification in these plants. Floral architecture is labile in basal angiosperms, with substantial variability in merosity and phyllotaxis. The flowers of many basal angiosperms do not have a differentiated perianth but are composed instead of tepals. The basalmost angiosperms have ascidiate carpels that are not postgenitally fused. Several lineages of basal angiosperms have laminar or leaf-like stamens that are not differentiated into filament and anther. To understand the genetic underpinning of floral development, including the early diversification of the flower, data are needed for both basal angiosperm lineages and basal eudicots (e.g., Kramer and Irish 1999, 2000; D. Soltis et al. 2002b; Kanno et al. 2003; Kramer et al. 2003). Such data will provide a critical link between what is known about derived model taxa such as *Arabidopsis, Antirrhinum,* and maize with other angiosperms. Research at the interface of phylogenetics, development, and developmental genetics is required to elucidate the genetic architecture of floral development.

Monocots

Introduction

Although the phylogenetic placement of the monocotyledons (often shortened to monocots) relative to eudicots, magnoliids, and Chloranthaceae is unclear (see Chapters 2 and 3), the monocots certainly represent one of the oldest lineages of angiosperms. The origin of the monocot clade must have occurred no later than the origin of the eudicot clade (see Chapter 5). Hence, given that eudicot pollen dates to at least 125 million years ago (Mya), the monocot lineage must also be that old, and molecular-based age estimates support this inference. Using *rbcL* sequence data and the mean-path lengths method for estimating divergence times, Bremer (2000, 2002) dated the origin of the monocot clade at 134 Mya. Using the nonparametric rate smoothing approach (Sanderson 1997), Wikström et al. (2001) estimated the age of the crown group of monocots to be in the range of 127 to 141 Mya (see Sanderson et al. 2004). There are about 52,000 species of monocots (Mabberly 1993), representing 22% of all angiosperms. Half of the monocots can be found in the two largest families, Orchidaceae and Poaceae, which include 34% and 17%, respectively, of all monocots. These two families are also among the largest families of angiosperms and dominant members of many plant communities. Although monocots are largely herbaceous, some, particularly agaves, palms, pandans, and bamboos, can reach great heights, lengths, and masses. Monocots provide most of the world's staple foods (e.g., grains and starchy root crops), abundant building materials, and a great number of medicines.

The single cotyledon is just one of several putative synapomorphies for the monocots. Phylogenetic studies of nonmolecular data (Donoghue and Doyle 1989a, 1989b; Loconte and Stevenson 1991; Doyle and Donoghue 1992) have identified many putative synapomorphies for the monocots (Table 4.1), including a single cotyledon, parallel-veined leaves, sieve-cell plastids with several cuneate protein crystals, scattered vascular bundles in the stem, and an adventitious root system.

Recognition of the monocots as a distinct group within the angiosperms dates from Ray (1703) and was based largely on their having a single cotyledon instead of the two cotyledons typical of the dicotyledons or "dicots." (As reviewed in Chapter 2, the dicot grouping is now known to be an unnatural, or nonmonophyletic, group that is not a relevant point of comparison.) Monocot seedlings display a great diversity of form (Tillich 1995), however, and not all possess an obvious single cotyledon. In the grasses, for example, the single cotyledon is thought to have been modified and become an absorptive organ within the seed.

Another major distinctive trait of the monocots is their vascular system, which is characterized by vascular bundles that are scattered throughout the medulla and cortex and are closed (i.e., they do not contain an active cambium). In contrast, most basal angiosperms formerly considered dicots (e.g., members of the magnoliid clade, Amborellaceae, Austrobaileyales) and eudicots possess vascular bundles in a ring that is open. Tomlinson (1995) argued that the vascular system of monocots is unique among the seed plants and so different that there is "no homology of organization between the primary vascular system of monocotyledons and dicotyledons." This statement would lead one to believe that monocots might not be angiosperms at all or might have an origin separate from other angiosperms, but that idea can be discarded on the basis of both DNA and morphological evidence. For example, scattered vascular bundles occur in other angiosperms outside of the monocots (e.g., Nymphaeaceae and some Piperaceae). Tomlinson's statement does, however, emphasize the distinctiveness of monocot habit and vegetative organization, despite the fact that the floral features of monocots are similar to many other basal angiosperms, particularly members of the magnoliid clade such as Aristolochiaceae and Annonaceae.

The fact that the vascular bundles of monocots are scattered rather than arranged in a ring as in the woody basal angiosperms and eudicots means that orderly, bifacial growth of the vascular cambium is not possible. That is, secondary growth does not occur in monocots as it does in eudicots and other angiosperms such as the magnoliids. As a result, many arborescent monocots achieve their stature by primary gigantism. For example, palm "trees" (Arecaceae) and members of Dasypogonaceae (i.e., *Kingia*) are overgrown herbs with

TABLE 4.1	Putative synapomorphies for the monocots from Donoghue and Doyle (1989b), Loconte and Stevenson (1991), and Doyle and Donoghue (1992)

1. Presence of calcium oxalate raphides
2. Absence of vessels in the leaves
3. Monocotyledonous anther wall formation*
4. Successive microsporogenesis
5. Syncarpous gynoecium
6. Parietal placentation
7. Monocotyledonous seedling
8. Persistent radicle
9. Haustorial cotyledon tip
10. An open cotyledon sheath
11. Presence of steroidal saponins*
12. Fly pollination*
13. Diffuse vascular bundles (lack of secondary growth)

* indicates characters that are synapomorphies for topologies in which *Acorus* (Acoraceae) is not sister to the rest of the monocots. *Acorus*, for example, lacks steroidal saponins and has dicotyledonous anther formation, so if *Acorus* is sister to the rest of the monocots, then these are not synapomorphies for the monocots as a whole.

primarily "woody" stems; once laid down, the stems cannot increase in diameter through secondary accretion. The stems of palms and others such as the pandans (Pandanaceae) do become stiffer with age by means of increased cell-wall thickness and lignin deposits, but this is an entirely different process from that occurring in plants with a bifacial cambium.

Primary gigantism is not the only method by which monocots can achieve a tree growth form. Other arborescent monocots, which are almost all members of Asparagales (see below) such as *Agave, Cordyline, Dracaena* (all Asparagaceae s.l. or Agavaceae, Laxmanniaceae, and Ruscaceae, respectively), and *Aloe* (Xanthorrhoeaceae s.l. or Asphodelaceae), achieve their large stature by means of an "anomalous" secondary growth produced by an etagen or tiered cambium. This growth is unidirectional and therefore still not like that of other angiosperms. Outside of Asparagales, an etagen cambium is reported only in *Dioscorea* (Dioscoreales). The inability to produce a well-organized bifacial cambium has limited the evolution of growth form in the monocots, but they nevertheless exhibit considerable habit diversity. Understanding how this radical reorganization of the monocot stem occurred and from what sort of ancestral state it evolved is a major unanswered problem in angiosperm evolution.

Another widely cited character of the monocots is their particular form of sieve cell plastids (Behnke 1969), which are triangular and have cuneate proteinaceous inclusions. Similar sieve cell plastids are found in Aristolochiaceae (Dahlgren et al. 1985). We now assume that this similarity between monocots and Aristolochiaceae represents convergence, not shared ancestry, because phylogenetic studies of DNA sequences from all three genomes (Qiu et al. 1999; Zanis et al. 2002, 2003; Chase et al. 2005) have demonstrated a strongly supported relationship of Aristolochiaceae to other Piperales within the magnoliid clade.

Trimerous flowers have long been considered a uniting characteristic of the monocots, but it is not an exclusive feature because many other basal angiosperms, including Nymphaeaceae and magnoliids, also exhibit trimery. In fact, character-state reconstructions indicate that trimery arose early in the angiosperms; it may be ancestral for basal angiosperms above *Amborella* (see Chapter 3). Trimery appears therefore to be a symplesiomorphic feature for monocots and other angiosperms and is not a "monocot character."

An often-overlooked synapomorphy for monocots is their sympodial growth. There are other sympodial angiosperms, but monocots are nearly exclusively so. Even branched, arborescent genera, such as *Aloe* (Xanthorrhoeaceae), are sympodial; new sympodia arise near the apex of the previous one and "displace" the terminal inflorescence into a lateral position, but these plants are nonetheless sympodial. Branching in arborescent monocots is achieved by production of more than one terminal sympodium, but such branching is limited by the demands it makes on the vascular system of older sympodia, which cannot expand to meet increased requirements (see Tomlinson 1995).

Phylogenetic Analyses Based on Morphology

Analysis of morphological characters to estimate relationships within monocots has been limited to those of Stevenson and Loconte (1995; 101 characters) and Chase et al. (1995b; 103 characters, compiled collectively by the participants of the 1993 Monocot Symposium at the Royal Botanic Gardens, Kew). These two sets of analyses reached similar conclusions about higher-level relationships in the monocotyledons, which is not surprising because they relied on a great deal of the same information. Both analyses used the same outgroups (Nymphaeales, Piperales, Lactoridales, Aristolochiales, *sensu* Takhtajan), but arranged them in different ways. Stevenson and Loconte (1995) specified *Lactoris* (as Lactoridales) as the ultimate outgroup; Chase et al. (1995b) used a polytomy. As a result of this difference in rooting, Aristolochiales alone appeared as the sister group of the monocots in Stevenson and Loconte (1995). Stevenson and Loconte (1995) then claimed that their analysis demonstrated that Aristolochiales were the sister group of the monocots, but this conclusion was not a result of their analysis but a product of assumptions made before the analysis about how to arrange the outgroups. Chase et al.'s (1995b) results could have been arranged in the same way.

The aquatic alismatids have most commonly been considered as the most "primitive" monocots (Hallier 1905; Arber 1925; Hutchinson 1934; Cronquist 1968, 1981; Takhtajan 1969, 1991; Stebbins 1974; Thorne 1976). Net-veined groups, such as Dioscoreales (Dahlgren et al. 1985) and Melanthiales (Thorne 1992a, 1992b), have also been considered most primitive among the monocots. As we stress throughout this book, the issues of which extant group is the most primitive and which is the sister of all others are not equivalent, although these statements are often confounded. We can infer primitive or ancestral character states via character-state reconstruction using the best estimate of phylogeny, as we have done throughout this book. Primitive characters for monocots could be present in some derived groups. Concomitantly, basal taxa can exhibit many morphological autapomorphies (derived traits; see Chapter 3). Hence, although *Acorus* has now been demonstrated in several DNA-based analyses to be sister to all other monocots (Duvall et al. 1993a, 1993b; Chase et al. 1993, 1995a, 1995b, 2000a, 2005; Davis et al. 1998; Fuse and Tamura 2000; Tamura et al. 2004; Givnish et al. 2005), this result does not imply that *Acorus* is "the most primitive monocot" in terms of its characters. Similarly, *Amborella* is not the most primitive angiosperm but the sister group of all other angiosperms (see Chapter 3). In fact, character-state reconstruction across tree topologies indicates that *Acorus* in general is highly derived in most morphological characters. Many of the characters of *Acorus* are autapomorphies.

A major effect of using basal angiosperms such as Aristolochiaceae as outgroups in strictly morphological analyses is a rooting among the monocots with net-veined leaves. The net-veined monocots were lumped in taxonomic treatments from the mid-1980s to mid-1990s (e.g., Cronquist 1981) under several higher categories, but they were largely referable to Dioscoreales *sensu* Dahlgren et al. (1985). The net-veined monocots include members of Dioscoreaceae, Smilacaceae, Stemonaceae, Taccaceae, and Trilliaceae, as well as several smaller families. Any other rooting would have been difficult to imagine because most of the other major groups of monocots, including *Acorus* and the alismatids, have no counterpart in any other group of basal angiosperms; their habits are unique to monocots, a feature that is likely to polarize characters in such a manner as to make these groups occupy a derived place in cladograms based solely on morphology. Thus, it should have been no surprise to see *Acorus* and the alismatids occupy relatively derived positions in the trees produced by Chase et al. (1995b) and Stevenson and Loconte (1995).

One advantage of using characters totally divorced from morphology in phylogenetic studies is that this approach can lead to new perspectives on old problems. Such is the case for the hypotheses that the alismatids are primitive monocots and that monocots were primitively aquatic (Henslow 1893). We would argue that the position occupied by the alismatids in the DNA analyses (see "Molecular Analyses of Monocot Relationships," below) is entirely consistent with them exhibiting a preponderance of traits that are relatively primitive for monocots. This contradicts the conclusion of Les and Haynes (1995) and Les et al. (1997a) that the *rbcL* results (Chase et al. 1993) did not support the idea that the alismatids were "basal within monocots," a conclusion that we take to mean that the alismatids were not sister to the rest (they are positioned as sister to the rest, exclusive of *Acorus*). As pointed out above, the position of *Acorus* in the *rbcL* tree does not preclude the possibility that the alismatids may still be "the most primitive monocots" in terms of their characters. We conclude that the DNA studies support such a conclusion, but not in a robust manner because tree topologies alone do not permit us to reach such conclusions. In this case, the conclusion that alismatids have the most primitive characters of all extant monocots is based on the same information that earlier authors (Hallier 1905; Arber 1925; Hutchinson 1934; Cronquist 1968, 1981; Takhtajan 1969, 1991; Stebbins 1974; Thorne 1976) who supported this idea had at their disposal—an aquatic scenario of monocot origins.

Plesiomorphic characters in both the Chase et al. (1995b) and Stevenson and Loconte (1995) analyses included stem vessels present, leaf petiole of the Dioscoreales type, primary venation palmate, secondary venation reticulate, and a eustele. These are all traits associated with plants occurring in forest understory conditions. Although not used in these analyses, a vining habit could be added to this list; Dioscoreales *sensu* Dahlgren et al. (1985) are mostly understory vines. Reproductive traits were also highly influenced in these analyses by the choice of outgroups (i.e., Nymphaeales, Piperales, Lactoridales, Aristolochiales *sensu* Takhtajan), and the following traits were plesiomorphic: differentiated perianth, stamen connectives protruding, lack of septal nectaries, parietal placentation, and septicidal capsules (i.e., a syncarpous fruit). These are not characters that would be considered primitive within monocots in general and are either absent in, or not applicable to, the alismatids. Stevenson and Loconte (1995) posited Stemonaceae and Trilliaceae as the most primitive monocots. Taccaceae exhibited more primitive states in the Chase et al. (1995b) study.

Both sets of morphological analyses then followed with either a grade (Chase et al. 1995b) or clade (Stevenson and Loconte 1995) composed of the lilioid monocots (treated as Asparagales and Liliales below; and including Zingiberales in the trees of Stevenson and Loconte 1995) sister to a clade composed of alismatids and commelinids. Characters for the lilioid clade were not specified, but the short branch shown in the figure in Stevenson and Loconte (1995) would indicate that there were probably only two synapomorphies. In the clade sister to the lilioids of Stevenson and Loconte (1995) was a grade of taxa associated with either Velloziaceae or Bromeliaceae followed by clades composed of alismatids, including Araceae, and commelinids (see below), except for Zingiberales, but including *Acorus* as sister to Sparganiaceae and Typhaceae. In the Chase et al. (1995b) results, the lilioid grade was followed by a clade of Bromeliaceae/Velloziaceae, Zingiberales, and alismatids and commelinids; Araceae were again associated with alismatids and *Acorus* with Sparganiaceae and Typhaceae.

The morphological cladistic analyses of Chase et al. (1995b) and Stevenson and Loconte (1995) produced highly congruent patterns of relationships, as would be expected from the similarity of the characters included. Neither morphological analysis surmounted the issues of how to address the unique characters of the monocots, particularly those of *Acorus* and the alismatids, and the problems this posed for determining the root by outgroup comparison. The many authors who supported the hypothesis of the primitiveness of the alismatids did not use an outgroup approach to arrive at this conclusion. Tomlinson (1995) likewise had problems with the use of an extant "dicot" as a model starting point for addressing the issue of the vegetative architecture of the monocots, particularly their unique vascular system.

Molecular Analyses of Monocot Relationships

Due to intensive molecular study and frequent collaboration, the monocots have now emerged as perhaps the best characterized major clade of angiosperms (Chase et al. 2005). The earliest well-sampled DNA analyses of monocots were the *rbcL* analyses of Duvall et al. (1993a, 1993b) and Chase et al. (1993). These analyses included enough monocots to conclude that *Acorus* was sister to the rest of the monocots, but several critical taxa were still missing (e.g., Triuridaceae) or underrepresented (e.g., only three alismatids were included) in these studies. Duvall et al. (1993a, 1993b) and Chase et al. (1993) did include at least some representatives of each of the three main contenders for "the most primitive monocot": alismatids, *Veratrum* (Melanthiaceae), and Dioscoreales. Bharathan and Zimmer (1995) found that *Acorus* did not fall with the other monocots in their study of partial 18S rDNA sequences, but they did not sample many monocots (only 24). A large-scale sampling of 18S rDNA sequences for monocots did not emerge until the studies of D. Soltis et al. (2000) and Chase et al. (2000a); the latter used 159 species of monocots. For comparison, Chase et al. (1995a) sampled 172 monocots but did not use an outgroup approach to root their trees.

In contrast to the results of the morphological analyses of angiosperm relationships (Donoghue and Doyle 1989a, 1989b; Loconte and Stevenson 1991; Doyle and Donoghue 1992), the *rbcL* analyses produced radically different ideas about the relationships of monocots. Morphological analyses alternated between two positions for the monocots (e.g., Donoghue and Doyle 1989a, 1989b): sister to Nymphaeaceae or to Aristolochiaceae. These studies coded monocots as a single terminal (as was generally done in studies of that kind at that time; see Chapter 2) and so presumed ideas of what constituted primitive monocot traits undoubtedly played a major role in where the monocots were positioned. For the reasons stated above, these sorts of studies have a degree of circularity about them that is difficult to avoid, but certainly such problems are exacerbated by the use of single, synthetic terminals coded with putatively primitive traits. The work with *rbcL* avoided such circularity and arrived at novel conclusions, not only for the position of monocots within the angiosperms but also for relationships within the monocots. In the *rbcL* results (Chase et al. 1993), the monocots were positioned with some relationship to the magnoliids Laurales, Magnoliales, and Piperales (at that time termed the "paleoherb I clade"), but none of these relationships received bootstrap support above 50% (Chase and Albert 1998).

Since the Chase et al. (1993) analysis, several studies of single, as well as combined, gene datasets have been analyzed (D. Soltis et al. 1998, 2000; Qiu et al. 1999; P. Soltis et al. 1999b; Savolainen et al. 2000a; Zanis et al. 2002, 2003; Hilu et al. 2003; Duvall et al. 2005), and only the last of these studies has provided clear evidence for monocot relationships (see also Chapter 3). The Duvall et al. paper placed the monocots as sister to the magnoliid clade (Magnoliales, Laurales, Canellales, and Piperales), but only with support from high Bayesian posterior probabilities, which have been shown to be over-estimates of support (Suzuki et al. 2002). What is clear from all of these studies is that an exclusive relationship to either Nymphaeaceae or Aristolochiaceae is excluded because there are intervening nodes that are strongly supported. The remaining possibilities for higher-level relationships are to Chloranthaceae, Ceratophyllaceae, the magnoliids collectively, eudicots, or to some combination of these. However, this problem is made difficult by the long branch of the monocots relative to several of these taxa (e.g., magnoliids) and the similar-

ly long branch of Ceratophyllaceae that have resulted in unstable estimates of relationships for these taxa. The most consistent answer thus far is that the monocots and Ceratophyllaceae are sister taxa (see Chapter 3). Model-based methods of analysis (e.g., maximum likelihood and Bayesian approaches) have also produced this pattern, but then the relationships of these two groups (monocots + Ceratophyllaceae) with the other lineages remain unclear (e.g., P. Soltis et al. 2000; Zanis et al. 2002, 2003). The Duvall et al. (2005) result needs to be considered further, but its reliance on Bayesian posterior probabilities renders it suspect.

Current Understanding of Relationships Within Monocots

Monocot phylogenetics have made immense strides over the past ten years due primarily to the foci provided by the international monocot symposia held in 1993, 1998, and 2003 (at the Royal Botanic Gardens, Kew, Rudall et al. 1995; the Royal Botanic Garden, Sydney, Wilson and Morrison 2000; and Rancho Santa Ana Botanical Garden, Columbus et al. 2005, respectively). These meetings have focused attention both on what was known and, more importantly, on which groups needed additional research. As a result, we now know more about monocots than about any other major group of angiosperms, a remarkable achievement given the paucity of information available in 1985 (Dahlgren et al. 1985). This model should now be adopted for the other large groups of angiosperms so that attention is likewise focused on integration of research programs and on gaps in the database (Chapters 3 and 8). Even the relatively well-studied asterid orders have new members that desperately need integration into the overall picture of eudicot evolution.

The overview of relationships among the monocots presented here is based in large part on the three-gene analyses of Chase et al. (2000a) and D. Soltis et al. (2000) and the seven-gene analysis of Chase et al. (2005). The first two studies were based on the same three genes (*rbcL, atpB,* 18S rDNA), but Chase et al. (2000a) focused just on the monocots and used more monocots than D. Soltis et al. (2000). The third large-scale analysis included those three genes, plus plastid *matK* and *ndhF*, mitochondrial *atpA,* and partial 26S rDNA (a seven-gene dataset; Chase et al. 2005). The patterns obtained in Chase et al. (2005) are generally like those in the three-gene analyses, but with stronger internal support. In the discussion that follows, we mention these results only when they offer new patterns or much stronger support than previous studies.

All but two molecular phylogenetic analyses of monocots have placed *Acorus* alone as sister to all other monocots. The first exception to this statement was the

18S rDNA analysis of Bharathan and Zimmer (1995) in which *Acorus* was placed outside of the monocots altogether, a result that must be considered spurious. Combining the 18S rDNA sequence data with sequences from *rbcL* and *atpB* (Chase et al. 2000a; D. Soltis et al. 2000) resulted in strong support for the monophyly of monocots, as well as strong support for the monophyly of all monocots excluding *Acorus*. A recent analysis of two of the seven genes used in Chase et al. (2005), *rbcL* and *atpA* (Davis et al. 2004), recovered an alismatid clade that contained *Acorus*. This deviating result is perplexing because neither *rbcL* (Chase et al. 1993, 1999a; Duvall et al. 1993a, 1993b) nor *atpA* (Davis et al. 1998) analyzed alone produced such a position for *Acorus*. In contrast, studies of basal angiosperm relationships that have used more genes (5 to 11) have consistently found *Acorus* sister to the remaining monocots (e.g., Qiu et al. 1999, 2000; Zanis et al. 2002, 2003; see Chapter 2). As reviewed here, a seven-gene analysis of monocots also found strong support for the placement of *Acorus* as sister to all other monocots (Figure 4.1).

The placement of *Acorus* as sister to all other monocots is therefore well supported by molecular data (Figure 4.1; see also Fuse and Tamura 2000). However, this rooting strongly contradicts the morphological analyses of Chase et al. (1995b) and Stevenson and Loconte (1995), but is consistent with the combined morphology/*rbcL* analysis of Chase et al. (1995b), in which *Acorus* was again well supported (Bremer support >3) as sister to the rest of the monocots. Chase et al. (1995b) noted that the ingroup networks of the morphological

and molecular trees were highly similar and that they differed most significantly in the point at which they were rooted by the outgroups. The potentially misleading effects of outgroup rooting from morphological-based data are discussed above. Chase et al. (1995b) also pointed out that the aquatic alismatids and the largely aquatic (helophytic) aroids and *Acorus* were nested in the morphological trees among other aquatic or semiaquatic taxa such as Pontederiaceae, Philydraceae, Sparganiaceae, and Typhaceae, a result likely due to convergence in characters associated with habitat conditions. Discounting these two differences, all other patterns of relationship were similar.

Following *Acorus*, the monophyly of the remaining monocots is again strongly supported (99% to 100% jackknife/bootstrap support). A strongly supported (99% to 100%) Alismatales clade appears as sister to the remaining monocots (Figures 4.1 and 4.2). Following Alismatales, the remaining monocots again form a well-supported clade (99% to 100%). Within this remaining large clade are several component subclades: commelinids, Dioscoreales, Petrosaviaceae, Pandanales, Liliales, and Asparagales. Although many of these component subclades receive moderate to strong support, relationships among these subclades are generally more poorly supported (Figure 4.1). In the strict consensus of Chase et al. (2000a), the branching order above Alismatales is Dioscoreales, Pandanales, Liliales, and Asparagales/commelinids. The seven-gene analysis is consistent with this pattern, except that Dioscoreales and Pandanales are sister taxa. Most of these relationships received at least moderate bootstrap support.

FIGURE 4.1 Strict consensus tree from Chase et al. (2005; see also Chase 2004) based on a combined analysis of seven genes: mitochondrial *atpA*, plastid *rbcL*, *atpB*, *ndhF*, *matK*, and nuclear 18S and 26S rDNA sequences. Numbers indicate bootstrap support. Due to a mistake in the the 18S rDNA exclusion set in an earlier version of the seven-gene matrix (Chase et al. 2005), Arecaceae (Arecales) were sister to Commelinales/Zingiberales (with low bootstrap support as shown here and in Figure 4.8), but when this mistake was corrected, Arecaceae then fell as sister to the rest of the commelinids, but again with low bootstrap support. This change in the exclusion set also affected some bootstrap percentages for other clades, so there are some minor deviations here from those reported in the final version of the Chase et al. (2005) paper. Results reported in Chase (2004) for this analysis also differ from Chase et al. (2005). Relationships of Arecaceae should thus be considered unclearly resolved within the commelinids.

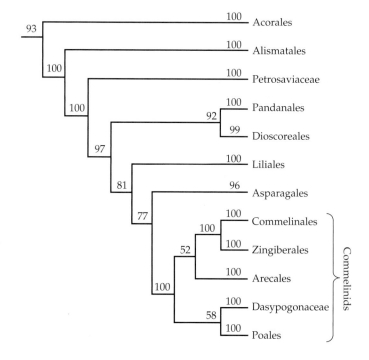

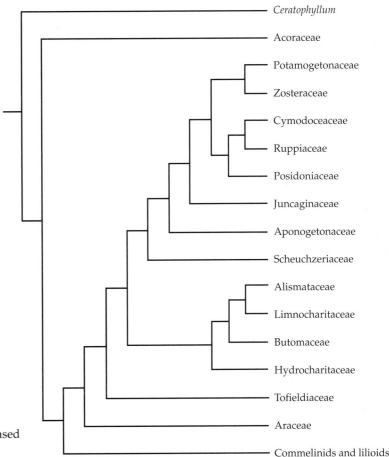

FIGURE 4.2 Summary tree of Alismatales. (Based on Les et al. 1995 and Chase et al. 2005.)

Acorales

Acorales consist of the single family Acoraceae, which, in turn, consist of the single genus *Acorus*. *Acorus* has typically been considered an aberrant member of the aroid family, Araceae (reviewed in Grayum 1987). On the basis of its highly divergent seedling morphology, Tillich (1985) stated that *Acorus* should be excluded from Araceae (Acoraceae was proposed by Martynov in 1820; see APG II 2003). The treatment followed here, placing *Acorus* in its own family and monofamilial order, Acorales, has now been widely accepted. In morphological terms, *Acorus* has many atypical features for a basal monocot, but it also has some that might be expected: dicotyledonous anther wall formation and ethereal oils, both of which are unique to *Acorus* among basal monocots. Its habit with unifacial, ensiform leaves, inflorescence structure (a spadix-like structure of flowers with an undifferentiated perianth), syncarpous ovary, and berry-like fruits, are not what would be expected for a "primitive" monocot based on traditional views (Figure 4.3, A to C).

Alismatales

Alismatales are composed of 14 families in the APG II (2003) circumscription: Alismataceae, Aponogetonaceae, Araceae, Butomaceae, Cymodoceaceae, Hydrocharitaceae, Juncaginaceae, Limnocharitaceae, Posidoniaceae, Potamogetonaceae, Ruppiaceae, Scheuchzeriaceae, Tofieldiaceae, and Zosteraceae (several of these are illustrated in Figure 4.3). The monophyly of the order is supported by molecular analyses (e.g., Chase et al. 1993, 1995a, 2000a; Duvall et al. 1993a; D. Soltis et al. 2000). This circumscription of Alismatales includes those families that most authors have previously placed in the order (e.g., Cronquist 1981; Thorne 1992a, 1992b; Takhtajan 1997), as well as Araceae and Tofieldiaceae. At least some authors have also referred Araceae to their own order with a possible close relationship to Arecales (e.g., Cronquist 1981); hence, the results of all molecular studies that placed Araceae with Alismatales were unanticipated (Figure 4.2).

The families of Alismatales all share scales or glandular hairs at the nodes within the sheathing leaf bases, and the embryos are uniquely green. Araceae (Figure 4.3, F to I) and Tofieldiaceae (3 genera, 27 species; North and South America, as well as Eurasia) are otherwise different from the remaining families in their inflorescence, floral structure, and habit (they are not aquatic and lack lacunae in their stems, etc.), but Tofieldiaceae have been poorly studied with respect to a relationship

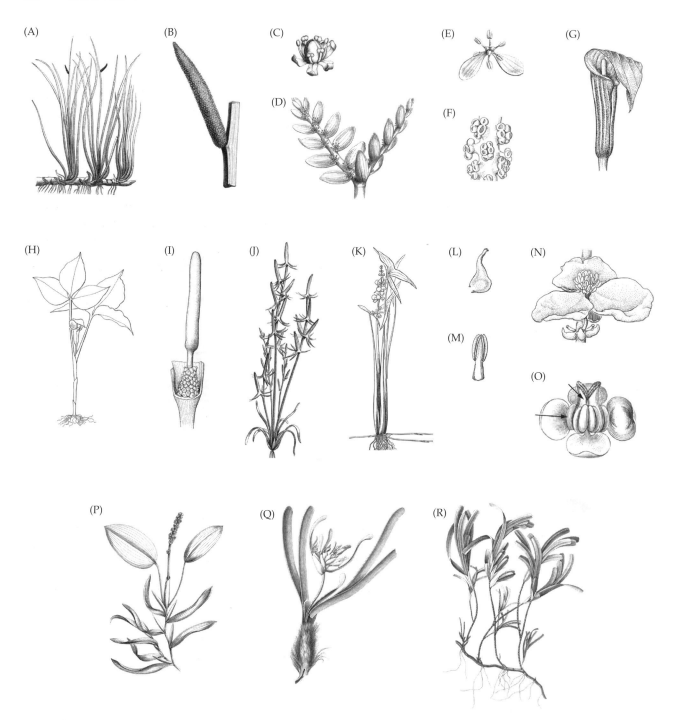

FIGURE 4.3 Flowers and plants representing Acorales (A–C) and Alismatales (D–R). (A) *Acorus calamus* (Acoraceae), plants showing habit; two inflorescences are visible. (B) *Acorus calamus* (Acoraceae), inflorescence. (C) *Acorus calamus* (Acoraceae), flower. (D) *Aponogeton monostachyus* (Aponogetonaceae), plant in flower. (E) *Aponogeton monostachyus* (Aponogetonaceae), single flower with two tepals, six stamens and three carpels. (F) *Arisaema triphyllum* (Araceae), portion of staminate spadix. (G) *Arisaema triphyllum* (Araceae), inflorescence. (H) *Arisaema triphyllum* (Araceae), plant in flower. (I) *Arisaema triphyllum* (Araceae), carpellate inflorescence. (J) *Triglochin calcitrapa* (Juncaginaceae), plant in flower and fruit. (K) *Sagittaria longifolia* (Alismataceae), plant in flower. (L) *Sagittaria longifolia* (Alismataceae), carpel. (M) *Sagittaria longifolia* (Alismataceae), stamen. (N) *Sagittaria longifolia* (Alismataceae), staminate flower. (O) *Potamogeton gramineus* (Potamogetonaceae), single flower, upper arrow indicates one of two styles, lower arrow indicates anther. (P) *Potamogeton gramineus* (Potamogetonaceae), plant in flower. (Q) *Posidonia oceanica* (Posidoniaceae), plant in flower. (R) *Cymodocea ciliata* (Cymodoceaceae), whole plant. (A to E, from Engler in Engler and Prantl 1889. F to I, from Wilson 1960b; K to N, from Rogers 1983. J, from Buchenau and Hieronymus in Engler and Prantl 1889. O to R, from Ascherson in Engler and Prantl 1889.)

to the other alismatid families. The flowers of Tofield-iaceae are nondescript and typical of monocots in general, except that some species have up to 12 stamens, and the fruit is a septicidal capsule. A "calyculus" (a simulated calyx of bracts or bractlets) subtending flowers is present in most species.

Araceae are by far the largest family of the order (106 genera, 4,025 species; cosmopolitan). This family is particularly diversified in the wet tropics, but some temperate genera (e.g., *Arisaema*) are also species-rich. Typically Araceae are herbs with a clear petiole and an expanded lamina; their most distinctive feature is their inflorescence, which has an often petaloid spathe and a spadix of sessile flowers with an expanded sterile terminal portion that emits odors (Figure 4.3, F to I). Many genera have separate male and female zones on the spadix. The pondweeds, five genera formerly placed in Lemnaceae (Les et al. 1997b; Rothwell et al. 2003), are embedded in Araceae, so they have been included there; these are the smallest angiosperms.

Molecular phylogenetic analyses have indicated that Araceae are sister to the remaining members of Alismatales (Figure 4.2), the monophyly of which is supported by DNA sequence data as well as by non-DNA characters. All families other than Araceae have seeds that lack endosperm and have root hair cells that are shorter than other epidermal cells (Judd et al. 2002). Members of this large subclade occur in wetland or aquatic habitats, and Judd et al. (2002) referred to these families (Alismataceae, Aponogetonaceae, Butomaceae, Cymodoceaceae, Hydrocharitaceae, Juncaginaceae, Limnocharitaceae, Posidoniaceae, Potamogetonaceae, Ruppiaceae, Scheuchzeriaceae, and Zosteraceae) as the aquatic clade. The large aquatic clade is divided into two subclades (Figure 4.2). The first subclade is smaller and comprises Alismataceae (Figure 4.3, K to N), Butomaceae, Hydrocharitaceae, and Limnocharitaceae; these have a scapose inflorescence, a differentiated calyx and corolla (in most), and exotestal seeds. Alismataceae (12 genera; 81 species) and Limnocharitaceae (3 genera; 7 species) are closely related and should perhaps be considered a single family (Soros and Les 2002). The families are linked by having latex, leaves with a pseudopetiole with a midvein and cross veins, spinose-pantoporate pollen, and a follicular fruit. Butomaceae (monospecific) are monopodial with two-ranked, three-angled leaves and an umbel-like inflorescence. Like Araceae and Alismataceae, members of Butomaceae have a differentiated perianth and a follicular fruit. Hydrocharitaceae are the largest family of this subclade (18 genera; 116 species). They are submersed aquatics, some of which are marine (e.g., *Halophila*, *Enhalus*, and *Thalassia*), with petiolate, usually undifferentiated leaves, an inflorescence subtended by two bracts, and an inferior ovary (the rest of this subclade has superior ovaries).

The second subclade within the aquatic clade of Alismatales consists of Aponogetonaceae, Cymodoceaceae, Juncaginaceae, Posidoniaceae, Potamogetonaceae, Ruppiaceae, Scheuchzeriaceae, and Zosteraceae (Figure 4.2; representatives are illustrated in Figure 4.3). Scheuchzeriaceae are highly apomorphic and not clearly linked by morphological characters to the rest of the second subclade. The family is monospecific and has two-ranked, auriculate leaves, and a racemose inflorescence with several bracts; as in the other subclade, the fruit is a follicle. The rest of the subclade contains Aponogetonaceae, Cymodoceaceae, Juncaginaceae, Posidoniaceae, Potamogetonaceae, Ruppiaceae, and Zosteraceae, which share more or less linear leaves and (except for Aponogetonaceae; see Figure 4.3D) united carpels and an unusual condition in which the perianth appears to be an "outgrowth" of the stamens. Aponogetonaceae are monogeneric (43 species) and typically have petiolate leaves, parallel venation and cross veins, and small spicate flowers with tepals (Figure 4.3D). Juncaginaceae (4 genera; 15 species) are emergent aquatic to terrestrial plants (but still of wet sites) and have imperfect flowers often lacking a perianth or with the perianth internal to the stamens (Figure 4.3J). The rest of the families are all rhizomatous, water-pollinated aquatics, with ascidiate carpels. Posidoniaceae (monogeneric; 5 species), Ruppiaceae (perhaps monospecific), and Cymodoceaceae (5 genera; 16 species) are largely marine with two-ranked leaves and filiform pollen (Figure 4.3, Q and R). Ruppiaceae might better be included in Posidoniaceae; these families are highly similar in most respects. Zosteraceae (2 genera; 14 species) and Potamogetonaceae (7 genera; 122 species) share a leaf with an apical pore (i.e., their sheaths are closed), although the former family is marine and the latter is found in freshwater (Figure 4.3P). Zosteraceae are seagrasses with linear leaves and leaf-opposed branches and spadix-like inflorescences enclosed in a spathe (see recent analysis by Les et al. 2002). Potamogetonaceae have petiolate leaves with a midvein and a spicate inflorescence densely packed with flowers in which there is a tepal opposing each stamen. Pollination is commonly by wind or on the water surface or in the water.

Several authors have suggested that the "flowers" of Alismatales, except for Araceae and Tofieldiaceae, are various types of modified inflorescences (Buzgo 2001; Rudall 2003), which could explain the wide variety of floral types found in these families. For a relatively small clade, floral diversity is exceptionally high in Alismatales, and their pollination mechanisms are among the most bizarre in the angiosperms (Cox and Humphries 1993).

Alismatales contain several families adapted to marine habitats. This feature has apparently evolved independently several times in the order, once in Hydrocharitaceae, and perhaps several additional times in the seagrass families Zosteraceae, Cymodoceaceae, Ruppiaceae, and Posidoniaceae (Les et al. 1997a). Until a detailed generic phylogenetic analysis is completed, it is not possible to say how many times adaptation to marine conditions occurred. However, marine angiosperms are confined to this order.

Petrosaviales

The relationships of Petrosaviaceae (*Petrosavia* and *Japonolirion*; four species) were uncertain until recently (Chase et al. 2005). They are now strongly supported as the sister group of all monocots except Acorales and Alismatales (Figure 4.1), which necessitates their elevation to an order (but APG II 2003 still treated their position as uncertain). Until DNA placed the two genera together (Chase et al. 2000a), *Petrosavia* and *Japonolirion* had not been considered closely related. Dahlgren et al. (1985) placed *Petrosavia* in Melanthiales, but did not list *Japonolirion* at all. Takhtajan (1997) placed Petrosaviaceae in their own order in Triurididae and Japonoliriaceae in Melanthiales in Liliidae. Cameron et al. (2003) reviewed the morphology of *Petrosavia* and *Japonolirion* and found several similarities, including a lack of completely fused carpels. *Japonolirion* is photosynthetic and has been poorly studied, whereas *Petrosavia* is achlorophyllous and mycotrophic. Both are rhizomatous herbs with spirally arranged leaves or bracts on racemes; microsporogenesis is simultaneous.

Pandanales

In Chase et al. (2005), a clade of Dioscoreales + Pandanales was found to be sister to all remaining monocots (Figure 4.1). A close relationship of Pandanales to Dioscoreales was not observed until the seven-gene analysis (Chase et al. 2005), in which this pair of orders received 92% bootstrap support. However, non-DNA characters linking these two orders are unknown.

Pandanales comprise five families, Cyclanthaceae (12 genera, 225 species; tropical America), Pandanaceae (3 genera, 805 species; West Africa to the Pacific), Stemonaceae (3 genera, 25 species; tropical East Asia to Australia with a disjunct genus in eastern North America), Triuridaceae (8 genera, 48 species; pantropical), and Velloziaceae (9 genera, 240 species; nearly pantropical; Figure 4.4). Within Pandanales, Cyclanthaceae and Pandanaceae form a well-supported clade to which Stemonaceae are sister. The closest relatives of Triuridaceae and Velloziaceae within Pandanales remain unclear (see below).

A close relationship of Cyclanthaceae (Figure 4.5D) and Pandanaceae (Figure 4.5, A to C) had long been supported (e.g., Dahlgren et al. 1985), and the two families are sister taxa in the morphological analyses of Chase et al. (1995b) and Stevenson and Loconte (1995). Cyclanthaceae and Pandanaceae share compound bundles in their stems, imperfect flowers, an indehiscent syncarpous fruit, and a nonphotosynthetic cotyledon. These two families have several instances of gigantism and can be either long vines (Cyclanthaceae and Pandanaceae) or large tree-like herbs (Pandanaceae; Figure 4.5A).

Stemonaceae, a small family of mostly vines, have previously been associated with Dioscoreaceae (e.g., Cronquist 1981; Dahlgren et al. 1985); their association with Cyclanthaceae and Pandanaceae in molecular analyses was completely unexpected. Stemonaceae were only weakly supported as related to Cyclanthaceae and Pandanaceae in Chase et al. (1995a), but in the three-gene analyses of Chase et al. (2000a) and D. Soltis et al. (2000), Pandanales received very strong (99% bootstrap or jackknife) support. With Cyclanthaceae and Pandanaceae, they share tetramerous (or at least not trimerous) flowers and parietal placentation; they differ in having broad leaves with a petiole and net venation and a dehiscent fruit. Likewise, Velloziaceae having a close relationship to any of the taxa noted above was a surprise. Velloziaceae were frequently placed with Bromeliaceae, as in Dahlgren et al. (1985). In D. Soltis et al. (2000), Velloziaceae are sister to Stemonaceae, a relationship that received low (75% jackknife) support, but in Chase et al. (2005), the former is well supported (100% bootstrap) as sister to all other Pandanales if Triuridaceae (represented only by 18S rDNA sequence data; see next paragraph) are excluded from the analysis. Characters supporting the position of Velloziaceae in Pandanales are unknown; all Pandanales have small embryos and a nucellar cap, but those are not particularly impressive synapomorphies. Velloziaceae can become woody, but they achieve this in the same manner as in Pandanaceae, by being giant herbs with toughened stems (Figure 4.5E). Some authors (e.g., Behnke et al. 2000) have considered *Acanthochlamys* to be too different from the rest of Velloziaceae to be included in the family, but its position as sister to Velloziaceae s.s. is well supported so it seems better to keep it as a divergent genus in the family.

The position of Triuridaceae in the monocots is currently based solely on 18S rDNA (Chase et al. 2000a), and this relationship was also unsuspected; most authors considered Triuridaceae to be related to Alismatales because they shared several characters with those families (and were placed there in the morphological analyses of Chase et al. 1995b, but not in Stevenson and Loconte 1995). The position of the anthers inside the carpel whorl in *Lacandonia* and the complexity and diversity of flowers in Triuridaceae have led some authors to speculate that, like several other families of Pandanales (e.g., Cyclanthaceae and Pandanaceae), the "flowers" of Triuridaceae are really inflorescences (Rudall 2003). Flowers of two genera of Triuridaceae are among the oldest fossils known for monocots, 90 Mya (Gandolfo et al. 2002; see "Fossil History of Monocots," below).

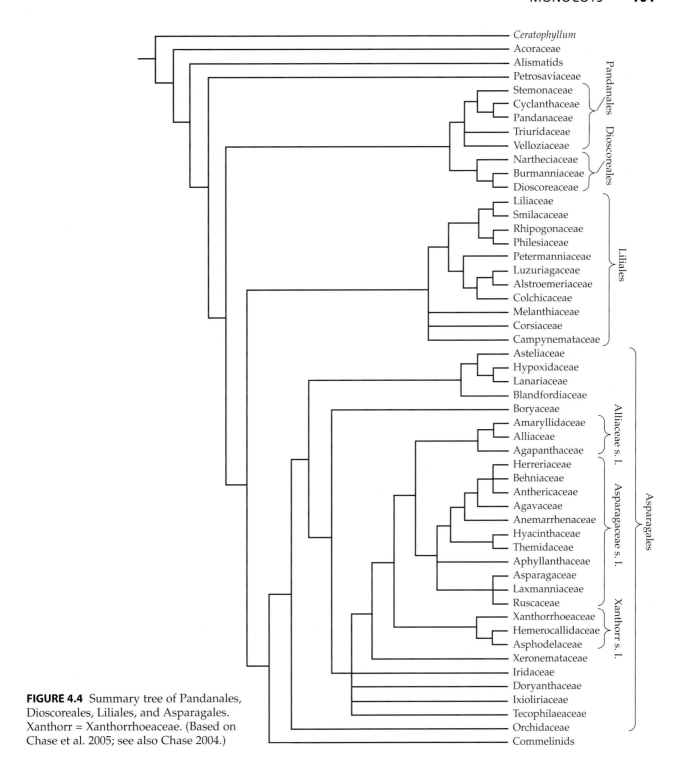

FIGURE 4.4 Summary tree of Pandanales, Dioscoreales, Liliales, and Asparagales. Xanthorr = Xanthorrhoeaceae. (Based on Chase et al. 2005; see also Chase 2004.)

Dioscoreales

Dioscoreales have a long history of recognition, but not in the circumscription given in APG (1998, 2003). In APG II (2003), the order included only three families, Burmanniaceae, Dioscoreaceae, and Nartheciaceae. Most previous circumscriptions (e.g., Dahlgren at al. 1985) included in Dioscoreales nearly all lilioid monocots with net-veined leaves, so in this sense, Dioscoreales comprised a set of families now known to be unrelated, such as Smilacaceae, Stemonaceae, and Trilliaceae.

Nartheciaceae were not included in Dioscoreales by Chase et al. (1993, 1995a, 1995b), but more recent analyses have consistently placed them as sister to a clade of Burmanniaceae and Dioscoreaceae (Chase et al. 2000a; Caddick et al. 2002a), although not with bootstrap support above 70% until the seven-gene analysis put them

FIGURE 4.5 Representatives of Pandanales (A–E), Dioscoreales (F–J), and Liliales (K–P). (A) *Pandanus dubius* (Pandanaceae), general habit, plant in flower. (B) *Pandanus racemosus* (Pandanaceae), carpellate inflorescence. (C) *Pandanus racemosus* (Pandanaceae), staminate inflorescence. (D) *Cyclanthus gracilis* (Cyclanthaceae), plant in flower, leaves removed. (E) *Vellozia brevifolia* (Velloziaceae), plant in flower. (F) *Dioscorea quaternata* (Dioscoreaceae), carpellate flower. (G) *Dioscorea quaternata* (Dioscoreaceae), staminate flower. (H) *Dioscorea quaternata* (Dioscoreaceae), plant in fruit. (I) *Tacca pinnatifida* (Dioscoreaceae), plant in flower. (J) *Burmannia longifolia* (Burmanniaceae), plant in flower. (K) *Paris polyphylla* (Melanthiaceae), plant in flower. (L) *Tricyrtis pilosa* (Liliaceae), flower. (M) *Luzuriaga erecta* (Luzuriagaceae), portion of shoot in flower. (N) *Smilax medica* (Smilacaceae), portion of plant showing habit, in fruit. (O) *Smilax herbacea* (Smilacaceae), carpellate flower. (P) *Smilax herbacea* (Smilacaceae), staminate flower. (A to C, from Solms in Engler and Prantl 1889. D, from Drude in Engler and Prantl 1889. E and I, from Pax in Engler and Prantl 1889. F to H, from Al-Shehbaz and Schubert 1989. J to N, from Engler in Engler and Prantl 1889. O and P, from Judd 1998.)

there with 99% bootstrap support. Nartheciaceae are a small family (four or five genera) that deviates in morphology from the general patterns found in the other two families of the order, but they do share with Burmanniaceae and Dioscoreaceae a perianth persistent on the fruit, as well as micromorphological characters (e.g., a tanniferous exotegmen and a short embryo). Like Tofieldiaceae (see above), Nartheciaceae are relatively poorly studied. Burmanniaceae (13 genera; 126 species) and Dioscoreaceae (4 genera; 870 species) have reflexed stamens, inferior, winged ovaries, and simultaneous microsporogenesis (the last trait is rare in monocots), but otherwise these two families are dissimilar. Burmanniaceae are rhizomatous mycoparasites, but some of them are green and have small leaves (Figure 4.5J), whereas Dioscoreaceae are free-living and are often vines (Figure 4.5, F to H), some with large underground storage organs. On the basis of characters that are convergent due to the mycoparasitic syndrome, many authors had suggested that Burmanniaceae were related to the orchids (e.g., Cronquist 1981), but this is now completely discredited. Most authors have separated out several additional families from Dioscoreaceae (e.g., Taccaceae; Figure 4.5I), but such narrow circumscriptions ignore their shared characters (Caddick et al. 2002a, 2002b). Taccaceae are now included within Dioscoreaceae (Caddick et al. 2002a, 2002b; APG II 2003). Likewise, several authors (e.g., Dahlgren et al. 1985) separated Thismiaceae from Burmanniaceae, but because they fall together and share most of their characters (*Thismia* and its relatives, such as *Afrothismia, Triscyphus,* and *Geomitra,* being only somewhat more reduced vegetatively than many Burmanniaceae s.s.), APG II (2003) placed these two families together in Burmanniaceae.

Liliales

Liliales (*sensu* APG II 2003) now appear to occupy a position as sister to Asparagales plus the large commelinid clade (Figure 4.1). Liliales (*sensu* APG II 2003) comprise 11 families and about 1,300 species. The concept of Liliales as distinct from other lilioid taxa such as Asparagales and Dioscoreales originated with Huber (1969, 1977), whose ideas were then adopted by Dahlgren et al. (1985). Nearly all of these taxa were considered to be closely related, perhaps even members of a single family, Liliaceae. Cronquist (1981) lumped these and other families such as Velloziaceae and Pontederiaceae into his subclass Liliidae, but if they were not arborescent or did not have net-veined leaves then he included them in Liliaceae. The broad concepts of Liliidae and Liliaceae were formulated because the patterns of characters did not indicate clear-cut groupings, and those authors who did segregate some genera into other families never did so consistently. For example,

Cronquist (1981) segregated *Aloe* and some of its close relatives into Aloaceae, largely because *Aloe* had some arborescent species and the other genera were highly succulent (e.g., *Haworthia* and its relatives), but retained the closely related *Bulbine* in Liliaceae because it was strictly herbaceous. However, *Kniphofia* is herbaceous and not succulent but was retained in Aloaceae because its flowers and inflorescences were "aloeoid." Dahlgren et al. (1985) kept both *Bulbine* and *Kniphofia* in Asphodelaceae, whereas he placed the others in Aloaceae (and both families into Asparagales due to their phytomelanous seeds; see below) because of their shared strongly bimodal and nearly identical karyotypes.

The *rbcL* analyses of Chase et al. (1993) and Duvall et al. (1993a) identified a Liliales clade containing Alstroemeriaceae, Colchicaceae, Liliaceae, Melanthiaceae, and Smilacaceae. Huber (1977) and Dahlgren et al. (1985) also recognized Liliales, but they included in this order Iridaceae and Orchidaceae, whereas in the DNA studies these two families appeared as part of Asparagales (see below). Huber (1977) and Dahlgren et al. (1985) excluded Melanthiaceae and Campynemataceae (as Melanthiales), which the DNA studies have consistently indicated to be members of Liliales. Chase et al. (1995a) added Luzuriagaceae, Philesiaceae, Smilacaceae, and Rhipogonaceae to the order, but they also placed here genera that previously had been referred to Calochortaceae (now in Liliaceae), Trilliaceae (now in Melanthiaceae), and Uvulariaceae (genera now in Colchicaceae or Liliaceae).

Liliales (APG 1998; APG II 2003) are mostly geophytes bearing elliptical leaves with the fine venation reticulate; their flowers have various forms of tepalar nectaries, extrorsely dehiscent anthers, and, often, spotted tepals. The morphological analyses of Chase et al. (1995b) and Stevenson and Loconte (1995) variously placed Alstroemeriaceae, Colchicaceae, Liliaceae, and Melanthiaceae in a clade, but the other families of Liliales *sensu* APG fell elsewhere; Iridaceae were also included in Liliales in both morphological studies. Relationships of the families within this clade are discussed in Rudall et al. (2000) and Vinnersten and Bremer (2001). Alstroemeriaceae (3 genera, 165 species; tropical America) are sister to Luzuriagaceae (2 genera; 5 species), which are native to South America (*Luzuriaga;* Figure 4.5M) and Australia/New Zealand (*Drymophila*); the two families share vegetative features such as being vines with the leaves twisted such that the developmentally upper surface is lowermost at maturity, although the ovary is superior in the former. They perhaps should be combined into a single family. Sister to these is Colchicaceae (18 genera, 224 species), which are native principally to the Old World, the only exception being *Uvularia* (North Temperate). Some genera of Colchicaceae have the twisted leaves of Alstroemeriaceae/Luzuriagaceae. Colchicine alkaloids (which are used to inhibit spindle formation and cause nondisjunction at meiosis, leading to polyploid offspring) are found in all members of the family. *Petermannia* was included in Colchicaceae in APG (1998) and APG II (2003), but it is now known that the DNA used came from a plant of *Tripladenia cunninghamii* that had been misidentified as *Petermannia* (M.W. Chase, unpubl. data.). Real material of *Petermannia* is sister to the above three families, so it would appear appropriate to reinstate Petermanniaceae (Figure 4.4).

Melanthiaceae (Figure 4.5K; 16 genera; 170 species) were studied in detail by Zomlefer et al. (2001); the family now includes the members of former Trilliaceae (APG II 2003; see Chapter 10). All genera have a North Temperate distribution; a single genus, *Schoenocaulon,* occurs in Peru, but it may have been taken there by humans (they are medicinal plants). The family contains potent alkaloids that provide them with some of their common names (e.g., fly poison, death camas). *Xerophyllum* (two species in North America) is sister to the genera of the former Trilliaceae (Chase et al. 1995a, 1995b; Rudall et al. 2000). Species of *Xerophyllum* are xerophytically adapted (narrow, grass-like leaves and tufted rosette habit), which contrasts sharply with *Trillium* and its relatives, taxa that are adapted to forest understories and have net-veined leaves. Some authors (Thorne 1992a, 1992b) considered *Veratrum* (Melanthiaceae) to be one of the most primitive monocots because of its plicate, net-veined leaves and mostly unfused carpels, but it is deeply embedded within Liliales and not at all close to the root of the monocots, meaning that it is unlikely to have retained these as primarily primitive features.

Liliaceae *sensu* APG (see Chapter 10) are composed of many fewer genera than in most previous circumscriptions of the family (e.g., Cronquist 1981). However, the APG concept is broader than that of others who would limit the family to just the core genera related to *Lilium* (Figure 4.5L; Tamura 1998) by excluding *Calochortus, Prosartes, Tricyrtis,* and so on, and placing these in Calochortaceae and/or Tricyrtidaceae. Liliaceae are exclusively North Temperate and are composed of geophytes with often large spotted flowers, extrorsely dehiscing anthers, and a superior ovary. Related to Liliaceae are Smilacaceae (Figure 4.5, N to P; monogeneric, 315 species; nearly cosmopolitan), Philesiaceae (2 monospecific genera; southern South America), and Rhipogonaceae (monogeneric, 6 species; Australasia), but non-DNA characters that reflect this pattern are unknown. Smilacaceae, Philesiaceae, and Rhipogonaceae all have unique spiny pollen (Rudall et al. 2000), but they have never formed a clade in any molecular analysis. *Rhipogonum* is often recovered as sister to *Philesia/Lapageria,* so it could be combined with them, but *Smilax* is generally sister to Liliaceae (but never higher than 80% support; Figure 4.4).

Melanthiaceae, Campynemataceae (2 genera, 4 species; Australasia), and Corsiaceae (3 genera, 30 species; China, South America, and Australasia) have an unclear pattern of relationships to the other members of Liliales (Figure 4.4). Campynemataceae were previously considered to be related to Melanthiaceae (Dahlgren et al. 1985) because of their mostly unfused carpels, whereas Corsiaceae were considered to be related to Burmanniaceae because of their shared mycoparasitic life history. However, both suites of characters are unreliable; unfused (free) carpels are potentially a symplesiomorphy, whereas the syndrome of traits associated with mycoheterotrophy is even convergent between eudicots and monocots with this life history. Campynemataceae and Corsiaceae generally fit the pattern of characters observed among the families of Liliales.

Asparagales

The other major clade of the lilioid monocots is Asparagales, a group that had originally been suggested by Huber (1977) and later adopted by Dahlgren et al. (1985). Asparagales are sister to the commelinid clade (Figure 4.1). As currently circumscribed, the order comprises 14 families: Alliaceae (including Agapanthaceae and Amaryllidaceae; 60 genera, 1,605 species; nearly cosmopolitan, but rarely tropical), Asparagaceae (including Agavaceae, Aphyllanthaceae, Hyacinthaceae, Laxmanniaceae, Ruscaceae, and Themidaceae; 120 genera, 2,640 species; nearly cosmopolitan, but infrequent in the tropics), Asteliaceae (4 genera, 36 species; mostly Australasia), Blandfordiaceae (monogeneric with 4 species; southeastern Australia), Boryaceae (2 genera, 12 species; Australia), Doryanthaceae (monogeneric with 2 species; southeastern Australia), Hypoxidaceae (9 genera, 200 species; nearly cosmopolitan but not in Europe), Iridaceae (67 genera, 1,800 species; cosmopolitan), Ixioliriaceae (probably monospecific; Central Asia), Lanariaceae (monospecific; South Africa), Orchidaceae (788 genera, 20,000 species; cosmopolitan, but particularly diverse in the wet tropics), Tecophilaeaceae (8 genera, 23 species; Africa, Madagascar, Chile, and California), Xanthorrhoeaceae (including Asphodelaceae and Hemerocallidaceae; 35 genera, 900 species; mostly Old World, including Australia, mostly temperate), and Xeronemataceae (monogeneric with 2 species; an island off New Zealand and New Caledonia; see Figure 4.4).

The preeminent characters uniting members of Asparagales are their phytomelanous (a dark, noncellular material) seed coat and collapsed/obliterated outer epidermis. In addition, nearly all Asparagales have septal nectaries and are geophytes, although in this case their leaves are mostly linear without reticulate fine-scale venation. Anomalous secondary growth with a tiered (etagen) meristem is nearly confined to the genera of this clade (outside this clade, it is known only in

Dioscorea of Dioscoreaceae, Dioscoreales). Some of these species, in *Agave, Aloe, Cordyline, Dracaena, Nolina,* and *Yucca,* become massive, short trees, generally, but not always, with limited branching (e.g., *Aloe, Dracaena,* and *Yucca*). Other genera, such as *Aphyllanthes* and *Lomandra,* have this same type of secondary growth confined to their underground stems. In the morphological analyses of Chase et al. (1995b) and Stevenson and Loconte (1995), at least some of these families of Asparagales formed clades, although members of Liliales (Chase et al. 1995b) and Zingiberales and Pandanales

(Stevenson and Loconte 1995) also fell among them. Seed anatomy would appear to be the best set of characters for the order, the members of which are otherwise heterogeneous, particularly if Orchidaceae are members of this clade (see below).

Orchidaceae (Figure 4.6K) are one of the largest families of the angiosperms, and their infrafamilial relationships are still under active investigation (e.g., Dressler 1983, 1993; Chase 1986, 1988; Chase and Hills 1992; Chase and Palmer 1992; Cameron et al. 1999; Kores et al. 2000; Whitten et al. 2000; Salazar et al. 2003). Some

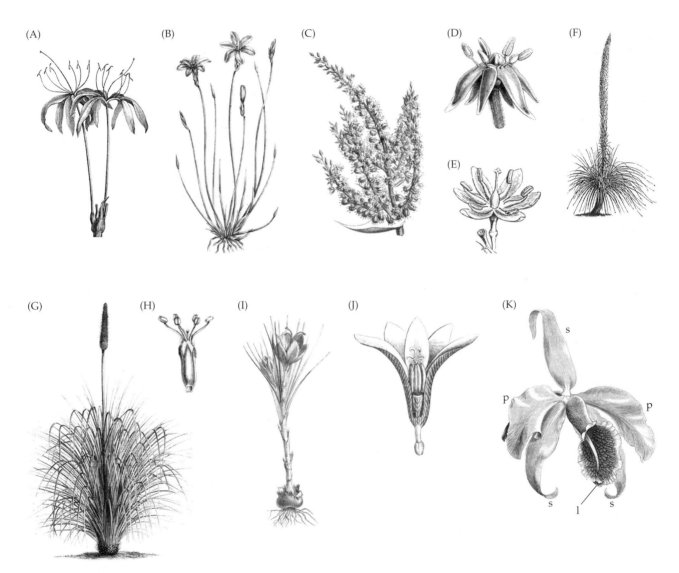

FIGURE 4.6 Flowers and plants representing Asparagales. (A) *Crinum purpurascens* (Alliaceae), flowers. (B) *Aphyllanthes monspeliensis* (Asparagaceae, formerly Aphyllanthaceae), whole plant in flower. (C) *Astelia cunninghamii* (Asteliaceae), whole plant in flower. (D) *Astelia cunninghamii* (Asteliaceae), single flower. (E) *Dracaena draco* (Asparagaceae, formerly Ruscaceae), flower. (F) *Dasylirion acrotrichum* (Asparagaceae, formerly Ruscaceae), flowering plant. (G) *Xanthorrhoea hastile* (Xanthorrhoeaceae), plant in flower. (H) *Xanthorrhoea hastile* (Xanthorrhoeaceae), flower. (I) *Crocus sativus* (Iridaceae), plant in flower. (J) *Romulea purpurascens* (Iridaceae), flower with portion cut away to show inferior ovary and three stamens. (K) *Cattleya maxima* (Orchidaceae), flower, s = sepal; p = petal; l = labellum, a modified petal characteristic of many orchids. (A, from Pax in Engler and Prantl 1889. B to J, from Engler in Engler and Prantl 1889. K, from Pfitzer in Engler and Prantl 1889.)

authors (e.g., Dahlgren et al. 1985) recognized three families of orchids, on the basis of the differing anther conditions present: Apostasiaceae (2 genera; 2 or 3 anthers only partially fused to the gynoecium), Cypripediaceae (5 genera; 2 anthers fused to the gynoecium), and Orchidaceae (all the rest of the orchids, all with a single anther fused to the gynoecium). All molecular evidence to date (summarized in Chase et al. 2003) shows that Orchidaceae *sensu* Dahlgren et al. (1985) are polyphyletic. The data indicate instead a broadly defined Orchidaceae; within this family there should be five subfamilies recognized with the following set of relationships: Apostasioideae (Vanilloideae (Cypripedioideae (Orchidoideae, Epidendroideae))). Thus, reduction from three anthers to one occurred at least twice in the evolution of the orchids. A new phylogenetic classification of Orchidaceae was published by Chase et al. (2003).

Orchidaceae appear to be sister to the rest of Asparagales (Figure 4.4; Chase et al. 2005); it is clear that no single family of Asparagales is closely related to orchids. The node separating the orchids from the rest of Asparagales has 91% bootstrap support (Chase et al. 2005). Interestingly, orchids lack most of the non-DNA synapomorphies of the order. The seeds of Orchidaceae lack phytomelan because of their dust-like nature (like most groups of mycoparasites), and their nectaries are only rarely septal. Orchids do have simultaneous microsporogenesis and inferior ovaries, characters that are typical of the clades at the basal nodes of Asparagales (Figure 4.7; the "lower Asparagales" of Rudall 1997), but it is not clear if these characters can be considered strict synapomorphies of the entire order. The only apparent synapomorphy for Orchidaceae is their protocorm (the structure produced by the growth of their undifferentiated embryos before roots and shoots develop). The hallmark of orchids is the fusion of androecium and gynoecium, but this feature is present only to varying degrees in members of subfamily Apostasioideae. Dust seeds distributed by the wind occur in most subfamilies of Orchidaceae, but there are crustose seeds in some members of subfamilies Cypripedioideae and Vanilloideae (probably an association with seed dispersal, which in these cases is due to their fleshy fruit fermenting *in situ* and releasing fragrant compounds attractive to birds and mammals, i.e., vanillin).

Asteliaceae (Figure 4.6, C and D), Blandfordiaceae, Hypoxidaceae, and Lanariaceae form a well-supported (94% bootstrap) clade (Figure 4.4) in molecular phylogenetic analyses. Within this clade, Blandfordiaceae are sister to a well-supported (100%) clade of Asteliaceae, Hypoxidaceae, and Lanariaceae. Morphology provides some support for these relationships. Of these four families, Asteliaceae and Hypoxidaceae are rosette-forming, covered with branched, multicellular hairs, and with root canals filled with mucilage; the former is also true for *La-*

naria (Lanariaceae), but *Blandfordia* (Blandfordiaceae) does not share these traits. A potential synapomorphy for these four families is ovule structure: these families all have a chalazal constriction and a nucellar cap. If, after more study, bootstrap support increases for this clade, including *Blandfordia,* then it should be possible to propose a merger of the four families into Hypoxidaceae (which is the only conserved name). *Lanaria* (Lanariaceae) had in the past been considered to be a member of Haemodoraceae (Hutchinson 1967) or Tecophilaeaceae (Dahlgren et al. 1985).

The relationship of Boryaceae to other Asparagales is still unclear (Figure 4.4). One of the two genera of Boryaceae (*Borya* and *Alania*) is typically a "resurrection plant" found on rocky slopes; during the dry season, plants of *Borya* fall to a fraction of their normal water content and turn rusty-orange, but they quickly become green and active again once it rains. Boryaceae are known to be mycorrhizal, but this is of the standard vesicular-arbuscular (VA) type and not like that of the orchids (their own unique type). Boryaceae were previously thought to be members of Anthericaceae (Dahlgren et al. 1985), a family shown to be grossly polyphyletic (Chase et al. 1996).

The next node in Asparagales (Figure 4.4; all families other than Orchidaceae, Boryaceae, and the clade of Asteliaceae, Blandfordiaceae, Hypoxidaceae, and Lanariaceae) has previously been demonstrated to be robust (81% jackknife support in D. Soltis et al. 2000; 100% bootstrap in Chase et al. 2005), but there are no known non-DNA characters that unite this large clade. Within this large subclade of Asparagales, some analyses have found that a clade of Ixioliriaceae and Tecophilaeaceae are sister to the rest. These two families share corms, a leafy inflorescence, and, often, a nearly capitate inflorescence. However, the relationship of Doryanthaceae to Ixioliriaceae and Tecophilaeaceae and also to Iridaceae is unclear from both morphological and molecular analyses, but all of these families are excluded from the next node in Asparagales (Fay et al. 2000a; Chase et al. 2005). The two species of *Doryanthes* are monstrous rosette-forming herbs; they are a conspicuous element of the flora around Sydney, Australia, and their flowering is difficult to miss. Iridaceae (Figure 4.6, I and J) are one of the largest and best studied families of Asparagales (e.g., Goldblatt 1990, 1991; Goldblatt et al. 1998, 2002; Reeves et al. 2001). Iridaceae consist of three subfamilies, Crocoideae (including Nivenioideae), Iridoideae, and Isophysidoideae, with the last sister to the other two. Iridaceae are distinctive among Asparagales because of their unique inflorescence structure (a rhipidium) and their combination of inferior ovaries and three stamens (Figure 4.6, I and J); unifacial leaves are also common in the family, whereas bifacial leaves are the norm in other Asparagales.

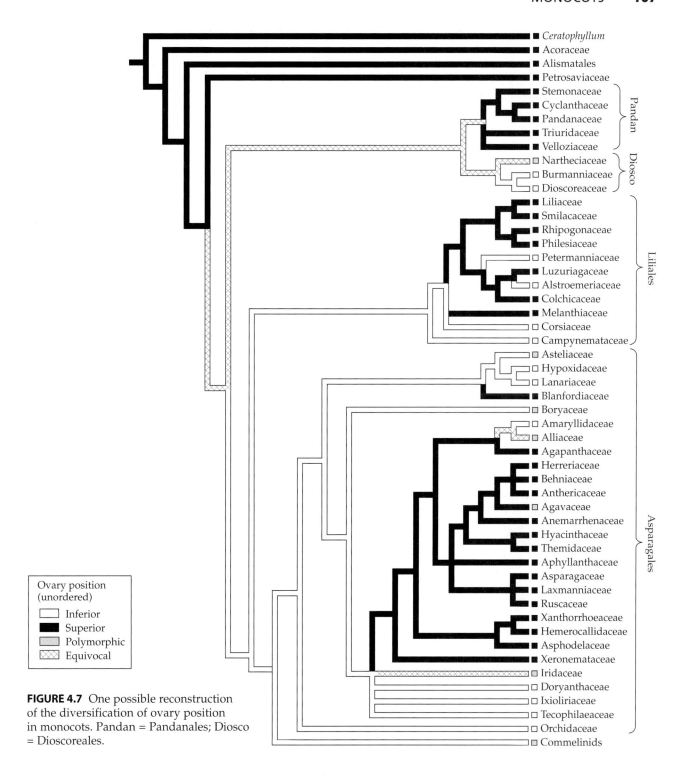

FIGURE 4.7 One possible reconstruction of the diversification of ovary position in monocots. Pandan = Pandanales; Diosco = Dioscoreales.

CHARACTER EVOLUTION IN ASPARAGALES For many years, it was known that *Allium* (Alliaceae) lacked the standard plant telomeric repeats, termed the *Arabidopsis*-type, that cap the ends of chromosomes. Then it was discovered (Adams et al. 2001) that *Aloe* (Xanthorrhoeaceae s.l.) also lacked this standard repeat type. Comparing the position of these two genera on the plastid DNA tree of Fay et al. (2000a) led to the hypothesis that the intervening genera of Asparagales should also lack these standard sequences if the phylogenetic tree were predictive. Representatives of all Asparagales families were then examined by a combination of fluorescent *in situ* hybridization (FISH) and slot blot Southern hybridization, and it was found that all but a single genus, *Ornithogalum*

(Asparagaceae s.l.), of those predicted to lack the *Arabidopsis*-type repeats, did indeed lack them. On further study (Sykorova et al. 2003), it was discovered that *Ornithogalum* and several other related genera had replaced the standard plant repeat with a human-type repeat and, with more sensitive probing, that nearly all genera of Asparagales have at least some copies of the human-type repeat, as well as a variety of others located at their chromosome tips. Thus, it would appear that Asparagales represent a dynamic situation in which a variety of telomeric cap motifs are maintained in low copy, in spite of the presence of the dominant *Arabidopsis*-type repeat. However, in some cases, one of the low-copy types is amplified and replaces the standard type. The presence and composition of the various repeat types provide a great deal of phylogenetic structure within Asparagales.

CHARACTER EVOLUTION IN IRIDACEAE   Asparagales offer many examples of the elucidation of character evolution in a phylogenetic context. Orchidaceae and pollination are discussed in Chapter 12; Orchidaceae and chromosomal evolution are noted in Chapter 13. Iridaceae have been well studied through the efforts of Goldblatt and coworkers. Goldblatt et al. (2002) used a molecular clock approach to date the radiation of *Moraea* (Iridaceae; nearly 200 species in Africa and Eurasia, but particularly diverse in the Cape Area of South Africa), and evaluated patterns of pollination and chromosomal evolution. *Moraea* and its sister genus *Ferraria* split from each other about 25 Mya, and the early radiation of *Moraea* took place against a background of increasing aridification and the spread of desert, shrublands, and fynbos (a diverse, shrubby community found in South Africa). An *Iris*-type flower (one with three separate pollination units) is the ancestral type in *Moraea*. This type of flower is pollinated by long-tongued bees foraging for nectar; pollen deposition is passive. In multiple, unrelated lineages, this flower type has been replaced by open, *Homeria*-type flowers, in which pollen is actively sought by short-tongued bees; pollen is taken to their nests and fed to the developing larval bees. In addition, some lineages have become adapted to nectar-seeking flies or scarab beetles, the latter using the flowers as sites for mate selection and mating.

Chromosome number change in *Moraea* took place relatively late in the evolution of the genus; most nodes near the base of the genus were *n* = 8. Some clades experienced increases and others decreases around the same time, so that lineage diversification over a short period fixed these differences in separate clades, after which little further change took place. Species-rich clades, mostly containing taxa endemic to the Cape, also contain species with ranges outside southern Africa. Some species managed to reach Europe and western Asia, but this occurred relatively late in the evolution of *Moraea*. Species-level analyses, such as that of Goldblatt et al.

(2002), demonstrate how a much-improved understanding of evolutionary patterns and processes, as well as molecular dating, can provide insights into why particular geographic areas contain so much species diversity.

Asparagales: remaining families

The next node of Asparagales—a clade of Xeronemataceae, Xanthorrhoeaceae s.l. (including Asphodelaceae and Hemerocallidaceae), Alliaceae s.l. (including Agapanthaceae and Amaryllidaceae), and Asparagaceae s.l. (including Agavaceae, Anthericaceae, Anemarrhenaceae, Aphyllanthaceae, Behniaceae, Herreriaceae, Hyacinthaceae, Laxmanniaceae, Ruscaceae, and Themidaceae—is also well supported (Fay et al. 2000a). If an inferior ovary is a synapomorphy of Asparagales (Chase et al. 1995b), then this node would mark the transition to a superior ovary (although there are inferior ovaries in Amaryllidoideae of Alliaceae and *Yucca* of Asparagaceae s.l., but these are obviously embedded clades). *Xeronema* (Xeronemataceae) was previously considered a close relative of *Phormium* (usually Phormiaceae; Dahlgren at al. 1985), but although there are superficial similarities (e.g., unifacial leaves), they are not identical in detail. Chase et al. (2000a) recently described Xeronemataceae, and it falls within the general patterns of variation within Asparagales.

The clades at this node all have infralocular septal nectaries, which Rudall (2000) interpreted to represent secondarily superior ovaries (see also Chapters 6 and 12). Since nearly all lower nodes in Asparagales are characterized by inferior ovaries, this seems a reasonable interpretation and demonstrates the reversibility of this trait that has often been emphasized in monocot systematics (Figure 4.7; Cronquist 1981; Dahlgren et al. 1985). Xanthorrhoeaceae s.l. (Figure 4.6, G and H) are characterized by anthraquinones. The value of this chemical character is seen by considering Asphodelaceae, a family now considered part of a broadly defined Xanthorrhoeaceae s.l., but historically placed with Anthericaceae (to which they are now known from DNA sequence studies to be only distantly related). Because of the presence of anthraquinones, Asphodelaceae were kept separate from Anthericaceae by Dahlgren et al. (1985; in addition, Asphodelaceae have simultaneous microsporogenesis, whereas Anthericaceae have successive). Within Xanthorrhoeaceae s.l., secondary growth occurs in *Aloe, Phormium,* and *Xanthorrhoea* (Figure 4.6G); some species of *Aloe* can attain enormous size. Included in Xanthorrhoeaceae s.s. in the past have been several other arborescent taxa such as *Kingia* and *Cordyline,* but the former does not have secondary growth and is now placed in Dasypogonaceae (see below) and the latter is only distantly related (it is a member of Asparagaceae-Laxmanniaceae).

Hemerocallidaceae have often included only *Hemerocallis* (Dahlgren et al. 1985), but even in the APG

(1998) circumscription, Hemerocallidaceae included a larger number of genera, such as *Dianella, Johnsonia,* and *Phormium.* All of these genera share trichotomosulcate pollen, a rare condition in monocots, and the early molecular studies (Chase et al. 1995a) were critical in determining if this was an important character for these genera (Rudall et al. 1997). Asphodelaceae have many genera with distinctive bimodal karyotypes (Chase et al. 2000c), and this trait has been argued as the basis for separating some genera as the distinct family Aloaceae from the rest (Dahlgren et al. 1985). Molecular studies have played an important role in identifying the members of this clade and changing its circumscription.

The remaining Asparagales (broadly defined families Alliaceae and Asparagaceae) are characterized by successive microsporogenesis (the "higher asparagoids" of Rudall et al. 1997), but steroidal saponins are also a potential character for them. In the APG II sense, there are two families: Asparagaceae and Alliaceae (Figure 4.6, A and B). These two families can be distinguished by umbelloid inflorescences with generally two fused enclosing bracts in Alliaceae versus racemes in Asparagaceae (although Themidaceae have umbelloid inflorescences; in this case, they have bracts subtending all flowers, which are taken to indicate that they are condensed racemes). There are several instances of arborescent genera in Asparagaceae (e.g., *Yucca, Agave, Cordyline*), as well as some with secondary growth in their roots (e.g., *Aphyllanthes, Lomandra*); others are vines (e.g., *Herreria, Thysanotus*), and many are geophytes. Alliaceae are much less diverse, with most being geophytes, although a few are rhizomatous (e.g., *Agapanthus, Tulbaghia, Clivia*). Recent papers dealing with phylogenetic relationships within these expanded concepts of Alliaceae, Asparagaceae, and Xanthorrhoeaceae include Meerow et al. (1999), Chase et al. (2000a), Yamashita and Tamura (2000), Reeves et al. (2001), and Pires and Sytsma (2002).

Commelinid monocots

The existence of a commelinid clade (Figures 4.1 and 4.8) had long been suspected (Dahlgren et al. 1985) because several characters occur exclusively in these taxa: cell walls with ultraviolet-fluorescent ferulic (Figure 4.9) and coumaric acids, silicon dioxide bodies in leaves, and epicuticular waxes of the *Strelitzia* type. Many authors confused the monophyly of this group relative to these characters because of the gross morphological similarity of Pandanaceae/Cyclanthaceae (which are not commelinids but rather part of Pandanales) to Arecaceae (which are commelinids). These three families share an arborescent or vining herbaceous habit, tetracytic stomata, fleshy, indehiscent fruits, and similar embryology. However, Pandanaceae/Cyclanthaceae lack all of the other commelinid characters mentioned above, so it is not too surprising that

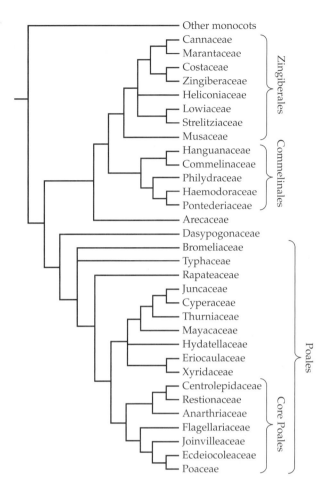

FIGURE 4.8 Summary tree of commelinids. (Based on Chase et al. 2005; see also Chase 2004.)

they should fall elsewhere in the monocots (see Figure 4.1). Some authors (e.g., Cronquist 1981) have also linked the aroids to Pandanaceae/Cyclanthaceae because of their shared inflorescence morphology (presence of a subtending spathe), but this character occurs in a wide range of unrelated taxa in the monocots (e.g., some of the alismatid families, aroids, cyclanths, pandans, palms). On the basis of their commelinid characters, Dasypogonaceae also were placed here (Rudall and Chase 1996), whereas earlier authors (Dahlgren et al. 1985) had associated them with other groups, such as Xanthorrhoeaceae because of similarities in gross habit.

The orders of commelinids (Figure 4.8) have typically been associated in various ways by previous authors. Cronquist (1981), for example, associated his Commelinidae (which approximates the APG order Poales, but without Bromeliaceae, which he placed in Zingiberidae, and with the addition of Commelinaceae) and Zingiberidae because of their shared starchy endosperm with compound starch grains and typically herbaceous perianth with distinct sepals and petals,

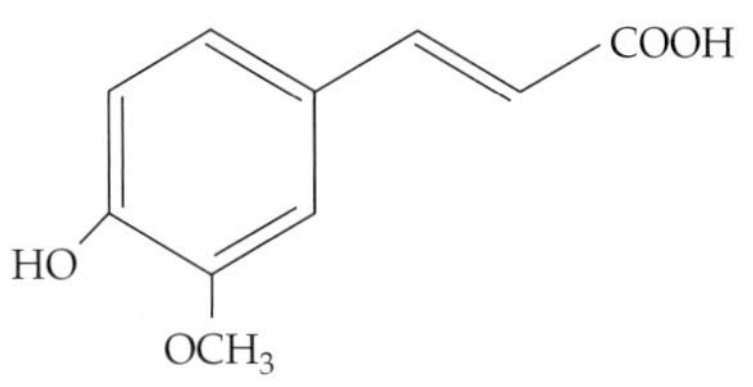

FIGURE 4.9 Chemical structure of ferulic acid, a synapomorphy of the commelinid monocots.

as opposed to the tepals (two indistinct whorls) of what he viewed as Liliidae. According to Cronquist (1981), Arecidae included Pandanales (Pandanaceae/Cyclanthaceae) and Arales (Araceae s.l., including Lemnaceae) and were not related to Commelinidae/Zingiberidae. The other families of the commelinids that Cronquist (1981) did not associate with Commelinidae/Zingiberidae were Dasypogonaceae (see above) and Hanguanaceae, Philydraceae, Pontederiaceae, and Haemodoraceae, the last three of which he considered to be members of Liliidae because they have large tepals and look like "lilies."

ARECALES Arecales consist of a single family, Arecaceae (or Palmae; 190 genera, 2,000 species), a family that is well supported as monophyletic (Chase et al. 1995a, 1995b; Asmussen and Chase 2001). Members of Arecaceae are easily recognized as trees, shrubs, or lianas with unbranched or rarely branched trunks with the stem apex consisting of a large apical meristem (Figure

4.10A). The shortest trees in Chase et al. (2005) placed Arecaceae as sister to Commelinales/Zingiberales, but with low bootstrap support. Flowers are trimerous in Arecaceae with three sepals, usually three petals, and three or six to numerous stamens; carpels are typically three, but sometimes there are as many as 10. The family is also well known from the fossil record, with fossils from Europe and North America dating to the Late Cretaceous (90 Mya).

Relationships within Arecaceae have also been investigated (e.g., Asmussen et al. 2000; Asmussen and Chase 2001; Hahn 2002), but some other published studies have used noncoding plastid DNA regions that cannot be aligned well against the outgroup taxa, so a robust evaluation of where the root should be placed within the family has been lacking. *Nypa* had been arranged as sister to the rest of Arecaceae because the genus had been used that way by Uhl et al. (1995) on the basis of plastid DNA restriction site studies. Plastid DNA has been shown to evolve slowly in the palms (Wilson et al. 1990), such that most researchers gave up using conserved regions, such as *rbcL*, early in the history of the use of this gene in angiosperm systematics. Asmussen and Chase (2001) used *rbcL* sequence data and found that, although it did have somewhat fewer variable sites than noncoding regions, it nonetheless had more changes at each site; hence, *rbcL* performed better in the family than had been predicted. Asmussen and Chase also found that Calamoideae were sister to the rest of Arecaceae, but this placement did not obtain bootstrap support above 50%. Following Calamoideae, *Nypa* (the sole member of Nypoideae) was then sister

FIGURE 4.10 Members of commelinid clade; representing Arecales (A–C), Commelinales (D–F), Zingiberales (G–L), Poales (M–Y). (A) *Roystonea regia* (Arecaceae), whole plant. (B) *Acoelorhaphe wrightii* (Arecaceae), portion of inflorescence. (C) *Acoelorhaphe wrightii* (Arecaceae), flower with corolla and androecium spread out. (D) *Callisia cordifolia* (Commelinaceae), flower. (E) *Pontederia cordata* (Pontederiaceae), flower of long-styled form. (F) *Pontederia cordata* (Pontederiaceae), leaf blade and portion of petiole behind flowering stem with leaf and bract subtending inflorescence. (G) *Strelitzia reginae* (Strelitziaceae), inflorescence with three flowers raised out of spathe, the flower on the right is at anthesis, the other two flowers are post-anthesis. (H) *Strelitzia reginae* (Strelitziaceae), a single flower showing the three more-or-less equal sepals and three unequal petals; the two lateral petals form a keel, the short median petal covers access to nectar. (I) *Canna flaccida* (Cannaceae), leaf blade and portion of petiole. (J) *Canna flaccida* (Cannaceae), fertile stamen with half-anther attached to the left margin. (K) *Canna flaccida* (Cannaceae), inflorescence, with open flower and flower bud; on the open flower, note the sepals (at base), reflexed petals, and five petaloid staminodes. (L) *Hedychium coronarium* (Zingiberaceae), inflorescence; note the bilaterally symmet-

ric flowers (calyx not visible), narrow petals, two broad lateral staminodes and lip-like staminodes with bifid apex, and the single stamen grasping the gynoecium. (M) *Tillandsia recurvata* (Bromeliaceae), flower. (N) *Tillandsia recurvata* (Bromeliaceae), cross-section of ovary. (O) *Tillandsia recurvata* (Bromeliaceae), plant in fruit. (P) *Tillandsia recurvata* (Bromeliaceae), scale hair from leaf. (Q) *Xyris fimbriata* (Xyridaceae), plant in flower. (R) *Xyris fimbriata* (Xyridaceae), inflorescence. (S) *Xyris fimbriata* (Xyridaceae), flower, showing the two persistent fimbriate lateral sepals and subtending bract (behind flower). (T) *Scirpus tabernaemontani* (Cyperaceae), single spikelet. (U) *Trichophorum caespitosum* (Cyperaceae), achene with bristles. (V) *Scirpus tabernaemontani* (Cyperaceae), flower and subtending bract. (W) *Typha latifolia* (Typhaceae), cluster of three carpellate flowers with many trichomes omitted to show stalked ovaries. (X) *Typha latifolia* (Typhaceae), cluster of staminate flowers. (Y) *Typha latifolia* (Typhaceae), inflorescence, staminate portion above, carpellate below. (A to C, from Zona 1997. D, from Tucker 1989. E and F, from Rosatti 1987. G and H, from Endress 1994c. I to L, from Rogers 1984. M to P, from Smith and Wood 1975. Q to S and V, from Kral 1983. T and U, from Tucker 1987. W to Y, from Thieret and Luken 1996.)

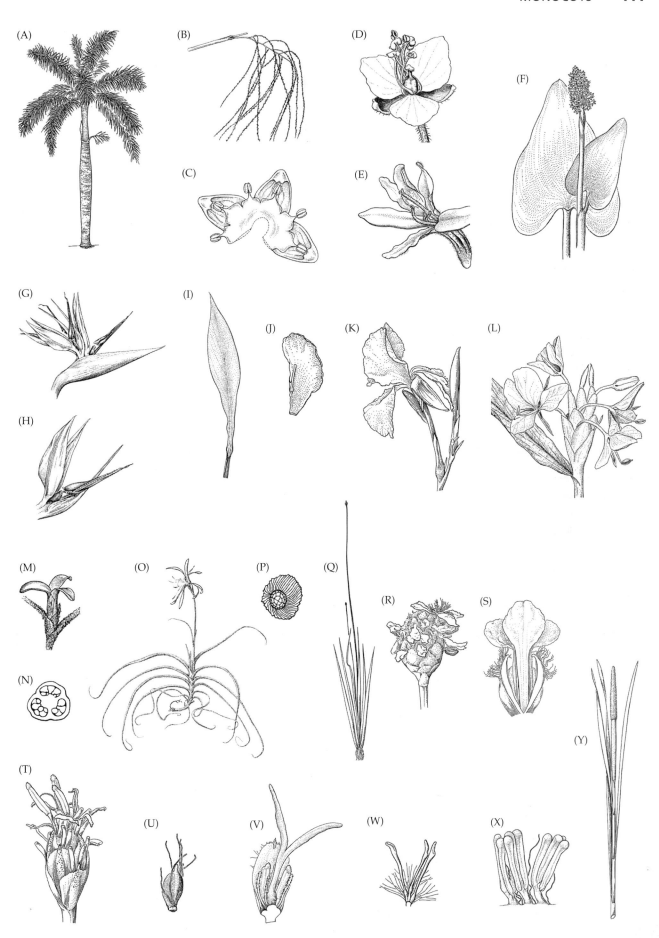

to the remaining taxa (71% bootstrap). In a combined analysis of three DNA regions, resolution and bootstrap support for large portions of the rest of the topology were poor, so little can be said with confidence regarding relationships within Arecaceae.

DASYPOGONACEAE Dasypogonaceae (4 genera, 16 species) are a family of stout shrubs or herbs, restricted to Australia (mostly Western Australia). They were previously linked to other, similar-looking groups, such as Xanthorrhoeaceae and Lomandraceae (the latter Laxmanniaceae in APG 1998; reviewed in Rudall and Chase 1996), families now known to be part of Asparagales (see above). The genus *Kingia* is a small tree, but this growth form is achieved just through primary growth, whereas the other three genera (*Baxteria*, *Dasypogon*, and *Calectasia*) are rhizomatous herbs or shrublike herbs (*Calectasia*). Floral characters of the family are like those of the lilioid monocots, with which they have always been placed previously, but Dasypogonaceae were found to have the diagnostic characters of the commelinids, which were reviewed above (Rudall and Chase 1996). In the shortest trees of Chase et al. (2005), Dasypogonaceae are sister to Poales, but with only 58% bootstrap support. If Dasypogonaceae continue in this position, and internal support increases as more genes are added, then they could ultimately be included in Poales.

ZINGIBERALES Zingiberales are a well-defined clade of eight families (Figure 4.8): Cannaceae (monogeneric with 19 species), Costaceae (4 genera, 110 species), Heliconiaceae (monogeneric with 100 to 200 species), Lowiaceae (monogeneric with 15 species), Marantaceae (31 genera, 550 species), Musaceae (2 genera, 35 species), Strelitziaceae (3 genera, 7 species), and Zingiberaceae (50 genera, 1,300 species).

A close relationship among these eight families has long been recognized from their morphology (Tomlinson 1962; Dahlgren and Rasmussen 1983; Kress 1990; Stevenson and Loconte 1995). Molecular phylogenetic studies involving single genes recovered a Zingiberales clade (Chase et al. 1993, 1995a, Duvall et al. 1993a; D. Soltis et al. 1997a), and combined molecular datasets provided strong support for this clade (Chase et al. 2000a; D. Soltis et al. 2000). These families are well known for their androecial modifications, such that what appears to be the perianth is often petaloid stamens (Figure 4.10, G to L). The flowers are thus difficult to interpret and are infamous among students for not being what they appear to be.

There are several non-DNA synapomorphies for Zingiberales, including presence of silica cells in the bundle sheath; leaves well differentiated into a petiole and blade; leaf blade with pinnate venation, often tearing between the secondary veins; leaf blade rolled into a tube in bud; petiole with enlarged air canals; flowers often with bilateral symmetry; ovary inferior (Figure 4.7); and arillate seeds with endosperm (Figure 4.10, G to L; Judd et al. 2002). Almost all of these plants are herbs, although some achieve large size, such as *Musa, Ravenala,* and *Strelitzia,* which are monstrous herbs.

Relationships within Zingiberales have been partially resolved in molecular phylogenetic studies (Chase et al. 2000a; D. Soltis et al. 2000; Kress et al. 2001). A tetrachotomy is present that consists of Lowiaceae + Strelitziaceae, Musaceae, Heliconiaceae, and a clade of Cannaceae, Marantaceae, Costaceae, and Zingiberaceae. The Cannaceae, Marantaceae, Costaceae, Zingiberaceae clade is supported by several putative synapomorphies, including androecium of a single functional stamen (Figure 4.10J), showy staminodes (Figure 4.10, K and L), seeds with more perisperm than endosperm, and absence of raphides in vegetative tissue (Tomlinson 1962, 1969; Kress et al. 2001; Judd et al. 2002). Only Marantaceae and Zingiberaceae have many species, and it is only by convention and history that some of the smaller families are still recognized. Some condensation of sister families into a single family would be desirable.

Musaceae are often large herbs with spirally arranged leaves in which the secondary veins are at right angles to the midvein; their inflorescences have large, deciduous bracts that subtend fascicles of flowers with five fused tepals (plus one free) and six stamens. Their stamen number is higher than other members of Zingiberales and is responsible for the family being viewed as relatively primitive within the order. Heliconiaceae are similar to Musaceae, except that the former have distichously arranged leaves. Like Musaceae, Heliconiaceae have large bracts subtending fascicles of flowers, but in Heliconiaceae the bracts persist and are highly colored.

Strelitziaceae (Figure 4.10, G and H) and Lowiaceae are sister families and share a petiole with air canals, a floral column that is formed from the sterile apex of the ovary, and a capsular fruit. Strelitziaceae have two-ranked, long-petiolate leaves, inflorescences that are fibrous, boat-shaped, and nonshowy, and flowers in which the lateral petals are connate. Lowiaceae have shorter-petiolate leaves with prominent veins and flowers with the median sepal uppermost and the median petal lowermost, forming a lip.

Cannaceae and Marantaceae form a clade that is sister to Costaceae + Zingiberaceae. Cannaceae and Marantaceae share flowers that lack any plane of symmetry (asymmetrical flowers; Figure 4.10, K and L); the single fertile stamen is modified with half of the structure expanded and staminodial (Figure 4.10, J and K). Cannaceae differ from the rest in their flattened style and muricate ovary. Marantaceae differ in having clearly petiolate leaves with a pulvinus (which facilitates their closing at night) and flowers in mirrored pairs. Several morphological characters unite Zingiberaceae and Costaceae. Both families share connate sepals, fused staminodes, reduction of two of the three stigmas, and a ligule at the apex of the leaf sheath. In addition, in Costaceae and Zingiberaceae the single functional stamen grasps the style (Figure 4.10L). Costaceae differ from Zingiberaceae in having spiromonostichous leaves, whereas Zingiberaceae have more or less distichously arranged leaves and often a branched inflorescence.

COMMELINALES The order consists of five families: Commelinaceae (38 genera, 640 species; worldwide except for Europe), Haemodoraceae (14 genera, 116 species; pantropical to warm temperate regions), Hanguanaceae (monogeneric, 6 species; Southeast Asia and northern Australia), Philydraceae (4 genera, 5 species; Australia to Southeast Asia), and Pontederiaceae (9 genera, 33 species; nearly cosmopolitan). The monophyly of the order has been well supported in recent molecular phylogenetic studies (e.g., Kellogg and Linder 1995; Chase et al. 2000a; D. Soltis et al. 2000). However, non-DNA synapomorphies for Commelinales are cryptic and thus far confined to chemistry (phenylphenalenones) and seed characters (e.g., abundant endosperm formed helobially). The families share the presence of tannin cells in the perianth and sclereids in the placentae (Judd et al. 2002).

This association of the families of Commelinales was another completely unpredicted phylogenetic result. Dahlgren et al. (1985) allied not only Haemodoraceae, Philydraceae, and Pontederiaceae but also Bromeliaceae, Typhaceae, and Velloziaceae (families of Poales). Within their concept of Bromeliiflorae, Dahlgren et al. placed each of these families in its own monofamilial order, signifying the degree to which they thought these families to be isolated from each other. Cronquist (1981) and others considered the tepalar perianth of Haemodoraceae, Philydraceae, and Pontederiaceae to indicate a relationship to Liliidae. Dahlgren et al. (1985) placed Commelinaceae in Commeliniflorae together with Eriocaulaceae, Mayacaceae, Rapateaceae, and Xyridaceae, which are members of Poales (*sensu* APG 1998; APG II 2003). Hanguanaceae were considered part of Asparagales by Dahlgren et al. (1985) and Cronquist (1981), and related to *Lomandra* (now considered Asparagaceae or Laxmanniaceae; APG 1998; APG II 2003).

Commelinaceae and Hanguanaceae are sister taxa (74% bootstrap support in Chase et al. 2005), but synapomorphies for this pair are thus far unknown. Commelinaceae are generally soft and fleshy herbs, some with a degree of succulence, and their inflorescences are cymose; flowers have differentiated perianth whorls (Figure 4.10D) and are typically fugacious (short lived); their hairy anthers are well known. *Hanguana* is

an often massive, coarse herb with petiolate leaves with many parallel secondary veins; its inflorescence is a panicle with sessile, nondescript flowers; its most distinctive trait is that the fruits are berries with a single, bowl-shaped seed. In many characters, *Hanguana* is a better match for Zingiberales (Rudall et al. 1999) than Commelinales, but the molecular data are clear about its position.

The other three families in the order (Haemodoraceae, Philydraceae, Pontederiaceae) share a few not particularly noteworthy characters, such as styloids and tanniniferous tepals that are typically persistent in fruit. As mentioned, these three families are relatively unusual in the commelinids due to their tepalar perianth (Figure 4.10E). Philydraceae are sister to the other two and are characterized by distichous phyllotaxis (except in *Philydrella*), pilose inflorescences, and monosymmetric flowers with a single stamen. The outer perianth whorl is petaloid, but much smaller than the inner whorl. In Haemodoraceae and Pontederiaceae, the outer whorl is nearly the same as the inner. Haemodoraceae grow typically in dry sites and have two-ranked, unifacial leaves and balanced cymes; most genera have mostly inferior ovaries. Roots of Haemodoraceae have a red color that is the basis for the family name. Pontederiaceae are emergent aquatics or wetland plants with sheathing leaf bases and a distinct petiole and blade; flowers exhibit tristyly, enantiostyly, and monosymmetry (Figure 4.10, E and F).

POALES Poales (sensu APG) are a large order of 17 families: Anarthriaceae (3 genera, 11 species; Western Australia), Bromeliaceae (57 genera, 1,400 species; mostly Neotropical, but 1 species in West Africa), Centrolepidaceae (3 genera, 35 species; mostly East Asia to Australia and New Zealand, but 1 genus in South America), Cyperaceae (98 genera, 4,350 species; cosmopolitan), Ecdeiocoleaceae (2 genera, 2 species; Western Australia), Eriocaulaceae (10 genera, 1,160 species; pantropical to temperate), Flagellariaceae (monogeneric, 4 species; paleotropics), Hydatellaceae (bigeneric, 10 species; India to New Zealand), Joinvilleaceae (monogeneric, 2 species; Malay Peninsula to the Pacific), Juncaceae (7 genera, 430 species; cosmopolitan), Mayacaceae (monogeneric, about 4 species; Neotropics, but 1 species in southeastern United States and 1 in West Africa), Poaceae (650 genera, 9,700 species; cosmopolitan), Rapateaceae (16 genera, 94 species; Neotropical with 1 genus in West Africa), Restionaceae (58 genera, 520 species; China, Indomalaysia to New Zealand, but 1 genus in Chile). Thurniaceae (bigeneric, 4 species; South Africa and eastern South America), Typhaceae (2 genera, about 24 species; nearly cosmopolitan; now includes Sparganiaceae), and Xyridaceae (5 genera, 260 species; pantropical with a few species in the temperate zone).

A broadly defined Poales are well supported by DNA sequence data (e.g., Chase et al. 2000a; D. Soltis et al. 2000). In addition, several non-DNA characters support a broadly defined Poales, including silica bodies in the epidermis and strongly branched styles. The order is also united by the loss of raphide crystals and sepal nectaries. Some authors have objected to the inclusion of so many families in Poales (Judd et al. 1999, but not Judd et al. 2002), but until recently, relationships among the families were too uncertain to justify recognition of additional orders carved from Poales. Chase et al. (2005) found at least some support for two subclades that Judd et al. (1999) recognized as orders: Juncales (100% bootstrap; Cyperaceae, Juncaceae, and Thurniaceae) and Poales (73% bootstrap; Anarthriaceae, Centrolepidaceae, Flagellariaceae, Joinvilleaceae, Ecdeiocoleaceae, and Poaceae). However, the presence in Judd et al. (1999) of several monofamilial orders (e.g., Bromeliales, Typhales) argues in favor of the broader circumscription. Certainly the argument that the order is too large and diverse does not hold up well against the background of the comparable levels of diversity in Asparagales and even the much smaller Commelinales (*sensu* APG). Poales s.l. also compare well with other recognized monocot orders if one considers naming clades of equal ages (Bremer 2002).

Within Poales in the broad sense (APG II 2003), Bromeliaceae and Typhaceae (including Sparganiaceae) are unresolved relative to the rest of the order, the monophyly of which is strongly (97% bootstrap) supported. Following Bromeliaceae and Typhaceae, molecular analyses place Rapateaceae as sister (81% bootstrap) to the remaining families.

Bromeliaceae have long been viewed as an isolated family of unclear relationships (Dahlgren et al. 1985), with possible links to the lilioids (through Velloziaceae), Commelinales (through Haemodoraceae and Pontederiaceae), and Poales (through Eriocaulaceae and Xyridaceae). As a member of the sister group to the rest of Poales, it is clear that Bromeliaceae are isolated and could share characters with these groups (those shared with the lilioids and Commelinales most likely being symplesiomorphies). Bromeliaceae are mostly rosette-forming, usually epiphytic herbs with lepidote hairs and a bracteate inflorescence (Figure 4.10, M to P). That Bromeliaceae may be related exclusively to Typhaceae is reflected in that both lack the mitochondrial gene *sdh4* (Adams et al. 2002) and possess helobial endosperm formation and an amoeboid tapetum; they differ principally in that Typhaceae are wind-pollinated as opposed to the entomophilous Bromeliaceae. Bremer (2002) placed Typhaceae (and Sparganiaceae) as sister to Bromeliaceae.

Typhaceae (Figure 4.10, W, X, Y) and Sparganiaceae are both small monogeneric families and could be combined (Judd et al. 2002). Both are rhizomatous emergent or marsh-loving herbs with two-ranked leaves and small, chaffy flowers arranged in complex inflorescences with the female flowers subtending the male flowers.

Rapateaceae are rosette-forming herbs that can reach large sizes; their flowers are arranged in scapose inflorescences and are large with distinct perianth whorls and six poricidally dehiscing anthers. Many authors have compared them to Xyridaceae (Cronquist 1981; Dahlgren et al. 1985), but their position as sister to a large clade containing Cyperaceae, Poaceae, Xyridaceae, and related families is clear, indicating that the similarities are parallelisms. The peculiar distribution of Rapateaceae in isolated areas of South America and West Africa compares well with that of both Bromeliaceae and Xyridaceae.

Following these basal members of Poales, a large clade of Juncaceae, Cyperaceae, Thurniaceae, Hydatellaceae, Eriocaulaceae, Xyridaceae, Flagellariaceae, Centrolepidaceae, Restionaceae, Anarthriaceae, Ecdeiocoleaceae, Joinvilleaceae, and Poaceae is supported by DNA sequence data. Non-DNA characters uniting this large clade are unclear, however. Within this large clade, several subclades are apparent; Xyridaceae and Eriocaulaceae are strongly supported as a clade. Several non-DNA characters also unite them—a perianth of both calyx and corolla and ovules with thin-walled megasporangia (Dahlgren et al. 1985).

Cyperaceae, Juncaceae, and Thurniaceae form a well-supported clade with molecular data (Figure 4.8). The genus *Prionium* had been placed in Juncaceae, but molecular data indicate that it should be considered a member of Thurniaceae. Thurniaceae are sister to Juncaceae and Cyperaceae (Plunkett et al. 1995; Munro and Linder 1997). A suite of non-DNA features also unites the three families: solid stems with three-ranked leaves and pollen in tetrads. The families share an unusual feature of chromosomes with diffuse centromeres (Plunkett et al. 1995; Simpson 1995; Munro and Linder 1998). Juncaceae were found in two studies not to be monophyletic. *Oxychloe* (Juncaceae) was found to be embedded in or sister to Cyperaceae (Plunkett et al. 1995; Muasya et al. 1998). The former study may have used a mixed leaf collection of *Oxychloe* and a sedge and almost certainly sequenced the sedge, whereas the latter study was based on a contaminated sample. True *Oxychloe* has now been sequenced, and it goes with the other genera of Juncaceae, so both families are monophyletic (Jones et al. 2005) and should be retained as separate.

A relationship of Xyridaceae to Eriocaulaceae has been widely accepted in the past (Dahlgren et al. 1985). The two families share a similar habit (rosette-forming herbs), strictly basal leaves with the same type of stomata (paracytic), capitate inflorescences with dimerous flowers in which the anthers are adnate to the corolla (Figure 4.10, Q to S), and spinulate/echinate pollen. However, the relationship of this pair to Mayacaceae and Hydatellaceae is less clear. Bremer (2002) had difficulties obtaining sequences from both Mayacaceae and Hydatellaceae, as did Chase et al. (2005) for the latter. Both families often appear near Eriocaulaceae/Xyridaceae, but the patterns are unclear at this point. Mayacaceae are morphologically unusual, looking much like a clubmoss, except that the plants are aquatic. Members of Mayacaceae have spirally arranged leaves with apical teeth and axillary flowers with clearly distinct calyx and corolla and three stamens. Hydatellaceae are also anomalous in many characters—inflorescence with minute flowers lacking a perianth emerges from basal, thin, filiform leaves; endosperm formation is cellular, a character known elsewhere in the monocots in Araceae; and the seeds are small and with storage function performed by a starchy perisperm. The two genera of Hydatellaceae were previously included in Centrolepidaceae, which are similarly small and adapted to seasonally inundated conditions. Using morphological characters, Dahlgren et al. (1985) concluded that their "inclusion even in any superorder will be most strained." All that can be said with much confidence at this stage is that Hydatellaceae belong to Poales.

The core Poales consist of the Restionaceae clade (see below) as sister to Flagellariaceae, Joinvilleaceae, Ecdeiocoleaceae, and Poaceae (Figure 4.8). The monophyly of core Poales is well supported by DNA analyses (e.g., Chase et al. 2000a; D. Soltis et al; 2000). However, a close relationship among these families had long been recognized (e.g., Dahlgren and Rasmussen 1983; Dahlgren et al. 1985).

Most families of core Poales are herbaceous plants native to the Southern Hemisphere and are small in terms of number of genera and species. The two largest families are Restionaceae and Poaceae. Non-DNA synapomorphies of core Poales include two-ranked leaves with an open sheath around the stem, stomata with dumbbell-shaped guard cells, a single apical orthotropous ovule per carpel, monoporate pollen with a rim around the pore, a complex (pinnately-branched) stigma, and nuclear endosperm development (Endress 1995a; Kellogg and Linder 1995; Soreng and Davis 1998).

The Restionaceae clade is composed of Anarthriaceae as sister to Centrolepidaceae + Restionaceae; these families share dioecy, peg cells in their chlorenchyma, and dorsifixed anthers. Linder et al. (2000) suggested that all of these are Restionaceae because they have the distinctive culm anatomy of that family; *Lyginia* also has starch in the embryo sac, as does Restionaceae. Certainly recognition of Lyginiaceae and Hopkinsiaceae as well as Anarthriaceae (all monogeneric) introduces redundan-

cy into the classification. Centrolepidaceae could well be paedomorphic Restionaceae (Linder et al. 2000); they are clearly sister to the latter (100% bootstrap; Chase et al. 2005).

The Restionaceae clade is sister to the remainder of core Poales, which includes Flagellariaceae, Joinvilleaceae, and Ecdeiocoleaceae as successive sister groups to Poaceae (Figure 4.8). Poaceae (Gramineae) are one of the largest and most important angiosperm families. Recently, a clear picture of the phylogenetic relationships has emerged for Poaceae from the combination of molecular datasets. These data have been combined and analyzed by a consortium of researchers—the Grass Phylogeny Working Group (GPWG 2000; www.virtualherbarium.org/GPWG/; see Figure 11.10 in Chapter 11).

The origin of Poaceae can be dated by the appearance of grass pollen—which is distinctive—in the fossil record. The earliest unequivocal records are from the Paleocene of South America and Africa (60–55 Mya; Thomasson 1987).

CHARACTER EVOLUTION IN POACEAE Because Poaceae have enormous ecological and economic significance, it is important to consider their origin and evolution, and Kellogg (2000) provided an excellent overview. By using the GPWG phylogeny (see Figure 11.10) and by comparing grasses with their closest relatives, Kellogg attempted to reconstruct the types of changes that likely occurred in the early evolution of Poaceae. A major change occurred in the timing of embryo development. Most monocotyledonous plants have undifferentiated embryos; seed maturation begins after the embryo has formed a shoot apical meristem, but the differentiation of cotyledon, leaves, root meristem, and vasculature largely occurs after the seed is shed from the parent plant. In the grasses, however, embryo development is accelerated relative to seed maturation (Kellogg 2000).

The immediate ancestors of the grasses had ovaries of three united carpels; each carpel possessed one locule with one ovule (Kellogg and Linder 1995). In the grasses, only one locule and one ovule form. As the ovule develops, the outer integument fuses with the inner ovary wall to form the distinctive fruit of the grasses, known as the grain, or caryopsis. The structure is unique among the flowering plants.

Perhaps the most striking characteristic of grasses today is their floral and inflorescence structure. Grass flowers are generally arranged in spikelets; each spikelet consists of one or more flowers (termed florets in Poaceae) and associated bracts (Figure 4.11). Phylogenetic trees reveal that the spikelet must have originated in several steps (GPWG 2000). The earliest grasses had three stigmas, a relict of the three fused carpels they inherited from their ancestors; this number was reduced to two

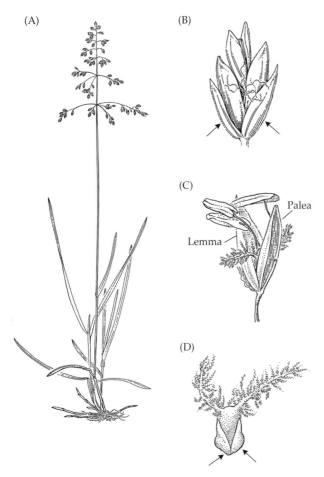

(A)

(B)

(C)

Palea

Lemma

(D)

FIGURE 4.11 Plants and flowers of Poaceae. (A) *Poa pratensis*, flowering plant with rhizomes at base. (B) *Poa pratensis*, spikelet with four florets; arrows indicate glumes. (C) *Poa pratensis*, a single floret, lemma and palea indicated. (D) *Poa pratensis*, lodicules (indicated by arrows) and gynoecium. (From Campbell 1985.)

after the speciation event that led to *Pharus*. The earliest species also had, like their nongrass ancestors, six stamens. It is not clear from the tree precisely when the shift from six to three stamens occurred, but it must have been after the divergence of the *Guaduella/Puelia* group.

Poaceae also provide a model for the evolution of C_4 photosynthesis. This topic is covered in Chapter 11.

Fossil History of Monocots

At least some putative monocot fossils have been found in strata as old as the eudicots (reviewed in Herendeen et al. 1995), and what we know of the branching order of the angiosperm tree (see Chapter 2) would lead us to conclude that monocots should be among the oldest lineages of flowering plants. The Wikström et al. (2001) effort to date all lineages of angiosperms produced dates ranging from 158 to 141 Mya for the stem node

of the monocots and from 141 to 127 Mya for the crown group (the split of *Acorus* and the rest of the monocots). Bremer (2000) put the split of *Acorus* and the rest of the monocots at 134 Mya, which then leads to the conclusion that the stem lineage of the monocots is even older. The discrepancy between these two estimates probably has to do with the highly conservative calibration point (the split between Fagales and Cucurbitales in the Late Santonian) used in the Wikström et al. (2001) study.

For some time, fossils of the palms were thought to be the oldest reliable monocots; they first appeared around 90 Mya. Fossil Araceae have been described that date from 97 Mya, but this date may not be entirely trustworthy (reviewed in Herendeen and Crane 1995). Friis et al. (2004) described a fossil attributed to Araceae that dates from 110 to 120 Mya. Fossil flowers of Triuridaceae (Pandanales) have been reported from Upper Cretaceous rocks of New Jersey (Gandolfo et al. 2002), making these the earliest known occurrence of the saprophytic/mycotrophic habit in flowering plants and among the oldest well-characterized monocot fossils. If we assume that Triuridaceae are members of Pandanales, then it becomes clear that the age of the crown group of the monocots could, accordingly, be closer to the estimates of Bremer (2000).

Future Research

Monocot relationships are now relatively well characterized at all levels, from genus to interordinal, making them one of the best understood of all major groups of angiosperms. Except for orchids and grasses, nearly every genus has been included in at least one published study, most involving plastid *rbcL*, and we anticipate that within five years it will be possible to complete a generic-level analysis of the monocots. Such an analysis will provide a framework for a wide range of other studies, including dating the origins of all major clades and facilitating analyses of the factors associated with this major radiation of angiosperms. This phylogenetic framework will also facilitate studies of floral evolution. Recent molecular investigations make it clear, for example, that the ABC model of floral initiation developed for eudicots must be modified to accommodate monocots and other basal angiosperms (see Chapter 3). Although homologues of the ABC genes function in monocots, the details of their expression differ from those of eudicot models (e.g., Kanno et al. 2003). There is also evidence from several sources that "flowers" in some monocots (e.g., Alismatales) may actually be modified inflorescences. Additional research in floral developmental genetics is needed to elucidate the genetic architecture of floral structures and floral diversification in the monocots.

The two largest families of monocots, the orchids and grasses, offer immensely contrasting pictures of how evolutionary success can be attained. Orchids are an old family (110 Mya; Bremer 2002), whereas the grasses are much more recent (approximately 60 Mya; Wikström et al. 2001; Bremer 2002). The orchids include some of the most complex biotic pollination syndromes known for the angiosperms, whereas the grasses are wind-pollinated, and the flowers of the two groups could not be more different in appearance. Understanding how such diverse forms have evolved will require a clear understanding of the relationships of each family, and that framework has now been laid. If an understanding can be achieved of how similar genes result in such different organisms, then we will be well on our way to understanding the diversity of flowering plants as a whole.

5

Early-Diverging Eudicots

Introduction

The eudicots, representing about 75% of all angiosperm species (Drinnan et al. 1994), are conspicuously marked by the presence of triaperturate pollen (Figure 5.1; Doyle and Hotton 1991; see Chapter 2). However, although triaperturate pollen is a synapomorphy for eudicots, not all eudicots have three-grooved pollen grains. Some other pollen types (e.g., inaperturate, polyporate, and polycolporate) also occur due to subsequent evolution of pollen structure. Analyses of three or more genes supported the monophyly of the eudicots with jackknife or bootstrap values of 100% (D. Soltis et al. 1998, 2000; Hoot et al. 1999; Qiu et al. 1999, 2000; P. Soltis et al. 1999b). Within eudicots, phylogenetic studies of individual genes and combined DNA datasets (e.g., Chase et al. 1993; D. Soltis et al. 1997a, 1997b, 2000; Hoot et al. 1999; Savolainen et al. 2000a, 2000b) all identified a basal grade of eudicots, followed by a clade of core eudicots (Figure 5.2). Because triaperturate pollen is so distinctive (Figure 5.1), the estimated age of the eudicots is perhaps the firmest date in the paleobotanical record. Fossilized pollen documents the presence of the eudicots at the Early Cretaceous (Barremian–Aptian boundary, 125 Mya; Crane et al. 1995; Magallón et al. 1999). Examination of the fossil record indicates an uneven distribution of species diversity across the major clades of eudicots, with the most species-rich groups known only from the relatively recent fossil record, suggesting that most of the diversity of eudicot species is the result of relatively recent (e.g., 50–70 Mya) radiations (Magallón et al. 1999).

FIGURE 5.1 SEM micrograph of triaperturate pollen. *Rumex* (Chenopodiaceae = Amaranthaceae, APG II 2003). (Courtesy of M. Sundberg.)

In this chapter, we focus on the basal branches of eudicots. Some of the smaller lineages of core eudicots (e.g., Gunnerales, Saxifragales, Santalales) are introduced in Chapter 6; the major groups (Caryophyllales, rosids, asterids) are discussed in more detail in Chapters 7, 8, and 9 (see also Judd and Olmstead 2004; Stevens 2004). The basal eudicot lineages (Figure 5.2) deserve special attention because of their critical phy-

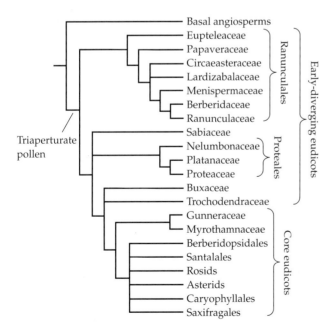

FIGURE 5.2 Phylogenetic summary tree depicting relationships among early-diverging eudicots, based on analyses of four-gene datasets. (From D. Soltis et al. 2003a)

logenetic positions and morphological diversity. The term "early-diverging eudicots" has been used to refer to this basal grade (e.g., D. Soltis et al. 2000), and we will primarily use this term for consistency; however, others may prefer "basal eudicots" for consistency with the "basal angiosperm grade." Of course, we do not advocate applying names to grades and do so here only as a way to collect these lineages into a convenient group for discussion.

The early-diverging eudicots are Ranunculales, Proteales, Sabiaceae, Buxaceae (including Didymelaceae; APG II 2003), and Trochodendraceae (including Tetracentraceae *sensu* APG 1998 and APG II 2003). (See Figure 5.3 for illustrations of some early-diverging eudicots.) Most of these lineages are small; only Ranunculales, Buxaceae, and Proteaceae contain more than a few genera and species. The oldest fossils of early-diverging eudicots are of Trochodendraceae and date from the Aptian (118 Mya); floral remains of both Platanaceae and Buxaceae have been reported from the Albian (108 Mya; Magallón et al. 1999), and fossils of Proteaceae are known from the mid-Cretaceous (97 Mya). Fossils of Ranunculales and Sabiaceae are considerably younger than those of other early-diverging eudicots; both have been reported from the Maastrichtian (69.5 Mya; see Figure 2.6; reviewed in Magallón et al. 1999).

Some groups of early-diverging eudicots have figured prominently in discussions of the early evolution of the angiosperms (e.g., Bessey 1915; Cronquist 1968, 1981, 1988; Takhtajan 1969, 1991). Ranunculales were typically placed in Magnoliidae in recent classifications because they were thought to possess ancestral floral features (e.g., Stebbins 1974; Cronquist 1981, 1988; Thorne 1992a, 1992b, 2001; Heywood 1993, 1998). Takhtajan (1997), in contrast, recognized a distinct subclass Ranunculidae that he considered to be closely related to Magnoliidae. Ranunculaceae and close relatives, such as Menispermaceae, Berberidaceae, and Papaveraceae, were central members of Bessey's (1915) ranalean complex, which contained those angiosperms he considered most primitive. Ranunculales include taxa with numerous stamens and carpels, which are often spirally arranged; numerous spirally arranged parts were considered primitive traits. Cronquist (1968, 1981, 1988), for example, considered members of Ranunculales to represent the herbaceous equivalent of his woody Magnoliales. Cronquist (1968, 1981) and Takhtajan (1987, 1991) suggested that Ranunculales originated from Magnoliales via Illiciales or some close relative of the latter. There has been renewed interest in Ranunculales (e.g., Drinnan et al. 1994; Albert et al. 1998) in light of their phylogenetic placement as sister to all other eudicots. *Trochodendron* and *Tetracentron* have attracted special attention because they possess several "magnoliid" features (e.g., the production of ethereal oil cells, chloranthoid leaf teeth, absence of ves-

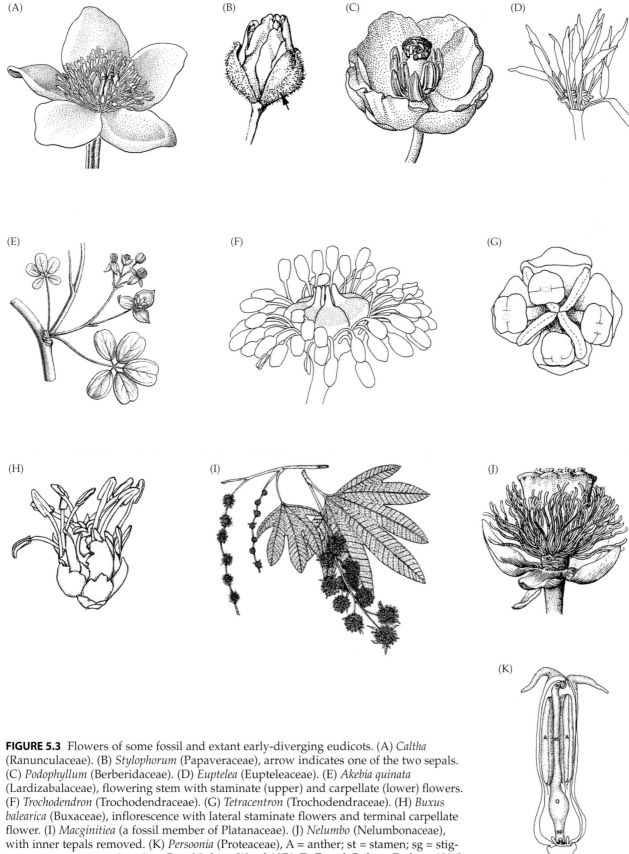

FIGURE 5.3 Flowers of some fossil and extant early-diverging eudicots. (A) *Caltha* (Ranunculaceae). (B) *Stylophorum* (Papaveraceae), arrow indicates one of the two sepals. (C) *Podophyllum* (Berberidaceae). (D) *Euptelea* (Eupteleaceae). (E) *Akebia quinata* (Lardizabalaceae), flowering stem with staminate (upper) and carpellate (lower) flowers. (F) *Trochodendron* (Trochodendraceae). (G) *Tetracentron* (Trochodendraceae). (H) *Buxus balearica* (Buxaceae), inflorescence with lateral staminate flowers and terminal carpellate flower. (I) *Macginitiea* (a fossil member of Platanaceae). (J) *Nelumbo* (Nelumbonaceae), with inner tepals removed. (K) *Persoonia* (Proteaceae), A = anther; st = stamen; sg = stigma; o = ovary; sp = stipe. A to C and J, from Wood 1974. D, F, and G, from Endress 1986b. E, from Engler and Prantl 1891. H, from von Balthazar and Endress 2002a. I, from Manchester 1986. K, from Douglas and Tucker 1996b.

sels, and valvate anther dehiscence) long considered primitive or plesiomorphic. These and other features were thought to have been retained from magnoliid ancestors (Cronquist 1981; Endress 1986b; Crane 1989).

Molecular as well as morphological (e.g., Hufford 1992) phylogenetic studies have identified the same families as successive sisters to all other eudicots. Ranunculales have consistently appeared as sister to all other eudicots in DNA analyses (e.g., Chase et al. 1993; D. Soltis et al. 1997a, 1997b, 2000, 2003a; Hoot et al. 1999; Savolainen et al. 2000a, 2000b; Hilu et al. 2003; Kim et al. 2004). Molecular phylogenetic studies have also concurred in recovering Proteales, Sabiaceae, Buxaceae, and Trochodendraceae as the remaining lineages of early-diverging eudicots. However, the placement of Ranunculales as sister to other eudicots does not receive high bootstrap support in analyses of combined three- (59% jackknife) and four-gene (87%) datasets (D. Soltis et al. 2000, 2003a). A recent analysis of the rapidly evolving plastid gene *matK* provided moderate (82%) jackknife support for this placement of Ranunculales (Hilu et al. 2003). When *matK* is ultimately combined with other genes (in progress), support for Ranunculales as sister to all other eudicots should be strong.

Following Ranunculales, relationships among Proteales, Sabiaceae, Buxaceae (including Didymelaceae), Trochodendraceae, and the core eudicots are not resolved. The best current estimates of phylogeny for the early-diverging eudicots are those based on three (Hoot et al. 1999; P. Soltis et al. 1999b; D. Soltis et al. 2000) or four genes (D. Soltis et al. 2003a; Kim et al. 2004). In these trees, Proteales and Sabiaceae follow Ranunculales as successive sisters to all other eudicots. However, the relationship between Proteales and Sabiaceae is unclear. In analyses of three genes, they formed a grade (Proteales followed by Sabiaceae), but without support above 50%. However, in parsimony analyses of a four-gene dataset, they formed a clade, but without support above 50%; in maximum likelihood analyses, they again form a grade (Kim et al. 2004). Analyses of single genes were similarly inconsistent. In some analyses of *rbcL* sequences, Proteales and Sabiaceae formed a clade (e.g., Chase et al. 1993; Savolainen et al. 2000b), but they did not form a clade in analyses of 18S rDNA or *atpB* sequences (D. Soltis et al. 1997a, 1997b; Savolainen et al. 2000a).

Following Proteales and Sabiaceae, there is strong support for Trochodendraceae and Buxaceae as sister groups to all other eudicots (100% in D. Soltis et al. 2000, 2003a), but the relationship between these two families is unclear. In the strict consensus of shortest trees inferred from three-gene datasets (Hoot et al. 1999; D. Soltis et al. 2000), they form a grade (with Trochodendraceae sister to the core eudicots), whereas in trees resulting from analysis of a four-gene dataset, Buxaceae and Trochodendraceae form a clade sister to

the core eudicots (Kim et al. 2004), although neither investigation provides support above 50% for these relationships. Similarly conflicting results were obtained in analyses that used individual genes (e.g., Chase et al. 1993; D. Soltis et al. 1997a, 1997b; Savolainen et al. 2000a, 2000b). Following Buxaceae and Trochodendraceae, there is strong support (100%) for the monophyly of the core eudicots (e.g., Hoot et al. 1999; P. Soltis et al. 1999b; Savolainen et al. 2000a; D. Soltis et al. 2000, 2003a; Hilu et al. 2003; see Chapter 6).

The early-diverging eudicots share some putatively plesiomorphic features with basal angiosperms (Chapter 3). Ellagic acid and gallic acid (Figure 5.4) are absent from the early-diverging eudicot lineages, as they are from most basal angiosperms, but these acids appear to be prevalent throughout the core eudicots (e.g., Chase et al. 1993; Savolainen et al. 2000a; D. Soltis et al. 2000). Using data from Nandi et al. (1998), we more critically examined the phylogenetic distribution of these chemical compounds. Ellagic acid is absent from basal angiosperms, with the exception of Nymphaeaceae (data are not available for *Amborella*), and from all early-diverging eudicot lineages (Figure 5.5). In contrast, ellagic acid is common throughout the core eudicots and is reconstructed as a putative synapomorphy for that large

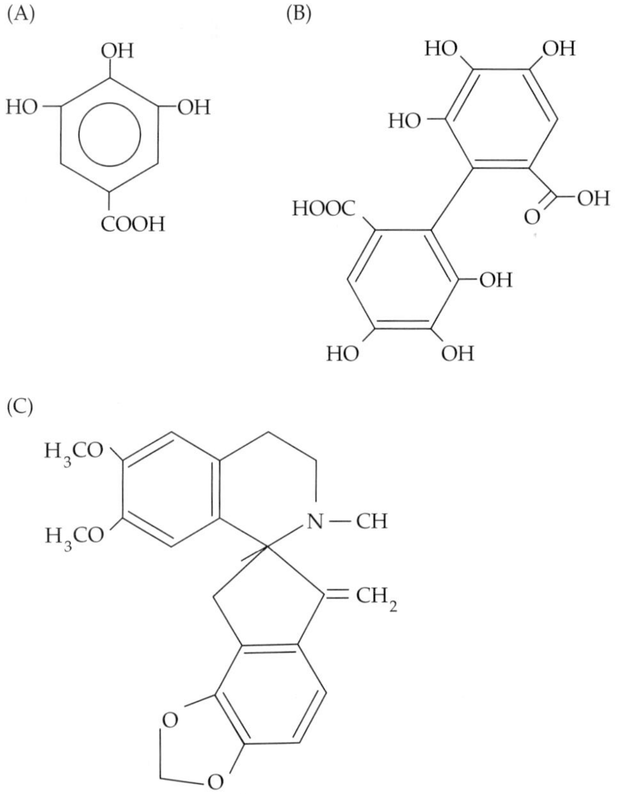

FIGURE 5.4 Chemical structure of (A) gallic acid, (B) ellagic acid, and (C) ochotensimine, a representative isoquinoline alkaloid.

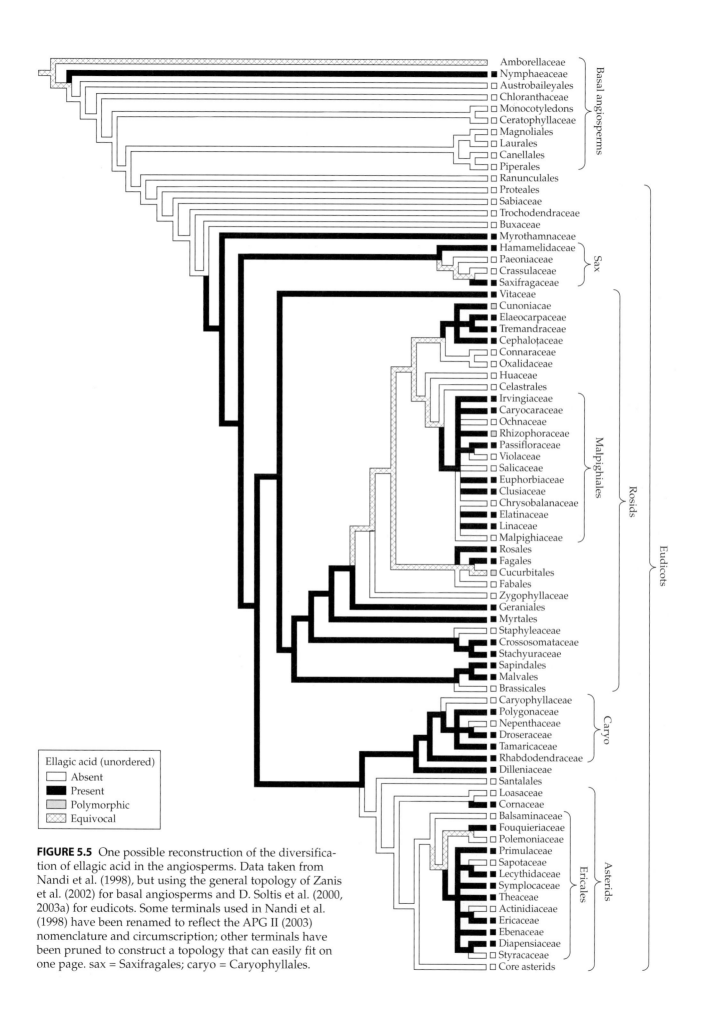

FIGURE 5.5 One possible reconstruction of the diversification of ellagic acid in the angiosperms. Data taken from Nandi et al. (1998), but using the general topology of Zanis et al. (2002) for basal angiosperms and D. Soltis et al. (2000, 2003a) for eudicots. Some terminals used in Nandi et al. (1998) have been renamed to reflect the APG II (2003) nomenclature and circumscription; other terminals have been pruned to construct a topology that can easily fit on one page. sax = Saxifragales; caryo = Caryophyllales.

clade. However, the diversification of ellagic acid production is complex, with several putative losses and secondary gains in the core eudicots (Figure 5.5). In fact, the absence of ellagic acid is reconstructed as ancestral for the large asterid clade. Our reconstructions indicate that the evolution of ellagic acid production in some large clades (e.g., Malpighiales and Ericales) is particularly dynamic. The production of gallic acid displays a similarly complex pattern of evolutionary diversification (reconstruction not shown; Nandi et al. 1998). Some aspects of floral structure, such as merosity, are labile in these early-branching eudicots, a situation comparable to that observed in basal angiosperms (see Chapter 3).

Ranunculales

Ranunculales are strongly supported as monophyletic and appear as the sister to all other eudicots. Ranunculales consist of seven families according to the APG (1998) and APG II (2003) classifications (Figure 5.2)—Berberidaceae, Circaeasteraceae (including *Kingdonia*, placed in Kingdoniaceae in some classifications; see Takhtajan 1997), Eupteleaceae (previously placed in subclass Hamamelidae), Lardizabalaceae (including *Sargentodoxa*), Menispermaceae, Papaveraceae, and Ranunculaceae. Papaveraceae are broadly defined here to include Fumariaceae, *Hypecoum* (treated as a distinct family, Hypecoaceae, in some classifications; Takhtajan 1997), and the monotypic *Pteridophyllum* (placed in its own family, Pteridophyllaceae, in some classifications; Takhtajan 1997). Ranunculaceae are also broadly defined to include *Hydrastis* and *Glaucidium*, monotypic genera sometimes placed in their own families, Hydrastidaceae and Glaucidiaceae (Takhtajan 1997). Ranunculales have been the focus of several phylogenetic investigations, both morphological and molecular (e.g., Drinnan et al. 1994; Hoot and Crane 1995; Loconte et al. 1995).

With the exception of the monotypic Eupteleaceae, most families now placed in Ranunculales have been considered closely related in recent morphology-based classifications (e.g., Cronquist 1981; Takhtajan 1987, 1997). In addition to Berberidaceae, Circaeasteraceae, Lardizabalaceae, Sargentodoxaceae, Menispermaceae, Papaveraceae, and Ranunculaceae, Cronquist also included Coriariaceae and Sabiaceae in his Ranunculales. Coriariaceae are now placed in the rosid order Cucurbitales (see Chapter 8), and Sabiaceae represent a distinct lineage of early-branching eudicots. Takhtajan (1997) also included Paeoniaceae in his Ranunculidae; Paeoniaceae are now recognized as a member of Saxifragales (see Chapter 6).

Families of Ranunculales share several non-DNA features, including the presence of benzyl-isoquinoline alkaloids of the berberine and morphine type (Figure 5.4; Jensen 1995), primarily herbaceous habit, hypogynous flowers, unusually large and homogeneous S-type sieve element plastids (Behnke 1995), epicuticular wax tubules (also found in some families outside of Ranunculales, such as Nelumbonaceae; Barthlott and Theissen 1995), and seeds with small embryos and copious endosperm (see also Hoot et al. 1999). However, no obvious synapomorphies are indicated for Ranunculales by floral morphology (Endress 1995b).

Eupteleaceae represent an unexpected addition to Ranunculales. Because its anemophilous (i.e., wind-pollinated) or partly entomophilous (insect-pollinated) flowers lack a perianth (Figure 5.3D), *Euptelea* has usually been placed in Hamamelidae close to Trochodendraceae, Cercidiphyllaceae, and Platanaceae (Cronquist 1981, 1988; Endress 1986b, 1993; Takhtajan 1987, 1997). Molecular phylogenetic studies clearly support *Euptelea* as a member of Ranunculales (e.g., Chase et al. 1993; Qiu et al. 1993; Hoot et al. 1999; D. Soltis et al. 2000), with Trochodendraceae, Platanaceae, and Cercidiphyllaceae representing more distantly related lineages (see below, Trochodendraceae; Proteales; and Chapter 6).

Molecular studies have also confirmed that *Sargentodoxa* (Lardizabalaceae) belongs to Ranunculales as sister to all other Lardizabalaceae (Hoot et al. 1995b; D. Soltis et al. 1997a, 2000). *Sargentodoxa* had typically been placed in its own family close to Lardizabalaceae in recent morphology-based classifications (Cronquist 1981, 1988; Takhtajan 1987, 1997).

With few exceptions, relationships within Ranunculales are well resolved and supported in trees inferred from multiple genes (e.g., Hoot et al. 1999; D. Soltis et al. 2000; Kim et al. 2004). In three-gene analyses, Papaveraceae are sister to all remaining Ranunculales; Eupteleaceae then appear as sister to the remaining members of the clade. However, internal support (as measured by the bootstrap or jackknife) for these positions of Papaveraceae and Eupteleaceae is very low (Hoot et al. 1999; D. Soltis et al. 2000). The monophyly of all Ranunculales excluding Papaveraceae receives only 53% jackknife support (D. Soltis et al. 2000). With a four-gene dataset that also includes 26S rDNA sequences (Kim et al. 2004), positions are reversed, with Eupteleaceae, followed by Papaveraceae, sisters to the remaining members of Ranunculales. Bootstrap support for this placement of Eupteleaceae is also fairly low (78%), but higher than the three-gene result. Thus, at this point the placements of Papaveraceae and Eupteleaceae within Ranunculales remain uncertain. Because of the stronger support, we illustrate the four-gene result (Figure 5.2).

Following Papaveraceae and Eupteleaceae, the remaining Ranunculales form a well-supported clade of Berberidaceae, Circaeasteraceae, Lardizabalaceae, Menispermaceae, and Ranunculaceae (Figure 5.2). Within this clade, with four genes the relationships of Lardizabalaceae and Circaeasteraceae remain unclear; there is moderate support for a clade in which Menispermaceae are sister to Ranunculaceae + Berberidaceae (Figure 5.2).

Detailed phylogenetic analyses have also been conducted for some of the families of Ranunculales. Papaveraceae, as well as genera within the family (e.g., *Papaver*), have been the subject of phylogenetic analyses using morphological and molecular characters (Kadereit and Sytsma 1992; Kadereit 1993; Kadereit et al. 1994, 1995; Loconte et al. 1995; Hoot et al. 1997). Several phylogenetic studies have focused on Ranunculaceae (Hoot 1991, 1995; Johansson and Jansen 1993; Tamura 1993; Hoot and Crane 1995; Jensen et al. 1995; Johansson 1995). Other studies have focused on large genera, such as *Anemone* (e.g., Hoot et al. 1994; Schuettpelz et al. 2002; S.B. Hoot, pers. comm.) and *Caltha* (S.B. Hoot, pers. comm.). Because Ranunculaceae are relatively large (50 genera, 1,800 species), a comprehensive phylogenetic analysis of the family is still needed. Berberidaceae have also been the focus of phylogenetic study (Loconte and Estes 1989; Kim and Jansen 1995, 1998). Phylogenetic relationships in Lardizabalaceae (including *Sargentodoxa*) have been analyzed using DNA sequence data (Hoot et al. 1995a, 1995b).

Proteales

Proteales comprise three families (Proteaceae, Platanaceae, and Nelumbonaceae) that have nearly always been considered distantly related (Figure 5.2). Although the composition of Proteales has to be one of the major surprises of molecular phylogenetics given the diverse habits and morphologies of its members, analyses of combined DNA datasets provide strong support for this clade (e.g., Hoot et al. 1999; Savolainen et al. 2000a; D. Soltis et al. 2000). For example, jackknife support of 85% for this clade was obtained in the three-gene analysis (D. Soltis et al. 2000). A relationship among these three families was previously indicated by separate analyses of *rbcL*, *atpB*, and 18S rDNA sequences, although without support above 50% (Chase et al. 1993; D. Soltis et al. 1997a; Savolainen et al. 2000a, 2000b). Within Proteales, Nelumbonaceae are sister to the strongly supported (93%) clade of Platanaceae and Proteaceae.

Although Proteales are an eclectic assemblage, there are possible non-DNA synapomorphies. These plants possess rod- or tube-shaped epicuticular waxes, seeds with scanty or no endosperm, and alternate vessel pitting (Nandi et al. 1998; Savolainen et al. 2000a). Other features indicating a close relationship between Proteaceae and Platanaceae include carpel closure by complete postgenital fusion, one or two ovules per carpel, and orthotropous ovules (Endress and Igersheim 1999). A close relationship between Proteaceae and Platanaceae is also in agreement with similarities in fossils. The two families are among the most common fossils in mid-Cretaceous floras, and it has been suggested that Proteaceae may be a Southern Hemisphere variant of Platanaceae that diverged early in angiosperm evolution (Drinnan et

al. 1994). This close relationship between Proteaceae and Platanaceae could, for example, explain the platanoid tooth of many leaf taxa in Late Cretaceous floras from the Southern Hemisphere and highlights the need for more studies of morphology (Drinnan et al. 1994). Several fossils attributed to Platanaceae have flowers with relatively conspicuous perianth parts and some are tetramerous rather than pentamerous, as in modern representatives (Magallón-Puebla et al. 1997). This tetramerous organization could provide another link with Proteaceae, although, as reviewed in more detail in "Merosity (Merism)," below, recent developmental studies indicate that Proteaceae are dimerous, rather than tetramerous (Douglas and Tucker 1996a, 1996b). Phylogenetic relationships within Proteaceae have been examined using plastid *atpB* and *atpB-rbcL* intergenic spacer region sequences (Hoot and Douglas 1998).

Sabiaceae

Sabiaceae are a small tropical and subtropical family of three genera, two of which (*Meliosma* and *Sabia*) have been included in molecular phylogenetic studies. Takhtajan (1997) recognized two separate families, Meliosmaceae and Sabiaceae, within Sabiales. The monophyly of Sabiaceae is strongly supported in phylogenetic studies of combined DNA datasets (D. Soltis et al. 2000), but is also evident in studies using single genes, such as *rbcL* alone (Savolainen et al. 2000b).

The proposed relationships of Sabiaceae have varied greatly in recent morphology-based classifications. Heimsch (1942) and Heywood (1993, 1998) placed the family in Sapindales (subclass Rosidae), whereas Takhtajan (1997) placed the family in its own order within Rosidae. Cronquist (1981, 1988), however, considered Sabiaceae a member of Ranunculales.

The precise placement of Sabiaceae among early-diverging eudicot lineages is not yet known (Figure 5.2). As noted above, in some analyses, Sabiaceae follow Proteales as sister to all other eudicots (e.g., Savolainen et al. 2000a; D. Soltis et al. 2000). However, in other analyses Sabiaceae and Proteales form a clade (e.g., Chase et al. 1993; Savolainen et al. 2000b; Kim et al. 2004). It is noteworthy that Sabiaceae and Proteaceae both possess wedge-shaped phloem rays (Metcalfe and Chalk 1950; see Nandi et al. 1998). These two families also share a nectar disk (Haber 1966; Beusekom 1971; Nandi et al. 1998), a rare feature among early eudicot lineages. However, these features are not apparent in Nelumbonaceae and Platanaceae, so it is unclear if these characters represent potential synapomorphies of Proteales and Sabiaceae that were lost in some members of Proteales, or if they simply evolved in parallel in Proteaceae and Sabiaceae. However, a nectar disk would not be expected in a wind-pollinated group such as Platanaceae.

Buxaceae

Molecular phylogenetic studies have consistently indicated a well-supported sister-group relationship between Buxaceae as traditionally recognized and Didymelaceae (Hoot et al. 1999; Savolainen et al. 2000a; D. Soltis et al. 2000; von Balthazar et al. 2000). As a result of this close relationship (see below), it has been proposed that the two families be combined into a single family, Buxaceae (APG II 2003).

Buxaceae in the narrow sense have been variously placed. Cronquist (1981, 1988) and Heywood (1993, 1998) placed the family in Euphorbiales, whereas Takhtajan (1987, 1997) placed Buxaceae with the hamamelids close to Myrothamnaceae, the latter family now known to be sister to Gunneraceae. Flowers of Buxaceae are characterized by considerable variability in organ number and differentiation (von Balthazar and Endress 2002a, 2002b). Flowers are unisexual with a perianth of small, inconspicuous tepals (Figure 5.3H). Most staminate Buxaceae are dimerous with two stamen whorls, although pistillate flowers are either trimerous or dimerous. The weakly differentiated perianth is spiral in pistillate flowers, whereas two whorls are present in staminate flowers (von Balthazar and Endress 2002a, 2002b).

Didymelaceae consist of the single genus *Didymeles*, with two species from Madagascar, and have long been considered taxonomically isolated. The plants are dioecious, and the staminate flowers lack a perianth; pistillate flowers sometimes have one to four minute scales. It is unclear, however, whether these scales represent perianth parts (von Balthazar et al. 2003). Cronquist (1981) placed the family in its own order in Hamamelidae. Stebbins (1974) also placed it in Hamamelidae, near Hamamelidaceae. Thorne (1976) considered it part of Euphorbiales, whereas Takhtajan (1997) placed it in its own order, together with Buxales and Simmondsiales, in his superorder Buxanae.

Buxaceae and Didymelaceae share distinctive steroid alkaloids, similar wood and leaf anatomy (Sutton 1989; Takhtajan 1997), and a simple bract-like perianth (Figure 5.3H; Nandi et al. 1998; Hoot et al. 1999; von Balthazar and Endress 2002a, 2002b; von Balthazar et al. 2003). Both families also have encyclocytic stomata, a rare feature in the angiosperms (see Metcalfe and Chalk 1988/1989) that also unites Aextoxicaceae and Berberidopsidaceae (see Chapter 6).

Trochodendraceae

Trochodendron and *Tetracentron* have long been considered to occupy a pivotal position in angiosperm evolution. They possess features long considered ancestral such as ethereal oil cells and chloranthoid leaf teeth. However, although the idioblasts in *Tetracentron* look like oil cells, they have not been studied in detail (Endress 1986b), and these oil cells may not be homologous to those of certain basal angiosperms. Cronquist (1981, 1988) considered the two genera (each placed in its own family, Trochodendraceae and Tetracentraceae, respectively) to be basal in Hamamelidae and connecting links between Magnoliidae and the rest of his Hamamelidae. The two genera also lack vessels; although usually considered a plesiomorphic feature, this trait is now interpreted as secondarily derived in these genera (Young 1981; Doyle and Endress 2000; see Chapter 3). Phylogenetic analyses that used non-DNA characters provided support for *Trochodendron* and *Tetracentron* as sister taxa (Crane 1989; Hufford and Crane 1989); Crane (1989) placed the families together in one family, Trochodendraceae. These same studies also indicated that Trochodendraceae are sister to all other eudicots (Crane 1989; Hufford and Crane 1989). Molecular data similarly strongly supported a sister-group relationship between *Trochodendron* and *Tetracentron* (e.g., Chase et al. 1993; Drinnan et al. 1994; D. Soltis et al. 1997a, 2000; Hoot et al. 1999; Savolainen et al. 2000a, 2000b), and the single family Trochodendraceae has been recognized (APG 1998; APG II 2003). These same molecular studies placed Trochodendraceae either alone as sister to the core eudicots or in a clade with Buxaceae as sister to the core eudicots (see above). Thus, the relationship of Trochodendraceae to Buxaceae remains uncertain (Figure 5.2).

Character Evolution in Early-Diverging Eudicots

Families now recognized as early-diverging eudicots were long considered to represent some of the most primitive angiosperms (e.g., Cronquist 1968, 1981,1988; Takhtajan 1969; Stebbins 1974; Dahlgren 1980; Thorne 1992a, 1992b). The trends in floral evolution observed in these plants were considered important in linking floral characters in basal angiosperms (Magnoliidae) with those of the remaining subclasses of dicots (Cronquist 1981, 1988).

Floral organization and development are considered "open" and highly labile in basal angiosperms (Endress 1987a, 1987b, 1990a, 1990b, 1994a, 1994c; see Chapter 3). In contrast, in most core eudicots, numbers of floral parts are low (i.e., 4 or 5) and fixed, with floral organs arranged in whorls, indicating that the basic floral bauplan became canalized during the early diversification of the eudicots (Albert et al. 1998; Zanis et al. 2003). The early-diverging eudicots have also been considered labile in their basic floral development and more similar to basal angiosperms than to core eudicots in their arrangement and number of floral parts (Endress 1987a, 1987b, 1994a, 1994c). For example, dimerous, trimerous, tetramerous, and pentamerous floral organizations are all found in the early-diverging eudicots, as are both spiral and whorled phyllotaxis (see Figure 5.3 and "Merosity [Merism]," below; see also Chapter 12).

In contrast to intensive efforts to infer angiosperm phylogeny, surprisingly little effort has been made to use DNA-based topologies to examine floral evolution in an explicitly phylogenetic context. This is especially true of early-diverging lineages of angiosperms, which might be expected to yield critical insights into the significance of the early radiation and success of flowering plants (but see Albert et al. 1998; Doyle and Endress 2000; Zanis et al. 2003). Several authors addressed character evolution in the context of the phylogeny (e.g., Drinnan et al. 1994; Hoot et al. 1999) but did not conduct explicit character mapping. Drinnan et al. (1994) suggested that (1) there was plasticity in floral form in the early-diverging eudicots, with frequent transitions between dimerous and trimerous forms; (2) a cyclic (whorled) floral plan is not only common in the early-diverging eudicots but also represents the basic formula for the eudicots as a whole; and (3) the helical (spiral) arrangement observed in some early-diverging eudicots (e.g., some Ranunculaceae) is "almost certainly secondary."

The mapping of floral traits by Albert et al. (1998) and Zanis et al. (2003) provided some support for the hypothesized floral lability of early-diverging eudicots, but these studies focused primarily on basal angiosperms and each family of eudicots was represented by a summary floral description rather than by individual genera. To evaluate hypotheses of floral evolution in the early diversification of the eudicots, we mapped floral characters onto a phylogenetic tree for eudicots. We focused on perianth, androecial, and gynoecial phyllotaxis and merosity, and on perianth differentiation in an effort to provide more explicit hypotheses regarding floral evolution in these plants. We constructed phylogenetic trees using the three- and four-gene topologies of D. Soltis et al. (2000) and Kim et al. (2004). Additional representatives of Ranunculaceae, Papaveraceae, Lardizabalaceae, and other Ranunculales were added to the backbone of the early-diverging eudicot topology to reflect relationships within these subclades (e.g., Drinnan et al. 1994; Hoot 1995; Hoot and Crane 1995; Hoot et al. 1995a, 1995b, 1997). We experimented with different reconstructions of relationships between Proteales and Sabiaceae (either as sister taxa or as successive sisters to remaining eudicots; see above). We conducted similar alternative reconstructions for Trochodendraceae and Buxaceae. Furthermore, although we typically illustrate Eupteleaceae followed by Papaveraceae as sister to other Ranunculales, we also explored the alternative of Papaveraceae followed by Eupteleaceae as successive sisters to other Ranunculales. A list of genera used as placeholders, as well as the family and order (if placed to order) designations for each genus (*sensu* APG II 2003), is provided in Table 5.1.

TABLE 5.1 Taxa used in character-state reconstructions (arranged by family in alphabetical order), with family and ordinal designations (following APG II 2003; see also Chapter 10)

Order (if assigned)	Family	Genus
—	Amborellaceae	*Amborella*
Magnoliales	Annonaceae	*Asimina*
Piperales	Aristolochiaceae	*Asarum*
Austrobaileyales	Austrobaileyaceae	*Austrobaileya*
—	Buxaceae	*Buxus*
—	Buxaceae	*Didymeles*
—	Buxaceae	*Pachysandra*
Laurales	Calycanthaceae	*Calycanthus*
Canellales	Canellaceae	*Cinnamodendron*
—	Ceratophyllaceae	*Ceratophyllum*
—	Chloranthaceae	*Chloranthus*
—	Chloranthaceae	*Hedyosmum*
Ranunculales	Eupteleaceae	*Euptelea*
Gunnerales	Gunneraceae	*Gunnera*
Laurales	Lauraceae	*Cinnamomum*
Magnoliales	Magnoliaceae	*Magnolia*
Ranunculales	Menispermaceae	*Menispermum*
Ranunculales	Menispermaceae	*Tinospora*
Magnoliales	Myristicaceae	*Myristica*
Gunnerales	Myrothamnaceae	*Myrothamnus*
Proteales	Nelumbonaceae	*Nelumbo*
—	Nymphaeaceae	*Cabomba*
—	Nymphaeaceae	*Nymphaea*
Ranunculales	Papaveraceae	*Dicentra*
Ranunculales	Papaveraceae	*Eschscholzia*
Ranunculales	Papaveraceae	*Hypecoum*
Ranunculales	Papaveraceae	*Platystemon*
Ranunculales	Papaveraceae	*Pteridophyllum*
Ranunculales	Papaveraceae	*Sanguinaria*
Piperales	Piperaceae	*Piper*
Proteales	Proteaceae	*Placospermum*
Proteales	Proteaceae	*Roupala*
Proteales	Platanaceae	*Platanus*
Ranunculales	Ranunculaceae	*Caltha*
Ranunculales	Ranunculaceae	*Coptis*
Ranunculales	Ranunculaceae	*Glaucidium*
Ranunculales	Ranunculaceae	*Ranunculus*
Ranunculales	Ranunculaceae	*Xanthorhiza*
—	Sabiaceae	*Meliosma*
—	Sabiaceae	*Sabia*
Austrobaileyales	Schisandraceae	*Illicium*
—	Trochodendraceae	*Tetracentron*
—	Trochodendraceae	*Trochodendron*
Canellales	Winteraceae	*Takhtajania*

We obtained floral data from Drinnan et al. (1994), Zanis et al. (2003), Ronse De Craene et al. (2003), and primary references therein. We reconstructed evolutionary history under the criterion of maximum parsimony as implemented in MacClade (v. 3.05, 3.07, or 4.0; Maddison and Maddison 1992). We standardly depict reconstructions using the "all most parsimonious states" trace option in MacClade, but we also used ACC-TRAN optimization in which evolutionary changes are "accelerated" or hypothesized to occur early in evolutionary history and DELTRAN optimizations in which evolutionary changes are "delayed."

Perianth phyllotaxis

In the analysis reported here, we scored perianth phyllotaxis as a binary character—spiral versus whorled. Considering all angiosperms, most of the earliest-branching lineages (Amborellaceae and Austrobaileyales, but not Nymphaeaceae; see Chapter 3) exhibit a spiral perianth. Above Amborellaceae, Nymphaeaceae, and Austrobaileyales, the whorled perianth is reconstructed as ancestral for all other angiosperms (see Chapter 3). Multiple reversals to a spiral perianth occurred in basal angiosperm lineages (e.g., Calycanthaceae, Monimiaceae, Gomortegaceae, some Magnoliaceae). A whorled perianth is reconstructed as ancestral for the eudicots, and a reversal to a spiral perianth has potentially occurred in *Nelumbo* (Williamson and Schneider 1993b; Hayes et al. 2000) and in some Ranunculales (Endress 1995b; reconstruction not shown; see Chapter 3). We coded perianth phyllotaxis in *Nelumbo* as spiral according to current interpretations (Williamson and Schneider 1993b; Hayes et al. 2000). Several early-diverging eudicots have lost a perianth, including *Trochodendron, Euptelea,* and *Didymeles*. It is noteworthy, however, that in *Trochodendron* very young flowers have, in addition to two lateral prophylls, minute rudiments of organs outside of the androecium in apparently irregular arrangement; these organs may represent perianth remnants (see "Merosity (Merism)," below; Endress 1986b). We also conducted multistate codings of perianth phyllotaxis (e.g., one whorl, two whorls, three whorls, multiple whorls), and these reconstructions (not shown) similarly show a dynamic picture of perianth evolution in basal angiosperms and early-diverging eudicots (Albert et al. 1998; Ronse De Craene et al. 2003).

Androecial and gynoecial phyllotaxis

As with perianth phyllotaxis (see preceding section), we coded androecial and gynoecial phyllotaxis as both binary and multistate characters. Stamen phyllotaxis can be complex and difficult to interpret (see also Chapters 3 and 12). For example, in Ranunculaceae and *Trochodendron,* stamen phyllotaxis may be both spiral and whorled (see Chapter 12; Endress 1987a, 1987b, 1990b, 1995b).

A whorled androecium (Figure 5.6) and whorled gynoecium are ancestral for the eudicots. The reconstruction of gynoecial evolution is not shown, but it mirrors that provided for androecial diversification. Our mapping studies indicate (Figure 5.6) that spirally arranged stamens and carpels have been derived independently within the early-diverging eudicot lineages Ranunculaceae, *Kingdonia* (Circaeasteraceae), and Nelumbonaceae. Our codings for *Kingdonia* follow Kosuge and Tamura (1989; see also Drinnan et al. 1994), who considered the floral architecture to be based on a "helical" plan. The taxa of Menispermaceae included here have whorled phyllotaxis, but some members of the family (e.g., *Hypserpa*) have spiral androecial phyllotaxis (Endress 1995b). Flowers in Ranunculaceae exhibit great variety in form (see also Chapter 12). It has been suggested that numerous spirally arranged parts may be ancestral for the family (Tamura 1965; see also Cronquist 1981, 1988), but our reconstructions are equivocal as to the ancestral state of the family. Almost all other members of Ranunculales have a whorled androecium and gynoecium, and in binary character-state reconstructions, the ancestral state throughout Ranunculales is whorled, with the ancestral state of Ranunculaceae ambiguous. *Hydrastis* and *Glaucidium* are sisters to the remainder of Ranunculaceae, and both have whorled phyllotaxis. Flowers of *Glaucidium* have numerous stamens that are sometimes misinterpreted as spirally arranged, although they are actually irregular (C. Wagner and P.K. Endress, unpubl.). Hence, whorled phyllotaxis could be ancestral for the family, as suggested by Drinnan et al. (1994). There may be multiple derivations of spiral stamen and gynoecial phyllotaxis just within Ranunculaceae, but more detailed phylogenetic analyses of the family are required to evaluate this hypothesis (Endress 1987a, 1987b, 1995b).

Ranunculales have been considered to represent an early lineage of angiosperms with features also seen in Magnoliales, such as numerous stamens and carpels (Cronquist 1968, 1988; Takhtajan 1969, 1991, 1997; see also Chapter 3). Coding androecial merosity as a multistate character allows us to evaluate whether numerous stamens in these taxa are ancestral or derived. We follow the definitions of monocycly and polycycly provided by Ronse De Craene and Smets (1998a, 1998b). Our reconstructions indicate that polycyclic androecia are derived, not ancestral, in the early-diverging eudicots. Above the monocot-*Ceratophyllum* clade, a monocyclic androecium is reconstructed as ancestral for the remaining angiosperms, but this reconstruction is somewhat tenuous. Chloranthaceae were scored as monocyclic (flowers of the family possess only a single stamen); the ancestral state for the magnoliid clade was reconstructed as monocyclic, although few members of the clade actually possess such androecia (*Piper* was scored as

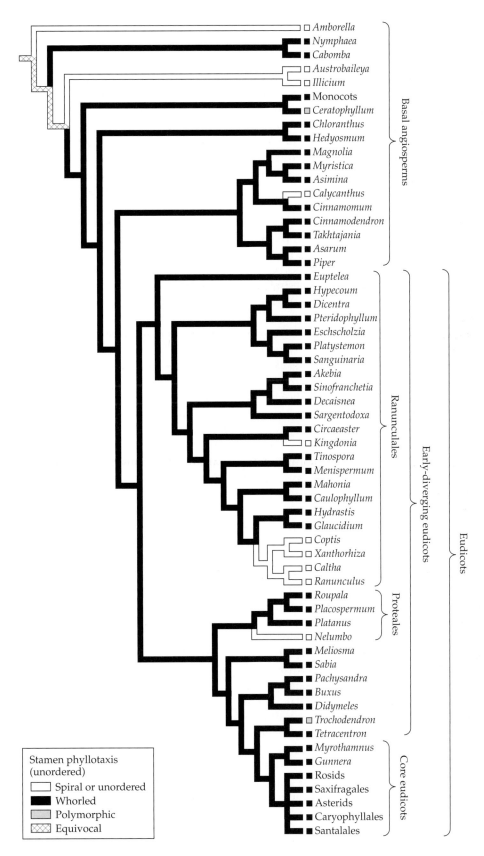

FIGURE 5.6 Reconstruction of the diversification of stamen phyllotaxis in early-diverging eudicots.

polymorphic, having both dicycly and monocycly). When we scored *Pteridophyllum* as equivocal for stamen phyllotaxis (it has not been studied in detail developmentally), the ancestral state for Ranunculales is unclear. However, if *Pteridophyllum* is scored as having a monocyclic androecium (Drinnan et al. 1994), then monocycly is reconstructed as ancestral for Ranunculales.

Within Ranunculales, a polycyclic androecium has been derived many times, at least once each within Ranunculaceae and Papaveraceae. The derivation of a polycyclic androecium from a dicyclic ancestor within Papaveraceae is more clearly seen by adding representatives according to the topology of Hoot et al. (1997; tree not shown). The polymerous androecia and gynoecia of Papaveraceae reflect secondary multiplication (Drinnan et al. 1994; Kadereit et al. 1995; Hoot et al. 1997). Within Ranunculaceae, a polycyclic androecium, rather than spiral phyllotaxis, may be the ancestral state, although more representatives of the family are needed to evaluate this hypothesis. Furthermore, stamen phyllotaxis is so labile in the family that it may be extremely difficult to tease apart "spiral" versus "whorled" within Ranunculaceae. For example, in *Actaea alba*, stamen phyllotaxis is normally whorled but sometimes is spiral; both spiral and whorled stamen phyllotaxis are also present within *Anemone* (Schöffel 1932).

Merosity (merism)

Merosity, or merism, is highly labile in early-diverging eudicots (Figures 5.7 and 5.8), a result similar to the lability in merosity reconstructed for basal angiosperms (see Chapter 3). A dimerous perianth has played a major role in floral evolution within the early-diverging eudicots (Figures 5.7 and 5.8, A to C). The extent of this role depends in part on the coding of Proteaceae (see following paragraphs) and the optimization method used. In ACCTRAN optimizations, the dimerous perianth is ancestral for all eudicots (not shown). With other approaches to tracing character evolution (DELTRAN, "all most parsimonious states"), dimery still plays a prominent role in the evolution of early-diverging eudicots, but is not reconstructed as ancestral for the entire eudicot clade; instead, it is ancestral for eudicots above the node to Ranunculales (Figure 5.7). A dimerous perianth is prevalent throughout Papaveraceae, with an apparent transition to trimery occurring in *Platystemon*. A dimerous perianth may also be ancestral for Ranunculaceae (Drinnan et al. 1994), although reconstruction of the ancestral state of Ranunculaceae is equivocal. *Hydrastis* and *Glaucidium* are both dimerous genera that are sister to the remainder of the family, which in our analyses is a clade (*Coptis*, *Xanthorhiza*, *Caltha*, and *Ranunculus*) of predominantly pentamerous taxa. This same result is obtained in other mapping studies (not shown) that use additional genera of Ranunculaceae.

The perianth in Proteaceae has been interpreted as tetramerous by some (Drinnan et al. 1994), but phyllotactically the four tepals "could represent two dimerous whorls of successive primordia" (Douglas and Tucker 1996a, 1996b). The patterns of perianth phyllotaxis in the family appear similar to those observed in Buxaceae and Papaveraceae, which have dimerous (opposite/decussate) arrangements of primordia (Douglas and Tucker 1996a, 1996b). We conducted reconstructions coding Proteaceae as either dimerous or tetramerous. If Proteaceae are tetramerous, then they represent one of the few instances in which this arrangement has evolved in the early-diverging eudicots. Some Ranunculaceae are tetramerous. Platanaceae, the sister group to Proteaceae, may also be tetramerous (Figure 5.7). Early fossils of Platanaceae are tetramerous, whereas other reports have indicated pentamerous merosity (Manchester 1986; Friis et al. 1988; Crane et al. 1993; Magallón-Puebla et al. 1997); we used both conditions in various reconstructions.

If Proteaceae are scored as tetramerous (e.g., Drinnan et al. 1994), then a dimerous perianth is reconstructed as the ancestral condition for Trochodendraceae. *Tetracentron* possesses a dimerous perianth (Figure 5.8C), and a perianth appears to be absent from its sister group, *Trochodendron*. However, rudiments of what may be a perianth are present in young flowers of *Trochodendron* (Figure 5.8D), although it is still unclear if these structures represent perianth parts (Endress 1986b). A dimerous perianth may also be ancestral for Buxaceae (Figure 5.8A). Reconstruction of the ancestral state for Buxaceae is confounded by the uncertain perianth condition in *Didymeles* (see above) (von Balthazar et al. 2003) and female flowers in some Buxaceae, which are not dimerous (von Balthazar and Endress 2002a, 2002b). We scored the merosity of *Didymeles* as uncertain; in our reconstruction the ancestral condition for the entire family is dimerous (Figure 5.7). Interpreting the merosity of Proteaceae as dimerous (Figure 5.8B; Douglas and Tucker 1996a, 1996b) has a major impact; the ancestral state of the branch to all eudicots except Ranunculales is then reconstructed as dimerous (using the "all most parsimonious states" optimization option; Figure 5.7). However, these reconstructions are hampered by equivocal interpretations of "perianth" in some lineages, such as Buxaceae, Trochodendraceae, and Myrothamnaceae. It is unclear whether the apparent perianth in these groups is actually a perianth or an assemblage of bracts, and the relationship between bracts and tepals is uncertain (see von Balthazar and Endress 2002a for discussion of Buxaceae).

Regardless of the coding of Proteaceae, our reconstructions indicate that a dimerous perianth may be the immediate precursor to the pentamerous condition characteristic of core eudicots. Significantly, a dimerous

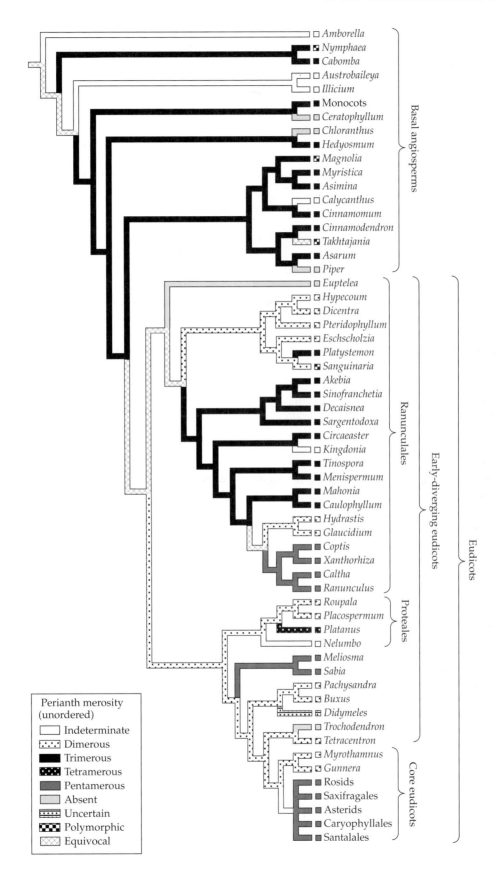

FIGURE 5.7 Reconstruction of the diversification of perianth merosity (merism) in early-diverging eudicots.

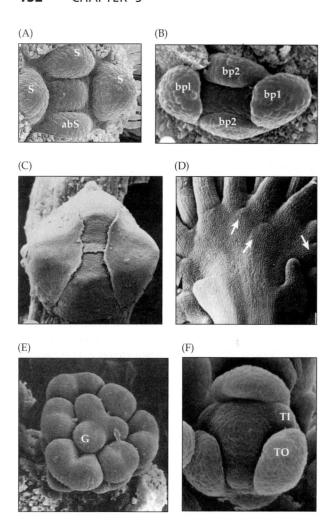

(A)

(B)

(C)

(D)

(E)

(F)

FIGURE 5.8 SEM micrographs illustrating perianth merosity of selected early-diverging eudicots. (A) *Persoonia myrtilloides* (Proteaceae), dimerous; S = stamen; abS = abaxial stamen (from Douglas and Tucker 1996b). (B) *Buxus balearica* (Buxaceae), dimerous; bp = bract-like phyllome (from von Balthazar and Endress 2002b). (C) *Tetracentron sinense* (Trochodendraceae), dimerous (from Endress 1986b). (D) *Trochodendron aralioides*, lateral flower from abaxial side showing two prophylls and rudimentary tepals (arrows; from Endress 1986b). (E) *Cocculus laurifolius* (Menispermaceae), trimerous; G = gynoecium (from Ronse De Craene and Smets 1993); (F) *Decaisnea fargesii* (Lardizabalaceae), trimerous; TI = inner tepal; TO = outer tepal (from Ronse De Craene and Smets 1993).

perianth is also found in Gunneraceae and perhaps Myrothamnaceae (Gunnerales), a clade of core eudicots (D. Soltis et al. 2000; Savolainen et al. 2000a). Recent phylogenetic analyses involving four genes indicated with moderate support that Gunnerales are sister to all core eudicots (see discussion of Gunnerales in Chapter 6). The canalization of merosity that yielded the pentamerous perianth typical of core eudicots occurred just after the node to Gunnerales.

As in the basal angiosperms (see Chapter 3), a trimerous perianth has also played a prominent evolutionary role in the early-diverging eudicots. Our reconstructions indicate that a trimerous perianth is the ancestral state for most members of Ranunculales (Figure 5.7). In these reconstructions, a trimerous perianth evolved in the ancestor of the clade containing Lardizabalaceae, Circaeasteraceae, Menispermaceae, Berberidaceae, and Ranunculaceae. Trimery characterizes all of these families (Figure 5.8, E and F) except Ranunculaceae and Circaeasteraceae. Ranunculaceae are highly variable in merosity, and trimerous, pentamerous, or dimerous flowers are common and occur in different genera (Salisbury 1919; Schöffel 1932; Endress 1987b, 1995b; Ronse De Craene and Smets 1993, 1995; Drinnan et al. 1994). Following the evolution of trimery in this clade, a transition to a dimerous or pentamerous perianth subsequently occurred in the ancestor of Ranunculaceae (Figure 5.7). Some members of Ranunculaceae (e.g., *Anemone* and *Clematis*) are trimerous, but this appears to be derived rather than ancestral within the family (see also Endress 1995b). An additional modification of the perianth occurred in *Kingdonia* (Circaeasteraceae), which has a variable number of tepals and indeterminate merosity. When the "all most parsimonious states" option is used in optimizations, the ancestral state for Ranunculales is equivocal, with a dimerous perianth appearing in both Papaveraceae and *Hydrastis* + *Glaucidium* (Ranunculaceae). However, trimery appears to be ancestral for Lardizabalaceae, Circaeasteraceae, Menispermaceae, Berberidaceae, and possibly Ranunculaceae (see also Hoot et al. 1999).

Our reconstructions of floral evolution support Drinnan et al.'s (1994) suggestion that plasticity in floral form was prevalent in the early-diverging eudicots. They suggested that dimerous (opposite/decussate) and trimerous (ternate) arrangements are widespread in early eudicots and "likely primitive" and that the transition from one to the other may have occurred multiple times. The pentamerous perianth has arisen at least four times within the eudicots: once in Sabiaceae, once in the ancestor of all core eudicots following Gunnerales, at least twice within Ranunculaceae (S.B. Hoot, pers. comm.), and possibly in Platanaceae. It is noteworthy that the origin of pentamery in both the core eudicots and Ranunculaceae appears to have involved dimerous ancestors (Figure 5.7); the origin of pentamery in Sabiaceae is unclear. These results provide some support for Kubitzki (1987), who maintained that the origin of the pentamerous condition from the trimerous condition was unlikely. Ronse De Craene and Smets (1994) advocated that tetramery is derived from pentamery. Although the interpretation of merosity (tetramerous versus dimerous) in Proteaceae and Platanaceae has been debated, our results provide some support for this suggestion in that families of Proteales potentially have as their sister Sabiaceae, which are pentamerous.

These examples of character reconstructions again illustrate the importance of detailed ontogenetic data in the study of angiosperm evolution and diversification (see also Chapter 3 and the section on Saxifragales in Chapter 6). Such data are critical for differentiating among dimerous, trimerous, and pentamerous flowers that may have arisen in different ways (Drinnan et al. 1994). For example, whereas Buxaceae may superficially appear tetramerous, developmental studies demonstrate that the flowers consist of two dimerous whorls of tepals with the position of the stamens (opposite the tepals) arising via the subsequent production of two decussate pairs of stamens (Drinnan et al. 1994; von Balthazar and Endress 2002a). Proteaceae, in contrast, are more difficult to interpret. Drinnan et al. (1994) suggested that Proteaceae may be truly tetramerous and only superficially similar to Buxaceae. However, rigorous developmental studies indicate that flowers of members of Proteaceae may also consist of two dimerous whorls of tepals (Douglas and Tucker 1996a, 1996b); we have followed this dimerous interpretation (Figures 5.7 and 5.8A). Similarly, trimery may arise in several ways (Ronse De Craene and Smets 1994).

Developmental data also provide insights into the polymerous androecia and gynoecia of some Papaveraceae and Ranunculaceae. These data appear to reflect secondary multiplication (e.g., Endress 1987a, 1987b; Ronse De Craene and Smets 1993). The presence of high petal numbers also appears to be secondarily derived in these families. In *Sanguinaria* (Papaveraceae), for example, polypetaly appears to be a specialized condition that has arisen through the modification of stamen primordia (Lehman and Sattler 1993). The basic developmental plan for the genus is one of dimery, and we coded it as such (Figure 5.7).

Perianth differentiation

Our analyses, and those of others (Albert et al. 1998; Zanis et al. 2003), suggest that ancestral eudicots lacked differentiated perianth whorls (Figure 5.9). Within Ranunculales, the evolution of the perianth is highly dynamic. Using the MacClade "all most parsimonious states" optimization option, it is unclear whether the ancestor of Ranunculales possessed a differentiated or undifferentiated perianth. There are clear transitions between the two states within the clade, with Papaveraceae, Menispermaceae, and some Ranunculaceae having a differentiated perianth, whereas some Lardizabalaceae and Ranunculaceae have an undifferentiated perianth (Figure 5.3). With ACCTRAN optimization (not shown) the ancestor of Ranunculales is reconstructed as having a differentiated perianth, with multiple transitions to an undifferentiated perianth. Multiple tran-

sitions between a differentiated and undifferentiated perianth apparently occurred even within Lardizabalaceae. Using the "all most parsimonious states" option in MacClade, it is unclear whether the ancestor of Lardizabalaceae possessed a differentiated or undifferentiated perianth. However, with ACCTRAN optimization, the ancestor of the family is reconstructed as having a differentiated perianth with a subsequent transition to an undifferentiated perianth occurring in *Decaisnea* and *Akebia,* followed by a return to a differentiated perianth in *Sinofranchetia*. Thus, this family may be of particular interest for more detailed analyses of perianth diversification.

The common ancestor of Proteales, Sabiaceae, Trochodendraceae, Buxaceae, and the core eudicots lacked a differentiated perianth (Figure 5.9); our reconstructions indicate that a differentiated perianth evolved independently in Sabiaceae, Platanaceae, and the core eudicots. Thus, within the early-diverging eudicots, the evolution of a differentiated perianth was highly dynamic. When the basal angiosperms are also considered, many additional transitions from an undifferentiated to a differentiated perianth have occurred. This is illustrated in more detail in Chapter 3. The petals of angiosperms almost certainly reflect different developmental origins and are likely not homologous in all cases. In Ranunculaceae, petals are extremely similar to stamens in morphology and ontogeny (Hiepko 1965a; Erbar et al. 1999); petals may be derived from stamens in these cases. In other instances, petals may have evolved from undifferentiated tepals or from sepals (see Chapter 3; Endress 1994a, 1994c; Takhtajan 1997; Albert et al. 1998).

Habit

Ranunculales have been considered to be derived from Magnoliales and "primitively herbaceous" (Cronquist 1968; 1988; Takhtajan 1987, 1991). However, reconstructions of habit, coded as woody versus herbaceous, indicate that the woody habit is ancestral not only for basal angiosperms but also for the eudicots (Kim et al. 2004b). Furthermore, Ranunculales may also not be ancestrally herbaceous. The ancestral condition for Ranunculales depends on the topology used, as well as the trace option used with MacClade. In the three-gene topology (Hoot et al. 1999; D. Soltis et al. 2000), Papaveraceae are sister to the remainder of Ranunculales. With this topology and the "all most parsimonious states" option, the ancestral state of Ranunculales is equivocal; with ACCTRAN optimization, the ancestral state is the herbaceous habit. However, with the four-gene topology (Kim et al. 2004b), Eupteleaceae are sister to all other Ranunculales, and the ancestral condition for Ranunculales is the woody habit (Figure 5.10). Internal support for Papaveraceae as sister to other Ranunculales is low (53% jack-

FIGURE 5.9 Reconstruction of the evolution of perianth differentiation in early-diverging eudicots.

FIGURE 5.10 Reconstruction of the evolution of habit in early-diverging eudicots

knife), whereas the support for Eupteleaceae in this position is higher (78%). More sequence data are needed to discern with more confidence the root of Ranunculales and thus evolutionary patterns both across this clade and among early-diverging eudicots.

Floral Developmental Genetics

Floral initiation and development in the model eudicots *Arabidopsis* and *Antirrhinum* are controlled by many genes with specific functions. During flower development in *Arabidopsis*, the identity of floral organs is specified by at least three classes of homeotic genes, including the well-known A-, B-, and C-function genes (Bowman et al. 1989; Carpenter and Coen 1990; Schwarz-Sommer et al. 1990; Coen and Meyerowitz 1991; Bradley et al. 1993; see Chapter 3). Genetic and molecular evidence for the conservation of portions of the ABC model indicates that this is an ancient regulatory network, perhaps applicable to most angiosperms (Ma and dePamphilis 2000). However, modifications of components of the ABC model may have occurred in different lineages of angiosperms (Baum 1998; Ma and dePamphilis 2000). For example, the function of *APETALA3* (*AP3*) and *PISTILLATA* (*PI*) as B-class organ identity genes is not rigidly conserved across all flowering plants (Heard and Dunn 1995; Kramer and Irish 1999; Yu et al. 1999; D. Soltis et al. 2002b). Expression of *AP3* and *PI* orthologues in stamens in *Ranunculus* (Ranunculaceae), *Papaver,* and *Dicentra* (Papaveraceae) is conserved, but expression patterns in petals differ from those in core eudicots such as *Arabidopsis* (Kramer and Irish 1999). Furthermore, the expression of *AP3* and *PI* homologues extends into both whorls of the perianth in *Aquilegia,* and duplicate *AP3* and *PI* genes are differentially expressed in perianth and stamens of both *Aquilegia* and *Clematis* (Kramer et al. 2003).

AP3 has apparently undergone duplication in the eudicots. Basal angiosperms and early-diverging eudicots examined possess one type of *AP3*-like genes, whereas some core eudicots possess two types of *AP3*-like genes that belong to different clades. The gene in basal angiosperms and early-diverging eudicots has been referred to as "paleo*AP3*." The two genes from core eudicots are referred to by Kramer and Irish (1999, 2000) as eu*AP3* and *TM6,* with members of the latter clade exhibiting a conserved motif most like that found in paleo*AP3* (Figure 5.11). The duplication event that resulted in the eu*AP3* and *TM6* clades may have occurred somewhere in the early-diverging eudicots or at the base of the core eudicots (Kramer and Irish 2000; P. Soltis and Soltis 2004; Kim et al. 2004a; E.M. Kramer, pers. comm.). The duplication is absent from the basal angiosperms (e.g., Magnoliaceae, Piperaceae, Poaceae) and the first few nodes in the eudicots (Ranunculaceae, Papaveraceae, Buxaceae). However, the putative dupli-

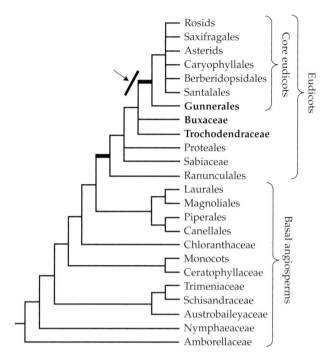

FIGURE 5.11 Possible phylogenetic location of putative *AP3* duplication in early-diverging eudicots. The branches leading to eudicots and core edicots, respectively, are in bold. Trochodendraceae, Buxaceae, and Gunnerales are indicated in bold font because the duplication may have occurred between the divergence of the basal eudicots Buxaceae and Trochodendraceae and the core eudicot Gunnerales (see text).

cation is present in the core eudicot lineages studied, including members of Caryophyllales (e.g., Caryophyllaceae) and several rosid (e.g., Brassicaceae, Rosaceae) and asterid (e.g., Solanaceae and Asteraceae) families. Both eu*AP3* and *TM6* were recently recovered from Saxifragales (Stellari et al. 2004; Kim et al. 2004a). Thus, the duplication may have occurred between Buxaceae and these major core eudicot clades. Gunneraceae, the sister to all other core eudicots (D. Soltis et al. 2003a), with plesiomorphic merosity possesses only a *TM6*-like sequence; eu*AP3* was not detected, raising the possibility of gene loss. Saxifragales, which may follow Gunnerales as sister to all other core eudicots, possess both *TM6* and eu*AP3* (Stellari et al. 2004; Kim et al. 2004a). If a duplication of *AP3* did occur in the early-diverging eudicots (or in the ancestor of core eudicots), then it may have been lost repeatedly.

Kramer and Irish (1999, 2000) proposed that this duplication may be of major evolutionary importance, reflecting "a de novo evolution of petals" at the base of the core eudicots. Thus, if this gene duplication has functional significance, then determining the timing of this duplication event in angiosperm phylogeny is critical for understanding the structure and evolutionary history of the eudicot flower. The presence or absence of this gene duplication should be determined in Trochodendraceae, a possible sister to the core eudicots.

A similar pattern of gene duplication has been demonstrated in another floral gene, *APETALA1* (*AP1*) (Litt and Irish 2003). The *AGAMOUS* subfamily of MADS box genes has also experienced duplication during the diversification of eudicots (Kramer et al. 2004). These duplications of floral genes may document a genome-wide duplication event that occurred during the origin or early diversification of the core eudicots (P. Soltis and Soltis 2004).

Future Research

The overall picture of relationships within the early-branching lineages of eudicots has been clarified. However, major questions remain, including the relationships between Sabiaceae and Proteales and also between Trochodendraceae and Buxaceae. Given that these relationships have not been clearly resolved with sequences from four genes and approximately 8,000 bp of sequence data, this will not be a simple task and may require the sequencing of many additional genes. Additional non-DNA characters would also be useful.

Perhaps the most important challenge offered by the early-diverging eudicots is the elucidation of the developmental genetics of floral diversification in these plants. Floral architecture remains labile in the early-diverging eudicots, with considerable variability in merosity as well as some variation in phyllotaxis. In addition, a differentiated perianth has arisen several times within the early-diverging eudicots, and the petals of these lineages do not appear to be homologous to each other or to those of core eudicots. To understand the genetic architecture of floral development, including the early diversification of the flower, data are needed for basal angiosperm lineages (see Chapter 3) and early-diverging eudicots. Such data will provide a critical link between what is known about derived model taxa such as *Arabidopsis*, *Antirrhinum*, and *Zea* relative to other angiosperms. Research at the interface of phylogenetics, developmental morphology, and developmental genetics is required to elucidate the nature of floral diversity in these plants.

6

Core Eudicots: Introduction and Smaller Lineages

Introduction

Within eudicots, Ranunculales, Proteales, Sabiaceae, Trochodendraceae, and Buxaceae are successive sisters (although their precise branching order is still unclear; Chapter 5) to a strongly supported core eudicot clade, which includes most angiosperms (Figure 6.1; P. Soltis et al. 1999b; Savolainen et al. 2000a, 2000b; D. Soltis et al. 2000). The circumscription of the core eudicots is not based on any morphological or other nonmolecular character, but on the strong internal support obtained for this clade in analyses of DNA datasets. Nonetheless, several key events seem to correspond fairly closely to the origin of the core eudicots, including the evolution of a highly synorganized flower (see Chapter 12), production of ellagic and gallic acids, and perhaps the duplication of several floral organ identity genes (see Chapter 5).

Strong internal support for the core eudicots was apparent in multigene analyses of two or three genes (Hoot et al. 1999; P. Soltis et al. 1999b; Savolainen et al. 2000a; D. Soltis et al. 2000). With two-gene analyses (*atpB* plus *rbcL*), support for the monophyly of the core eudicots was 91%; with three-gene analyses, jackknife support was 100% (D. Soltis et al. 2000). Even in phylogenetic analyses involving *rbcL* alone (Albert and Chase 1998; Savolainen et al. 2000a), weak jackknife support (58% and 72%, respectively) was apparent for the core eudicot clade.

mately requiring the sequencing of several additional genes for hundreds of taxa. Great success was realized in resolving relationships among basal angiosperms by constructing datasets of five or more genes for numerous taxa (Qiu et al. 1999; Zanis et al. 2002). The addition of the rapidly evolving plastid gene *matK* (Hilu et al. 2003) to the existing four-gene dataset may improve our understanding of eudicot relationships. On the other hand, a starburst may accurately depict relationships among major lineages of core eudicots—that is, they might in fact represent an ancient, rapid radiation. Fossil evidence provides some support for a rapid radiation of core eudicots in that the minimum age of mesofossils reported for Saxifragales, Gunnerales, rosids, and asterids are comparable (89.5 Mya; Magallón et al. 1999; see Table 2.1).

This chapter reviews the smaller lineages of core eudicots—Gunnerales, "Berberidopsidales," Saxifragales, and Santalales. This circumscription is for convenience only and in no way implies that these clades are more closely related to each other than to the other lineages of core eudicots. Subsequent chapters provide coverage of the larger clades of core eudicots: Caryophyllales (and Dilleniaceae) (Chapter 7), rosids (Chapter 8), and asterids (Chapter 9).

Gunnerales

Several molecular analyses have indicated that *Gunnera* (Gunneraceae) and *Myrothamnus* (Myrothamnaceae) form a clade. Weak support for this relationship first appeared in analyses that used *rbcL* and *rbcL* plus *atpB* sequences (Savolainen et al. 2000a, 2000b). Somewhat stronger support (75% jackknife) was provided for this clade in the three-gene analysis (D. Soltis et al. 2000). Stronger support (85%) was provided for Gunneraceae + Myrothamnaceae in analyses of a combined four-gene dataset of more than 200 eudicots (D. Soltis et al. 2003a). Recent analysis of a large *matK* dataset provided strong (100%) bootstrap support for Gunneraceae + Myrothamnaceae (Hilu et al. 2003). This Gunneraceae + Myrothamnaceae family pair was also strongly excluded from all other eudicot clades, providing support for the isolated status of this clade (see D. Soltis et al. 2000). As a result, it seemed most appropriate to treat these two families as an order, Gunnerales (APG II 2003).

Although molecular data indicate a close relationship between Gunneraceae and Myrothamnaceae, the two families exhibit different morphologies and have not previously been considered closely related. We have therefore retained them as distinct families rather than including *Gunnera* and *Myrothamnus* in an expanded Gunneraceae—a possible treatment suggested by APG II (2003). In fact, all modern classifications have considered the two families to be distantly related (e.g.,

Dahlgren 1980; Cronquist 1981; Thorne 1992a, 1992b, 2001; Takhtajan 1997). Cronquist (1981) and Takhtajan (1997) both placed Myrothamnaceae in Hamamelidae and Gunneraceae in Rosidae, although their respective placements of Gunneraceae within the rosids differed. Phylogenetic analysis of non-DNA characters (Nandi et al. 1998) did not indicate a close relationship between the two families, but instead placed Myrothamnaceae with early-diverging eudicots (i.e., Buxaceae, Proteaceae, and Platanaceae), whereas Gunneraceae appeared as part of the asterid clade.

The two species of *Myrothamnus* are small xerophytic shrubs (Kubitzki 1993). *Gunnera* (40 species) consists of herbs with short, upright stems (illustrated in Figure 6.4A). The flowers of *Myrothamnus* are unisexual and apetalous, with up to four scales interpreted variously as sepals or tepals or bracts (Jäger-Zürn 1966; Cronquist 1981; Drinnan et al. 1994; Takhtajan 1997). Staminate flowers have three to eight stamens, and pistillate flowers have a gynoecium of three or four united carpels. In *Gunnera* the flowers are often unisexual. Typically, the basal flowers of the inflorescence are pistillate, the upper flowers staminate, and the middle flowers bisexual, although all flowers of an inflorescence may be either unisexual or bisexual. Floral parts are in twos, with two sepals, two petals, one or two stamens, and two carpels. *Gunnera* is also of interest in that plants host nitrogen-fixing cyanobacterial symbionts in vegetative parts; *Gunnera* therefore represents an independent evolution of symbiotic nitrogen fixation in the angiosperms (see Chapters 8 and 11). Relationships within *Gunnera* have been resolved using gene sequence data (Wanntorp et al. 2002).

The relationships of *Gunnera* and *Myrothamnus* to other angiosperms were unclear in initial molecular analyses. Some single-gene studies indicated a sister-group relationship of *Gunnera,* or of *Gunnera* + *Myrothamnus* (*Myrothamnus* was not included in some of the initial *rbcL* analyses), to all other core eudicots (e.g., one of the two analyses of Chase et al. 1993; D. Soltis et al. 1997a; Savolainen et al. 2000a). Analysis of *atpB* sequence data alone (Savolainen et al. 2000a) placed (*Gunnera* + *Myrothamnus*) + Dilleniaceae as sister to all core eudicots. Analysis of combined two- and three-gene datasets placed *Gunnera* + *Myrothamnus* as sister to all core eudicots, but without support of 50% or higher (D. Soltis et al. 1998, 2000; Hoot et al. 1999; Savolainen et al. 2000a). Moderate support (84%) for the position of Gunnerales as sister to all other core eudicots was realized in analyses of a four-gene dataset for eudicots (D. Soltis et al. 2003a). Analysis of *matK* sequences also placed Gunnerales as sister to other core eudicots, albeit without support above 50% (Hilu et al. 2003). When *matK* sequences are combined with the existing four-gene dataset for eudicots, support for this position of Gunnerales will likely increase.

The position of Gunnerales as sister to all other eudicots has important implications for floral evolution. As reviewed in Chapter 5, a dimerous or trimerous perianth is frequently encountered in early-diverging eudicots. In contrast, the pentamerous condition predominates in core eudicots. However, a dimerous perianth also is present in *Gunnera;* it is not clear whether *Myrothamnus* possesses a perianth (von Balthazar and Endress 2002a; see Chapter 5). All reconstructions indicate that the canalization of merosity that yielded the pentamerous perianth typical of core eudicots occurred following the divergence of Gunnerales from the remaining core eudicots (D. Soltis et al. 2003a; P. Soltis and Soltis 2004; Chapter 5).

"Berberidopsidales"

This clade consists of two small families, Berberidopsidaceae and Aextoxicaceae, considered distantly related in recent classifications, but strongly supported as sister taxa in molecular phylogenetic investigations (Savolainen et al. 2000a; D. Soltis et al. 2000, 2003a). The order Berberidopsidales was not recognized in APG II (2003) pending additional investigation. The strong support (100% jackknife) for the clade, combined with its phylogenetic isolation from other eudicots in a recent four-gene analysis (D. Soltis et al. 2003a) indicates, however, that the clade merits designation as a distinct order. As noted in APG II (2003), the name Berberidopsidales is available, and we will use this name in this chapter.

Berberidopsidaceae are treated either as a single genus, *Berberidopsis,* or two genera, *Berberidopsis* and *Streptothamnus* (Takhtajan 1997). This family of scandent vines and shrubs displays a transPacific disjunction: *Berberidopsis* from temperate South America (Chile) and Australia and *Streptothamnus* from southeastern Australia. Some authors placed *Berberidopsis* (including *Streptothamnus*) in Flacourtiaceae (e.g., Gilg 1925; Cronquist 1981; see Chapter 8). Takhtajan (1987, 1997) removed *Berberidopsis* and *Streptothamnus* from Flacourtiaceae on the basis of wood anatomy (Miller 1975; Baas 1984) and floral morphology (van Heel 1984), but still considered Berberidopsidaceae closely allied with the formerly recognized Flacourtiaceae (see Chapter 8).

Aextoxicaceae consist of a single species of tree, also from Chile. The family was considered closely related to Euphorbiaceae (e.g., Bentham 1878, 1880). Although Euphorbiaceae and *Aextoxicon* do share similar pollen features, they differ in many other respects. *Aextoxicon* is distinctive in having apotropous ovules and ruminate endosperm. Aextoxicaceae have often been placed in Celastrales (Hutchinson 1973; Cronquist 1981, 1988; Takhtajan 1997). However, Takhtajan stressed that *Aextoxicon* is a poorly known plant in need of more intensive study, and Thorne (1992a, 1992b) could not

place the family using morphology and went so far as to include *Aextoxicon* in a list of "taxa incertae sedis."

Only one obvious non-DNA feature unites Aextoxicaceae and Berberidopsidaceae. *Aextoxicon* differs from Berberidopsidaceae in aspects of wood anatomy (but see Carlquist 2003), its unisexual flowers with clearly distinct sepals and petals, and a drupaceous fruit. In contrast, *Berberidopsis* flowers are bisexual, sepals and petals are spirally arranged with a gradual transition from sepals to petals, and the fruit is a berry. Although the close relationship of Berberidopsidaceae and Aextoxicaceae is considered a "surprise" result of molecular systematics, the two families have encyclocytic stomata, a rare feature in the angiosperms (Metcalfe and Chalk 1988, 1999) and an apparent synapomorphy for them. This same feature also unites Buxaceae and Didymelaceae (now Buxaceae; APG II 2003), two families of early-diverging eudicots that are sisters in DNA studies, but that were considered distantly related in morphological classifications (see Chapter 5). This stomatal feature may therefore be a good indicator of relationship.

Berberidopsidales may represent an old eudicot lineage. Although the exact placement of the order remains unclear, the clade often appears near the base of the core eudicots. In some of the shortest trees obtained in the three-gene analysis (D. Soltis et al. 2000), Berberidopsidales were sister to the Saxifragales + rosid clade. In Savolainen et al. (2000b), the order appeared as sister to the rosids. In a four-gene analysis of eudicots, Berberidopsidales followed Gunnerales as sister to the rest of the core eudicot lineages (Santalales, Caryophyllales, Saxifragales, asterids, rosids; Figure 6.2); however, this placement received only weak support (54%; D. Soltis et al. 2003a). Clearly, additional gene sequence data will be needed to determine the placement of Berberidopsidales with more confidence.

Saxifragales

The composition of Saxifragales is one of the major surprises of molecular phylogenetic analyses of the angiosperms because this set of families had never before been associated with one another (Chase et al. 1993; Morgan and Soltis 1993; D. Soltis and Soltis 1997; D. Soltis et al. 1997a, 2000; Qiu et al. 1998; Hoot et al. 1999; P. Soltis et al. 1999b; Savolainen et al. 2000a, 2000b). Saxifragales include Altingiaceae, Aphanopetalaceae (included here in Haloragaceae), Cercidiphyllaceae, Crassulaceae, Daphniphyllaceae, Grossulariaceae, Haloragaceae, Hamamelidaceae, Iteaceae, Paeoniaceae, Peridiscaceae, Pterostemonaceae (which optionally could be included in Iteaceae; APG II 2003), and Saxifragaceae (Figure 6.3). This unexpected assemblage (representatives are illustrated in Figure 6.4, B to N) consists of taxa placed in three subclasses in recent classifica-

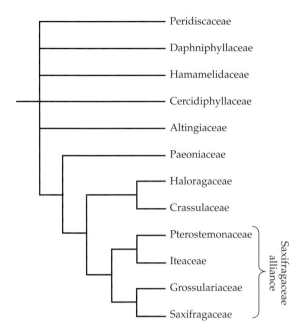

FIGURE 6.3 Summary tree of relationships in Saxifragales based in part on parsimony analyses of a combined five-gene dataset (Fishbein et al. 2001; Fishbein and Soltis 2004). Only relationships receiving bootstrap support of 50% and above are depicted. *Aphanopetalum* is included here in Haloragaceae.

tions. The circumscription of Saxifragales based on DNA studies therefore departs markedly from morphology-based classifications (e.g., Cronquist 1981; Takhtajan 1997; reviewed in Morgan and Soltis 1993; Qiu et al. 1998). Altingiaceae, Hamamelidaceae, Cercidiphyllaceae, and Daphniphyllaceae were previously placed in subclass Hamamelidae. Saxifragaceae, Iteaceae, Pterostemonaceae, Peridiscaceae, Grossulariaceae, Crassulaceae, and Haloragaceae, which now include Tetracarpaeaceae, Penthoraceae, and *Aphanopetalum* (the latter formerly placed in Cunoniaceae), were treated in Rosidae. Paeoniaceae were variously placed in subclasses Magnoliidae or Dilleniidae (reviewed in Cronquist 1981).

Analysis of *rbcL* sequences first revealed the Saxifragales clade (Chase et al. 1993; Morgan and Soltis 1993). Investigation of *atpB* alone, as well as *atpB* plus *rbcL*, also indicated a monophyletic Saxifragales, but none of these studies provided support of 50% or higher (Chase and Albert 1998; Savolainen et al. 2000a, 2000b). 18S rDNA alone and 18S rDNA plus *rbcL* revealed a Saxifragales clade with moderate bootstrap support (e.g., D. Soltis et al. 1997a, 1998; D. Soltis and Soltis 1997). The monophyly of Saxifragales received strong support in analyses of three-gene datasets (*rbcL*, 18S rDNA, *atpB*; Hoot et al. 1999; D. Soltis et al. 2000).

Molecular data also demonstrated that the initial diversification of Saxifragales was rapid (Fishbein et al. 2001) and further indicated that the early diversifica-

tion of Saxifragales was contemporaneous with the initial radiation of the major lineages of eudicots and of noneudicot (magnoliid and monocot) angiosperms. Molecular data therefore agree with the fossil record in that fossils attributed to Saxifragales are present in Turonian–Campanian strata (89.5 Mya; Table 2.1), which is comparable in age to the oldest fossils of core eudicots (Magallón et al. 1999).

Several members of Saxifragales were considered closely related in some previous classifications. Saxifragaceae, Grossulariaceae, Iteaceae, Pterostemonaceae, Penthoraceae, and Tetracarpaeaceae were previously considered part of a much more broadly defined Saxifragaceae s.l. (Engler 1930; reviewed in Morgan and Soltis 1993). A close relationship of these core families of Saxifragaceae s.l. to Crassulaceae was also proposed (e.g., Cronquist 1981; Takhtajan 1987, 1997). Hamamelidaceae, Altingiaceae, Cercidiphyllaceae, and Daphniphyllaceae were also considered closely related in modern classifications (e.g., the "lower hamamelids" *sensu* Walker and Doyle 1975). However, Saxifragales also include Haloragaceae and Paeoniaceae, families that have never been placed with any members of Saxifragales in previous classifications. Haloragaceae have been placed in or near the rosid order Myrtales (Cronquist 1981), whereas Paeoniaceae have been considered closely related to Magnoliaceae (e.g., Worsdell 1908), Ranunculaceae (e.g., Takhtajan 1997), or Dilleniaceae (e.g., Cronquist 1981). The inclusion of *Aphanopetalum* (previously placed in the rosid family Cunoniaceae) in Saxifragales initially seemed surprising but had been suggested previously based on anatomical data (Dickison et al. 1994). Davis and Chase (2004) demonstrated that Peridiscaceae, previously included in Flacourtiaceae in the rosid order Malpighiales (see Chapter 8), are part of Saxifragales, although the relationship of the family to other Saxifragales is unclear. Peridiscaceae consist of the small genera *Peridiscus, Soyauxia,* and *Whittonia* from tropical South America and West Africa.

The inability of classical systematics to reveal the close relationship among members of Saxifragales reflects the great morphological diversity encompassed by this small clade. Although including far fewer species than some of the other core eudicot clades, such as asterids, rosids, and Caryophyllales, Saxifragales exhibit similar levels of diversity in vegetative and reproductive morphology. Saxifragales include trees, shrubs, lianas, annual and perennial herbs, succulents, and aquatics. Flowers vary considerably in arrangement, merosity, degree of fusion of perianth parts, stamen and carpel number, ovary position, and degree of syncarpy (Figure 6.4, B to N; Cronquist 1981; Takhtajan 1997).

Because of the morphological diversity encompassed by Saxifragales, synapomorphies for the clade are not yet clear. The woody members are similar in their anatomy: solitary vessels, scalariform perfora-

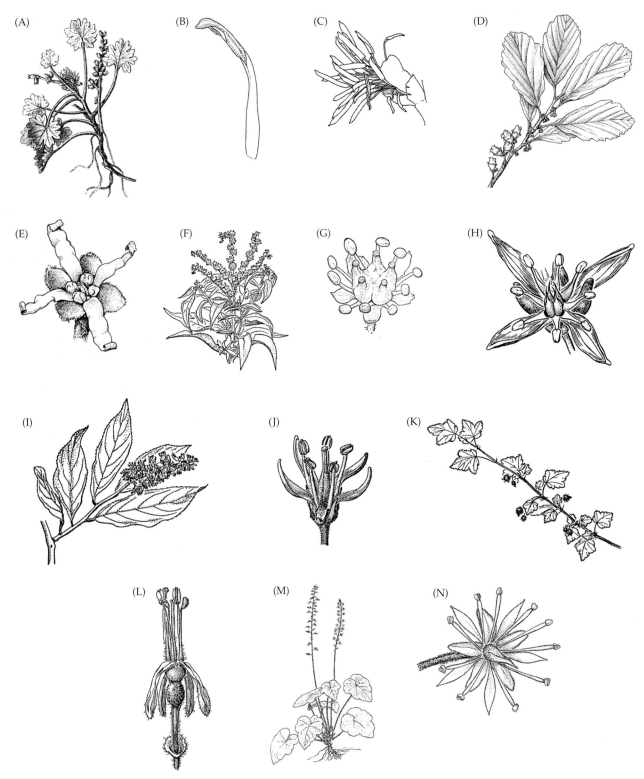

FIGURE 6.4 Flowers and plants representing Gunnerales (A) and Saxifragales (B–N). (A) *Gunnera monoica* (Gunneraceae). (B) *Cercidiphyllum japonicum* (Cercidiphyllaceae), pistillate flower. (C) *Cercidiphyllum japonicum* (Cercidiphyllaceae), staminate inflorescence. (D) *Hamamelis virginiana* (Hamamelidaceae), branch with flower buds and fruit. (E) *Hamamelis xvirginiana* (Hamamelidaceae), flower. (F) *Penthorum sedoides* (Haloragaceae; formerly Penthoraceae), plant in flower. (G) *Penthorum sedoides* (Haloragaceae; formerly Penthoraceae), flower. (H) *Sedum pulchellum* (Crassulaceae), flower. (I) *Itea virginica* (Iteaceae), branch in flower. (J) *Itea virginica* (Iteaceae), flower. (K) *Ribes cynosbati* (Grossulariaceae), branch in flower. (L) *Ribes rotundifolia* (Grossulariaceae), flower. (M) *Tiarella cordifolia* (Saxifragaceae), whole plant. (N) *Tiarella cordifolia* (Saxifragaceae), flower. (A, from Engler and Prantl 1898; B and C, from Endress 1986; D and E, from Wood 1974; F and G, from Spongberg 1972; H to N, from Spongberg 1972 and Wood 1974.)

tions, opposite to scalariform intervessel pits, nonseptate fibers with distinctly bordered pits, and apotracheal parenchyma (Baas et al. 2000). Exceptions to this suite of features occur, however, in Crassulaceae and Saxifragaceae, but in these primarily herbaceous families, woodiness probably arose secondarily (Carlquist 1988; Mort et al. 2001). A seed character represents a possible synapomorphy for a portion of the clade. An exotestal palisade with thickened outer walls (Nandi et al. 1998) is found in Saxifragaceae, *Ribes* (Grossulariaceae), Crassulaceae, and Paeoniaceae. However, Hamamelidaceae are characterized by mesotestal seeds (see Corner 1976 for a review of angiosperm seeds; see also Chapter 8, Box 1).

Whereas the composition of Saxifragales is now clear, the position of the clade among the core eudicots remains uncertain (Figures 6.1 and 6.2). Initial analyses of *rbcL* sequences (Chase et al. 1993) placed the order as sister to all rosids, whereas analyses of *atpB* sequences placed Saxifragales as sister to a large clade containing most core eudicots (Savolainen et al. 2000a). None of these placements received support above 50%. The three-gene analyses (P. Soltis et al. 1999b; D. Soltis et al. 2000) placed Saxifragales as sister to rosids, but with only weak jackknife support (60%). Analyses of a four-gene dataset for eudicots revealed a placement of Saxifragales as sister to all other core eudicots except Gunnerales and Berberidopsidales (D. Soltis et al. 2003a), but also without support above 50%. Analysis of *matK* placed Gunnerales sister to other core eudicots, followed by a trichotomy of Saxifragales, rosids, and remaining core eudicots (Santalales, asterids, Caryophyllales, Berberidopsidales; Hilu et al. 2003). Thus, Saxifragales may occupy a pivotal position near the base of the core eudicot radiation.

Relationships within Saxifragales

Phylogenetic analyses of a five-gene dataset identified two well-supported clades of families in Saxifragales (Figure 6.3; Fishbein et al. 2001; Fishbein and Soltis 2004). However, neither of these analyses included Peridiscaceae, which were only recently shown to be a member of Saxifragales (Davis and Chase 2004). Hence, our discussion of relationships within Saxifragales will focus on the remaining members of the clade. Peridiscaceae have been depicted as part of a polytomy that also includes Hamamelidaceae, Altingiaceae, Daphniphyllaceae, Cercidiphyllaceae, and a clade of Paeoniaceae as sister to (Haloragaceae + Crassulaceae) + the Saxifragaceae alliance (Figure 6.3).

Strong support was found for an alliance of Saxifragaceae and several woody members of the former Saxifragaceae s.l. (i.e., Grossulariaceae, Iteaceae, and Pterostemonaceae). Within this clade, *Ribes* (Grossulariaceae; see Senters and Soltis 2002 and Weigend et al. 2002 for phylogenetic analyses of *Ribes*) was the immediate sister to

Saxifragaceae (see Soltis et al. 2001b), which, in turn, is sister to a clade of Iteaceae + Pterostemonaceae.

A second strongly supported clade within Saxifragales consists of a well-supported Crassulaceae (Mort et al. 2001) as sister to a clade of *Tetracarpaea* (formerly Tetracarpaeaceae), *Penthorum* (formerly Penthoraceae), Haloragaceae, and *Aphanopetalum* (formerly of Cunoniaceae). *Tetracarpaea, Penthorum,* and *Aphanopetalum* are all small genera that should be included within an expanded family Haloragaceae (APG II 2003), and we have followed that treatment here (Figure 6.3). The Saxifragaceae alliance and Crassulaceae + Haloragaceae form a clade in the strict consensus of shortest trees, but this clade does not receive support higher than 50% in parsimony analyses and is only weakly supported (68% bootstrap) in maximum likelihood analyses (Fishbein et al. 2001; Figure 6.3). However, a more recent mixed-model Bayesian analysis has provided greater support (1.00 posterior probability) for this sister-group relationship (Fishbein and Soltis 2004).

The monophyly of Paeoniaceae, consisting of the single genus *Paeonia*, is strongly supported (e.g., Sang et al. 1997; Fishbein et al. 2001), but the position of Paeoniaceae within Saxifragales remains uncertain. In some parsimony analyses of multigene datasets, Paeoniaceae appear as sister to Haloragaceae + Crassulaceae. However, in maximum likelihood and Bayesian analyses, Paeoniaceae are sister, with low to sometimes strong support, to (Haloragaceae + Crassulaceae) + the Saxifragaceae alliance; this latter relationship is the one depicted in the summary tree (Figure 6.3). The relationships of Paeoniaceae have long been problematic. Cronquist (1981) placed Paeoniaceae in Dilleniales, considering the family transitional between magnoliids and the remainder of Dilleniidae (see also Corner 1946; Melchior 1964; Keefe and Moseley 1978). Takhtajan (1997), in contrast, placed the family near Ranunculaceae. Others considered *Paeonia* closely related to Magnoliaceae (e.g., Sawada 1971). The morphological characters of Paeoniaceae do not indicate a close relationship to most other Saxifragales. The family is polystemonous (i.e., having many stamens), a feature also found in some Hamamelidaceae (see below); Cercidiphyllaceae have perhaps 1 to 13 stamens, although delimitation of flowers is difficult. Paeoniaceae also possess an apocarpous gynoecium of more than two carpels, a feature shared only with Crassulaceae and *Tetracarpaea* of the Crassulaceae + Haloragaceae clade.

The relationships of Hamamelidaceae, Altingiaceae, Daphniphyllaceae, and Cercidiphyllaceae have not been clarified with strong support in analyses that used five genes (Fishbein et al. 2001; Fishbein and Soltis 2004). Hence, these families are depicted as part of a polytomy (Figure 6.3). Bayesian analyses provide varying levels of support (from low to strong) that these four families form a clade. Sequence data provided strong support for a monophyletic Altingiaceae (*Altingia* + *Liquidambar*) and also supported removal of Altingiaceae from Hamamelidaceae (Hoot et al. 1999; D. Soltis et al. 2000; Fishbein et al. 2001), as proposed in studies of morphology, anatomy, cytology, and chemistry (Doweld 1998). Molecular data provided weak support for the inclusion in Hamamelidaceae of *Exbucklandia* and *Rhodoleia,* two genera sometimes excluded from Hamamelidaceae (Doweld 1998). Relationships within Hamamelidaceae have been the subject of several phylogenetic analyses (e.g., Li et al. 1999a, 1999b).

The relationships of Cercidiphyllaceae and Daphniphyllaceae have long been contentious (reviewed in Cronquist 1981; Takhtajan 1997; Fishbein et al. 2001). *Cercidiphyllum* has been considered to be closely related to *Myrothamnus* (Hufford and Crane 1989; Hufford 1992) or to the early-diverging eudicot family Trochodendraceae and viewed as a derivative of the magnoliids, either distinct from hamamelids (Swamy and Bailey 1949) or transitional between magnoliids and hamamelids (Cronquist 1981). *Daphniphyllum* has been considered to be closely related to Euphorbiaceae (Scholz 1964), *Cercidiphyllum* (Croizat 1941), and various hamamelids, including Hamamelidaceae (Bhatnagar and Garg 1977; Cronquist 1981; Zavada and Dilcher 1986; Sutton 1989; Takhtajan 1997).

Character evolution in Saxifragales

Saxifragales are an important clade in which to examine floral evolution and diversification for several reasons. First, although the exact phylogenetic position of the clade remains unclear, it appears to be an early-diverging group of core eudicots (see above). In addition, although small in terms of number of species, Saxifragales exhibit levels of diversity in floral morphology comparable to the major eudicot clades. Thus, understanding floral diversification in Saxifragales may elucidate general evolutionary processes that operated early in eudicot diversification.

Using the best estimate of phylogeny (Figure 6.3, but without Peridiscaceae), we examined the evolution of several floral characters in Saxifragales. In the next sections, we consider perianth merosity, stamen number, anther dehiscence, and several gynoecial features: carpel union, carpel number, and ovary position (M. Fishbein, D. E. Soltis, L. Hufford, unpubl.). These character reconstructions provided support for some longheld ideas about evolutionary trends in angiosperms, but in other cases, they suggested opposing trends or labile evolution. These reconstructions were conducted before it was realized that Peridiscaceae are a member of Saxifragales; hence, this family does not appear in the figures and text that follow.

PERIANTH A longstanding view of floral evolution is a trend toward reduction in the number of perianth parts

(Bessey 1915; Stebbins 1974; Cronquist 1981, 1988; Takhtajan 1987, 1997). In agreement with these proposed "reduction trends," analyses of perianth evolution in Saxifragales revealed a reduction in number of perianth parts. In addition, the complete loss of perianth appears irreversible in this clade (perianth loss, which is not shown, has occurred independently in Altingiaceae, Cercidiphyllaceae, and some Hamamelidaceae; for the latter see Endress 1978).

Most resolutions of relationships at the base of Saxifragales are consistent with pentamery as ancestral for the clade (Zanis et al. 2003; M. Fishbein, D. E. Soltis, L. Hufford, unpubl.). From this ancestral state of pentamery, most transitions within the clade (trees not shown) are putative reductions in merosity (e.g., to tetramerous in some Crassulaceae and Haloragaceae) or a loss of stability in merosity (Paeoniaceae, Hamamelidaceae, male *Cercidiphyllum*).

ANDROECIUM Another longstanding view of floral evolution is that angiosperms were ancestrally polystemonous (numerous stamens) with a trend toward reduction in the number of stamens (Bessey 1915; Sprague 1925; Cronquist 1988; Leins and Erbar 1991, 1994; Takhtajan 1991; Ronse De Craene and Smets 1993). Others proposed that both increases and decreases in stamen number have occurred commonly (Stebbins 1974; Takhtajan 1991; Ronse De Craene and Smets 1992), although both Stebbins (1974) and Cronquist (1988) suggested that reductions have been much more common than increases.

In Saxifragales, polystemony in *Paeonia* is clearly derived from an ancestral state of diplostemony (10 stamens; Figure 6.5). Due in large part to its numerous stamens, *Paeonia* was often considered closely allied to magnoliid groups (e.g., Leins and Erbar 1994; Takhtajan 1997). Two additional independent origins of polystemony from diplostemony appear to have occurred within Hamamelidaceae, in which *Fothergilla* and *Matudaea* both exhibit polystemony (reconstruction not shown), but with two different patterns of androecial development, centrifugal and centripetal (Endress 1976). On the basis of this and other structural features, Endress assumed that polystemony evolved separately in these two genera, a hypothesis supported by molecular studies (Li et al. 1999a). Optimization of stamen number also indicated several evolutionary transitions from diplostemony to haplostemony in Saxifragales (Figure 6.5). However, uncertainty regarding relationships makes interpretation of some transitions between diplostemony and haplostemony equivocal.

ANTHER DEHISCENCE Despite the wide occurrence of valvate anther dehiscence among basal angiosperms and early-diverging eudicots, as well as the presence of this pattern in the earliest fossil stamens that are well pre-

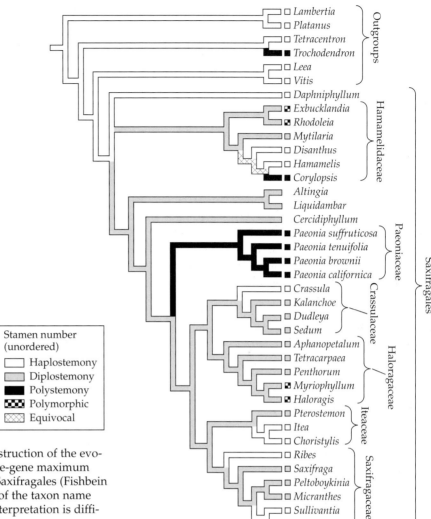

FIGURE 6.5 Most parsimonious reconstruction of the evolution of stamen number using the five-gene maximum likelihood estimate of phylogeny for Saxifragales (Fishbein et al. 2001). A missing square in front of the taxon name indicates missing data (in this case, interpretation is difficult or states not applicable).

served (Crane and Blackmore 1985; Crane et al. 1995), this condition is not necessarily a plesiomorphic feature for the angiosperms. Possibly, only the predisposition for easy development of valvate dehiscence was present in the earliest angiosperms, which may have possessed anthers that dehisced via simple longitudinal slits. This predisposition would have been lost in more derived angiosperms (Endress 1986a, 1986b, 1989a; Endress and Hufford 1989; Hufford and Endress 1989).

Our reconstructions and those of M. Fishbein, D. E. Soltis, L. Hufford (unpubl.) for Saxifragales placed longitudinal anther dehiscence as ancestral for Saxifragales with the derivation of valvate dehiscence occurring at least once, and possibly many times, within the clade (in Hamamelidaceae, *Altingia,* and *Cercidiphyllum;* reconstruction not shown). These results challenge Hufford and Endress (1989), who concluded that valvate dehiscence was plesiomorphic in Hamamelidaceae, Cercidiphyllaceae, and Altingiaceae. Firm conclusions

regarding the number of origins of valvate dehiscence in Saxifragales are not possible because of phylogenetic uncertainty at the base of the clade (Figure 6.3).

Valvate anther dehiscence in Saxifragales does not appear to be homologous with that found among basal angiosperms or early-diverging eudicots (e.g., Trochodendraceae, Platanaceae). In addition, the presence of longitudinal dehiscence in *Disanthus* (Hamamelidaceae) is noteworthy in that, because of its derived phylogenetic position within Hamamelidaceae, *Disanthus* clearly represents a reversal within the family from valvate back to longitudinal dehiscence. Thus, anther dehiscence within Saxifragales may be more labile than was previously envisioned (M. Fishbein, D. E. Soltis, L. Hufford, unpubl.).

CARPEL NUMBER AND UNION   Evolutionary trends toward reduction in number of carpels and increasing syncarpy have been widely accepted in the angiosperms, with reversals considered unlikely (e.g., Bessey 1915; Sprague

1925; Grant 1950; Cronquist 1968, 1981, 1988; Takhtajan 1969, 1987, 1991; Stebbins 1974). It has been proposed that syncarpy offers a selective advantage over apocarpy due to increased pollination efficiency (Carr and Carr 1961; Stebbins 1974; Endress 1982; Takhtajan 1991) and centralized selection of male gametophytes (Endress 1982); it is also a precondition for fruit diversification (Endress 1982; Takhtajan 1991). Intermediate stages have also been proposed for the evolution of syncarpy. Cronquist (1988), for example, suggested that partial syncarpy, in which the apical portion of the gynoecium remains free (as in many Saxifragales), is an intermediate between apocarpy and complete syncarpy.

Character mapping studies of Saxifragales provided strong evidence against the longstanding view of reduction in carpel number (not shown). Saxifragales were inferred to be ancestrally bicarpellate with at least two independent increases in carpel number. The actual number of independent increases in carpel number (reconstruction not shown) is unclear, but may be two or three.

Analyses of carpel union in Saxifragales also argued against the widely accepted trend of increased syncarpy. The ancestral condition for Saxifragales is carpels that are free apically (Figure 6.6; the variation in extent of carpel union in Saxifragales is shown in Figure 6.7). Several independent derivations of apocarpy have occurred in Saxifragales, in Paeoniaceae, *Tetracarpaea*, and within Crassulaceae (Figures 6.6 and 6.7). These represent the first documentations of the derivation of apocarpy from syncarpous ancestors outside of Apocynaceae, Malvaceae, Rutaceae, and Simaroubaceae (Endress et al. 1983; Fallen 1986; Sennblad and Bremer 1996). No transitions from apocarpy to complete syncarpy were reconstructed within Saxifragales (M. Fishbein, D. E. Soltis, L. Hufford, unpubl.), although this transformation was the traditionally accepted trend of evolution.

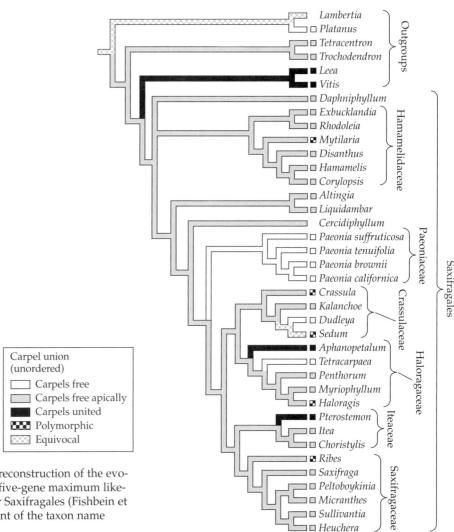

FIGURE 6.6 Most parsimonious reconstruction of the evolution of carpel union using the five-gene maximum likelihood estimate of phylogeny for Saxifragales (Fishbein et al. 2001). A missing square in front of the taxon name indicates missing data.

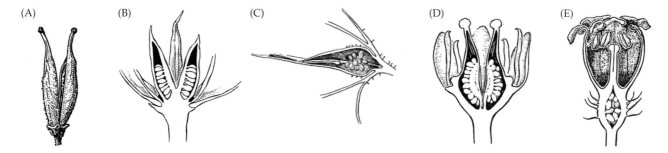

FIGURE 6.7 Carpel union and variation in ovary position in Saxifragales illustrating various degrees of carpel union and ovary inferiority. (A) *Itea virginica* (Iteaceae) in fruit, carpels united only at base; ovary with only a short inferior portion (from Spongberg 1972). (B) *Sedum pusillum* (Crassulaceae) in flower, carpels nearly free (slight union at base); ovary superior in appearance (from Wood 1974). (C) *Tiarella cordifolia* (Saxifragaceae) in flower, carpels fused for about one-third their length; ovary with only a short inferior region (perhaps best termed pseudosuperior; see section on ovary position) (from Spongberg 1972; Wood 1974). (D) *Micranthes virginiensis* (Saxifragaceae) in flower, carpels fused only at base; ovary perhaps one-half inferior (from Spongberg 1972; Wood 1974). (E) *Ribes cynosbati* (Grossulariaceae) in flower, carpels completely united; ovary completely inferior (from Spongberg 1972).

OVARY POSITION The position of the ovary (Figure 6.8) has a major impact on floral architecture, concomitantly influencing various interactions with animals, such as predation, pollination, and seed dispersal (e.g., Grant 1950; Stebbins 1974; Thompson 1994). In general, ovary positions have been treated as either superior or inferior according to the point of attachment of the perianth and androecium relative to the ovary in an anthetic (i.e., open or mature) flower (Figure 6.8). A superior ovary is one that is situated above the point of attachment of the perianth and androecium to form a hypogynous flower; an inferior ovary has the perianth and androecium attached above the ovary to form an epigynous flower (Figure 6.8). Flowers in which the outer floral appendages are basally fused to form a floral cup, or hypanthium, that surrounds the ovary are often called perigynous. Ovary position has long been used as a key descriptive feature of families and has been considered a relatively stable character (Grant 1950; Stebbins 1974) that is sometimes used to distinguish closely related families.

A common view is that ovary position has evolved in a unidirectional manner from superior to greater inferiority, generally via congenital fusion of the outer floral appendages to the ovary wall (e.g., Langdon 1939; Douglas 1944; Gauthier 1950; Eames 1961); reversals were considered rare or impossible (Bessey 1915; Grant 1950; Cronquist 1968, 1988; Takhtajan 1969, 1991; Stebbins 1974). The putative selective advantage of epigyny is protection of ovules (Grant 1950; Stebbins 1974; Takhtajan 1991). Epigyny also brings the perianth whorl into more contact with the androecium, permitting more interactions between these whorls in synorganized flowers. Reversals to superior ovaries were also thought to be constrained by the complex alterations of developmental pathways presumed to be associated with the formation of inferior ovaries (Grant 1950; Stebbins 1974).

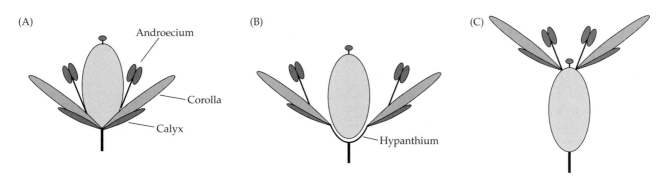

FIGURE 6.8 Ovary positions in angiosperms. (A) In a flower with a superior ovary the perianth and androecium are inserted beneath the ovary; the flower is hypogynous. (B) Alternatively, these outer appendages can be united to form a hypanthium; the flower is perigynous. (C) A flower with an inferior ovary has the floral appendages inserted at the apex of the ovary; the flower is epigynous.

Gustafsson and Albert (1999) analyzed these purported trends in ovary position diversification across the angiosperms. Their mapping studies indicated that the evolution of ovary position at this broad scale was dynamic, with reversals from epigyny to hypogyny occurring on many occasions. Gustafsson and Albert's results demonstrated that change in ovary position is not unidirectional. Importantly, the apparent reversals observed are not evenly distributed; whereas some groups are essentially fixed for epigyny, others, including Saxifragales, appeared more labile.

Given that ovary position has been viewed as a fairly stable character, families and orders that display extensive variation in ovary position have long been of evolutionary interest (e.g., Klopfer 1973; Stebbins 1974; Cronquist 1988; Gustafsson and Albert 1999). Saxifragales exhibit a remarkable diversity in ovary position (Figures 6.7 and 6.8). Ovary position is highly variable in the small family Saxifragaceae (about 30 genera), ranging from inferior to what has been termed superior with a complete range of intermediate positions also present. This entire range of ovary positions is present within single genera, such as *Lithophragma* (Figure 6.9), *Saxifraga* s.s., *Micranthes*, and *Chrysosplenium* (e.g., Hitchcock et al. 1961; Taylor 1965; Elvander 1984; Webb and Gornall 1989; D. Soltis et al. 1993; D. Soltis and Hufford 2002). Even at this level, unidirectional ovary evolution from superior to increasing inferiority was invoked to explain the remarkable range of ovary positions (e.g., Taylor 1965; Stebbins 1974).

Mapping ovary position onto phylogenetic trees for Saxifragaceae revealed repeated reversals from inferior ovaries to those that appear to be superior. This dynamic nature of ovary position evolution was detected across the entire family, within well-defined clades of genera such as the *Heuchera* group (Kuzoff et al. 1998), and also within genera, including *Chrysosplenium* (D. Soltis et al. 2001a), *Lithophragma* (Kuzoff et al. 1999, 2001), and *Micranthes* (Mort and Soltis 1999). For example, ovary position evolution in *Lithophragma* has been highly labile, moving toward greater superiority in some lineages and greater inferiority in others, contrary to the longstanding view of a unidirectional trend in ovary position evolution (Kuzoff et al. 2001; reviewed in D. Soltis et al. 2003c; Figures 6.9 and 6.10).

Studies across Saxifragales also revealed several reversals from inferior ovaries to those that appear to be superior (Figure 6.11). Inferior (or partially inferior) ovaries characterize most Saxifragales, but several reversals to ovaries that appear superior have occurred (M. Fishbein, D. E. Soltis, L. Hufford, unpubl.). The actual number of reversals is unclear, however, due to uncertainty in phylogenetic relationships at the base of the Saxifragales clade (D. Soltis et al. 2003c; M. Fishbein, D. E. Soltis, L. Hufford, unpubl.). In *Daphniphyllum*, retention of a superior ovary and a reversal to a superior

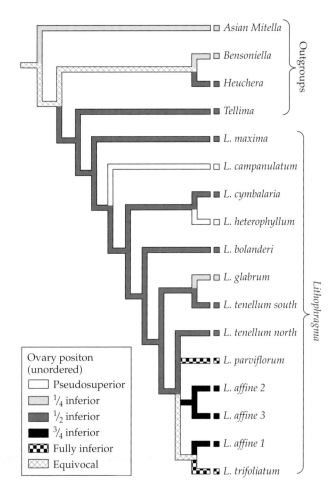

FIGURE 6.9 Diversification of ovary position in *Lithophragma* (Saxifragaceae). Strict consensus for species of *Lithophragma*. From an ancestral condition that was half inferior, there has been evolution toward increasing inferiority in some species (*L. affine, L. parviflorum, L. trifoliatum*) and also toward increasing superiority in several other species (*L. heterophyllum, L. campanulatum,* and *L. glabrum*). (Modified from Kuzoff et al. 1999.)

ovary are equally parsimonious. Unambiguous reversals to superior ovaries are seen in Crassulaceae and in *Tetracarpaea* (Figure 6.11; D. Soltis et al. 2003c; M. Fishbein, D. E. Soltis, L. Hufford, unpubl.).

GYNOECIAL DIVERSIFICATION AND DEVELOPMENT Studies of phylogenetic relationships and floral development in Saxifragales have indicated a dynamic picture of floral diversification (Kuzoff et al. 2001; D. Soltis et al. 2003c; M. Fishbein, D. E. Soltis, L. Hufford, unpubl.). Typically, ovary positions have been described and decisions concerning structural homology made exclusively on the basis of anthetic (i.e., mature) floral structure. Superior and inferior ovaries typically are distinguished by the point of attachment of other floral organs relative to the ovary in anthetic flowers (Fig-

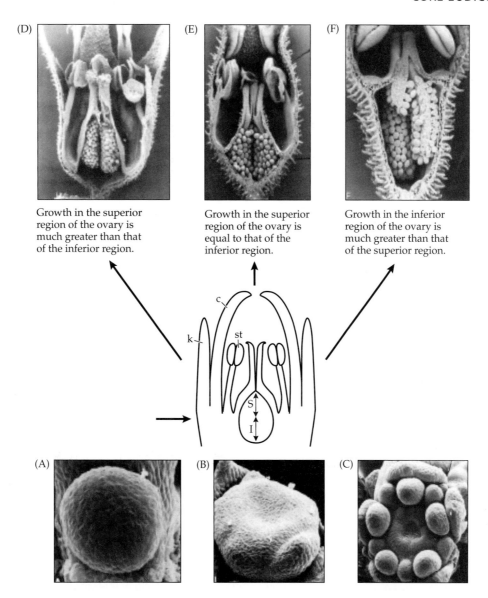

FIGURE 6.10 Gynoecial development in *Lithophragma* (Saxifragaceae). Ontogenetic studies have indicated that all species of *Lithophragma* and relatives have an appendicular epigynous developmental ground plan (see text and Figure 6.12). Early floral ontogeny (A, B, and C) is identical in species of *Lithophragma* having very different ovary positions at anthesis (D, E, and F). The arrow on the flower section indicates the point of attachment of the hypanthium to the ovary (note superior, S, and inferior, I, portions of the ovary); k = calyx, c = corolla, st = stamen. In *Lithophragma*, the ovary position at anthesis (D, E, and F) is the result of the amount of vertical extension of the inferior versus superior region of the ovary. Differences in ovary position are the direct result of allometric shifts in the growth proportions of the superior versus inferior regions. (Photographs from Kuzoff et al. 2001.)

ure 6.8). Early floral development serves as an additional line of evidence to distinguish between flowers that are hypogynous and those that are epigynous (Kuzoff et al. 2001; D. Soltis and Hufford 2002). Boke (1963, 1964) and Kaplan (1967) demonstrated that hypogynous and epigynous flowers actually differ from the time of organogenesis, but the significance of their research has often been overlooked. There are two basic ground plans of floral development, hypogynous and appendicular epigynous (reviewed in Figure 6.12).

Studies of floral development in Saxifragaceae have indicated that species reported to have superior ovaries actually have an appendicular epigynous ground plan (Figures 6.10 and 6.12; e.g., Kuzoff et al. 2001; D. Soltis and Hufford 2002; D. Soltis et al. 2003c). Thus, these ovaries are not truly superior but represent "superior mimics" or "pseudosuperior" ovaries. Pseudosuperior ovaries should not be considered homologous to truly superior ovaries derived from an hypogynous ground plan. Importantly, in Saxifragaceae,

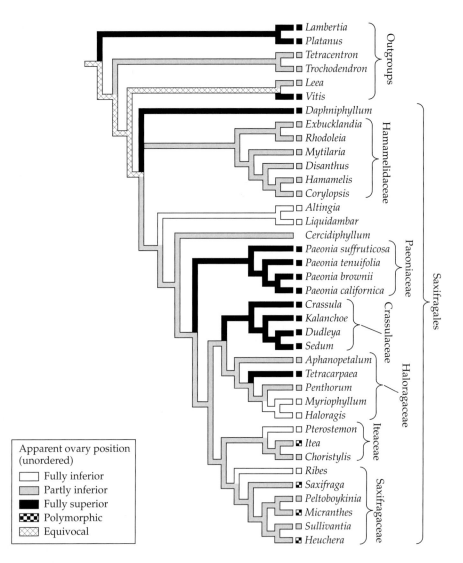

FIGURE 6.11 Most parsimonious reconstruction of diversification of ovary position using the maximum likelihood (and Bayesian) estimate of phylogeny for Saxifragales (Fishbein et al. 2001; Fishbein and Soltis 2004). A missing square in front of the taxon name indicates missing data. Reversals to ovaries that appear to be superior have occurred in Crassulaceae, *Tetracarpaea*, and *Paeonia*. Ovaries that appear superior in Crassulaceae and Daphniphyllaceae are noteworthy because each develops by a distinctive process in which the carpels are initiated on a flattened meristem that is less conical than that found in angiosperms characterized by a hypogynous ground plan. In *Daphniphyllum* this switch is associated with the loss of floral organs. It is unclear at this point if floral development in Crassulaceae and *Daphniphyllum* is truly homologous with a hypogynous ground plan. Although the ovary of Cercidiphyllaceae has been described as superior (Endress 1993; Endress and Igersheim 1999), *Cercidiphyllum* is a special case because the flowers are unisexual, and each female flower is essentially no more than a carpel (there is no perianth); hence, the terms superior and inferior ovary are not applicable (M. Fishbein, D. E. Soltis, L. Hufford, unpubl.).

the ovary position at anthesis is a consequence of the amount of vertical extension of the inferior versus superior region of the ovary. Differences in ovary position are a direct result of allometric shifts in the growth proportions of the superior versus inferior regions (Kuzoff et al. 2001), as observed for *Lithophragma* (Figure 6.10). The relative ease with which allometric shifts can occur in the course of evolution explains the wide range of ovary positions in Saxifragaceae, as well as the numerous reversals that have occurred. These data also indicate that gynoecial evolution may be more rapid than has been generally understood, an important result in light of studies in Saxifragaceae that ovary position affects pollinator preferences (e.g., Thompson and Pellmyr 1992; Thompson 1994; Segraves and Thompson 1999).

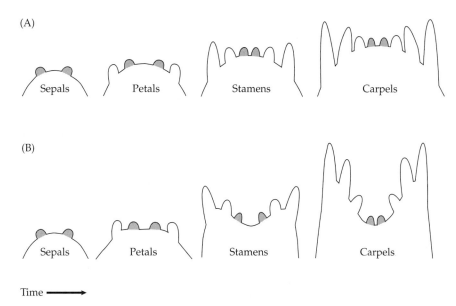

Time ⟶

FIGURE 6.12 Patterns of ovary development in angiosperms. The early floral conformation of most hypogynous flowers represents what is termed a hypogynous ground plan (A). Hypogynous flowers generally exhibit a convex floral apex throughout organogenesis (e.g., Endress 1994c). This ground plan is common in the angiosperms and likely accounts for many flowers with ovaries described as "superior" based on the examination of anthetic flowers. An hypogynous ground plan character-izes many basal angiosperm lineages, such as Piperaceae and Alismataceae, as well as some early-branching eudicots, such as members of Ranunculaceae and Papaveraceae (Sattler 1973). Most inferior ovaries are the result of an appendicular epigynous pattern of floral development (B). Appendicular epigyny also begins with a convex floral apex, but during, or just after, perianth initiation the floral apex becomes concave (Boke 1964, 1966; Kaplan 1967; Leins 1972; Magin 1977; reviewed in D. Soltis et al. 2003c).

Gynoecial diversification across Saxifragales is more complex than in Saxifragaceae (M. Fishbein, D. E. Soltis, L. Hufford, unpubl.). Phylogenetic data indicate that reversals to ovaries that appear to be superior have occurred in Crassulaceae, *Tetracarpaea*, and *Cercidiphyllum* (Figure 6.11). Appendicular epigyny characterizes most Saxifragales and is also plesiomorphic for the clade. However, in *Tetracarpaea*, the floral apex remains convex throughout organogenesis—hypogyny in *Tetracarpaea* is therefore ontogenetically homologous to that of magnoliids (Figure 6.12). This is the first unambiguous documentation of a true reversal from an appendicular epigynous ground plan to a hypogynous ground plan (M. Fishbein, D. E. Soltis, L. Hufford, unpubl.). In contrast, ovaries termed "superior" in Iteaceae and Paeoniaceae are pseudosuperior (for floral development in Paeoniaceae see Hiepko 1965b), derived via an appendicular epigynous ground plan, and are comparable to those observed in Saxifragaceae (Figures 6.10 and 6.12).

Data for Saxifragaceae and Saxifragales have important implications for floral evolution across the angiosperms. The same processes described for gynoecia in Saxifragales are likely operating throughout the rosids (D. Soltis and Hufford 2002). In fact, truly inferior ovaries derived from an epigynous ground plan may actually be the plesiomorphic state for many rosids. In this regard, data for the rosid family Vochysiaceae (Myrtales) suggest that the same developmental processes occurring in Saxifragaceae also occur in this family (Litt 1999; Litt and Stevenson 2003). Richardson et al. (2000) similarly proposed that the Saxifragaceae model of ovary position diversification also occurs in Rhamnaceae. Similar allometric shifts in development may explain the transition to a superficially superior ovary observed in *Gaertnera* (Rubiaceae; Igersheim et al. 1994). Although recent studies have greatly improved our understanding of the evolution of epigyny, new complexities have emerged. For example, developmental complexities are present in Rosaceae, in which species of *Physocarpus* and *Oemlera* have flowers that exhibit a concave floral apex before gynoecial initiation, yet have clearly superior ovaries at maturity (Evans and Dickinson 1999a, 1999b).

Santalales

The composition of Santalales has differed among recent morphology-based classifications (Cronquist 1981; Thorne 1992a, 1992b, 2001; Takhtajan 1997). In recent systems, the order has included Santalaceae, Eremolepidaceae, Loranthaceae, Opiliaceae, Olacaceae, Misodendraceae, and Viscaceae, as well as Balanophoraceae, Medusandraceae, and in some treatments Dipentodont-

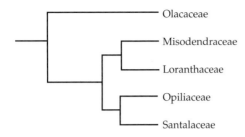

FIGURE 6.13 Summary tree for Santalales. Olacaceae are shown as a clade, but may be paraphyletic (see text).

aceae (see Cronquist 1981). The monophyly of Santalales was well supported even in initial broad analyses of *rbcL* and 18S rDNA alone (Nickrent and Soltis 1995; Chase et al. 1993; D. Soltis et al. 1997a; Chase and Albert 1998) with 100% support in analyses of combined datasets (e.g., Nickrent et al. 1998; P. Soltis et al. 1999b; Savolainen et al. 2000a; D. Soltis et al. 2000). Available molecular evidence also indicates that other families sometimes included in Santalales (such as Balanophoraceae, Medusandraceae, and Dipentodontaceae) should be excluded from the order (Nickrent et al. 1998; APG II 2003).

Thus, a major contribution of molecular phylogenetic studies has been to circumscribe Santalales as a small order that consists of the families Santalaceae s.l. (which includes Eremolepidaceae and Viscaceae), Loranthaceae, Opiliaceae, Olacaceae, and Misodendraceae (Figure 6.13; several are illustrated in Figure 6.14; Nickrent et al. 1998; Nickrent and Malécot 2001; APG II 2003). The APG II classification includes Viscaceae within Santalaceae, otherwise the latter family is paraphyletic. Although Olacaceae are tentatively retained in the traditional sense, as a large family (Figure 6.13), the circumscription of the family remains unclear; recent molecular analyses (see below) suggest that Olacaceae may be paraphyletic relative to the remainder of the Santalales (see also Chapter 11).

Members of Santalales are considered hemiparasites (Figure 6.14). As defined in Nickrent et al. (1998), these plants photosynthesize during part of their life cycle and obtain mainly water and dissolved minerals from their hosts. Holoparasites, in contrast, are nonphotosynthetic (some holoparasites are discussed in the chapters on rosids and asterids; see Chapters 8 and 9). Ascertaining the relationships of many parasitic plants has long been problematic because evolutionary modifications, such as the loss of leaves, perianth parts, integuments, and chlorophyll, associated with the parasitic habit may result in extreme morphologies and may be subject to convergence (see Nickrent et al. 1998). The relationships of many holoparasites remain particularly problematic because the plants are nonphotosynthetic, and the plastid genomes of these plants are highly

modified with many genes absent or with accelerated rates of molecular evolution (Nickrent et al. 1998; see Chapters 8, 9, and 11).

Relationships of Santalales

Although the circumscription of Santalales emerged early in molecular phylogenetic studies, the relationship of Santalales to other core eudicots remains problematic (P. Soltis et al. 1999b; D. Soltis et al. 2000, 2003a). The position of Santalales has differed among the broad phylogenetic analyses conducted to date. Analysis of *atpB* plus *rbcL* sequences (Savolainen et al. 2000a) placed Santalales as sister to Dilleniaceae, and these two clades together were, in turn, sister to the asterids. In the shortest trees found in a three-gene analysis (D. Soltis et al. 2000), Santalales appeared as sister to a clade of Caryophyllales + Dilleniaceae; this large clade was, in turn, sister to asterids. The shortest trees found in a four-gene analysis of eudicots (D. Soltis et al. 2003a) placed Santalales sister to the asterids, but without support above 50%. Analyses of combined DNA datasets therefore consistently associate Santalales, Caryophyllales, and asterids, but without support above 50%. Additional characters will be needed to resolve these relationships adequately (see "Future Research").

Relationships within Santalales

The best-supported topology for Santalales is that of Nickrent and Malécot (2001), which is based on a combined 18S rDNA plus *rbcL* dataset with broad representation of the families within Santalales. (See Nickrent et al. 1998 for an earlier analysis of Santalales using 18S rDNA plus *rbcL* data, but fewer taxa.) Olacaceae are paraphyletic to the remainder of Santalales, a result in general agreement with longstanding views of relationships. Olacaceae had been considered ancestral in the clade because it contains both parasitic and nonparasitic members (reviewed in Nickrent et al. 1998.) However, neither 18S rDNA nor *rbcL*, nor the two genes combined, recovered a monophyletic Olacaceae. Instead, a series of subclades of Olacaceae appeared as successive sisters to the remainder of the family (Figures 6.13 and 11.3; Nickrent and Malécot 2001). However, Olacaceae have been retained as a distinct family in APG II (2003) until the relationships are better resolved and supported.

One member of Olacaceae, *Schoepfia*, emerged as sister to *Misodendron*, the only genus of Misodendraceae (Figures 6.13 and 11.3). Anatomical evidence, including features of the parenchyma, tracheids, and vessels (reviewed in Nickrent et al. 1998), supports the distinctiveness of *Schoepfia* from other Olacaceae. Nonetheless, the close relationship between *Schoepfia* and *Misodendron* indicated by molecular data was unanticipated, and non-DNA synapomorphies for this small clade remain unclear. The *Schoepfia* + *Misodendron* sister group (which should ultimately be named Misodendraceae)

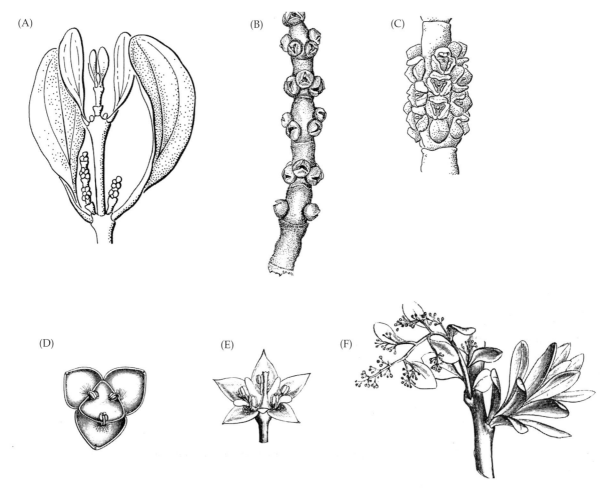

FIGURE 6.14 Flowers and plants of Santalales. (A) *Phora-dendron rubrum* (formerly Viscaceae, now included in Santalaceae), tip of staminate plant. (B) *Phoradendron leucocarpon*, staminate inflorescence. (C) *Phoradendron leucocarpon*, carpellate inflorescence. (D) *Osyris alba* (Santalaceae), staminate flower. (E) *Acanthosyris spinescens* (Santalaceae), bisexual flower. (F) *Misodendron brachystachyum* (Misodendraceae), flowering branch. (A to C, from Kuijt 1982. D to F, from Pilger in Engler and Prantl 1935.)

appeared, in turn, as a well-supported sister to a well-supported Loranthaceae (Nickrent and Malécot 2001). The relationship of *Schoepfia* to Loranthaceae was noted as early as 1830 by deCandolle (see Nickrent and Malécot 2001). *Schoepfia* and Loranthaceae share several features; for example, both have a reduced calyx.

Following Olacaceae, Misodendraceae (including *Schoepfia*), and Loranthaceae, Opiliaceae form a well-supported clade sister to the remainder of Santalales (Nickrent and Malécot 2001; Figure 6.13). The remainder of the order consists of a paraphyletic Santalaceae within which are nested a well-supported Viscaceae and Eremolepidaceae (the latter family represented in these analyses by *Eubrachion* and *Lepidoceras*), both families of aerial parasites. In contrast, many members of Santalaceae are terrestrial, although some are "mistletoes," such as *Phacellaria* and some species of *Exocarpos*. The topologies of Nickrent et al. (1998) and Nickrent and Malécot (2001) supported the morphology- and

karyology-based hypothesis (Wiens and Barlow 1971) that Eremolepidaceae belong to Santalaceae. The APG (1998) and APG II (2003) classifications both included Eremolepidaceae within Santalaceae (see Chapter 10). In addition, given the well-supported position of Viscaceae within Santalaceae, Santalaceae were further expanded to include Viscaceae (APG II 2003).

Character evolution in Santalales

Santalales provide an excellent model for the study of the evolution of parasitism (see Chapter 11). The early-branching members of the clade (some members of the paraphyletic Olacaceae) are nonparasitic (Nickrent and Malécot 2001), whereas most members of Santalales are hemiparasites. Parasitism appears to have arisen late in the diversification of the paraphyletic Olacaceae. Many of the parasitic taxa in Santalales are terrestrial. Aerial parasites (those confined to trees) are also present, and this strategy arose many times (see Figure 11.3).

Many nonphotosynthetic plants exhibit increased rates of molecular evolution (reviewed in Nickrent et al. 1998). Because Santalales include hemiparasites as well as nonparasites, this clade has been the focus of several studies of rates of molecular evolution (Nickrent and Starr 1994; Nickrent et al. 1998). 18S rDNA in the hemiparasite *Arceuthobium* (Viscaceae) and in several unrelated holoparasites (*Balanophora, Prosopanche, Rafflesia,* and *Rhinanthes*) evolves approximately three times faster than in nonparasites, consistent with a possible association between accelerated rates of molecular evolution and unusual life histories (see Chapter 2). However, accelerated rates of molecular evolution are not observed in all Santalales—other hemiparasites, as well as nonparasites, in the clade do not exhibit accelerated rates of molecular evolution. *Arceuthobium* also exhibits an accelerated rate of molecular evolution for some plastid genes, but not for the mitochondrial 16S rDNA. Thus, these studies of *Arceuthobium* and other Santalales have revealed that not all parasitic plants are associated with accelerated rates of molecular evolution. Furthermore, some plants from other clades that have lost photosynthesis do not have increased rates of molecular evolution in their plastid genomes (e.g., *Monotropa,* Ericaceae; *Corallorhiza,* Orchidaceae; Nickrent et al. 1998).

Future Research

Resolving relationships among the major lineages of core eudicots, which include most of the extant angiosperms, should be a major priority of angiosperm phylogenetic analyses. An understanding of core eudicot relationships is crucial not only for angiosperm systematics but also for an improved understanding of character evolution across the eudicots. Considerable success was realized in resolving relationships among basal angiosperms by constructing datasets of five or more genes for numerous taxa (Qiu et al. 1999; Zanis et al. 2002); similar, large datasets are needed for core eudicots.

The smaller lineages discussed in this chapter (Gunnerales, Berberidopsidales, Saxifragales, Santalales) may occupy significant phylogenetic positions among the core eudicots. For example, Gunnerales appear to be sister to all other core eudicots. Berberidopsidales may follow Gunnerales as sister to all remaining core eudicots, but additional data are needed to evaluate this relationship. One important avenue of research, therefore, is to clarify the position of each of these smaller clades among the core eudicots.

Both Saxifragales and Santalales require additional investigation. Deep-level relationships remain unclear within Saxifragales despite an enormous sequencing effort. Additional molecular analyses may help to resolve deep-level relationships, but the analyses of Fishbein et al. (2001) indicated that Saxifragales represent a rapid, ancient radiation, and the resolution of deep-level relationships will require a monumental effort (perhaps 10^7 base pairs). Relationships within Santalales also remain unclear. Although Olacaceae have been tentatively retained as a distinct family, recent molecular studies found Olacaceae paraphyletic to the remainder of the clade. Additional taxa and sequence data are needed to resolve relationships within Santalales.

Caryophyllales

Introduction

Since the 1980s, the name "Caryophyllales" has been applied to an increasingly broad, and correspondingly diverse, circumscription of families. Before that, for 50 years or more, the concept of Caryophyllales generally corresponded to that of Centrospermae (Harms 1934), a group long recognized as an assemblage of closely related families (e.g., Braun 1864; Eichler 1875–1878). Morphological and embryological characters were used to unite the core families of Centrospermae, but Centrospermae gained particular attention as one of the earliest groups for which circumscription was modified on the basis of chemical characters. The discovery that all but two families of Centrospermae—Caryophyllaceae and Molluginaceae—produced betalain pigments instead of the anthocyanin pigments produced in other angiosperms supported a close relationship among the core group of families. Furthermore, Cactaceae and Didiereaceae, not previously considered closely related to Centrospermae, were also discovered to produce betalains. Revised classifications (e.g., Dahlgren 1975, 1980; Takhtajan 1980; Cronquist 1981; Thorne 1983, 1992a, 1992b), applying the name Caryophyllales and incorporating chemical, morphological, embryological, and anatomical characters, included 12 families. These circumscriptions have remained largely intact for the past 30 years. Cronquist's (1981) Caryophyllidae, comprising Caryophyllales, Polygonaceae, and Plumbaginaceae (the last two placed in his monofamilial Polygonales and Plumbaginales, respectively), also recognized similarities among these three groups. However, emphasis on certain sets of characters over others has resulted in the inclusion or exclusion of additional families and genera and has altered views on the closest relatives of Caryophyllidae.

A series of molecular phylogenetic analyses has reshaped concepts of Caryophyllales by identifying the closest relatives of the traditional order and resolving patterns of relationship within the clade. Most notable is the discovery that certain carnivorous plants—the sundews and Venus flytrap (Droseraceae) and Old World pitcher plants (Nepenthaceae)—are closely related to Cronquist's Caryophyllidae (Albert et al. 1992; Chase et al. 1993; Williams et al. 1994; Meimberg et al. 2000; Cuénoud et al. 2002). In addition, many families previously considered distantly related to Caryophyllales have been included in a large clade with Caryophyllales (e.g., Asteropeiaceae and Physenaceae, Morton et al. 1997; Rhabdodendraceae, Fay et al. 1997a; Simmondsiaceae, e.g., D. Soltis et al. 2000). The strong support for this clade in recent multigene analyses (e.g., D. Soltis et al. 2000; Cuénoud et al. 2002) has led to a revised—and broader—circumscription of Caryophyllales by APG (1998) and APG II (2003). Caryophyllales *sensu* APG II (2003) comprise 29 families; two other families have been proposed ("Agdestidaceae" and "Petiveriaceae," discussed below, in Phylogeny of Caryophyllales; Figure 7.1; Table 7.1). Because of the

TABLE 7.1	**Comparison of treatments of Caryophyllales. Cronquist's (1981) treatment is conceptually similar to other modern treatments and represents those classifications**
Cronquist (1981)	**APG II (2003)**
Caryophyllidae	Caryophyllales
Caryophyllales	Achatocarpaceae
Phytolaccaceae	Aizoaceae
Achatocarpaceae	Amaranthaceae
Nyctaginaceae	Asteropeiaceae[†]
Aizoaceae	Barbeuiaceae[†]
Didiereaceae	Basellaceae
Cactaceae	Cactaceae
Chenopodiaceae	Caryophyllaceae*
Amaranthaceae	Didiereaceae
Portulacaceae	Gisekiaceae
Basellaceae	Halophytaceae
Molluginaceae*	Molluginaceae*
Caryophyllaceae*	Nyctaginaceae
Polygonales	Physenaceae[†]
Polygonaceae*	Phytolaccaceae
Plumbaginales	Portulacaceae
Plumbaginaceae*	Rhabdodendraceae*
	Sarcobataceae
	Simmondsiaceae*
	Stegnospermataceae
	Ancistrocladaceae*
	Dioncophyllaceae*
	Droseraceae*
	Drosophyllaceae*
	Frankeniaceae*
	Nepenthaceae*
	Plumbaginaceae*
	Polygonaceae*
	Tamaricaceae*

* indicates anthocyanin production.
no * indicates betalain production.
[†] indicates type of pigment unknown.

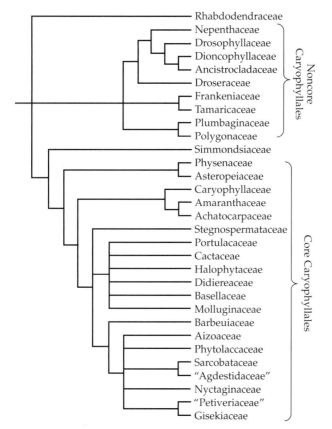

FIGURE 7.1 Phylogenetic relationships among families of Caryophyllales *sensu* APG II (2003), taken mostly from the *rbcL* + *matK* tree of Cuénoud et al. (2002). Branches receiving less than 60% bootstrap support in their tree have been collapsed. Names in quotation marks have been recommended for familial status.

potential confusion introduced by applying the name Caryophyllales to a large clade that includes the traditional Caryophyllales and several other families, not all investigators have accepted this name change. Judd et al. (2002), for example, continue to use Caryophyllales in a narrow sense and refer to the larger, more inclusive clade as the caryophyllid clade. In this book, we follow the broader APG II (2003) system. Although morphological and anatomical characters are generally consistent with these relationships, non-DNA synapomorphies for this newly recognized expanded Caryophyllales have not yet been discovered.

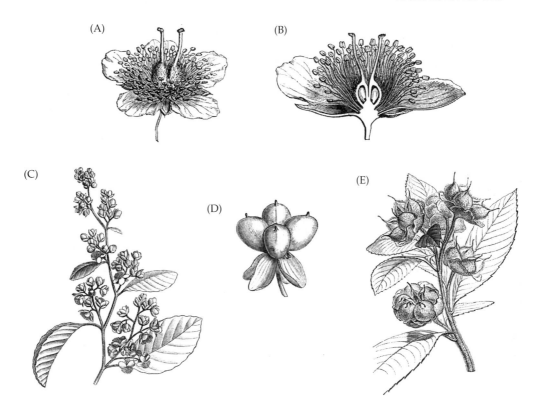

FIGURE 7.2 Flowers and plants of Dilleniaceae. (A) *Davilla villosa*, flower. (B) *Tetracera radula*, flower in longitudinal section. (C) *Tetracera radula*, plant in flower. (D) *Tetracera assa*, fruit. (E) *Tetracera boiviniana*, plant in fruit. (From Gilg in Engler and Prantl 1895.)

Relationships of Caryophyllales to other major clades of eudicots remain unclear. Cronquist (1981) and Takhtajan (1980, 1997), on the basis of floral characters, viewed Caryophyllidae as being derived from Ranunculales-type ancestors. However, phylogenetic analyses using many gene sequences place Caryophyllales firmly within the core eudicots. They have alternatively been considered to be close relatives of rosids, asterids, or Santalales and are best regarded at this time simply as one of the major clades of core eudicots (see Chapter 6; e.g., D. Soltis et al. 2000). The sister group of Caryophyllales may be Dilleniaceae (e.g., D. Soltis et al. 2000), several representatives of which are shown in Figure 7.2. Although a close relationship between Dilleniaceae and Caryophyllales (*sensu* APG) was also detected by Chase and Albert (1998; using the "search 2" dataset of Chase et al. 1993), support was less than 50%, and analyses of *rbcL* and *atpB* did not find this relationship. A relationship between Dilleniaceae and members of Caryophyllales had not previously been suggested. Cronquist (1981) considered Dilleniaceae, together with Paeoniaceae, as occupying a basal position within subclass Dilleniidae, whereas Takhtajan (1997) viewed Dilleniaceae alone as the "most archaic family in Dilleniidae." Further investigation is needed to assess this potential relationship and to identify the sister group(s) of Caryophyllales.

Despite the small size of Caryophyllales, relative to the rosids or asterids, for example, this clade encompasses tremendous diversity in morphology, physiology, and biochemistry. Caryophyllales exemplify spectacular adaptations to extreme environments, including deserts, saline soils, and nutrient-poor substrates. These environmental challenges have been met through a variety of mechanisms. In addition, breeding systems range from hermaphroditic outcrossing flowers to complex systems of dioecy and subdioecy. Floral morphology also follows developmental patterns different from those in other eudicots. Rates of diversification in Caryophyllales indicate a bimodal distribution in which some families have only one or a few genera (e.g., Rhabdodendraceae, Drosophyllaceae) and other families consist of 100 or more genera (e.g., Cactaceae). This disparity may be, at least in part, an artifact of classification, but the pattern of sister groups of disparate size is real. Although these patterns have not yet been correlated with habitat or presumed adaptations, this differential species richness in sister clades may suggest key innovations in some members of Caryophyllales.

Phylogeny of Caryophyllales

The most comprehensive phylogenetic study of Caryophyllales to date is that of Cuénoud et al. (2002), in which 127 taxa of Caryophyllales, representing all 29 families included in the order by APG II (2003), were sampled for *matK* sequence analysis. A subset of this matrix was combined with sequences from *rbcL, atpB,* and 18S rDNA to provide a four-gene matrix for 26 genera of Caryophyllales. This combination of analyses strongly supported the monophyly of the expanded Caryophyllales *sensu* APG II. Within Caryophyllales are two large clades (core and noncore Caryophyllales, which Judd et al. 2002 referred to as Caryophyllales and Polygonales, respectively), with *Rhabdodendron* (of the monogeneric Rhabdodendraceae) and *Simmondsia* (of the monospecific Simmondsiaceae) unplaced; the last two genera appear in different positions in different analyses, always with low support (Figure 7.1). *Rhabdodendron,* with three species from tropical South America, and *Simmondsia,* with a single species from the southwestern United States, are possibly sisters to each other and then to core Caryophyllales, or *Rhabdodendron* is sister to the entire Caryophyllales and *Simmondsia* is sister to the core Caryophyllales. Several representatives of Caryophyllales are depicted in Figure 7.3.

Neither *Rhabdodendron* nor *Simmondsia* was included in Cronquist's (1981) Caryophyllidae or their counterparts. *Simmondsia* (Simmondsiaceae) is similar to core Caryophyllales in having normal secondary growth and seeds without endosperm, and Airy-Shaw (1966) therefore suggested a close relationship between the two. However, few authors have followed this suggestion, with Simmondsiaceae being variously viewed as closely related to Buxaceae (Hutchinson 1959; Scholz 1964), Euphorbiaceae, or Pandaceae (Takhtajan 1969; Cronquist 1981). (Note that Buxaceae, now considered an early-diverging eudicot, was itself variously placed by these authors in Hamamelidales [Hutchinson 1959], Celastrales [Scholz 1964], and Euphorbiales [Takhtajan 1969; Cronquist 1981]). *Rhabdodendron* was originally placed in Rutaceae (Gilg and Pilger 1905), and it has been associated with species that were placed in Chrysobalanaceae by Bentham (1853) and in Phytolaccaceae by Record (1933), Chalk and Chattaway (1937), and Metcalfe and Chalk (1950). Prance (1968) raised

Rhabdodendron to familial level and considered it to be close to Phytolaccaceae and other Centrospermae. On the basis of different characters, it was subsequently viewed as similar to Rutaceae (Puff and Weber 1976; Behnke 1977) or to various other groups of rosids (e.g., Cronquist 1983; Wolter-Filho et al. 1985, 1989; Tobe and Raven 1989). Both *Simmondsia* and *Rhabdodendron* appear firmly placed within Caryophyllales in molecular phylogenetic analyses (Figure 7.1).

The core Caryophyllales clade has long been recognized; it corresponds generally to the Caryophyllales of recent classifications (e.g., Cronquist 1981; Cronquist and Thorne 1994; Takhtajan 1997). The core Caryophyllales (or Caryophyllales *sensu* Judd et al. 2002) comprise 18 families, although several currently recognized families are clearly poly- or paraphyletic and require recircumscription (Cuénoud et al. 2002). *Asteropeia* (Asteropeiaceae) is sister to all other core Caryophyllales; when included, *Physena* (Physenaceae) is sister to *Asteropeia,* and this pair is sister to all other core Caryophyllales (Figure 7.1).

Asteropeia and *Physena,* with five and two species, respectively, all of which are restricted to Madagascar, have been difficult to place taxonomically. In recent classifications, *Asteropeia* has been considered a member of Theales, either within Theaceae (Cronquist 1988) or in its own family, Asteropeiaceae (Takhtajan 1987; Thorne 1992a, 1992b). Monogeneric Physenaceae have been placed with Urticales (Cronquist 1988) or Sapindales (Takhtajan 1987); Thorne (1992a, 1992b) listed Physenaceae as *incertae sedis.* Neither genus has been considered closely related to Caryophyllales in either the narrow or broad sense. The sister-group relationship of *Physena* and *Asteropeia* was demonstrated by Morton et al. (1997), and the sister relationship of this clade to the remaining core Caryophyllales has been confirmed in several studies (e.g., Morton et al. 1997; D. Soltis et al. 2000; Cuénoud et al. 2002).

Within core Caryophyllales, several clades are well supported (Figure 7.1). Caryophyllaceae (86 genera, 2,200 species; mostly with a temperate distribution), Achatocarpaceae (3 genera, 7 species; western North

FIGURE 7.3 Flowers and plants of Caryophyllales. (A) *Opuntia pusilla* (Cactaceae), entire flower. (B) *Opuntia pusilla* (Cactaceae), flower in longitudinal section showing inferior ovary. (C) *Opuntia pusilla* (Cactaceae), areole with spines (1) and glochids (2). (D) *Arenaria patula* (Caryophyllaceae), flower in longitudinal section showing free central placentation. (E) *Silene virginica* (Caryophyllaceae), flower. (F) *Chenopodium album* (Amaranthaceae), tip of flowering shoot. (G) *Chenopodium album* (Amaranthaceae), bisexual flower. (H) *Plumbago europaea* (Plumbaginaceae), flower. (I) *Plumbago europaea* (Plumbaginaceae), inflorescence. (J) *Phytolacca americana* (Phytolaccaceae), inflorescence. (K)

Polygonum scandens (Polygonaceae), flower. (L) *Polygonum scandens* (Polygonaceae), stem in fruit, arrow indicates ochrea. (M) *Portulaca oleracea* (Portulacaceae), flower in longitudinal section, upper arrow (1) shows perigynous insertion of perianth; lower arrow (2) indicates basal placentation. (N) *Portulaca oleracea* (Portulacaceae), withered perianth attached to upper part of pyxis (circumscissile capsule). (O) *Portulaca oleracea* (Portulacaceae), base of pyxis. (P) *Drosera tracyi* (Droseraceae), plant. (Q) *Drosera tracyi* (Droseraceae), flower. (A to G, K, L, P, and Q, from Wood 1974; H and I, from Pax in Engler and Prantl 1897; J from Rogers 1985; M to O, from Bogle 1969.) ▶

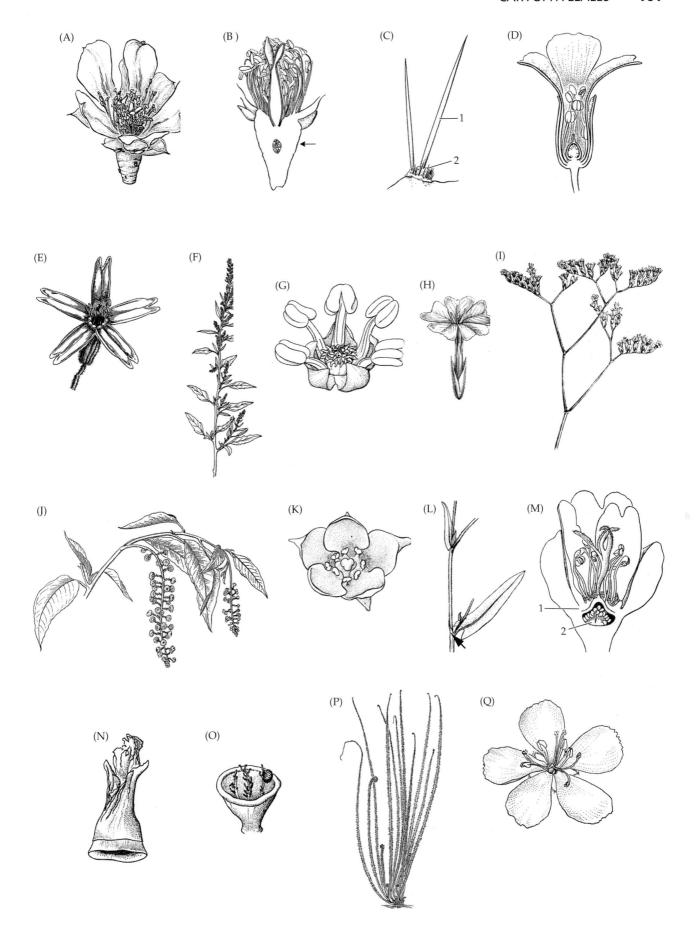

America to South America), and Amaranthaceae (including Chenopodiaceae; 174 genera, 2,050 species; warm temperate and subtropical regions) form a clade, with the last two as sisters. Amaranthaceae and Chenopodiaceae have appeared as sisters in several phylogenetic analyses (e.g., Giannasi et al. 1992; Rettig et al. 1992), but further sampling of both families has indicated that Amaranthaceae are derived from within Chenopodiaceae, making Chenopodiaceae paraphyletic. The monophyly of a broadly defined Amaranthaceae is supported by plastid DNA restriction site data, sequence data, and morphology (e.g., Rodman et al. 1984; Rodman 1990, 1994; Downie and Palmer 1994; Manhart and Rettig 1994; Downie et al. 1997). The conserved name Amaranthaceae has been applied to this clade. In the *rbcL* plus *matK* tree of Cuénoud et al. (2002), this Caryophyllaceae + (Achatocarpaceae + Amaranthaceae) clade is sister to all other core Caryophyllales; however, in the *matK*-only tree, this clade collapsed.

A second strongly supported clade within core Caryophyllales consists of Basellaceae (4 genera, 20 species; pantropical), Didiereaceae (4 genera, 11 species; Madagascar), Portulacaceae (32 genera, 385 species; cosmopolitan), Cactaceae (100 genera, 1,500 species, mostly New World deserts), and Halophytaceae (monospecific; Argentina). This succulent group largely corresponds to the "portulacaceous alliance" of Hershkovitz (1993), the suborder Portulacineae of Cronquist and Thorne (1994), or the "portulacaceous cohort" of Applequist and Wallace (2001) and reflects relationships long recognized, such as that between Basellaceae and Portulacaceae (e.g., Franz 1908; Bogle 1969; Sperling and Bittrich 1993) and Cactaceae and Didiereaceae or Portulacaceae (Chorinsky 1931; Gibson and Nobel 1986). The monophyly of this group was previously supported by plastid DNA sequence data (e.g., Rettig et al. 1992; Downie et al. 1997), and phylogenetic analyses of both morphology (Hershkovitz 1993) and ITS sequences (Hershkovitz and Zimmer 1997) indicated that Basellaceae, Cactaceae, and Didiereaceae are all embedded in Portulacaceae.

A third clade within core Caryophyllales is composed of Aizoaceae (125 genera, 2,020 species; mostly warm southern Africa and Australia) and Nyctaginaceae (30 genera, 395 species; pantropical)—each of which is monophyletic—and Phytolaccaceae (18 genera, 65 species; pantropical to warm temperate) and its segregates. Within this clade, Phytolaccaceae are polyphyletic (with an additional lineage outside this clade), and several segregate families are recommended by Cuénoud et al. (2002): Agdestidaceae (monospecific; warm Americas; not recognized by APG II), Petiveriaceae (monospecific; warm Americas; not recognized by APG II), Barbeuiaceae (monospecific; Madagascar),

and Sarcobataceae (monogeneric, two species; southwestern North America; segregated from Chenopodiaceae but clearly within this clade). Molluginaceae (13 genera, 120 species; pantropical and warm temperate) are also polyphyletic, and the precise relationships of these lineages are not clear. However, Molluginaceae seem to be part of the clade of core Caryophyllales that excludes Caryophyllaceae, Achatocarpaceae, Amaranthaceae, and Asteropeiaceae + Physenaceae. Gisekiaceae (monogeneric, 7 species; Africa, central and eastern Asia) have been segregated from Molluginaceae and appear closely related to Petiveriaceae (Cuénoud et al. 2002). *Stegnosperma* (3 species from Central America and the Antilles; Stegnospermataceae; or Stegnospermaceae Nakai 1942) is sister to this clade in the *rbcL* plus *matK* tree (Cuénoud et al. 2002).

Stegnosperma was segregated from Phytolaccaceae, in which it was originally placed, by Nakai (1942). It has since been viewed either as a monogeneric family, Stegnospermaceae, related to Phytolaccaceae or Pittosporales (Hutchinson 1959, 1973) or as a tribe or subfamily of Phytolaccaceae (e.g., Eckardt 1964; Cronquist 1968; Nowicke 1968; Thorne 1968;Takhtajan 1969). Hofmann's (1977) and Bedell's (1980) analyses of morphology and vegetative anatomy of *Stegnosperma*, combined with other available evidence, supported recognition of Stegnospermaceae and found it to be more similar to Caryophyllaceae than to Phytolaccaceae.

Synapomorphies for core Caryophyllales are unilacunar nodes, stems often with concentric rings of xylem and phloem or of vascular bundles, phloem sieve tubes with plastids with a peripheral ring of proteinaceous filaments and a central protein crystal (rather than the sieve tube plastids with starch grains of most other core eudicots), betalains (rather than anthocyanins, although anthocyanins are present in Caryophyllaceae and Molluginaceae), loss of the intron in the plastid gene *rpl2,* a single perianth whorl, "secondarily" free central to (primarily) basal placentation (Eckardt 1976), embryo curved around the seed, and presence of perisperm with scanty or no endosperm (per Judd et al. 2002). Placentation in core Caryophyllales merits special attention. Gynoecial development in those Caryophyllales considered to have free central placentation is completely different from the free central placentation found in other taxa (e.g., the asterid family Primulaceae). Caryophyllales have "secondarily" free central placentation; the ovary has normal septa in young flowers, and these disintegrate in later development, either before or after anthesis. In contrast, the gynoecia in other groups with free central placentation never develop septa (Eckardt 1976).

Sieve-element plastids of subtype PIII characterize core Caryophyllales. Different forms of PIII plastids, classified by the type of crystal in the plastid, correspond to families or groups of families within core Caryophyllales. For example, form PIIIc'f plastids, characterized by a polygonal protein crystal (Behnke 1976a, 1976b, 1982), occur in Caryophyllaceae, Stegnospermataceae, and Achatocarpaceae (Behnke 1981, 1993; Behnke and Barthlott 1983), whereas most Caryophyllales have form PIIIcf plastids characterized by a globular crystal (Behnke 1981), and taxa classified in Amaranthaceae (including the former Chenopodiaceae) lack the protein crystal altogether (Behnke 1991). The presence of form PIIIcf plastids in *Sarcobatus* (Behnke 1993) supports its distinctness from those genera formerly classified in Chenopodiaceae, in which it had been placed; *Sarcobatus* (Sarcobataceae) is well removed from Chenopodiaceae (now included in Amaranthaceae) in the phylogenetic analyses of Cuénoud et al. (2002) (Figure 7.1). Some members of Polygonaceae (Polygonales) also have P-type plastids, although most Polygonaceae and all Plumbaginaceae studied to date have S-type plastids (Behnke 1999). The presence of form Pfs plastids, which lack crystals and contain starch, in at least three genera of Polygonaceae indicates multiple derivations of this plastid form.

Despite extensive study from many perspectives for several decades, relationships and patterns of evolution within core Caryophyllales are not entirely clear. Phytolaccaceae and Molluginaceae are polyphyletic, and Portulacaceae are paraphyletic. In addition, several new families have been proposed to accommodate the placement of single genera or small groups of genera as sister to larger clades. For example, Cuénoud et al. (2002) recommended the use of the names Agdestidaceae, Barbeuiaceae, Petiveriaceae, and Sarcobataceae, some of which were previously suggested by Nakai (1942), Behnke (1997), and Takhtajan (1997). Barbeuiaceae and Sarcobataceae have been used in APG II. Agdestidaceae and Petiveriaceae are proposed and therefore given in quotation marks in Figure 7.1.

The noncore Caryophyllales clade (*sensu* Cuénoud et al. 2002), which corresponds to the Polygonales of Judd et al. (2002), has only recently been identified through molecular phylogenetic analysis and includes several families that were previously classified in Cronquist's (1981) Rosidae or Dilleniidae. Within noncore Caryophyllales, Polygonaceae (43 genera, 1,100 species; cosmopolitan) and Plumbaginaceae (22 genera, 800 species; cosmopolitan, but centered in the Mediterranean region) form a strongly supported clade, as they have in many previous analyses (e.g., Giannasi et al. 1992; Rettig et al. 1992), with Tamaricaceae (5 genera, 90 species; Eurasia and Africa) and Frankeniaceae (monogeneric with 90 species; cosmopolitan) weakly supported as their sister. The sister to this clade comprises the carnivorous plants classified in Droseraceae (3 genera; 115 species; cosmopolitan), Drosophyllaceae (monospecific; Spain and Mo-

rocco), Dioncophyllaceae (3 monospecific genera; western Africa), and Nepenthaceae (monogeneric with 90 species; Madagascar to New Caledonia), together with Ancistrocladaceae (monogeneric with 12 species; tropical Africa to Borneo and Taiwan; see also Chapter 11). The relationship between Droseraceae and Nepenthaceae, and their relationship to Caryophyllidae, was first recognized on the basis of *rbcL* sequences (Albert et al. 1992; Chase et al. 1993; Williams et al. 1994). Analyses of *matK* sequences alone for several species of *Nepenthes, Drosera,* and *Ancistrocladus,* plus *Drosophyllum* and all three genera of Dioncophyllaceae, resulted in a well-supported tree with Droseraceae sister to a clade of Nepenthaceae + (*Drosophyllum* + (Dioncophyllaceae + Ancistrocladaceae)) (Meimberg et al. 2000). Judd et al. (2002) refer to the noncore Caryophyllales as Polygonales, reserving "Caryophyllales" for the expanded traditional order or core Caryophyllales *sensu* Cuénoud et al. (2002).

Possible synapomorphies for the noncore Caryophyllales are scattered secretory cells containing plumbagin, naphthaquinone (which has been lost in several clades), an indumentum of stalked, gland-headed hairs, basal placentation (with shifts to parietal in some Droseraceae and axile in Nepenthaceae), and starchy endosperm (Judd et al. 2002). Polygonaceae and Plumbaginaceae are united by ovaries with a single basal ovule and generally indehiscent fruits such as achenes or nuts. The clade of Droseraceae, Nepenthaceae, and relatives is probably supported by carnivory (see Chapter 11), circinate leaves, and pollen grains in tetrads (Judd et al. 2002).

Character Evolution in Caryophyllales

Chemical evolution: anthocyanins versus betalains

Anthocyanin and betalain production are mutually exclusive chemical pathways. In all core Caryophyllales examined except Caryophyllaceae and Molluginaceae, anthocyanin pigments have been replaced by betalains, which have taken on the functions of anthocyanins in both reproductive and vegetative tissues and are produced upon wounding, pathogenic infection, and senescence. These two pigment classes are apparently regulated in a similar, if not identical, manner (see e.g., Stafford 1994; Clement and Mabry 1996).

The biochemical and evolutionary mechanisms responsible for the complete replacement of anthocyanins by betalains in this single group of angiosperms remain a mystery. Anthocyanin production occurs via the flavonoid biosynthetic pathway, but, although anthocyanins are absent from betalain-producing plants, the precursors of anthocyanins are present. It appears that a single enzymatic step in the formation of anthocyanins is missing in betalain producers—the step that converts leucoanthocyanidin into anthocyanidin, the sugarless backbone of an anthocyanin pigment (Clement and Mabry 1996).

Molecular biology may hold the answers to the mystery of betalain versus anthocyanin production (Stafford 1994; Clement and Mabry 1996). As the anthocyanin pathway and its underlying genes become better understood, it may be possible to identify the crucial anthocyanidin-producing gene in Caryophyllaceae and Molluginaceae and compare its sequence to those of anthocyanin producers outside Caryophyllales. Furthermore, analysis of expression patterns of betalains and flavonoids after wounding could indicate whether the genes controlling these pathways are coregulated. The evolutionary costs and benefits of anthocyanin versus betalain production also deserve further attention.

From our reconstruction of the evolution of anthocyanins and betalains in Caryophyllales (Figure 7.4), it appears that either betalains were acquired independently in Amaranthaceae + Achatocarpaceae and the ancestor of the remainder of the betalain-producing members of core Caryophyllales or betalain production evolved once in the ancestor of the core Caryophyllales (minus Simmondsiaceae, Physenaceae, and Aster-

opeiaceae), with independent losses in Caryophyllaceae and Molluginaceae. Pigment type is still unknown for Barbeuiaceae, Asteropeiaceae, and Physenaceae.

Pollen

No single pollen type characterizes all members of Caryophyllales. However, all core Caryophyllales examined possess pollen with a spinulose and tubuliferous/punctate ektexine and a spinulose aperture membrane or operculum (Nowicke 1975). Within core Caryophyllales, most families exhibit multiple pollen morphologies, indicating that pollen structure is labile in this clade. Three basic types of pollen have been observed in core Caryophyllales: tricolpate, pantoporate, and pantocolpate, all with a spinulose and tubuliferous/punctate ektexine (Nowicke 1975). Tricolpate pollen of similar morphology (type I) is found in Aizoaceae and some Cactaceae, Caryophyllaceae, Molluginaceae, Nyctaginaceae, Phytolaccaceae, and Portulacaceae. Pantoporate pollen (type II) occurs in Amaranthaceae (including Chenopodiaceae) and some genera of Caryophyllaceae, Nyctaginaceae, Phytolaccaceae, and Portulacaceae. The pantocolpate pollen (type III) of core Caryophyllales typically has 12 to 15 colpi and is found in Basellaceae and some genera of Cactaceae, Molluginaceae, Nyctaginaceae, Phytolaccaceae, and Portulacaceae. In addition to these common types of pollen, two narrowly distributed pollen types have been reported for core Caryophyllaceae (Nowicke 1975): tricolpate with a reticulate ektexine (type IV) in Nyctaginaceae and pantoporate with a reticulate ektexine (type V) in Caryophyllaceae. A divergent form of pantoporate grains with reticulate ektexine is found in some Amaranthaceae and Cactaceae.

Certain aspects of structural characteristics of the pollen (e.g., partially fused columellae) unite some members of Basellaceae, Didiereaceae, Portulacaceae, and Cactaceae (Nowicke 1994, 1996; see also Erdtman 1966) in Cronquist and Thorne's (1994) suborder Portulacineae. However, at least some of the exine characteristics that unite these groups are also found elsewhere in core Caryophyllales (Nowicke 1975, 1996). Despite their small size (fewer than 20 species in 4 genera), Basellaceae are highly polymorphic in pollen morphology (Nowicke 1994, 1996). All pollen of Didiereaceae is 5-7-zonocolpate, a pollen type that is apparently unique in core Caryophyllales, with a spinulose and annular perforate or punctate tectum (Nowicke 1996). Cactaceae and Portulacaceae are polymorphic, with pollen types I and III and types II and III, respectively (e.g., Nowicke 1975, 1994).

Variation in pollen type within Amaranthaceae generally supports the subfamilies (Amaranthoideae and Gomphrenoideae) but not the tribes and subtribes recognized by Schinz (1893) (e.g., Erdtman 1952, 1966; Vishnu-Mittre 1963; Nowicke 1975; Eliasson 1988; Borsch

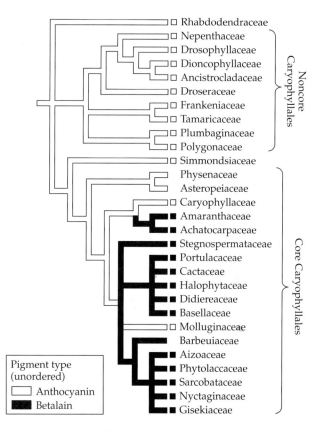

FIGURE 7.4 Reconstruction of evolution of anthocyanin and betalain pigments in Caryophyllales.

1998). The "*Amaranthus*-type" pollen (Erdtman 1952, 1966), corresponding to Nowicke's (1975) type II pollen, occurs in most Amaranthaceae and some Caryophyllaceae and Phytolaccaceae (Nowicke 1975). Erdtman's (1952, 1966) "*Gomphrena*-type" pollen occurs in only a handful of genera (and corresponds to a "specialized" category in Nowicke 1975). Variation within both categories is extensive (Borsch 1998).

Pollen of Achatocarpaceae (*Achatocarpus* and *Phaulothamnus*) differs from that of all core Caryophyllales in being hexaporate with a scabrate ektexine (Bortenschlager et al. 1972; Nowicke 1975; Skvarla and Nowicke 1976, 1977, 1979, 1982). Pollen of *Simmondsia* is triporate with an irregularly scabrate tectum (Nowicke and Skvarla 1984), distinct from that of all core Caryophyllales and from Buxaceae, Euphorbiaceae, and Pandaceae with which it had previously been placed (e.g., Hutchinson 1959; Scholz 1964; Takhtajan 1969; Cronquist 1981).

A comprehensive comparative analysis of pollen in noncore Caryophyllales has not yet been conducted. Given the many gaps in pollen data across Caryophyllales, reconstruction of pollen evolution is premature.

Wood anatomy

Some features of wood anatomy appear to be synapomorphies for clades within Caryophyllales. Successive cambia are widespread in core Caryophyllales (Metcalfe and Chalk 1950) and have also been reported from *Simmondsia, Rhabdodendron* (Carlquist 2001b), *Stegnosperma* (Horak 1981; Carlquist 1999a), and at least one genus each of Plumbaginaceae (Carlquist and Boggs 1999) and Polygonaceae (see Carlquist 2001b). Successive cambia are missing from Achatocarpaceae, Cactaceae, Didiereaceae, Portulacaceae, and some Phytolaccaceae. Carlquist (2001b) suggested that the genetic mechanisms for the production of successive cambia may have arisen at the base of the core Caryophyllales and were then lost in those families that lack successive cambia. However, it may be that initiation of subsequent cambial layers is delayed, as in *Stegnosperma* (Horak 1981), rather than absent. Successive cambia have not been reported in the woody members of the carnivorous clade, but their presence in Polygonaceae and Plumbaginaceae indicates that the origin of this feature may predate the entire Caryophyllales, with multiple losses of the trait. Alternatively, it could have arisen independently in Polygonaceae and Plumbaginaceae and core Caryophyllales.

Several wood features could be synapomorphies for the entire Caryophyllales and were later lost in all or most of the core Caryophyllales (depending on the placement of *Rhabdodendron*). Alternatively, these wood features could represent cases of parallel evolution or possibly synapomorphies of Judd et al.'s (2002) Polygonales. The vestured pits in vessels and tracheids of *Rhabdodendron* also occur in Polygonaceae (Ter Welle 1976; Carlquist 2001b). Rhabdodendraceae, Polygonaceae, Plumbaginaceae, Ancistrocladaceae, and Dioncophyllaceae all have silica bodies (Gottwald and Parameswaran 1968; Ter Welle 1976; Carlquist 1988, 2001b; Carlquist and Boggs 1996). Nonbordered perforation plates are also present in Rhabdodendraceae and are common in much of Caryophyllales (Carlquist 1999b, 2000, 2001b). Many wood characters, such as simple perforation plates, small pits on lateral vessel walls, libriform fibers, paratracheal axial parenchyma, storying, silica bodies, and dark-staining amorphous deposits, also indicate a close relationship between Polygonaceae and Plumbaginaceae (Carlquist and Boggs 1996).

The carnivorous clade of Droseraceae, Nepenthaceae, Drosophyllaceae, Dioncophyllaceae, and Ancistrocladaceae is also apparently supported by wood anatomical features. Although secondary growth is limited in members of this clade, vessel elements with simple perforation plates, fibriform vessel elements, tracheids with large, fully bordered pits, diffuse (and variously grouped) axial parenchyma, and paedomorphic rays (*sensu* Carlquist 1988) one to two cells wide are shared by those members of this clade that have been analyzed to date (Gottwald and Parameswaran 1968; Carlquist 1981; Carlquist and Wilson 1995). Some features, such as the simple perforation plates, are also common to other members of Caryophyllales and may not be synapomorphies for the carnivorous clade.

Despite the many investigations of various aspects of wood anatomy for members of Caryophyllales, gaps remain for most characters. Even a character such as successive cambia, the presence of which has been investigated in many taxa, has not been scored in several other groups, leaving potentially significant gaps in our knowledge. Furthermore, scoring of successive cambia as "present or absent" may introduce additional problems because evidence of successive cambia may arise at different stages of development and in different tissues in different taxa. Thus, further analysis of the evolution of wood characters requires additional comparative data before optimization of the observed variation across a phylogenetic tree should be attempted.

Extreme environments

Many members of Caryophyllales are adapted structurally or physiologically to extreme environments such as deserts, high-alkaline soils, high-saline substrates, and nutrient-poor soils. They have conquered these habitats through a variety of adaptations such as unusual photosynthetic pathways (crassulacean acid metabolism, CAM, and C_4 as opposed to C_3 photosynthesis), unusual morphologies (e.g., succulence), secretion of excessive salt by special glands, and unusual methods of nutrient uptake (e.g., carnivory). Given the distributions of these adaptations across Caryophyllales, it appears that most of these adaptations have arisen many times.

PHOTOSYNTHETIC PATHWAYS C_4 photosynthesis has apparently evolved independently in several lineages of Caryophyllales. However, the predisposition to evolve this suite of traits may have arisen in an ancestor common to those lineages that eventually developed this adaptation to high-light, high-temperature environments. For example, the lability of photosynthetic pathways is clear in members of *Salsola* and relatives (formerly classified in tribe Salsoleae of Chenopodiaceae but now considered part of Amaranthaceae). This group exhibits in microcosm the patterns of photosynthetic variation present in the family as a whole. A highly resolved phylogenetic tree of *Salsola* based on ITS (internal transcribed spacer) sequences largely agrees with the photosynthetic type and anatomy of leaves and cotyledons and provides strong support for the origin and evolution of two main lineages of plants in "tribe Salsoleae"—the NAD-ME and NADP-ME types of C_4 photosynthesis, respectively (Pyankov et al. 2001). Reconstruction of photosynthetic characters on the ITS phylogenetic tree demonstrates a single origin of C_4 photosynthesis, with subsequent divergence into the NAD-ME and NADP-ME lineages and two reversions to C_3 photosynthesis.

The topology for core Caryophyllales indicates that CAM has also arisen multiple times within this clade, including independently in Cactaceae and Aizoaceae. The similar selection pressures exerted by the arid habitats occupied by members of these clades have resulted in a spectacular convergence in morphology and photosynthetic pathway.

CARNIVORY Multiple mechanisms of carnivory have evolved in Caryophyllales (see discussion of carnivory in Chapter 11). Insects are trapped by pitfall traps in Nepenthaceae, flypaper traps in *Drosera*, *Drosophyllum*, and *Triphyophyllum*, and snap-traps in *Dionaea* and *Aldrovanda*. Although previous classifications considered only *Drosera* and *Dionaea* to be closely related to each other (both in Droseraceae, e.g., Cronquist 1981) and none of these groups to be close to Caryophyllales, all phylogenetic analyses of the past decade have indicated both their relationship to each other and to Caryophyllidae *sensu* Cronquist (1981) (e.g., Albert et al. 1992; Chase et al. 1993; Williams et al. 1994; D. Soltis et al. 1997a, 2000; Meimberg et al. 2000). Although the relationships among these carnivorous genera have varied with taxon sampling and gene(s) analyzed (e.g., Albert et al. 1992; Williams et al. 1994; Fay et al. 1997a; Lledó et al. 1998; Meimberg et al. 2000; Cameron et al. 2002), their phylogenetic relationships now seem clear: Droseraceae are sister to a clade of Nepenthaceae +

(Drosophyllaceae + (Ancistrocladaceae + Dioncophyllaceae)) (Meimberg et al. 2000; Cameron et al. 2002). This topology implies that "carnivory" was gained a single time in Caryophyllales and lost independently in Ancistrocladaceae and some Dioncophyllaceae (see Meimberg et al. 2000). Carnivory was achieved through several different mechanisms (pitchers, flypaper traps, snap-traps), involving extreme modifications of leaves and glands (see Williams 1976; Juniper et al. 1989; Albert et al. 1992; Chapter 11); each method apparently evolved only once in this clade (Cameron et al. 2002 for snap-traps, and inferences from trees of Meimberg et al. 2000 and Cameron et al. 2002 for pitchers and flypaper traps; see Chapter 11). The absence of carnivory in Ancistrocladaceae and some Dioncophyllaceae results from the loss of the flypaper trap.

Floral development

The perianth of Caryophyllales is highly variable, often unicyclic and sepaloid or sepal-derived, but in other cases dicyclic with both sepaloid and petaloid whorls. However, the nature of the petals in Caryophyllales has long been debated: are they true petals, or staminodes, or appendices of stamens (see review by Ronse De Craene et al. 1998a, 1998b)? Inferences of the perianth require analyses of the androecium as well; stamen number in Caryophyllales ranges from one to more than 100. Analyses of floral development in many members of Caryophyllaceae and additional members of other families may indicate that the apparent "diplostemonous" condition of Caryophyllales (i.e., 5 sepals, 5 petals, 10 stamens, 5-carpellate gynoecium) differs developmentally from diplostemony in other core eudicots (Ronse De Craene and Smets 1993, 1995; Vanvinckenroye and Smets 1996; Ronse De Craene et al. 1997, 1998). The five "petals" were interpreted as being derived from a hexamerous whorl of three stamen pairs, followed by loss of a stamen. A similar process was suggested to account for the outer (antesepalous) stamens, which likely correspond to one whorl of three stamens and a second whorl of two; loss of one stamen from a hexamerous whorl (or two whorls of three stamens) resulted in five inner (antepetalous) stamens. The result of these modifications is an apparently diplostemonous flower; however, this final form would not be homologous to that of other core eudicots because it was achieved independently via a different mechanism (as interpreted by Ronse De Craene and Smets 1993). It has therefore been referred to as "pseudodiplostemony" and corresponds to a floral formula of 5 sepals, 5 "petals" (sterile stamens), 3 plus 2 fertile antesepalous stamens, 5 antepetalous stamens, 5-carpellate gynoecium. Analyses of expression of organ-initiation genes could evaluate these morphologically and developmentally based hypotheses of floral structure.

Future Research

Although many aspects of core Caryophyllales have been thoroughly investigated during the past several decades (e.g., chemistry by Mabry and colleagues, pollen by Nowicke and colleagues, wood anatomy by Carlquist and colleagues), revised views on the phylogenetic relationships of the former Centrospermae and its relatives, plus the noncore Caryophyllales clade, require renewed efforts to identify synapomorphies and to reconstruct patterns of morphological, chemical, and anatomical evolution. Comparative datasets for many characters have many gaps because several families are poorly studied or were not previously considered closely related to other Caryophyllales. Furthermore, the poly- and paraphyly of several important families of core Caryophyllales demonstrate that additional phylogenetic study, incorporating increased taxon sampling above previous analyses, is needed to resolve relationships and develop a truly phylogenetic classification. Finally, the placement of Caryophyllales relative to other core eudicots is required for understanding the origins of many features found only, or mostly, in this clade, such as betalains, P-type sieve tube plastids, and pseudodiplostemony. Caryophyllales therefore continue to offer exciting research possibilities in systematics and evolutionary biology.

Rosids

Introduction

Initial analyses of the angiosperms using single genes (*rbcL*, Chase et al. 1993; *atpB*, Savolainen et al. 2000a; 18S rDNA, Soltis et al. 1997a) recovered a rosid clade that included more families than belonged to rosid groups (e.g., Rosidae, Rosanae) of modern classifications (Cronquist 1981, 1988; Dahlgren 1983; Takhtajan 1987, 1997; Thorne 1992a, 1992b, 2001). Analysis of non-DNA characters by Hufford (1992) recovered a similar rosid clade, although far fewer taxa were sampled. The larger non-DNA analysis of Nandi et al. (1998) recovered a large rosid clade, but the clade also included Caryophyllales, and not all taxa now considered rosids from DNA-based analyses were part of this rosid clade. However, none of these analyses provided internal support (e.g., bootstrap or jackknife) above 50% for this expanded rosid clade. As datasets were combined, *atpB* plus *rbcL* provided weak internal support (61%) for this large assemblage of angiosperms, and with the combined three-gene data (P. Soltis et al. 1999b; D. Soltis et al. 2000), the monophyly of the rosid clade, excluding Vitaceae, received strong jackknife support (99%).

The rosid clade is enormous, encompassing approximately 140 families and perhaps one-third of all angiosperm species and roughly 39% of eudicot species diversity (Magallón et al. 1999). The oldest fossils representing the clade are from the late Santonian to Turonian (about 84–89.5 Mya; Crepet and Nixon 1998; Magallón et al. 1999). A rosid flower, as yet unnamed, from the Dakota Formation in Nebraska is reported to be 94 Mya (Basinger and Dilcher 1984). In addition to most of the families previously placed in Rosidae, many families of Dilleniidae and Hamamelidae (groups that are now known to be grossly polyphyletic—see Chapter 2) are also placed in the rosid clade by molecular characters. For example, Casuarinaceae, Fagaceae, Juglandaceae, Moraceae, Myricaceae, Ulmaceae, and Urticaceae are all members of the former subclass Hamamelidae now placed in the rosid clade. Similarly, Brassicaceae, Caricaceae, Datiscaceae, Dipterocarpaceae, Malvaceae, Ochnaceae, Passifloraceae, and Violaceae are some of the families of Dilleniidae now part of the rosid clade. Even a few families formerly placed in Magnoliidae (e.g., Coriariaceae, Corynocarpaceae) are now considered rosids. A few families of Rosidae of recent classifications do not appear in the rosid clade, but instead have been shown to be asterids. Examples include Apiaceae, Araliaceae, Cornaceae, Hydrangeaceae, and Pittosporaceae (see Chapter 9).

Although great progress has been made in defining and resolving relationships within the rosids, many relationships within this clade remain poorly understood. Non-DNA synapomorphies are still unknown for the rosid clade, as well as for many subclades within the rosids. Although several morphological and anatomical features are shared by many rosids, at this point none of these traits represent clear synapomorphies. Noteworthy shared non-DNA features of rosids include nuclear endosperm development, reticulate pollen exine, generally simple perforations of vessel end-walls, alternate intervessel pitting, mucilaginous leaf epidermis, ellagic acid (see Chapter 5), and often two (diplostemony) or more whorls of stamens (Hufford 1992; Nandi et al. 1998).

One factor contributing to the complexity of the rosids is the sheer size of the group. However, aspects other than size contribute to the poor understanding of rosid relationships. The asterid clade, which is of comparable size, is much better understood in terms of relationships, possible non-DNA synapomorphies, and character evolution (see Chapter 9). Similarly, enormous advances have been made in elucidating monocot relationships (see Chapter 4). One reason the rosid clade still needs intensive study is that it contains many morphologically disparate families that were long considered so distinctive they were placed in four different subclasses of recent classifications—Rosidae, Hamamelidae, Dilleniidae, and Magnoliidae. These morphologically diverse families now occur interdigitated within the rosid clade. As is evident from the overview of rosid orders provided in this chapter, some families long placed in subclass Rosidae are now known to have as their closest relatives members of the former subclasses Dilleniidae and Hamamelidae. Hence in many cases, families now considered closely related in the rosid clade were not considered closely related in the past; as a result, the uniting morphological, chemical, and other non-DNA characters have not yet been ascertained. Complicating the problem, many

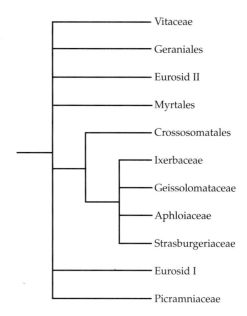

FIGURE 8.1 Summary tree providing a conservative estimate of phylogeny for rosids from recent molecular phylogenetic analyses, including Savolainen et al. (2000a, 2000b) and D. Soltis et al. (2000, 2003a).

non-DNA characters of interest have not been surveyed throughout the rosids. There are so many gaps in taxonomic coverage of these characters that the potential utility of many traits cannot be meaningfully assessed until more basic survey work is conducted. In contrast, although the asterid clade is larger than the Asteridae of recent classifications, most of the additions (e.g., Cornales, Ericales) have been placed at the basalmost nodes and are outside the core of the asterid clade. The core asterids remain highly similar to the Asteridae recognized before DNA-based phylogenetic analyses.

Phylogenetic studies using *rbcL* and *atpB* alone revealed two large subclades within the rosids, but neither of these subclades obtained bootstrap or jackknife support above 50% (e.g., Chase et al. 1993; Savolainen et al. 2000a, 2000b). These two large subclades were termed rosid I and II (see Chase et al. 1993) or eurosid I and II (Savolainen et al. 2000a, 2000b; D. Soltis et al. 2000; Figure 8.1). Analyses of combined molecular datasets provided the first internal support for these two large subclades of rosids (Figure 8.1). The three-gene analysis (P. Soltis et al. 1999b; D. Soltis et al. 2000) provided moderate internal support (77%) for a eurosid I clade of Celastrales, Cucurbitales, Fabales, Fagales, Zygophyllaceae + Krameriaceae, Malpighiales, Oxalidales, and Rosales, and strong jackknife support (97%) for a eurosid II clade of Brassicales, Malvales, and Sapindales, plus Tapisciaceae—a family not included in the Chase et al. (1993) or Nandi et al. (1998) analyses.

The three-gene analysis, which sampled more taxa than earlier investigations, also provided evidence for a new subclade, Crossosomatales (APG II 2003; see "Other Rosids," below), but its placement, as well as those of Myrtales and Geraniales, did not receive support above 50%. In the strict consensus of shortest trees in the three-gene analysis, Crossosomatales and Geraniales appeared as successive sisters to the eurosid II clade, but the placement of Myrtales was uncertain. In analyses using only *rbcL* or *rbcL* plus *atpB,* Myrtales also appeared as a sister group to eurosid II (e.g., Savolainen et al. 2000a, 2000b). In a four-gene analysis of eudicots (including 26S rDNA sequences), relationships of these rosid lineages remained unclear (D. Soltis et al. 2003a; Figure 8.1).

Thus, the placements of Myrtales, Geraniales, Crossosomatales, and Picramniaceae (Figure 8.1) remain among the major deep-level questions of relationships within the rosids. At this point, the eurosid I clade comprises Celastrales, Cucurbitales, Fabales, Fagales, Malpighiales, Oxalidales, Rosales, and Zygophyllaceae + Krameriaceae; eurosid II should be limited to Brassicales, Malvales, Sapindales, and Tapisciaceae. Once their circumscriptions are clear, eurosid I and eurosid II should be given more meaningful names that are more easily remembered; fabids (for Fabaceae) and malvids (for Malvaceae) have been proposed to refer to the eurosid I and eurosid II clades, respectively.

In the next sections, we discuss in turn the composition of the rosid clade, beginning with Vitaceae, a family that appeared as sister to all other rosids in several molecular phylogenetic analyses. We then discuss the two large rosid subclades, eurosid I (fabids) and eurosid II (malvids). Lastly, we present an overview of the three groups of rosids of uncertain placement, Myrtales, Geraniales, and Crossosomatales. The placements of several rosid families also remain uncertain: Picramniaceae (including Alvaradoaceae), Huaceae, Aphloiaceae, Geissolomataceae, Ixerbaceae, and Strasburgeriaceae. Picramniaceae and Huaceae are discussed briefly at the end of the chapter. Aphloiaceae, Geissolomataceae, Ixerbaceae, and Strasburgeriaceae may be members of Crossosomatales and are covered there.

Vitaceae

In modern classifications, Vitaceae have been placed within Rosidae, typically within Rhamnales (Cronquist 1981). However, the placement of Vitaceae, which include *Leea* (see Ingrouille et al. 2002; APG II 2003), differs in molecular phylogenetic analyses. In analyses of *rbcL,* Vitaceae appeared either as sister to Caryophyllales or as sister to the asterid clade (Chase et al. 1993). With *atpB* alone, Vitaceae were sister to Saxifragales (Savolainen et al. 2000a). In analyses of *rbcL* plus *atpB* (Savolainen et al. 2000a) and in analyses of the three-gene dataset, Vitaceae appeared as sister to the large, well-supported rosid clade, although with only 73%

jackknife support (D. Soltis et al. 2000). In a four-gene analysis of eudicots, Vitaceae again appeared as sister to the rosids, but without support above 50% (D. Soltis et al. 2003a; Figure 8.1). However, a recent analysis of a large *matK* dataset for angiosperms did not place Vitaceae sister to the rosids, but rather as sister to Dilleniaceae, albeit with low support (Hilu et al. 2003). In our own analyses, conducted with a larger three-gene dataset (with more taxa than in D. Soltis et al. 2000), Vitaceae again appear as sister to the remaining rosids. Vitaceae have been placed in the rosid clade (see APG II 2003; Chapter 10), but analyses involving additional genes are needed to evaluate this placement further.

The inclusion of *Leea* (Leeaceae) in Vitaceae is not a great surprise. Bentham and Hooker (1862–1883) recognized only three genera in Vitaceae: *Leea, Vitis,* and *Pterisanthes.* Thorne (1992a, 1992b) also placed *Leea* in Vitaceae as a distinct subfamily. Most modern classifications, however, have followed Planchon (1887) in placing *Leea* in its own family, Leeaceae, and recognizing 10 genera in Vitaceae. Vitaceae, *sensu* Planchon, are a family of mainly woody climbers characterized by the presence of leaf-opposed tendrils or inflorescences. *Leea* (a genus of about 70 species) is a large shrub or small tree lacking tendrils. Although most authors have treated it as a distinct family, Leeaceae, they have also recognized a close relationship to Vitaceae (e.g., Cronquist 1981; Takhtajan 1997; Judd et al. 2002).

Vitaceae, including *Leea,* have been extensively analyzed using *rbcL* sequences (Ingrouille et al. 2002). *Leea,* followed by *Ampelopsis,* appeared as sister to the remainder of the family (Ingrouille et al. 2002). Vegetative morphology is less specialized in *Leea* and *Ampelopsis* than in other Vitaceae. There is also a trend in leaf and tendril evolution: *Leea* lacks tendrils and has pinnate to multiply compound leaves; *Ampelopsis* is weakly tendrillate, but has tendrillate inflorescences and multiply compound leaves; other more derived genera are more regularly ternate or have simple, often tripartite leaves and clearly distinct tendrils and inflorescences. Putative synapomorphies of Vitaceae include raphide sacs, specialized stalked, gland-headed hairs, diminutive calyx, stamens opposite the petals, and berries (Judd et al. 2002).

Eurosid I (Fabids)

From the phylogenetic analyses of the three-gene dataset (D. Soltis et al. 2000), the eurosid I clade (fabids) comprises three subclades: (1) a weakly supported nitrogen-fixing clade (see also D. Soltis et al. 1995) consisting of Rosales, Fabales, Cucurbitales, and Fagales; (2) a well-supported clade of Krameriaceae and Zygophyllaceae; and (3) a weakly supported large clade of Malpighiales, Oxalidales, and Celastrales. Each of these groups is discussed in the next sections.

Nitrogen-fixing clade

Molecular phylogenetic analyses support a clade of Rosales, Fabales, Cucurbitales, and Fagales (APG 1998; APG II 2003; Figure 8.2). Representatives of these orders are illustrated in Figure 8.3. As reviewed in the next section ("Character Evolution: Nitrogen-Fixing Symbiosis"), these orders share a genetic predisposition for nitrogen fixation via root nodules, and this condition represents a possible synapomorphy for this group of four orders.

Only 10 of the approximately 400 families (APG II) of angiosperms form symbiotic associations with nitrogen-fixing bacteria involving root nodules (Table 8.1): Betulaceae, Casuarinaceae, Cannabaceae (including the nitrogen-fixing genus *Parasponia* of the former Celtidaceae; APG II 2003), Coriariaceae, Datiscaceae, Elaeagnaceae, Fabaceae, Myricaceae, Rhamnaceae, and Rosaceae (Akkermans and van Dijk 1981; Torrey and Berg 1988). In recent morphology-based classifications (e.g., Dahlgren 1980; Cronquist 1981, 1988; Takhtajan 1987, 1997; Thorne 1992a, 1992b, 2001), these families were considered distantly related. Cronquist (1981, 1988), for example, distributed these families among four of his six subclasses of dicots. Thus, the presence of these 10 families in one well-defined clade is a remarkable result that contrasts sharply with all modern morphology-based classifications.

The monophyly of those angiosperms forming such symbiotic associations was initially indicated by analyses of *rbcL* sequences (D. Soltis et al. 1995), but without bootstrap support above 50%. The clade containing these 10 families, distributed among what are now the orders Rosales, Cucurbitales, Fabales, and Fagales, has been referred to since that time as the nitrogen-fixing clade. However, many of the families in the nitrogen-fixing clade do not form symbiotic associations with nitrogen-fixing bacteria; furthermore, many of the genera and species within the 10 nitrogen-fixing families also lack this symbiosis (Table 8.1).

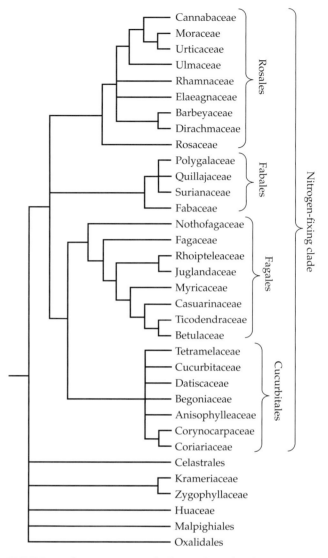

FIGURE 8.2 Summary tree of relationships for the nitrogen-fixing clade of rosids using the three-gene analysis of D. Soltis et al. (2000) as a backbone with portions of the tree updated following Sytsma et al. (2002), Li et al. (2004), and analyses conducted for this book.

FIGURE 8.3 Flowers and flowering stems representing the nitrogen-fixing clade: Rosales (A–H), Fagales (I–O), Fabales (P–R), and Cucurbitales (S–W). (A) *Cannabis sativa* (Cannabaceae), tip of flowering stem of staminate plant. (B) *Cannabis sativa* (Cannabaceae), carpellate flower; arrow indicates perianth. (C) *Cannabis sativa* (Cannabaceae), staminate flower. (D) *Potentilla canadensis* (Rosaceae), flower. (E) *Ulmus rubra* (Ulmaceae), branch with maturing fruit. (F) *Ulmus rubra* (Ulmaceae), flower. (G) *Maclura pomifera* (Moraceae), shoot with staminate inflorescences. (H) *Maclura pomifera* (Moraceae), staminate flower. (I) *Juglans nigra* (Juglandaceae), branch with staminate inflorescences. (J) *Juglans nigra* (Juglandaceae), carpellate flower; arrow indicates one of the two large stigmatic regions. (K) *Ostrya virginiana* (Betulaceae), flowering branch, with staminate catkins at tip; line indicates location of carpellate catkin. (L)

Fagus grandifolia (Fagaceae), branch with two infructescences ▶ (cupules enclosing fruits). (M) *Fagus grandifolia* (Fagaceae), staminate flower. (N) *Casuarina equisetifolia* (Casuarinaceae), staminate flower; 1 = outer bracteole; 2 = inner bracteole. (O) *Casuarina equisetifolia* (Casuarinaceae), branch with young infructescences. (P) *Vicia ludoviciana* (Fabaceae), flower. (Q) *Vicia ludoviciana* (Fabaceae), diadelphous stamens. (R) *Polygala pauciflora* (Polygalaceae), flower. (S) *Echinocystis lobatus* (Cucurbitaceae), staminate flower, line indicates androecium. (T) *Echinocystis lobatus* (Cucurbitaceae), node with leaf, tendrils, flowers, and fruit. (U) *Datisca cannabina* (Datiscaceae), carpellate flower. (V) *Datisca cannabina* (Datiscaceae), staminate flower. (W) *Datisca cannabina* (Datiscaceae), staminate plant in flower. (A to T, from Wood 1974. U to W, from Warburg in Engler and Prantl 1895.)

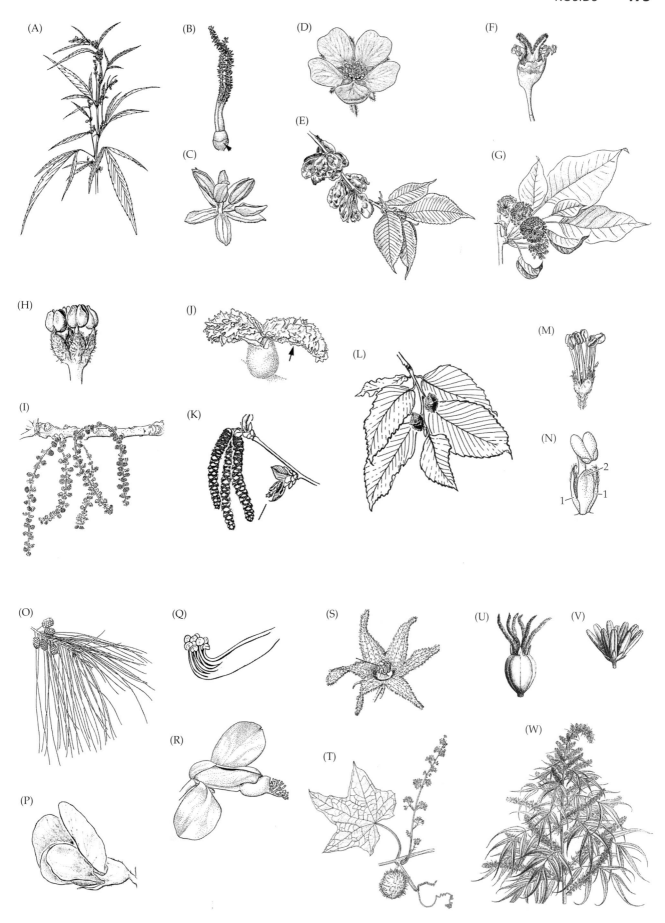

TABLE 8.1 Angiosperm families with nodular nitrogen-fixing symbioses and the frequency of this association in each family

Prokaryote	Angiosperm Family	Genera Having Root Nodules / Approx. Total Genera
Rhizobiaceae	Fabaceae	530 / 730
	Cannabaceae (including Celtidaceae)	1 / 11
Frankia	Betulaceae	1 / 6
	Casuarinaceae	4 / 4
	Elaeagnaceae	3 / 3
	Myricaceae	2 / 3
	Rhamnaceae	7 / 55
	Rosaceae	5 / 100
	Datiscaceae	1 / 1
	Coriariaceae	1 / 1

Although a clade of Rosales, Cucurbitales, and Fagales was also apparent with 18S rDNA sequence data, inclusion of Fabales in the same clade was less certain (D. Soltis et al. 1995, 1997a). Phylogenetic analyses of *atpB* and *atpB* plus *rbcL* also indicated a monophyletic nitrogen-fixing clade (Savolainen et al. 2000a), but again without strong internal support. The combination of three genes (*rbcL, atpB,* 18S rDNA) provided the strongest evidence yet (68%) for a monophyletic group containing all angiosperms engaged in nitrogen-fixing symbioses in root nodules. Although the internal support for this clade is still low, the evidence reinforces the hypothesis of a single origin of the predisposition of nitrogen-fixing symbiosis (D. Soltis et al. 1995). Some of the nitrogen-fixing families (Betulaceae, Casuarinaceae, Fabaceae, Elaeagnaceae, Myricaceae, and Ulmaceae) also share the ability to produce nodular hemoglobin (Landsmann et al. 1986).

The morphologically diverse mix of families present in the nitrogen-fixing clade illustrates well the problem of diagnosing DNA-based clades within the rosids. Families in the nitrogen-fixing clade were once placed in four different subclasses (e.g., Cronquist 1981, 1988). Families of Fagales (Betulaceae, Casuarinaceae, Fagaceae, Juglandaceae, and Myricaceae) represent former Hamamelidae. Other former hamamelids are found in Rosales, including Cannabaceae, Ulmaceae, Moraceae, and Urticaceae; Rosales also include members of the formerly recognized subclass Rosidae, such as Elaeagnaceae, Fabaceae, Rhamnaceae, and Rosaceae. Many Cucurbitales were formerly placed in Dilleniidae (e.g., Begoniaceae, Cucurbitaceae, and Datiscaceae),

with one family in the order, Coriariaceae, placed in Magnoliidae as a member of Ranunculales.

Within the nitrogen-fixing clade, the monophyly of each of the four subclades (Rosales, Cucurbitales, Fagales, and Fabales) received jackknife support of 100% or nearly so in analyses based on three genes (D. Soltis et al. 2000). However, relationships among these orders are unclear, with only a sister-group relationship between Cucurbitales and Fagales obtaining weak support (60%) (Figure 8.2).

CHARACTER EVOLUTION: NITROGEN-FIXING SYMBIOSIS Root nodules are induced and inhabited by one of two groups of distantly related bacteria. Species of Rhizobiaceae (gram-negative motile rods) nodulate the legumes and *Parasponia* (Cannabaceae; formerly of Celtidaceae; Trinick and Galbraith 1980). In the legumes, the symbiont is either *Rhizobium, Bradyrhizobium,* or *Azorhizobium;* only *Bradyrhizobium* is symbiotic with *Parasponia.* Actinomycetes of the genus *Frankia* (gram-positive, non-endospore–forming, mycelial bacteria) nodulate hosts in the remaining families (Table 8.1); these plants are referred to as actinorhizal (Akkermans and van Dijk 1981). Gunneraceae, a family that hosts nitrogen-fixing cyanobacterial symbionts in shoots (rather than in root nodules), were not found to be a member of the nitrogen-fixing clade, but instead appeared (with Myrothamnaceae) as sister to all other core eudicots (Savolainen et al. 2000a; D. Soltis et al. 2000, 2003a; see Chapter 6). Gunneraceae therefore represent an independent evolution of nitrogen-fixing symbiosis in angiosperms.

The putative gains and losses of nitrogen-fixing symbioses within the nitrogen-fixing clade were discussed by Swensen (1996) with a primary focus on the actinorhizal families. Doyle et al. (1997) considered the origins and losses of nitrogen-fixing symbioses in the legumes. We addressed the pattern of gain and loss of symbiosis by constructing a phylogenetic tree based on multiple genes and more taxa than were used in earlier analyses (Figure 8.4).

Because the bacterial symbionts involved (Rhizobiaceae versus *Frankia*) are different, those taxa involving Rhizobiaceae (Fabaceae and *Parasponia*) may represent separate origins from those using the genus *Frankia.* Furthermore, Fabaceae and *Parasponia* are distantly related (Figure 8.4) within the nitrogen-fixing clade, implying a minimum of two separate origins of nitrogen-fixing symbiosis with Rhizobiaceae.

There is evidence for multiple origins of nodulation involving Rhizobiaceae just within Fabaceae (J.J. Doyle 1994; Sprent 1994; Doyle et al. 1997). Most members of subfamily Mimosoideae (92% of those examined) nodulate, and *rbcL* data indicate that Mimosoideae are monophyletic. Similarly, most members of Papilionoideae nodulate (97% of those taxa investigated); the

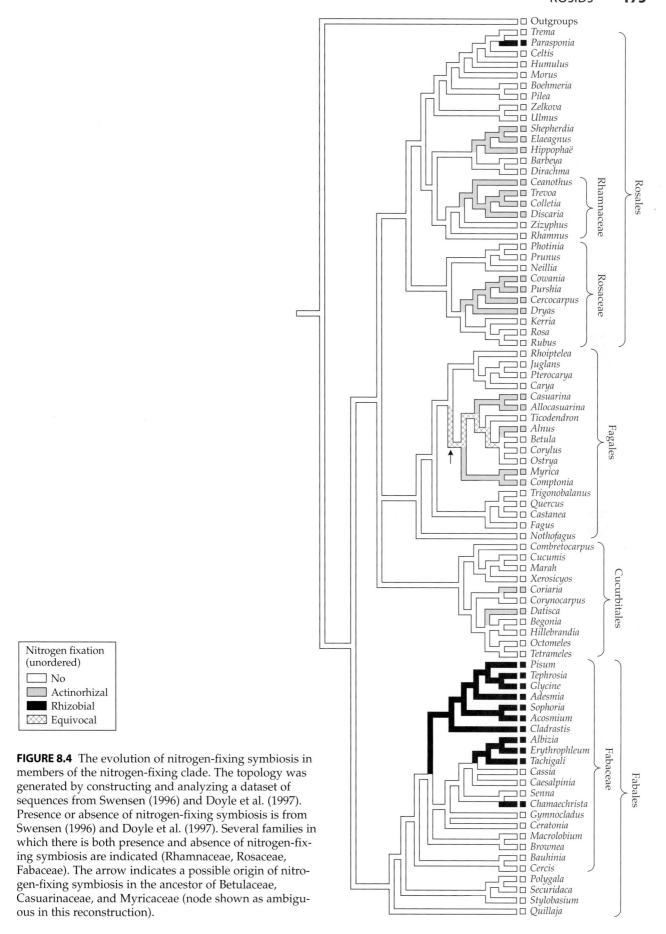

FIGURE 8.4 The evolution of nitrogen-fixing symbiosis in members of the nitrogen-fixing clade. The topology was generated by constructing and analyzing a dataset of sequences from Swensen (1996) and Doyle et al. (1997). Presence or absence of nitrogen-fixing symbiosis is from Swensen (1996) and Doyle et al. (1997). Several families in which there is both presence and absence of nitrogen-fixing symbiosis are indicated (Rhamnaceae, Rosaceae, Fabaceae). The arrow indicates a possible origin of nitrogen-fixing symbiosis in the ancestor of Betulaceae, Casuarinaceae, and Myricaceae (node shown as ambiguous in this reconstruction).

subfamily is also monophyletic. Those Papilionoideae that do not nodulate mostly represent early-branching lineages within the subfamily. Caesalpinioideae are paraphyletic to the other branches of the family, and few genera nodulate (approximately 5%). There is no evidence of nodulation at the base of the phylogenetic tree for Fabaceae; it now is clear that nodulation arose only after the earliest lineages had diverged (Doyle and Luckow 2003). Caesalpinioids confirmed as nodulating are either sister to mimosoids or belong to the more distantly related genus *Chamaecrista*. The most parsimonious explanation is that there are three origins of nitrogen-fixing symbiosis in the legumes—(1) the papilionoids; (2) the clade consisting of some caesalpinioids (e.g., the *Sclerolobium* and *Dimorphandra* groups of the highly polyphyletic tribe Caesalpinieae) and Mimosoideae; and (3) *Chamaecrista* (Doyle et al. 1997; Doyle and Luckow 2003).

Our analysis of the nitrogen-fixing clade indicates a minimum of six gains with two subsequent losses of actinorhizal symbiosis (Figure 8.4). Evolution of this symbiosis apparently occurred in: (1) Elaeagnaceae; (2) some Rhamnaceae; (3) some Rosaceae; (4) the common ancestor of Betulaceae, Casuarinaceae, and Myricaceae (see arrow on Figure 8.4); (5) *Coriaria*; and (6) *Datisca*. Our studies actually indicate that the ancestral state for the subclade (within Fagales) of Betulaceae, Casuarinaceae, and Myricaceae is ambiguous. If actinorhizal symbiosis is ancestral for this subclade, then two losses occurred, one in *Ticodendron* and a second in Betulaceae. Alternatively, Casuarinaceae (*Casuarina* and *Allocasuarina*), *Alnus,* and Myricaceae (*Myrica* and *Comptonia*) could represent three separate gains of actinorhizal symbiosis.

Swensen (1996) argued that comparative morphological data provide additional evidence for multiple origins of this symbiotic relationship within the actinorhizal lineages. For example, only host plants in the Fagales clade (Betulaceae, Casuarinaceae, and Myricaceae) are infected via root hairs (all others are infected by intercellular penetration) and possess nodules with a specialized oxygen diffusion-limiting cell layer; these same taxa also have higher levels of hemoglobin than hosts in other clades (reviewed in Swensen 1996). This provides evidence that Fagales represent a separate origin of actinorhizal nitrogen-fixing symbiosis. There is some evidence for multiple origins of nitrogen-fixing symbiosis within Fagales alone. *Alnus* differs from other members of this clade in several respects involving nodulation.

Coriaria and *Datisca* (Cucurbitales) are united by several anatomical features of the nodules. The infected cells are multinucleate, and the arrangement of vesicles within the cells is unique (reviewed in Swensen 1996). Thus, although the most parsimonious reconstruction

is two independent gains, one in *Coriaria* and one in *Datisca*, an alternative, slightly less parsimonious explanation is one gain, followed by two losses (one in *Corynocarpus* and another in the clade of *Begonia, Hillebrandia, Octomeles,* and *Tetrameles*).

CHARACTER EVOLUTION: WIND POLLINATION Wind pollination (anemophily) occurs in approximately 18% of flowering plant families and has clearly evolved many times (Linder 1998; Ackerman 2000; Culley et al. 2002), most likely in response to pollinator limitation and environmental variation. Wind pollination would be facilitated by movement of individuals into a geographic area with a distinct dry season in which conditions are unfavorable for insect pollination (Whitehead 1969, 1983; Weller et al. 1998; Culley et al. 2002). A phylogenetic analysis focused on the nitrogen-fixing clade indicates that wind pollination is more likely to evolve in those groups that have small, simple flowers and dry pollen (Culley et al. 2002). Culley et al. (2002) mapped perianth condition onto the strict consensus tree for the nitrogen-fixing clade and observed that wind pollination occurs only in those families in which the perianth is absent or reduced, whereas all the families having a well-developed perianth are biotically pollinated. Fagaceae illustrate this evolutionary tendency well. The family is characterized by small flowers, and ancestral members were apparently insect-pollinated; molecular analyses indicate multiple derivations of wind pollination within this single family (see "Fagales," below).

ROSALES Within the nitrogen-fixing clade, Rosales are strongly supported (100%) as monophyletic in molecular phylogenetic analyses (e.g., D. Soltis et al. 2000), consisting of nine families (Figure 8.2): Barbeyaceae, Cannabaceae (including Celtidaceae), Dirachmaceae, Elaeagnaceae, Moraceae, Rhamnaceae, Rosaceae, Ulmaceae, and Urticaceae (including Cecropiaceae). Rosales are only one of several examples of an order consisting of families of the former subclasses Hamamelidae (e.g., Ulmaceae, Urticaceae, Moraceae) and Rosidae (e.g., Rosaceae, Rhamnaceae, and Elaeagnaceae). Judd et al. (2002) suggested that a reduction (or lack) of endosperm and the presence of a hypanthium (a character absent in some wind-pollinated members, however) may be synapomorphies for the clade. Other possible uniting features are craspedodromous leaf venation and anthraquinones. As with many other rosid clades, critical analyses are needed to evaluate these and other potential synapomorphies.

Within Rosales (Figure 8.2) some relationships remain unclear, particularly within the Cannabaceae–Moraceae–Ulmaceae–Urticaceae complex, although recent molecular phylogenetic studies (Sytsma et al. 2002) have clarified many of these problems (e.g., Savolainen et al.

2000a, 2000b; D. Soltis et al. 2000). Rosaceae are sister to the remainder of the order, which comprises several sub-groups: Rhamnaceae; Barbeyaceae + Dirachmaceae (see Thulin et al. 1998 for discussion of these two families); Elaeagnaceae; and a clade of Ulmaceae, Cannabaceae, Urticaceae, and Moraceae (e.g., Savolainen et al. 2000a, 2000b; D. Soltis et al. 2000). Within the latter clade, Ulmaceae are sister to the remaining families; Cannabaceae form a well-supported clade that is sister to a well-supported Urticaceae + Moraceae.

Celtidaceae should be subsumed within Cannabaceae, a result supported by other lines of evidence, including ultrastructure and chromosome number (Sytsma et al. 2002). Additional molecular evidence for the recognition of Cannabaceae and Ulmaceae as distinct families was provided by Wiegrefe et al. (1998); morphological data supporting this distinction were presented by Grudzinskaja (1967).

Monophyly of Rhamnaceae and relationships within the family were discussed by Richardson et al. (2000). Several studies have clarified relationships within Rosaceae or major subclades within this large family (e.g., Morgan et al. 1994; Lee and Wen 2001). *Quillaja*, a genus placed in Rosaceae in most classifications, is distinct from this family and a member of Fabales (discussed below).

CHARACTER EVOLUTION IN ROSALES Rosaceae represent one of several examples in which base chromosome number agrees well with clades recovered in phylogenetic analyses, indicating that base chromosome number may be a good predictor of relationships in many groups. Other examples include Crassulaceae (Mort et al. 2001), Rutaceae (Chase et al. 1999b), and Onagraceae (Sytsma and Smith 1988; Levin et al. 2003; see Chapter 13).

Phylogenetic analyses based on sequence data indicated that there are clades within Rosaceae similar to the traditionally recognized subfamilies Maloideae, Prunoideae (Amygdaloideae), and Rosoideae; however, the clades Maloideae and Prunoideae each contain genera not traditionally placed in these subfamilies. Spiraeoideae are not monophyletic but consist of several distinct lineages (Morgan et al. 1994). Although both chromosome number and fruit type have been used to define groups within Rosaceae, chromosome number agrees well with the circumscription of clades within the family and is a more reliable indictor of relationships than fruit type (Figure 8.5). Phylogenetic analysis revealed a well-supported Rosoideae consisting primarily of taxa with $x = 7$ and some with $x = 8$. Those genera previously placed in Rosoideae with $x = 9$ were shown to be distantly related to this clade. Members of this Rosoideae clade possess achenes or fruits derived from achenes; however, the achene is not restricted to this clade but evolved more than once in the family (Figure 8.5). A Prunoideae clade (*Prunus, Prinsepia, Osmaronia* (= *Oemleria*), and *Exochorda*) is composed of taxa with $x = 9$; all of these genera except *Exochorda* have drupes. Drupes also occur in genera not part of Prunoideae.

Subfamily Maloideae has been defined by fruit type (a fleshy fruit, the pome; e.g., apple or pear). However, sequence data placed several problematic genera (*Vauquelinia, Lindleya,* and *Kageneckia,* usually placed in Spiraeoideae) with capsules and follicles in the maloid clade. These three genera possess high chromosome numbers ($x = 15, 17$) that agree with those in the traditionally recognized Maloideae ($x = 17$); these high numbers are the result of an ancient polyploid event that occurred in the ancestor of the maloid clade (e.g., Stebbins 1976; Evans and Campbell 2002). These results also indicate that the distinctive pome fruit was derived from dry follicular or capsular fruits (Figure 8.6). More genetic clues to the origin of the apple are provided by Harris et al. (2002).

FABALES The circumscription of Fabales recovered by phylogenetic analyses includes only four families, Fabaceae, Polygalaceae, Quillajaceae, and Surianaceae (Figure 8.2), although phylogenetic analyses of non-DNA characters did not find a close relationship between Fabaceae and Polygalaceae (Surianaceae and Quillajaceae were not included; Nandi et al. 1998). The composition of this clade is important in that the closest relatives of Fabaceae have long been debated (see J. J. Doyle 1994; Doyle and Luckow 2003). Non-DNA synapomorphies for Fabales are not clear, but may include vestured pits, vessel elements with single perforations, and a large, green embryo (Judd et al. 2002).

Fabales represent another surprising result of molecular phylogenetic analyses—the four constituent families (Fabaceae, Surianaceae, Polygalaceae, and Quillajaceae) have not been placed together in any modern classifications. Cronquist (1981, 1988) and Takhtajan (1997) both considered Fabaceae to be close to Connaraceae (see also Dickison 1981), the latter sister to Oxalidaceae (Oxalidales; discussed below). Polygalaceae have also been variously placed because of their unusual, bilaterally symmetric flowers (Figure 8.3R). Typically they have been associated with Malpighiaceae and Krameriaceae, two other families exhibiting bilateral floral symmetry (Cronquist 1981). Takhtajan (1997) considered Polygalaceae to be closely related to Xanthophyllaceae, Vochysiaceae, and Emblingiaceae, but only the first of these families is closely related to Polygalaceae and has been submerged within it (APG II 2003).

Quillaja had typically been placed in Rosaceae with *Kageneckia*, sometimes as a separate subfamily (Takhtajan 1987, 1997; Thorne 1992a, 1992b), because of several

(A)

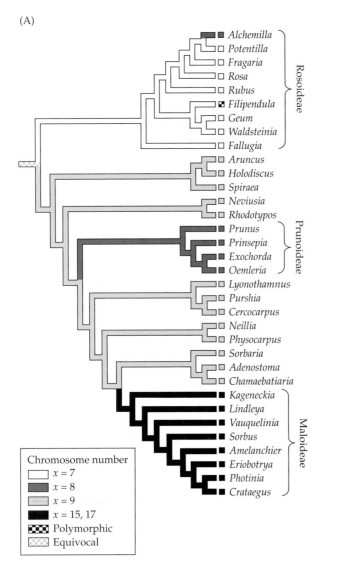

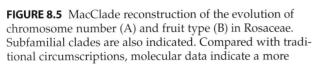

(B)

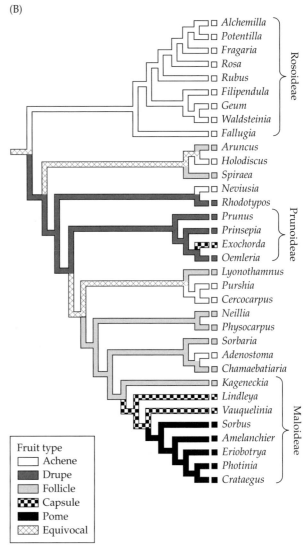

FIGURE 8.5 MacClade reconstruction of the evolution of chromosome number (A) and fruit type (B) in Rosaceae. Subfamilial clades are also indicated. Compared with traditional circumscriptions, molecular data indicate a more narrowly defined Rosoideae and more broadly defined Prunoideae and Maloideae; Spiraeoideae are polyphyletic. (Modified from Morgan et al. 1994.)

shared features including follicular fruits with numerous pleurotropic ovules that mature into winged seeds. Other features, however, distinguish *Quillaja* from Rosaceae, such as its base chromosome number of $x = 14$, which is not found elsewhere in Rosaceae. Furthermore, cytological studies indicated that $x = 14$ in *Quillaja* does not represent a tetraploid derivative of $x = 7$, a common number in Rosaceae (Goldblatt 1976). Bate-Smith (1965) noted chemical characters of *Quillaja* that are discordant with Rosaceae, including the presence of saponins and trihydroxy-substituted flavonoids.

The placement within Fabales of Surianaceae, a small, enigmatic family of five genera from Australia and Mexico, was also unexpected. The family had been placed with Simaroubaceae and Rutaceae in Sapindales (Takhtajan 1997), and Cronquist (1981) placed it in Rosales.

Relationships within Fabales are still unclear; resolving these relationships will require the sequencing of additional genes. Sequences from three genes (D. Soltis et al. 2000) provided weak support for Surianaceae + Polygalaceae as sister to Fabaceae, but the monotypic Quillajaceae were not included in the analysis. In analyses based on *rbcL* sequence data, *Quillaja* appeared as sister to Polygalaceae, with Quillajaceae + Polygalaceae sister to Surianaceae, but these relationships were not strongly supported (Savolainen et al. 2000b).

Both Fabaceae and Polygalaceae have been the focus of recent phylogenetic analyses (see Persson 2001 for a

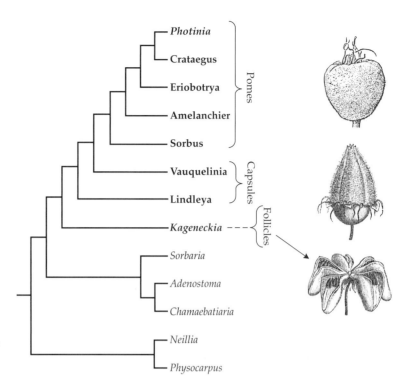

FIGURE 8.6 Proposed evolution of the pome from ancestral taxa having capsules in Rosaceae. Taxa in bold are ancient polyploids. Topology is modified from Morgan et al. (1994). (Illustration courtesy of D. Morgan.)

phylogenetic analysis of Polygalaceae). Phylogenetic analyses of Fabaceae based on *rbcL* sequences provided insights into the broad picture of relationships within the family. Both subfamilies Mimosoideae and Papilionoideae are monophyletic; however, Caesalpinioideae are paraphyletic to the other two (Doyle et al. 1997; Doyle and Luckow 2003).

Considerable progress has been made in recent years in clarifying relationships across Fabaceae as a whole and also within subclades within the family (e.g., Doyle et al. 1997; Bruneau et al. 2001; Kajita et al. 2001; Pennington et al. 2001; Hu et al. 2002; Doyle and Luckow 2003). Nonetheless, considering that Fabaceae are the third largest angiosperm family, relationships within the family are not as clearly understood as are those of the two largest angiosperm families, Orchidaceae and Asteraceae, both of which have been more intensively studied.

Cucurbitales

Cucurbitales as defined by APG (1998) and APG II (2003) consist of seven families: Anisophylleaceae, Begoniaceae, Coriariaceae, Corynocarpaceae, Cucurbitaceae, Datiscaceae, and Tetramelaceae (Figure 8.2). These families represent three subclasses of recent classifications. Cucurbitaceae, Begoniaceae, Tetramelaceae, and Datiscaceae were placed in Dilleniidae; Coriariaceae and Corynocarpaceae were variously placed within Rosidae or Magnoliidae; and Anisophylleaceae were often placed in Rosidae (Cronquist 1981, 1988; Takhtajan 1987, 1997).

A close relationship of Cucurbitaceae, Begoniaceae, Tetramelaceae, and Datiscaceae has been recognized by most authors (e.g., Cronquist 1981; Takhtajan 1997). However, Takhtajan (1997) considered Cucurbitaceae to be more closely related to Passifloraceae and Achariaceae (now in Malpighiales) than to Begoniaceae, Tetramelaceae, or Datiscaceae. The placement of Anisophylleaceae, Coriariaceae, and Corynocarpaceae as closely related to the other families of Cucurbitales was surprising, however, because a close relationship among these families had not been proposed previously. Corynocarpaceae, Coriariaceae, and Anisophylleaceae are all small, problematic families for which affinities were long considered uncertain, and a diverse array of placements had been proposed (e.g., Cronquist 1981, 1988; Takhtajan 1987, 1997; Thorne 1992a, 1992b, 2001).

As with many other orders of rosids, non-DNA synapomorphies for Cucurbitales are unclear. Cucurbitaceae, Begoniaceae, Tetramelaceae, and Datiscaceae share some distinctive features, including an inferior ovary, imperfect flowers, strongly protruding placentae, and a distinctive cucurbitoid tooth (Nandi et al. 1998). Comparisons of floral structure (Matthews and Endress 2004) indicate a close relationship between Coriariaceae and Corynocarpaceae, as well as between Tetramelaceae and Datiscaceae; these same comparisons also support a relationship of Tetramelaceae and Datiscaceae to Begoniaceae.

Most interfamilial relationships within Cucurbitales remain poorly resolved (see also Swensen et al. 1994,

1998), and the order requires additional study using more gene sequences and nonmolecular characters. Within the order, only the sister relationship of Coriariaceae and Corynocarpaceae (94%) receives support above 50% in analyses using three genes (D. Soltis et al. 2000). The sister-group relationship of Coriariaceae and Corynocarpaceae is noteworthy in that the relationships of both families have long been uncertain. This close relationship was first recovered in the combined *rbcL* and morphology analysis of Nandi et al. (1998). Corynocarpaceae, a family of one genus and perhaps seven species, have been placed in diverse orders including Ranunculales, Sapindales, and Celastrales (Takhtajan 1997; Cronquist 1981, 1988). Coriariaceae are also monogeneric, comprising five species; the placement of the family has been extremely controversial (Sharma 1968; Takhtajan 1997). Whereas some authors proposed relationships with Sapindales (Takhtajan 1997), Cronquist (1981) placed the family in Ranunculales. Coriariaceae and Corynocarpaceae share several features: fibers with simple pits, vascicentric axial parenchyma, scanty or lacking endosperm, and bitter chemical compounds (identified as sesquiterpenoids in *Coriaria*). Phylogenetic analyses using *rbcL* sequences indicated that Anisophylleaceae, an enigmatic tropical family of four genera, were part of Cucurbitales (Schwarzbach and Ricklefs 2000), but the closest relatives of the family were unclear. Anisophylleaceae have often been considered either closely related to, or part of, Rhizophoraceae (reviewed in Schwarzbach and Ricklefs 2000; see also "Malpighiales," below). However, molecular data indicated that Anisophylleaceae are distantly related to Rhizophoraceae. Matthews et al. (2001) and Schönenberger et al. (2001) showed striking similarities in floral structure of Anisophylleaceae with the distantly related Cunoniaceae (Oxalidales) and a Cretaceous fossil exhibiting features of both families.

FAGALES Fagales consist of eight woody families formerly considered part of subclass Hamamelidae (Figure 8.2): Betulaceae, Casuarinaceae, Fagaceae, Juglandaceae, Myricaceae, Nothofagaceae, Rhoipteleaceae, and Ticodendraceae (*Ticodendron* and Ticodendraceae were only recently described by Gómez-Laurito and Gómez 1989). Several of these families are illustrated in Figure 8.3. Molecular and nonmolecular analyses have demonstrated, however, that Fagales are rosids and not related to other former Hamamelidae such as Platanaceae, Cercidiphyllaceae, and Daphniphyllaceae (see chapters 2, 5, and 6; see also Hufford 1992; Nandi et al. 1998). Members of Fagales share several non-DNA characters, including an indumentum of gland-headed and/or stellate hairs, an inferior ovary (except Rhoipteleaceae), one or two ovules per locule, unisexual flowers with tepals reduced or lacking, nectaries lacking, a pollen tube that enters the ovule via the chalaza (cha-lazogamy), and indehiscent fruits with only one seed (by abortion of additional ovules; Hufford 1992; Manos and Steele 1997; Judd et al. 1999, 2002).

Manos and Steele (1997) provided an excellent overview of relationships for Fagales, and broader analyses (D. Soltis et al. 2000; Savolainen et al. 2000a, 2000b) agreed with their results. Topologies derived from sequence data recovered four clades within Fagales (Figure 8.2): Nothofagaceae, followed by Fagaceae, were shown to be well supported as successive sisters to the remainder of the clade. The rest of Fagales consists of two subclades: (1) a strongly supported Rhoipteleaceae + Juglandaceae; and (2) a weakly supported clade in which Myricaceae are sister to a clade of Casuarinaceae and Ticodendraceae + Betulaceae (Manos and Steele 1997). A recent multigene analysis of Fagales provided some support for Myricaceae as sister to Casuarinaceae and Ticodendraceae + Betulaceae (Li et al. 2004). However, Myricaceae share several morphological features with Rhoipteleaceae and Juglandaceae, including aromatic glandular hairs.

Molecular phylogenetic results for Fagales conflicted with relationships indicated by phylogenetic analyses of non-DNA characters (Hufford 1992; Loconte 1996). However, several characters provided additional support for the DNA-based topology. The distribution of the three types of pollen in the order (colporate, porate, stephanoporate) is in general agreement with the four clades of families within the order (see Manos and Steele 1997). In addition, Fagales are well represented in the fossil record (e.g., Jones 1986; Crepet and Nixon 1989; Herendeen et al. 1995), and the position of Nothofagaceae as sister to the remainder of the order is in agreement with the appearance of this family in the fossil record before other Fagales (Manos and Steele 1997). Manos (1997) examined phylogenetic relationships and biogeographic patterns within *Nothofagus*.

The sister-group relationship of Ticodendraceae and Betulaceae indicated by DNA data (Figure 8.2; see also Conti et al. 1996) is also supported by several non-DNA characters, such as similarities in floral morphology (Tobe 1991), wood and bark anatomy (Carlquist 1991), sieve element plastids (Behnke 1991), pollen (Feuer 1991), and leaf architecture (Hickey and Taylor 1991).

Molecular data also provided strong support for a clade of Juglandaceae and Rhoipteleaceae (Figure 8.2; Manos and Steele 1997; Chen et al. 1998; Li et al. 2004). Rhoipteleaceae are an enigmatic monogeneric family that could be included in Juglandaceae (APG II 2003; see Chapter 10), with several features that are unusual in Fagales (e.g., a superior ovary). However, Juglandaceae and Rhoipteleaceae are distinctive in Fagales in possessing compound leaves; a close relationship is also suggested by stem anatomy, aspects of floral morphology (Manning 1938, 1940; Withner 1941), and chromosomal data (Oginuma et al. 1995).

(A)

(B)

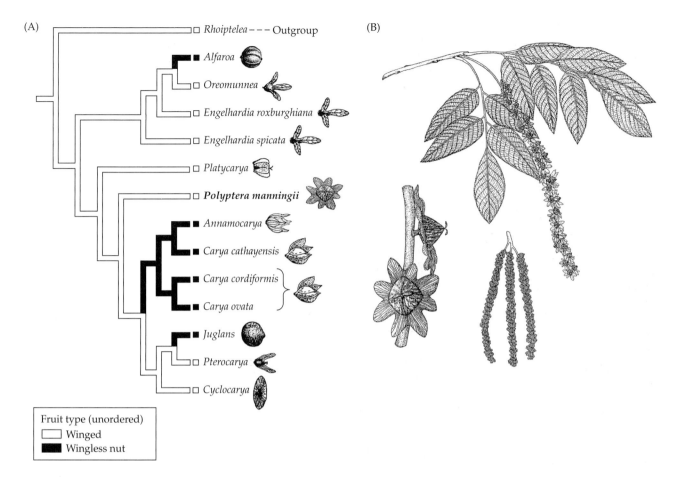

Fruit type (unordered)
☐ Winged
■ Wingless nut

FIGURE 8.7 Fruit diversification and fossils in Juglandaceae. The topology is modified from P. Manos (unpubl. data). The tree is based on both gene sequence and morphological data; the placement of the fossil *Polyptera manningii* (indicated in bold) is based on a total evidence analysis. (A) Fruit type. *Carya cordiformis* and *Carya ovata* are characterized by the same general fruit type. Fruit illustrations are from Manos and Stone (2001). (B) *Polyptera manningii*, a fossil member of Juglandaceae (from Manchester and Dilcher 1997). (Top) Twig with infructescence. (Bottom left) Detail of fruits. (Bottom right) Staminate catkins.

CHARACTER EVOLUTION IN FAGALES Molecular analyses of Fagales have provided new insights into character evolution within both Juglandaceae and Fagaceae. Analysis of fruit evolution in Juglandaceae indicated a dynamic picture of diversification (Manos and Stone 2001). Although wind dispersal appears to be ancestral within the family, four independent origins of wings from a combination of floral and accessory structures are suggested (Figure 8.7): (1) *Engelhardia* and *Oreomunnea* have wings formed from a trilobed bract; (2) *Platycarya* has two small wings that result from the fusion of bracteoles and lateral sepals; (3) *Pterocarya* has wings derived from bracteoles; and (4) *Cyclocarya* has a distinctive circular wing that is the result of fusion of bracts and bracteoles. Our reconstruction indicated that animal dispersal evolved independently three times, resulting in nuts formed uniquely by the fusion of different organs: (1) fusion of four sepals in *Alfaroa*; (2) fusion of bract, bracteoles, and sepals in *Juglans*; and (3) fusion of the bract and bracteoles in *Carya* and *Annamocarya*. In general, wind-dispersed seeds have epigeal germination, and those that are animal-dispersed are hypogeal, but *Oreomunnea* and *Cyclocarya* are exceptions in their respective clades in having wind-dispersed seeds with hypogeal germination.

Juglandaceae have been thoroughly investigated using DNA characters (Manos and Stone 2001). In addition, they have an excellent fossil record, and these fossils have been carefully investigated (e.g., Manchester and Dilcher 1997). Hence, the family affords the opportunity to integrate fossils into the excellent phylogenetic framework for extant taxa (P. Manos, unpubl. data). These studies have indicated placements for well-characterized fossil Juglandaceae, such as the genus *Polyptera* (Figure 8.7; for more examples of placing fossils within a phylogenetic framework see the Deep Time Website, www.flmnh.ufl.edu/deeptime/).

Fagaceae (about 1,000 species), the largest family within Fagales, have also been the focus of recent phylogenetic analyses (Manos and Steele 1997; Manos and Stanford 2001; Manos et al. 2001). If we assume that ancestral Fagaceae were insect-pollinated, as indicated by the fossil record (Herendeen et al. 1995; Sims et al. 1998), then the DNA-based reconstructions support at least three origins of wind pollination.

Other Eurosid I

The remaining members of the fabids (eurosid I) are Zygophyllaceae and a large, weakly supported subclade consisting of Malpighiales, Oxalidales, and Celastrales (D. Soltis et al. 2000). Because of the weak support in this portion of the eurosid I clade, relationships among these remaining groups are depicted as a polytomy (Figure 8.2). In the next sections, we cover these remaining eurosid I lineages—Zygophyllaceae and Krameriaceae, and then Malpighiales, Oxalidales, and Celastrales.

ZYGOPHYLLACEAE AND KRAMERIACEAE *Krameria* (Krameriaceae) and Zygophyllaceae s.s. formed a strongly supported (100%) clade within eurosid I (Savolainen et al. 2000a; D. Soltis et al. 2000; Figure 8.2). However, a close relationship between these two families had not been suggested in any previous classifications. *Krameria* had been variously placed in Fabaceae (Caesalpinioideae; reviewed in Cronquist 1981), Polygalales (Cronquist 1981), or Vochysiales (Heywood 1993; Takhtajan 1997). Zygophyllaceae, a heterogeneous assemblage of 25 genera, have been variously placed in five different orders, but have most frequently been classified in Sapindales or Geraniales (e.g., Cronquist 1981; Takhtajan 1997).

The strongly supported relationship between Krameriaceae and Zygophyllaceae indicated by molecular data prompted inclusion of *Krameria* in Zygophyllaceae as an optional treatment in APG II (2003). However, synapomorphies for this expanded Zygophyllaceae remain unclear. *Krameria* shares few features with Zygophyllaceae, probably because both have become so highly specialized that "little remains to indicate affinities" (Sheahan and Chase 2000). Hence, it may be justified to maintain Zygophyllaceae and Krameriaceae as distinct families rather than combining them into a single broadly defined Zygophyllaceae. Relationships within Zygophyllaceae are discussed in more detail by Sheahan and Chase (1996, 2000).

Zygophyllaceae and Krameriaceae share few non-DNA traits with any other rosid lineage. Anthraquinones are present in Zygophyllaceae, and these are also common in the nitrogen-fixing clade (Nandi et al. 1998), but rare elsewhere, so this could be a synapomorphy (Sheahan and Chase 2000). However, there is no evidence from molecular analyses to support a close relationship

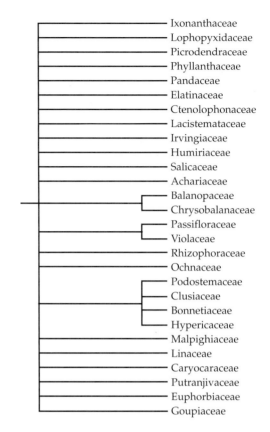

FIGURE 8.8 Summary tree for Malpighiales. General topology is from D. Soltis et al. (2000) with modifications from Savolainen et al. (2000b).

with the nitrogen-fixing clade (D. Soltis et al. 2000); Zygophyllaceae + Krameriaceae appeared in all shortest trees as the sister to the remainder of eurosid I, but without support above 50% (Figure 8.2).

MALPIGHIALES Molecular phylogenetic analyses provided strong support for a Malpighiales clade of 28 families not recognized in any previous classification (Figure 8.8): Achariaceae, Balanopaceae, Bonnetiaceae, Caryocaraceae, Chrysobalanaceae (including Dichapetalaceae, Euphroniaceae, and Trigoniaceae), Clusiaceae, Ctenolophonaceae, Elatinaceae, Euphorbiaceae, Goupiaceae, Humiriaceae, Hypericaceae, Irvingiaceae, Ixonanthaceae, Lacistemataceae, Linaceae (including Hugoniaceae), Lophopyxidaceae, Malpighiaceae, Ochnaceae (expanded to include Medusagynaceae and Quiinaceae; see APG II 2003), Pandaceae, Passifloraceae (expanded to include Malesherbiaceae, Medusandraceae, and Turneraceae; see APG II), Phyllanthaceae, Podostemaceae (including Tristichaceae), Picrodendraceae, Putranjivaceae, Rhizophoraceae (including Erythroxylaceae), Salicaceae (including part of Flacourtiaceae and Scyphostegiaceae), and Violaceae. Several of these families are illustrated in Figure 8.9.

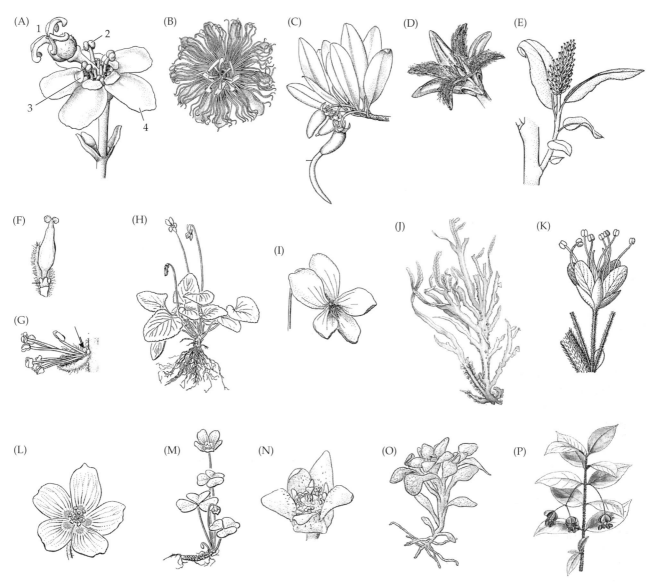

FIGURE 8.9 Flowers representing Malpighiales (A–J), Oxalidales (K–M), and Celastrales (N–P). (A) *Euphorbia corollata* (Euphorbiaceae), cyathium; 1 = carpellate flower; 2 = staminate flower; 3 = nectar gland; 4 = appendage of nectar gland. (B) *Passiflora incarnata* (Passifloraceae), flower. (C) *Rhizophora mangle* (Rhizophoraceae), branch in fruit; line indicates hypocotyl of embryo. (D) *Rhizophora mangle* (Rhizophoraceae), flower. (E) *Salix caroliniana* (Salicaceae), carpellate inflorescence. (F) *Salix caroliniana* (Salicaceae), carpellate flower; line indicates nectar gland. (G) *Salix caroliniana* (Salicaceae), staminate flower; arrow indicates nectar gland. (H) *Viola primulifolia* (Violaceae), plant in flower. (I) *Viola primulifolia* (Violaceae), flower. (J) *Dicraeia stylosa* (Podostemaceae), plant in flower. (K) *Weinmannia guyanensis* (Cunoniaceae), flower. (L) *Oxalis acetosella* (Oxalidaceae), flower. (M) *Oxalis acetosella* (Oxalidaceae), plant with flower and immature fruit from cleistogamous flower. (N) *Lepuropetalon spathulatum* (Parnassiaceae), flower. (O) *Lepuropetalon spathulatum* (Parnassiaceae), whole plant. (P) *Euonymus verrucosa* (Celastra-ceae), branch in fruit. (A–I, from Wood 1974. J, from Warming in Engler and Prantl 1891. K, from Engler in Engler and Prantl 1890. L and M, from Robertson 1975. N and O, from Spongberg 1972. P, from Loesener in Engler and Prantl 1897.)

Monophyly of Malpighiales received strong support (99%) in analyses based on three genes (D. Soltis et al. 2000), but this is a morphologically diverse clade. Many families of the order were previously placed in Violales (subclass Dilleniidae): Flacourtiaceae, Lacistemataceae, Passifloraceae, Malesherbiaceae and Turneraceae (both now included in Passifloraceae; APG II), and Violaceae. Ochnaceae represent another family formerly placed in Dilleniidae in recent classifications (Dilleniales). Malpighiales also include families previously attributed to a diverse array of orders in Rosidae, such as Linales (Humiriaceae, Linaceae, and Erythroxylaceae (now included in Rhizophoraceae), Polygalales (Malpighiaceae, Trigoniaceae, and Euphroniaceae, the last two now in Chrysobalanaceae), Euphorbiales (Euphorbiaceae), and Rosales (Chrysobalanaceae).

Relationships within Malpighiales are poorly understood, and the topology within the clade is essentially a polytomy (D. Soltis et al. 2000; Figure 8.8); Malpighiales may represent an excellent example of a rapid radiation (see Chapter 3). Because the order is such a heterogeneous assemblage, non-DNA characters uniting the constituent taxa are also poorly understood. In addition, phylogenetic studies indicated that several families of Malpighiales are polyphyletic (Flacourtiaceae, Euphorbiaceae). It will likely require a large amount of additional data, and perhaps additional taxa, to resolve relationships within this order. Malpighiales are the clade of rosids most in need of extensive study, in terms of both molecular phylogenetics and critical examination of non-DNA characters. We provide an overview of the subclades and other major features of Malpighiales below.

One of the few well-supported subclades within Malpighiales includes Balanopaceae and Chrysobalanaceae (including Trigoniaceae, Dichapetalaceae, and Euphroniaceae; Figure 8.8). This clade illustrates well a problem encountered throughout Malpighiales: it contains families not considered closely related in any modern classifications. For example, according to Cronquist (1981) the monogeneric Balanopaceae had been placed in Hamamelidae, with the remaining families placed in different orders of Rosidae (Dichapetalaceae in Sapindales; Chrysobalanaceae in Rosales; Trigoniaceae and Euphroniaceae in Polygalales). Trigoniaceae, Dichapetalaceae, Chrysobalanaceae, and Euphroniaceae, as well as several other Malpighiales, all have tenuinucellate ovules, a feature rare outside of asterids (Endress 2003a; APG II 2003; see Chapter 9). However, *Balanops* has crassinucellate ovules (Merino Sutter and Endress 2003).

Another well-supported subclade within Malpighiales is Rhizophoraceae, including Erythroxylaceae (see also Schwarzbach and Ricklefs 2000). A phylogenetic analysis of non-DNA characters first recovered the sister group of Rhizophoraceae and Erythroxylaceae s.s. (Hufford 1992), and their close relationship is supported by several non-DNA characters, including embryological features, the presence of the alkaloid hygroline, and a unique sieve-tube plastid type (Behnke 1988b; Nandi et al. 1998; reviewed in Schwarzbach and Ricklefs 2000). Therefore, these two families were combined into a single family, Rhizophoraceae (APG II 2003; see Chapter 10). The large number of features shared by these families was also recognized by Dahlgren (1988), who nonetheless did not consider the families to be closely related. The mangrove habit associated with Rhizophoraceae is restricted to a few members of that family (Cronquist 1988). Relationships

and character evolution within Rhizophoraceae have been analyzed by Schwarzbach and Ricklefs (2000). In addition, the poorly known African genus *Aneulophus* (Erythroxylaceae) is morphologically similar to some members of Rhizophoraceae.

Molecular phylogenetic studies provided support for an expanded Ochnaceae that includes Quiinaceae and Medusagynaceae (Fay et al. 1997a, 1997b). Based on morphological data, there is no clear reason for separating these families. In addition to similar overall floral morphologies, these families are united by contorted petal aestivation, stratified phloem, mucilage cells, and fusion of the styles (Fay et al. 1997a, 1997b; Nandi et al. 1998).

Another subclade of Malpighiales receiving support consists of several families formerly placed together in Violales (Dilleniidae): Passifloraceae s.s. are sister to Malesherbiaceae + Turneraceae. A close relationship among these three families has long been recognized (e.g., Cronquist 1981; Takhtajan 1997), and they produce structurally related cyanogenic glycosides with a cyclopentenoid ring system (see also Achariaceae) and similar hydrolytic enzymes (Spencer and Seigler 1985a, 1985b; Takhtajan 1997). Turneraceae and Passifloraceae possess foliar glands, and both have biparental or paternal transmission of plastids (Malesherbiaceae have not been examined); in contrast, in most angiosperms plastids are maternally inherited (e.g., Shore et al. 1994). Malesherbiaceae, some Turneraceae, and Passifloraceae also possess an extrastaminal corona, and all three families possess a hypanthium-like structure that does not bear the stamens. Furthermore, unpublished results indicate that Paropsieae (two arborescent genera, *Barteria* and *Paropsia*) of Passifloraceae are sister to a clade composed of Malesherbiaceae, the rest of Passifloraceae, Turneraceae, and *Medusandra* (M. W. Chase and L. W. Chatrou, unpubl. data). It may therefore be appropriate to combine Passifloraceae, Turneraceae, Malesherbiaceae, and Medusandraceae into a single family, Passifloraceae. This is an option presented in the APG II (2003) reclassification of flowering plants (see Chapter 10).

Both Euphorbiaceae and Flacourtiaceae as previously recognized are polyphyletic according to the results of phylogenetic studies. Phylogenetic analyses (Savolainen et al. 2000b; D. Soltis et al. 2000) revealed a narrowly defined Euphorbiaceae and indicated the presence of several other well-separated clades of former euphorbs that have closest relatives elsewhere in Malpighiales. In APG II, Euphorbiaceae were divided into four families: Euphorbiaceae s.s., Phyllanthaceae, Picrodendraceae, and Putranjivaceae. In this treatment, Euphorbiaceae s.s. consist of the uniovulate Euphorbioideae, Crotonoideae, and Acalyphoideae. Phyllanthaceae include the biovulate Phyllanthoideae, whereas Picrodendraceae include the biovulate Oldfield-ioideae. *Drypetes* and *Putranjiva* (Putranjivaceae) represent another distinct clade of former euphorbs (see APG 1998; Rodman et al. 1998; D. Soltis et al. 2000; APG II 2003). There is evidence from seed-coat morphology (structure of the exotegmen; seeds exotegmic versus endotestal) to support some of these groups (Corner 1976; Tokuoka and Tobe 2001), but additional investigations are needed. Genera of Pandaceae have also been placed within Euphorbiaceae (Thorne 1992a, 1992b) or close to the family in Euphorbiales (Cronquist 1981; Takhtajan 1997). However, sequence data reveal that Pandaceae, a family of three genera, may be sister to Lophopyxidaceae, a family of a single genus placed in or near Celastraceae in some modern classifications (e.g., Cronquist 1981). Euphorbiaceae, in the narrow sense, have only recently been the subject of a broad molecular phylogenetic analysis (Wurdack et al. 2004). Phyllanthaceae are the subject of two reports, one using a nearly complete set of genera with just *rbcL,* and another with somewhat fewer genera using *matK* and *PHYC* (Samuel et al. 2005; Wurdack et al. 2004).

Flacourtiaceae, as previously defined (see Gilg 1925; Sleumer 1954, 1980; Lemke 1988; Judd 1997), were shown to be polyphyletic in molecular phylogenetic studies (Chase et al. 2002). The former, broadly defined family is composed of two major subclades that are more closely related to other families in Malpighiales than to each other. Thus, Chase et al. (2002) recognized two families: a broadly defined Salicaceae (in addition to Salicaceae and some members of Flacourtiaceae, this clade also includes the tribe Abatieae of Passifloraceae and the monogeneric Scyphostegiaceae; see also Bernhard 1999) and the newly expanded family Achariaceae (which include other former Flacourtiaceae, such as *Kiggelaria,* as well as *Acharia, Ceratiosicyos,* and *Guthriea* of Achariaceae). The name Achariaceae was used because members of this poorly known family appeared as sister to *Kiggelaria* (D. Soltis et al. 2000; Chase et al. 2002), and so the name of the family becomes the older conserved name, Achariaceae (not Kiggelariaceae as in several recent papers).

Salicaceae as now defined (APG II 2003) are much expanded to include taxa, formerly placed in Flacourtiaceae, that possess salicoid teeth (Nandi et al. 1998), have centrifugal stamen initiation (Bernhard and Endress 1999), lack cyanogenic glycosides of the gynocardin type, usually have small globose anthers, and possess generally small flowers in which the sepals and petals, if both are present, are equal in number (Figure 8.9, E to G; in many taxa of Salicaceae s.l., petals are absent). In contrast, the situation is reversed in Achariaceae; most taxa lack salicoid teeth, have centripetal stamen initiation (Bernhard and Endress 1999), produce cyclopentenoid cyanogenic glycosides, usually have

elongate anthers, and have large flowers typically with sepals and petals not equal in number. In addition, members of the lepidopteran genus *Cupha* feed only on members of the expanded Salicaceae (Nandi et al. 1998). Although the expansion of Salicaceae to include the largely tropical tribes of Flacourtiaceae would appear to challenge the concept of the family as one of cool, north-temperate regions, the taxonomic diversity of *Populus* (Salicaceae s.s.) actually reaches its peak in warm temperate to subtropical latitudes, and temperate members of Flacourtiaceae occur in China and Chile. Furthermore, those species of *Populus* most resembling Flacourtiaceae in morphology are tropical (Eckenwalder 1996; Chase et al. 2002). Also, whereas *Salix* is most diverse in the Northern Hemisphere, its distribution includes Africa, Malaysia, and Central and South America (Argus 1997). *Salix* has also been the focus of molecular phylogenetic study (Azuma et al. 2000).

Although Lacistemataceae had been placed in Flacourtiaceae by some investigators (e.g., Sleumer 1980), authors of modern classifications, including Cronquist (1981) and Takhtajan (1997), recognized Lacistemataceae as a distinct family. The family consists of *Lacistema* and *Lozania,* two small woody genera from Central and South America. In analyses using the sequences of three genes, the placement of Lacistemataceae within Malpighiales was unclear (D. Soltis et al. 2000), although Chase et al. (2002) suggested a close relationship to the expanded Salicaceae.

Peridiscaceae consist of the monotypic genera *Peridiscus* and *Whittonia* from tropical South America, and *Soyauxia,* which has several species from West Africa. Although some recent classifications (e.g., Cronquist 1981) included these genera in Flacourtiaceae, Takhtajan (1997) recognized the distinct family Peridiscaceae. The family was not included in the three-gene survey (D. Soltis et al. 2000), but based on the *rbcL* analysis of eudicots (Savolainen et al. 2000b), *Whittonia* appeared closely related to Malpighiaceae. However, Davis and Chase (2004) demonstrated that the sequence reported for *Whittonia* was chimeric and consisted of two parts from different taxa. Sequencing and phylogenetic analysis of *Peridiscus* demonstrated that Peridiscaceae are part of Saxifragales (see Chapter 6).

The extent of the taxonomic confusion surrounding the former broadly defined Flacourtiaceae is further illustrated by the placements of several other genera formerly placed in the family by at least some authors (e.g., *Aphloia, Asteropeia, Berberidopsis, Muntingia,* and *Plagiopteron*). Molecular analyses indicated that these and other genera must be excluded not only from Flacourtiaceae but also from Malpighiales. For example, *Berberidopsis* (Berberidopsidaceae) appeared with strong support as sister to *Aextoxicon* (Aextoxicaceae) in several multigene analyses (Savolainen et al. 2000a; D. Soltis et al. 2000, 2003a; "Berberidopsidales," see Chapter 6). "Berberidopsidales" were placed outside of the rosids in these analyses, although their exact position among the core eudicots is unclear (see Chapter 6). Similarly, *Aphloia* (Aphloiaceae), formerly placed in Flacourtiaceae, appeared in an isolated position near the newly recognized rosid order Crossosomatales (discussed below). Another former member of Flacourtiaceae, *Plagiopteron* (sometimes also recognized as a distinct family, Plagiopteridaceae; Airy Shaw 1964; Takhtajan 1997), appeared in a broadly defined Celastraceae (discussed below). *Lethedon* was referred to Flacourtiaceae because it possesses cyanogenic glycosides (Spencer and Seigler 1985b), but the genus is well supported as a member of Thymelaeaceae in Malvales (Savolainen et al. 2000b). Alzateaceae were included by Hutchinson (1967) in Flacourtiaceae, but the family has been shown to be a member of Myrtales (Graham 1984; Conti et al. 1996). Both *Asteropeia* and *Dioncophyllum* have been suggested as members of Flacourtiaceae (e.g., Hutchinson 1967). In contrast, Schmid (1964) proposed that *Dioncophyllum* was related to Droseraceae, and both *Asteropeia* and *Dioncophyllum* (with Droseraceae) appeared as members of Caryophyllales in molecular analyses (Morton et al. 1997; Lledó et al. 1998; Savolainen et al. 2000a, 2000b; D. Soltis et al. 2000). *Muntingia* (included in Flacourtiaceae only by Cronquist 1981), together with *Dicraspidia*, were placed in Malvales as a distinct family, Muntingiaceae (Bayer et al. 1998).

Linaceae have been expanded to include Hugoniaceae, and a close relationship of the two has long been suggested (see Takhtajan 1997). In part, the separation of the families was geographical. Whereas Linaceae in the strict sense are widely distributed, especially in temperate areas and the subtropics, members of Hugoniaceae are largely found in the tropical areas of the Southern Hemisphere.

Malpighiales also illustrate the value of broad phylogenetic analyses in placing enigmatic taxa. The relationship of *Medusagyne* (formerly Medusagynaceae) is a clear example (see Fay et al. 1997b). The single species, *M. oppositifolia*, was thought to be extinct and is presently considered endangered (Melville 1971; Fay et al. 1997b). Molecular phylogenetic analyses not only placed the species firmly within Malpighiales but also indicated a position within Ochnaceae (Melville 1971; Fay et al. 1997b). In fact, Ochnaceae, Medusagynaceae, and Quiinaceae form a distinctive and monophyletic group (Nandi et al. 1998; Savolainen et al. 2000a) in which leaves have well-defined secondary and tertiary venation. It now seems appropriate to combine all of these into a single family, Ochnaceae (APG II 2003).

Another noteworthy example of the placement of a problematic family within Malpighiales involves Podostemaceae, a family of highly modified submerged aquatics variously associated with Piperaceae, Ne-

penthaceae, Polygonaceae, Caryophyllaceae, Scrophulariaceae, Rosaceae, Crassulaceae, or Saxifragaceae (reviewed in Les et al. 1997c; Ueda et al. 1997; D. Soltis et al. 1999). Molecular phylogenetic analyses using multiple genes placed Podostemaceae in Malpighiales (D. Soltis et al. 1999, 2000) as sister to *Hypericum* (Hypericaceae) with strong support. Another small enigmatic family, Bonnetiaceae (segregated from Theaceae, see Chapter 9), was found to be closely related to, if not part of, Clusiaceae (Savolainen et al. 2000b). In a phylogenetic analysis of Clusiaceae using *rbcL* sequences, Gustafsson et al. (2002) found Podostemaceae to be nested within Clusiaceae.

Relationships within the Hypericaceae-Clusiaceae-Bonnetiaceae-Podostemaceae clade are, however, still unclear (Figure 8.8). Elatinaceae were also shown to be closely related to these families (Savolainen et al. 2000b; Stevens, in press), although support for this was weak; Davis and Chase (2004) showed that Elatinaceae were instead sister to Malpighiaceae with strong bootstrap support. Bonnetiaceae share a distinctive exotegmen with Clusiaceae in which the palisade cells are radially elongate, but with stellate-undulate or lobate facets (Corner 1976). In addition, Bonnetiaceae and Clusiaceae have distinctive xanthones in common. Xanthones have also been reported from some Podostemaceae, and tenuinucellate ovules are known from Clusiaceae and Podostemaceae (e.g., Contreras et al. 1993; Jäger-Zürn 1997). Thus, although considerable progress has been made, additional phylogenetic analyses (with an emphasis on increased taxon sampling) of Hypericaceae and Clusiaceae are imperative to sort out relationships. Until relationships are better resolved in this portion of Malpighiales, these enigmatic families can be treated either as part of a broadly defined Clusiaceae or as separate families (as in APG II).

Other similar examples of small, enigmatic families that appeared as part of a well-supported Malpighiales in molecular phylogenetic studies include Goupiaceae, Irvingiaceae, Balanopaceae, Lophopyxidaceae, Pandaceae, Ctenolophonaceae, and Caryocaraceae (Savolainen et al. 2000b). One intriguing example involves a newly discovered and as-yet-unnamed genus that is sister to the equally enigmatic *Medusandra*. Another exciting discovery is molecular evidence for a close relationship of the parasitic genus *Rafflesia* (Rafflesiaceae) with Malpighiales (Barkman et al. 2004; see "Parasitism in the Rosids," below, and Chapter 11).

Some families of Malpighiales have been the focus of detailed phylogenetic analyses. Recent studies have clarified phylogenetic relationships within Malpighiaceae and also provided insights into biogeography, as well as fruit and pollen diversification within the family (Cameron et al. 2001; Davis et al. 2001, 2002). Relationships within Salicaceae and Achariaceae (*sensu* APG II 2003) have also been investigated (Chase et al. 2002).

Rhizophoraceae have been the focus of considerable study using non-DNA characters (an entire issue of the *Annals of the Missouri Botanical Garden* was devoted to the family; e.g., Tobe and Raven 1988; Dahlgren 1988), as well as molecular phylogenetic analysis (Schwarzbach and Ricklefs 2000).

Oxalidales

Phylogenetic analyses (e.g., Savolainen et al. 2000a, 2000b; D. Soltis et al. 2000) indicated that Oxalidales consist of six families (Figure 8.10): Brunelliaceae, Cephalotaceae, Connaraceae, Cunoniaceae (including Eucryphiaceae and Davidsoniaceae), Elaeocarpaceae (including Tremandraceae), and Oxalidaceae (several of these are illustrated in Figure 8.9). Brunelliaceae, a monogeneric family from tropical America, were included in Cunoniaceae in APG (1998; see also Hufford and Dickison 1992), but their position in recent analyses as sister to the subclade of Elaeocarpaceae–Cunoniaceae–Cephalotaceae, rather than as part of Cunoniaceae justifies continued familial status (APG II 2003).

Oxalidales represent another rosid clade composed of families previously placed in several different orders in morphology-based classifications. Most families were placed in Rosidae, with Cephalotaceae, Brunelliaceae, and Cunoniaceae all considered part of Rosales (*sensu* Cronquist 1981); Oxalidaceae were placed in Geraniales; Connaraceae were considered part of Sapindales, and Tremandraceae were placed in Polygalales. Elaeocarpaceae were placed in Malvales of subclass Dilleniidae. Thus, this one small clade again illustrates the pattern seen throughout the rosids—clades now recognized as orders (*sensu* APG 1998; APG II 2003) are composed of taxa previously considered distantly related. As with many other orders of rosids, non-DNA synapomorphies for Oxalidales are unclear.

The analysis of three genes (D. Soltis et al. 2000) and the more inclusive eudicot analysis of Savolainen et al. (2000b) showed that Oxalidales consist of Oxalidaceae + Connaraceae as sister to a strongly supported clade of Brunelliaceae, Cephalotaceae, Cunoniaceae, and Elaeocarpaceae (Figure 8.10). Tremandraceae (a family of three genera, *Platytheca, Tetratheca,* and *Tremandra*) were embedded within Elaeocarpaceae and are now considered members of that family (APG II 2003). Brunelliaceae appeared as sister to a clade of Cephalotaceae, Cunoniaceae, and Elaeocarpaceae. However, relationships among these four families were uncertain and did not receive support above 50%.

Molecular phylogenetic studies indicated that Cunoniaceae should be broadly defined to include both *Bauera* and *Davidsonia*, genera sometimes placed in Baueraceae and Davidsoniaceae, respectively (see also Hufford and Dickison 1992). However, *Aphanopetalum*, previously placed in Cunoniaceae, has been shown to be a member of Saxifragales (Fishbein et al. 2001; see

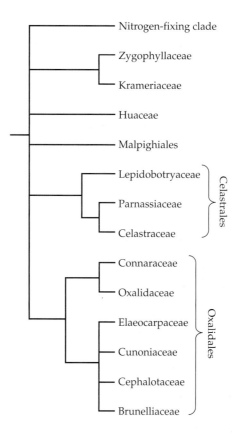

FIGURE 8.10 Summary tree for Oxalidales and Celastrales. General topology is from D. Soltis et al. (2000) with modifications from Savolainen et al. (1994, 1997, 2000b) and Simmons et al. (2001).

Chapter 6). *Cephalotus* (Cephalotaceae), a monotypic genus of carnivorous plants (see Chapter 11), was not included in the combined molecular analyses (Savolainen et al. 2000a; D. Soltis et al. 2000), but *rbcL* sequence data indicated that this enigmatic family is sister to Cunoniaceae (Savolainen et al. 2000b). Both of these families had been placed in Rosales in recent classifications (e.g., Cronquist 1981), share several floral features, and occur or are centered in Australia.

The close relationship between Elaeocarpaceae and Tremandraceae was an unexpected outcome of phylogenetic analyses. Elaeocarpaceae had been placed in Malvales in recent classifications, but they lack many of the non-DNA synapomorphies for the latter group (see "Malvales," below). Tremandraceae had been placed in Polygalales (Rosidae) in recent classifications and therefore were not considered a close relative of Elaeocarpaceae. Oxalidales are one of the first newly recognized rosid orders to be the subject of comparative studies in floral architecture (Matthews and Endress 2002). The clade of Oxalidaceae + Connaraceae is morphologically well separated from the remainder of the order, possessing tristyly, postgenital union of petals into a basal tube, congenitally united stamens, hemianatropous to almost orthotropous ovules, and a special type of sieve tube plastids. Elaeocarpaceae are also well supported, characterized by buzz-pollination and a syndrome of structural features functionally connected with it. In addition, flowers of both families have involute petals that are longer than the sepals in advanced buds, three vascular traces, and stamens partly wrapped by adjacent petals; ovules have a chalazal appendage and a thick inner integument. In contrast, Cephalotaceae are morphologically isolated with their pitcher-trap leaves (see Chapter 11) and unusual floral structure.

Celastrales

Celastrales were not recognized as an order in the reclassification of angiosperms proposed by APG (1998), but were recognized as an order in Judd et al. (1999). More recent phylogenetic analyses provided additional support for recognition of Lepidobotryaceae, Parnassiaceae (including *Lepuropetalon*), and an expanded Celastraceae (further discussed below) as an order, Celastrales (APG II 2003; Figure 8.10; several families are illustrated in Figure 8.9). This circumscription of Celastrales differs substantially from those provided in recent classifications. The delimitation of the order has been highly variable, with several families, including Rhamnaceae, Salvadoraceae, Vitaceae, Aquifoliaceae, and Icacinaceae, placed in Celastrales in various classifications (Hutchinson 1973; Cronquist 1981, 1988; Takhtajan 1987, 1997; Heywood 1993; see Savolainen et al. 1994). However, molecular data demonstrated that most families previously placed in Celastrales should be assigned to other orders (see Savolainen et al. 1994, 1997; Simmons et al. 2000). As with many of the rosid clades now recognized as orders based on DNA sequence data, nonmolecular synapomorphies for Celastrales are not clear.

Within Celastrales, molecular data placed Lepidobotryaceae as sister to Celastraceae and Parnassiaceae (Savolainen et al. 2000b). Both Parnassiaceae and Lepidobotryaceae are small, enigmatic families. Parnassiaceae comprise *Parnassia* and the monotypic *Lepuropetalon*, the smallest terrestrial angiosperm (the latter genus has often been placed in its own family; see Spongberg 1972; Gastony and Soltis 1977). Parnassiaceae have been variously placed in Saxifragaceae (e.g., Cronquist 1981) and Celastrales (Takhtajan 1991, 1997). Lepidobotryaceae consist of two disjunct genera, *Rutiliocarpon* and *Lepidobotrys* in Central America and tropical Africa, respectively. Cronquist (1981) placed these genera in Oxalidaceae, whereas Takhtajan (1997) considered Lepidobotryaceae and Oxalidaceae closely related. However, Oxalidaceae have been shown to be related to families discussed in the preceding section (under Oxalidales).

Molecular phylogenetic analyses also indicated that Celastraceae should be greatly expanded to include several small families, such as Brexiaceae, Hippocrateaceae, Stackhousiaceae, Plagiopteridaceae, and Canotiaceae (see Savolainen et al. 1994, 1997, 2000b; Simmons et al. 2001). Molecular results indicating an expanded Celastraceae were not totally unexpected. For example, questions about the distinctiveness of Celastraceae and Hippocrateaceae have existed since the two families were first described, so it is not surprising that the latter were found to be nested within Celastraceae. Similarly, although *Brexia* has been variously placed in Saxifragaceae s.l., Escalloniaceae, Grossulariaceae, and Brexiaceae (reviewed in Morgan and Soltis 1993), a close relationship between *Brexia* and Celastraceae has also been proposed (Perrier de la Bâthie 1933; see Takhtajan 1991, 1997). *Plagiopteron* has been placed in several families, including Tiliaceae and Flacourtiaceae, and is sometimes recognized as a distinct family, Plagiopteridaceae (Airy Shaw 1965; Hutchinson 1967). However, the genus also fits well in a broadly defined Celastraceae based on leaf and wood anatomy (Baas et al. 1979) and embryology (Tang 1994). Stackhousiaceae differ from Celastraceae, as previously recognized, only in being mostly herbaceous (rather than woody, although most species of Stackhousiaceae have a woody rootstock) and in having a better-developed floral tube (Cronquist 1981). Hence, the molecular placement of Stackhousiaceae within Celastraceae was not surprising. *Canotia* has been variously placed in Rutaceae, Koeberliniaceae, Canotiaceae, and Celastraceae, but both molecular and embryological data (Tobe and Raven 1993) indicated a close relationship of *Canotia* to *Acanthothamnus* within Celastraceae (Simmons et al. 2001).

Four morphological synapomorphies, all of which show reversals, support a broadly defined Celastraceae (Simmons et al. 2001): stamen and staminode number each equal petal number; filaments are inserted at the outer border of, or within, the conspicuous nectar disk; styles are connate; and two to four ovules per locule are present. Other possible synapomorphies include the presence of crystals in leaf epidermis (Baas et al. 1979) and an integumentary tapetum (Johri et al. 1992; Tang 1994; Nandi et al. 1998).

CHARACTER EVOLUTION IN CELASTRALES Simmons et al. (2001) showed the complex nature of fruit and aril diversification within Celastraceae, with multiple origins of most fruit and aril forms. There have been many transitions from dehiscent to indehiscent fruit types. Nuts, drupes, samaras, and berries have arisen many times independently, most often from capsules; similarly, arils (fleshy covering surrounding seed) apparently arose once or twice in the family and have been lost in five or six lineages (Simmons et al. 2001; Figure 8.11). Other recent studies have similarly indicated the lability of fruit evolution within families (e.g., Manos and Stone 2001—see "Fagales," above; Richardson et al. 2000; Davis et al. 2001).

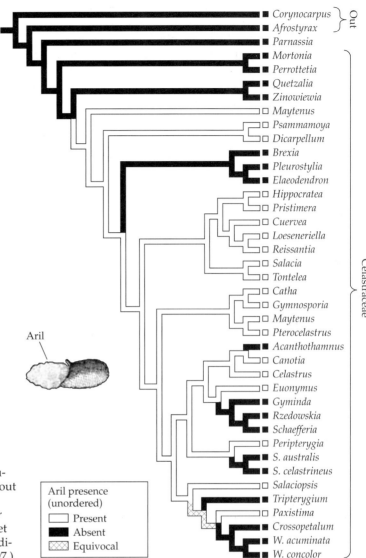

FIGURE 8.11 MacClade reconstruction of the evolution of arils (presence or absence) in Celastraceae. out = outgroup taxa. *S. celastrineus* and *S. australis* are species of *Siphonodon*; *W. acuminata* and *W. concolor* are species of *Wimmeria*. Modified from Simmons et al. (2001). Seed of *Catha edulis* is illustrated; line indicates aril. (From Loesener in Engler and Prantl 1897.)

Eurosid II (Malvids)

Molecular analyses provided strong support for a core clade of eurosid II (malvids) that includes the well-supported groups Brassicales, Malvales, Sapindales, and Tapisciaceae (e.g., D. Soltis et al. 2000; Figure 8.12). It is still unclear, however, if other orders may also be part of eurosid II. With the sampling of taxa used in the three-gene analysis, Myrtales appeared as sister in all shortest trees to the eurosid I clade, but Geraniales and Crossosomatales were related to the well-supported eurosid II clade. However, at this point there is no support above 50% for the relationships of these three problematic orders. Therefore, eurosid II is restricted to Brassicales, Malvales, Sapindales, and Tapisciaceae (APG II 2003). As with eurosid I, there are no known non-DNA synapomorphies for eurosid II.

Although studies have consistently recovered a eurosid II clade (e.g., Chase et al. 1993; Savolainen et al. 2000a, 2000b; D. Soltis et al. 2000), relationships among the members of this clade are still uncertain. Savolainen et al. (2000a) found weak support (62%) for a sister-group relationship of Brassicales and Malvales with combined *rbcL* plus *atpB* sequence data, and *matK* also provided support for Malvales + Brassicales (Hilu et al. 2003). However, the pair Malvales + Sapindales had weak support (51% in D. Soltis et al. 2000; 78% in a four-gene analysis, D. Soltis et al. 2003a; Figure 8.12). In these analyses, Tapisciaceae appeared as sister to these three orders, but this relationship did not receive support above 50% (Figure 8.12); Tapisciaceae were not included in Hilu et al. (2003). Additional work is clearly needed to determine relationships among the subclades of eurosid II.

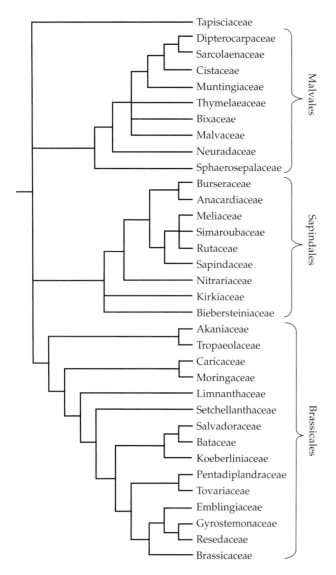

Malvales

Sapindales

Brassicales

FIGURE 8.12 Summary tree for the eurosid II clade. General topology is from D. Soltis et al. (2000) with modifications from Savolainen et al. (2000b) and D. Soltis et al. (2003a).

Tapisciaceae

The three-gene analysis (D. Soltis et al. 2000) placed Tapisciaceae in eurosid II with strong support; however, the relationships of the family to other members of this clade were unclear. The family consists of two genera, *Tapiscia* and *Huertea,* and only one species of the former has so far been included in published phylogenetic analyses. In classifications, the two genera have consistently been placed in or near Staphyleaceae. Cronquist (1981), for example, considered *Tapiscia* and *Huertea* to represent a distinct subfamily within Staphyleaceae, following the treatment of Pax (in Engler and Prantl 1893). Based on distinctive features of seed structure, Spongberg (1971) suggested the separation of these two genera into their own family, Tapisciaceae. Takhtajan (1980, 1997) similarly recognized a separate family Tapisciaceae, considering it closely allied to Staphyleaceae. Thus, the placement of Tapisciaceae with Brassicales, Sapindales, and Malvales, rather than near Staphyleaceae in Crossosomatales would have to be considered another unexpected result of phylogenetic analyses. *Dipentodon* (Dipentodontaceae), placed in Santalales by Cronquist (1981) and unplaced in APG II (2003), is also related to Tapisciaceae (M.W. Chase, unpubl. data).

Brassicales

Molecular phylogenetic studies revealed a well-defined and strongly supported Brassicales consisting of 15 families (Figure 8.12): Akaniaceae (including Bretschneideraceae), Bataceae, Brassicaceae (including Capparaceae), Caricaceae, Emblingiaceae, Gyrostemonaceae, Koeberliniaceae, Limnanthaceae, Moringaceae, Pentadiplandraceae, Resedaceae, Salvadoraceae, Setchellanthaceae, Tovariaceae, and Tropaeolaceae (some of these families are illustrated in Figure 8.13). This circumscription of Brassicales is significant because the order includes all of the families that produce mustard-oil glucosides (also called glucosinolates), except for Putranjivaceae of

FIGURE 8.13 Flowers representing Sapindales (A–I), Brassicales (J–Q), and Malvales (R–U). (A) *Acer saccharum* (Sapindaceae), branch with flowers and expanding leaves. (B) *Acer saccharum* (Sapindaceae, representing samaroid clade), staminate flower with portion of perianth removed; 1 = perianth part; 2 = nectar disc; 3 = rudimentary gynoecium. (C) *Exothea paniculata* (Sapindaceae), staminate flower with prominent nectar disc and rudimentary gynoecium. (D) *Exothea paniculata* (Sapindaceae), branch with developing fruits. (E) *Poncirus trifoliata* (Rutaceae), flowering branch. (F) *Poncirus trifoliata* (Rutaceae), flower with one sepal and two petals and several stamens removed; 1 = nectar disc; 2 = stamens removed; 3 = one sepal removed. (G) *Commifera kataf* (Burseraceae), stem in flower. (H) *Commifera kataf* (Burseraceae), flower. (I) *Rhus coriaria* (Anacardiaceae), plant in fruit. (J) *Carica papaya* (Caricaceae), carpellate flower. (K) *Carica papaya*

(Caricaceae), staminate flower. (L) *Moringa oleifera* (Moringaceae), flower. (M) *Moringa oleifera* (Moringaceae), plant in flower. (N) *Capsella bursa-pastoris* (Brassicaceae), flower with perianth partially removed to show four long and two short stamens. (O) *Capsella bursa-pastoris* (Brassicaceae), fruit. (P) *Cleome spinosa* (Brassicaceae), flower. (Q) *Cleome spinosa* (Brassicaceae), plant in flower. (R) *Anisoptera curtisii* (Dipterocarpaceae), distinctive winged fruit. (S) *Shorea robusta* (Dipterocarpaceae), flower. (T) *Kosteletzkya virginica* (Malvaceae), plant in flower. (U) *Helianthemum atriplicifolium* (Cistaceae), plant in flower. (A, B, N, O, and T, from Wood 1974. C to F, from Brizicky 1962, 1963. G to I, from Engler in Engler and Prantl 1897. J and K, from Graf zu Solms in Engler and Prantl 1895. L. M, P, and Q, from Pax in Engler and Prantl 1891. R and S, from Brandis and Gilg in Engler and Prantl 1895. U, from Reiche in Engler and Prantl 1895.)

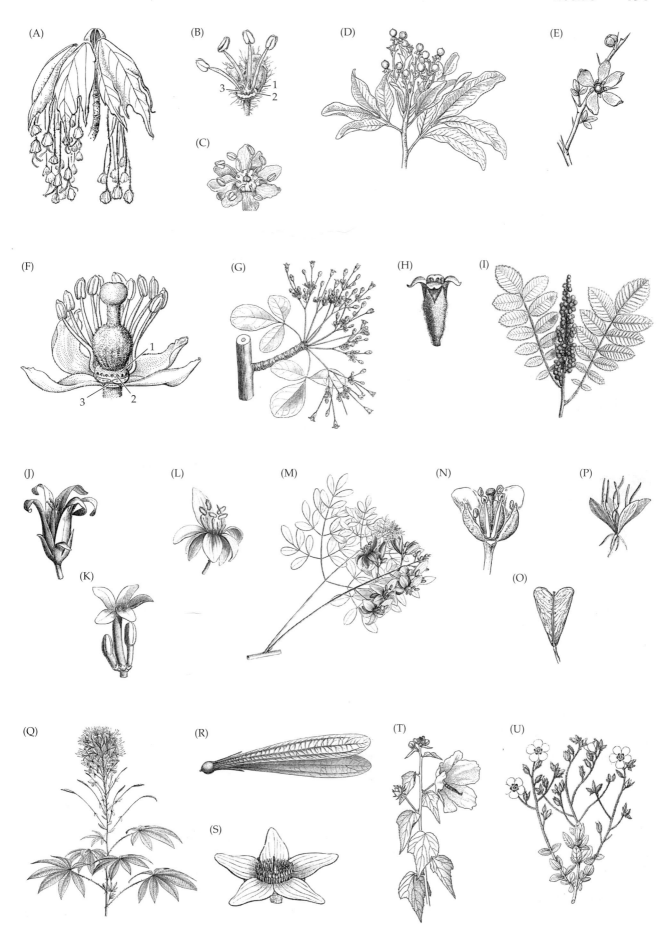

Malpighiales (Rodman 1991a, 1991b; Gadek et al. 1992; Rodman et al. 1993, 1998). Thus, glucosinolate production represents a non-DNA synapomorphy for the Brassicales clade and is a feature that evolved twice in the angiosperms (see paragraphs below in this section).

Recent morphology-based classifications placed members of Brassicales in many different orders (e.g., Cronquist 1981, 1988; Takhtajan 1987, 1997; Thorne 1992a, 1992b, 2001). However, Dahlgren (1975, 1977) challenged traditional approaches to the classification of these taxa and placed in his order Capparales nearly all plant families producing glucosinolates. He later backed off from this "radical" classification (Dahlgren 1980, 1983) and, like other authors, emphasized the striking morphological differences among these families.

The concept of a broadly defined "glucosinolate" clade (*sensu* Dahlgren 1975) reemerged after Rodman's (1991a, 1991b) phenetic and cladistic analyses of morphological and phytochemical characters. A similar glucosinolate-producing clade (now termed Brassicales rather than Capparales) was recovered in analyses using either *rbcL* or 18S rDNA sequence data (Gadek et al. 1992; Chase et al. 1993; Rodman et al. 1993, 1994; D. Soltis et al. 1997a, 1997b). Strong support ultimately emerged not only for the monophyly of this clade but also for relationships within this clade as gene sequences were combined (Rodman et al. 1998; D. Soltis et al. 2000). *Drypetes* (Putranjivaceae; formerly in Euphorbiaceae), another glucosinolate-producing taxon, is not part of Brassicales, but instead is part of the well-supported Malpighiales in the eurosid I clade (see above).

Three strongly supported subclades were detected within Brassicales (Rodman et al. 1998): (1) Tropaeolaceae + Akaniaceae (including Bretschneideraceae); (2) Moringaceae + Caricaceae; and (3) a large clade in which Limnanthaceae followed by *Setchellanthus* (Setchellanthaceae) were shown to be subsequent sisters to the remainder of the order. Following Limnanthaceae and Setchellanthaceae, a clade of Koeberliniaceae + (Bataceae + Salvadoraceae) emerged as sister to two subclades: Pentadiplandraceae + Tovariaceae and Brassicaceae + (Resedaceae + Gyrostemonaceae). Emblingiaceae, not included in Rodman et al. (1998), appear to be sister to Gyrostemonaceae + Resedaceae (Figure 8.12).

Chandler and Bayer (2000) suggested that *Emblingia* (Emblingiaceae), a problematic monotypic genus from Australia, belonged in Brassicales, as part of the third large subclade of the order. Their analysis of *rbcL* sequences placed *Emblingia* as sister to Resedaceae. In contrast, however, the *rbcL* sequence obtained for *Emblingia* by Savolainen et al. (2000b), using different material, resulted in a placement within Gentianales. An unpublished sequence for *Emblingia* using another source of material also indicated a placement within Brassicales (J.E. Rodman, pers. comm.); hence, the sequence of Savolainen et al. is suspect. Our own analy-

ses using the sequence of Chandler and Bayer (2000) placed *Emblingia* in Brassicales (Figure 8.12).

Morphology has failed to provide a clear answer to the phylogenetic position of *Emblingia*. The single species was originally placed in Capparaceae based on the presence of an androgynophore (a stalk supporting the androecium and gynoecium above the insertion of the corolla), a structure also seen in Brassicaceae (*sensu* APG II) and Resedaceae. However, pollen morphology and floral features indicated affinities with Polygalaceae and Sapindaceae, respectively. In contrast, leaf and stem anatomy supported a relationship to the asterid family Goodeniaceae (Erdtman et al. 1969; reviewed in Chandler and Bayer 2000). The presence of glucosinolates has apparently not been investigated in *Emblingia*, although initial field observations suggested that these compounds are present in the genus (J.E. Rodman, pers. comm.). *Emblingia* is noteworthy in having several features also observed in taxa of subclade 3 of Brassicales, including curved embryos, scanty endosperm, and some floral features that are tetramerous, whereas others are pentamerous. The genus has two petals and four stamens, but five sepals (Takhtajan 1997; Chandler and Bayer 2000). We have placed the family in Brassicales following Chandler and Bayer (2000), pending additional study (see also APG II 2003; Chapter 10).

More detailed analyses have been conducted on several families within Brassicales. Olson (2002a, 2002b) analyzed phylogenetic relationships among the 13 species of *Moringa*, the only genus in Moringaceae. Brassicaceae were expanded to include Capparaceae in the initial APG (1998) classifications because Capparaceae, long considered the closest relative of Brassicaceae, are paraphyletic relative to Brassicaceae (see Judd et al. 1994). A broadly defined Brassicaceae including Capparaceae are also indicated by non-DNA characters such as a long gynophore, vacuolar or utricular cisternae of the endoplasmic reticulum, a similar method of glucosinolate production involving synthesis from methionine via long chain-extensions, and the presence of sinapine (Rodman 1991a, 1991b; Judd et al. 1994, 1999). However, using a broad sampling of taxa, J. Hall et al. (2002) found evidence for three well-supported, primary clades within Brassicaceae s.l.: (1) Capparaceae subfamily Capparoideae; (2) Capparaceae subfamily Cleomoideae; and (3) Brassicaceae s.s. In their analyses, Brassicaceae and Capparaceae subfamily Cleomoideae are sister taxa. These authors suggest that three families be recognized corresponding to each of these clades: Capparaceae, Cleomaceae, and Brassicaceae. Morphological characters corresponding to each of these three clades were also provided (J. Hall et al. 2002). Capparaceae typically have fleshy fruits whereas the fruits of Cleomaceae and Brassicaceae are typically dry. Cleomaceae possess bracteate inflorescences as a synapomorphy; Brassicaceae s.s. have a false septum and six stamens (4 long and 2 short). We favor the broader circumscription of Brassicaceae in APG II.

Members of Brassicaceae have served as model systems for studies of floral development and genome evolution (e.g., species of *Arabidopsis* and *Brassica*). The importance of understanding the phylogenetic relationships of these organisms has been repeatedly emphasized (see Chapter 3; see also D. Soltis and Soltis 2000, 2003). For example, phylogenetic inferences for Brassicaceae and relatives will allow the *Arabidopsis* model of floral development to be extended to encompass more floral diversity (reviewed in J. Hall et al. 2002; Mitchell-Olds and Clauss 2002). However, even with three-gene analyses, the closest relatives of Brassicaceae s.l. remained unclear. In the shortest trees obtained, a clade of Resedaceae and Gyrostemonaceae emerged as sister to Brassicaceae, but without internal support above 50%. Within Brassicaceae, phylogenetic studies have provided a general framework of generic relationships and elucidated the closest relatives of *Arabidopsis* (e.g., Koch et al. 1999, 2001, 2003; Koch 2003; O'Kane and Al-Shehbaz 2003).

CHARACTER EVOLUTION IN BRASSICALES The core members of Brassicales (e.g., Koeberliniaceae, Bataceae, Salvadoraceae, Pentadiplandraceae, Tovariaceae, Emblingiaceae, Gyrostemonaceae, Resedaceae, and Brassicaceae) share several features typically associated only with Brassicaceae, such as tetramerous flowers (Table 8.2), seeds with curved or folded embryos and campylotropous ovules, endosperm lacking or nearly absent, vessels with vestured pits, and protein-rich, vascular or utricular cisternae of the endoplasmic reticulum (see Rodman 1991a, 1991b; Judd et al. 1999). There are important exceptions, however. Pentadiplandraceae and Emblingiaceae have perianth parts in five (the latter having two petals and five sepals); Resedaceae and Tovariaceae have six- to eight-merous flowers; Salvadoraceae have a straight rather than curved embryo; both Gyrostemonaceae and Pentadiplandraceae also have nonvestured pits. Other families illustrate the difficulty of some character codings: in Bataceae, the embryo is reported to be "slightly curved," not fitting neatly into the categories of "straight" or "curved."

To examine character evolution in Brassicales, we added the *rbcL* sequence of *Emblingia* (Chandler and Bayer 2000) to the existing data matrix of three genes and conducted additional phylogenetic analyses. Using the shortest tree, we conducted character-state reconstructions using MacClade (Maddison and Maddison 1992) and the characters and character states of Rodman (1991b), with modifications (*sensu* Cronquist 1981; Takhtajan 1997) for some characters, such as sepal and petal merosity and embryo curvature.

The topology we generated for Brassicales (see also Rodman et al. 1998; Chandler and Bayer 2000) indicated that the sister group of Brassicaceae is Emblingiaceae +

TABLE 8.2   Characters and character states in Brassicales[a]

Family	Floral Merosity	Embryo
Tropaeolaceae	5	Straight
Akaniaceae		
Bretschneidera	5	Curved
Akania	5	Straight
Caricaceae	5	Straight
Moringaceae	5	Straight
Limnanthaceae		
Limnanthes	5	Straight
Floerkea	3	Straight
Setchellanthaceae	4	Curved
Koeberliniaceae	4 (5)	Curved
Bataceae	4	Slightly curved
Salvadoraceae	4 (5)	Straight
Pentadiplandraceae	5	Curved
Tovariaceae	(8) 4	Curved
Emblingiaceae	5[b]	Curved
Gyrostemonaceae	uncertain	Curved
Resedaceae	4–8[c]	Curved
Brassicaceae		
Cleome	4	Curved
Capparis	4	Curved
Brassica	4	Curved
Arabidopsis	4	Curved

[a]Character states generally follow Cronquist (1981) and Takhtajan (1997).

[b]Reports for the number of sepals and petals vary. Some reports indicate that there are five sepals and two petals (reviewed in Chandler and Bayer 2000).

[c]Variously considered 4- or 6-merous.

(Gyrostemonaceae + Resedaceae). Slightly more distantly related are the clades of (1) Pentadiplandraceae + Tovariaceae and (2) Koeberliniaceae + (Bataceae + Salvadoraceae), followed by Setchellanthaceae. Some features associated with Brassicaceae, such as embryo morphology, actually evolved in the common ancestor of Brassicaceae and its close relatives. A curved embryo apparently evolved multiple times in Brassicales (Figure 8.14A), once in *Bretschneidera* (Akaniaceae) and again in the common ancestor of Setchellanthaceae, Koeberliniaceae, Bataceae, Salvadoraceae, Pentadiplandraceae, Tovariaceae, Emblingiaceae, Gyrostemonaceae, Resedaceae, and Brassicaceae; this feature was then lost in Salvadoraceae.

Character-state mapping indicated that perianth evolution in Brassicales is similar to that observed for embryo diversification. A pentamerous perianth was reconstructed as ancestral in Brassicales with a tetramerous perianth evolving in the common ancestor of Setchellanthaceae, Koeberliniaceae, Bataceae, Salvadoraceae, and taxa above these nodes (Figure 8.14B). A reversal to a

(A)

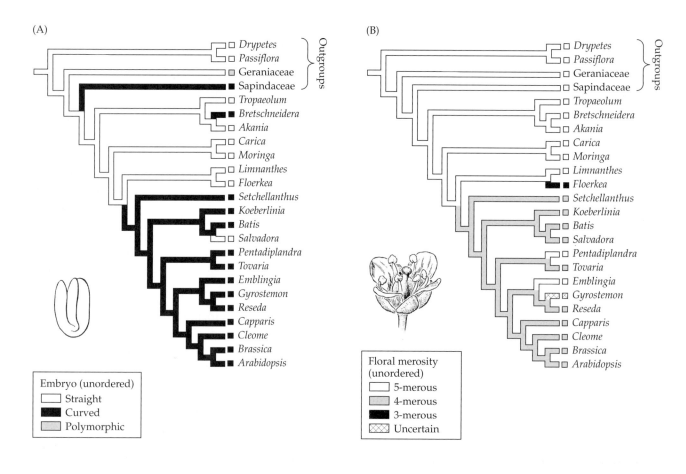

(B)

(C)

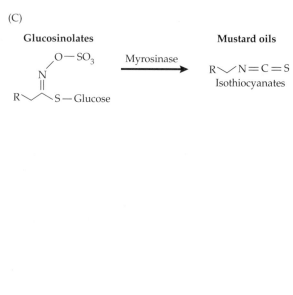

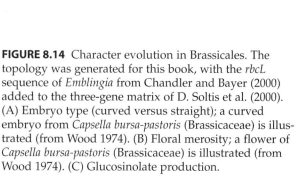

Glucosinolates

$$O\!-\!SO_3$$

Myrosinase

Mustard oils

$R\!\smallsmile\!N\!=\!C\!=\!S$
Isothiocyanates

R S—Glucose

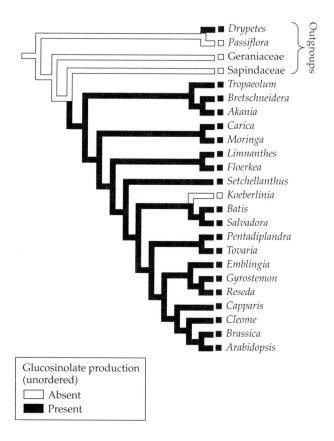

FIGURE 8.14 Character evolution in Brassicales. The topology was generated for this book, with the *rbcL* sequence of *Emblingia* from Chandler and Bayer (2000) added to the three-gene matrix of D. Soltis et al. (2000). (A) Embryo type (curved versus straight); a curved embryo from *Capsella bursa-pastoris* (Brassicaceae) is illustrated (from Wood 1974). (B) Floral merosity; a flower of *Capsella bursa-pastoris* (Brassicaceae) is illustrated (from Wood 1974). (C) Glucosinolate production.

pentamerous perianth occurred in Pentadiplandraceae and Emblingiaceae. However, the coding of perianth merism is not always straightforward. Families such as Emblingiaceae and Resedaceae are complicated. *Emblingia* has five sepals and two petals. Similarly, although Resedaceae were scored here as tetramerous, the number of sepals and petals each range from four to eight; the family has been considered hexamerous by some (e.g., Cronquist 1981).

Mustard-oil glucosides (Ettlinger and Kjaer 1968) are oxime-derived sulfur-containing compounds for which breakdown products include the familiar pungent principles of mustard, radish, and capers. The compounds are usually accompanied in the plant by a hydrolytic enzyme, myrosinase (a β-thioglucoside glucohydrolase, E.C. 3.2.3.1), which may be compartmentalized in special myrosin cells (reviewed in Rodman et al. 1998). This "mustard oil bomb" may deter herbivores and pathogens (Lüthy and Matile 1984).

Phylogenetic analyses indicate that the phytochemical system of glucosinolates with myrosinase enzyme, the latter in special myrosin cells, evolved twice, once in the ancestor of Brassicales and again, independently, in *Drypetes* (Putranjivaceae; Malpighiales; Figure 8.14C). *Drypetes* is a large pantropical genus of about 150 species, and the few species that have been screened for mustard-oil glucosides have yielded positive results (Ettlinger and Kjaer 1968). Additional species of *Drypetes*, as well as the related genus *Putranjiva*, should be analyzed for the presence of glucosinolates. Ultrastructural studies of *Drypetes* revealed protein-accumulating phloem cells similar to developing myrosin cells, although tests for ascorbate-activated myrosinase enzymes proved negative (reviewed in Rodman et al. 1998). However, no studies have been reported on the enzymology of mustard-oil glucoside biosynthesis in *Drypetes;* in contrast, several members of the Brassicales have been analyzed in detail (Du et al. 1995). At this point, the available data indicate that glucosinolate biosynthesis and myrosin cell formation in *Drypetes* are remarkably similar to those in Brassicales. Thus, the mustard oil bomb was invented twice (Rodman et al. 1993, 1998).

Within Brassicales, 14 families are known to produce glucosinolates. Only *Koeberlinia* clearly does not produce these compounds. The deeply embedded position of *Koeberlinia* within Brassicales indicates that it has lost critical gene functions or otherwise repressed the capacity to synthesize glucosinolates. Significantly, *Koeberlinia* possesses myrosin cells (Gibson 1979), which is consistent with this hypothesis of glucosinolate loss or repression (Rodman et al. 1993, 1998). Also, *Koeberlinia* is thorny, with small, deciduous leaves, so physical means of protection may have been associated with the loss of glucosinolate production. *Emblingia* represents a special situation in that the presence of glucosinolates in this monotypic genus has not been critically investigated, but the observations of several investigators indicate the presence of glucosinolates (C. Quinn, pers. comm.; J. E. Rodman, pers. comm.).

The documentation of two evolutionary origins of glucosinolate production poses an intriguing question. What biochemical pathway was modified for glucosinolate production, and is it the same pathway in both *Drypetes* and Brassicales? Kjaer (1973) and Harborne (1982) stressed the similarities in biosynthesis between mustard-oil glucosides and cyanogenic glycosides. The latter are cyanide-releasing compounds widespread in the angiosperms. Kjaer (1973) suggested that glucosinolate biosynthesis evolved via the recruitment and modification of cyanogen biosynthesis. Both glucosinolates and cyanogenic glucosides are groups of natural plant products derived from amino acids and have oximes as intermediates. It is noteworthy that Brassicales are in a clade with both Malvales and Sapindales, lineages with cyanogenic taxa. Similarly, *Drypetes* is part of Malpighiales, which also contain cyanogenic taxa such as *Euphorbia*, *Passiflora*, and Violaceae (see above).

More recently, molecular genetic analyses of *Arabidopsis* have provided insights into the evolution of glucosinolate production. Hansen et al. (2000a, 2000b) showed that cytochromes P450 (belonging to the CYP79 family) are involved in oxime formation in the biosynthetic pathway of both cyanogenic glucosides and glucosinolates. This is consistent with the hypothesis that glucosinolate production evolved from the cyanogenic glucoside pathway and indicates that the oxime-metabolizing enzyme is a possible branch point between the cyanogenic glucoside and glucosinolate pathway (Hansen et al. 2000a, 2000b; for more information on glucosinolates see http://www.plbio.kvl.dk/plbio/glucosinolate.htm).

Sapindales

Molecular phylogenetic investigations demonstrated the presence of a well-supported Sapindales clade consisting of nine families (Figure 8.12): Anacardiaceae, Biebersteiniaceae, Burseraceae, Kirkiaceae, Meliaceae, Nitrariaceae (including Peganaceae and Tetradiclidaceae), Rutaceae, Simaroubaceae, and Sapindaceae (the last broadly defined to include Aceraceae and Hippocastanaceae; D. Soltis et al. 2000; APG 1998; APG II 2003; several of these families are illustrated in Figure 8.13). This composition of Sapindales corresponds closely to Cronquist's (1968, 1981, 1988) broad view of the order, although he included families in Sapindales now considered only distantly related to the group (e.g., Akaniaceae, including Bretschneideraceae, of Brassicales). In contrast, other authors divided the constituent families of Sapindales between two orders, Rutales and Sapindales s.s. (e.g., Scholz 1964; Takhtajan 1987, 1997; G. Dahlgren 1989; Thorne 1992a, 1992b).

Although Sapindales were well supported in recent molecular phylogenetic analyses (Gadek et al. 1996; D. Soltis et al. 2000; Savolainen et al. 2000a), non-DNA synapomorphies for the clade are less clear. Several characters are widespread in the order, including woodiness, exstipulate, pinnately compound leaves (sometimes becoming palmately compound, trifoliate, or unifoliate), and generally small, tetra- or pentamerous flowers with a distinct nectar disc and imbricate perianth parts.

Molecular phylogenetic studies (Gadek et al. 1996; Savolainen et al. 2000b) indicated the presence of several major subclades in Sapindales (Figure 8.12): (1) Biebersteiniaceae, (2) Kirkiaceae, (3) Nitrariaceae, (4) Sapindaceae, (5) Burseraceae + Anacardiaceae, and (6) Rutaceae + Meliaceae + Simaroubaceae. Phylogenetic analyses provided support for a broadly defined Nitrariaceae that also included Peganaceae and Tetradiclidaceae (APG II 2003). In molecular analyses, Nitrariaceae (consisting of the single genus *Nitraria*) emerged as sister to Peganaceae (a family first recognized by Takhtajan 1969; see also Dahlgren 1980), which also included *Tetradiclis* (Tetradiclidaceae). Both *Nitraria* and *Tetradiclis* had previously been included in Zygophyllaceae (Sheahan and Chase 1996). Takhtajan (1987, 1997) recognized Kirkiaceae as a distinct family of two genera, *Kirkia* and *Pleiokirkia* (the latter has not been included in phylogenetic analyses), from Africa and Madagascar, respectively. Molecular evidence confirmed that the enigmatic Kirkiaceae represent a distinct family within Sapindales, but their relationships were unclear in these analyses (Gadek et al. 1996; Savolainen et al. 2000b). The position of Biebersteiniaceae within Sapindales also was uncertain in molecular phylogenetic studies. *Biebersteinia* comprises about five species previously placed in Geraniaceae (Heywood 1993). Takhtajan (1997), however, placed the genus in its own family and order close to Geraniales. Biebersteiniaceae appeared as sister to the remaining members of Sapindales in analyses based on *rbcL* sequences (Savolainen et al. 2000b). In our analyses conducted for this book, the genus appeared within Sapindales as part of an unresolved basal polytomy.

Molecular data have provided strong support for a broadly defined Sapindaceae that include Hippocastanaceae and Aceraceae, a definition of the family similar to that of Thorne (1992a, 1992b). Both Cronquist (1981) and Takhtajan (1987, 1997) considered Aceraceae and Hippocastanaceae to be closely related to Sapindaceae. Judd et al. (1994, 1999, 2002) similarly provided evidence of morphological support for a broadly defined Sapindaceae. Non-DNA characters uniting Aceraceae and Hippocastanaceae with Sapindaceae include the production of triterpenoid saponins in secretory cells, cyclopropane amino acids, usually eight stamens, pubescent or papillose filaments, ovules that lack a funiculus and are broadly attached close to a pro-

truding portion of the placenta (the obturator), the presence of a nectar disc, and an embryo with the radicle separated from the rest of the embryo by a deep fold or pocket in the seed coat. The presence of hypoglycin, an unusual amino acid, may be another synapomorphy for the clade (Gadek et al. 1996). These data supported the view that Aceraceae and Hippocastanaceae are temperate specializations within Sapindaceae, a family that has both temperate and tropical members. Phylogenetic relationships within a broadly defined Sapindaceae have also been investigated, and several well-defined clades are indicated (Judd et al. 1994).

Molecular data also indicated a strongly supported sister group of Anacardiaceae and Burseraceae (D. Soltis et al. 2000). These two families display some noteworthy non-DNA similarities, including the presence of vertical intercellular secretory canals in the primary and secondary phloem (Gadek et al. 1996); they also possess resin canals in the leaves and are the only two families of Sapindales from which biflavones are reported (Wannan et al. 1985). Anacardiaceae as defined here and by APG (1998) and APG II (2003) also include Podoaceae, Julianiaceae, and Blepharocaryaceae; these families have typically been included within Anacardiaceae in recent classifications.

A final well-supported clade in Sapindales is that of Meliaceae, Simaroubaceae, and Rutaceae. Molecular phylogenetic studies indicated that the last family also includes *Cneorum* (Cneoraceae) and *Ptaeroxylon* (Ptaeroxylaceae) (Gadek et al. 1996; Chase et al. 1999b; Savolainen et al. 2000a, 2000b). Relationships within Rutaceae are discussed in detail by Chase et al. (1999b), and Meliaceae have been analyzed phylogenetically by Muellner et al. (2003). Members of this clade of three families are united by the presence of biosynthetically related triterpenoid derivatives, limonoids and quassinoids (Hegnauer 1962–1994). Cronquist (1988) also considered the families closely related because of the absence of resin ducts in the bark, rays, and veins.

Molecular phylogenetic analyses also demonstrated that some taxa classified in Simaroubaceae are only distantly related to that family. *Picramnia* and *Alvaradoa* have usually been included within Simaroubaceae, but phylogenetic studies indicated a placement of these two genera well removed from Sapindales (Savolainen et al. 2000b). However, the exact placement of this small clade within the rosids is still unclear (see "Picramniaceae (including Alvaradoaceae)," below).

Savolainen et al. (2000b) reported that *Lissocarpa* (Lissocarpaceae) falls within Rutaceae. However, because the sequence was obtained from degraded herbarium DNA, this anomalous placement was reexamined; further study indicated that *Lissocarpa* is sister to *Diospyros* (Ebenaceae), as would be expected based on morphology (Berry et al. 2001). APG II included *Lissocarpa* in Ebenaceae.

Malvales

Malvales consist of nine families (Figure 8.12): Bixaceae (including Cochlospermaceae and Diegodendraceae), Cistaceae, Dipterocarpaceae, Malvaceae (including Bombacaceae, Tiliaceae, and Sterculiaceae), Muntingiaceae, Neuradaceae, Sarcolaenaceae, Sphaerosepalaceae, and Thymelaeaceae (several of these families are illustrated in Figure 8.13). Thymelaeaceae have been expanded (APG II 2003) to include the enigmatic genus *Tepuianthus,* a genus of five species from Venezuela and Colombia that had been placed in Celastraceae, but transferred to its own family (Cronquist 1981). Takhtajan (1997) considered Tepuianthaceae closely related to Simaroubaceae. However, molecular phylogenetic studies revealed that *Tepuianthus* was the well-supported sister to the remainder of Thymelaeaceae (Savolainen et al. 2000b; Wurdack and Horn 2001). Several molecular phylogenetic studies have focused on Malvales, and the topology we generated is in agreement with the results of Alverson et al. (1998), Bayer et al. (1998, 1999), Dayanandan et al. (1999), and van der Bank et al. (2002).

Malvales have been variously defined in recent classifications. A narrow view was provided by Cronquist (1981), who restricted the order to Bombacaceae, Tiliaceae, Malvaceae, Sterculiaceae, and Elaeocarpaceae. However, other authors have included additional families such as Bixaceae, Cistaceae, Cochlospermaceae, Diegodendraceae, Dipterocarpaceae, Dirachmaceae, Huaceae, Peridiscaceae, Plagiopteridaceae, Sarcolaenaceae, Sphaerosepalaceae, and Thymelaeaceae (Dahlgren 1983; Thorne 1992a, 1992b; Huber 1993; Takhtajan 1997).

A suite of morphological and other nonmolecular characters unites members of the Malvales clade, although in many cases not all families in the order have been examined for these features. In most cases, only one or a few representatives of a family have been examined for these characters, so it is not known if exceptions to the occurrence of these features exist. With these caveats, potential synapomorphies for Malvales include the presence of cyclopropenoid fatty acids in the seeds, complex chalazal anatomy, and dilated phloem rays; in addition, all members of Malvales have simple perforations in the secondary xylem (Metcalfe and Chalk 1950; Nandi et al. 1998). Most families (except Cistaceae) have representatives with mucilage cells or mucilage cavities (Metcalfe and Chalk 1950; Cronquist 1981). All taxa for which the character is known (all but Sarcolaenaceae and Sphaerosepalaceae) are also characterized by centrifugal polyandry, or rarely, as in Thymelaeaceae, lateral polyandry (e.g., Cronquist 1981, 1988; Ronse De Craene and Smets 1992; Nandi 1998b; Nandi et al. 1998). Seed anatomy provides another useful character because families of Malvales have the exotegmen differentiated as a palisade layer (Corner 1976; Nandi 1998a). This seed feature occurs only rarely outside of Malvales (e.g., in Trochodendraceae, Huaceae, and some Euphorbiaceae; Nandi et al. 1998).

Most Malvales (except Thymelaeaceae) possess a stratified phloem (Metcalfe and Chalk 1950), another feature that is uncommon outside of the order (Nandi et al. 1998). Other characters common in Malvales, but also found outside the clade, include stellate hairs, peltate scales, strong phloem fibers, valvate sepals, and the presence of an epicalyx (see Alverson et al. 1998; Nandi et al. 1998).

Of those families previously placed in Malvales but now excluded by phylogenetic evidence, Elaeocarpaceae (now placed in Oxalidales) are the most notable because they were frequently associated with families now considered to be the single family Malvaceae (see Cronquist 1981). However, Elaeocarpaceae lack most of the putative nonmolecular synapomorphies of Malvales as now circumscribed. For example, Elaeocarpaceae lack fatty acids, stellate hairs, peltate trichomes, and mucilage cavities; they have unstratified phloem, opposite lateral pitting of the vessels, and a fibrous exotegmen. As a result, Takhtajan (1997) noted that Elaeocarpaceae "differ markedly from the major families of traditional Malvales."

Neuradaceae, now firmly placed in Malvales, have often been included in Rosaceae or considered a close relative of Rosaceae. The family shares two potential synapomorphies with the expanded Malvales, an exotegmic seed coat with a palisade layer and cyclopropenoid fatty acids (Huber 1993). Neuradaceae also possess valvate sepals, lysigenous mucilage canals, stellate hairs, and an epicalyx. In the three-gene analysis of D. Soltis et al. (2000) and the *rbcL* analysis of Alverson et al. (1998), Neuradaceae emerged as sister to the remainder of the clade. However, support for this placement is less than 50% and, in other *rbcL* analyses, Neuradaceae appeared elsewhere in Malvales (Bayer et al. 1999; Savolainen et al. 2000b). Hence, the position of this family within the order is still unclear.

Analyses of *rbcL* sequences provided support for a clade of Bixaceae, Diegodendraceae, and Cochlospermaceae (Alverson et al. 1998; Bayer et al. 1998; Fay et al. 1998b). The last two families have been included in an expanded Bixaceae (APG II 2003). Muntingiaceae are weakly supported as sister to a well-supported group of Cistaceae and Sarcolaenaceae + Dipterocarpaceae (D. Soltis et al. 2000; Savolainen et al. 2000a, 2000b). Neotropical *Muntingia* is noteworthy in that the genus had historically been placed in Elaeocarpaceae, Tiliaceae, or Flacourtiaceae (see Cronquist 1981), but molecular data clearly indicate that it, and the related genus *Dicraspidia,* are a distinct lineage of Malvales (Bayer et al. 1998; Savolainen et al. 2000b). Bayer et al. (1998) recognized these two genera as a new family, Muntingiaceae, and later (Bayer et al. 1999) demonstrated that *Petenaea,* a monotypic genus from Central America, is allied with (if not part of) Muntingiaceae. Dipterocarpaceae are not monophyletic in the analyses conducted to date (Savolainen et al. 2000b; Dayanandan et al. 1999). Cistaceae and Sarcolaenaceae should probably be combined with Dipterocarpaceae; the oldest conserved name is Cistaceae. Hence, several interfamilial relationships within Malvales remain uncertain. In the *rbcL* analysis of Savolainen et al. (2000b), *Ploiarium,* the only Old World member of Bonnetiaceae, appeared as a member of Thymelaeaceae. This unexpected finding requires corroboration.

Within Malvales is a broadly defined Malvaceae that received strong support in molecular analyses (Alverson et al. 1998, 1999; Bayer et al. 1999; Savolainen et al. 2000a, 2000b; D. Soltis et al. 2000). In addition to the traditionally recognized Malvaceae, Malvaceae now include former members of Bombacaceae, Sterculiaceae, and Tiliaceae (Judd and Manchester 1997; Bayer et al. 1999). A close relationship among these four families has been recognized since Linnaeus, although relationships among them have long been problematic. Phylogenetic analyses of DNA sequences, as well as of non-DNA characters, demonstrated the polyphyly of both Sterculiaceae and Tiliaceae, as well as the paraphyly of Bombacaceae, and indicated that these four families should be merged into the single family Malvaceae (Judd and Manchester 1997; Bayer et al. 1999; see also Judd et al. 1999, 2002). *Theobroma* of Sterculiaceae received strong support as sister to *Grewia* of Tiliaceae, and *Durio* and its close relatives in Bombacaceae were found to be more closely related to some members of Sterculiaceae than to other Bombacaceae. Several genera treated as Bombacaceae and other genera treated as Sterculiaceae emerged as sister to traditional Malvaceae. These inter-digitating relationships among genera assigned to different families in traditional classifications (all recognizing four families) indicated that a broadly defined Malvaceae is required (Judd and Manchester 1997; Alverson et al. 1999; Bayer et al. 1999).

Likely synapomorphies for this broad Malvaceae include distinctive nectaries composed of tightly packed, multicellular hairs, normally found on the adaxial surface of the sepals (Judd and Manchester 1997; Vogel 2000) and the distinctive upright "tile" cells in wood rays (Manchester and Miller 1978; Figure 8.15). In addition, the inflorescences of all members of a broadly defined Malvaceae are composed of special modules called "bicolor units" (Figure 8.15, B to D; Bayer and Kubitzki 1996). In contrast, there are no clear synapomorphies for Malvaceae s.s.

CHARACTER EVOLUTION IN MALVALES   Molecular analyses indicate that Malvaceae have experienced many gains and losses of characters, including staminal tubes, palmately compound leaves, spiny pollen, winged seeds or fruits, and non-dithecate anthers. Considering the last feature, it has long been assumed that monothecate anthers ("half-anthers") were a derived character uniting Malvaceae and Bombacaceae (Cronquist 1981, 1988; Judd et al. 1994; Judd and Manchester 1997). However, DNA-based trees indicated instead that non-dithecate anthers have been derived at least twice independently (Figure 8.16): once in core Malvaceae ("/Malvoideae" of Alverson et al. 1999) and again in members of tribe

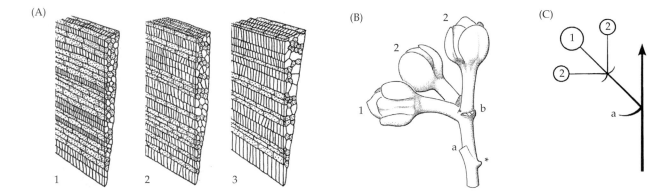
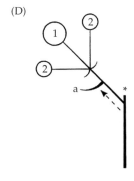

FIGURE 8.15 Diagnostic morphological features of Malvaceae. (A) The distinctive upright "tile" cells in wood rays. Cutaway diagrams of multiseriate rays showing different types of tile cells. Procumbent cells stippled, tile cells not stippled. 1. Tile cells of the *Durio* type, approximately the same height as the procumbent cells. 2. Tile cells of the intermediate type, slightly higher than the procumbent cells. 3. Tile cells of the *Pterospermum* type, two to several times higher than the procumbent cells (from Manchester and Miller 1978). B–D. Bicolor unit (from Bayer 1999). (B) Inflorescence of *Monotes kerstingii* axillary to a displaced subtending bract (a = scar). Asterisk designates what is interpreted as a rudiment of the apical bud; b = the scar of the subtending bract of a second-order lateral flower. C, D. Possible derivation of the bicolor unit from a lateral cymose inflorescence. (C) The lateral cymose inflorescence is subtended by a bract (a). (D) The bract (a) is shifted to a more distal position (arrow). With the suppression of the apex of the main axis (indicated by asterisk), the bicolor unit appears to occupy a terminal position.

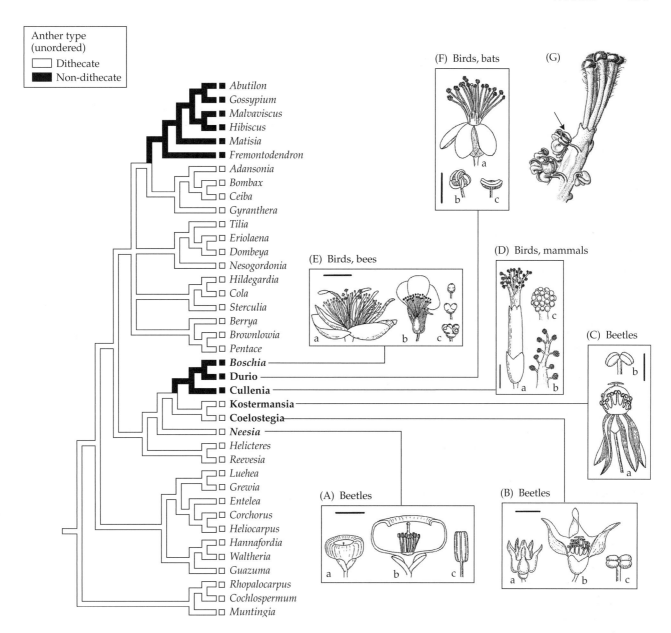

FIGURE 8.16 Evolution of non-dithecate (monothecate and non-thecal) anthers in Malvaceae. The topology shown is modified from the strict consensus of shortest trees of Alverson et al. (1998). To make the topology easier to visualize, some genera have been removed without altering the general pattern of relationships. Additional genera were added to tribe Durioneae, which contains *Neesia, Coelostegia, Kostermansia, Cullenia, Boschia,* and *Durio;* the relationships depicted among members of Durioneae (genera are indicated in bold) follow Nyffeler and Baum (2000). The molecular phylogeny indicates that non-dithecate anthers evolved twice: once in a clade of several genera of Durioneae (see E, F, D) and again in a clade that largely corresponds to Malvaceae in the traditional or narrow sense (G). Floral structure and pollinators are indicated for members of Durioneae. Within this tribe, the molecular phylogeny suggests that beetle pollination was ancestral in the tribe, with subsequent evolution of bat and bird pollination. Furthermore, the transition to bat and bird pollination was accompanied by extensive changes in the androecium. The early-branching genera *Neesia, Coelostegia,* and *Kostermansia* all have bithecate, tetrasporangiate anthers. However, in *Cullenia, Boschia,* and *Durio,* the anthers are highly modified (D, E, F). Flowers and stamens of Durioneae (A–F; modified from Nyffeler and Baum 2000). (A) *Neesia glabra:* (a) flower at anthesis; (b) longitudinal section of the flower at anthesis; (c) bithecate stamen. (B) *Coelostegia griffithii:* (a) flower before anthesis; (b) flower at anthesis with part of calyx removed; (c) bithecate stamen. (C) *Kostermansia malayana:* (a) flower at anthesis; (b) bithecate stamen. (D) *Cullenia exarillata:* (a) flower at anthesis; (b) terminal segment of the floral tube with nine clusters of pollen chambers; (c) detail of cluster of locules. (E) *Boschia griffithii:* (a) flower at anthesis; (b) flower at anthesis; (c) anthers, showing three different forms of poricidal locules; *Boschia grandiflora.* (F) *Durio zibethinus:* (a) flower at anthesis; (b) polylocular anther. (G) Stamens characteristic of the monothecate stamen clade, which largely corresponds to narrowly defined Malvaceae. *Kosteletzkya virginica:* arrow designates monothecate anther (from Wood 1974).

Durioneae of former Bombacaceae ("/Helicteroideae" of Alverson et al. 1999). Furthermore, in Helicteroideae, modifications in the androecium may be related to pollinator syndromes (Figure 8.16). Helicteroideae contain *Neesia, Coelostegia, Kostermansia, Cullenia, Boschia*, and *Durio* (the durians). The molecular results indicate that beetle pollination is ancestral in the tribe, with subsequent evolution of bat and bird pollination. Furthermore, the transition to bat and bird pollination was accompanied by extensive changes in the androecium. For example, the genera attached to the basalmost nodes, *Neesia, Coelostegia*, and *Kostermansia*, all have dithecate, tetrasporangiate anthers, the typical condition in the angiosperms, and their flowers are beetle-pollinated. However, in *Cullenia, Boschia*, and *Durio*, the anthers are nondithecate, and the flowers are pollinated primarily by birds and bats (Alverson et al. 1999; Nyffeler and Baum 2000).

A molecular study provided similar insights regarding floral evolution and pollinators in the famous baobabs (*Adansonia*) (Baum et al. 1998). In *Adansonia*, character-state mapping using a molecular tree indicated that hawkmoth pollination is ancestral in the genus and that there were two independent switches to pollination by fruit bats (Baum et al. 1998). Flowers in the two clades with bat pollination differ greatly in morphology.

Other Rosids

Although Myrtales, Crossosomatales, Geraniales, and Picramniaceae (including Alvaradoaceae) are clearly rosids, molecular phylogenetic studies have been unable to place them relative to other rosid groups with confidence. Each group is discussed in more detail below.

Myrtales

Myrtales consist of 13 families (Figure 8.17): Alzateaceae, Combretaceae, Crypteroniaceae, Heteropyxidaceae, Lythraceae (including Punicaceae and Trapaceae), Melastomataceae (including Memecylaceae; or these two can be treated as sister families), Myrtaceae, Oliniaceae, Onagraceae, Penaeaceae, Psiloxylaceae, Vochysiaceae, and Rhynchocalycaceae (several of these are illustrated in Figure 8.18). Several molecular phylogenetic studies have focused on the group (Conti et al. 1993, 1996; Savolainen et al. 2000b; Clausing and Renner 2001).

Myrtales represent one of the most thoroughly investigated orders of angiosperms. An entire issue of the *Annals of the Missouri Botanical Garden* (1984; vol. 71, no. 3) was dedicated to the order, summarizing aspects of morphology, chemistry, anatomy, palynology, and embryology (e.g., Behnke 1984; Cronquist 1984; Dahlgren and Thorne 1984; Johnson and Briggs 1984; Patel et al. 1984; Tobe and Raven 1984; van Vliet and Baas 1984).

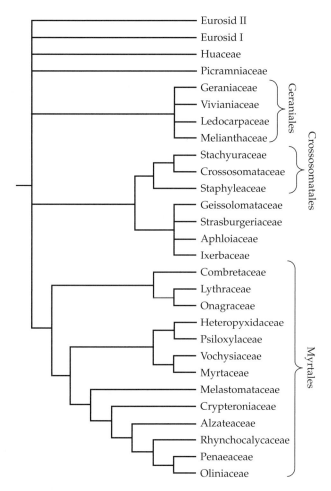

FIGURE 8.17 Summary tree for Myrtales, Crossosomatales, and Geraniales. General topology is from D. Soltis et al. (2000) with modifications from Savolainen et al. (2000b) and Clausing and Renner (2001).

Despite intensive study, disagreement persisted regarding the composition of, as well as relationships within, the order (reviewed in Dahlgren and Thorne 1984; Conti et al. 1996). Treatments agreed in placing 14 families in the order: Myrtaceae, Heteropyxidaceae, Psiloxylaceae, Lythraceae, Punicaceae, Sonneratiaceae, Combretaceae, Melastomataceae, Memecylaceae, Onagraceae, Trapaceae, Crypteroniaceae, Alzateaceae, and Rhynchocalycaceae. Most authors also included the small South African families Oliniaceae and Penaeaceae. However, additional families were sometimes placed in Myrtales, including Thymelaeaceae, Lecythidaceae, Rhizophoraceae, Haloragaceae, and Gunneraceae (e.g., Cronquist 1981, 1988; Heywood 1993).

Phylogenetic analyses based on sequence data clarified the composition of Myrtales, as well as relationships within the order (e.g., Conti et al. 1993; Clausing and Renner 2001; Schönenberger and Conti 2003; Fig-

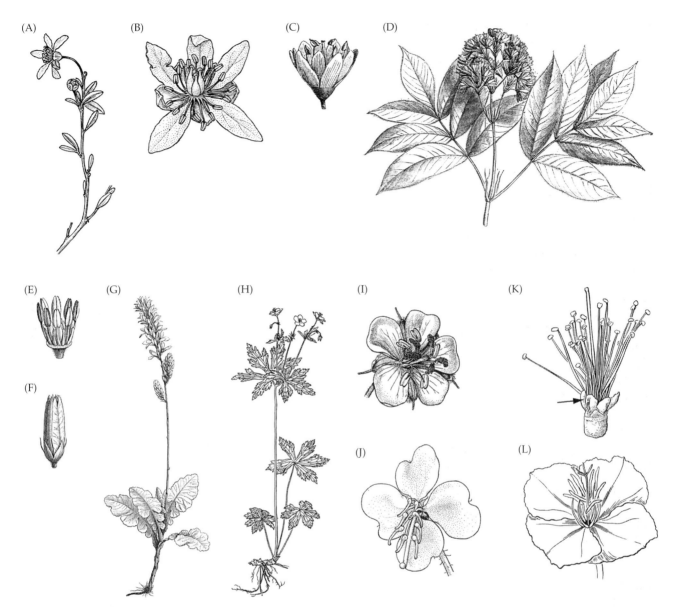

FIGURE 8.18 Flowers and flowering plants representing Crossosomatales (A–D), Geraniales (E–I), and Myrtales (J–L). (A) *Crossosoma bigelovii* (Crossosomataceae), flowering branch. (B) *Crossosoma bigelovii* (Crossosomataceae), flower. (C) *Staphylea pinnata* (Staphyleaceae), flower. (D) *Staphylea pinnata* (Staphyleaceae), flowering stem. (E) *Francoa sonchifolia* (Melianthaceae), flower, with perianth removed to show androecium and gynoecium (the latter partially obscured by stamens). (F) *Francoa sonchifolia* (Melianthaceae), fruit. (G) *Francoa sonchifolia* (Melianthaceae), plant in flower. (H) *Geranium maculatum* (Geraniaceae), plant in flower. (I) *Geranium maculatum* (Geraniaceae), flower. (J) *Rhexia virginica* (Melastomataceae), flower. (K) *Melaleuca quinquenervia* (Myrtaceae), flower; arrow indicates petal. (L) *Oenothera macrocarpa* (Onagraceae), flower. (A and B, from Sosa and Chase 2003. C and D, from Pax in Engler and Prantl 1897. E to G, from Engler in Engler and Prantl 1891. H to L, from Wood 1974.)

ure 8.17). The DNA-based circumscription of Myrtales agreed closely with that of Dahlgren and Thorne (1984), except that molecular phylogenetic studies indicated that Vochysiaceae also belong in Myrtales, a noteworthy result in that this family had never been associated with the order, but was placed in Polygalales (e.g., Cronquist 1981, 1988; Heywood 1993) or recognized as Vochysiales close to Polygalales (Takhtajan 1997). The unique floral morphology of Vochysiaceae (strongly asymmetric flowers, one fertile stamen, and the number of petals reduced to three, one, or none; Cronquist 1981) had made it difficult to place the family using morphology. Other families sometimes placed in Myrtales (e.g., Thymelaeaceae, Lecythidaceae, Rhizophoraceae, Haloragaceae, and Gunneraceae) were clearly excluded from the order (Conti et al. 1993; D. Soltis et al. 2000; Savolainen et al. 2000a, 2000b).

Myrtales are characterized in large part by two distinctive features of wood anatomy, presence of bicollateral vascular bundles in the primary stem and vestures in bordered pits of secondary xylem. Significantly, this combination is found in Vochysiaceae, supporting their placement in Myrtales. The occurrence of both features outside of Myrtales (e.g., Apocynaceae, Thymelaeaceae) is rare. Myrtales also possess methylated ellagic acids and intraxylary phloem (Nandi et al. 1998). Other synapomorphies include stamens incurved in bud and a single style (carpels completely connate). Still other features are present in most Myrtales; these include (Tobe and Raven 1983, 1984) glandular anther tapetum, an inner integument with two layers, micropyle formed by both integuments, ephemeral antipodal cells, and exalbuminous seeds.

DNA sequence data revealed that Myrtales are composed of two major clades (Figure 8.17; Conti et al. 1993; Savolainen et al. 2000b; D. Soltis et al. 2000; Clausing and Renner 2001). One of these clades consists of Combretaceae as sister to the strongly supported (95%) sister group of Lythraceae (broadly defined to include *Trapa* and *Punica*) + Onagraceae. The remaining families formed a second clade, which, in turn, consisted of two subclades. In one of these subclades, Myrtaceae + Vochysiaceae were sister to Heteropyxidaceae + Psiloxylaceae. The second subclade consisted of Melastomataceae (including Memecylaceae) as the sister group to a strongly supported "CAROP" clade (Clausing and Renner 2001) consisting of a grade of Crypteroniaceae, Alzateaceae, and Rhynchocalycaceae, as successive sisters to Oliniaceae + Penaeaceae. These relationships are in agreement with the results of Savolainen et al. (2000a) based on *atpB* plus *rbcL*, as well as those of Conti et al. (1996) and Savolainen et al. (2000b) based on *rbcL* alone.

Johnson and Briggs (1984) provided a cladistic analysis of interfamilial relationships within Myrtales based on morphological characters, using the circumscription of the order proposed by Dahlgren and Thorne (1984). Some of their results compare favorably with the DNA-based topology. For example, Johnson and Briggs recovered a clade of Myrtaceae, Heteropyxidaceae, and Psiloxylaceae (they did not include Vochysiaceae in their non-DNA analysis; see above) as sister to Penaeaceae and Oliniaceae (Schönenberger and Conti 2003). Several analyses have also focused on phylogenetic relationships within families of Myrtales. Onagraceae, for example, are one of the best-studied angiosperm families (e.g., Raven 1979; Sytsma et al. 1990; Conti et al. 1993; Levin et al. 2003). Other families of the order also have been the focus of phyloge-

netic analyses, including Myrtaceae (Briggs and Johnson 1979; Wilson et al. 2001), Vochysiaceae (Litt 1999; Litt and Stevenson 2003), and Melastomataceae (Clausing et al. 2000; Clausing and Renner 2001; Renner and Meyer 2001; Renner et al. 2001; Michelangeli et al. 2004; see also www.flmnh.ufl.edu/natsci/herbarium/melastomes/default.htm), and Lythraceae (Graham et al. 1993; K. Sytsma et al. unpubl. data).

CHARACTER EVOLUTION IN MYRTALES Fruit characters have been evolutionarily labile in Myrtales; this lability in fruit types is seen throughout the angiosperms (see above, Fagales, Celastrales). In Myrtaceae, the plesiomorphic state is a dehiscent fruit; indehiscent fruits have evolved in four lineages, with three origins of fleshy fruit (Wilson et al. 2000). In Melastomataceae, berries evolved from capsules at least four times (Clausing and Renner 2001). Renner et al. (2001) also examined the historical biogeography of Melastomataceae in light of existing molecular phylogenetic hypotheses. The family has a pantropical distribution and, in contrast to earlier hypotheses, Renner et al.'s results indicated that the current distribution of Melastomataceae is the result of long-distance dispersal during the Neogene, rather than fragmentation of Gondwana.

Geraniales

Molecular phylogenetic studies indicated that Geraniales comprise Geraniaceae (including Hypseocharitaceae), Melianthaceae (including Francoaceae and Greyiaceae), Vivianiaceae, and Ledocarpaceae (Figure 8.17; several of these families are illustrated in Figure 8.18). This circumscription of Geraniales represents another unexpected result of molecular phylogenetic studies (see also Price and Palmer 1993). Geraniaceae, Hypseocharitaceae (now part of Geraniaceae), and Vivianiaceae have been considered closely related in some treatments (e.g., Takhtajan 1997); Cronquist (1981) also placed them together in his Geraniaceae. They share several features, including vessels with simple perforations, fibers with simple pores, extrastaminal nectar glands, two ovules per locule, and five connate carpels (usually). However, Geraniaceae have often been associated with Oxalidaceae, a family now known to be a member of Oxalidales in eurosid I. Furthermore, the remaining members of the molecular-based Geraniales clade (i.e., Melianthaceae, including Francoaceae and Greyiaceae) have not been closely associated with Geraniaceae in morphology-based classifications. For example, Francoaceae and Greyiaceae represent former members of Saxifragaceae s.l. (Morgan and Soltis 1993), whereas Melianthaceae in the traditional sense were placed in Sapindales (Cronquist 1981; Takhtajan 1997). At this point, there are no known non-DNA synapomorphies for Geraniales.

Relationships within Geraniales remain unclear. The *rbcL* sequence analysis of Savolainen et al. (2000b) provided weak support for Geraniaceae as sister to a clade of Melianthaceae and Vivianiaceae. However, in three-gene analyses, Vivianiaceae emerged as sister to Geraniaceae, but without support above 50%. The expansion of Melianthaceae to include Greyiaceae and Francoaceae is noteworthy. Both morphological and molecular evidence indicates that these families are closely related (Danilova 1996; Ronse De Craene and Smets 1999; Savolainen et al. 2000b; D. Soltis et al. 2000). As a result, in APG II (2003), Greyiaceae were synonymized under Melianthaceae, with Francoaceae (comprising *Francoa* and *Tetilla*) an optional further synonym under Melianthaceae. However, *Francoa* and *Tetilla* (both from Chile and the only members of Melianthaceae s.l. outside eastern and southern Africa) are both herbaceous and otherwise dissimilar to the other woody taxa of this group. Only *Francoa* has been included in molecular phylogenetic analyses; it would be worthwhile to add *Tetilla* in future investigations.

The relationships of the small family Ledocarpaceae (consisting of *Balbisia* and *Wendtia*) remain obscure. With *rbcL*, *Wendtia* (*Balbisia* was not included) was sister to *Viviania* (Savolainen et al. 2000b). *Wendtia* was not included in the three-gene analysis of D. Soltis et al. (2000). However, in our larger three-gene analysis conducted for this book, *Wendtia* again appeared as sister to *Viviania*. Ledocarpaceae are provisionally maintained as a distinct family pending more data.

Crossosomatales

Phylogenetic analyses of gene sequence data have indicated the presence of a previously unrecognized assemblage of closely related taxa designated Crossosomatales (APG II 2003). Circumscription of this order was conservative in APG II, consisting of only three families, Crossosomataceae, Stachyuraceae, and Staphyleaceae (Figure 8.17). Analyses of *rbcL* sequences that included samples from all eudicot families indicated the presence of this Crossosomatales clade (Savolainen et al. 2000b), as did the analysis of three genes (D. Soltis et al. 2000). In contrast, earlier analyses of *rbcL* (Chase et al. 1993) and *rbcL* plus *atpB* (Savolainen et al. 2000a) did not contain adequate taxon sampling to demonstrate the presence of this clade. Relationships within Crossosomatales are well understood. Molecular analyses revealed that Crossosomatales (*sensu* APG II 2003) consist of a strongly supported clade of Crossosomataceae + Stachyuraceae as sister to Staphyleaceae (Savolainen et al. 2000b; D. Soltis et al. 2000). Phylogenetic relationships within Crossosomataceae have recently been investigated (Sosa and Chase 2003).

In addition to the currently recognized members (Staphyleaceae, Stachyuraceae, Crossosomataceae; APG II 2003), molecular studies indicated, albeit with weak support, that several other families may be related to Crossosomatales: Ixerbaceae, Aphloiaceae, Geissolomataceae, and Strasburgeriaceae. In the three-gene analysis (D. Soltis et al. 2000), *Aphloia* and *Ixerba* were sister taxa, and this clade, in turn, appeared as the weakly supported (56%) sister to Crossosomatales. However, Strasburgeriaceae and Geissolomataceae were not included in that study. In the *rbcL* analysis of Savolainen et al. (2000b), a clade of *Ixerba*, *Strasburgeria*, and *Geissoloma* was sister to Crossosomatales, but this relationship did not receive support above 50%. In our phylogenetic analysis conducted for this book, *Aphloia* was sister to a clade of *Geissoloma* + (*Ixerba* + *Strasburgeria*). These four genera then formed the sister group of Crossosomatales (Figure 8.17). Using a small *rbcL* dataset, Sosa and Chase (2003) found the same results, with very strong support (100%) for the sister group of *Ixerba* + *Strasburgeria*. Much weaker support (64%) was apparent for the clade of *Geissoloma* + (*Ixerba* + *Strasburgeria*) and the clade of *Aphloia* + (*Geissoloma* + (*Ixerba* + *Strasburgeria*)). Bootstrap support above 50% was not obtained in Sosa and Chase (2003) for the sister-group relationship of these four genera to Crossosomatales. If additional data strengthen the support for a close relationship of these families (Ixerbaceae, Aphloiaceae, Geissolomataceae, and Strasburgeriaceae) to Crossosomatales, the order should ultimately be expanded.

Crossosomatales represent another unexpected assemblage of taxa revealed by molecular phylogenetic analyses. Members of Crossosomatales are enigmatic taxa, previously placed in distantly related orders. Both Crossosomataceae and Stachyuraceae were placed by Heywood (1993) in Dilleniidae, in Dilleniales and Theales, respectively. However, other authors have placed Crossosomataceae in or near Rosales (Cronquist 1981; Takhtajan 1997). Staphyleaceae have been placed in Sapindales (Rosidae) (Cronquist 1981; Heywood 1993, 1998; Takhtajan 1997).

Geissolomataceae and Stachyuraceae have at times been placed together in Celastrales (Heywood 1993), but in other treatments have been variously placed. *Ixerba* has been placed in its own family, with *Brexia* and *Roussea* in Brexiaceae, or the three genera have each been placed in separate families within Brexiales (Takhtajan 1997). However, these three genera are only distantly related (Koontz and Soltis 1999). The monotypic *Aphloia* (Aphloiaceae) has been treated as a member of Flacourtiaceae, to which it is clearly only distantly related (see Malpighiales, above). Takhtajan (1997) considered the monotypic *Strasburgeria* to represent "a New Caledonian relict" with "primitive characters"; he placed Strasburgeriaceae close to Ochnaceae in his Ochnales (Dilleniidae), whereas Cronquist (1981) considered *Strasburgeria* to be a member of Ochnaceae.

The three families of this clade, Crossosomataceae, Stachyuraceae, and Staphyleaceae, appear to share a seed in which the cell walls of the many-layered testa are all, or mostly, lignified. This same seed morphology also occurs in *Geissoloma, Ixerba, Strasburgeria,* and *Aphloia,* strengthening support for a close relationship of these taxa to Crossosomatales and perhaps ultimately a broader circumscription of the order. This finding further demonstrates that seed anatomy may be a valuable source of new systematic information; seed characters often appear highly congruent with phylogenetic relationships inferred from analyses of molecular data (Nandi et al. 1998; APG II 2003). Despite the diverse taxonomic placements, most members of Crossosomatales are similar in floral morphology, possessing four or five sepals and petals, four or five stamens (numerous stamens in Aphloiaceae and some Crossosomataceae), four or five carpels, and a superior ovary, but these characters are, of course, widespread in the eudicots. A comparative study of floral characters in all seven families considered here (Crossosomataceae, Stachyuraceae, Staphyleaceae; Ixerbaceae, Aphloiaceae, Geissolomataceae, and Strasburgeriaceae) reveals potential new synapomorphies for subclades recovered with molecular data within this extended Crossosomatales clade (Matthews and Endress, 2005). For example, floral structure supports a close relationship among the core members of Crossosomatales (Crossosomataceae, Stach-yuraceae, and Staphyleaceae). In addition, floral data support a particularly close relationship of Ixerbaceae and Strasburgeriaceae, as well as a close relationship of Aphloiaceae and Geissolomataceae to these two families (Matthews and Endress 2005).

Crossosomatales are much in need of additional investigation. More molecular data are needed to resolve the composition of the clade and determine whether Crossosomatales should be more broadly defined. Relationships within the clade also require more investigation.

Picramniaceae (including Alvaradoaceae)

Alvaradoa and *Picramnia* are two poorly understood genera from Central and tropical America. Takhtajan (1997) and Cronquist (1981) placed both *Alvaradoa* and *Picramnia* in Simaroubaceae; Takhtajan (1997) treated each as a distinct subfamily within Simaroubaceae. The two genera appeared as sisters with strong support in analyses based on *rbcL* and also *rbcL* plus *atpB* (Savolainen et al. 2000a, 2000b) but were not included in the three-gene study of angiosperms (D. Soltis et al. 2000). The exact placement of *Alvaradoa* and *Picramnia* within the rosids is uncertain. However, the two genera do not appear to be closely related to Simaroubaceae. With *rbcL* alone, *Alvaradoa + Picramnia* appeared as sister to all other rosids (Savolainen et al. 2000b), but without support above 50%. Analysis of a combined *atpB* plus *rbcL* dataset placed *Alvaradoa* and *Picramnia* as sister to Zygophyllaceae (including Krameriaceae) in eurosid I, but also without support above 50%. *Alvaradoa* has been included in Picramniaceae, but at this point the family remains unassigned to order within the rosid clade (APG II 2003), and their relationships are unknown (Figures 8.1 and 8.17).

Huaceae

The affinities of Huaceae have long been obscure. The two genera (*Hua* and *Afrostyrax*) have been placed in or near Styracaceae (Heywood 1993; a family now placed in Ericales), in Violales (Cronquist 1981), or considered closely related to Sterculiaceae in Malvales (Takhtajan 1997).

With three genes (D. Soltis et al. 2000), the problematic Huaceae appeared as sister to the strongly supported Celastrales clade, but with only weak support (62%), indicating a tentative association of Huaceae with Celastrales (Savolainen et al. 2000a; D. Soltis et al. 2000). However, other analyses have not placed Huaceae with Celastrales. Our analysis of a large three-gene dataset for eudicots conducted for this book placed Huaceae as sister to Oxalidales and Malpighiales. A four-gene analysis of eudicots indicated a relationship of *Afrostyrax* to Oxalidales, but without support above 50% (D. Soltis et al. 2003a). Thus, at this point, the relationships of Huaceae remain unclear (Figure 8.17).

Parasitism in the Rosids

Phylogenetic placement of many parasitic plants has remained obscure, despite the general success of molecular phylogenetics in clarifying the relationships of taxonomically troublesome plants. The reasons for these difficulties are several, but generally involve the fact that these plants have degenerate plastid genomes. As a result, plastid genes such as *rbcL* and *atpB*, which have been so widely used to infer angiosperm phylogeny, are absent or have accelerated evolutionary rates (reviewed in Nickrent et al. 1998). However, these taxa can be placed by using nuclear genes, such as 18S rDNA, or mitochondrial DNA sequences.

Sequence data have indicated that several groups of parasitic plants are nested within the asterids (e.g., Mitrastemonaceae, Orobanchaceae, *Cuscuta;* see Chapter 9). Several other lineages of parasitic plants that have been particularly difficult to place using morphological data now appear to be nested within the rosids. Using DNA sequence data for the mitochondrial gene *matR*, Barkman et al. (2004) found evidence for the placement of *Rafflesia* and *Rhizanthes* (Rafflesiaceae) with Malpighiales. These genera are noteworthy in that the plants lack leaves, stems, and roots and rely entirely on the host plants for nutrients (see Chapter 11); the genus *Rafflesia* also contains the world's largest flowers

(*R. arnoldii*). Barkman et al. (2004) also showed that *Mitrastema*, another genus sometimes placed in Rafflesiaceae, is distantly related to *Rafflesia* and *Rhizanthes*, exhibiting a close relationship to Ericales in the asterid clade (see Chapter 9). Additional analyses of mtDNA sequence data suggest that Balanophoraceae may be a member of Santalales and that Cytinaceae are embedded within Malvales (T.J. Barkman and C.W. dePamphilis, pers. comm.). Apodanthaceae may also be a member of Malvales. Recent morphological studies indicate that Apodanthaceae share with Malvales special features of floral structure, including an androecial tube and a trend from normal stamens to synandria without a thecal anther organization (Blarer et al. 2004). A possible close relationship of Apodanthaceae and Malvales is also suggested by preliminary molecular data (Nickrent 2002). One caveat of these molecular analyses is that branch lengths to the parasitic taxa are often long; hence, spurious attraction could be occurring. Molecular data have not yet placed Cynomoriaceae with confidence (T. J. Barkman and C. W. dePamphilis, pers. comm.). These parasitic lineages are discussed in more detail in Chapter 11.

Future Research

The rosid clade remains the most poorly understood major subgroup of angiosperms. Other large clades of roughly comparable size, such as the monocots and asterids, are much better understood in terms of phylogenetic relationships, non-DNA synapomorphies for clades, and patterns of character evolution. Several major questions of relationship remain, such as the exact placement of Myrtales, Geraniales, Crossosomatales, and several families (Picramniaceae, Aphloiaceae, Geissolomataceae, Ixerbaceae, and Strasburgeriaceae). Relationships within several orders, particularly Malpighiales, also remain poorly understood and require intensive investigation. One of the major challenges is the discovery of non-DNA synapomorphies for the entire rosid clade, eurosid I and II, and many of the orders within the rosids. However, identifying such characters will not be a simple task because many characters have not been surveyed completely throughout the rosids. There are so many gaps in taxonomic coverage of these non-DNA characters that the potential utility of many traits cannot be meaningfully assessed until more basic survey work is conducted.

9

Asterids

Introduction

The taxonomic recognition of a group of angiosperms that corresponds in large part to the asterids traces back to de Jussieu's (1789) Monopetalae, those angiosperms with fused petals. Subsequently referred to as Sympetalae (Reichenbach 1827–1829), this group was divided into Pentacyclicae and Tetracyclicae by Warming (1879). Tetracyclicae, with a single series of stamens, formed the core of modern treatments of subclass Asteridae (e.g., Takhtajan 1969, 1980, 1997; Cronquist 1981).

Asteridae *sensu* Takhtajan and Cronquist have been the focus of phylogenetic analysis for more than 15 years, with data from morphology (e.g., Hufford 1992), restriction site analysis of the plastid genome (e.g., Jansen and Palmer 1987, 1988; Downie and Palmer 1992), gene sequences (e.g., Olmstead et al. 1992, 1993, 2000; D. Soltis et al. 2000; Albach et al. 2001a; B. Bremer et al. 2002), as well as molecular data plus morphology (K. Bremer et al. 2001) used to improve understanding of relationships within Asteridae and between Asteridae and other groups. The most significant result of these phylogenetic analyses is the firm conclusion that Asteridae *sensu* Takhtajan and Cronquist are not monophyletic; instead, they form the core of a large clade that also includes members of Cronquist's (1981) Hamamelidae (Eucommiales), Dilleniidae (most Theales, Lecythidales, Ericales, Ebenales, Diapensiales, Primulales, Sarraceniaceae, Fouquieriaceae, Loasaceae), and Rosidae (Hydrangeaceae, Cornales, Pittosporaceae, Apiales, Byblidaceae, Columelliaceae, Alseuosmiaceae, Aquifoliaceae, Icacinaceae, Balsaminaceae, *Escallonia* and *Montinia* from Grossulariaceae, and *Eremosyne* and *Vahlia* from

Saxifragaceae; Olmstead et al. 1992, 1993; Chase et al. 1993). The asterids *sensu* APG (1998; APG II 2003) include approximately one-third of all angiosperm species, with almost 80,000 species in nearly 4,700 genera and 114 families (Thorne 1992a, 1992b; Albach et al. 2001a).

In contrast to the large rosid clade, for which no unifying non-DNA characters are known (see Chapter 8), several features unite all, or most, asterids. The asterid clade includes nearly all species of angiosperms that produce iridoids (Jensen 1992) and tropane alkaloids (Romeike 1978) and most angiosperms that produce caffeic acid (Mølgaard 1985; Grayer et al. 1999). Dahlgren (1975) noted that the occurrence of iridoids is strongly correlated with sympetalous flowers and embryological characters such as unitegmic-tenuinucellate ovules and cellular endosperm development. However, although several non-DNA characters are prevalent throughout the asterids, circumscription of the clade based on morphological data alone has been difficult because of parallelisms (Hufford 1992) and losses of characters (Olmstead et al. 1992; Albach et al. 2001b). Patterns of character evolution are discussed in more detail later in this chapter.

The circumscription of and relationships within the asterids *sensu* APG have been clarified through both broad analyses of the angiosperms (e.g., Chase et al. 1993; D. Soltis et al. 1997a, 1997b, 2000; P. Soltis et al. 1999b; Savolainen et al. 2000a, 2000b) and more focused analyses of specific clades. The most comprehensive recent molecular analyses of the asterids themselves (Albach et al. 2001a; B. Bremer et al. 2002) have built on the earlier work of Olmstead et al. (1992, 1993, 2000). Most asterids fall into one of four major clades—Cornales, Ericales, euasterid I (or lamiids), and euasterid II (or campanulids) (Figure 9.1). With combined datasets of two (Savolainen et al. 2000a), three (D. Soltis et al. 2000), four (Albach et al. 2001a), and six (B. Bremer et al. 2002) genes, internal support is generally high (bootstrap values >95%) for these clades, as well as for the relationships among them. An exception is euasterid II, which received only weak support (56% jackknife) with three genes (D. Soltis et al. 2000) and very strong support (99% jackknife) with six DNA regions, but with fewer taxa used in the analysis (B. Bremer et al. 2002). The four-gene (*rbcL, atpB,* 18S rDNA, and *ndhF*) topology (Albach et al. 2001a) and six-gene (*rbcL, ndhF, matK,* and three noncoding plastid regions) topology (B. Bremer et al. 2002) are the best supported and will form the framework for our discussion in this chapter. Cornales are sister to all remaining asterids, followed by Ericales, which are sister to euasterid I (lamiids) + euasterid II (campanulids) (Figure 9.1). These placements of Cornales and Ericales received strong support, however, only in the analysis of B. Bremer et al. (2002). The composition and internal phylogenetic structure of these clades are discussed in more detail below.

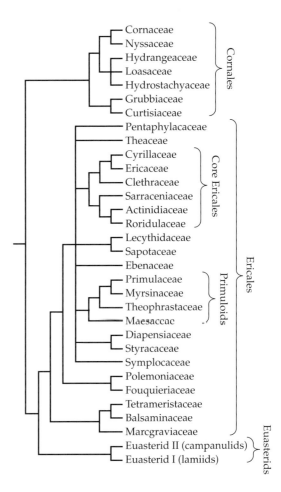

FIGURE 9.1 Summary tree of phylogenetic relationships in the asterids, with emphasis on familial relationships within Cornales and Ericales.

Cornales

Cornales (asterid IV *sensu* Chase et al. 1993; Cornales *sensu* Olmstead et al. 1993) consist of six or seven families (*sensu* APG II 2003; Figure 9.1): Cornaceae (including *Alangium,* which was sometimes recognized as a distinct family, Alangiaceae), Nyssaceae, Loasaceae, Hydrangeaceae, Grubbiaceae, Curtisiaceae, and Hydrostachyaceae (several of these families are illustrated in Figure 9.2). This circumscription expands Cronquist's (1981) Cornales with the addition of Hydrangeaceae (Cronquist's Rosidae-Rosales), Hydrostachyaceae (Asteridae-Callitrichales), Loasaceae (Dilleniidae-Violales), and *Grubbia* (Dilleniidae-Ericales). Phylogenetic analyses of molecular data also indicated that several families should be excluded from Cornales *sensu* Cronquist: Garryaceae, Aucubaceae (now included in Garryaceae), and Helwingiaceae. These three families have affinities elsewhere in the asterids (see "Garryales and Aquifoliales," below). The DNA-based circumscription of Cornales resembles more closely the Cornales of Takhtajan (1997), although he placed Hydrostachyaceae and Loasaceae in Lamiidae and Grubbiaceae in Bruniales of his Dilleniidae. Thorne's (1992a, 1992b) Cornineae resemble Cronquist's Cornales, but also include Eucommiaceae and Icacinaceae, both of which are now known to have relationships elsewhere in the asterids.

The circumscription of Cornales and relationships within the clade have been the subject of several investigations. Cladistic analyses using restriction sites (Downie and Palmer 1992), morphological and chemical data (Hufford 1992), *rbcL* or *matK* sequences (Olmstead et al. 1992, 1993; Chase et al. 1993; Xiang et al. 1993), and combined DNA datasets involving multiple genes (e.g., Albach et al. 2001a, 2001c; Xiang et al. 2002) have revealed a similar Cornales clade. Cornales include 40 genera and approximately 660 species.

Possible non-DNA synapomorphies for Cornales include an ovary that is at least partially inferior, an epigynous nectar disk, small sepals, and drupaceous fruits (Figure 9.2); however, there are deviations from these features, particularly in the highly modified aquatic *Hydrostachys* (see Albach et al. 2001c; Hufford et al. 2001).

Cornales comprise three subclades: Curtisiaceae + Grubbiaceae; Nyssaceae + Cornaceae; and Hydrangeaceae, Loasaceae, Hydrostachyaceae. However, relationships among these groups are not completely resolved. The Curtisiaceae + Grubbiaceae clade (each family consisting of a single genus) is sister to the remainder of Cornales. The southern African genus *Curtisia* was included in Cornaceae by Cronquist (1981) and APG (1998), but based on more recent analyses, Curtisiaceae were reinstated as a distinct family in APG II (2003). The relationships of *Grubbia*, a genus of three species from southern Africa, have long been problematic. It has variously been referred to Santalales, Ericales, and Rosales (reviewed in Cronquist 1981; see also Carlquist 1978; Xiang et al. 2002).

Cornaceae (including *Alangium*) form a strongly supported subclade. *Cornus* and *Alangium* resemble each other in floral morphology and embryology, but differ in biochemistry and some morphological features (Eyde 1988). Nyssaceae, including *Camptotheca, Mastixia,* and *Diplopanax,* also form a well-supported clade. Some analyses suggest that Nyssaceae and Cornaceae are sisters; however, bootstrap support for this sister-group relationship is less than 50% (Xiang et al. 2002). Although it has been proposed that Nyssaceae be included in Cornaceae (APG II 2003), this may be premature (Fan and Xiang 2003). Shared characters possibly indicating a sister-group relationship between Cornaceae and Nyssaceae are germination valves in their seeds, a base chromosome number of 11, and transseptal vascular bundles to the ovules. Cornaceae and most Nyssaceae also share the presence of ellagitannins (Bate-Smith et al. 1975).

Loasaceae and Hydrangeaceae appear as well-supported sister families in many molecular analyses (e.g., Hempel et al. 1995; D. Soltis et al. 1995, 2000; Savolainen et al. 2000a, 2000b; Hufford et al. 2001; Xiang et al. 2002),

a relationship not indicated before Hufford's (1992) cladistic analysis of morphological and chemical data. Both families have rare iridoids (deutzioside and derivatives, C_{10}-decarboxylated iridoids (see "Patterns of Character Evolution in Asterids," below); El-Naggar and Beal 1980) and share tuberculate trichomes with basal cell pedestals (Hufford 1992). *Fendlera* of Hydrangeaceae and Loasaceae share very similar vessel elements (Quibell 1972). The relationship between Hydrangeaceae and Loasaceae is also supported by floral morphology (Roels et al. 1997; Albach et al. 2001a). Relationships within Hydrangeaceae and Loasaceae have been the subject of several morphological and molecular analyses (D. Soltis et al. 1995; Hufford et al. 2001, 2003; Moody et al. 2001).

The enigmatic submerged aquatic genus *Hydrostachys,* the only member of Hydrostachyaceae, appears to be a member of Cornales, perhaps derived from within the Hydrangeaceae–Loasaceae clade. However, the exact position of *Hydrostachys* remains uncertain despite intensive study. Most single-gene analyses have indicated a close relationship with Hydrangeaceae, but without bootstrap support above 50% (Xiang et al. 1993; Hempel et al. 1995; Olmstead et al. 2000); multigene analyses using both parsimony and maximum likelihood indicated the derivation of *Hydrostachys* from within Hydrangeaceae (D. Soltis et al. 2000; Albach et al. 2001c; Xiang et al. 2002), but the sister group of *Hydrostachys* varies among analyses and is not well supported in any study. In the three-gene analysis (D. Soltis et al. 2000), *Hydrostachys* appears with *Decumaria* and *Hydrangea* within Hydrangeaceae; the monophyly of Hydrangeaceae including *Hydrostachys* received jackknife support of 71%. The four-gene study by Albach et al. (2001a) indicated that *Hydrostachys* is sister to *Philadelphus* and *Carpenteria,* but without bootstrap support above 50%. In maximum likelihood and parsimony analyses of a two-gene dataset, the placement of *Hydrostachys* within Hydrangeaceae also varies (Xiang et al. 2002). The placement of *Hydrostachys* will be difficult to resolve with confidence given the long branch to this taxon in most analyses.

Ericales

Based on phylogenetic studies (e.g., Chase et al. 1993; Kron and Chase 1993; Olmstead et al. 1993; Johnson and Soltis 1995; Johnson et al. 1996, 1999; Kron 1996, 1997; Morton et al. 1996; D. Soltis et al. 1997a, 2000; Anderberg et al. 1998, 2002; Savolainen et al. 2000a, 2000b; B. Bremer et al. 2002), Ericales consist of 23 families (Figure 9.1): Actinidiaceae, Balsaminaceae, Clethraceae, Cyrillaceae, Diapensiaceae, Ebenaceae (including Lissocarpaceae), Ericaceae, Fouquieriaceae, Lecythidaceae (including Scytopetalaceae), Maesaceae, Marcgraviaceae, Myrsinaceae, Pentaphylacaceae (including Ternstroemiaceae and Sladeniaceae; Anderberg et al. 2002),

FIGURE 9.2 Flowers representing Cornales (A–E) and Ericales (F–T). (A) *Cornus florida* (Cornaceae), line indicates flowers. (B) *Cornus amomum* (Cornaceae), stone (endocarp), top view. (C) *Cornus amomum* (Cornaceae), single flower. (D) *Nyssa multiflora* (Nyssaceae), branch with inflorescences. (E) *Philadelphus inodorus* (Hydrangeaceae), flower at anthesis and flower in bud. (F) *Chimaphila maculata* (Ericaceae), plant in flower. (G) *Chimaphila maculata* (Ericaceae), flower; line indicates inverted anther. (H) *Gaultheria procumbens* (Ericaceae), flower. (I) *Gaultheria procumbens* (Ericaceae), stamen: 1 = awns; 2 = anther opening via pores. (J) *Sarracenia oreophila* (Sarraceniaceae), flower. (K) *Dodecatheon meadia* (Primulaceae), flower with corolla lobes reflexed. (L) *Dodecatheon meadia* (Primulaceae), plant in flower. (M) *Dodecatheon meadia* (Primulaceae), flower in vertical section; arrow indicates connective of anther; line indicates free central placenta. (N) *Couroupita guianensis* (Lecythidaceae), flower in frontal view. (O) *Couroupita guianensis* (Lecythidaceae), androecium consisting of a ring of smaller fertile stamens, surrounding the gynoecium, and basal extension of united sterile stamens. (P) *Diapensia lapponica* (Diapensiaceae), plant in flower. (Q) *Halesia hispida* (Styracaceae), flower. (R) *Phlox subulata* (Polemoniaceae), plant in flower. (S) *Gilia squarrosa* (Polemoniaceae), flower. (T) *Symplocos phacoclados* (Symplocaceae), flower. (A to C, and E to M, from Wood 1974. D, from Harms in Engler and Prantl 1898. N and O, from Endress 1994c. P, from Drude in Engler and Prantl 1897. Q and T, from Gürke in Engler and Prantl 1897. R and S, from Peter in Engler and Prantl 1897.)

Polemoniaceae, Primulaceae, Roridulaceae, Sapotaceae, Sarraceniaceae, Styracaceae (including Halesiaceae), Symplocaceae, Tetrameristaceae (including Pellicieraceae), Theaceae, and Theophrastaceae (several of these families are illustrated in Figure 9.2). Phylogenetic analysis of mtDNA sequences indicates that the enigmatic parasite *Mitrastema* (Mitrastemonaceae) may also be part of Ericales, although the exact placement of the genus within the clade remains unclear (Barkman et al. 2004; see Chapter 11).

Ericales consist mainly of taxa previously placed in Dilleniidae (*sensu* Cronquist 1981); the clade includes many former members of the dilleniid orders Theales, Lecythidales, Ericales, Diapensiales, Ebenales, and Primulales, and families Fouquieriaceae and Sarraceniaceae, as well as Polemoniaceae (Asteridae) and Balsaminaceae (Rosidae). The monophyly of the clade is strongly supported in analyses of combined datasets of three, four, and six genes (D. Soltis et al. 2000; Albach et al. 2001a; B. Bremer et al. 2002). Weak support for the monophyly of Ericales is also apparent with *rbcL* alone (e.g., Savolainen et al. 2000b; Källersjö et al. 1998).

Ericales differ from most other asterids in the common occurrence of ellagic acid, a compound frequently found outside the asterids, but not within it (see Chapter 5, Figure 5.5). In general, morphological synapomorphies for the clade are few. A possible synapomorphy for the clade is the presence of protruding diffuse placentae (Nandi et al. 1998). Another uniting character is theoid leaf teeth in which a single vein enters the tooth and ends in an opaque deciduous cap or gland (Judd et al. 2002).

Several relationships within Ericales remain unclear (Figure 9.1). Even with three, four, or six genes combined, the backbone of the tree for Ericales is a polytomy, with few interfamilial relationships receiving strong support. Anderberg et al. (2002) used sequences of the plastid genes *rbcL, ndhF, atpB,* and the mitochondrial genes *atp1* and *matR.* A combination of all gene sequences did not fully resolve the relationships among all major clades in Ericales. Marcgraviaceae, Balsaminaceae, Tetrameristaceae, and Pellicieraceae (now included in Tetrameristaceae; APG II 2003) form a well-supported (100%) monophyletic group that is sister to the remaining families. The remaining Ericales are well supported (89%) as a clade. Fouquieriaceae and Polemoniaceae then form a weakly supported (72%) clade that is sister to the remaining families. Among the remaining families, the interrelationships of the eight supported clades are unresolved: (1) subfamily Ternstroemioideae of Theaceae, with *Ficalhoa, Sladenia,* and Pentaphylacaceae (this clade now recognized as Pentaphylacaceae; APG II 2003); (2) subfamily Theoideae of Theaceae (now recognized as Theaceae; APG II 2003); (3) Ebenaceae and Lissocarpaceae (the latter now included in Ebenaceae; APG II 2003); (4) Symplocaceae; (5) Maesaceae, Theophrastaceae, Primulaceae, and Myrsinaceae; (6) Diapensiaceae and Styracaceae; (7) Lecythidaceae and Sapotaceae; and (8) Actinidiaceae, Roridulaceae, Sarraceniaceae, Clethraceae, Cyrillaceae, and Ericaceae. The results of B. Bremer et al. (2002) generally agreed with those of Anderberg et al. (2002), although in the former Sapotaceae are placed with the clade of Maesaceae, Theophrastaceae, Primulaceae, and Myrsinaceae. Descriptions of these subclades of families within Ericales follow.

The clade of Tetrameristaceae (including Pellicieraceae), Marcgraviaceae, and Balsaminaceae is strongly supported (Morton et al. 1996; Savolainen et al. 2000b; D. Soltis et al. 2000; Albach et al. 2001a; Anderberg et al. 2002; B. Bremer et al. 2002) and occupies a pivotal position as sister to the rest of Ericales (Albach et al. 2001a), a placement that is strongly supported in B. Bremer et al. (2002). Within this clade, Marcgraviaceae are well supported as sister to Balsaminaceae + Tetrameristaceae (including Pellicieraceae) in B. Bremer et al. (2002; Figure 9.1), whereas in Anderberg et al. (2002) relationships among these families are unresolved.

A close relationship among Tetrameristaceae (including Pellicieraceae), Marcgraviaceae, and Balsaminaceae (Figure 9.1) had never been proposed. These families are different in habit, varying from herbaceous plants in Balsaminaceae to climbing shrubs in Marcgraviaceae.

(A)

(B)

(C)

(D)

(E)

(F)

(G)

(H)

(I) 1 2

(J) Bract Sepal Lobe of style Petal

(K)

(L)

(M)

(N)

(O)

(P)

(Q)

(R)

(S)

(T)

However, they do share several non-DNA characters, including specialized nectaries or glands on leaves, petioles, sepals, or petals (Morton et al. 1996), bitegmictenuinucellate ovules with cellular endosperm and endothelium, mostly simple perforation plates in vessels, and hypogynous flowers (Albach et al. 2001a). Oxalate druses are absent from the clade; members of the clade instead form oxalate raphides (Nandi et al. 1998).

Polemoniaceae and Fouquieriaceae may form another small clade within Ericales (Figure 9.1). Anderberg et al. (2002) found weak support (72%) for this clade; B. Bremer et al. (2002) found strong support (88%) for this sister group, but their taxon sampling throughout the asterids is more limited. Nonetheless, a sister-group relationship of Polemoniaceae and Fouquieriaceae is consistent with the similarity observed between *Fouquieria* (Fouquieriaceae) and *Acanthogilia,* attached at one of the basal-most nodes within Polemoniaceae (Johnson et al. 1996), a similarity indicated on the basis of morphology (Nash 1903; Henrickson 1967; reviewed by Hufford 1992) and biochemistry (Scogin 1977, 1978). However, there was no support for a clade of Polemoniaceae and Fouquieriaceae in earlier multigene analyses, and the position of Polemoniaceae has varied in other molecular analyses (e.g., Morton et al. 1996; Johnson et al. 1999; Savolainen et al. 2000b; D. Soltis et al. 2000; Albach et al. 2001a).

The relationships of Fouquieriaceae and Polemoniaceae to the primuloid clade (Primulaceae, Myrsinaceae, Theophrastaceae, and Maesaceae; discussed below) remain unresolved with molecular data. There is some evidence, however, for a close relationship of these families. Polemoniaceae are sister to the primuloid clade in the multigene analyses of both D. Soltis et al. (2000) and Albach et al. (2001a); this relationship appears in the strict consensus tree of the former study and received weak (61%) bootstrap support in the latter analysis. However, in both studies, Fouquieriaceae appear elsewhere in Ericales. Importantly, non-DNA characters also indicate a close relationship of Polemoniaceae to the primuloid group. Two characters shared by the primuloids and Polemoniaceae are nuclear endosperm and simple perforation plates in vessels (see Albach et al. 2001a). Neither character is restricted to these two groups within Ericales, but they occur in combination elsewhere only in Sapotaceae. Another character shared by at least some members of the primuloid clade and Polemoniaceae is the occurrence of separate sepals in *Cobaea, Polemonium,* and some genera of Theophrastaceae and Myrsinaceae, generally a rare condition among sympetalous plants (Stebbins 1974). Additionally, Primulaceae and Polemoniaceae both initiate stamen primordia earlier than petal primordia (Nishino 1978, 1983). Polemoniaceae and Primulaceae share many biochemical characters, such as the presence of cucurbitacins, and plants in both families

excrete methylated 6- or 8-oxygenated flavonols from glandular hairs (Hegnauer 1962–1994). More analyses of molecular and morphological data are needed to resolve relationships here and elsewhere in Ericales.

The polyphyly of Theaceae was noted earlier in this book with the placement of *Asteropeia* (now Asteropeiaceae) within Caryophyllales (Chapter 7) and the placement of subfamily Bonnetioideae (now Bonnetiaceae) within Malpighiales (Chapter 8; Savolainen et al. 2000b). The two remaining former subfamilies of Theaceae (Theoideae, Ternstroemioideae) have not formed a clade in any molecular analyses (e.g., Morton et al. 1996; Savolainen et al. 2000a, 2000b; D. Soltis et al. 2000; Albach et al. 2001a; B. Bremer et al. 2002), supporting the recognition of two separate families (APG 1998), Theaceae and Ternstroemiaceae (the latter included in Pentaphylacaceae; APG II 2003). Theaceae and Ternstroemiaceae differ in several respects, including embryological characters (curved rather than straight embryo in the latter; Tsou 1995) and floral characters, including fruit and stamen morphology. The phylogenetic positions of the two families within Ericales are not clear, but they repeatedly appear as part of different clades. Savolainen et al. (2000b) first showed that the monogeneric Sladeniaceae should be added to Ericales, as a member of Ternstroemiaceae (now Pentaphylacaceae). The DNA-based position of *Sladenia* in Ericales is in general agreement with the treatments of Dahlgren (1980) and Cronquist (1981), who placed Sladeniaceae in Theales/Theaceae, the component members of which are found in Ericales as circumscribed here. In a more recent study, Anderberg et al. (2002) found *Sladenia* and *Filcalhoa* (Theaceae) to be a weakly supported (51%) sister group to a well-supported (96%) clade of Pentaphylacaceae + Ternstroemiaceae. B. Bremer et al. (2002) found stronger support (72%) for a clade of *Sladenia* as sister to Pentaphylacaceae + Ternstroemiaceae, but they did not include *Filcalhoa* in their analysis. In K. Bremer et al. (2001), Ternstroemiaceae appear with Sladeniaceae and Pentaphylacaceae in a weakly supported (72% jackknife) clade now referred to as a broadly defined Pentaphylacaceae (*sensu* APG II 2003). In D. Soltis et al. (2000), Albach et al. (2001a), and B. Bremer et al. (2002), the narrowly defined Theaceae are sister to the clade of Symplocaceae (Diapensiaceae + Styracaceae), but this relationship received support above 50% (52% jackknife) only in K. Bremer et al. (2001).

Diospyros (Ebenaceae) and *Lissocarpa* (Lissocarpaceae) are strongly supported sisters (D. Soltis et al. 2000; Berry et al. 2001; Anderberg et al. 2002). Savolainen et al. (2000b) suggested that *Lissocarpa* might be a member of Sapindales, but this placement was based on an *rbcL* sequence obtained from degraded DNA. The placement of Lissocarpaceae as a close relative of Ebenaceae agrees with morphological data; the two families share similar floral morphology and wood anatomy, differing chiefly in that *Lissocarpa* has an inferior ovary (Cronquist 1981;

Berry et al. 2001; Anderberg et al. 2002). On the basis of molecular and morphological evidence, Lissocarpaceae have been included within Ebenaceae (APG II 2003).

Molecular analyses revealed a well-supported primuloid clade (Figure 9.1) consisting of Primulaceae, Myrsinaceae, and Theophrastaceae, all three of which have been recircumscribed, and Maesaceae, a new family consisting only of *Maesa* (formerly of Myrsinaceae; Anderberg et al. 2000, 2002; Källersjö et al. 2000). This primuloid clade corresponds to Cronquist's (1981) Primulales and is strongly supported by both molecular (D. Soltis et al. 2000; Albach et al. 2001a) and morphological (Cronquist 1981; Anderberg and Ståhl 1995) data. This clade is united by free central placentation and by stamens equal in number and opposite the corolla lobes (Figure 9.2M); free central placentation is an unusual feature, occurring also in Santalales and Lentibulariaceae (as reviewed in Chapter 7, Caryophyllales have secondarily free central placentation). Maesaceae are sister to the remainder of the primuloid clade, followed by Theophrastaceae as sister to Myrsinaceae + Primulaceae (B. Bremer et al. 2002; Anderberg et al. 2002). Recent analyses of these families included studies of Primulaceae (Mast et al. 2001) and Myrsinaceae (Anderberg and Ståhl 1995).

Diapensiaceae and Styracaceae (including Halesiaceae) form a clade with low to moderate support (D. Soltis et al. 2000; Albach et al. 2001a; B. Bremer et al. 2002). The strongest support (82% jackknife) came from the six-gene analysis of B. Bremer et al. (2002). In contrast to previous analyses based on *rbcL* (Morton et al. 1996) that showed *Halesia* and other genera of Styracaceae to be only distantly related to *Styrax*, B. Bremer et al.'s multigene analyses found Styracaceae (including *Halesia*) to form a well-supported monophyletic group with 100% bootstrap support (see also Kron 1996). Characters that unite Styracaceae and Diapensiaceae include crystals of calcium oxalate, unitegmic ovules (but not in all Styracaceae), unilacunar nodes, cellular endosperm, and binucleate pollen (Albach et al. 2001a; see also Morton et al. 1996). The two families also share the *Chenopodium* variant of polygonad type embryo formation (Yamazaki 1974). These characters (except polygonad embryo formation) are also present in Symplocaceae. Symplocaceae appear as sister to Styracaceae + Diapensiaceae in the six-gene analysis (B. Bremer et al. 2002), albeit with weak (50% jackknife) support. Styracaceae have been the focus of a recent study (e.g., Fritsch 2001).

Several studies, both molecular and morphological, have provided compelling evidence for a broadly defined Ericaceae that also includes Pyrolaceae, Monotropaceae, Epacridaceae, and Empetraceae (Anderberg 1993; Judd and Kron 1993; Kron and Chase 1993; Kron 1996; Kron et al. 2002). Ericaceae are part of a larger clade (sometimes referred to as the core Ericales; see Judd et al. 2002) that also includes Actinidiaceae, Sarraceniaceae, Roridulaceae, Cyrillaceae, and Clethraceae (Figure 9.1; Albach et al. 2000a; Savolainen et al. 2000a, 2000b; D. Soltis et al. 2000; B. Bremer et al. 2002). Within the core Ericales, Sarraceniaceae, Actinidiaceae, and Roridulaceae form one weakly supported subclade (B. Bremer et al. 2002), with the second subclade consisting of Clethraceae as sister to a clade of Cyrillaceae and a broadly defined Ericaceae. This core Ericales clade is characterized by inverted anthers, a usually hollow style that emerges from an apical depression in the ovary, and endosperm with haustoria at both ends (Anderberg 1992, 1993; Judd and Kron 1993; Kron and Chase 1993; Kron 1996); this group also contains all of the iridoid-producing genera of Ericales except *Fouquieria* and *Symplocos* (Albach et al. 2001a).

Several families of Ericales have been the focus of more detailed phylogenetic studies, including Ericaceae s.l. (Kron and Chase 1993; Kron 1996, 1997; Kron et al. 2002), Primulaceae and related families (Anderberg et al. 1998, 2002; Trift et al. 2002), Polemoniaceae (Johnson and Soltis 1995; Johnson et al. 1996, 1999), Diapensiaceae (Rönblom and Anderberg 2002), Actinidiaceae (Li et al. 2002), and Ebenaceae and related families (Morton et al. 1996).

Euasterids

Following Cornales and Ericales, which are successive sisters to all other asterids, the remaining asterids are referred to as the euasterids (Figure 9.1). These euasterid families have flowers with epipetalous stamens that equal the corolla lobes in number and a gynoecium of usually two fused carpels. Molecular analyses have provided strong support for this clade and also indicated the presence of two subclades of euasterids—euasterid I (lamiids) and euasterid II (campanulids) (Olmstead et al. 1993; D. Soltis et al. 2000; Albach et al. 2001a; B. Bremer et al. 2002). The names euasterid I and II are still used in the recent APG II (2003) classification, and we therefore use them here (together with the names lamiids and campanulids per B. Bremer et al. 2002), but the names euasterid I and II bear no obvious connection to their constituent families and should ultimately be replaced with the clade names lamiids and campanulids.

An analysis conducted by K. Bremer et al. of 142 genera of euasterids using 143 morphological, anatomical, embryological, palynological, chemical, and RFLP (restriction fragment length polymorphisms) characters recovered some groups (mostly families and small groups of families) supported by DNA sequence data and recognized by APG (1998), but the analysis identified only one order, Dipsacales, of the eight APG orders of euasterids. Furthermore, euasterid I and II were not distinguished until *rbcL* and *ndhF* sequences were added to the analysis (K. Bremer et al. 2001). Thus, although morphological and other nonmolecular data contain historical signal, it is not sufficiently strong to discern all of the groups identified by molecular data.

Some possible morphological synapomorphies for euasterid I and II (lamiids and campanulids) can be inferred from the analysis of K. Bremer et al. (2001). Members of euasterid I (lamiids) are generally characterized by opposite leaves, entire leaf margins, hypogynous flowers, "early sympetaly" with a ring-shaped corolla primordium, fusion of stamen filaments with the corolla tube, and capsular fruits (K. Bremer et al. 2001). Taxa appearing in euasterid II (campanulids) typically have alternate leaves, serrate-dentate leaf margins, epigynous flowers, free stamen filaments, indehiscent fruits, and "late sympetaly" with distinct petal primordia (K. Bremer et al. 2001). Although these morphological features are useful as general descriptions, it is unclear which are truly synapomorphies and which are symplesiomorphies, and both reversals and parallelisms have generated many exceptions to these general patterns.

Euasterid I (Lamiids)

The euasterid I clade (asterid I *sensu* Chase et al. 1993; lamiids *sensu* B. Bremer et al. 2002) received only weak (56%) support with three genes (D. Soltis et al. 2000), but support increased greatly (86%) in Albach et al.'s (2001a) four-gene analysis; with six genes the clade received 100% jackknife support (B. Bremer et al. 2002). Within euasterid I, Garryales, Icacinaceae, Metteniusaceae, and Oncothecaceae are sisters to the rest of the clade, but relationships among these lineages are still unclear (Figure 9.3). The remainder of euasterid I is strongly supported as monophyletic in analyses of both three and four genes and consists of several large subclades, Gentianales, Solanales, and Lamiales (*sensu* APG 1998; APG II 2003), as well as *Vahlia* (Vahliaceae) and Boraginaceae (Figure 9.3; several representatives are illustrated in Figure 9.4).

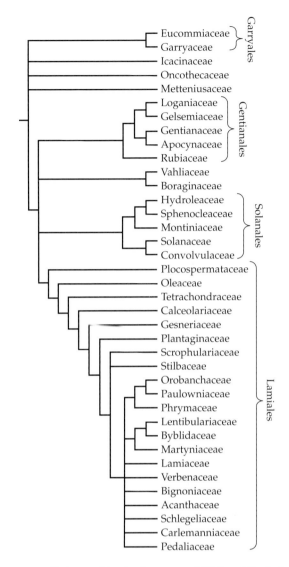

FIGURE 9.3 Relationships in the euasterid I (lamiid) clade.

FIGURE 9.4 Flowers of the euasterid I (lamiid) clade. Representatives of Garryales (A–C), Gentianales (D–H), Vahliaceae (I–K), Boraginaceae (L), Solanales (M, N), and Lamiales (O–Z). (A) *Garrya elliptica* (Garryaceae), male flowering branch. (B) *Aucuba japonica* (Garryaceae), staminate flower with one stamen removed. (C) *Aucuba japonica* (Garryaceae), carpellate flower. (D) *Sabatia kennedyana* (Gentianaceae), inflorescence with flowers and partially mature fruit. (E) *Sabatia kennedyana* (Gentianaceae), flower. (F) *Diodia teres* (Rubiaceae), flower. (G) *Asclepias syriaca* (Apocynaceae), lateral view of flower; 1 = clip; 2 = corona horn; 3, 4 = anther. (H) *Amsonia tabernaemontana* (Apocynaceae), flower in long section; 1 = united style; 2 = free (separate; not fused) ovaries. (I) *Vahlia capensis* (Vahliaceae), lateral view of flower. (J) *Vahlia capensis* (Vahliaceae), gynoecium in longitudinal section. (K) *Vahlia capensis* (Vahliaceae), plant in flower. (L) *Phacelia tanacetifolia* (Boraginaceae), plant in flower showing scorpioid cymes. (M) *Calystegia spithamaea* (Convolvulaceae), plant in flower. (N) *Physalis heterophylla* (Solanaceae), flower. (O) *Chionanthus pygmaeus* (Oleaceae), functionally staminate flower. (P) *Chionanthus virginicus* (Oleaceae), flowering branchlet. (Q) *Plantago lanceolata* (Plantaginaceae), flower. (R) *Plantago rugellii* (Plantaginaceae), plant in flower. (S) *Penstemon canescens* (Scrophulariaceae), side view of flower. (T) *Penstemon canescens* (Scrophulariaceae), flower in long section. (U) *Salvia urticifolia* (Lamiaceae), tip of plant in flower. (V) *Salvia urticifolia* (Lamiaceae), side view of flower. (W) *Salvia urticifolia* (Lamiaceae), gynoecium; 1 = gynobasic style; 2 = nectar disc. (X) *Verbena bipinnatifida* (Verbenaceae), inflorescence. (Y) *Justicia ovata* (Acanthaceae), side view of flower. (Z) *Catalpa bignonioides* (Bignoniaceae), flower. (A to C, from Harms in Engler and Prantl 1898. D to H, and M to Z, from Wood 1974. I to K, from Engler in Engler and Prantl 1891. L, from Peter in Engler and Prantl 1897.)

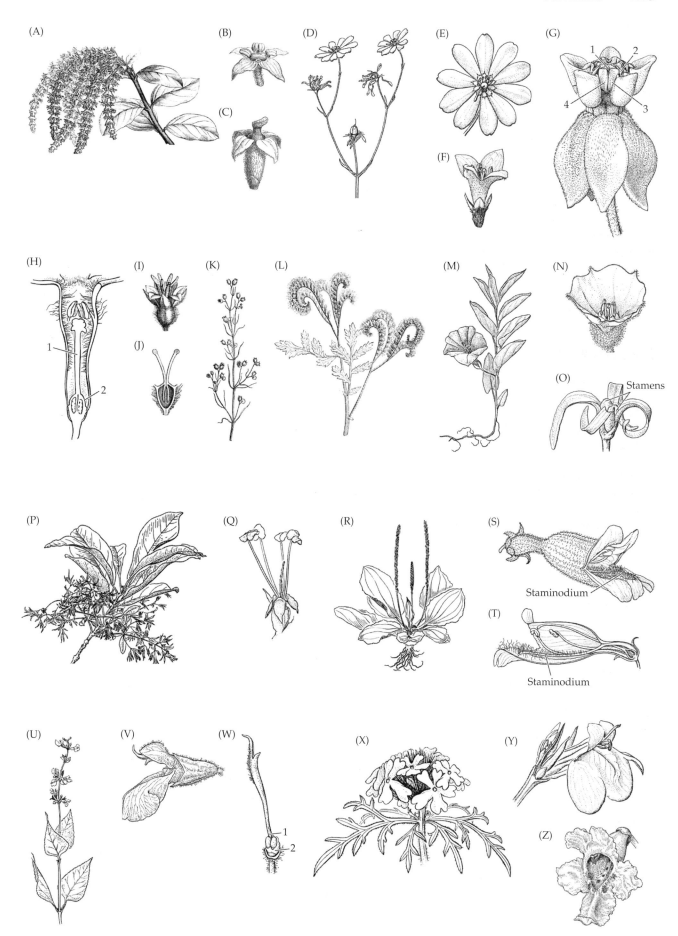

Oncothecaceae, Metteniusaceae, and Icacinaceae

The relationships of Oncothecaceae, Metteniusaceae, and Icacinaceae to each other and other euasterid I families are not clear, although they, with Garryales (below), consistently appear as the basal-most lineages of the clade (D. Soltis et al. 2000; B. Bremer et al. 2002). In some analyses, Icacinaceae appeared (with weak support) as sister to Garryales (D. Soltis et al. 2000). In the strict consensus of shortest trees obtained in B. Bremer et al. (2002), Oncothecaceae and Icacinaceae form a clade that is sister to the remainder of euasterid I, although neither relationship received support above 50%. Metteniusaceae are poorly known and only recently have been shown to belong here (M. W. Chase and F. Gonzales, unpubl.).

Icacinaceae as previously circumscribed are polyphyletic. *Icacina* and related genera such as *Cassinopsis* and *Pyrenacantha* form a well-supported clade and are part of euasterid I (Figure 9.3); several studies point to a close relationship of these genera to Garryales, but without strong support (D. Soltis et al. 2000; Kårehed 2002). With the addition of more characters, this relationship may receive strong support, and Icacinaceae should then be included in Garryales. Several other genera previously attributed to Icacinaceae are part of either Cardiopteridaceae or a new family, Stemonuraceae, both placed in Aquifoliales of euasterid II (Kårehed 2002).

The relationships of Oncothecaceae remain obscure. The family consists of two species (evergreen trees or shrubs) native to New Caledonia. Cronquist (1981) placed the family in his Theales (Dilleniidae), noting that it was poorly studied and distinctive in morphology. Additional data are required to confirm its placement, but analyses conducted to date concur in placing it at or near the base of euasterid I (Figure 9.3).

Garryales

Garryales consist of two families: Eucommiaceae and Garryaceae (including *Aucuba*) (Figure 9.3). In APG (1998), the order also included Oncothecaceae, but that placement was premature; it did not receive support above 50% in D. Soltis et al. (2000). Eucommiaceae and Garryaceae were not considered closely related in morphology-based classifications, but this clade has been recognized in analyses of *rbcL* alone (Olmstead et al. 1993; Xiang et al. 1993; Savolainen et al. 2000b), as well as in combined analyses (e.g., D. Soltis et al. 2000; Albach et al. 2001a; B. Bremer et al. 2002). Possible morphological synapomorphies for Garryales are unisexual flowers and apical placentation, although there are parallelisms in Aquifoliales and Apiales of euasterid II (K. Bremer et al. 2001).

Eucommiaceae, which have a wide fossil distribution, today consist of a single species from China that has been placed in Hamamelidae in recent morphology-based classifications because of the simple nature of the flowers (Cronquist 1981). A close relationship of the family to Garryaceae is another example of an unexpected result of molecular systematic studies. Garryaceae are now broadly defined to include *Aucuba*, as well as *Garrya* (APG II 2003). Although the sister-group relationship between *Garrya* and *Aucuba* received strong support in molecular phylogenetic studies (e.g., D. Soltis et al. 2000; Savolainen et al. 2000a, 2000b; Albach et al. 2001a; B. Bremer et al. 2002), the two have not been considered closely related in recent classifications (e.g., Cronquist 1981). However, a close relationship between *Garrya* and *Aucuba* was first proposed by Baillon (1879–1895), and this relationship is supported by several non-DNA characters, including palynological data (Ferguson 1977), the presence of petroselinic acid in seed oils (Breuer et al. 1987), two-seeded fruits lacking a septum, and a similar gynoecial vasculature (Eyde 1964). In addition, both *Garrya* and *Aucuba* are dioecious, evergreen shrubs or small trees with tetramerous flowers (Figure 9.4, B and C) and opposite leaves. Non-DNA characters also unite Eucommiaceae with *Garrya* and *Aucuba*. All three are dioecious and woody, and all three contain similar iridoid compounds; aucubin has been detected in all three genera (Jensen et al. 1975), and eucommioside has been found in *Eucommia* and *Aucuba* (Boros and Stermitz 1991).

Boraginaceae

The position of Boraginaceae (including Hydrophyllaceae) in euasterid I (lamiids) remains uncertain despite multigene analyses involving as many as six genes (Figure 9.3). Boraginaceae have usually been placed in or close to Solanales in morphology-based classifications (Dahlgren 1980; Thorne 1992a, 1992b, 2001; Takhtajan 1997) and in some molecular-based trees (e.g., Chase et al. 1993; Olmstead et al. 2000), but most molecular analyses have indicated that Boraginaceae are an isolated family with no clear or close relatives. With *ndhF* sequences, Boraginaceae are sister to Solanales, but without support above 50% (Olmstead et al. 2000). In other studies, Boraginaceae are sister to Lamiales (*rbcL*; Olmstead et al. 1992) or to Vahliaceae + Lamiales (multigene analyses; D. Soltis et al. 2000; Albach et al. 2001a). However, this close relationship of Boraginaceae to Lamiales did not receive support above 50%. In B. Bremer et al. (2002), Boraginaceae + Vahliaceae are sister to Solanales + Lamiales, but again without support above 50%. Development of the endosperm may provide synapomorphies for Boraginaceae and Lamiales; Boraginaceae and all Lamiales (except Oleaceae) share terminal endosperm haustoria.

Boraginaceae are now circumscribed broadly to include Hydrophyllaceae (excluding *Hydrolea*, which is in Solanales). In contrast, Cronquist (1981, 1988) split Boraginaceae and Hydrophyllaceae. He placed Hydrophyllaceae in Solanales and Boraginaceae in Lami-

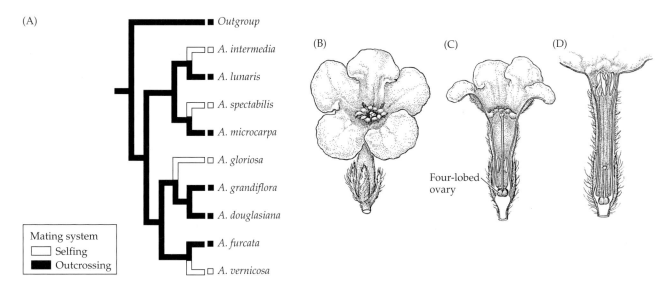

FIGURE 9.5 (A) Reconstruction of evolution of mating system in *Amsinckia* (Boraginaceae; modified from Schoen et al. 1997). Flowers are from *Lithospermum canescens* (Boraginaceae) and illustrate a distylous condition (from Wood 1974). (B) Entire flower. (C) Flower in long section; long-style form. (D) Flower in long section, short-style form.

ales on the basis of similarities between tropical, woody Boraginaceae and tropical, woody Verbenaceae. A morphological character uniting Boraginaceae as now circumscribed (APG II 2003) is a distinctive inflorescence, usually composed of helicoid or scorpioid cymes, which straighten as the flowers mature (see also Judd et al. 2002; Figure 9.4L).

Boraginaceae offer the opportunity to examine facets of mating system evolution. Although there are obvious genetic benefits to outcrossing, many (20%) angiosperms have evolved predominant selfing (autogamy; reviewed in Barrett 2002). Multiple origins of small-flowered predominantly autogamous species from outcrossing species is a common feature of herbaceous flowering plant families (Barrett 2002). An excellent example is provided by *Amsinckia* (Boraginaceae). Phylogenetic analysis and character-state mapping indicated that predominant selfing has evolved from outcrossing at least four times in the genus (Schoen et al. 1997; Figure 9.5). The outcrossing species are distylous with large flowers; the selfing species are homostylous and have small flowers. In addition, the repeated occurrence of short branches separating selfers from related outcrossers indicates that selfing may be of recent origin (Schoen et al. 1997).

Vahliaceae

Vahliaceae contain five species in the genus *Vahlia* (= *Bistella*) from tropical and southern Africa to northwestern India. *Vahlia* usually has been considered close to Saxifragaceae because of their free sepals and petals (Figure 9.4, I to K), alternatively included there (Cronquist 1981) or treated as a separate related family (e.g., Dahlgren 1980; Thorne 1992a, 1992b; Takhtajan 1997). However, the placement of *Vahlia* in the asterids is supported by the presence of iridoids (Al-Shammary 1991) and tenuinucellate ovules. Nonetheless, *Vahlia* differs from all other taxa in euasterid I in having bitegmic rather than unitegmic ovules.

The position of Vahliaceae in the euasterid I clade remains somewhat uncertain. Based on *rbcL* sequence data, Morgan and Soltis (1993) first suggested that *Vahlia* occupies an isolated position as sister to Lamiales, a position it also occupies in other analyses of *rbcL* alone (e.g. Savolainen et al. 2000a) and in some multigene analyses (D. Soltis et al. 2000; Albach et al. 2001a). However, even in three- and four-gene analyses, this placement did not receive internal support above 50%. In the analysis by B. Bremer et al. (2002), *Vahlia* is weakly supported (69%) as sister to Boraginaceae. We illustrate this latter placement of Vahliaceae (Figure 9.3), but it should be considered tentative.

Lamiales

Lamiales, with almost 18,000 species in about 1,100 genera, are among the most taxonomically difficult clades of asterids. Based on molecular phylogenetic studies, Lamiales include the former orders Lamiales, Scrophulariales, Callitrichales, and Plantaginales of Cronquist (1981). Lamiales are strongly supported, comprising 21 families (APG II 2003; Figure 9.3): Acanthaceae (including Avicenniaceae), Bignoniaceae, Byblidaceae, Calceolariaceae, Carlemanniaceae, Gesneriaceae, Lamiaceae, Lentibulariaceae, Martyniaceae, Oleaceae, Orobanchaceae (including Nesogenaceae), Paulowniaceae, Pedaliaceae, Phrymaceae, Plantaginaceae, Plocospermataceae, Schlegeliaceae, Scrophulariaceae (including Buddlejaceae and Myoporaceae), Stilbaceae, Tetrachondraceae, and Verbenaceae (several of these families are illustrated in Figure 9.4). Most Lamiales share several characters that are extremely rare outside of this clade: verbascoside; anthraquinones derived from the shikimic acid pathway; C_{11}-decarboxylated iridoids (Jensen 1991, 1992); oligosaccharides; 6-oxygenated flavones; protein inclusions in nuclei of mesophyll cells; often diacytic stomates; hairs with a stalk of one or more uniseriate cells and a head of two or more vertical cells; young anthers with pollen sac placentoids; and fewer than five fertile stamens (Rahn 1996).

A large part of the taxonomic complexity of Lamiales involves compelling molecular phylogenetic evidence for the polyphyly of Scrophulariaceae in the traditional sense (see paragraphs below; reviewed in Olmstead et al. 2001). The demonstrated polyphyly of Scrophulariaceae and other recent phylogenetic results have resulted in several newly named families in Lamiales. Families added (see APG II 2003) include Calceolariaceae, Carlemanniaceae, Plocospermataceae, and Tetrachondraceae (Figure 9.3). Other families have been redefined. For example, phylogenetic studies dictate that Buddlejaceae and Myoporaceae be included in Scrophulariaceae (see Olmstead et al. 1992, 1999, 2001; Oxelman et al. 1999). *Avicennia* (Avicenniaceae) appears to be part of Acanthaceae and has been included in that family (APG II 2003). Orobanchaceae have been expanded to contain parasitic Scrophulariaceae (plus the nonparasitic *Lindenbergia*, formerly of Scrophulariaceae; see Chapter 11 for evolution of parasitism in the angiosperms). Orobanchaceae also include core genera of the former Cyclocheilaceae. B. Bremer et al. (2002) found *Cyclocheilon* to be embedded within Orobanchaceae with strong support; *Asepalum* is also nested within Orobanchaceae (R.G. Olmstead, pers. comm.).

Some confusion has been introduced in the naming of clades in this disintegrated Scrophulariaceae. Veronicaceae have been resurrected for one clade of the former, broadly defined (and polyphyletic) Scrophulariaceae (Olmstead and Reeves 1995; Olmstead et al. 1999, 2001), but the appropriate name for this family is Plantaginaceae (APG II 2003). The newly recognized Calceolariaceae (slipper flowers) are distinct from Scrophulariaceae s.s. and include *Calceolaria* (250 species from South America), *Jovellana* (six species from New Zealand and Chile), and *Porodittia* (=*Stemotria*) from Peru (Olmstead et al. 1999, 2001). The relationships of some genera formerly placed in Scrophulariaceae (e.g., *Mimulus*) remain somewhat uncertain. It appears, however, that *Mimulus* may be best placed in Phrymaceae, although the bootstrap support for this relationship is low (Beardsley et al. 2001; Beardsley and Olmstead 2002). Carlemanniaceae consist of five species and two genera (*Carlemannia* and *Silvianthus*) from Southeast

Asia to Sumatra; they were formerly included in Caprifoliaceae/Dipsacales by Cronquist (1981), but their connivent anthers are consistent with their placement in Lamiales. Martyniaceae consist of four genera from tropical and subtropical America; they were included in Pedaliaceae in APG (1998), but recent multigene analyses (Olmstead et al. 1999, 2001; Albach et al. 2001a; B. Bremer et al. 2002) indicated that their familial status should be maintained. Plocospermataceae are a monogeneric family comprising three species from Central America; they were included in Loganiaceae by Cronquist (1981) and not assigned to order in APG (1998). Tetrachondraceae are a new family comprising *Tetrachondra* from Patagonia and New Zealand and *Polypremum procumbens* from warm temperate and tropical America (Oxelman et al. 1999; B. Bremer et al. 2002).

Within Lamiales, relationships among families in general remain poorly understood, despite the use of combined datasets of multiple genes (Figure 9.3). In B. Bremer et al. (2002), Plocospermataceae are sister to other Lamiales, followed by Oleaceae. In other analyses in which *Plocospermum* was not included, a well-supported Oleaceae appeared as sister to all other Lamiales (e.g., D. Soltis et al. 2000; Albach et al. 2001a; Olmstead et al. 2001). In contrast to these molecular results placing Oleaceae firmly in Lamiales, Oleaceae were often considered closely related to Gentianales in morphology-based classifications (e.g., Dahlgren et al. 1980). Cronquist (1981) placed the family in his Santalales within Rosidae. Oleaceae are distinct from the remaining Lamiales in their embryogeny, which is of the polygonad type, in contrast to the onagrad type found in the remainder of the order. Phylogenetic relationships within Oleaceae have recently been studied (e.g., Wallander and Albert 2000).

Following Oleaceae, Tetrachondraceae are sister to the remaining Lamiales (Figure 9.3), a placement receiving strong (100%) jackknife support in B. Bremer et al. (2002). Following Tetrachondraceae, the remaining members of Lamiales form a well-supported clade that is also supported by two non-DNA characters: (1) flowers having bilateral symmetry (a two plus three pattern); and (2) four stamens in a two-long and two-short configuration (e.g., Judd et al. 2002). Within this latter clade, Calceolariaceae, followed by Gesneriaceae and Plantaginaceae, are subsequent sisters to the rest of Lamiales (Olmstead et al. 2001). Calceolariaceae were not included in the analyses of either Albach et al. (2001a) or B. Bremer et al. (2002). The placement of Gesneriaceae received only low (60% bootstrap) support with four genes (Albach et al. 2001a), but strong (95% jackknife) support in B. Bremer et al. (2002). Gesneriaceae are distinct from most other Lamiales in their parietal placentation and lack of iridoid compounds. Relationships within Gesneriaceae have been the subject of several phylogenetic investigations (e.g., Smith and Carroll 1997; Smith and Atkinson 1998; Smith 2000a, 2000b; Zimmer et al. 2002; Mayer et al. 2003). Plantaginaceae are morphologically diverse; they include the wind-pollinated genus *Plantago* (placed by some in a separate family and order; Cronquist 1981), as well as Callitrichaceae and Hippuridaceae, two families of aquatic herbs. *Antirrhinum* (snapdragon), a model organism for the study of floral developmental genetics, is also a member of Plantaginaceae (see "Floral Symmetry," below).

Within the remaining Lamiales, relationships remain poorly understood (Figure 9.3). Only a few clades with support above 50% have been identified. Scrophulariaceae s.s., including *Buddleja* (formerly Buddlejaceae) and *Myoporum* (formerly Myoporaceae), constitute one such clade (Olmstead and Reeves 1995; Oxelman et al. 1999; Albach et al. 2001a; B. Bremer et al. 2002). The clade receives 78% bootstrap support in Olmstead et al. (2001); support is 100% in B. Bremer et al. (2002), but taxon sampling is sparse in that analysis. A close relationship between Scrophulariaceae and Buddlejaceae was first noted by Dahlgren (1983). Olmstead et al. (2001) provided support for a close relationship of Buddlejaceae, Scrophulariaceae s.s., and Myoporaceae. The genus *Androya*, previously placed in Loganiaceae, is also part of the *Myoporum* clade of the newly circumscribed Scrophulariaceae (B. Bremer et al. 2002). Members of this redefined Scrophulariaceae are characterized by acylated rhamnosyl iridoids (Jensen et al. 1998).

Olmstead et al. (2001) obtained weak support (63%) for a clade of Martyniaceae, Verbenaceae, Schlegeliaceae, Bignoniaceae, Acanthaceae, Pedaliaceae, Lamiaceae, Phrymaceae (represented by *Mimulus*), Paulowniaceae, and Orobanchaceae; Lentibulariaceae (see Chapter 11; not included in Olmstead et al. 2001) are also part of this clade. Within this clade, relationships remain particularly problematic. Weak support was observed for a clade of the carnivorous families Martyniaceae, Byblidaceae, and Lentibulariaceae (see below; Chapter 11). Phrymaceae, Paulowniaceae, and Orobanchaceae form a well-supported subclade in the B. Bremer et al. (2002) analysis (90% jackknife support). Orobanchaceae (see Chapter 11) and Paulowniaceae appear as sister groups in Olmstead et al. (2001) with moderate (79% bootstrap) support. However, in the latter study, Phrymaceae (represented by *Mimulus*) and Lamiaceae appeared as sisters to Paulowniaceae + Orobanchaceae, but without support above 50%. Increased taxon sampling again revealed a clade of Phrymaceae, Paulowniaceae, and Orobanchaceae, with a close relationship of these families to Lamiaceae (R.G. Olmstead, pers. comm.). In contrast, in B. Bremer et al. (2002), Lamiaceae appear as sister to Verbenaceae with strong support, a result that may be attributable to low taxon sampling.

There is also some evidence for a close relationship between Acanthaceae (including *Avicennia*, previously Avicenniaceae) and *Sesamum* (Pedaliaceae). In Olmstead et al. (2001) and Albach et al. (2001a) they are sister taxa, but without support above 50%. Both Acanthaceae and Pedaliaceae contain amyloid, an oligosaccharide not found in other Lamiales (Hegnauer and Hegnauer 1962–1994), and share pollen morphology (Erdtman 1952). Phylogenetic relationships within Acanthaceae have been examined by several investigators (e.g., Scotland et al. 1995; McDade et al. 2000). Schwarzbach and McDade (2002) provided molecular evidence for the close relationship of the mangrove genus *Avicennia* (formerly Avicenniaceae) to subfamily Thunbergioideae of Acanthaceae.

In the analysis by B. Bremer et al. (2002), Acanthaceae and Pedaliaceae are part of a clade with Stilbaceae and Schlegeliaceae. However, with the greater taxon sampling of Olmstead et al. (2001), Stilbaceae and Schlegeliaceae appeared in different positions. Schlegeliaceae, a small family sometimes placed in Bignoniaceae (e.g., Cronquist 1981), are sister to Bignoniaceae; Stilbaceae are part of a large polytomy with no clear close relatives in the core Lamiales (Olmstead et al. 2001).

Lamiaceae include those taxa traditionally placed in Lamiaceae, as well as some taxa previously placed in Verbenaceae (e.g., *Callicarpa, Clerodendrum, Tectona, Vitex*; Olmstead and Reeves 1995; Wagstaff and Olmstead 1996; Wagstaff et al. 1998; Cantino 1992a). Previously, Verbenaceae were circumscribed much more broadly (e.g., Cronquist 1981), but phylogenetic analyses revealed this broadly defined family to be paraphyletic. The family now includes only the former subfamily Verbenoideae (with the exclusion of tribe Monochileae) (Cantino 1992a, 1992b; Chadwell et al. 1992; Judd et al. 1994; Wagstaff and Olmstead 1996). To make Lamiaceae monophyletic, roughly two-thirds of the genera of the former, broadly defined Verbenaceae were transferred to Lamiaceae.

In most studies, the Verbenaceae and Lamiaceae clades appear distantly related within Lamiales (Olmstead and Reeves 1995; Wagstaff and Olmstead 1997; Wagstaff et al. 1998; Savolainen et al. 2000a, 2000b; D. Soltis et al. 2000; Albach et al. 2001a). However, with the smaller taxon sampling of B. Bremer et al. (2002), the two families are strongly supported as sisters. Verbenaceae in the narrow sense differ from Lamiaceae (as redefined) in accumulation of 4-carboxy iridoids in the former versus C_4-decarboxylated iridoids in the latter (von Poser et al. 1997). Verbenaceae as now defined can be distinguished from Lamiaceae by several morphological characters, including the presence in the former of indeterminate racemes, spikes, or heads (versus inflorescences with an indeterminate main axis and cymosely branched lateral axes), simple style with conspicuous two-lobed stigma (versus apically forked style with inconspicuous stigmatic region), pollen exine thickened near apertures (versus not thickened), and nonglandular hairs exclusively unicellular (versus multicellular, uniseriate) (Judd et al. 2002).

The relationships of the carnivorous families, Byblidaceae, Martyniaceae, and Lentibulariaceae, also remain unclear (see also Chapter 11). In B. Bremer et al. (2002) they formed a clade with weak support, but Byblidaceae and Lentibulariaceae were not included in Olmstead et al. (2001). Some molecular phylogenetic analyses indicated a close relationship between Byblidaceae and Martyniaceae, although support for this sister group was not apparent in analyses of three genes (D. Soltis et al. 2000). *Byblis* was generally considered related to Pittosporaceae (e.g., Dahlgren 1980; Cronquist 1981; Dahlgren et al. 1981), whereas Martyniaceae were placed near Pedaliaceae and Bignoniaceae, and sometimes even included in Pedaliaceae (Cronquist 1981). More recently, B. Bremer et al. (2002) found weak support (51%) for a clade of *Byblis* (Byblidaceae), *Proboscidea* (Martyniaceae), and Lentibulariaceae. These results raise the possibility of a common ancestry of the carnivorous syndrome with considerable subsequent structural divergence in this part of the Lamiales, but these relationships are tentative and should be examined further (see Chapter 11).

Gentianales

Gentianales constitute a well-supported clade of five families (Figure 9.2): Apocynaceae (including Asclepiadaceae), Gelsemiaceae, Gentianaceae, Loganiaceae, and Rubiaceae (several of these are illustrated in Figure 9.4). The Gentianales clade comprises about 1,000 genera and 14,000 species. Although *Emblingia* (Emblingiaceae) was placed in Gentianales in the *rbcL* analysis of eudicots by Savolainen et al. (2000b), this appears to be the result of a DNA mixup. Other evidence indicates that *Emblingia* is a member of Brassicales (see Chapter 8). Dialypetalanthaceae, consisting of one species in eastern Brazil, should be included in Rubiaceae (Fay et al. 2000b; Savolainen et al. 2000b).

A close relationship among these families of Gentianales had long been proposed in classifications based solely on morphology. Cronquist (1981), for example, had a similar Gentianales but also recognized Retziaceae (= Stilbaceae, now included in Lamiales); like most authors at that time, Cronquist also included Asclepiadaceae as a family distinct from Apocynaceae. Morphological and anatomical synapomorphies for Gentianales include vestured pits, interpetiolar stipules with thick glandular hairs (termed colleters; these were lost in most Gentianaceae; Backlund et al. 2000), corollas convolute in bud (Bremer and Struwe 1992), intraxylary phloem (shared by all Gentianales except Rubiaceae), and possibly some types of complex indole alkaloids (K. Bremer et al. 2001; Judd et al. 2002).

Rubiaceae are sister to the remainder of the clade (Figures 9.3) in most molecular phylogenetic analyses

(Downie and Palmer 1992; Chase et al. 1993; Olmstead et al. 1993; B. Bremer et al. 1994; Backlund et al. 2000; Savolainen et al. 2000a, 2000b; D. Soltis et al. 2000; Albach et al. 2001a), a placement that is well supported in some analyses (B. Bremer et al. 2002). A morphological cladistic analysis (Struwe et al. 1994) differed, with Rubiaceae placed as sister to Gelsemiaceae and Loganiaceae s.s. as sister to the remaining Gentianales. Phylogenetic relationships within Rubiaceae have been analyzed using morphology (e.g., B. Bremer and Struwe 1992), DNA characters (e.g., Manen et al. 1994; B. Bremer et al. 1995, 1999; Anderrson and Rova 1999), and combined morphology and DNA datasets (Andreasen and Bremer 2000).

Following Rubiaceae, a clade of the remaining families (Apocynaceae, Gelsemiaceae, Gentianaceae, Loganiaceae) received moderate to strong support in analyses based on three or four genes (D. Soltis et al. 2000; Albach et al. 2001a) and strong support in B. Bremer et al. (2002), but in the last study each family is represented by a single placeholder. This clade of four families also shares a feature of wood anatomy; all Gentianales except Rubiaceae are characterized by intraxylary phloem (Carlquist 1992). Relationships among these four families, however, are not completely resolved. There is weak support (< 65%) for Loganiaceae as sister to Gentianaceae and for Apocynaceae as sister to Gelsemiaceae in D. Soltis et al. (2000), but taxon sampling was low. In contrast, in B. Bremer et al. (2002), Apocynaceae are sister to Gentianaceae (95% jackknife support), and Gelsemiaceae are sister to Loganiaceae (85% support; as depicted in Figure 9.3). Adding to the uncertainty in relationships is the polyphyly or paraphyly of Loganiaceae as previously defined (e.g., Chase et al. 1993; Olmstead et al. 1993; B. Bremer et al. 1994; Struwe et al. 1994; Backlund et al. 2000). *Fagraea* and *Potalia* (typically placed in Loganiaceae) should be placed in Gentianaceae; *Gelsemium* and *Mostuea* are well separated from other Loganiaceae and have been treated as a separate family, Gelsemiaceae (reviewed in Judd et al. 2002, APG II 2003).

Phylogenetic analyses of both molecular and non-DNA characters indicated that Asclepiadaceae and Apocynaceae should be combined to form a single family, Apocynaceae (Judd et al. 1994; M. Endress and Bruyns 2000). Apocynaceae in this broad sense are united by several non-DNA characters, including tissues with laticifers and usually milky sap; carpels united by styles and/or stigmas with ovaries usually distinct (Figure 9.4H); and apical portion of the style expanded and modified to form a secretory head (Figure 9.4H; Judd et al. 1994). Relationships within Apocynaceae have been the subject of phylogenetic analyses (Judd et al. 1994; M. Endress et al. 1996; Sennblad and Bremer 1996; Civeyrel et al. 1998; M. Endress and Stevens 2001; Potgieter and Albert 2001). The family exhibits a stepwise series of morphological modifications to the androecium and gynoecium that appear to be associated with increased specialization and efficiency of pollination (Schick 1980, 1982; Judd et al. 1994, 2002; M. Endress 2002).

Solanales

Solanales consist of five families (APG 1998; APG II 2003; Figure 9.3): Convolvulaceae, Hydroleaceae, Montiniaceae, Solanaceae (including Duckeodendraceae, Goetzeaceae, and Nolanaceae *sensu* Cronquist 1981), and Sphenocleaceae (several of these are illustrated in Figure 9.4). Solanales include about 140 genera and 7,000 species.

This molecular-based circumscription of Solanales differs from concepts of the order in recent classifications. Cronquist (1981), for example, also included Polemoniaceae, Hydrophyllaceae, and Menyanthaceae in his Solanales, but he did not include Montiniaceae. Montiniaceae comprise three genera (*Montinia, Kaliphora*, and *Grevea*), and the relationships of the family have long been problematic. Some authors have included the family in a broadly defined Saxifragaceae (reviewed in Morgan and Soltis 1993). Members of Montiniaceae are characterized by unisexual flowers, an unusual feature in asterids, but found also in Garryales and Aquifoliales. Molecular studies also clearly placed *Hydrolea* in Solanales. The split of Hydrophyllaceae, with *Hydrolea* assigned to Solanales and *Hydrophyllum* and *Phacelia* to Boraginaceae, was first proposed by de Jussieu (1789). However, this inference was not accepted by Gray (1875), and modern morphology-based classifications included *Hydrolea* in Hydrophyllaceae (e.g., Cronquist 1981). Characters separating *Hydrolea* from other Hydrophyllaceae include a bilocular capsule and axile placentation in the former.

Few non-DNA characters have been identified as possible synapomorphies for Solanales. The families have radially symmetric flowers with a plicate, sympetalous corolla (Figure 9.4), and the number of stamens equals the number of petals; they also lack iridoids (discussed below; Albach et al. 2001b; K. Bremer et al. 2001; Judd et al. 2002). Alternate leaves may also be a synapomorphy; however, a reversal to opposite leaves has occurred in *Grevea* (Montiniaceae). Solanaceae and Convolvulaceae are united by similar alkaloids and wood with inter- and intraxylary phloem.

Within Solanales are two major clades (Figure 9.3). One clade consists of Montiniaceae as sister to Sphenocleaceae + Hydroleaceae, and the second clade consists of Solanaceae and Convolvulaceae (D. Soltis et al. 2000; B. Bremer et al. 2002). Solanaceae and Convolvulaceae are chemically distinct from other members of euasterid I in the replacement of iridoids by tropane alkaloids (Jensen et al. 1975; Romeike 1978). Circumscription of the major lineages within Convolvulaceae is provided by Stefanovic et al. (2002), and a molecular phylogenetic framework and provisional reclassification are available for Solanaceae (Olmstead et al. 1999). Solanaceae have been shown to include Duckeodendraceae (Fay et al. 1998b; APG II 2003). At lower levels within Solanaceae, a series of phylogenetic analyses have been conducted (e.g., Spooner et al. 1993; Olmstead and Sweere 1994; Olmstead et al. 1999). Tomato (formerly *Lycopersicon*) is clearly embedded within the large genus *Solanum* (e.g., Spooner et al. 1993; Bohs and Olmstead 1997).

Euasterid II (Campanulids)

The euasterid II clade (asterid II *sensu* Chase et al. 1993; campanulids of B. Bremer et al. 2002) comprises four major subclades (Figure 9.6A): Dipsacales, Aquifoliales, Asterales, and Apiales. Also part of euasterid II are Bruniaceae + Columelliaceae and a small clade of Tribelaceae, Polyosmaceae, Escalloniaceae, and Eremosynaceae (several of these families are illustrated in Figure 9.7). Paracryphiaceae may also be part of the euasterid II clade. Although the compositions of Dipsacales, Asterales (Asteranae *sensu* Thorne), and Apiales correspond well to the orders described by Cronquist (1981) and Thorne (1992a, 1992b), the euasterid II clade does not agree with any group of orders suggested in recent classifications. The clade combines members of traditional Asteridae with several groups of Rosidae (e.g., Aquifoliales, Apiales, Escalloniaceae; Cronquist 1981, 1988).

Monophyly of the euasterid II clade was indicated by analyses of single genes, such as *rbcL* (Chase et al. 1993; Olmstead et al. 1993; Cosner et al. 1994; Backlund and Bremer 1997), *atpB* (Savolainen et al. 2000a), and 18S rDNA (D. Soltis et al. 1997a), and is also supported by several non-DNA characters, including floral ontogeny (Erbar and Leins 1996; Roels and Smets 1996). In addition, euasterid II mostly have alternate leaves (with exceptions mainly in Dipsacales). This clade includes many taxa characterized by either secoiridoids or the apparent loss of iridoid production (the latter in Campanulaceae, Aquifoliaceae, and Apiales except *Griselinia* and *Melanophylla*; Albach et al. 2001b). Carbocyclic iridoids have been found in only five genera of euasterid II: three genera of Icacinaceae (Kaplan et al. 1991), *Stylidium* (Stylidiaceae), and *Escallonia* (Plouvier and Favre-Bonvin 1971). Iridoids and polyacetylenes seem to be complementary in distribution among euasterid II members; they compete for acetyl-CoA and mevalonic acid as biosynthetic precursors (Stuhlfauth et al. 1985; reviewed in Albach et al. 2001b).

Aquifoliales are strongly supported as sister to the rest of euasterid II (e.g., D. Soltis et al. 2000; B. Bremer et al. 2002), but the relationships among the remaining members of euasterid II (Dipsacales, Asterales, Apiales, Bruniaceae + Columelliaceae, Sphenostemonaceae, Paracryphiaceae; a clade of Tribelaceae, Polyosmaceae, Escalloniaceae, and Eremosynaceae) are

(A)

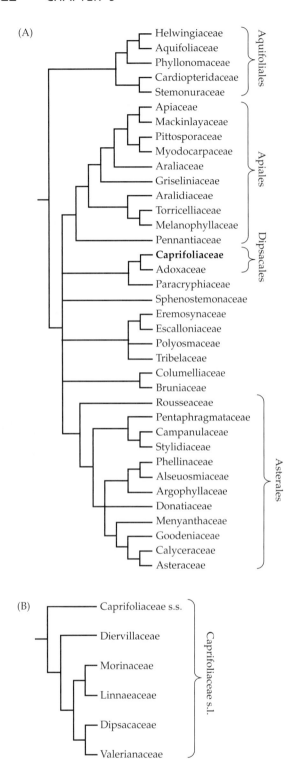

(B)

FIGURE 9.6 Relationships in the euasterid II (campanulid) clade. (A) Summary tree of ordinal and familial relationships. Caprifoliaceae indicated in bold (see B for details of relationships). Phylogenetic relationships of several families remain highly uncertain: Paracryphiaceae, Polyosmaceae, Tribelaceae, Columelliaceae, Bruniaceae, and Sphenostemonaceae. (B) Detail of relationships within Caprifoliaceae when component clades are treated as distinct families (e.g., Backlund and Bremer 1997; Donoghue et al. 2001; see text).

FIGURE 9.7 Flowers of the euasterid II (campanulid) clade. Representatives of Aquifoliales (A–C), Apiales (D–F), Dipsacales (G–J), Eremosynaceae (K–M), Asterales (N–U). (A) *Phyllonoma laticuspis* (Phyllonomaceae), branchlet with staminate flowers on leaf blades. (B) *Helwingia ruscifolia* (Helwingiaceae), branchlet, with flowers on leaf blades. (C) *Ilex glabra* (Aquifoliaceae), carpellate flower. (D) *Daucus carota* (Apiaceae), plant in flower. (E) *Daucus carota* (Apiaceae), flower. (F) *Heracleum maximum* (Apiaceae), fruit. (G) *Dipsacus laciniatus* (Caprifoliaceae s.l.; or Dipsacaceae), plant in flower. (H) *Sambucus canadensis* (Adoxaceae), fruit (drupe). (I) *Sambucus canadensis* (Adoxaceae), tip of branch in flower. (J) *Sambucus canadensis* (Adoxaceae), flower. (K) *Eremosyne pectinata* (Eremosynaceae), top view of flower. (L) *Eremosyne pectinata* (Eremosynaceae), plant in flower. (M) *Eremosyne pectinata* (Eremosynaceae), side view of flower. (N) *Lobelia cardinalis* (Campanulaceae), flower; petals not connate at base. (O) *Triodanis perfoliata* (Campanulaceae), flower. (P) *Roussea simplex* (Rousseaceae), branchlet in flower. (Q) *Donatia fasicularis* (Donatiaceae), side view of flower. (R) *Donatia fasicularis* (Donatiaceae), plant in flower. (S) *Cichorium intybus* (Asteraceae), head of flowers. (T) *Cichorium intybus* (Asteraceae), flower. (U) *Cichorium intybus* (Asteraceae), androecium and gynoecium. (A, from Harms in Engler and Prantl 1898. B, K to M, and P to R, from Engler in Engler and Prantl 1891. C to F, H to J, N, O, and S to U, from Wood 1974. G, from Höck in Engler and Prantl 1897.)

unclear. None of the relationships among these clades received support above 50% in the combined analyses of three, four, or six genes.

Aquifoliales

Aquifoliales consist of five families (APG II 2003; Figure 9.6A): Aquifoliaceae, Cardiopteridaceae, Stemonuraceae, Helwingiaceae, and Phyllonomaceae (several of these families are illustrated in Figure 9.7). The placement of these families together in the asterid clade was not anticipated from their treatments in recent morphology-based classifications. Cronquist (1981) placed Aquifoliaceae and Icacinaceae in Celastrales, Helwingiaceae within Cornaceae, and Phyllonomaceae within Grossulariaceae. Morphological synapomorphies for Aquifoliales are stipules (which are rare in asterids), unisexual flowers, and fleshy fruits, which evolved in parallel in other orders (K. Bremer et al. 2001).

The phylogenetic analysis of Kårehed (2001) indicated that Icacinaceae are polyphyletic. Some members (Icacinaceae s.s.) are part of euasterid I (unassigned to order), with remaining members part of Aquifoliales. Some of these taxa constitute a new family, Stemonuraceae, with other genera part of an enlarged Cardiopteridaceae. Within the order, Stemonuraceae and Cardiopteridaceae form a clade that is sister to the remaining families. The second subclade of Aquifoliales consists of Phyllonomaceae as sister to Helwingiaceae

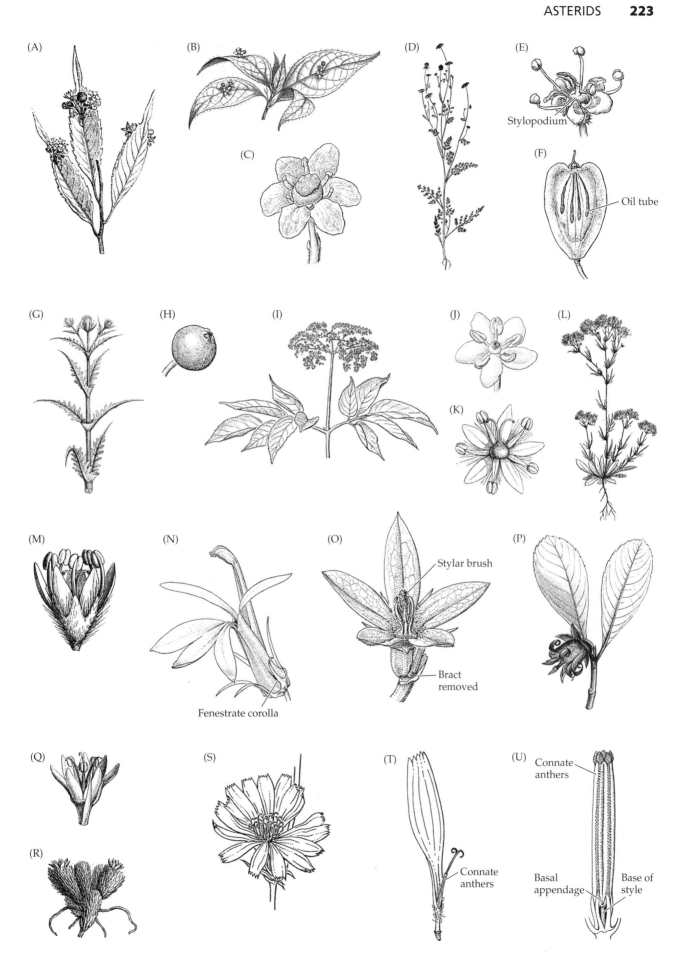

(A)

(B)

(C)

(D)

(E)

Stylopodium

(F)

Oil tube

(G)

(H)

(I)

(J)

(K)

(L)

(M)

(N)

Fenestrate corolla

(O)

Stylar brush

Bract removed

(P)

(Q)

(R)

(S)

(T)

Connate anthers

(U)

Connate anthers

Basal appendage

Base of style

+ Aquifoliaceae. These familial relationships are strongly supported (above 95%) in Kårehed (2001). Helwingiaceae and Aquifoliaceae were weakly supported sisters in D. Soltis et al. (2000). However, other analyses indicated with strong support that Aquifoliaceae are sister to Phyllonomaceae + Helwingiaceae (Savolainen et al. 2000b; Albach et al. 2001a). The association of *Helwingia* and *Phyllonoma* is interesting, given that they share the rare character of flowers borne directly on the leaf blades (Figure 9.7, A and B; Dickinson and Sattler 1974; Morgan and Soltis 1993). *Ilex* (Aquifoliaceae) has been the focus of phylogenetic study (Cuénoud et al. 2000).

Asterales

Asterales are strongly supported as a clade and consist of 12 families (Figure 9.6A): Alseuosmiaceae, Argophyllaceae, Asteraceae, Calyceraceae, Campanulaceae (including Lobeliaceae), Donatiaceae, Goodeniaceae, Menyanthaceae, Pentaphragmataceae, Phellinaceae, Rousseaceae (including *Carpodetus*), and Stylidiaceae (several of these families are illustrated in Figure 9.7; see Backlund and Bremer 1997; Gustafsson and Bremer 1997; Kårehed et al. 1999). The option of retaining Lobeliaceae as a family distinct from Campanulaceae has been noted (see also Lundberg and Bremer 2002; APG II 2003). Donatiaceae and Stylidiaceae may be closely related (Lundberg and Bremer 2002); in APG II the former is placed in optional synonymy under the latter. Asterales in this circumscription include approximately 1,700 genera and 26,000 species, dominated by Asteraceae. This circumscription differs from that of Lammers (1992) in the addition of Stylidiaceae and Donatiaceae and the inclusion of *Alseuosmia* and *Corokia*, which Lammers did not consider. Lammers (1992) did not include Stylidiaceae because of the presence of carbocyclic iridoids in this family but not in the other families of this order. However, the production of carbocyclic iridoids also occurs in Cornales and early-diverging members of euasterid II; the presence of carbocyclic iridoids in Stylidiaceae may represent a separate origin from these lineages.

Possible nonmolecular synapomorphies for Asterales are a base chromosome number of $x = 9$, the production of the oligosaccharide inulin as a storage compound (reported from Asteraceae, Calyceraceae, Goodeniaceae, Campanulaceae, and Stylidiaceae), valvate corolla aestivation (with a reversal to imbricate aestivation in Donatiaceae and Stylidiaceae and parallel derivation of valvate aestivation in other euasterids), and secondary pollen presentation (K. Bremer et al. 2001), sometimes referred to as "plunger pollination" (Judd et al. 2002). However, because this is really a pollen presentation mechanism rather than a means of pollination per se, "plunger pollen presentation" may be more appropriate (for the

diversity of this mechanism, see Erbar and Leins 1995b). A possible synapomorphy for Campanulaceae and Stylidiaceae, which are supported as sisters in both Albach et al. (2001a) and B. Bremer et al. (2002), is the filamentous suspensor of the proembryo (Tobe and Morin 1996). Petal venation (Gustafsson 1995) and loss of micropylar endosperm haustoria (Cosner et al. 1994) are synapomorphies for the well-supported clade of Asteraceae, Calyceraceae, Goodeniaceae, and Menyanthaceae. Other characters shared by these four families are multinucleate tapetal cells and the production of secoiridoids (Lammers 1992), although in Asteraceae, the production of secoiridoids is mostly replaced by the production of sesquiterpene lactones (but see Wang and Yu 1997).

There have been several recent comprehensive phylogenetic analyses of Asterales (e.g., Cosner et al. 1994; Gustafsson et al. 1996; Backlund and Bremer 1997; Gustafsson and Bremer 1997; Kårehed et al. 1999; Olmstead et al. 2000; Albach et al. 2001a; Lundberg and Bremer 2002; B. Bremer et al. 2002). Despite differences in taxon sampling, these studies reported many congruent relationships, although some differences are also apparent. Furthermore, interrelationships among many families of Asterales remain poorly understood. For example, the basal relationships are not well supported, even in four-gene (Albach et al. 2001a) and six-gene (B. Bremer et al. 2002) analyses. Within Asterales, *Roussea* is sister to *Carpodetus* with strong support (Savolainen et al. 2000b; B. Bremer et al. 2002). *Roussea*, which contains a single species, *Roussea simplex*, was recognized as part of the Asterales clade by D. Soltis et al. (1997a, 1997b) and Koontz and Soltis (1999). The genus had often been placed with *Brexia* and *Ixerba* in Saxifragaceae s.l., Grossulariaceae, or Brexiaceae in Rosidae. *Carpodetus* comprises 10 species and was formerly included in Escalloniaceae. *Roussea* and *Carpodetus* have been included in an expanded Rousseaceae (APG II 2003). Rousseaceae (including *Carpodetus*) form a weakly supported sister group to Campanulaceae (D. Soltis et al. 2000), to Campanulaceae + Stylidiaceae (Albach et al. 2001a), or to all remaining Asterales (Savolainen et al. 2000b; B. Bremer et al. 2002) (the latter relationship is depicted in Figure 9.6A). The monophyly of all Asterales except Rousseaceae received only 61% jackknife support in B. Bremer et al. (2002); however, with the inclusion of Rousseaceae, Asterales received strong (100%) support.

There is weak support for a sister-group relationship of Stylidiaceae and Campanulaceae in the four-gene analysis of Albach et al. (2001a); support for this relationship increased (82%, but with only one Campanulaceae sampled) in B. Bremer et al. (2002). The relationships of Pentaphragmataceae also remain unclear. In B. Bremer et al. (2002), the family appeared as sister to Campanulaceae + Stylidiaceae with weak (61%) jackknife support (Figure 9.6A). Also uncertain is the relationship of Donatiaceae. In B. Bremer et al. (2002), *Donatia* is sister to Argophyllaceae, Alseuosmiaceae, and Phellinaceae, but without support above 50%. *Donatia* was sister to *Stylidium* in D. Soltis et al. (2000), and this pair was sister to a clade of Alseuosmiaceae, Phellinaceae, Argophyllaceae, Menyanthaceae, Goodeniaceae, Asteraceae, and Calyceraceae, but without strong support.

There is strong support in some multigene analyses for a clade of Argophyllaceae, Alseuosmiaceae, and Phellinaceae, with Argophyllaceae sister to Alseuosmiaceae + Phellinaceae (B. Bremer et al. 2002). However, the relationships of this clade are still unclear (Figure 9.6A). In several analyses (Savolainen et al. 2000b; Albach et al. 2001a; B. Bremer et al. 2002), Argophyllaceae, Alseuosmiaceae, and Phellinaceae (sometimes with Donatiaceae) were sister to the clade of Menyanthaceae, Goodeniaceae, Asteraceae, and Calyceraceae, but support for this placement was low (only 55% jackknife support in B. Bremer et al. 2002). In the analyses of D. Soltis et al. (2000), Argophyllaceae, Phellinaceae, and Alseuosmiaceae formed a clade in the shortest trees, but without support above 50%, and these families formed a weakly supported (55%) tetrachotomy with a clade of Menyanthaceae, Goodeniaceae, Calyceraceae, and Asteraceae.

A well-supported clade within Asterales consists of Menyanthaceae, Goodeniaceae, Calyceraceae, and Asteraceae (Savolainen et al. 2000a, 2000b; D. Soltis et al. 2000; Albach et al. 2001a; B. Bremer et al. 2002; Lundberg and Bremer 2002). Menyanthaceae are well supported as sister to the remaining three families (Figure 9.6A). The sister group of Asteraceae appears to be either Calyceraceae (Gustafsson and Bremer 1995; Albach et al. 2001a; B. Bremer et al. 2002; Lundberg and Bremer 2002) or Calyceraceae + Goodeniaceae (Cosner et al. 1994; Gustafsson et al. 1996; Savolainen et al. 2000a; D. Soltis et al. 2000; Albach et al. 2001a). The conflict among these analyses is of interest. For example, B. Bremer et al. (2002) obtained strong support (88%) for Calyceraceae as the immediate sister to Asteraceae; morphology plus DNA sequences also supported this relationship (Lundberg and Bremer 2002). However, D. Soltis et al. (2000) found strong support (88%) for Calyceraceae + Goodeniaceae. Characters supporting the sister-group relationship of Calyceraceae and Asteraceae are pollen morphology (Skvarla et al. 1977), petal venation, and morphology of the receptacles and receptacle bracts (Hansen 1992). However, biochemical evidence seems to support a relationship between Goodeniaceae and Calyceraceae: both produce bis-secoiridoids of the sylvestroside type (Jensen et al. 1979; Capasso et al. 1996; Albach et al. 2001b).

The large family Asteraceae (approximately 23,000 species) has been the focus of several phylogenetic analyses (K. Bremer 1987, 1994, 1996; Jansen et al. 1991,

1992; Kim et al. 1992; Kim and Jansen 1995). Barnadesioideae, a small South American group comprising mostly trees and shrubs, are sister to the remainder of the family (Jansen and Palmer 1987). Jansen et al. (1991) also examined character evolution in the family.

Dipsacales

Dipsacales consist of two families (Figure 9.6A), Adoxaceae (including *Sambucus* and *Viburnum*, genera formerly placed in Caprifoliaceae) and a broadly defined Caprifoliaceae (Judd et al. 2002; APG II 2003; several representatives are illustrated in Figure 9.7). The order comprises about 40 genera and more than 100 species. This treatment of the order is broader than APG (1998) with the inclusion of Adoxaceae (K. Bremer et al. 2001; B. Bremer et al. 2002). Rather than recognizing a broadly defined Caprifoliaceae, some systematists prefer subdividing Caprifoliaceae into several families (Figure 9.6B): Caprifoliaceae s.s., Diervillaceae, Dipsacaceae, Linnaeaceae, Morinaceae, and Valerianaceae, a treatment that is also provided as an option in the APG II (2003) classification (see Backlund and Pyck 1998; B. Bremer et al. 2002). Savolainen et al. (2000a, 2000b) suggested that Desfontainiaceae (now included in Columelliaceae), Polyosmaceae, Paracryphiaceae, and Sphenostemonaceae might also be part of Dipsacales, but bootstrap support for this placement was lacking; other multigene analyses placed these families outside of Dipsacales. B. Bremer et al. (2002) found weak support (55%) for a sister-group relationship of Paracryphiaceae and Dipsacales (Figure 9.3).

Dipsacales as now circumscribed (Figure 9.6A) correspond closely to the Dipsacales of Cronquist (1981) and Thorne (1992a, 1992b) and Dipsacanae *sensu* Takhtajan (1997). Dahlgren's (1980) treatment of Dipsacales differed in that he also included Calyceraceae in the order, a family firmly placed in Asterales (see above). Takhtajan (1997) placed several genera of Dipsacales (i.e., *Sambucus, Viburnum, Triplostegia,* and *Morina*) in monogeneric families.

Morphological synapomorphies for Dipsacales include pericyclic cork and a gynoecium of three or more carpels (with some reversals to two; K. Bremer et al. 2001). The relationships of Dipsacales remain unclear. Analyses based on three (D. Soltis et al. 2000) and four (Albach et al. 2001a) genes indicated that Dipsacales and Apiales form a clade with Bruniaceae and Escalloniaceae + Eremosynaceae, although this clade did not receive support of 50% or higher and relationships among these groups are not clear. A possible close relationship between Apiales and Dipsacales is supported by the shedding of trinucleate pollen (Albach et al. 2001a).

Dipsacales have been the subject of a series of molecular phylogenetic investigations (e.g., Donoghue et al. 1992, 2001; Backlund and Bremer 1997). Within Dipsacales, Adoxaceae received moderate to strong support as sister to the remainder of the order (D. Soltis et al. 2000; Albach et al. 2001a; Donoghue et al. 2001; B. Bremer et al. 2002). Caprifoliaceae s.l. are easily separated from Adoxaceae by bilateral floral symmetry in the former versus radial in the latter; flowers with an elongate versus short style, capitate versus lobed stigma, spiny versus reticulate pollen exine, and differences in the structure of the nectary (Judd et al. 2002; for the nectary, see also Wagenitz and Laing 1984).

A series of morphological and molecular phylogenetic analyses has focused on Caprifoliaceae s.l., some as part of more comprehensive analyses (e.g., Donoghue et al. 1992, 2001; Downie and Palmer 1992; Judd et al. 1994; Backlund and Bremer 1998; Backlund and Pyck 1998; Källersjö et al. 1998; Savolainen et al. 2000b; D. Soltis et al. 2000; Albach et al. 2001a; K. Bremer et al. 2001; B. Bremer et al. 2002). The multigene analysis of B. Bremer et al. (2002) strongly supported relationships among the component families of Caprifoliaceae s.l. found in a cladistic analysis of non-DNA characters (Judd et al. 1994). Donoghue et al. (2001) provided a comprehensive overview of the order.

Caprifoliaceae in the strict sense (represented by genera such as *Lonicera*) are sister to the remaining taxa, a placement that received strong (100% jackknife) support in B. Bremer et al. (2002). Caprifoliaceae in the strict sense are united by 4 or 5 carpels and an indeterminate inflorescence (Judd et al. 1994, 2002). Diervillaceae are then sister to the remaining families, which form two subclades: Linnaeaceae + Morinaceae and Dipsacaceae + Valerianaceae (Figure 9.6B); these relationships all received strong support (Backlund and Pyck 1998; Albach et al. 2001a; B. Bremer et al. 2002). Diervillaceae, Linnaeaceae, Valerianaceae, and Dipsacaceae formed a well-supported clade; synapomorphies for this clade include presence of an achene fruit (rather than a berry or drupe), a base chromosome number of $x = 8$, and abortion of two or three carpels, leaving a single-ovuled carpel occupying half of the ovary (Judd et al. 1994). Biochemistry may also link these families: they produce bis-secoiridoids, such as sylvestroside and related compounds (Jensen et al. 1979; Plouvier 1992; Capasso et al. 1996).

Removal of *Abelia* and tribe Linnaeeae from the remaining Caprifoliaceae was proposed based on morphological features (Donoghue 1983), including a putative reduction in the number of seeds, stamens, and fertile locules in the ovary in the former. Morphological and molecular evidence ultimately resulted in the description of the family Linnaeaceae (Backlund and Pyck 1998). Strong support for a sister-group relationship between Dipsacaceae and Valerianaceae has been provided in several molecular analyses (e.g., Albach et al. 2001a; B. Bremer et al. 2002). The two families are united by several non-DNA characters, including reduced or lacking endosperm and a modified calyx.

Apiales

Apiales, with approximately 425 genera and 3,250 species, consist of 10 families following APG II (2003; Figure 9.6A): Apiaceae, Araliaceae, Aralidiaceae, Griseliniaceae, Mackinlayaceae, Melanophyllaceae, Myodocarpaceae, Pennantiaceae, Pittosporaceae, and Torricelliaceae (several are illustrated in Figure 9.7; see also Plunkett et al. 1997a, 1997b, 2001, 2004; Plunkett 2001; Plunkett and Lowry 2001; Lowry 2001; Chandler and Plunkett 2004a). In recent classifications (e.g., Cronquist 1981), Apiales have been included in Rosidae; their placement within asterids, as evident in all molecular phylogenetic studies, was not anticipated. In recent morphology-based classifications, the order was often limited to Apiaceae and Araliaceae. Aralidiaceae, Griseliniaceae, Melanophyllaceae, and Torricelliaceae were often included in Cornaceae (Rosidae); Pittosporaceae were also considered part of Rosidae, and *Pennantia* was placed in Icacinaceae. *Mackinlaya* and *Myodocarpus*, both formerly placed in Araliaceae, represent distinct lineages within Apiales and require recognition at the family level.

Apiales consist of taxa with mostly choripetalous flowers, whereas most euasterids are characterized by sympetaly. However, Erbar and Leins (1996) showed that members of Apiales exhibit early sympetaly. Thus, choripetaly is a synapomorphy for Apiales (assuming that sympetaly is a synapomorphy for all asterids), together with sheathing petioles (K. Bremer et al. 2001).

The relationships of Apiales within euasterid II are not clear (Figure 9.6A). In some analyses, they formed a clade with Dipsacales and a clade of Tribelaceae + (Escalloniaceae + Eremosynaceae), but without support above 50% (see above). Pittosporaceae, Araliaceae, and *Torricellia* share a base chromosome number of $x = 12$ with *Escallonia*. *Griselinia* has $x = 9$, a number found in some Apiaceae and common in euasterid II (e.g., many Asterales and Adoxaceae).

Pennantia (Pennantiaceae), previously placed in Icacinaceae, appears to be sister to the remainder of the order (Kårehed 2002; Figure 9.6A). B. Bremer et al. (2002) found strong support for two major clades within Apiales: (1) a clade of Aralidiaceae as sister to Melanophyllaceae + Torricelliaceae; and (2) Griseliniaceae followed by Araliaceae as subsequent sisters to Apiaceae + Pittosporaceae. However, the taxon sampling in B. Bremer et al. (2002) was sparse. Plunkett and Lowry (2001) stressed that within Apiales, some relationships remain uncertain. However, recent analyses (Plunkett et al. 2004) have clarified the positions of several "ancient araliads" and enigmatic hydrocotyloids of Apiaceae, resulting in the recognition of Myodocarpaceae. Myodocarpaceae contain *Myodocarpus*, *Delarbrea*, and *Pseudosciadium* (Plunkett and Lowry 2001; Plunkett et al. 2004). A distinct family Mackinlayaceae could be recognized (APG II 2003) as sister to Apiaceae. Additional phylogenetic analyses of Araliales, Apiaceae, and Araliaceae include Plunkett et al. (1996a, 1996b, 2004), Downie et al. (1998, 2000a, 2000b), Plunkett and Downie (1999), Henwood and Hart (2001), and Chandler and Plunkett (2004).

CHARACTER EVOLUTION IN APIALES The relationships of Apiales had long been enigmatic. Resolution of relationships within the order and clarification of appropriate outgroups for the order afforded the opportunity to reexamine longstanding hypotheses of character evolution in the clade. For example, pluricarpellate (many-carpellate) flowers were considered "primitive" in Apiales by some (Philipson 1970; Eyde and Tseng 1971), whereas Cronquist (1981) considered pentacarpellate flowers primitive. Bicarpellate flowers, typical of many Apiaceae, were considered derived. However, the placement of *Aralidium* (unicarpellate), *Mel-anophylla* (bicarpellate), *Torricellia* (tricarpellate), and *Griselinia* (tricarpellate) as closest relatives of Apiaceae indicates that a lower carpel number was probably the ancestral state for Apiales (reconstruction not shown) (Plunkett et al. 1997a).

Similarly, compound leaves are prevalent throughout Apiales and were also considered "primitive" by some authors (e.g., Cronquist 1981). However, all of the closest relatives of Apiales revealed by molecular phylogenetic analyses have simple leaves. Simple leaves are also scattered throughout Apiales, but they have typically been considered to represent examples of reduction. However, character mapping indicates that simple leaves are ancestral for Apiales (Figure 9.8). Additional character-state mapping (not shown) further indicates that the ancestor of Apiales was woody with a tropical or Southern Hemisphere geographic distribution (Plunkett et al. 1996a).

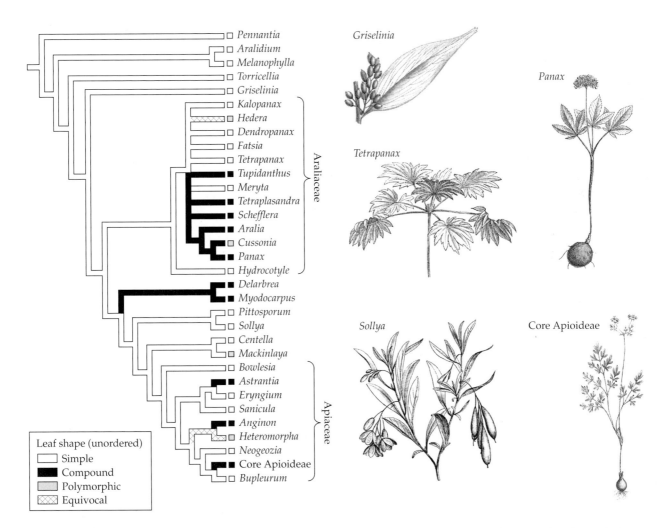

FIGURE 9.8 Reconstruction of evolution of leaf shape in Apiales (modified from Plunkett et al. 1996). Illustrations of representative leaf shapes are depicted (from Harms in Engler and Prantl 1898). *Griselinia* (Griseliniaceae), simple; *Tetrapanax* (Araliaceae), simple; *Panax* (Araliaceae), compound; *Sollya* (Pittosporaceae), simple; core Apioideae, represented by *Erigenia* (Apiaceae), compound.

Thus, Plunkett et al. (1996a) suggested that bicarpellate flowers and simple leaves were ancestral within Apiales. Both characters also occur in Stylidiaceae and Donatiaceae, which occupy basal branches in Asterales. Escalloniaceae and Adoxaceae have two to five carpels and mostly simple leaves.

Unplaced Euasterid II

In APG II, seven families remain unplaced to order in euasterid II: Bruniaceae, Columelliaceae (including Desfontainiaceae), Eremosynaceae, Escalloniaceae, Polyosmaceae, Sphenostemonaceae, and Tribelaceae (Figure 9.6A). Many of these unplaced families have either been placed in, or considered close to, Escalloniaceae in recent classifications (e.g., Takhtajan 1997). The relationships of these families have long been enigmatic, and they remain somewhat obscure after analyses of multigene datasets. Escalloniaceae and Eremosynaceae consistently formed a clade in recent studies and are discussed together. The relationships of the remaining unplaced families to each other and to other members of euasterid II are unclear even with combined datasets of four (Albach et al. 2001a) and six (B. Bremer et al. 2002) genes; these five families are discussed after Escalloniaceae and Eremosynaceae.

Escalloniaceae and Eremosynaceae

Escalloniaceae and Eremosynaceae formed a clade (Figure 9.6A) that is well supported in most multigene molecular phylogenetic analyses (e.g., D. Soltis et al. 2000; Albach et al. 2001a). The close relationship of these two families is another somewhat unexpected finding, although both had been placed in Saxifragaceae s.l. by Engler (1930; reviewed in Morgan and Soltis 1993). Treatments of these two families have been diverse. Cronquist (1981), for example, included Escalloniaceae in his Grossulariaceae and Eremosynaceae within Saxifragaceae (both in his Rosales, Rosidae). Takhtajan (1997), in contrast, placed Escalloniaceae in his Cornidae, while similarly considering *Eremosyne* (as Eremosynaceae) close to Saxifragaceae in his Saxifragales (Rosidae).

Escalloniaceae as previously circumscribed are polyphyletic, with members having diverse relationships within the euasterid II clade. Other genera that formerly were placed in the family and with relationships elsewhere in euasterid II include *Columellia* (unplaced to order), *Desfontainia* (included within Columelliaceae; unplaced to order), *Phelline* (Asterales), and *Sphenostemon* (unplaced to order) (Backlund and Bremer 1997; V. Savolainen, A. Backlund, and M. Chase, pers. comm.). *Quintinia*, another former Escalloniaceae, may be sister to *Paracryphia* (B. Bremer et al. 2002), although *Quintinia* was in an unresolved position in Gustafsson et al. (1996).

The relationships of *Escallonia* + *Eremosyne* in euasterid II remain unclear. They (with Bruniaceae) formed a clade that is sister to Apiales and Dipsacales in Albach et al.'s (2001a) four-gene analysis, but were sister to Dipsacales in the three-gene analysis (D. Soltis et al. 2000) and in analyses based on *rbcL* alone (Hibsch-Jetter et al. 1997). In B. Bremer et al. (2002), *Escallonia* + *Eremosyne* were part of a weakly supported clade with *Tribeles* and *Polyosma*, but again the relationship within euasterid II was unclear. Trichomes with radially arranged apical cells in the glandular head provide a synapomorphy for *Escallonia* and *Eremosyne* (Al-Shammary and Gornall 1994).

Bruniaceae

Bruniaceae are considered a well-defined family of 12 genera native to southern Africa. The family has at times been associated with Hamamelidaceae. However, most modern classifications have placed the family among what were considered the woody relatives of Saxifragaceae in Rosales (e.g., Greyiaceae, Escalloniaceae, Columelliaceae, Grossulariaceae of Cronquist 1981). However, Takhtajan (1997) placed the family with Grubbiaceae (now in Cornales) in Bruniales in Dilleniidae.

To date, two genera, *Berzelia* (sometimes considered a separate family, Berzeliaceae) and *Brunia*, have standardly been sampled in molecular analyses, with one or the other genus used as a placeholder for Bruniaceae, but not both. The position of Bruniaceae remains uncertain in molecular analyses. With three genes (D. Soltis et al. 2000), *Berzelia* was strongly supported (97% bootstrap value) as part of a major lineage of euasterid II that also included Dipsacales, Asterales, Apiales, Escalloniaceae, and Eremosynaceae, but its relationship to *Eremosyne* + *Escallonia*, Dipsacales, Apiales, and Asterales was unclear. In four-gene analyses (Albach et al. 2001a), *Berzelia* was sister to *Escallonia* + *Eremosyne*, but with internal support less than 50%. In studies based on *rbcL* alone (e.g., Olmstead et al. 1993; Savolainen et al. 2000b), *Berzelia* was sister to Dipsacales, but without support above 50%. In the six-gene analysis (B. Bremer et al. 2002), there was weak (56% jackknife) support for Bruniaceae as sister to Columelliaceae, a placement we have followed here (Figure 9.6A).

Columelliaceae (including Desfontainiaceae)

Columelliaceae as previously circumscribed consist of the single genus *Columellia* with four species native to the Andes. Cronquist (1981) considered the family taxonomically isolated, but placed them in his Rosales with several other genera and families now recognized as part of the asterid clade. Takhtajan (1997) envisioned a relationship of *Columellia* to some of the same "escallonioid" taxa, but placed these families in his Hydrangeales (Cornidae).

There was strong molecular support for a sister relationship of *Columellia* and *Desfontainia* (formerly Desfontainiaceae) (B. Bremer et al. 2002). *Desfontainia* is another enigmatic genus, usually included in Loganiaceae (Asteridae) (see Cronquist 1981). However, Takhtajan (1997) placed Desfontainiaceae in its own order near Hydrangeales and noted a possible close relationship between Desfontainiaceae and Columelliaceae, supported by similarities in wood anatomy (Mennega 1980). Given the close relationship between *Columellia* and *Desfontainia,* the two could be combined to form a single family, Columelliaceae (B. Bremer et al. 2002; APG II 2003); this is presented as an optional treatment in APG II (2003) and is followed here.

The relationship of Columelliaceae to other asterids is unknown. In analyses of *rbcL* alone, the family appeared as sister to most of euasterid I (Solanales, Lamiales, Gentianales), but this relationship lacked support above 50% (Savolainen et al. 2000b). The family was not included in three- and four-gene analyses, but in the six-gene analysis of B. Bremer et al. (2002), Columelliaceae were sister with weak (52%) support to Bruniaceae in euasterid II (Figure 9.6A), but this placement must be considered tentative.

Polyosmaceae

Polyosmaceae consist only of *Polyosma,* a genus of 60 species. The genus has been included in Escalloniaceae by some authors (Takhtajan 1987). Cronquist (1981) placed *Polyosma* in his Grossulariaceae with other putative woody relatives of Saxifragaceae, including Montiniaceae, Tribelaceae, and Phyllonomaceae, families now known to be part of the asterid clade. The relationships of the family remain unclear. In the B. Bremer et al. (2002) analysis, Polyosmaceae, as well as Tribelaceae, were weakly (69%) supported as members of a clade with Eremosynaceae + Escalloniaceae.

Tribelaceae

Tribelaceae represent still another asterid lineage that Cronquist (1981) placed in his Grossulariaceae with other putative woody relatives of Saxifragaceae. Takhtajan (1997) placed the family close to Escalloniaceae (Hydrangeales, Cornidae). In the B. Bremer et al. (2002) analysis, Tribelaceae, together with Polyosmaceae, were part of a weakly (69%) supported clade with Eremosynaceae + Escalloniaceae.

Sphenostemonaceae

Sphenostemon, comprising seven species, represents another enigmatic genus. Cronquist (1981) placed the genus in his Aquifoliaceae, as did Takhtajan (1997; his Icacinales + Aquifoliales). The genus has been included in few comprehensive molecular phylogenetic analyses. Based on *rbcL* sequence data, the genus appeared as sister to another problematic genus, *Paracryphia,* within Dipsacales of euasterid II (Savolainen et al. 2000b). *Sphenostemon* was not included, however, in the multigene analyses of D. Soltis et al. (2000), Albach et al. (2001a), and B. Bremer et al. (2002). More research is

needed to establish the relationships of this family. An important next step would be to obtain sequence data for genes other than *rbcL.*

Unplaced Asterids

Paracryphiaceae consist of the single woody species, *Paracryphia alticola,* native to New Caledonia. The relationships of this poorly understood species have been debated. Cronquist (1981) placed the family in his Theales (Dilleniidae). Takhtajan (1997) and Thorne (1992a, 1992b, 2000) also considered the species a member of Theales (see also Schmid 1978).

Paracryphiaceae are the only family considered unplaced within the asterids as a whole in APG II (2003). However, two analyses that included the family have placed Paracryphiaceae within euasterid II (campanulids). In Savolainen et al. (2000b), *Paracryphia* appeared as sister to *Sphenostemon* (Sphenostemonaceae), a family in euasterid II that is unplaced to order (discussed above). In the B. Bremer et al. (2002) six-gene analysis, *Paracryphia* is sister to *Quintinia* (Escalloniaceae) and well nested within euasterid II (*Sphenostemon* was not included in that study). *Paracryphia* and *Quintinia* are weakly (55%) supported as sister to Dipsacales. We have followed this placement of Paracryphiaceae here, depicting the family as sister to Dipsacales (Figure 9.6A), but this should be considered tentative. Available evidence therefore indicates that Paracryphiaceae are part of euasterid II, but the relationships of the family remain poorly understood. *Paracryphia* exhibits an unusual combination of characters (Dickison and Baas 1977). In vegetative anatomy, the species is unspecialized, whereas in some aspects of floral morphology, it appears highly specialized (Takhtajan 1997).

Patterns of Character Evolution in Asterids

The asterid clade affords an excellent opportunity to examine the evolution of several key morphological and chemical characters, including sympetaly, integument number, type of nucellus, floral symmetry, and iridoid production. Many of these non-DNA characters were examined in a phylogenetic context by Albach et al. (2001b) using the four-gene topology for this large clade (Albach et al. 2001a), and the evolution of floral symmetry was addressed by Donoghue et al. (1998). However, because phylogenetic relationships within the asterid clade are now better understood (e.g., Olmstead et al. 2001; B. Bremer et al. 2002; Kårehed 2002; Lundberg and Bremer 2002; Xiang et al. 2002; Plunkett et al. 2004a), in this section we evaluate the evolution of several key features in the asterid clade. We used a topology that closely reflects that in B. Bremer et al. (2002), but with specific subclades modified by increased taxon sampling to reflect recent phylogenetic analyses (e.g., Ericales, Kron et al. 2002; Apiales, Plunkett and Lowry 2001, Kårehed 2002, and Plunkett et al. 2004a; Cornales, Xiang et al. 2002; Asterales, Lundberg and Bremer 2002; Lamiales, Olmstead et al. 2001).

Sympetaly characterizes most of the asterids, and the core of this clade has been recognized as a natural group in classifications for more than 200 years on the basis of this character (e.g., de Jussieu 1789; de Candolle 1813; Reichenbach 1827–1829). A suite of embryological and chemical characters also appears to be highly but not perfectly correlated with sympetaly: unitegmy (Warming 1879), tenuinucellate ovules (cf. Warming 1879; Philipson 1974), cellular endosperm formation (Sporne 1954; Dahlgren 1975), terminal endosperm haustoria (Dahlgren 1977), pollen grains released at the tricellular stage (Dahlgren 1975), and presence of iridoids (Dahlgren et al. 1981).

Nearly all species of eudicots with sympetalous corollas are found in the asterids. Notable exceptions are Plumbaginaceae, Caricaceae, and Cucurbitaceae, three families representing Caryophyllales (the first) and different orders of rosids (the last two), in which sympetalous corollas undoubtedly evolved independently. However, even within asterids, homology of sympetaly is not certain because sympetalous corollas may arise through different ontogenetic pathways (Erbar 1991) and may have been derived many times (at least four or more; Olmstead et al. 1992; Albach 1998). Consequently, additional information, particularly on the development of sympetalous corollas, is needed for many groups before the evolutionary history of sympetaly can be clarified. Instead of examining the evolution of sympetaly itself, we therefore present hypotheses of the evolution of those characters typically associated with sympetaly in the asterids: unitegmy, tenuinucellate ovules, cellular endosperm formation, trinucleate pollen, and iridoids. The evolution of these characters in asterids was studied in detail by Albach et al. (2001b), using Albach et al.'s (2001a) phylogenetic tree inferred from sequences of four genes (see above); we provide an abbreviated discussion of these patterns in the following sections.

Unitegmy

Unitegmy, although typically associated with sympetaly (cf. Warming 1879; Olmstead et al. 1993), also occurs frequently outside asterids (Philipson 1974; Corner 1976). Multiple transitions to unitegmic ovules are supported by the discovery of multiple ontogenetic pathways to unitegmy (Bouman and Calis 1977; Bouman 1984), although a general lack of information on ovule ontogeny hinders attempts to relate the distribution of unitegmy to specific mechanisms of unitegmic ovule formation. Our assignment of unitegmic versus bitegmic ovules is based on Philipson (1974), Corner (1976), Cronquist (1981), and Takhtajan (1997).

The ancestral condition of asterids is not entirely clear because their sister group is not certain, even in angiosperm-wide studies using data from three or four genes (e.g., P. Soltis et al. 1999b; D. Soltis et al. 2000, 2003a). Most angiosperms outside the asterid clade have bitegmic ovules, and hence it is often assumed that the sister group of the asterids possesses this condition. However, several comprehensive phylogenetic analyses point to Santalales or Santalales + Caryophyllales as the possible sister group to the asterids, albeit without support above 50% (e.g., Chase et al. 1993; D. Soltis et al. 2000, 2003a). If Santalales are sister to asterids, then the ancestral state for the asterids is unambiguously reconstructed as unitegmy, regardless of the model of character optimization used. If the sister group of the asterids possesses bitegmic ovules, then the ancestral condition for asterids is ambiguous, depending on the optimization method used. If the "all most parsimonious states" or delayed transformation (DELTRAN) optimization is used, the ancestral condition is uncertain; if accelerated transformation (ACCTRAN) optimization is used, the ancestral condition is bitegmy. However, given the lack of jackknife/bootstrap support for a sister group of the asterid clade, the ancestral state of the asterids is best considered equivocal.

Within asterids, Cornales are sister to all other asterids, and these taxa have unitegmic ovules. However, the situation is more complex in Ericales, the immediate sister to euasterids. Bitegmy occurs frequently in Ericales (cf. Endress 2003a) and is reconstructed as the ancestral state for this clade regardless of the outgroup or optimization model (Figure 9.9). Within Ericales, independent shifts to unitegmy have occurred on many occasions (perhaps six times) in (1) some Balsaminaceae (reported to have both bitegmy and unitegmy; Boesewinkel and Bouman 1991; Albach et al. 2001b), (2) some Primulaceae (reported to have both bitegmy and unitegmy; Anderberg 1995; Albach et al. 2001b), (3) core Ericales (Ericaceae, Clethracae, Cyrillaceae, Actinidiaceae, Roridulaceae, Sarraceniaceae), (4) Polemoniaceae, (5) Sapotaceae, and (6) Diapensiaceae–Styracaceae–Symplocaceae. The last clade is of interest because there may have been a reversal to bitegmy in *Halesia* (Styracaceae) (Figure 9.9).

Four possible pathways from bitegmy to unitegmy have been proposed (Bouman and Calis 1977; Boesewinkel and Bouman 1991): (1) reduction of one integument, (2) fusion of two integument primordia, (3) shift of the inner integument along a subdermal outer integument, and (4) shift of a dermal outer integument along the inner integument. However, the developmental sequence for unitegmy has rarely been studied. Asterids, in general, appear to have lost one integument by integumentary shifting (pathway 3 or 4; Bouman and Schier 1979; Boesewinkel and Bouman 1991; see review by Albach et al. 2001b, for pathways in other angiosperms).

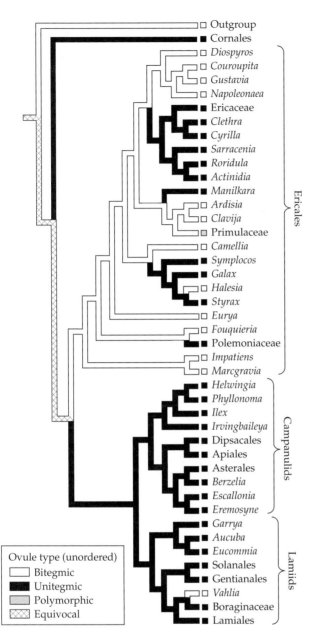

FIGURE 9.9 Reconstruction of the evolution of unitegmic and bitegmic ovules in asterids. Relationships among major clades are depicted as resolved (to facilitate character reconstruction) following the shortest trees of B. Bremer et al. (2002). (Modified from Albach et al. 2001b.)

Genes that cause loss or deformation of one integument have been identified in *Arabidopsis thaliana* (e.g., Robinson-Beers et al. 1992; Leon-Kloosterziel, et al. 1994; Gaiser et al. 1995; Baker et al. 1997) and provide possible mechanisms for the transition from bitegmy to unitegmy. However, given the number of genes involved in ovule development, it seems likely that integument formation and transition to unitegmy may require changes in several genes (Leon-Kloosterziel et al. 1994; Baker et al. 1997). Hence, there may be several genetic

mechanisms by which these transitions can be achieved. It remains uncertain whether unitegmic ovules in asterids are developmentally and genetically homologous.

Tenuinucellate ovules

Tenuinucellate ovules are defined by the absence of parietal tissue between the megaspore mother cell and the nucellar epidermis (see Figure 12.12 and Bouman 1984, and references therein). As with sympetaly and integument number, interpretation of tenuinucellate versus crassinucellate ovules is not always clear because different processes may lead to similar nucellar structures (Endress 2003a). However, although information on nucellar development is critical for determining ontogenetic homology of nucellar structure, this information is lacking for most angiosperms, and our analysis relies on reports of tenuinucellate versus crassinucellate ovules taken from Philipson (1974), Corner (1976), Cronquist (1981), and Takhtajan (1997).

Tenuinucellate ovules occur throughout asterids, with only a few exceptions (some Cornales, many Apiales, *Viburnum*, and Garryales; Figure 9.10). The ancestral state for the asterids is equivocal because of the presence of crassinucellate ovules in *Alangium* and most species of *Cornus* (see Xiang et al. 1993) of Cornales, which represent the sister to all other asterids. Furthermore, as noted above, the sister to the overall asterid clade is unclear. Most angiosperms outside of asterids have crassinucellate ovules. If the sister group of asterids is considered to be crassinucellate, then the ancestral state for asterids is reconstructed as equivocal when both the "all most parsimonious states" and DELTRAN optimizations are used. In contrast, with ACCTRAN optimization, the ancestral state is reconstructed as tenuinucellate (Figure 9.10). However, several molecular phylogenetic analyses have indicated that Santalales may be the immediate sister to asterids. If Santalales, with their tenuinucellate ovules, are considered the sister group of asterids, the ancestral state for asterids is tenuinucellate regardless of the optimization.

Although tenuinucellate ovules characterize asterids in general and some clades are exclusively tenuinucellate (Lamiales, Solanales, Gentianales, Asterales, and Ericales), crassinucellate ovules also occur in at least some Cornales, Apiales, Dipsacales, Aquifoliales, and Garryales. Multiple transitions between tenuinucellate and crassinucellate ovules are apparent (Figure 9.10), and both types have been reported within single species of *Viburnum* (Philipson 1974). However, the number and direction of such transitions depend on outgroup state and optimization. If the ancestral state for asterids as a whole is ambiguous (see above), then directions of transitions in Cornales are also unclear. Regardless of outgroup and optimization, however, transitions from tenuinucellate to crassinucellate ovules have occurred in *Berzelia* (Bruniaceae), *Griselinia* and

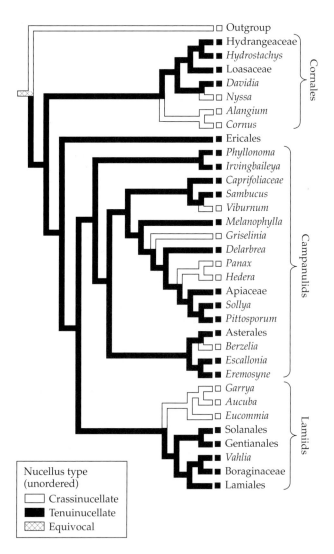

FIGURE 9.10 Reconstruction of the evolution of crassinu-cellate and tenuinucellate ovules in asterids using ACCTRAN optimization. Relationships among major clades are depicted as resolved (to facilitate character reconstruction) following the shortest trees of B. Bremer et al. (2002). (Modified from Albach et al. 2001b.)

Araliaceae (Apiales), *Viburnum* (Adoxaceae), and Gar-ryales. *Nyssa* (Nyssaceae), *Alangium,* and *Cornus* (Cor-naceae) in Cornales would represent additional transi-tions to crassinucellate ovules should the ancestral state for all asterids be tenuinucellate ovules.

A single loss of parietal tissue leading to production of tenuinucellate ovules has been inferred for the Ericales-euasterids clade (i.e., all asterids other than Cornales). A different ontogenetic pathway may, how-ever, be responsible for tenuinucellate ovules in Cor-nales and some rosid families (see Philipson 1974). Our assessment of evolutionary patterns in ovule structure will be improved by identifying the sister group of as-terids and through additional information on ovule de-velopment (Albach et al. 2001b).

Endosperm formation

Patterns of endosperm formation may be related to nu-cellus development because a nucellus without protec-tion from parietal tissue may favor a cellular endosperm, whereas a well-developed nucellus may experience a delay in cell wall formation, resulting in free nuclear en-dosperm (e.g., Wunderlich 1959; Dahlgren 1975). Thus, cellular endosperm formation is expected in asterids in which tenuinucellate ovules predominate. However, cel-lular endosperm formation is also common in Magno-liales, Laurales, and Piperales and is proposed to be an-cestral in the angiosperms (e.g., Coulter and Cham-berlain 1903; Schürhoff 1926; Glisic 1928; Swamy and Ganapathy 1957; Wunderlich 1959; Friedman 1992, 1994). Therefore, cellular endosperm formation in the as-terids may represent either the retention of the ancestral cellular endosperm or a reversal (e.g., Sporne 1954; Swamy and Ganapathy 1957; Dahlgren 1991). Howev-er, recent studies of endosperm development reveal that traditional categories of endosperm formation—that is, cellular, nuclear, helobial—do not capture the diversity of developmental pathways, and patterns considered "nuclear," for example, in many species may not be on-togenetically homologous (Floyd et al. 1999; Floyd and Friedman 2000). Because developmental data are avail-able for relatively few species, here we report analyses of the asterids that are based on the traditional designa-tions of endosperm formation as cellular versus nuclear, from Rao (1972), Corner (1976), Kamelina (1984), Cron-quist (1981), Dahlgren (1991), and Takhtajan (1997), but we recognize that at least some such designations may require reevaluation.

Cellular endosperm is reconstructed as ancestral for the asterids (see Albach et al. 2001b). Although rosids and Caryophyllales have nuclear endosperm, cellular endosperm is present in Santalales, which, as noted, ap-pear as sister to asterids in several molecular analyses. Furthermore, cellular endosperm is widespread in Cor-nales and Ericales and is reconstructed as ancestral for both of these clades. The reconstructions of Albach et al. (2001b), as well as our reconstructions (not shown) based on more recent topologies, indicated many independent derivations of nuclear endosperm formation: once in Cornales (in *Alangium*), perhaps twice in Ericales, and multiple times in euasterid I and euasterid II. In euas-terid I, there were derivations of nuclear endosperm in Gentianales (see Wagenitz 1959), *Garrya,* Convolvu-laceae, and Boraginaceae; in euasterid II, nuclear en-dosperm is present in *Irvingbaileya* and most Apiales. Several of these derivations match reversals to crassinu-cellate ovules, supporting Wunderlich's (1959) hypoth-esis of a close association between nucellar structure and patterns of endosperm development. Multiple deriva-tions of nuclear endosperm development from cellular have also been observed across the angiosperms as a whole (Floyd et al. 1999). Despite multiple transitions from cellular to nuclear endosperm in the asterids, no reversals were identified, indicating that perhaps transformations from nuclear to cellular are not possible in asterids. An evaluation of this hypothesis requires additional information on patterns and processes of endosperm development throughout the angiosperms.

Tricellular pollen

In most angiosperms, pollen grains are shed when the microgametophyte reaches the bicellular stage. In some angiosperms, however, including many asterids, pollen grains are shed at the tricellular stage. Release of pollen at the tricellular stage has been considered a selective advantage under certain circumstances (Stebbins 1974). However, this character is generally uniform within large clades recognized as orders or families. Data on bicellular versus tricellular pollen grains were obtained from Brewbaker (1967), Kamelina (1984), Eyde (1988), Takhtajan (1997), and Albach et al. (2001b).

Pollen shed in a tricellular state occurs mostly in asterids, Caryophyllales, some rosid families (e.g., Brassicaceae), Alismatales, and grasses. The ancestral state for asterids appears to be bicellular pollen (Albach et al. 2001b); tricellular pollen has not been reported in Cornales, and it occurs only sporadically in Ericales. Additional derivations of tricellular pollen are found in euasterid II, independently in *Irvingbaileya* (Cardiopteridaceae), Asteraceae, *Corokia* (Argophyllaceae), and the clade of Apiales and Dipsacales (reconstruction not shown). The pattern of pollen evolution is more complex in euasterid I, in which both bicellular and tricellular pollen are present in members of several large clades (i.e., Lamiales, Boraginaceae, and Gentianales) and occasionally even within a single genus (*Plantago, Ipomoea*; Brewbaker 1967). However, there are no apparent reversions from tricellular to bicellular pollen (see Albach et al. 2001b), supporting Brewbaker's (1967) suggestion of a unidirectional shift. Tricellular pollen

appears to be associated with efficient cross-pollination (Stebbins 1974) and sporophytic self-incompatibility, shorter storage longevity, and an aquatic habit with submersed flowers (Brewbaker 1957, 1967).

Iridoids

Iridoids are monoterpenes characterized by a cyclopentanopyran ring system with a double bond between C_3 and C_4 (Figure 9.11). Iridoids may have several functions in plants, such as anti-herbivory (e.g., Bernays and De Luca 1981; Bowers and Puttick 1988) or inhibition of

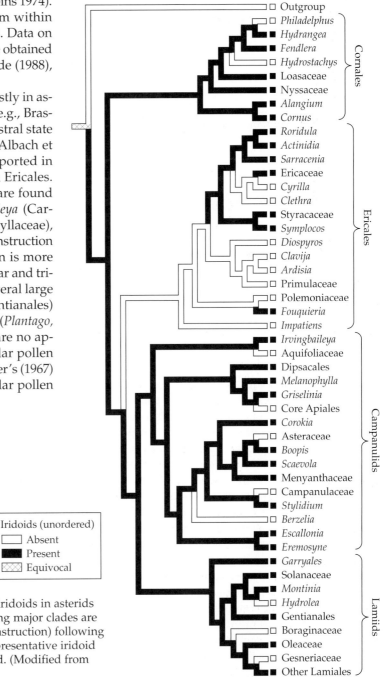

Sweroside, a secoiridoid
(*Swertia*, Gentianaceae)

Iridoids (unordered)
☐ Absent
■ Present
▨ Equivocal

FIGURE 9.11 Reconstruction of the evolution of iridoids in asterids using ACCTRAN optimization. Relationships among major clades are depicted as resolved (to facilitate character reconstruction) following the shortest trees of B. Bremer et al. (2002). A representative iridoid compound, sweroside (a secoiridoid), is depicted. (Modified from Albach et al. 2001b.)

seed germination (Pardo et al. 1998). The occurrence of iridoids is highly correlated with sympetaly (Jensen et al. 1975), with only a few occurrences of iridoids outside asterids, in *Stigmaphyllon* (Malpighiaceae; Davioud et al. 1985), *Homalium* (Salicaceae; Ekabo et al. 1993), *Bhesa* (Celastraceae; Ohashi et al. 1993), *Ailanthus* (Simaroubaceae; Kosuge et al. 1994), *Liquidambar* (Plouvier 1964) and other Altingiaceae (Jiang and Zhou 1992), *Daphniphyllum* (Daphniphyllaceae; Inouye et al. 1966), and *Paeonia* (Paeoniaceae; Kolpalova and Popov 1994). Several clades within asterids do not produce iridoids, such as Solanales, some Apiales, Asteraceae (but see Wang and Yu 1997), and several families of Ericales (reviewed in Albach et al. 2001b). Iridoids have been classified by their chemical properties, such as functional features or oxidation level (Kaplan and Gottlieb 1982), biosynthetic pathway (Jensen 1991), or a combination of the two (Jensen et al. 1975). However, the pathway followed in any given species is typically not known, and there may be several pathways to the same compound (Jensen 1991). Thus, an extensive classification based on biosynthetic pathway, although desirable, is not currently possible, although two main, presumably mutually exclusive, pathways have been identified (Jensen 1991): one leading to secoiridoids and a second to iridoids decarboxylated at the C_4 position. In this section, we report analyses of the presence or absence of (1) iridoids generally and (2) secoiridoids; additional information on the distribution of iridoids can be found in Albach et al. (2001b). Data on the distribution of iridoids are from Kostecka-Madalska and Rymkiewick (1971), Jensen et al. (1975), Degut and Fursa (1980), El-Naggar and Beal (1980), Lammel and Rimpler (1981), Makboul (1986), Boros and Stermitz (1990, 1991), Jensen (1991, 1992), Nicoletti et al. (1991), Drewes et al. (1996), and von Poser et al. (1997).

Iridoids are fairly widespread in asterids (Figure 9.11); however, evolution and diversification of iridoids and related alkaloids are too complex to be completely understood on the level of the entire asterids. Nevertheless, some patterns are evident. The presence of iridoids is reconstructed as the ancestral state in at least three of the four subclades of asterids. However, whether iridoid production is the ancestral state for all asterids, as suggested by Olmstead et al. (1993), cannot be inferred from Albach et al.'s (2001b) analysis (Figure 9.11). Even if iridoids were present in the ancestral asterid, the unsolved question of the immediate sister to the asterids (P. Soltis et al. 1999b; D. Soltis et al. 2000) prevents the designation of iridoids as a synapomorphy for asterids.

Most of the transitions in asterids between iridoid production and lack thereof involve losses of iridoids (Figure 9.11), as in *Hydrostachys* of Cornales, at least four clades in euasterid I, and several clades of euasterid II (with perhaps a subsequent derivation in *Aster*; Wang and Yu 1997). In Ericales, the ancestral state is difficult to infer (it is ambiguous with DELTRAN and "all most parsimonious states" optimizations), and therefore direction of change is less clear; however, iridoid evolution is clearly dynamic within this clade also, with numerous transitions between iridoid production and lack of iridoid production.

Secoiridoids, characterized by a ring fission between C_7 and C_8 and alkaloids derived from them, are mostly derived via loganin and secologanin (Cordell 1974) or their corresponding acids, but variants of this pathway have been recently reported (see Albach et al. 2001b). Secoiridoids and related alkaloids occur in Oleaceae, Gentianales, most Cornales, and many euasterid II (Figure 9.11) and have been reported for *Lamium album*, although these are formed via a different biosynthetic pathway (Damtoft et al. 1992). Aucubin and other iridoid compounds decarboxylated at C_4 seem also to have evolved independently in Lamiales, Garryales, and some Ericaceae (see Albach et al. 2001b). Thus, presence of aucubin alone is not characteristic of any particular clade.

Floral symmetry

Most asterids have zygomorphic (bilaterally symmetric, monosymmetric) flowers. However, actinomorphic (radially symmetric, polysymmetric) flowers are found in several lineages, including most Rubiaceae, Apocynaceae, Apiales, Cornales, and Ericales (see Figure 12.3). Given the wealth of floral developmental genetic data available for *Antirrhinum* (Plantaginaceae), asterids present ideal opportunities for investigations of evolutionary changes in floral symmetry (e.g., Endress 1992a, 1999; Donoghue et al. 1998; Reeves and Olmstead 1998). Donoghue et al. (1998) examined the evolution of floral symmetry in the asterids by grafting subclades of major subgroups of asterids onto a backbone tree based on *rbcL* sequences (Olmstead et al. 1993) and mapping character states using MacClade (Maddison and Maddison 1992). According to their analysis, the ancestral condition for asterids was an actinomorphic flower. Zygomorphic flowers originated independently at least eight times in asterids, in Asterales, Dipsacales, Solanales, Ericales, and Lamiales.

There are also clear examples of reversals from zygomorphy to actinomorphy within asterids. The greatest number of such shifts occurs in Lamiales, providing a model clade for more detailed examination of these reversals. Donoghue et al. (1998) therefore examined floral symmetry in more detail in Lamiales, a clade in which actinomorphic flowers are apparently derived from zygomorphic ancestors. Their most parsimonious reconstruction of ancestral states (with equal costs of going from zygomorphy to actinomorphy and vice-versa) indicates at least seven reversals from zygomorphy to actinomorphy. We reanalyzed floral symmetry in Lamiales in light of more recent phylogenetic hypotheses (e.g.,

D. Soltis et al. 2000; Olmstead et al. 2001; B. Bremer et al. 2002), including additional actinomorphic taxa (*Sibthorpia*, Plantaginaceae; *Ramonda*, Gesneriaceae), and obtained evidence for additional reversals to actinomorphy in these genera. Actinomorphic flowers occur in other Gesneriaceae, and more detailed character-state mappings of the family are needed to elucidate transitions between zygomorphy and actinomorphy. An additional reversal to near actinomorphy may have occurred in *Byblis* (Figure 9.12).

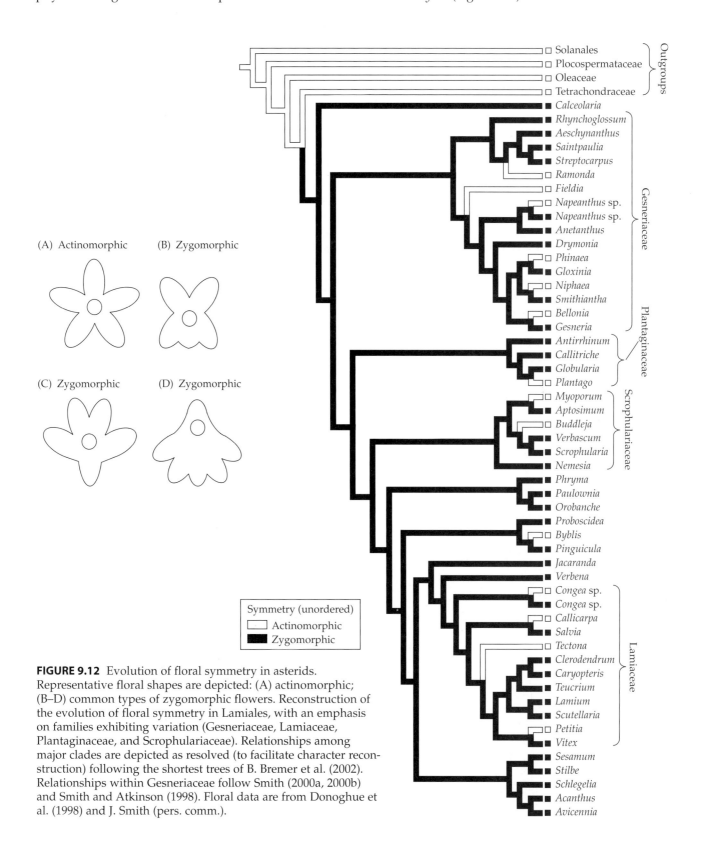

FIGURE 9.12 Evolution of floral symmetry in asterids. Representative floral shapes are depicted: (A) actinomorphic; (B–D) common types of zygomorphic flowers. Reconstruction of the evolution of floral symmetry in Lamiales, with an emphasis on families exhibiting variation (Gesneriaceae, Lamiaceae, Plantaginaceae, and Scrophulariaceae). Relationships among major clades are depicted as resolved (to facilitate character reconstruction) following the shortest trees of B. Bremer et al. (2002). Relationships within Gesneriaceae follow Smith (2000a, 2000b) and Smith and Atkinson (1998). Floral data are from Donoghue et al. (1998) and J. Smith (pers. comm.).

Donoghue et al. (1998) also explored possible explanations for these reversals to actinomorphy. *Plantago* is an excellent organism in which to examine these shifts in more detail because it is well nested within a zygomorphic clade. This and other reversals to actinomorphy often co-occur with shifts from pentamery to tetramery and reduction in flower size. Other examples include reversals to actinomorphy in *Callicarpa* and *Petitia*. A likely explanation is that actinomorphy arises in these taxa by union of the adaxial petals, accompanied by a shift in the orientation of the two lateral petals. Some derived actinomorphic flowers resemble mutants caused by loss-of-function mutations in genes such as *CYCLOIDEA* (*CYC*), a gene that is needed for normal development of zygomorphic flowers (Carpenter and Coen 1990). Actinomorphic flowers in several genera of Gesneriaceae (e.g., *Ramonda*) and *Sibthorpia* (Plantaginaceae) may have arisen via this mechanism (Figure 9.12). Detailed developmental and genetic investigations are needed at several levels in the asterid clade. First, such analyses may elucidate the different origins of zygomorphic flowers within the asterids. Second, investigations of different types of actinomorphic flowers in Lamiales and their close zygomorphic relatives are needed to elucidate pathways to actinomorphy.

Future Research

Recent studies have greatly clarified the overall pattern of asterid relationships; as a result, asterids are much better understood than rosids. However, phylogenetic studies are still needed at multiple levels to clarify several questions. At deep levels, relationships among Asterales, Apiales, Dipsacales, and smaller clades within euasterid II are poorly understood, as are those among Lamiales, Solanales, Gentianales, Garryales, Boraginaceae, Vahliaceae, Icacinaceae, and Oncothecaceae within euasterid I. On a broader scale, placement of Paracryphiaceae among asterids in general is uncertain, although it may belong to euasterid II (B. Bremer et al. 2002). Relationships within many clades recognized as orders require additional investigation, including Ericales and Lamiales. The extent of polyphyly and relationships of the constituent lineages of the former Icacinaceae also require detailed study. Further studies of nonmolecular characters are needed to ascertain synapomorphies for clades of asterids at several levels. Furthermore, nonmolecular synapomorphies uniting euasterid I and euasterid II, as well as some of the subclades recognized as orders, have not been identified.

10

Angiosperm Classification

Introduction

Until very recently, m ost investigators, systematists and nonsystematists alike, consulted one of a few recent classifications of angiosperms (e.g., Dahlgren 1980; Cronquist 1981; Takhtajan 1980, 1987, 1997; Thorne 1992a, 1992b, 2001). Dahlgren et al.'s (1985) classification of monocots remained the system of choice for more than a decade. Although the systems are generally similar, they differed in several respects. For example, Takhtajan established many small families and orders not recognized by other authors (e.g., Cronquist 1981; Thorne 1992a, 1992b, 2001; Heywood 1993). These recent classifications were based chiefly on morphological characters. Only Dahlgren (1980) used chemical data to a significant extent, and Thorne's (1992a, 1992b, 2001) updates incorporated information from molecular phylogenetics.

Relationships among major groups of taxa (e.g., orders, subclasses) were often depicted in terms of phylogenetic shrubs or "bubble" diagrams that represented the author's intuitive ideas of evolutionary relationships (see Chapter 2 and Figure 2.4; Judd et al. 2002). In this sense, these classifications, and those tracing back to Bessey (1915), are phylogenetic. However, the inferences of phylogeny were not based on topologies derived from the explicit phylogenetic analysis of either DNA or non-DNA characters. Irreconcilable differences among these classifications reflected subjective weights of characters imposed by the authors. Nonetheless, these modern classifications represented tremendous improvements over earlier classifications in many ways. They contained more information on obscure taxa, and more characters were described. The classifications of Cronquist (1981), Takhtajan (1997), Thorne (1992a, 1992b, 2001), and Heywood (1993, 1998) in particular continue to receive wide use, particularly among nonsystematists.

The past decade has seen enormous strides in reconstruction of angiosperm phylogenetic relationships at all taxonomic levels, from species to higher levels (e.g., Chase and Albert 1998; Chase 2001; Schaal and Leverich 2001; P. Soltis and Soltis 2001; Sytsma and Pires 2001; P. Soltis et al. 2004); in this chapter we focus primarily on developments at the family level and above. As reviewed in Chapter 2, broad molecular phylogenetic analyses provided the general framework of angiosperm relationships, and major clades of angiosperm families were also identified. Although many relationships inferred from molecular phylogenetic analyses were congruent with the modern classifications (e.g., Cronquist 1981; Thorne 1992a, 1992b, 2001; Takhtajan 1997), others conflicted with them.

The Angiosperm Phylogeny Group (APG) Classification

It became clear that no previous classification system adequately reflected the relationships of angiosperms as reconstructed by molecular phylogenetics and that a dramatic change was needed. The molecular trees became the stimulus for a new classification proposed by a consortium of angiosperm systematists, the Angiosperm Phylogeny Group (APG). As a result of the APG's efforts, angiosperms became the first large clade of organisms to be reclassified on the basis of explicit phylogenetic analysis of DNA sequence data. Furthermore, the first APG classification developed by 29 systematists (APG 1998) represented a landmark in the history of angiosperm systematics because for the first time a classification was provided not by one or a few individuals, as had been the case for several hundred years, but by a broad representation of the systematics community.

The first APG system recognized 462 families placed in 40 putatively monophyletic orders and a few monophyletic higher groups. The general strategy of the APG authors, as discussed in APG (1998), was to focus on orders and, to a lesser extent, on families. In general, orders were quite widely circumscribed, especially in comparison with those of Takhtajan (1997). Several formal clades above the level of order were also recognized, using informal names rather than Linnaean ranks: monocots, eudicots, core eudicots, rosids, and asterids, as well as two major subclades of the rosid clade (eurosid I, or fabids, and eurosid II, or malvids; see Chapter 8) and the asterid clade (euasterid I, or lamiids, and euasterid II, or campanulids; see Chapter

9). The APG authors conservatively recognized only those clades that received moderate to strong support in molecular phylogenetic analysis of either *rbcL* gene sequences (Källersjö et al. 1998; Chase and Albert 1998) or an initial analysis of a three-gene dataset (D. Soltis et al. 1998).

In APG (1998), many families were not placed in an order because their positions were unknown or not strongly supported. When possible, these families of uncertain position were simply placed in the larger clade in which they appeared (e.g., asterids, rosids). For example, Adoxaceae, Bruniaceae, Icacinaceae, Tribelaceae, and several other families were placed within the euasterid II clade, but the positions of these families within euasterid II were unclear.

Some families could not be reliably placed in any supraordinal group and were therefore placed at the end of the classification in a list of eudicot families of "uncertain position." In APG (1998), 25 families could not be placed with confidence within any major clade. Other families could be placed within major clades, but not within any well-supported subclade (designated as orders) within that clade.

Monophyletic groups were also recognized at the family level in APG (1998). In many cases, family circumscriptions were unchanged or modified only slightly from longstanding concepts. In other cases, however, dramatic restructuring of familial boundaries was required. The APG (1998) authors also recognized the limitations of molecular analyses available at that time. Reclassification of some families required further molecular phylogenetic study that used additional taxon sampling; still other families awaited phylogenetic analysis.

APG (1998) provided names for previously unrecognized clades and much-needed nomenclatural stability. With the explosion of molecular systematic endeavors, different authors were using different names for the same clade. For example, the DNA-based asterid clade was variously referred to as Asteridae, Asteridae s.l., asterids, and euasterids.

The APG (1998) authors recognized that changes in familial circumscription and recognition of new orders would ultimately be needed, requiring a revised APG classification. What was surprising, however, was how quickly such an updated APG classification was needed. After publication of APG (1998), rapid advances in angiosperm phylogenetics continued (see Chapter 2). Completion of multigene analyses for all major groups of angiosperms (e.g., Savolainen et al. 2000a; D. Soltis et al. 2000), as well as more focused studies of monocots (e.g., Chase et al. 2000a, 2004), asterids (Albach et al. 2001a; B. Bremer et al. 2002), and many orders and families (see other chapters of this book), necessitated an update of the APG (1998) classification, resulting in APG II (2003). APG II involved fewer authors (seven) than the initial APG (1998). With the changes provided in APG II, the number of angiosperm orders increased from 40 to 44, and the number of families decreased from 462 to 397, if all suggestions to combine families are followed. In many instances, APG II suggested that sister families be combined; for example, Platanaceae and Proteaceae (Proteaceae), Nymphaeaceae and Cabombaceae (Nymphaeaceae), and Illiciaceae and Schisandraceae (Schisandraceae). Many investigators may choose not to follow all these broader circumscriptions.

The APG II (2003) classification is provided at the end of this chapter, and the specific taxonomic changes are described and justified in detail in APG II (2003). Note that in APG II, some families still remain unplaced. For example, Aphloiaceae, Ixerbaceae, Geissolomataceae, Picramniaceae, Strasburgeriaceae, and Vitaceae remain unplaced to order in the rosids, and Bruniaceae, Columelliaceae, and Eremosynaceae are three of several families unplaced to order in euasterid II. In APG II, the placement of some taxa is considered uncertain across the angiosperms as a whole (e.g., Balanophoraceae, Rafflesiaceae, *Cytinus*). However, the molecular phylogenetic analysis of Barkman et al. (2004) has placed Rafflesiaceae with Malpighiales in the rosid clade. Other results indicate that Balanophoraceae and *Cytinus* may soon be placed within the angiosperms, as well (see Chapter 11). Hence, although additional updates of the APG classification will ultimately be required, modifications will likely be relatively minor, focusing on the placement of currently unplaced taxa.

Ranked versus Rank-Free Classification

For more than 200 years, the Linnaean hierarchy (e.g., kingdom, phylum, class, order, family, genus, species), published in 1753, has been the foundation of taxonomy. However, the Linnaean hierarchy was established a century before publication of Darwin's (1859) *Origin of Species*, which introduced the principle of descent with modification. During the past decade, use of the Linnaean hierarchy has been challenged by investigators who seek to apply the principle of descent to the field of nomenclature. In fact, it has been suggested that the Linnaean hierarchy be abandoned.

De Queiroz (1997) made a distinction between *taxonomic* and *nomenclatural* systems. A Linnaean system, whether of taxonomy or nomenclature, is based on the Linnaean hierarchy, whereas a phylogenetic system, whether of taxonomy or nomenclature, is based on the principle of descent (de Queiroz and Gauthier 1992; de Queiroz 1997). Following this terminology, *phylogenetic taxonomy* represents only one branch of *phylogenetic systematics*; it deals with representation (rather than reconstruction) of phylogenetic relationships. The goal of *phylogenetic taxonomy* is to produce classifications that reflect phylogenetic relationships accurately and efficiently. A *phylogenetic system of nomenclature* is a body of principles and rules (nomenclatural conventions) governing taxonomic practice, "the components of which are unified by their relation to the central tenet of evolutionary descent" (de Queiroz and Gauthier 1992).

Hennig, well-known for introducing principles of phylogenetic reconstruction (Hennig 1950, 1965, 1966), also initiated the development of a phylogenetic system of nomenclature (Hennig 1969, 1981, 1983), and many other authors subsequently contributed to the development of phylogenetic taxonomy. In a series of papers, de Queiroz and Gauthier (1990, 1992, 1994) attempted to formulate nomenclatural conventions based on the principle of descent. Specific rules and recommendations of phylogenetic taxonomy are still being developed (e.g., de Queiroz and Gauthier 1990, 1992, 1994; Bryant 1994, 1996; Sundberg and Pleijel 1994; de Queiroz 1996; Lee 1996; Cantino et al. 1997; www.ohio.edu/phylocode/).

Until recently, most phylogenetic systems of classification have attempted to use the basic conventions of the Linnaean hierarchy. As several authors have noted, the familiar set of nested categories offered by the Linnaean system of classification (kingdom, phylum, class, order, etc.) can be used in phylogenetic taxonomy to provide information on rank (that is, the relative nested positions of clades). However, Linnaean categories are not really needed to provide this information to the reader. Linnaean categories do not contain any information about ancestry that is not already present in a branching diagram or in an indented list of names. Furthermore, assignment of taxa to categories is, alone, insufficient information to specify relationships. That is, knowing that one taxon is a family and another an order does not indicate that the family in question is nested within that given order.

Although Linnaean categories can be used in such a way that they are consistent with phylogenetic taxonomy, several problems have been cited (reviewed in de Queiroz and Gauthier 1992). One problem is that of "mandatory categories." That is, certain categories (kingdom, phylum, class, order, family, and genus) are mandatory with the Linnaean system. Every named species must be assigned to a taxon at every one of these taxonomic levels, causing a proliferation of names in the case of monotypic taxa (see below). Ancestors cannot be assigned to monophyletic taxa less inclusive than those originating with them. These ancestors are not part of clades that are less inclusive than the one stemming from that ancestral node or population. For example, the stem species of the angiosperms would be included in Magnoliophyta, but not in any of the subgroups of Magnoliophyta.

Another difficulty with the Linnaean system is the proliferation of categories that occurs as a direct result of the attempt to reflect fine levels of systematic resolution. As an example, if we consider the current summary topology for angiosperms (Figure 10.1), *Amborella* appears as sister to all other angiosperms. Hence, if angiosperms are treated as a phylum (Magnoliophyta), *Amborella* belongs to class Amborellopsida, and all remaining angiosperms, which are the sister to *Amborella*, would be part of a second class (here labeled Euangiopsida). Within the class Euangiopsida, Nymphaeaceae are sister to all other angiosperms. Nymphaeaceae would

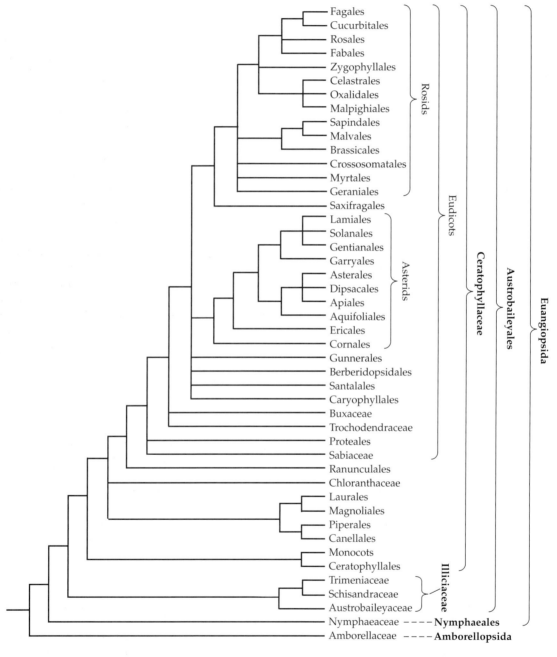

FIGURE 10.1 An example of the proliferation of categories that results from an attempt to reflect fine levels of systematic resolution. The topology used is one possible tree based on a summary of the three-gene analysis of angiosperms (D. Soltis et al. 2000). The names eudicots, rosids, asterids are common names given to clades. Names in bold (e.g., **Illiciaceae**) represent an attempt to place the topology in a Linnaean-based hierarchical classification. We exhaust the taxonomic categories afforded by the Linnaean system very quickly and would have to introduce many more taxonomic categories.

be placed in the order Nymphaeales and their sister group (all remaining angiosperms) would be a separate order, here named Austrobaileyales. As illustrated in Figure 10.1, because the early-branching angiosperms form a graded series, the problem of proliferation of categories would continue. In a short time, we would exhaust the taxonomic categories afforded by the Linnaean system and would need to introduce many additional taxonomic categories. Crane and Kenrick (1997a) similarly noted this problem of proliferation of hierarchical levels in their attempt to translate recent topologies for green plants into a classification that follows the Linnaean system of nomenclature.

Several modifications of the Linnaean system have been proposed to deal with this problem, but none is totally satisfactory. The addition of new taxonomic categories was proposed (e.g., McKenna 1975; Farris 1976; Gaffney and Meylan 1988). Farris, for example, proposed a method for producing new taxonomic categories that used rank-modifying prefixes. This method, although not widely used, could generate many taxonomic categories through the application of single-prefix as well as multi-prefix categories (reviewed in Kron 1997). Other investigators attempted to avoid the proliferation of names and categories by introducing systems that conveyed information regarding phylogenetic relationships without using Linnaean categories. Nelson (1972, 1973) noted that the sequence of taxon names in a list could be used to convey information about phylogenetic relationships. This method, which became known as the "sequencing convention," greatly reduced the number of taxa and categories, but it also left many clades unnamed (reviewed in de Queiroz 1997).

Because of these and other problems associated with the Linnaean system of classification, and because categories themselves are not needed for conveying phylogenetic relationships, some authors have proposed that the best way to implement a phylogenetic taxonomy is to abandon Linnaean categories. Phylogenetic nomenclature differs fundamentally from the Linnaean system (as governed by Botanical, Zoological, and Bacterial Codes of Nomenclature) in that it does not require ranks above the level of species (hence, it is often referred to as *"rank-free"* classification). The idea of "phylogenetic" nomenclature as an alternative to the traditional Linnaean system is not new. Hennig (1969, 1981, 1983), for example, used numerical prefixes rather than Linnaean categories in his taxonomies of insects and chordates. Other investigators subsequently contributed to development of the concept and application of a rank-free classification (Griffiths 1973, 1976; Farris 1976; Løvtrup 1977; Stevens 1984; Ax 1987; Ereshefsky 1994). More recently, de Queiroz and Gauthier (1990, 1992, 1994) further developed a phylogenetic system of classification by discussing specific rules and principles. Their papers sub-

sequently stimulated extensive discussion and debate (e.g., Bryant 1994, 1996; Wyss and Meng 1996; Cantino et al. 1997; de Queiroz 1997; Kron 1997; Cantino 1998, 2000; Hibbett and Donoghue 1998). There have also been strong counter-arguments against phylogenetic nomenclature (e.g., Liden and Oxelman 1996; Brummitt 1997; Stuessy 2000). It is important to stress, for example, that the problems noted with the Linnaean system can be addressed without abandoning the Linnaean hierarchy. Thus, many systematists who readily embrace phylogenetic principles continued to use the Linnaean hierarchy (see Wiley 1979, 1981; Eldredge and Cracraft 1980).

Phylogenetic nomenclature does not mandate that taxa be assigned to ranked categories such as genus, family, or order. Names such as Lamiaceae, Rosales, or Caryophyllales can be used, following phylogenetic nomenclature, without implying any taxonomic category; they are simply names of clades that do not have to correspond to ranks within a taxonomic hierarchy. Elimination of Linnaean ranks is not as dramatic a change as it might seem. For example, the APG classification of angiosperms is, in part, rank-free; above the category of order, the Linnaean hierarchy is not followed. The intent of the APG editors was not to use a rank-free system, but rather to name formally these higher levels only when their inter-relationships become clear. An advantage of phylogenetic nomenclature is that newly discovered clades can be named without changing the names of other taxa. A major problem with the Linnaean system is that ranks of taxa, and as a result names given to taxa, depend on their relative position to other groups of taxa. Thus, naming a new clade can require a cascading series of name changes (Kron 1997; Cantino 1998; Hibbett and Donoghue 1998). A code of phylogenetic nomenclature, referred to as the PhyloCode, has been developed, although it is still considered a work in progress (www.ohiou.edu/phylocode/).

Other authors have proposed a compromise method of classification that includes elements of both Linnaean and phylogenetic nomenclature. For example, Sennblad and Bremer (2002) proposed a nomenclatural system for Apocynaceae that includes "standard" Linnaean tribal names, but, above that level, uses a variant of the definitions from the PhyloCode (Figure 10.2) for an infrafamilial classification. Four rankless taxa are recognized, each ending in "ina" and each progressively more inclusive (Asclepiadoidina, Asclepiadacina, Euapocynoidina, Apocynoidina).

The APG (1998) and APG II (2003) classifications specifically assigned familial and ordinal names and hence are clearly Linnaean in intent. However, "families" and "orders" directly correspond to monophyletic groups and therefore could equally be incorporated into a rank-free system of classification. Above the level of order, a rank-free system of classification has quiet-

(A)

(B)

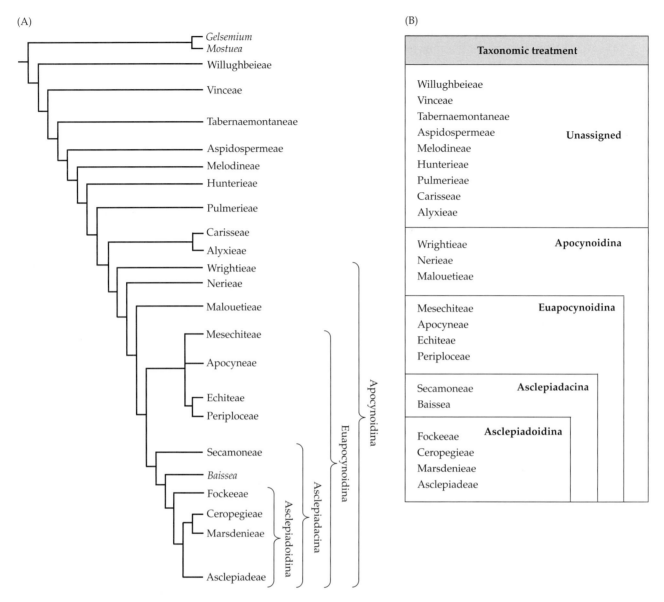

FIGURE 10.2 A "compromise" reclassification of the Apocynaceae. Standard tribal names (Asclepiadeae, etc.) are given and these follow the Linnaean system. However, four nonstandard taxa, Apocynoidina, Euapocynoidina, Asclepiadacina, and Asclepiadoidina, are also indicated (see text). (A) Nonstandard names superimposed on phylogenetic tree of Apocynaceae. (B) Phylogenetic treatment in table form. (Modified from Sennblad and Bremer 2002.)

ly slipped into wide usage in the angiosperms, even among ardent advocates of Linnaean classification. Thus, plant systematists readily use terms such as monocot, eudicot, core eudicot, rosid, asterid, and eu-asterid I, all of which appear in the APG classifications and represent informal, yet highly meaningful, names for well-defined and well-supported clades. Forcing a Linnaean system of ranks and names onto the existing phylogenetic framework for angiosperms above the level of order would not help but hinder angiosperm systematics. That is, following the Linnaean system of classification and naming the asterid or rosid clades As-teridae and Rosidae, respectively, would force a cas-

cade of name changes throughout the angiosperms, as well as establishment of new ranks to accommodate sister groups. Furthermore, many clade names, such as the term "eudicots," have become widely used, even among nonspecialists. What benefit would accrue, for example, from naming the eudicots as a class, and what implications would this change have for the nomenclature of the remaining angiosperms that form a grade basal to the eudicots? Thus, angiosperm systematics currently has a hybrid system with orders, families, genera, and species following the Linnaean system of classification, but a rank-free system above the level of order.

Although some investigators have been strong proponents of a rank-free system at even the species level (e.g., Mishler 1999), it seems unlikely that a rank-free system will be widely used by angiosperm systematists at the level of genus and below in the near future. The Linnaean system seems entrenched at lower levels (family and below), but rankless taxon names continue to be proposed and will likely continue to be used more and more frequently by plant systematists.

Classification of Angiosperms (Modified from APG II 2003)

This section lists the angiosperms as classified in APG II (2003). The symbols used in the classification are as follows: * new family placement; † newly recognized order for the APG system; § new family circumscription (described in APG II 2003).

Names in brackets are optional; for example, APG II suggested the use of a broadly defined Nymphaeaceae, but an alternative is to recognize a narrowly defined Nymphaeaceae and Cabombaceae. Similarly, under Austrobaileyales, APG II proposed a broadly defined Schisandraceae that includes Illiciaceae, but as an option, the recognition of Schisandraceae in a narrow sense and Illiciaceae is provided.

For more detail on ordinal names and synonyms, see APG II (2003); see also www.mobot.org/MOBOT/research/APweb/.

Amborellaceae Pichon (1948), nom. cons.
Chloranthaceae R.Br. ex Sims (1820), nom. cons.
Nymphaeaceae Salisb. (1805), nom. cons.
[+ Cabombaceae Rich. ex A.Rich. (1822), nom. cons.]

† Austrobaileyales Takht. ex Reveal (1992)
Austrobaileyaceae (Croizat) Croizat (1943), nom. cons.
§ Schisandraceae Blume (1830), nom. cons.
[+ Illiciaceae A.C.Sm. (1947), nom. cons.]
Trimeniaceae L.S.Gibbs (1917), nom. cons.

Ceratophyllales Bisch. (1839)
Ceratophyllaceae Gray (1821), nom. cons.

MAGNOLIIDS

† Canellales Cronquist (1957)
Canellaceae Mart. (1832), nom. cons.
Winteraceae R.Br. ex Lindl. (1830), nom. cons.

Laurales Perleb (1826)
Atherospermataceae R.Br. (1814)
Calycanthaceae Lindl. (1819), nom. cons.
Gomortegaceae Reiche (1896), nom. cons.

Hernandiaceae Bercht. & J.Presl (1820), nom. cons.
Lauraceae Juss. (1789), nom. cons.
Monimiaceae Juss. (1809), nom. cons.
Siparunaceae (A.DC.) Schodde (1970)

Magnoliales Bromhead (1838)
Annonaceae Juss. (1789), nom. cons.
Degeneriaceae I.W.Bailey & A.C.Sm. (1942), nom. cons.
Eupomatiaceae Endl. (1841), nom. cons.
Himantandraceae Diels (1917), nom. cons.
Magnoliaceae Juss. (1789), nom. cons.
Myristicaceae R.Br. (1810), nom. cons.

Piperales Dumort. (1829)
Aristolochiaceae Juss. (1789), nom. cons.
* Hydnoraceae C.Agardh (1821), nom. cons.
Lactoridaceae Engl. (1888), nom. cons.
Piperaceae Bercht. & J.Presl (1820), nom. cons.
Saururaceae Martynov (1820), nom. cons.

MONOCOTS

§ Petrosaviaceae Hutch. (1934), nom. cons.

Acorales Reveal (1996)
Acoraceae Martynov (1820)

Alismatales Dumort. (1829)
Alismataceae Vent. (1799), nom. cons.
Aponogetonaceae J.Agardh (1858), nom. cons.
Araceae Juss. (1789), nom. cons.
Butomaceae Mirb. (1804), nom. cons.
Cymodoceaceae N.Taylor (1909), nom. cons.
Hydrocharitaceae Juss. (1789), nom. cons.
Juncaginaceae Rich. (1808), nom. cons.
Limnocharitaceae Takht. ex Cronquist (1981)
Posidoniaceae Hutch. (1934), nom. cons.
Potamogetonaceae Rchb. (1828), nom. cons.
Ruppiaceae Horan. (1834), nom. cons.
Scheuchzeriaceae F.Rudolphi (1830), nom. cons.
Tofieldiaceae Takht. (1995)
Zosteraceae Dumort. (1829), nom. cons.

Asparagales Bromhead (1838)
§ Alliaceae Batsch ex Borkh. (1797), nom. cons.
[+ Agapanthaceae F.Voigt (1850)]
[+ Amaryllidaceae J.St.-Hil. (1805), nom. cons.]
§ Asparagaceae Juss. (1789), nom. cons.
[+ Agavaceae Dumort. (1829), nom. cons.]
[+ Aphyllanthaceae Burnett (1835)]
[+ Hesperocallidaceae Traub (1972)]
[+ Hyacinthaceae Batsch ex Borkh. (1797)]
[+ Laxmanniaceae Bubani (1901-1902)]
[+ Ruscaceae Spreng. (1826), nom. cons.]
[+ Themidaceae Salisb. (1866)]

Asteliaceae Dumort. (1829)
Blandfordiaceae R.Dahlgren & Clifford (1985)
Boryaceae (Baker) M.W.Chase, Rudall & Conran (1997)
Doryanthaceae R.Dahlgren & Clifford (1985)
Hypoxidaceae R.Br. (1814), nom. cons.
Iridaceae Juss. (1789), nom. cons.
Ixioliriaceae Nakai (1943)
Lanariaceae H. Huber ex R. Dahlgren & A.E. vanWyk(1988)
Orchidaceae Juss. (1789), nom. cons.
Tecophilaeaceae Leyb. (1862), nom. cons.
§ Xanthorrhoeaceae Dumort. (1829), nom. cons.
[+ Asphodelaceae Juss. (1789)]
[+ Hemerocallidaceae R.Br. (1810)]
Xeronemataceae M.W. Chase, Rudall & M.F. Fay (2001)

Dioscoreales Hook.f. (1873)
§ Burmanniaceae Blume (1827), nom. cons.
§ Dioscoreaceae R.Br. (1810), nom. cons.
* Nartheciaceae Fr. ex Bjurzon (1846)

Liliales Perleb (1826)
Alstroemeriaceae Dumort. (1829), nom. cons.
Campynemataceae Dumort. (1829)
Colchicaceae DC. (1804), nom. cons.
* Corsiaceae Becc. (1878), nom. cons.
Liliaceae Juss. (1789), nom. cons.
Luzuriagaceae Lotsy (1911)
Melanthiaceae Batsch ex Borkh. (1796), nom. cons.
Petermanniaceae Hutch. (1934), nom. cons.
Philesiaceae Dumort. (1829), nom. cons.
Rhipogonaceae Conran & Clifford (1985)
Smilacaceae Vent. (1799), nom. cons.

Pandanales Lindl. (1833)
Cyclanthaceae Poit. ex A.Rich. (1824), nom. cons.
Pandanaceae R.Br. (1810), nom. cons.
Stemonaceae Caruel (1878), nom. cons.
* Triuridaceae Gardner (1843), nom. cons.
Velloziaceae Hook. (1827), nom. cons.

COMMELINIDS
Dasypogonaceae Dumort. (1829)

Arecales Bromhead (1840)
Arecaceae Schultz Sch. (1832), nom. cons.

Commelinales Dumort. (1829)
Commelinaceae Mirb. (1804), nom. cons.
Haemodoraceae R.Br. (1810), nom. cons.
* Hanguanaceae Airy Shaw (1964)
Philydraceae Link (1821), nom. cons.

Pontederiaceae Kunth (1816), nom. cons.

Poales Small (1903)
Anarthriaceae D.F.Cutler & Airy Shaw (1965)
* Bromeliaceae Juss. (1789), nom. cons.
Centrolepidaceae Endl. (1836), nom. cons.
Cyperaceae Juss. (1789), nom. cons.
Ecdeiocoleaceae D.F.Cutler & Airy Shaw (1965)
Eriocaulaceae Martynov (1820), nom. cons.
Flagellariaceae Dumort. (1829), nom. cons.
Hydatellaceae U.Hamann (1976)
Joinvilleaceae Toml. & A.C.Sm. (1970)
Juncaceae Juss. (1789), nom. cons.
* Mayacaceae Kunth (1842), nom. cons.
Poaceae (R.Br.) Barnh. (1895), nom. cons.
* Rapateaceae Dumort. (1829), nom. cons.
Restionaceae R.Br. (1810), nom. cons.
§ Thurniaceae Engl. (1907), nom. cons.
Typhaceae Juss. (1789), nom. cons.
§ Xyridaceae C.Agardh (1823), nom. cons.

Zingiberales Griseb. (1854)
Cannaceae Juss. (1789), nom. cons.
Costaceae Nakai (1941)
Heliconiaceae Nakai (1941)
Lowiaceae Ridl. (1924), nom. cons.
Marantaceae R.Br. (1814), nom. cons.
Musaceae Juss. (1789), nom. cons.
Strelitziaceae Hutch. (1934), nom. cons.
Zingiberaceae Martynov (1820), nom. cons.

EUDICOTS
§ Buxaceae Dumort. (1822), nom. cons.
[+ Didymelaceae Leandri (1937)]
Sabiaceae Blume (1851), nom. cons.
Trochodendraceae Eichler (1865), nom. cons.
[+ Tetracentraceae A.C.Sm. (1945), nom. cons.]

Proteales Dumort. (1829)
Nelumbonaceae Bercht. & J.Presl (1820), nom. cons.
§ Proteaceae Juss. (1789), nom. cons.
[+ Platanaceae T.Lestib. (1826), nom. cons.]

Ranunculales Dumort. (1829)
Berberidaceae Juss. (1789), nom. cons.
Circaeasteraceae Hutch. (1926), nom. cons.
[+ Kingdoniaceae A.S.Foster ex Airy Shaw (1964)]
Eupteleaceae K.Wilh. (1910), nom. cons.
Lardizabalaceae R.Br. (1821), nom. cons.
Menispermaceae Juss. (1789), nom. cons.
Papaveraceae Juss. (1789), nom. cons.
[+ Fumariaceae Bercht. & J.Presl (1820), nom. cons.]

[+ Pteridophyllaceae (Murb.) Nakai ex Reveal & Hoogland (1991)]
Ranunculaceae Juss. (1789), nom. cons.

CORE EUDICOTS

Aextoxicaceae Engl. & Gilg (1920), nom. cons.
Berberidopsidaceae Takht. (1985)
Dilleniaceae Salisb. (1807), nom. cons.

† Gunnerales Takht. ex Reveal (1992)
§ Gunneraceae Meisn. (1842), nom. cons.
[+ Myrothamnaceae Nied. (1891), nom. cons.]

Caryophyllales Perleb (1826)
Achatocarpaceae Heimerl (1934), nom. cons.
Aizoaceae Martynov (1820), nom. cons.
Amaranthaceae Juss. (1789), nom. cons.
Ancistrocladaceae Planch. ex Walp. (1851), nom. cons.
Asteropeiaceae (Szyszyl.) Takht. ex Reveal & Hoogland (1990)
* Barbeuiaceae Nakai (1942)
Basellaceae Raf. (1837), nom. cons.
Cactaceae Juss. (1789), nom. cons.
Caryophyllaceae Juss. (1789), nom. cons.
Didiereaceae Radlk. (1896), nom. cons.
Dioncophyllaceae Airy Shaw (1952), nom. cons.
Droseraceae Salisb. (1808), nom. cons.
Drosophyllaceae Chrtek, Slavíková & Studnicka (1989)
Frankeniaceae Desv. (1817), nom. cons.
* Gisekiaceae Nakai (1942)
Halophytaceae A.Soriano (1984)
Molluginaceae Bartl. (1825), nom. cons.
Nepenthaceae Bercht. & J.Presl (1820), nom. cons.
Nyctaginaceae Juss. (1789), nom. cons.
Physenaceae Takht. (1985)
Phytolaccaceae R.Br. (1818), nom. cons.
Plumbaginaceae Juss. (1789), nom. cons.
Polygonaceae Juss. (1789), nom. cons.
Portulacaceae Juss. (1789), nom. cons.
Rhabdodendraceae Prance (1968)
Sarcobataceae Behnke (1997)
Simmondsiaceae Tiegh. (1899)
Stegnospermataceae Nakai (1942)
Tamaricaceae Bercht. & J.Presl (1820), nom. cons.

Santalales Dumort. (1829)
Olacaceae R.Br. (1818), nom. cons.
Opiliaceae Valeton (1886), nom. cons.
Loranthaceae Juss. (1808), nom. cons.
Misodendraceae J.Agardh (1858), nom. cons.
Santalaceae R.Br. (1810), nom. cons.

Saxifragales Dumort. (1829)
Altingiaceae Horan. (1843), nom. cons.

Aphanopetalaceae Doweld (2001)[1]
Cercidiphyllaceae Engl. (1907), nom. cons.
Crassulaceae J.St.-Hil. (1805), nom. cons.
Daphniphyllaceae Müll.-Arg. (1869), nom. cons.
Grossulariaceae DC. (1805), nom. cons.
§ Haloragaceae R.Br. (1814), nom. cons.
[+ Penthoraceae Rydb. ex Britt. (1901), nom. cons.]
[+ Tetracarpaeaceae Nakai (1943)]
Hamamelidaceae R.Br. (1818), nom. cons.
§ Iteaceae J.Agardh (1858), nom. cons.
[+ Pterostemonaceae Small (1905), nom. cons.]
Paeoniaceae Raf. (1815), nom. cons.
Peridiscaceae Kuhlm. (1950)[2]
Saxifragaceae Juss. (1789), nom. cons.

ROSIDS

Aphloiaceae Takht. (1985)
* Geissolomataceae Endl. (1841)
Ixerbaceae Griseb. (1854)
Picramniaceae Fernando & Quinn (1995)
* Strasburgeriaceae Soler. (1908), nom. cons.
* Vitaceae Juss. (1789), nom. cons.

† Crossosomatales Takht. ex Reveal (1993)
Crossosomataceae Engl. (1897), nom. cons.
Stachyuraceae J.Agardh (1858), nom. cons.
Staphyleaceae Martynov (1820), nom. cons.

Geraniales Dumort. (1829)
Geraniaceae Juss. (1789), nom. cons.
[+ Hypseocharitaceae Wedd. (1861)]
Ledocarpaceae Meyen (1834)
§ Melianthaceae Bercht. & J.Presl (1820), nom. cons.
[+ Francoaceae A.Juss. (1832), nom. cons.]
Vivianiaceae Klotzsch (1836)

Myrtales Rchb. (1828)
Alzateaceae S.A.Graham (1985)
Combretaceae R.Br. (1810), nom. cons.
Crypteroniaceae A.DC. (1868), nom. cons.
Heteropyxidaceae Engl. & Gilg (1920), nom. cons.
Lythraceae J.St.-Hil. (1805), nom. cons.
§ Melastomataceae Juss. (1789), nom. cons.
[+ Memecylaceae DC. (1827), nom. cons.]
Myrtaceae Juss. (1789), nom. cons.
Oliniaceae Arn. (1839), nom. cons.
Onagraceae Juss. (1789), nom. cons.
Penaeaceae Sweet ex Guill. (1828), nom. cons.
Psiloxylaceae Croizat (1960)
Rhynchocalycaceae L.A.S.Johnson & B.G. Briggs (1985)
Vochysiaceae A.St.-Hil. (1820), nom. cons.

EUROSIDS I (FABIDS)
   § * Zygophyllaceae R.Br. (1814), nom. cons.
   [+ Krameriaceae Dumort. (1829), nom. cons.]
   Huaceae A.Chev. (1947)
† Celastrales Baskerville (1839)
   § Celastraceae R.Br. (1814), nom. cons.
   † Lepidobotryaceae J.Léonard (1950), nom. cons.
   Parnassiaceae Martynov (1820), nom. cons.
   [+ Lepuropetalaceae Nakai (1943)]

Cucurbitales Dumort. (1829)
   Anisophylleaceae Ridl. (1922)
   Begoniaceae Bercht. & J.Presl (1820), nom. cons.
   Coriariaceae DC. (1824), nom. cons.
   Corynocarpaceae Engl. (1897), nom. cons.
   Cucurbitaceae Juss. (1789), nom. cons.
   Datiscaceae Bercht. & J.Presl (1820), nom. cons.
   Tetramelaceae Airy Shaw (1964)

Fabales Bromhead (1838)
   Fabaceae Lindl. (1836), nom. cons.
   Polygalaceae Hoffmanns. & Link (1809),
      nom. cons.
   Quillajaceae D.Don (1831)
   Surianaceae Arn. (1834), nom. cons.

Fagales Engl. (1892)
   Betulaceae Gray (1821), nom. cons.
   Casuarinaceae R.Br. (1814), nom. cons.
   Fagaceae Dumort. (1829), nom. cons.
   § Juglandaceae DC. ex Perleb (1818), nom. cons.
   [+ Rhoipteleaceae Hand.-Mazz. (1932)
      nom. cons.]
   Myricaceae A.Rich. ex Kunth (1817), nom. cons.
   Nothofagaceae Kuprian. (1962)
   Ticodendraceae Gómez-Laur. & L.D.Gómez
      (1991)

Malpighiales Mart. (1835)
   § Achariaceae Harms (1897), nom. cons.
   Balanopaceae Benth. & Hook.f. (1880),
      nom. cons.
   * Bonnetiaceae (Bartl.) L.Beauv. ex Nakai (1948)
   Caryocaraceae Voigt (1845), nom. cons.
   § Chrysobalanaceae R.Br. (1818), nom. cons.
   [+ Dichapetalaceae Baill. (1886), nom. cons.]
   [+ Euphroniaceae Marc.-Berti (1989)]
   [+ Trigoniaceae Endl. (1841), nom. cons.]
   § Clusiaceae Lindl. (1836), nom. cons.
   * Ctenolophonaceae (H.Winkl.) Exell &
      Mendonça (1951)
   * Elatinaceae Dumort. (1829), nom. cons.
   § Euphorbiaceae Juss. (1789), nom. cons.
   Goupiaceae Miers (1862)
   Humiriaceae A.Juss. (1829), nom. cons.
   § Hypericaceae Juss. (1789), nom. cons.

Irvingiaceae (Engl.) Exell & Mendonça (1951),
   nom. cons.
* Ixonanthaceae Planch. ex Miq. (1858), nom. cons.
Lacistemataceae Mart. (1826), nom. cons.
§ Linaceae DC. ex Perleb (1818), nom. cons.
* Lophopyxidaceae (Engl.) H.Pfeiff. (1951)
Malpighiaceae Juss. (1789), nom. cons.
§ Ochnaceae DC. (1811), nom. cons.
[+ Medusagynaceae Engl. & Gilg (1924),
   nom. cons.]
[+ Quiinaceae Choisy ex Engl. (1888), nom. cons.]
Pandaceae Engl. & Gilg (1912-1913), nom. cons.
§ Passifloraceae Juss. ex Roussel (1806),
   nom. cons.
[+ Malesherbiaceae D.Don (1827), nom. cons.]
[+ Turneraceae Kunth ex DC. (1828), nom. cons.]
§ Phyllanthaceae Martynov (1820)
§ Picrodendraceae Small (1917), nom. cons.
* Podostemaceae Rich. ex C.Agardh (1822),
   nom. cons
Putranjivaceae Endl. (1841)
§ Rhizophoraceae Pers. (1807), nom. cons.
[+ Erythroxylaceae Kunth (1822), nom. cons.]
§ Salicaceae Mirb. (1815), nom. cons.
Violaceae Batsch (1802), nom. cons.

Oxalidales Heintze (1927)
   § Brunelliaceae Engl. (1897), nom. cons.
   Cephalotaceae Dumort. (1829), nom. cons.
   Connaraceae R.Br. (1818), nom. cons.
   Cunoniaceae R.Br. (1814), nom. cons.
   § Elaeocarpaceae Juss. ex DC. (1816), nom. cons.
   Oxalidaceae R.Br. (1818), nom. cons.

Rosales Perleb (1826)
   Barbeyaceae Rendle (1916), nom. cons.
   § Cannabaceae Martynov (1820), nom. cons.
   Dirachmaceae Hutch. (1959)
   Elaeagnaceae Juss. (1789), nom. cons.
   Moraceae Link (1831), nom. cons.
   Rhamnaceae Juss. (1789), nom. cons.
   Rosaceae Juss. (1789), nom. cons.
   Ulmaceae Mirb. (1815), nom. cons.
   § Urticaceae Juss. (1789), nom. cons.

EUROSIDS II (MALVIDS)
   Tapisciaceae (Pax) Takht. (1987)

Brassicales Bromhead (1838)
   Akaniaceae Stapf (1912), nom. cons.
   [+ Bretschneideraceae Engl. & Gilg (1924),
      nom. cons.]
   Bataceae Perleb (1838), nom. cons.
   Brassicaceae Burnett (1835), nom. cons.
   Caricaceae Dumort. (1829), nom. cons.
   Emblingiaceae Airy Shaw (1964)

Gyrostemonaceae Endl. (1841), nom. cons.
Koeberliniaceae Engl. (1895), nom. cons.
Limnanthaceae R.Br. (1833), nom. cons.
Moringaceae Martynov (1820), nom. cons.
Pentadiplandraceae Hutch. & Dalziel (1928)
Resedaceae Bercht. & J.Presl (1820), nom. cons.
Salvadoraceae Lindl. (1836), nom. cons.
Setchellanthaceae Iltis (1999)
Tovariaceae Pax (1891), nom. cons.
Tropaeolaceae Bercht. & J.Presl (1820), nom. cons.

Malvales Dumort. (1829)
§ Bixaceae Kunth (1822), nom. cons.
[+ Diegodendraceae Capuron (1964)]
[+ Cochlospermaceae Planch. (1847), nom. cons.]
Cistaceae Juss. (1789), nom. cons.
Dipterocarpaceae Blume (1825), nom. cons.
Malvaceae Juss. (1789), nom. cons.
Muntingiaceae C.Bayer, M.W.Chase & M.F.Fay (1998)
Neuradaceae Link (1831), nom. cons.
Sarcolaenaceae Caruel (1881), nom. cons.
Sphaerosepalaceae (Warb.) Tiegh. ex Bullock (1959)
§ Thymelaeaceae Juss. (1789), nom. cons.

Sapindales Dumort. (1829)
Anacardiaceae R.Br. (1818), nom. cons.
Biebersteiniaceae Endl. (1841)
Burseraceae Kunth (1824), nom. cons.
Kirkiaceae (Engl.) Takht. (1967)
Meliaceae Juss. (1789), nom. cons.
§ Nitrariaceae Bercht. & J.Presl (1820), nom. cons.
[+ Peganaceae (Engl.) Tiegh. ex Takht. (1987)]
[+ Tetradiclidaceae (Engl.) Takht. (1986)]
Rutaceae Juss. (1789), nom. cons.
Sapindaceae Juss. (1789), nom. cons.
Simaroubaceae DC. (1811), nom. cons.

ASTERIDS

Cornales Dumort. (1829)
Cornaceae Dumort. (1829), nom. cons.
[+ Nyssaceae Juss. ex Dumort. (1829), nom. cons.]
Curtisiaceae (Engl.) Takht. (1987)
Grubbiaceae Endl. (1839), nom. cons.
Hydrangeaceae Dumort. (1829), nom. cons.
Hydrostachyaceae (Tul.) Engl. (1894), nom. cons.
Loasaceae Juss. (1804), nom. cons.

Ericales Dumort. (1829)
Actinidiaceae Gilg & Werderm. (1825), nom. cons.
Balsaminaceae Bercht. & J.Presl (1820), nom. cons.
Clethraceae Klotzsch (1851), nom. cons.

Cyrillaceae Endl. (1841), nom. cons.
Diapensiaceae Lindl. (1836), nom. cons.
§ Ebenaceae Gürke (1891), nom. cons.
Ericaceae Juss. (1789), nom. cons.
Fouquieriaceae DC. (1828), nom. cons.
Lecythidaceae A.Rich. (1825), nom. cons.
Maesaceae (A.DC.) Anderb., B.Ståhl & Källersjö (2000)
Marcgraviaceae Juss. ex DC. (1816), nom. cons.
§ Myrsinaceae R.Br. (1810), nom. cons.
Pentaphylacaceae Engl. (1897), nom. cons.
[+ Ternstroemiaceae Mirb. ex DC. (1816)]
[+ Sladeniaceae Airy Shaw (1964)]
Polemoniaceae Juss. (1789), nom. cons.
§ Primulaceae Batsch ex Borkh. (1797), nom. cons.
Roridulaceae Bercht. & J.Presl (1820), nom. cons.
Sapotaceae Juss. (1789), nom. cons.
Sarraceniaceae Dumort. (1829), nom. cons.
§ Styracaceae DC. & Spreng. (1821), nom. cons.
Symplocaceae Desf. (1820), nom. cons.
§ Tetrameristaceae Hutch. (1959)
[+ Pellicieraceae (Triana & Planch.) L.Beauvis. ex Bullock (1959)]
Theaceae Mirb. ex Ker Gawl. (1816), nom. cons.
§ Theophrastaceae Link (1829), nom. cons.

EUASTERIDS I (LAMIIDS)

Boraginaceae Juss. (1789), nom. cons.
§ * Icacinaceae (Benth.) Miers (1851), nom. cons.
* Oncothecaceae Kobuski ex Airy Shaw (1964)
Vahliaceae Dandy (1959)
Metteniusaceae H.Karst. ex Schnizl. (1860–1870)

Garryales Lindl. (1846)
Eucommiaceae Engl. (1909), nom. cons.
§ Garryaceae Lindl. (1834), nom. cons.
[+ Aucubaceae J.Agardh (1858)]

Gentianales Lindl. (1833)
Apocynaceae Juss. (1789), nom. cons.
Gelsemiaceae (G.Don) Struwe & V. Albert (1995)
Gentianaceae Juss. (1789), nom. cons.
Loganiaceae R.Br. (1814), nom. cons.
Rubiaceae Juss. (1789), nom. cons.

Lamiales Bromhead (1838)
§ Acanthaceae Juss. (1789), nom. cons.
Bignoniaceae Juss. (1789), nom. cons.
Byblidaceae (Engl. & Gilg) Domin (1922), nom. cons.
Calceolariaceae (D. Don) Olmstead (2001)
* Carlemanniaceae Airy Shaw (1964)
Gesneriaceae Rich. & Juss. ex DC. (1816), nom. cons.
Lamiaceae Martynov (1820), nom. cons.
Lentibulariaceae Rich. (1808), nom. cons.

* Martyniaceae Horan. (1847), nom. cons.
Oleaceae Hoffmanns. & Link (1809), nom. cons.
§ Orobanchaceae Vent. (1799), nom. cons.
Paulowniaceae Nakai (1949)
Pedaliaceae R.Br. (1810), nom. cons.
§ Phrymaceae Schauer (1847), nom. cons.
§ Plantaginaceae Juss. (1789), nom. cons.
* Plocospermataceae Hutch. (1973)
Schlegeliaceae (A.H.Gentry) Reveal (1996)
§ Scrophulariaceae Juss. (1789), nom. cons.
Stilbaceae Kunth (1831), nom. cons.
Tetrachondraceae Wettst. (1924)
Verbenaceae J.St.-Hil. (1805), nom. cons.

Solanales Dumort. (1829)
Convolvulaceae Juss. (1789), nom. cons.
Hydroleaceae Bercht. & J.Presl (1820)
§ Montiniaceae Nakai (1943), nom. cons.
Solanaceae Juss. (1789), nom. cons.
Sphenocleaceae (Lindl.) Baskerville (1839), nom. cons.

EUASTERIDS II (CAMPANULIDS)
Bruniaceae Bercht. & J.Presl (1820), nom. cons.
Columelliaceae D.Don (1828), nom. cons.
[+ Desfontainiaceae Endl. (1841), nom. cons.]
Eremosynaceae Dandy (1959)
Escalloniaceae R.Br. ex Dumort. (1829), nom. cons.
Paracryphiaceae Airy Shaw (1964)
Polyosmaceae Blume (1851)
Sphenostemonaceae P.Royen & Airy Shaw (1972)
Tribelaceae Airy Shaw (1964)

Apiales Nakai (1930)
Apiaceae Lindl. (1836), nom. cons.
Araliaceae Juss. (1789), nom. cons.
Aralidiaceae Philipson & B.C.Stone (1980)
Griseliniaceae J.R.Forst. & G.Forst. ex A.Cunn. (1839)
Mackinlayaceae Doweld (2001)
Melanophyllaceae Takht. ex Airy Shaw (1972)
Myodocarpaceae Doweld (2001)
Pennantiaceae J.Agardh (1858)
Pittosporaceae R.Br. (1814), nom. cons.
Torricelliaceae Hu (1934)

Aquifoliales Senft (1856)
Aquifoliaceae DC. ex A.Rich. (1828), nom. cons.
* § Cardiopteridaceae Blume (1847), nom. cons.
Helwingiaceae Decne. (1836)
Phyllonomaceae Small (1905)
Stemonuraceae (M. Roem.) Kårehed (2001)

Asterales Lindl. (1833)
Alseuosmiaceae Airy Shaw (1964)
Argophyllaceae (Engl.) Takht. (1987)

Asteraceae Martynov (1820), nom. cons.
Calyceraceae R.Br. ex Rich. (1820), nom. cons.
§ Campanulaceae Juss. (1789), nom. cons.
[+ Lobeliaceae Juss. ex Bonpl. (1813), nom. cons.]
Goodeniaceae R.Br. (1810), nom. cons.
Menyanthaceae Bercht. & J.Presl (1820), nom. cons.
Pentaphragmataceae J.Agardh (1858), nom. cons.
Phellinaceae (Loes.) Takht. (1967)
§ Rousseaceae DC. (1839)
Stylidiaceae R.Br. (1810), nom. cons.
[+ Donatiaceae B.Chandler (1911), nom. cons.]

Dipsacales Dumort. (1829)
* Adoxaceae E.Mey. (1839), nom. cons.
§ Caprifoliaceae Juss. (1789), nom. cons.
[+ Diervillaceae (Raf.) Pyck (1998)]
[+ Dipsacaceae Juss. (1789), nom. cons.]
[+ Linnaeaceae (Raf.) Backlund (1998)]
[+ Morinaceae Raf. (1820)]
[+ Valerianaceae Batsch (1802), nom. cons.]

TAXA OF UNCERTAIN POSITION
If an unplaced genus is the type of a family, the family name is given for information purposes.

Aneulophus Benth.
Apodanthaceae Tiegh. ex Takht. in Takht. (1997) [three genera]
Bdallophyton Eichl.
Balanophoraceae Rich. (1822), nom. cons.
Centroplacus Pierre
Cynomorium L. [Cynomoriaceae Lindl. (1833) nom. cons.]
Cytinus L. [Cytinaceae A.Rich. (1824)]—probably in Malvales (see Chapter 11)
Dipentodon Dunn [Dipentodontaceae Merr. (1941), nom. cons.]
Gumillea Ruiz & Pav.
Hoplestigma Pierre [Hoplestigmataceae Engl. & Gilg (1924), nom. cons.]
Leptaulus Benth.
Medusandra Brenan [Medusandraceae Brenan (1952), nom. cons.]
Mitrastema Makino [Mitrastemonaceae Makino (1911), nom. cons.]—in Ericales (see Chapter 11)
Pottingeria Prain [Pottingeriaceae (Engl.) Takht. (1987)]
Rafflesiaceae Dumort. (1829), nom. cons. [three genera included]—in Malpighiales (see Chapter 11)
Soyauxia Oliv.—in Peridiscaceae (Saxifragales)
Trichostephanus Gilg

---

[1] Aphanopetalaceae could be included in Haloragaceae (see Chapter 6).

[2] Peridiscaceae belong in Saxifragales (see Chapter 6).

11

Parallel and Convergent Evolution

Introduction

Clarification of phylogenetic relationships at all levels in the angiosperms has facilitated a more accurate assessment of character evolution. Several morphological and chemical features implied by recent classifications (e.g., Cronquist 1981; Takhtajan 1987, 1997) to have evolved many times are now established as having arisen only once or a few times. For example, glucosinolate production, a chemical defense long thought to have evolved recurrently (Cronquist 1981; but see Dahlgren 1975), was shown to have evolved only twice (once in Brassicales and again in Putranjivaceae of Malpighiales; see Chapter 8 and Rodman et al. 1998). Similarly, symbiosis with nitrogen-fixing bacteria was thought to have evolved many times, but molecular phylogenetic analyses revealed instead that the predisposition for this character evolved only once, in the ancestor of the nitrogen-fixing clade. However, within this one relatively small clade, nitrogen-fixing symbiosis may have evolved several times (see Chapter 8). Other features have clearly evolved repeatedly. In this chapter we provide an overview of three features that are highly prone to parallel or convergent evolution: parasitism, carnivory, and C_4 photosynthesis. Molecular phylogenetic investigations have helped to clarify the extent of parallel or convergent evolution of these traits and also have provided additional evolutionary insights. The possible parallel or convergent evolution of other features, including several chemical characters (e.g., ellagic acid, iridoids), is examined in other chapters (e.g., Chapters 5 and 9).

Parasitism

Not all angiosperms produce all of their own nutrients via photosynthesis (i.e., not all are autotrophic). Some angiosperms are heterotrophic, of which there are two categories: mycotrophs and haustorial parasites (reviewed in Nickrent et al. 1998). Mycotrophs derive nutrients through a symbiotic relationship with mycorrhizae and will not be discussed further here. Haustorial parasites directly penetrate host tissues by way of a haustorium, which is commonly a modified root. Our discussion of the evolution of parasites deals only with haustorial parasites. The degree of nutritional dependence on the host varies among haustorial parasites (Nickrent et al. 1998). Hemiparasites photosynthesize during part of their life cycle and obtain mainly water and dissolved minerals from their hosts. Holoparasites, in contrast, are nonphotosynthetic. Both holoparasites and hemiparasites are discussed to some extent in the chapters on rosids and asterids (see Chapters 8 and 9, respectively).

In our definition, "hemiparasite" as a category contains both facultative (those that can survive without being connected to a host during part of their life cycle, e.g., *Castilleja* of Orobanchaceae) and obligate hemiparasites (those that are incapable of survival if not connected to their host, e.g., *Viscum* of Santalaceae). This altered definition of hemiparasite is thus not completely satisfactory because it lumps together two different phenomena, but at least it is an improvement over the previous use of these terms, which were difficult to apply.

In the APG II (2003) classification, approximately 4,000 species in 265 genera and perhaps 17 families of angiosperms are considered to be parasites (see also Nickrent et al. 1998). Our estimate of the number of parasitic families (Table 11.1) is lower than other recent estimates because of changes in classification in the angiosperms (APG II 2003; some are illustrated in Figure 11.1). Most of these families of parasites are holoparasites. Members of Santalales, Krameriaceae, and one genus of Lauraceae are considered hemiparasites, al-

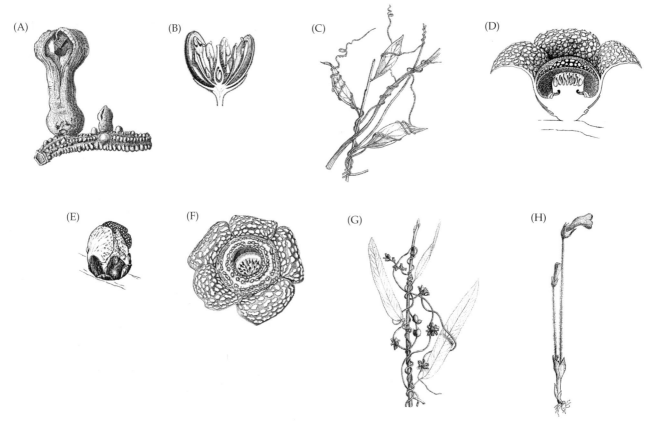

FIGURE 11.1 Flowers of parasitic taxa (members of Santalales are shown in Figure 6.14). (A) *Hydnora africana* (Hydnoraceae), flower and flower bud on host. (B) *Cassytha filiformis* (Lauraceae), flower. (C) *Cassytha americana* (Lauraceae), twining habit. (D) *Rafflesia arnoldii* (Rafflesiaceae), open flower in longitudinal section. (E) *Raffle-sia keithii* (Rafflesiaceae), floral bud, surrounded by bracts, emerging from host. (F) *Rafflesia* *arnoldii* (Rafflesiaceae), open flower. (G) *Cuscuta monogyna* (Convolvulaceae), plant in flower, illustrating twining habit. (H) *Orobanche ramosa* (Orobanchaceae), plant in flower. (A, from Solms in Engler and Prantl 1894. B and C, from Pax in Engler and Prantl 1891. D to F, from Endress 1994c. G, from Peter in Engler and Prantl 1897. H, from Beck in Engler and Prantl 1897.)

TABLE 11.1 Angiosperm families containing parasitic members, with degree of specialization, and general phylogenetic placement in the angiosperms

Family	Number of Parasitic Genera / Total Species	Specialization	Higher-Level Placement
Hydnoraceae	2 / 14–18	Holoparasites	Piperales
Lauraceae	1 / 20	Hemiparasites	Laurales
Cynomoriaceae	1 / 1–2	Holoparasites	Uncertain
Balanophoraceae[a]	11 / 32	Holoparasites	Uncertain
Cytinaceae	1 / 7	Holoparasites	Rosid—Malvales
Rafflesiaceae	3 / 19	Holoparasites	Rosid—Malpighiales
Apodanthaceae	2 / 19	Holoparasites	Uncertain
Krameriaceae	1 / 17	Hemiparasites	Rosid
Mitrastemonaceae	1 / 2	Holoparasites	Asterid—Ericales
Convolvulaceae	1 / 158	Hemiparasites	Asterid—Solanales
Boraginaceae	3 / 5	Holoparasites	Asterid
Orobanchaceae	78 / 1,950	Hemi- and holoparasites	Asterid—Lamiales
Olacaceae	27 / 190	Hemiparasites	Santalales
Misodendraceae	1 / 8	Hemiparasites	Santalales
Loranthaceae	80 / 1,000	Hemiparasites and 1 holoparasite	Santalales
Opiliaceae	9 / 29	Hemiparasites	Santalales
Santalaceae	37 / 480	Hemiparasites and 1 holoparasite	Santalales

[a]Balanophoraceae are broadly defined to include Dactylanthaceae, Sarcophytaceae, and Lophophytaceae.

though single genera in Santalaceae and Loranthaceae appear to be holoparasites (see below, Santalales parasites; see Chapter 6). Orobanchaceae (*sensu* APG II 2003; see Chapter 9) include both hemi- and holoparasites.

Ascertaining the phylogenetic relationships of many parasitic plants has long been problematic because morphological features associated with the parasitic habit are often extensively modified (Nickrent et al. 1998). Modifications include the loss of diverse organs, including leaves, perianth parts, and integuments, as well as the loss of chlorophyll. Some morphological modifications are subject to parallelism and convergence, and similar features have evolved in distantly related groups. It is somewhat surprising, therefore, that the parasitic habit was used by Cronquist (1981) to link Balanophoraceae, Hydnoraceae, and Rafflesiaceae with Santalales in Rosidae.

DNA systematics has not readily resolved the relationships of most holoparasites to other angiosperms, because molecular analyses of plants have, until recently, relied primarily on plastid genes for reconstructing phylogenies. Holoparasitic plants are nonphotosynthetic and often have highly modified plastid genomes that lack many genes or have accelerated rates of molecular evolution and deletions not in triplets (e.g., dePamphilis and Palmer 1990; Nickrent et al. 1998; see Chapters 8 and

9). As a result, plastid genes such as *rbcL* and *atpB,* which have been widely used to infer angiosperm phylogeny, have typically not been of value in placing holoparasites. There are exceptions; sequences of several plastid genes were used to resolve the relationships of Orobanchaceae, parasitic members of Lamiales (Olmstead et al. 2001) and of Burmanniaceae in Dioscoreales (Caddick et al. 2002; parasitic monocots are not covered by this discussion). Although parasitic taxa generally can be placed with nuclear genes, such as 18S and 26S rDNA, and with mtDNA sequences, the phylogenetic placement of many parasitic plants has remained obscure until recently, despite the enormous success of molecular phylogenetics in clarifying the relationships of other taxonomically troublesome plants.

According to results of molecular phylogenetic analyses, parasitism appears to have arisen independently many times in the angiosperms (Figure 11.2), with the following clades, or members thereof, each likely representing a putative separate origin: Convolvulaceae, Boraginaceae, Orobanchaceae, Cynomoriaceae, Cytinaceae, Balanophoraceae, Apodanthaceae, Mitrastemonaceae, Rafflesiaceae, Hydnoraceae, Lauraceae, Krameriaceae, and Santalales (see also Nickrent et al. 1998; Smith et al. 2001; Barkman et al. 2004, in prep.).

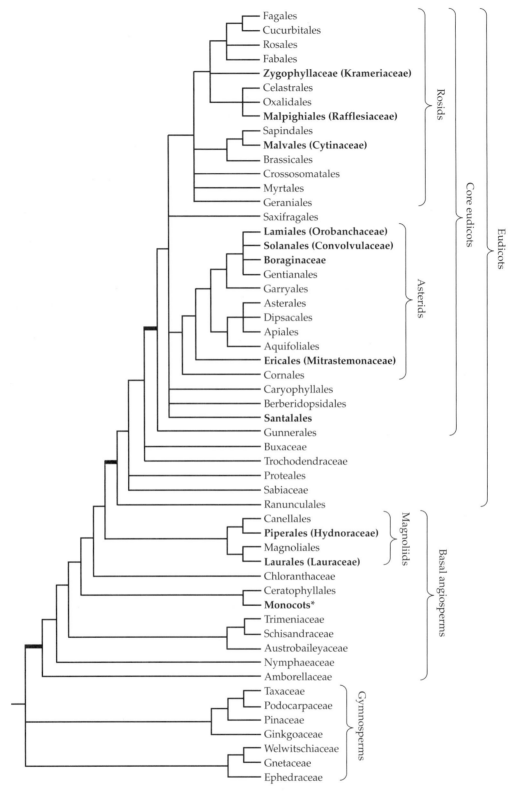

FIGURE 11.2 Major occurrences of parasitism in the angiosperms. Parasitic taxa are shown in bold; * designates parasitic monocots, which are not covered in this discussion. The phylogenetic tree is one possible topology based on the summary topology of D. Soltis et al. (2000) with modifications of the basal angiosperms from Zanis et al. (2002). Three key nodes are indicated with bold lines: angiosperms, eudicots, and core eudicots (all receive 100% support). The relationships of several parasitic families remain uncertain and are not depicted here (see text and Table 11.1).

Basal angiosperm parasites

Hydnoraceae represent a family of holoparasites that is firmly placed among the basal angiosperms (Figure 11.2). To address the question of phylogenetic position of Hydnoraceae among angiosperms, nuclear 18S and 26S rDNA and mitochondrial *atp1* and *matR* sequences were obtained for *Hydnora* and *Prosopanche* (Nickrent et al. 2002) and included in a dataset of these four genes, plus the plastid genes *rbcL* and *atpB* for many taxa. These plastid genes are absent in Hydnoraceae and were therefore coded as missing, but inclusion of these sequences helped to strengthen and resolve the structure of the overall molecular topology (see the review of combining molecular datasets in Chapter 2). Analyses of separate and combined data partitions supported the monophyly of Hydnoraceae and the association of the family with Aristolochiaceae and Lactoridaceae (Nickrent et al. 2002) as part of Piperales (see Chapter 3). The phylogenetic analysis permitted the general placement of Hydnoraceae, but, despite more than 5 kb of sequence data per taxon, the precise relationship of Hydnoraceae is still unclear. It cannot yet be determined whether Aristolochiaceae, Hydnoraceae, and Lactoridaceae should be considered distinct families, or whether Hydnoraceae or Lactoridaceae, or both, were derived from within Aristolochiaceae.

In contrast to most recent classifications, molecular phylogenetic analyses clearly do not suggest a close relationship between Hydnoraceae and Rafflesiaceae (e.g., Cronquist 1981). Molecular data indicate, instead, that Rafflesiaceae in the traditional sense represent a polyphyletic assemblage with constituent members associated with various asterid and rosid families (see below). Thus, several floral morphological features shared by Hydnoraceae and Rafflesiaceae represent independent acquisitions. These features include flowers or inflorescences arising from the host, fleshy flowers, a perianth of a single series (three to five tepals), and a highly modified androecium.

The genus *Cassytha* (Lauraceae) represents another independent origin of the parasitic habit among basal angiosperm lineages (Figure 11.2). The genus of about 20 species of hemiparasites is distinctive in Lauraceae because of its twining, herbaceous habit, whereas other members of Lauraceae are trees or shrubs. Although sometimes placed in its own family, Cassythaceae, most recent treatments have included *Cassytha* in Lauraceae (e.g., Cronquist 1981; Takhtajan 1987, 1997). Takhtajan (1997) placed the genus in its own subfamily within Lauraceae. *Cassytha* is typical of Lauraceae in features other than habit, including floral and fruit morphology, pollen structure, and alkaloid chemistry (Rohwer 1993). Hence, it has long been clear that *Cassytha* is allied with Lauraceae, and this genus therefore differs from many other parasitic lineages that are so highly modified morphologically that their relationships have been difficult, if not impossible, to ascertain from morphology alone.

Santalales parasites

Santalales (reviewed in Chapter 6; plants illustrated in Figure 6.13) provide an excellent model for studying the evolution of parasitism. The early-branching members of the order (some members of the paraphyletic Olacaceae) are nonparasitic (Nickrent and Malécot 2001), whereas other members of Santalales are hemiparasites; *Daenikera* (Santalaceae) and *Tristerix* (= *Macrosolen*; Loranthaceae) are apparently holoparasitic (Hürlimann and Stauffer 1957; Kraus et al. 1995). Parasitism appears to have arisen late in the diversification of the paraphyletic Olacaceae, apparently in the common ancestor of *Malania* and *Ximenia* and their large sister clade (i.e., the remainder of Olacaceae, Misodendraceae, Loranthaceae, Opiliaceae, and Santalaceae; Figure 11.3). Many of the parasitic taxa in Santalales are terrestrial. Aerial parasites (those confined to trees) are also present, and this strategy arose many times (Figure 11.3; Nickrent and Malécot 2001), in *Misodendron*, in Loranthaceae, and at least twice within a broadly defined Santalaceae (APG II 2003). In Santalaceae, aerial parasitism has evolved in those genera previously classified in Eremolepidaceae (i.e., *Antidaphne*, *Eubrachion*, *Lepidoceras*, and *Eremolepis*; the latter genus is not shown in Figure 11.3). Aerial parasitism also arose at least two additional times to account for the occurrence of aerial parasites in other Santalaceae, such as *Dendrotrophe*, *Dufrenoya*, some species of *Exocarpos*, and *Phacellaria* (the latter not shown in Figure 11.3), as well as in *Viscum*, *Arceuthobium*, *Phoradendron*, and *Ginalloa* (former Viscaceae). The number of origins of aerial parasites in Santalaceae will remain unclear until a phylogenetic tree that includes more taxa and more resolution is obtained for this portion of Santalales. For more information on Santalales, see http://www.science.siu.edu/parasitic-plants.

Rosid parasites

Several other lineages of parasitic plants that have been particularly difficult to place using morphological data now appear to be nested within the rosids (Barkman et al. 2004; T.J. Barkman and C.W. dePamphilis, pers. comm.; see also Chapter 8; Figure 11.2). These rosid parasites include some of the best known of the parasitic plants, such as *Rafflesia* and other genera placed in Rafflesiaceae. However, the rosid parasite that has proved to be one of the easiest to place in molecular phylogenetic analyses is *Krameria*. *Krameria* (Krameriaceae) comprises about 15 species of hemiparasitic shrubs and perennial herbs. In modern classifications there has been longstanding debate regarding the closest relatives of

FIGURE 11.3 Evolution of parasitism in Santalales. Topology modified from Nickrent and Malécot (2001).

Krameria. Some authors have allied the genus with Polygalaceae and others with the caesalpinioid legumes (reviewed in Cronquist 1981; Takhtajan 1997). The genus has long been considered a rosid, although its relationships within the rosids were unclear. Plastid sequences were easily obtained for *Krameria,* and the genus was therefore readily included in broad phylogenetic analyses. Molecular data indicate, with strong support, that *Krameria* is sister to Zygophyllaceae and a member of eurosid I (fabids). However, the relationships of this Krameriaceae + Zygophyllaceae clade to other members of eurosid I remain unclear (see Chapter 8).

Rafflesia has been of special interest because *Rafflesia arnoldii* from Sumatra is renowned for having the largest

flower in the world (Figure 11.1, D to F). *Cytinus,* considered to represent a separate family (Cytinaceae) by some investigators, has also been placed in Rafflesiaceae (e.g., Cronquist 1981). Cronquist (1981) also placed Mitrastemonaceae with Rafflesiaceae in his Rafflesiales, close to Santalales in Rosidae. All of these plants lack chlorophyll and are endoparasitic on the roots, or less often on the shoots, of other plants. Only the flowers or the short flowering shoot of the parasite emerge from the host plant (Figure 11.1, A, D, and E). The vegetative body of the plant is largely filamentous and resembles a fungal mycelium. For these reasons, these taxa have been extremely difficult to place. These parasites seem to have modified plastid genomes, and it has not been possible to place them with plastid genes. Barkman et al. (2004), using mtDNA sequence data, found evidence for placing *Rafflesia* and *Rhizanthes* (Rafflesiaceae) as close relatives of the rosid order Malpighiales, a finding reinforced by Davis and Wurdack (2005). This placement of *Rafflesia* and *Rhizanthes* with Malpighiales has not been suggested in previous classifications, although floral similarities between *Rafflesia* and Passifloraceae of Malpighiales had been noted (reviewed in Barkman et al. 2004). *Rafflesia* shares several morphological features with Passifloraceae, including a central column (termed an androgynophore in Passifloraceae) and a corona, characteristics that are unusual in flowering plants (Barkman et al. 2004).

Phylogenetic analysis of mtDNA sequence data has also helped to place Cytinaceae, another family traditionally considered closely related to Rafflesiaceae (e.g., Cronquist 1981; Takhtajan 1997). However, rather than a placement close to Rafflesiaceae, or Malpighiales (see above), molecular data suggest, instead, that *Cytinus* is embedded within Malvales (Barkman et al., in prep.).

The relationships of Apodanthaceae, Cynomoriaceae, and Balanophoraceae have been even more difficult to ascertain than the parasites noted above (T. Barkman and C.W. dePamphilis, pers. comm.). Because the placements of these three families remain tentative, they are not depicted in Figure 11.2. Apodanthaceae, considered part of either Rafflesiaceae (e.g., Cronquist 1981) or Rafflesiales (e.g., Takhtajan 1997) by some, may be a member of Malvales. Recent morphological studies indicate that Apodanthaceae share with Malvales several noteworthy floral structures, including an androecial tube and a trend from normal stamens to synandria without thecal organization (Blarer et al. 2004). If these data are correct, Rafflesiaceae and Apodanthaceae should be retained as distinct families. Balanophoraceae appear to be related to Santalales based on recent phylogenetic analyses of mtDNA sequences (Barkman et al., in prep.; T.J. Barkman and C.W. dePamphilis, pers. comm.). The phylogenetic relationships of Cynomoriaceae remain uncertain.

The Barkman et al. (2004; in prep.) analyses are major contributions to our understanding of the relationships of parasitic plants. One caveat, however, is that branch lengths to the parasitic taxa are often long, and some placements could be spurious as a result of long-branch attraction. Barkman et al.'s placement of some of these families with various rosid lineages was totally unexpected; modern morphology-based taxonomic treatments did not suggest the placement of parasites with orders of rosids. Of course, as noted, morphological characters are often problematic in the placement of parasites. Regardless of concerns about branch lengths and branch attraction, however, the molecular trees of Barkman et al. (2004) provide important hypotheses that can now be rigorously examined with other approaches, as well as with additional gene sequence data.

Asterid parasites

Several genera and families of parasites are clearly members of the asterid clade (Figure 11.2)—*Cuscuta* (previously of Cuscutaceae, but now placed in Convolvulaceae; Solanales), Orobanchaceae (Lamiales), and *Lennoa* (a genus of two species now included in Boraginaceae) (APG II 2003). These parasitic taxa have long been associated with asterids (Chapter 9) based on non-DNA evidence (i.e., obvious morphological synapomorphies derived from floral characters), but molecular data have provided additional important insights. *Cuscuta* has long been considered closely related to, or derived from within, Convolvulaceae (e.g., Cronquist 1981). *Cuscuta* contains more than 100 species of hemiparasites and holoparasites (Nickrent et al. 1998). The holoparasitic members of the genus lack thylakoids, chlorophyll, and RUBISCO, but interestingly retain the gene *rbcL*. Molecular data place *Cuscuta* with Convolvulaceae (e.g., D. Soltis et al. 1997a; Nickrent et al. 1998; Stefanovic et al. 2002, 2004), and the genus has now been included within Convolvulaceae (APG II 2003). In *Cuscuta*, most of the changes in the plastid genome previously attributed to its parasitic mode of life can be explained as either a plesiomorphic condition within Convolvulaceae or as autapomorphies of particular species of *Cuscuta* (Stefanovic et al. 2002).

Lennoa (formerly placed in Lennoaceae) represents another lineage of parasites long considered to be an asterid. Based on morphology, the family has been associated either with Boraginaceae (Lamiales) or Hydrophyllaceae (Solanales) (Cronquist 1981; Takhtajan 1987, 1997; Thorne 1992a, 1992b). Cronquist (1981) placed Lennoaceae in Lamiales with Boraginaceae, Verbenaceae, and Lamiaceae because of similarities in the structure of the gynoecium. Phylogenetic analysis of mtDNA sequence data place *Lennoa* with Boraginaceae (T.J. Barkman and C.W. dePamphilis, pers. comm.), and it now seems appropriate to include the genus in that family (APG II 2003).

Many genera of parasitic plants have been associated with Scrophulariaceae and Orobanchaceae. In fact, this group of parasites exhibits the greatest range in degree of specialization. In recent classifications (e.g., Cronquist 1981), the parasitic taxa have been distributed between Scrophulariaceae and Orobanchaceae (reviewed in Young et al. 1999) and have been considered to represent an evolutionary series beginning with hemiparasitism and culminating with holoparasitism. The genera *Lathraea, Harveya*, and *Hyobanche* have been viewed as links or transitional taxa between parasites placed in Scrophulariaceae and the holoparasitic Orobanchaceae (Young et al. 1999).

Molecular phylogenetic analyses of Scrophulariaceae and Orobanchaceae have revealed a single well-supported clade that contains all of the parasitic taxa (Young et al. 1999; Olmstead et al. 2001). This clade, containing taxa traditionally placed in either Scrophulariaceae or Orobanchaceae, is now considered a broadly defined Orobanchaceae (APG II 2003; for more information on the disintegration of the Scrophulariaceae, see Olmstead et al. 2001 and Chapter 9). All members of this broadly defined Orobanchaceae are parasitic except the genus *Lindenbergia,* which is nonparasitic and sister to the remainder of the family (Young et al. 1999; Olmstead et al. 2001). Young et al. (1999) further demonstrated that holoparasitism has evolved independently five times from hemiparasitic ancestors in the Orobanchaceae clade (not shown). Lastly, the three transitional genera *Lathraea, Harveya,* and *Hyobanche* are not supported as part of an evolutionary series between hemiparasites in Scrophulariaceae and holoparasites in Orobanchaceae. Instead, all three of these "transitional" genera are more closely related to green, hemiparasitic lineages. Other hemiparasitic members of Orobanchaceae, such as *Castilleja,* do not need to be attached to a host and can flower in a pot of regular soil, so it is clear that parasitism is a continuum, which makes applying these terms difficult.

Genera of Orobanchaceae also have been of interest in the analysis of plastid genome evolution. The plastid genome of the holoparasite *Epifagus virginiana* (Orobanchaceae, *sensu* APG II 2003) has been investigated in detail and has become a model for the evolution of a reduced plastome in a nonphotosynthetic angiosperm. Although the plant lacks chlorophyll and does not photosynthesize, the cells retain plastids, which in turn contain DNA (dePamphilis and Palmer 1990). The plastid genome of *Epifagus virginiana* is much smaller than that of other angiosperms (71 kb versus 156 kb for *Nicotiana* [tobacco], which is typical of most angiosperms) and contains only 42 intact genes (dePamphilis and Palmer 1990; Wolfe et al. 1992). Importantly, however, the greatly reduced plastid genome of *Epifagus* is nearly colinear with the asterid *Nicotiana* (Solanaceae) (dePamphilis and Palmer 1990).

Using mitochondrial DNA sequence data, Barkman et al. (2004) demonstrated that *Mitrastema* (Mitrastemonaceae), an enigmatic holoparasite sometimes placed in Rafflesiaceae (see "Rosid Parasites," above), is placed with the asterid order Ericales. A close relationship of *Mitrastema* to asterids had not been suggested previously. However, the genus exhibits a sympetalous corolla, which certainly supports a placement with the asterid clade (see Chapter 9). In addition, a relationship of *Mitrastema* with Ericales is also supported by morphological features found in various Ericales, including opposite and decussate leaves, parietal placentation, and circumscissile fruit dehiscence (Barkman et al. 2004).

Carnivory

Carnivory in angiosperms is a highly integrated system that involves morphological modifications to attract, trap, kill, and ultimately digest (and absorb nutrients from) animal prey (Juniper et al. 1989). Several highly divergent trap types have evolved, all relying on modified leaves. The most familiar traps are the pitcher, flypaper, and snap trap or steel trap (Figure 11.4; Table 11.2). The pitcher trap is found in such well-known families as Sarraceniaceae (Figure 11.4A) and Nepenthaceae, the New and Old World pitcher plants, respectively, as well as in the Australian pitcher plant family, Cephalotaceae (Figure 11.4, C and D). Flypaper traps are present in several families, including Droseraceae (*Drosera;* Figure 11.4F), Drosophyllaceae (*Drosophyllum*), Dioncophyllaceae (*Triphyophyllum;* Figure 11.8I), Lentibulariaceae (*Pinguicula*), Byblidaceae (*Byblis*), and Martyniaceae (*Proboscidea;* Figure 11.4G). Snap traps are found in *Dionaea* and *Aldrovanda* (Droseraceae; Figure 11.4, L to N).

Less well known are the intricate bladder traps found in the genus *Utricularia* (Lentibulariaceae) and the "lobster pot" traps of *Genlisea* (Figure 11.4K; also of Lentibulariaceae). The trap of *Utricularia* is highly sophisticated and involves negative pressure or suction within the "bladder," which is a thin-walled sac with an intricately constructed doorway that opens inwardly. The "lobster pot" traps of *Genlisea* have a unique configuration, consisting of a swollen utricle or bulb (thought to serve as a digestive cavity). A tubular channel (5–20 mm long) leads to this bulb area. Below this tubular channel are two helical arms (10–50 mm long). A slit, which extends into both arms, forms an extremely long and narrow mouth through which protozoa enter the trap by chemotaxis (for overviews of these and other traps, see Juniper et al. 1989 and Barthlott et al. 1998).

A mechanism approaching carnivory has also evolved in some monocots. Tank-forming members of Bromeliaceae, such as *Brocchinia* (Figure 11.4E), have absorptive trichomes that are viewed as a mechanism that has evolved to acquire nutrients from trapped animals or degrading plant debris (Givnish et al. 1984; Benzing et al. 1985). However, there is debate as to whether tank bromeliads are true carnivores; the plants lack true attractive zones of the leaf (for attracting prey) and do not actively digest prey.

Carnivory in angiosperms has long been considered as having several independent origins. Darwin (1875), for example, suggested that carnivory had evolved many times. Modern taxonomic treatments also indicate multiple origins of carnivory, but both the actual extent to which multiple origins had occurred, as well as the relationships of carnivorous plants to other angiosperms, were much debated (e.g., Takhtajan 1980, 1987, 1997; Cronquist 1981; Thorne 1992a, 1992b). Cron-

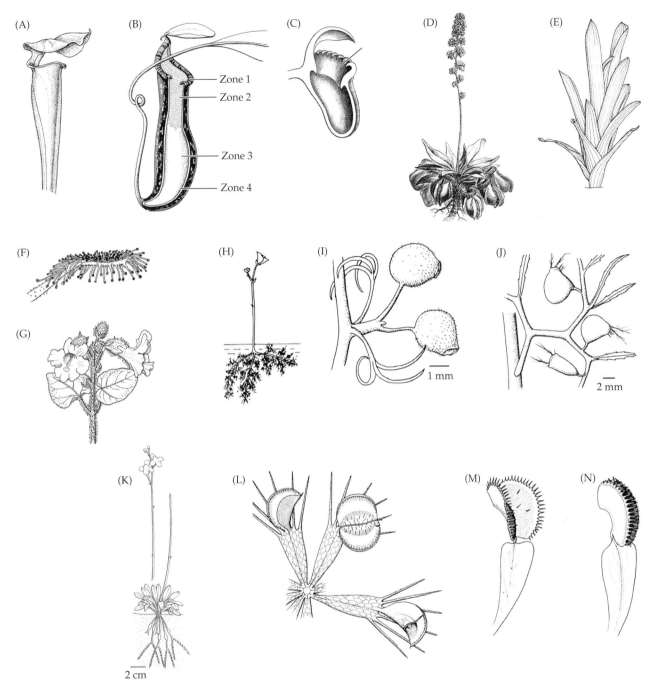

FIGURE 11.4 Trap types in carnivorous angiosperms. A–D, pitcher traps. E, tank trap. F and G, flypaper traps. H–J, bladder traps. K, lobster trap. L–N, snap traps. (A) *Sarracenia flava* (Sarraceniaceae), upper portion of trap. (B) Longitudinal section through a leaf of *Nepenthes* showing the four zones of a pitcher trap. 1. The attractive zone consisting of the lid and peristome with its associated nectaries. 2. The conductive or slide zone, which is wax-coated and slippery. 3. An apparently functionless zone in *Nepenthes*. 4. The principal absorptive zone. (C) Longitudinal section through a leaf of *Cephalotus follicularis*, a pitcher trap; line indicates slide zone. (D) Plant of *Cephalotus follicularis* (showing rosette of pitchers) in flower. (E) *Brocchinia reducta* (Bromeliaceae), a tank species; insects are retained, drown, and decay (not considered a true carnivorous plant by some). (F) *Drosera rotundifolia* (Droseraceae), leaf viewed lat- erally. (G) *Proboscidea lutea* (Martyniaceae), a flypaper-trap species. (H) *Utricularia purpurea* (Lentibulariaceae), whole plant in flower, showing submerged traps. (I) *Utricularia purpurea* (Lentibulariaceae), two traps. (J) *Utricularia vulgaris* (Lentibulariaceae), traps with conspicuous trigger hairs. (K) *Genlisea* (Lentibulariaceae), general habit of plant; the tall flowering stem has been cut so that the entire inflorescence can fit into the frame and remain in scale; traps project down beneath the basal leaves. (L) *Aldrovanda vesiculosa* (Droseraceae), with several leaves. (M) *Dionaea muscipula* (Droseraceae), leaf open, unstimulated. (N) *Dionaea muscipula* (Droseraceae), leaf immediately after mechanical stimula- tion. (A, from Wood 1974. B, E, and G to N, from Juniper et al. 1989. C and D, from Engler in Engler and Prantl 1891. F, original from Darwin 1875, in Juniper et al. 1989.)

TABLE 11.2 Conventionally recognized carnivorous plants, including family and higher-level placement

Family and Genera	Number of Species	Trap Type
Sarraceniaceae (asterids—Ericales)		
Heliamphora	5	Pitcher
Sarracenia	8	Pitcher
Darlingtonia	1	Pitcher
Roridulaceae (asterids—Ericales)		
Roridula	2	Flypaper
Byblidaceae (asterids—Lamiales)		
Byblis	2	Flypaper
Lentibulariaceae (asterids—Lamiales)		
Pinguicula	52	Flypaper
Utricularia	200	Bladder
Genlisea	15	Lobster pot
Martyniaceae (asterids—Lamiales)		
Proboscidea	9	Flypaper
Nepenthaceae (Caryophyllales)		
Nepenthes	68	Pitcher
Dioncophyllaceae (Caryophyllales)		
Triphyophyllum	1	Flypaper
Droseraceae (Caryophyllales)		
Aldrovanda	1	Snap trap
Drosera	110	Flypaper
Dionaea	1	Snap trap
Drosophyllaceae (Caryophyllales)		
Drosophyllum	1	Flypaper
Cephalotaceae (rosids—Oxalidales)		
Cephalotus	1	Pitcher
Bromeliaceae (monocots)		
Brocchinia reducta	–	Tank
Catopsis berteroniana	–	Tank

families were homologous (Markgraf 1955). Cronquist (1981) also noted that Droseraceae and Nepenthaceae have similar pollen (see Chapter 7). Takhtajan (1997) also considered these three families of carnivorous plants to be closely related members of Dilleniidae; he placed Nepenthaceae and Droseraceae together in superorder Nepenthanae and Sarraceniaceae in Sarracenianae next to Nepenthanae. Molecular phylogenetic results have confirmed a close relationship between Droseraceae and Nepenthaceae (both in Caryophyllales; see Chapter 7), but Sarraceniaceae (Ericales, asterid clade; Chapter 9) are distantly related to these two families.

Albert et al. (1992) addressed the number of origins of carnivory using *rbcL* sequence data and found evidence for at least six origins of carnivory among eudicots: in (1) *Cephalotus* (Cephalotaceae); (2) a single origin for the predisposition for carnivory in Caryophyllales (see below); (3) *Roridula* (Roridulaceae); (4) Sarraceniaceae; (5) *Proboscidea* (Martyniaceae); and (6) *Byblis* (Byblidaceae) and Lentibulariaceae. *Byblis* and Lentibulariaceae, although both appearing in the same part of the asterid clade, could also represent independent origins, raising to seven the total number of possible origins of carnivory in the eudicots. In some analyses, however, *Proboscidea* (Martyniaceae), *Byblis,* and Lentibulariaceae form a clade (see paragraphs below and asterid clade; Chapter 9). Members of Bromeliaceae (monocots), if truly carnivorous, would represent additional origins (Figure 11.5).

We reanalyzed the evolution of carnivory in light of new phylogenetic trees that have appeared since Albert et al.'s analysis. Recent multigene topologies for asterids (e.g., Albach et al. 2001a; Anderberg et al. 2002; B. Bremer et al. 2002) are particularly important for determining the likely number of origins of carnivory. Also of importance are multigene analyses of carnivorous plants in Caryophyllales (Meimberg et al. 2000; Cameron et al. 2002). The analyses of Albert et al. and our own analyses both indicate that, although there is clear evidence for multiple origins of carnivory, there are also clades that are evolutionary hotspots for the carnivorous habit (Figure 11.5). Caryophyllales (*sensu* APG II 2003), for example, contain a large proportion of the carnivorous taxa (five genera in four families: Droseraceae, Drosophyllaceae, Nepenthaceae, and Dioncophyllaceae). The asterid order Lamiales represent another hotspot of carnivory with five genera in three families: Martyniaceae, Lentibulariaceae, and Byblidaceae (Table 11.2). Two families of Ericales (Roridulaceae and Sarraceniaceae) are also carnivorous. Only one species of the large rosid clade is carnivorous, *Cephalotus follicularis* (Cephalotaceae; Oxalidales). Carnivory is not present in any basal eudicot lineages or in the basal angiosperms (other than the possible occurrence of carnivory in the monocot family Bromeliaceae; Figures 11.4 and 11.5).

quist (1981), for example, placed three prominent, and morphologically divergent, carnivorous families—Nepenthaceae, Sarraceniaceae, and Droseraceae—in his order Nepenthales. He based this treatment not only on "the obvious exomorphic characters" (i.e., the shared carnivorous habit) but also on the fact that some data suggested that the insect-catching leaves of all three

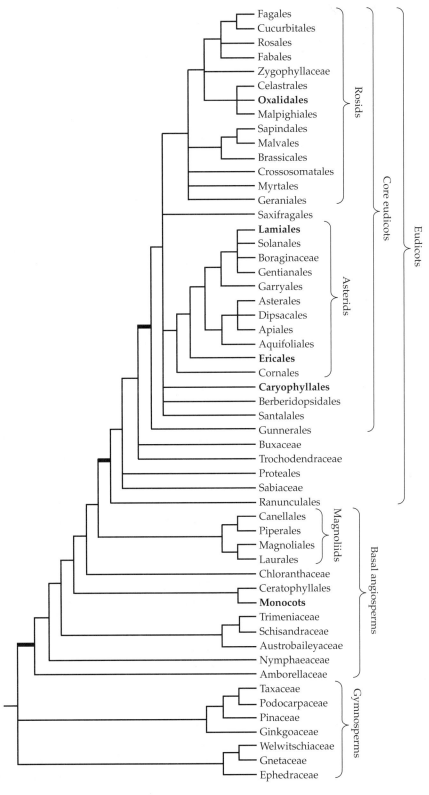

FIGURE 11.5 Occurrences of carnivory in the angiosperms. Clades containing carnivorous taxa are shown in bold. Monocots are indicated as carnivorous based on reports for Bromeliaceae (e.g., *Brocchinia*), but some authors consider these taxa not to be carnivorous in the strict sense (see text). The phylogenetic tree is one possible topology based on the summary topology of D. Soltis et al. (2000) with modifications of the basal angiosperms from Zanis et al. (2002). Three key nodes are indicated with bold lines: angiosperms, eudicots, core eudicots (all receive 100% support).

(A)

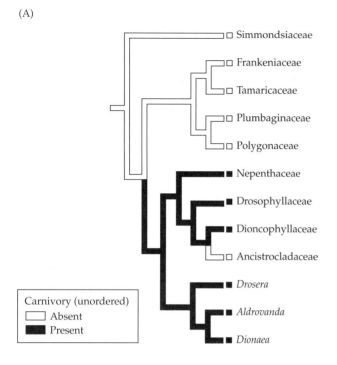

(B)

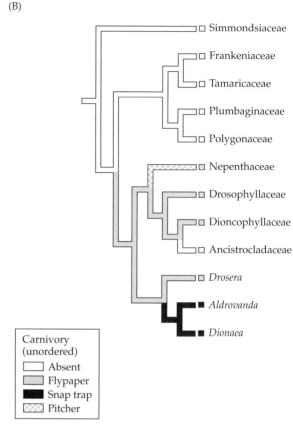

FIGURE 11.6 MacClade reconstruction of the evolution of carnivory in Caryophyllales. (A) Carnivory treated as either present or absent. (B) Different trap types considered as distinct character states.

Recent topologies suggest the possibility of fewer origins of carnivory than initially proposed by Albert et al. (1992) (Figures 11.6 and 11.7). Our data, coupled with the analysis of Cameron et al. (2002) for Caryophyllales, indicate that carnivory may have evolved only three times in the eudicots: once each in Caryophyllales and Ericales, and perhaps only once in Lamiales (Figures 11.6 and 11.7). Although *Roridula* and Sarraceniaceae are both in Ericales, it was initially thought likely that they represented separate origins of carnivory because the two lineages did not appear closely related in some molecular analyses (e.g., Albert et al. 1992; Chase et al. 1993; D. Soltis et al. 2000). However, even with three large datasets involving many taxa and three genes, relationships within Ericales were essentially unresolved (D. Soltis et al. 2000), including the relationship between *Roridula* and Sarraceniaceae. Analyses of asterids using four (Albach et al. 2001a), five (Anderberg et al. 2002), and six DNA regions (B. Bremer et al. 2002) all placed *Sarracenia* as sister to a clade of *Roridula* and the nonparasitic *Actinidia* (Figure 11.7). Support for this clade is very weak in two of these studies—less than 50% in Albach et al. (2001a) and only 50% in B. Bremer et al. (2002). However, Anderberg et al. (2002) found strong support (100%) for a sister group of *Actinidia* + *Roridula*; support was also strong (89%) for a clade of *Sarracenia* + (*Actinidia* + *Roridula*). These recent analyses indicate a single

possible origin of carnivory in Ericales, followed by a loss in Actinidiaceae (Figure 11.7). Interestingly, the trap types of *Roridula* and Sarraceniaceae differ, representing the flypaper and pitcher types, respectively.

The carnivorous members of Lamiales, *Proboscidea* (Martyniaceae), *Byblis* (Byblidaceae), and the genera of Lentibulariaceae, were considered to represent three distinct origins of carnivory (Albert et al. 1992). Furthermore, in the shortest trees obtained in some recent molecular analyses, these three families did not appear together as close relatives within Lamiales (e.g., Albert et al. 1992; Chase et al. 1993; Savolainen et al. 2000a, 2000b; D. Soltis et al. 2000), reinforcing the view of separate origins. However, relationships among these three lineages of Lamiales were poorly resolved and supported, even with three- and four-gene analyses (D. Soltis et al. 2000; Albach et al. 2001a; Olmstead et al. 2001). However, in the six-gene analysis of asterids (B. Bremer et al. 2002), *Proboscidea* (Martyniaceae) appeared as sister to a small clade of *Byblis* (Byblidaceae) + *Pinguicula* (Lentibulariaceae). Although this clade of the three carnivorous lineages in Lamiales was only weakly supported (51%), as was the sister group of *Byblis* + *Pinguicula* (65%), this analysis raised the possibility of a single origin of carnivory in Lamiales; this hypothesis should be tested with larger datasets involving more DNA regions. Again, different trap types are present in

(A)

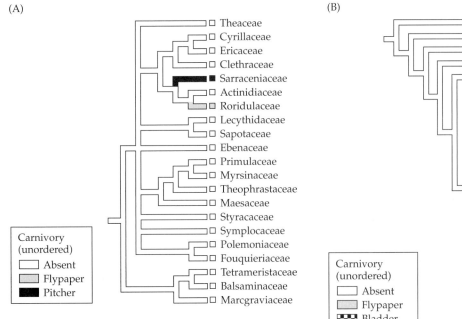

(B)

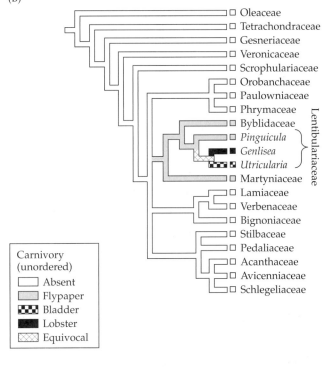

FIGURE 11.7 MacClade reconstruction of evolution of carnivory in Ericales (A) and Lamiales (B).

this small clade of carnivorous Lamiales. *Proboscidea* and *Byblis* have flypaper traps; Lentibulariaceae have flypaper (*Utricularia*), bladder (*Lentibularia*), and lobster-pot (*Genlisea*) trap types (Figure 11.4).

Cameron et al. (2002) provided strong evidence for a single origin of carnivory within Caryophyllales (Figure 11.6; see also Meimberg et al. 2000; Cuénoud et al. 2002). Within this carnivorous clade, two subclades are apparent. One subclade contains *Nepenthes* (Nepenthaceae) as sister to *Drosophyllum* (Drosophyllaceae) + (Dioncophyllaceae + Ancistrocladaceae). Dioncophyllaceae contain the carnivorous *Triphyophyllum* and two noncarnivorous genera; *Ancistrocladus* is not carnivorous (Figure 11.6). The second subclade contains the carnivorous taxa *Dionaea* + *Aldrovanda* as sister to *Drosera* (Droseraceae). These results indicate a single origin of carnivory, followed by a loss of carnivory in *Ancistrocladus* and some members of Dioncophyllaceae. As in the other carnivorous clades, there are several different trap types within the carnivorous clade of Caryophyllales: pitcher (*Nepenthes*), flypaper (*Drosophyllum, Triphyophyllum, Drosera*), and snap trap (*Aldrovandra* and *Dionaea*).

Phylogeny reconstruction has provided important insights into the patterns of structural evolution associated with carnivory (Albert et al. 1992; Williams et al. 1994; Cameron et al. 2002). *Byblis* (Lamiales) and *Drosophyllum* (Caryophyllales), although distantly related (Figure 11.5), have similar flypaper traps. The two genera share several characters, including woody habit, linear leaves with reverse circinate vernation (i.e., fiddle-

heads), and dimorphic stalked and sessile glands. In both genera, the stalked glands secrete mucilage that encoats prey, whereas the sessile glands secrete digestive enzymes. However, *Byblis* differs from *Drosophyllum* in that its mucilage-secreting glands have unicellular stalks that protrude from reservoir cells of the epidermis, whereas the mucilage-secreting cells of *Drosophyllum* are multicellular and vascularized with tracheids and phloem (Figure 11.8); the glands of *Drosophyllum* are similar to those of *Triphyophyllum* (Dioncophyllaceae; Figure 11.8, A and B) to which it is closely related (Figure 11.6). The mucilage-secreting glands of both *Drosophyllum* and *Triphyophyllum* are also similar to those of *Drosera* (Droseraceae), also of Caryophyllales (Figure 11.8, A to C). There are other differences between the mucilage-secreting glands of these members of Caryophyllales and those of *Byblis*. The sessile digestive glands of *Byblis* are simple with four to eight head cells, whereas those of *Drosophyllum, Triphyophyllum,* and *Drosera* are multicellular and multilayered (Figure 11.8, A to C, and E). The glands of *Byblis* are more similar, structurally, to those of *Pinguicula* (Lentibulariaceae; Figure 11.8, D and E). Therefore, the flypaper trapping mechanism is nonhomologous (as indicated by both phylogenetic and structural analyses), despite remarkable structural and functional similarities (Albert et al. 1992).

Different trap types occur together in Ericales, Lamiales, and Caryophyllales. Lentibulariaceae (Lamiales; Figure 11.7) and the carnivore subclade of Caryophyllales (Figure 11.6) can serve as models for the evolution

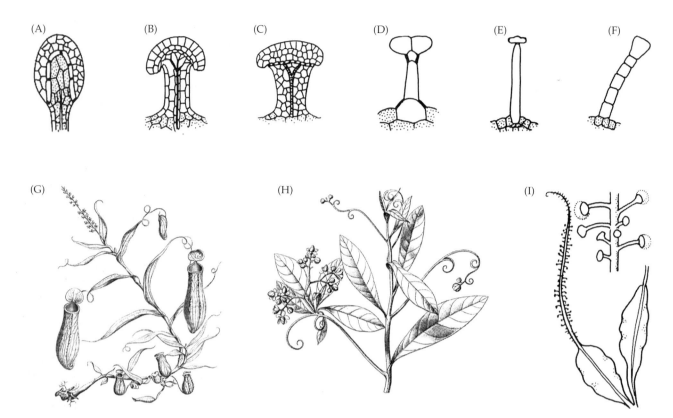

FIGURE 11.8 Glandular hairs in different genera having flypaper traps (A–F) and tendril morphology in three members of Caryophyllales (G–I). (A) *Drosera* (Droseraceae; Caryophyllales). (B) *Drosophyllum* (Drosophyllaceae; Caryophyllales). (C) *Triphyophyllum* (Dioncophyllaceae; Caryophyllales). (D) *Pinguicula* (Lentibulariaceae; Lamiales). (E) *Byblis* (Byblidaceae; Lamiales). (F) *Proboscidea* (Martyniaceae; Lamiales). (G) *Nepenthes* (Nepenthaceae); tendrils terminate in pitchers. (H) *Ancistrocladus heyneanus* (Ancistrocladaceae); tendrils present, but plant not carnivorous. (I) *Triphyophyllum* (Dioncophyllaceae); tendrils with glandular hairs (flypaper trap). (A to F, and I, from Juniper et al. 1989. G, from Wunschmann in Engler and Prantl 1891. H, from Gilg in Engler and Prantl 1895.)

and diversification of carnivory. In these two clades, which contain several different trap types, our data raise the possibility that the flypaper trap is the ancestral trap type. For example, three trap types occur together in a small subclade of Caryophyllales. Nepenthaceae have pitcher traps; Dioncophyllaceae (*Triphyophyllum*) and Drosophyllaceae (*Drosophyllum*) have flypaper traps; and Droseraceae have both flypaper traps (*Drosera*) and snap traps (*Dionaea* and *Aldrovanda*). Reconstructions of trap type indicate that the flypaper trap was the ancestral carnivorous type within Caryophyllales (Figure 11.6; Cameron et al. 2002). It is noteworthy that *Ancistrocladus* has retained leaf glands similar to those of the related carnivorous families Droseraceae and Dioncophyllaceae, despite having lost carnivory.

The carnivorous genus *Aldrovanda* is sister to *Dionaea* (Cameron et al. 2002). An earlier phylogenetic analysis based on both DNA (*rbcL*) and morphological data (Williams et al. 1994) indicated *Aldrovanda* as sister to *Drosera*, but there were no molecular data for *Aldrovanda* in that study. With DNA for *Aldrovanda* included,

Cameron et al. (2002) provided strong support for a clade of *Aldrovanda* and *Dionaea*. *Aldrovanda* (Figure 11.4L), a little-known carnivore from the Old World, is a rootless, submerged aquatic herb. Each leaf resembles a leaf of *Dionaea* (Figure 11.4, M and N), having two lobes with each lobe possessing roughly 20 trigger hairs. Darwin (1875), in fact, first noted the strong similarity of the leaves of *Aldrovanda* and *Dionaea*. *Aldrovanda* was later shown to have a rapidly closing snap-trap mechanism, somewhat similar to that of *Dionaea*. Cameron et al.'s (2002) results indicated a single origin of this snap-trap mechanism.

The topologies for the carnivorous clade of Caryophyllales (Meimberg et al. 2000; Cuénoud et al. 2000; Cameron et al. 2002) place *Nepenthes* as sister to a clade of *Drosophyllum* + (Dioncophyllaceae + *Ancistrocladus*). MacClade reconstructions (Figure 11.6) indicate that the pitcher trap of *Nepenthes* was derived from the flypaper trap type in this clade. Also of interest is the fact that the pitcher traps of *Nepenthes* are produced at the tips of tendrils; this contrasts with the pitcher traps of other families (e.g., Sarraceniaceae, Cephalotaceae) in which the pitcher trap is an entire leaf (Figure 11.4, A to D). Tendril formation is present throughout the subclade of carnivores to which *Nepenthes* belongs, with the possible exception of *Drosophyllum,* which has linear leaves. *Triphyophyllum,* which has a flypaper trap mechanism, has tendrils similar to those of *Nepenthes,* as does the noncarnivorous *Ancistrocladus* (Figure 11.8H). Hence, tendril formation may have been an ancestral trait for this subclade of carnivores.

Several different trap types are also present within Lamiales. In the B. Bremer et al. (2002) topology, Martyniaceae and Byblidaceae—both of which have flypaper traps—are successive sisters to Lentibulariaceae. Within Lentibulariaceae alone, three different trap types have evolved. *Pinguicula* (butterworts), which also possesses a flypaper trap, is sister to *Genlisea* (lobster pot) + *Utricularia* (bladder) (Jobson and Albert 2002; Jobson et al. 2003). If the B. Bremer et al. topology is correct in its reconstruction of relationships, the flypaper trap is unambiguously reconstructed as ancestral in Lamiales (Figure 11.7), regardless of the resolving option used (ACCTRAN, DELTRAN, "all most parsimonious states"), and both the lobster pot and bladder traps are derived.

The fact that *Utricularia* contains several different types of bladders adds to the structural and evolutionary complexity of Lentibulariaceae. Molecular phylogenetic data indicate that *Pinguicula* is sister to a clade of *Genlisea* and *Utricularia* (Jobson and Albert 2002; Jobson et al. 2003; Figure 11.7). Jobson and Albert (2002) argue that the clade of *Utricularia* and *Genlisea* is substantially more species-rich and morphologically divergent than *Pinguicula.* They suggest that bladderworts have a flexible vegetative development that could be viewed as a key innovation (e.g., Sanderson and Donoghue 1994; Givnish 1997; Hodges 1997a; see Chapter 12) that permits these plants to invade a broad range of nutrient-poor niches, ranging from water-saturated terrestrial to epiphytic and suspended aquatic. Jobson and Albert (2002) also found that bladderwort genomes evolve significantly faster across seven genes than do nonbladderwort genomes and proposed that increased cladogenesis in this instance is related to increased nucleotide substitution. Molecular phylogenetic investigations of Lentibulariaceae have also provided insights into character evolution, including habit (Jobson et al. 2004). The terrestrial habit is the most common for all three genera, whereas other species exhibit other habits, including affixed aquatic, suspended aquatic, and epiphytic. Jobson et al. (2004) found that the epiphytic habit evolved independently at least three times in the family.

C₄ Photosynthesis

Some chemical pathways seem to be restricted to only one or a few lineages and must therefore have originated only one or a few times. Betalain production, for example, is limited to the core Caryophyllales clade (see Chapter 7). Similarly, glucosinolate production has evolved only twice, once in the Brassicales and again in Putranjivaceae (Malpighiales) (see Chapter 8). Symbiosis with nodule-forming nitrogen-fixing bacteria appears to have evolved many times, but all within a single well-defined clade of rosids—the nitrogen-fixing clade—leading to the suggestion that the predisposition for nitrogen-fixing symbiosis evolved only once in the angiosperms (D. Soltis et al. 1995; see Chapter 8). In contrast, C_4 photosynthesis has evolved repeatedly in the angiosperms, in distantly related groups, but often within a small number of species in a given group (Kellogg 1999; Sage et al. 1999; Figure 11.9).

The C_4 pathway is considered to be an adaptation to high light intensities, high temperatures, and a dry climate. The optimal temperature range for C_4 photosynthesis is much higher than that for C_3 photosynthesis. The C_4 photosynthetic pathway is well understood and is developmentally and genetically complex (Edwards and Walker 1983). Typically, for example, C_4 plants have a characteristic leaf anatomy (Kranz anatomy) in which the leaves have an orderly arrangement of mesophyll cells around a layer of large bundle-sheath cells; together, the two cell types form two concentric layers or rings around the vascular bundle.

Attempts to explain the broad distribution of C_4 photosynthesis among the angiosperms have involved repeated loss and lateral transfer. However, the most likely explanation is repeated evolution. Using the *rbcL* topologies for angiosperms available at that time, Kellogg (1999) addressed the phylogenetic distribution of C_4 photosynthesis across the angiosperms and within particular clades, including the asterids and Poaceae, in which many C_4 species occur. Although the phylogenetic analysis of large, multigene datasets provided higher resolution and support for many relationships than the original *rbcL* topologies (e.g., Chase et al. 1993), Kellogg's (1999) general conclusions remain valid. Her analyses indicated at least 31 origins of C_4 photosynthesis and multiple origins within some large clades such as the monocots and asterids. Within the asterids, for example, C_4 photosynthesis has evolved in *Heliotropium* (Boraginaceae), *Blepharis* (Acanthaceae), *Anticharis* (Scrophulariaceae), and several additional times within Asteraceae (see below). Within the monocots, C_4 photosynthesis has evolved in the basal monocot Hydrocharitaceae (see Chapter 4), Cyperaceae, and several times in Poaceae (see below). Caryophyllales were not examined in any detail by Kellogg (1999) because the topologies available were not comprehensive and familial circumscriptions were uncertain. Hence, despite widespread variation in photosynthetic systems in Caryophyllales, the pattern of origins and losses of C_4 photosynthesis in Caryophyllales, although labile, remains unclear (e.g., Pyankov et al. 2001).

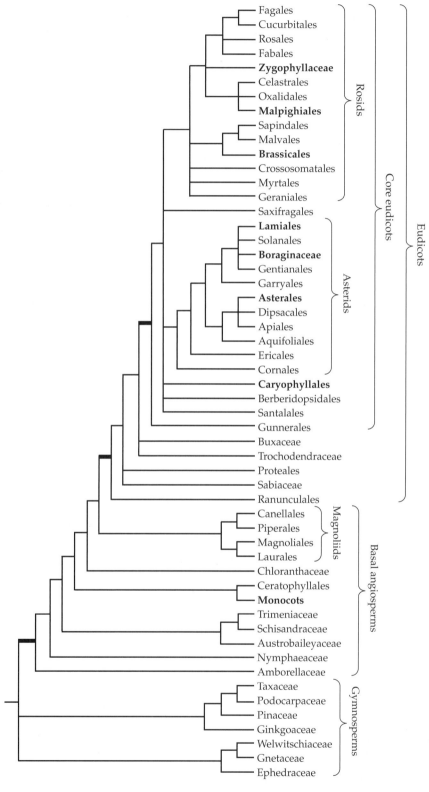

FIGURE 11.9 Major occurrences of C$_4$ photosynthesis in the angiosperms. Lineages with some members with C$_4$ photosynthesis are indicated in bold. The phylogenetic tree is one possible topology based on the summary topology of D. Soltis et al. (2000) with modifications of the basal angiosperms from Zanis et al. (2002). Three key nodes are indicated with bold lines: angiosperms, eudicots, core eudicots (all receive 100% support).

The C_4 pathway is not identical in all angiosperm lineages. For example, C_4 plants differ in their decarboxylating enzymes (Hatch et al. 1975), using either NAD-ME, NADP-ME, or phosphoenolpyruvate carboxykinase (PCK). In addition, Kranz anatomy is typically present in C_4 plants, but not always. Thus, the C_4 pathway is not a single pathway, but several different pathways that share phosphoenolpyruvate carboxylase (PEP carboxylase or PEPC) as a carbon acceptor. The use of PEPC as a carbon acceptor is common in plants, and the activation of a PEPC pathway for C_4 photosynthesis may be relatively easy. The required anatomical changes may be more difficult to achieve, and this may be more of a limiting factor, and ultimately the causal agent, in the scattered and sporadic distribution of C_4 photosynthesis among flowering plants (Kellogg 1999). The absence of Kranz anatomy in some members of Caryophyllales (Pyankov et al. 2001; Voznesenskaya et al. 2001) demonstrates, however, that biochemical and anatomical components of C_4 photosynthesis can be decoupled.

In addition to the many origins of C_4 photosynthesis on a broad scale, Kellogg (1999) also showed that multiple origins were frequent within several families, such as Asteraceae, Poaceae, Cyperaceae, and Zygophyllaceae. Within Zygophyllaceae, there is evidence for at least two origins, and there have been at least four separate origins in Cyperaceae. Within Asteraceae, there have been three origins of C_4 photosynthesis. One origin occurred in the common ancestor of the genus *Pectus* (six species have been examined and all have C_4 photosynthesis); once in the common ancestor of one subclade of four species within the genus *Flaveria*; and again in the common ancestor of one subclade within the Coreopsidae. *Flaveria* is of special interest because this single genus contains some species that are C_3 and others that are C_4, as well as several species that are considered C_3–C_4 intermediates. Based on phylogenetic data coupled with other evidence, Kopriva et al. (1996) favored a single origin of C_3–C_4 intermediate photosynthesis, with a subsequent origin of C_4.

Several possible origins of C_4 photosynthesis are also indicated for Poaceae. On a broad scale, two to four origins are suggested across Poaceae (Figure 11.10, A and B), with a combination of phylogenetic and morpho-

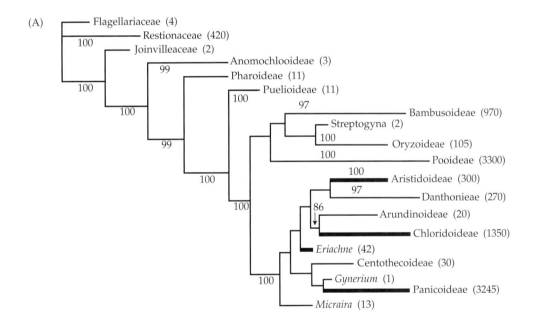

FIGURE 11.10 Evolution of C_4 photosynthesis (indicated by thickened lines) in Poaceae (courtesy of E. Kellogg) based on the work of the Grass Phylogeny Working Group. Numbers below or above branches indicate bootstrap support. Numbers in parentheses after taxon names indicate the total number of species. Phylogenetic tree is based on phylogenetic analyses of a large dataset of seven molecular datasets (*rbcL, rpoC2, ndhF,* cpDNA restriction sites, *phyB,* ITS, *waxy*), plus morphology. (A) Summary of grass phylogeny. (B) Additional detail of the clade (from A) that includes *Micraira,* Panicoideae, Centothecoideae, Chloridoideae, Arundinoideae, Danthonieae, and Aristidoideae.

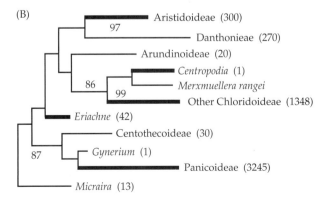

logical data favoring four major origins (reviewed in Sinha and Kellogg 1996): in Aristidoideae, Chloridoideae, *Eriachne,* and Panicoideae. However, several critical nodes do not receive bootstrap support above 50% (Figure 11.10). The precise number of origins depends on the placement of *Eriachne,* the relationships of which are not resolved. On a broad scale, there could be as few as two major origins, with the panicoids representing an origin separate from the other three C_4 lineages (all of which could have had a single origin).

Within the panicoid clade, the evolution of C_4 photosynthesis appears to be particularly complex. The data are compatible with multiple origins of C_4 photosynthesis (perhaps as many as eight) in the panicoid clade alone (Giussani et al. 2001). However, it is equally parsimonious to infer a single origin of C_4 photosynthesis in the ancestor of the clade, followed by multiple reversals from C_4 to C_3 photosynthesis (Giussani et al. 2001; E.A. Kellogg, pers. comm.). Additional phylogenetic analyses beyond those of Giussani et al. (2001) further illustrate the complexity of the problem. The number of origins is sensitive to taxon inclusion and also differs with the gene or genes that are analyzed phylogenetically (E.A. Kellogg, pers. comm.). Perhaps the safest conclusion is that the evolution of C_4 photosynthesis is particularly labile in the panicoids. Importantly, the data suggest that some genetic change occurred at the base of what is called the PACCAD clade (Panicoideae, Aristidoideae, Chloridoideae, Centothecoideae, Arundinoideae, Danthonieae; Davis and Soreng 1993) that facilitated the evolution of C_4 photosynthesis. In other words, a genetic predisposition for C_4 photosynthesis may have evolved at the base of this clade in perhaps much the same way that a predisposition for nitrogen-fixing symbiosis apparently evolved at the base of the nitrogen-fixing clade (D. Soltis et al. 1995; see Chapter 8).

Kellogg (1999) and coworkers (Sinha and Kellogg 1996; Giussani et al. 2001) also used Poaceae as an example of the variability in the C_4 pathway within a single lineage. The NAD-malic enzyme (ME) type is present in *Centropodia* and Chloridoideae. *Eriachne* exhibits the NADP-ME subtype. Considering the Panicoideae, both the PCK and NAD-ME subtypes of C_4 photosynthesis have each evolved only once; the other origins of C_4 photosynthesis are NADP-ME (Giussani et al. 2001).

The evolution of C_4 photosynthesis within Caryophyllales appears to be very complex. Phylogenetic analysis of tribe Salsoleae (of the former Chenopodiaceae, now in Amaranthaceae) reveals the complexities of this family. The molecular topology provides strong support for the origin and evolution of two main lineages, one having NAD-ME and the other NADP-ME C_4 photosynthesis (Figure 11.11; Pyankov et al. 2001). These NAD-ME and NADP-ME clades have not only

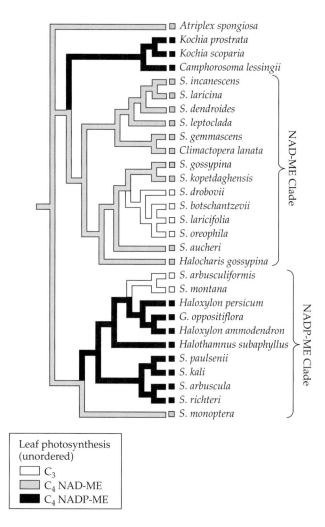

FIGURE 11.11 Parsimony reconstruction of leaf photosynthesis in tribe Salsoleae (Amaranthaceae). The molecular topology provides strong support for the origin and evolution of two main lineages, one having NAD-ME and the other NADP-ME C_4 photosynthesis. A reversal to C_3 photosynthesis occurred in the NAD-ME clade and another reversal occurred in the NADP-ME clade (modified from Pyankov et al. 2001). Names of genera are abbreviated: *S. = Salsola; G. = Girgensohnia.*

different photosynthetic types but also different structural and photosynthetic characteristics in cotyledons. The topology also reveals two independent reversions to C_3 photosynthesis and loss of Kranz anatomy (Figure 11.11; Pyankov et al. 2001).

Future Directions

In this chapter we reviewed current evidence for homoplasy in life history strategies and major metabolic processes across the angiosperms. Additional independent derivations of C_4 photosynthesis have undoubtedly occurred, although current information does not allow further assessment. Many other features of angiosperms—ranging from "simple" morphological structures to complex biosynthetic pathways—have undoubtedly arisen recurrently. As phylogeny estimations improve and encompass more botanical diversity, new opportunities arise for evaluating patterns of evolution across the tree. However, in addition to new phylogenetic information and documentation of the presence or absence of a trait, pathway, or other feature in species

not yet examined, new data are also needed on the details of the characteristic under study. For example, if a morphological feature is considered, does it develop in the same manner in different species? If a chemical compound, did it arise via the same pathway? This sort of data will ultimately be useful in evaluating the patterns of evolutionary change and perhaps the selective pressures under which a character has evolved.

12

Floral Diversification

Introduction

Flowers are exclusive to angiosperms. Although the flowers of all angiosperms are homologous in a general sense, floral organization shows quite different traits if the basalmost angiosperms (composed of many former members of Magnoliidae; see Chapter 3) are compared with more highly modified groups, such as Orchidaceae, a highly derived monocot family, or asterids, a highly derived major clade of eudicots. These different floral traits are not single characters, but entire sets of traits that make up the evolutionary "behavior" of flowers. Such major evolutionary changes are referred to as "key innovations" (e.g., Müller and Wagner 1991; Endress 2001b; Wagner 2001). Such innovations can also be seen as evolution of evolvability. In other words, these major evolutionary steps enable new radiations to occur, and the presence of many new forms is more likely to produce additional key innovations, which in turn can be seen as enhanced evolvability. In this chapter, we provide an overview of some of the key innovations in floral diversification.

In basal angiosperms, floral organ number and phyllotaxis are flexible. Thus a wide range of organ numbers is possible—from one to thousands—in flowers with spiral or whorled or irregular organ position (see Chapter 3). In the course of angiosperm evolution, key innovations have involved the following changes: (1) synorganization (cohesion of floral organs, either through fusion or close association) of organs of the same kind, and (2) synorganization between organs of different kinds (Endress 1990a, 1994a, 1994c, 2001b). A precondition for, or a consequence of, this synorganization was the restriction to whorled phyllotaxis and to a fixed and low number of organs. Examples of synorganization are syncarpy, sympetaly, the fusion of petals and stamens, and the fusion of stamens and carpels. The most complicated flowers, such as in members of Orchidaceae and Asclepiadoideae (Apocynaceae), are characterized by many and highly elaborated kinds of synorganization, and by a completely uniform organ number. Other key innovations include the differentiation of sepals and petals in the perianth, and the transition from crassinucellar (and bitegmic) to tenuinucellar (and unitegmic) ovules (Figure 12.12, and see Chapter 9, Figures 9.9 and 9.10).

Tepals (or sepals and petals), stamens, and carpels are the structural elements or modular structures that compose a flower. Their number and arrangement pattern can vary within wide limits. However, if a flower is bisexual, a gynoecium is always in the center of the flower surrounded by an androecium. This sequence is therefore part of the fixed floral organization of angiosperms. Only one exception is known in which an androecium is regularly surrounded by a gynoecium in bisexual flowers—*Lacandonia* (Triuridaceae; Márquez-Guzmán et al. 1989; Vergara-Silva et al. 2003).

Flowers with a whorl of sepals, a whorl of petals, a whorl of stamens, and a whorl of carpels have been used as a basis for the ABC model of flower development in molecular developmental genetics (Coen and Meyerowitz 1991; see Chapters 3 and 5). In unisexual flowers, the number of stamens or carpels is reduced to various degrees; they are either suppressed (initiated but not differentiated) or lost (not noticeably initiated) (cf. Tucker 1988). In unisexual flowers, development takes various pathways (Dellaporta and Calderon-Urrea 1993; Korpelainen 1998), and the way that male or female organs are suppressed during development even within a species may also be different (e.g., *Zea*; Dellaporta and Calderon-Urrea 1994).

The crown group of angiosperms has existed for at least 130 million years (see Chapters 1 and 2), and flowers with a basic structure similar to those of the basalmost extant clades have probably existed at least that long. Many evolutionary progressions and radiations have taken place; some floral traits have undergone many changes, and others have remained more conservative.

Because floral traits are so variable, it is practical to distinguish different levels of structure, such as organization (bauplan), construction (architecture), and mode (style), if the terms are applied loosely (cf. Vogel 1954; Endress 1994a, 1994c). Organization (bauplan) refers to features that are conservative (e.g., presence and pattern of disposition of floral organs: sepals, petals, stamens, carpels). Construction (architecture) refers to the shape of flowers (e.g., bowl-shaped flowers, tubular flowers, salverform flowers, lip flowers, revolver flowers). Mode (style) refers to adaptation of flowers to specific pollinators (e.g., bird-pollinated flowers, moth-pollinated flowers, wind-pollinated flowers; colors and scents are also attributes of mode). The levels of organization, construction, and mode are increasingly unstable in evolutionary terms—that is, mode varies at lower taxonomic levels than does organization.

That these concepts should be applied only in a loose way is shown by the fact that a feature referred to organization may suddenly become labile in a group, which is to be expected in an evolving group of organisms. For example, Brassicaceae consistently have flowers with four petals and six stamens, but in the genus *Lepidium*, this trait has become unstable, and flowers with four or two stamens and without petals have evolved several times (Endress 1992; Bowman et al. 1999; Karoly and Conner 2000; Mummenhoff et al. 2001).

Pathways of Evolutionary Innovations

Innovations in floral evolution can take any of several potential pathways.

1. Use of existing structures for new functions. An example is the odd numbered staminode in flowers of Lamiales. Concomitant with the evolution of monosymmetry, the upper median stamen has become reduced to a staminode or has been lost. In a few genera, such as *Penstemon* and *Jacaranda*, this staminode has become enlarged and functions as a nectar guide for pollinating bees (Endress 1994c; Walker-Larsen and Harder 2000, 2001). This type of evolutionary innovation was called "diverted development" by Crane and Kenrick (1997b).

2. Synorganization of organs of the same kind and/or organs of different kinds. An example is the formation of pollinia with transport organs (translators) in Orchidaceae and Apocynaceae–Asclepiadoideae, a precondition of which was the close synorganization of androecium and gynoecium by congenital fusion in orchids (gynostemium) or postgenital fusion in Apocynaceae (gynostegium) (Vogel 1959, 1969; Endress 1990a, 1994c, 1997b; Rudall and Bateman 2002).

3. Ectopic gene expression. A putative example is the proposed evolutionary origin of carpels by ectopic development of ovules on microsporophylls in a Mesozoic group of gymnosperms, as proposed by Frohlich and Parker (2000) (see "The Mostly Male Hypothesis," Chapter 1).

Evolutionary Radiations

Adaptive radiations are spectacular evolutionary events (see Givnish and Sytsma 1997). Within angiosperms, there are many examples of floral adaptive radiations triggered by different groups of pollinators, which have been studied in some detail. In those plant genera, distinctive features in flowers are used to distinguish species. Classic examples are provided by Vogel (1954) who studied various angiosperm genera in South Africa (e.g., *Bauhinia, Clerodendrum, Erica, Pelargonium*) and by Grant and Grant (1965) who studied Polemoniaceae. Both Vogel and Grant and Grant analyzed floral adaptation to different pollinators and the accompanying structural, visual, and olfactory diversification. Bees, butterflies, sphingids, flies, birds, and bats are all involved in the rich diversification of many of these genera. More recently, it has become possible to base such studies on a cladistic framework derived from molecular studies, an approach that facilitates the reconstruction of evolutionary change. The number of such studies involving putative adaptive radiations has increased dramatically

in just the past decade. A general work that marks this new situation is the book by Givnish and Sytsma (1997), in which Givnish (1997) provides a general introduction to the field, and several authors discuss case studies, among them several on flowers. It may suffice to mention a few studies of various angiosperm genera or higher groups, such as on *Adansonia* (Malvaceae; Baum et al. 1998; see also Chapter 8), *Aphelandra* (Acanthaceae; McDade 1992), *Aquilegia* (Ranunculaceae; Hodges 1997a, 1997b; Hodges et al. 2002), *Caesalpinia* (Fabaceae; Tucker et al. 1985; Vogel 1990; Lewis 1998), *Dalechampia* (Euphorbiaceae; Armbruster 1996; Armbruster and Baldwin 1998; Hansen et al. 2000), *Disa* (Orchidaceae; Johnson et al. 1998; Figure 12.1), *Gladiolus* (Iridaceae; Goldblatt et al. 1998), Fabaceae (Lewis et al. 2000), Oncidiinae (Orchidaceae; Chase and Palmer 1997), *Pelargonium* (Geraniaceae; Struck 1997), *Platanthera* (Orchidaceae; Hapeman and Inoue 1997), Pontederiaceae (Barrett and Graham 1997), and *Swertia / Halenia* (Gentianaceae; von Hagen and Kadereit 2002).

The general lesson for flower evolution derived from such studies is that many structural (and other) changes

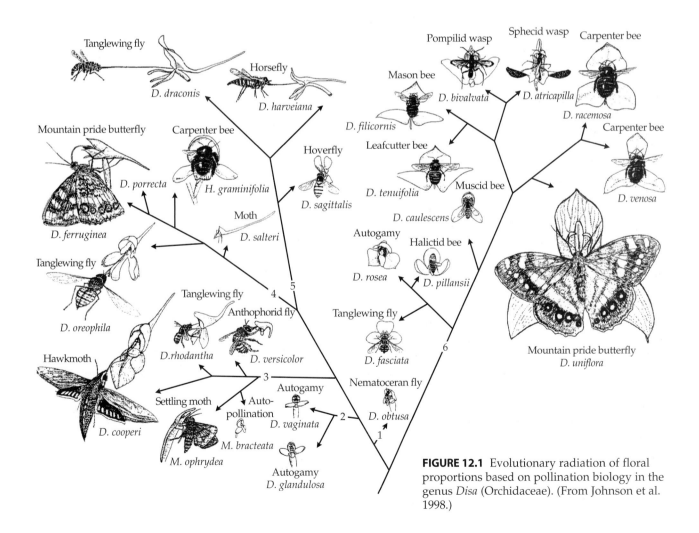

FIGURE 12.1 Evolutionary radiation of floral proportions based on pollination biology in the genus *Disa* (Orchidaceae). (From Johnson et al. 1998.)

arise very easily, whereas others are much more difficult. Floral features may differ greatly in detail, however, from group to group. Nonetheless, some previously overlooked patterns may later emerge from a comparative survey of many groups.

Floral Phyllotaxis

Floral phyllotaxis may be spiral (Figure 12.2A), whorled (Figure 12.2B), or unordered (Figure 12.2C; Endress 1987a, 1987b; Doust 2001). In spiral phyllotaxis, all subsequent organs are initiated with the same divergence angle (the angle between the lines drawn from the floral center to the center of each of two subsequent organs). In whorled phyllotaxis, a fixed number of organs is commonly arranged in a whorl, and subsequent whorls alternate with each other. Unordered phyllotaxis often occurs when the number of organs is greatly increased and the ratio of the size of the floral apex to the size of individual organs is increased, or when the floral apex has an irregular shape. However, these phyllotactic patterns may not be exclusive in an individual flower. Flowers of core eudicots are known to have whorled phyllotaxis (Chapter 5), but even in such flowers, phyllotaxis commonly begins spiral in the calyx and becomes whorled only in the corolla; or flowers with an unordered phyllotaxis in the polymerous androecium or gynoecium often have an ordered phyllotaxis in the perianth. Even in flowers with a completely whorled phyllotaxis, the organs may be initiated in a spiral sequence (Erbar and Leins 1985, 1997; Endress 1987a), but the plastochrons (the time lapse between the initiation of two subsequent organs) seem to be much shorter than in spiral phyllotaxis (see Chapter 3 on basal angiosperms). If a perianth is lacking, the phyllotaxis of the remaining organs is often irregular (En-

dress 1989b, 1990a; Tucker 1991; Doust 2001). Thus, the perianth may set the boundary conditions for phyllotaxis regularity of the inner floral region, that is, the androecium and gynoecium.

In flowers with spiral phyllotaxis, the divergence angle commonly approaches 137.5 degrees (following a Fibonacci pattern). Developmentally, it can be understood as an activator-inhibitor mechanism in the floral apex in which the inhibitor exponentially decays and results in this pattern (Meinhardt 1982). Ecologically, this pattern provides a favorable spacing effect if many organs are arranged on a short axis, such as the showy perianth organs in a flower or the leaves in a vegetative rosette—no two organs fall exactly on the same radius. In contrast, a whorled position is favorable if only a small number of organs that can be displayed in a single series is present. Thus, it is not surprising that in core eudicots and monocots, with generally low numbers of petals (or tepals), the whorled pattern is almost exclusively present, whereas in basal angiosperms and some families of basal eudicots, with greater variation in the number of perianth members, both patterns of phyllotaxis occur (Endress 1987a, 1987b). In basal angiosperms, in extreme cases, both spiral and whorled patterns may even occur in the same species (e.g., male flowers of *Ceratophyllum,* Endress 1994b; flowers of *Drimys,* Winteraceae, Doust 2001; see Chapter 3).

The developmental process is more complicated in flowers than in vegetative shoots because flowers, in contrast to vegetative shoots, have repeated switches from one kind of organ to another. In whorled flowers, these switches are abrupt. Because the width of different kinds of organs within a flower may differ considerably, there may also be a change in organ number from one whorl to the next. For instance, double positions may occur after the transition from the perianth

(A)

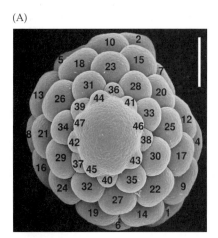

(B)

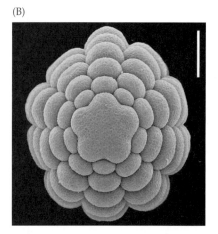

(C)

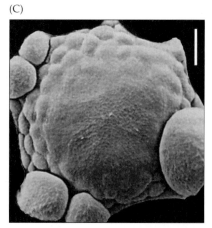

FIGURE 12.2 Floral phyllotaxis, as shown in floral buds. (A) Spiral phyllotaxis. *Ranunculus acris* (Ranunculaceae), young stamens and carpels, organs numbered according to their developmental sequence. (B) Whorled phyllotaxis. *Aquilegia* *vulgaris* (Ranunculaceae), young stamens and carpels. (C) Irregular phyllotaxis. *Zygogynum stipitatum* (Winteraceae), innermost tepals and young stamens. Bars = 0.1 mm. (Modified from Endress 1987a.)

to the androecium, that is, instead of one organ in an expected position, two are side by side (Endress 1994a, 1994c). Such cases are known from Nymphaeaceae, Aristolochiaceae, Butomaceae, Velloziaceae, Papaveraceae, Phytolaccaceae, Rosaceae, and Zygophyllaceae. In eudicots, toward the floral center the reverse takes place—the gynoecium generally has fewer organs than the inner whorl of the androecium (many rosids, most asterids; Endress 1987a, 1987b).

Due to synorganization, "phyllotaxis theories" (Lacroix and Sattler 1988) can be applied to flowers only with limitations. There seem to be cases in which the association of different organs of two different whorls is so intimate that the two types of organs always occur together, even if not expected in a normal phyllotactic situation. An example is the pentamerous terminal flowers of *Berberis vulgaris,* in which the five petals and five stamens are in exactly the same sector, although they are in spiral phyllotaxis (Endress 1987a). The petal–stamen complex forms a functional apparatus in which the nectaries (on the petal base) are close to the sensitive stamen base, where stamen movement is triggered by touching. Flower organization has changed in some respect during evolution. Floral development became more "closed," and the organs more integrated (Endress 1987a, 1987b, 1990a). Progressive limitations in the application of "phyllotaxis theories" accompany increasing integration of floral organs.

Floral Merism

Floral merism (mery, merosity) designates the number of organs in a whorl or series (Eichler 1875–1878; Endress 1987a, 1987b, 1994a; Ronse De Craene and Smets 1994). This number is related to floral phyllotaxis and also depends on the ratio of organ primordium size (width) to the floral apex size (diameter) at the time the organs are formed. The smaller the ratio, the more organs are formed in a whorl or series. In basal angiosperms and some early-diverging eudicots (especially Ranunculaceae, e.g., Schöffel 1932), floral merism may be highly variable within a species or variable even within an individual (Chapters 3 and 5). In contrast, in more derived angiosperms, especially in monocots and asterids, floral merism may be constant (or nearly so), in some cases even for entire families (e.g., Apocynaceae) or orders (Lamiales). The genetic canalization of floral organ number (e.g., Huether 1969) allows increased synorganization of organs (Endress 1990a). Thus, in more derived angiosperms, diversity is less in organ number than in organ shape plasticity and evolution of various precision mechanisms and specialization for a wide variety of pollinators (Stebbins 1970, 1974; Endress 1994a, 1994c). Variation in floral merism is discussed in more detail in the chapters on basal angiosperms and early-diverging eudicots.

(A)  (B)

FIGURE 12.3 Floral symmetry. (A) Polysymmetric flower. *Anagallis arvensis* (Myrsinaceae). (B) Monosymmetric flower. *Columnea* (Gesneriaceae). The two stigmatic lobes and the four anthers (by postgenital fusion) are exactly in the median plane.

Floral Symmetry

Most flowers are polysymmetric (actinomorphic, with several symmetry planes; Figure 12.3A) or monosymmetric (zygomorphic, with one symmetry plane; Figure 12.3B). In monosymmetric flowers, the symmetry plane is almost always vertical. Commonly, this position is present from the beginning of development. Only in a few taxa is the plane of symmetry horizontal or oblique at the beginning of development and secondarily vertical only at anthesis by some bending or rotation of the flower. The flowers of basal angiosperms are polysymmetric. Floral monosymmetry has evolved many times from polysymmetry. A general trend is that in more synorganized flowers, monosymmetry appears more often as a key innovation than in groups with less synorganized flowers. Particularly species-rich groups with exclusively or predominantly monosymmetric flowers are Fabaceae, Lamiales, Asteraceae, Orchidaceae, and Zingiberales; among these, Asteraceae and Orchidaceae are the two most species-rich families of the angiosperms (Endress 1999, 2001a; Ree and Donoghue 2000; Rudall and Bateman 2002). There have also been multiple reversals to approximate secondary polysymmetry from monosymmetry, as in *Veronica* and *Plantago* (Plantaginaceae), as well as in several other members of Lamiales (Donoghue et al. 1998, Endress 1999; see Chapter 9, Figure 9.12) and Melianthaceae (Ronse De Craene et al. 2001).

The method by which monosymmetry evolved from polysymmetry has not been elucidated. To address this question, the old observations of plant groups with "constitutionally" monosymmetric flowers (e.g., Lamiales) and others with "positionally" monosymmetric flowers (e.g., *Epilobium, Cleome, Hemerocallis*) (Vöchting 1886) may be of interest. In the first type, flowers are always monosymmetric and cannot be experimentally converted into polysymmetric ones. In the second type, flowers can be turned into polysymmetric ones by experimental intervention during flower development. An interplay between internal, organizational factors and external, selective factors (by pollinators) must be assumed, the details of which are unknown.

The establishment of floral symmetry is to a great extent independent of floral phyllotaxis (Endress 1999). As mentioned above, monosymmetric flowers do not occur in basal angiosperms (except for monosymmetry by organ reduction, e.g., unistaminate flowers in Chloranthaceae or unilaterally bistaminate flowers in Piperaceae). Especially noteworthy are relatively highly elaborated monosymmetric flowers in early-diverging eudicots (Ranunculaceae: *Aconitum, Delphinium*), which are based on spiral phyllotaxis (Schöffel 1932; Hiepko 1965a). Probably all other elaborate monosymmetric flowers have whorled phyllotaxis. Unelaborate monosymmetric flowers occur in Proteaceae, among Caryophyllales (some Cactaceae, some Caryophyllaceae); among rosids in some Capparaceae, some Malvaceae, some Rutaceae; among asterids in some Rubiaceae; among monocots in some Iridaceae and Amaryllidaceae (Vöchting 1886; Douglas and Tucker 1996b; Endress 1999). More elaborate monosymmetric flowers occur, in addition to the most prominent ones already mentioned, in some other Asterales (Campanulaceae, Goodeniaceae, Stylidiaceae).

Perianth

In most angiosperms, flowers have a perianth that commonly consists of one or two whorls or series of organs with protective and attractive functions. If there is only one kind of organ, these structures are referred to as tepals; if the perianth is differentiated into two kinds of organs, they are commonly called sepals and petals (for a review, see Hiepko 1965a). Sometimes all perianth organs of monocots are referred to as tepals. All petals are probably not homologous, and that may also be true for the other categories of perianth organs. However, this problem has not yet been resolved (see below), and it should be approached using a combination of phylogenetic and developmental (comparative and molecular) studies (Chapter 3).

Stamens and carpels are easy to define because of their specific functions. Stamens contain pollen sacs in which meiosis takes place and microspores are formed; carpels contain ovules in which meiosis takes places and megaspores are formed. In contrast, tepals, or sepals and petals, are more difficult to define because their

functions are less specific. Commonly, tepals or sepals have a mostly protective function for the young inner floral organs, and petals have a mostly attractive function for pollinators during anthesis. However, these functions can also be performed by other organs. In contrast, there is much transference of function between sepals and petals, and also between sepals or petals and organs more inside the flower (stamens, carpels) or outside the flower (bracts) (Baum and Donoghue 2002b). This is in contrast to stamens and carpels, which are strictly defined by their functions (and the complicated specific structures that underlie their functions). In most eudicots, the perianth organs are present in two whorls or series. In these cases, sepals and petals are best defined by their position. Sepals are the organs of the first whorl (or series), and petals are the organs of the second whorl (or series). No single character will always differentiate between sepals and petals, a realization that prompted Albert et al. (1998) to propose that floral organs be defined by the expression of floral organ identity genes (Chapter 3). Nonetheless, several traits tend to be more common in sepals or more common in petals, allowing a loose characterization. These traits can be grouped into four categories:

1. Position: In a flower that contains sepals and petals, each organ type is commonly in one series. Sepals are always in the outer series and petals in the inner. However, if only one series or more than two series are present, the distinction may be more problematic.

2. Structure: Sepals tend to be robust, persistent, green, with a broad base and more narrow upper part, acute, with three vascular strands in the floral base (Figure 12.4A). In contrast, petals tend to be delicate, ephemeral, colored, with a narrow

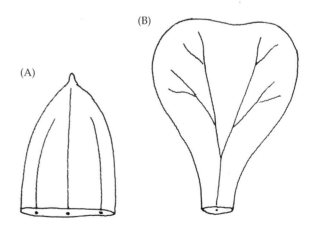

FIGURE 12.4 Typical perianth elements in a choripetalous flower in core eudicots, with vascular system (schematic diagrams). (A) Sepal. (B) Petal. (From Endress 1994c.)

(A)

(B)

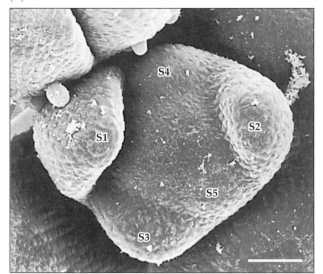

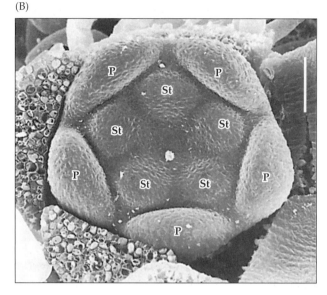

FIGURE 12.5 Typical perianth development in core eudicots. *Asclepias curassavica* (Apocynaceae). (A) Sepals (S1–S5) with successive initiation in spiral sequence. (B) Petals (P) and stamens (St) with simultaneous initiation. Bars = 50 µm. (Modified from Endress 1994c.)

base and broader upper part, obtuse or emarginate, one vascular strand in the floral base, large intercellular spaces, and papillate (Figure 12.4B).

3. Development: Sepals tend to be initiated in a spiral sequence (resulting in quincuncial aestivation) and differentiated early in flower development (Figure 12.5A). In contrast, petals tend to be initiated almost simultaneously or in a rapid spiral, and differentiated late, often being retarded until shortly before anthesis (Figure 12.5B).

4. Function: Sepals are commonly protective organs for inner floral organs in bud and the young fruit, sometimes they are also involved in fruit dispersal, and they tend to be photosynthetic. Petals are commonly attractive organs for pollinators at anthesis (shape, color, scent).

Petals that are characterized by the features listed above are common in core eudicots, especially in rosids (e.g., Geraniaceae) and Caryophyllales (e.g., Caryophyllaceae; Rohweder 1967; Rohweder and Endress 1983; Endress 2003a.). Among early-diverging eudicots, similar petals are present only in some Ranunculales, especially in Ranunculaceae and Berberidaceae (Hiepko 1965a; Erbar et al. 1999); however, the petals of these families are mostly characterized by the presence of nectaries, whereas nectaries occur at other sites in the flowers of core eudicots (Chapter 5). In other early-diverging eudicots, the perianth is simple (uniseriate), and the tepals are well differentiated (Proteaceae), weakly differentiated (Buxaceae, Myrothamnaceae, *Tetracentron*), rudimentary (*Trochodendron*), or lacking

(Eupteleaceae, some Buxaceae) (Jäger-Zürn 1967; Endress 1986a, 1986b; Douglas and Tucker 1996a, 1996b; von Balthazar and Endress 2002a, 2002b; see also D. Soltis et al. 2003a; and Chapter 5). Tepal differentiation was addressed in a comparative developmental study by von Balthazar and Endress (2002a). However, the distinction between bracts and tepals in this entire group of early-diverging eudicots has not yet been critically investigated in a comparative manner.

Even in some core eudicots the perianth remains diverse. In Saxifragales, for example, the perianth ranges from that typical of core eudicots to no perianth at all (e.g., *Cercidiphyllum;* Chapter 6). Among Santalales, Olacaceae—which are sister to all other members of the order—have a double perianth, with the sepals small and the petals exerting both protective and attractive functions. In more derived lineages of Santalales, the calyx is completely reduced, and only the petals are left (Endress 1994c).

In basal angiosperms, several groups exhibit some features of petals in the inner perianth members (Chapter 3). However, the collection of such features is much smaller than in eudicots. Therefore, one hesitates to speak of sepals and petals in this assemblage (Hiepko 1965a). Among basal angiosperms, *Cabomba* (Nymphaeaceae) is perhaps the genus with inner perianth parts that come closest to the characterization of petals (Endress 2001a). Character-state reconstructions of basal angiosperms (Chapter 3) indicate that a differentiated perianth of what have been termed sepals and petals has evolved multiple times in these plants (Figure 3.14).

In monocots, outer and inner perianth members are often less different from each other than they are in eudicots, and there may also be fluctuation between more petal-like or more sepal-like features in both whorls (Weber 1980). Therefore, one also hesitates to apply the terms sepals and petals in monocots. Although in some groups this distinction is commonly made (e.g., orchids), this is more for convenience than it is based on critical comparative research. The organs of the outer whorl are then called sepals, those of the inner whorl petals, irrespective of how similar they are.

It is hoped that molecular developmental genetic studies, in concert with new comparative developmental studies, will provide more insight into the evolution of petals and the homology of perianth organs (e.g., Albert et al. 1998; Irish and Kramer 1998; Kramer and Irish 2000; D. Soltis et al. 2002; Kanno et al. 2003; Kramer et al. 2003; Zanis et al. 2003).

Sympetaly and floral tubes

In asterids, petals are commonly congenitally united (sympetalous) (Figure 12.3B; Chapter 9, Figures 9.4 and 9.7). This fusion has repercussions for petal structure, and some features are different from those of free petals. Petals that are united do not have narrow bases, but are more often acute and have three main vascular bundles. They are also less delicate than those in flowers with free petals. In addition, united petals are less retarded in development than free petals and often take over a protective function of the inner organs in later bud stages by overtopping the sepals. Thus, in asterids, petals have several features that are more characteristic of sepals (Endress 1994c; Figure 12.6, A and B).

Sympetaly is a prominent key innovation in angiosperm evolution. It is widespread in the asterids, the most species-rich major clade of angiosperms and the clade in which some of the most complicated flowers (Apocynaceae) and inflorescences (Asteraceae) have evolved (Chapter 9). Sympetaly has also evolved in several other groups, with fewer genera and species than asterids, such as some Crassulaceae and Malvaceae (Chapter 8). The union of petals allows the formation of floral tubes, which in turn facilitates adaptive radiations correlated with different pollinators because the length and width of floral tubes can easily be changed (Alexandersson and Johnson 2002). Another, less often considered aspect is that petal union allows the construction of larger flowers because sympetaly provides stability to the floral architecture. Thus, the world's largest flowers are in groups with united petals (or tepals). Examples are *Brugmansia* (Solanaceae) and *Fagraea* (Gentianaceae), both asterids, which attain flower lengths of more than 50 cm. *Aristolochia* (Piperales) and *Rafflesia* (a eudicot that now appears to belong to Malpighiales; see Chapters 8 and 11; Barkman et al.

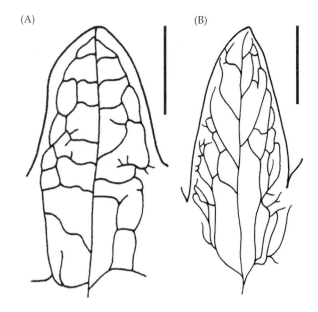

FIGURE 12.6 Typical perianth elements in a sympetalous flower in core eudicots, with vascular system. *Solanum luteum* (Solanaceae). (A) Sepal. (B) Petal. Bars = 1 mm. (From Rohweder and Endress 1983.)

2004) have united tepals. This architecture extends the upper end of a spectrum of diversification that is narrower in plants with free perianth parts (see also Endress 1994c).

In asterids, two kinds of developmental patterns of petals have been distinguished, "late" and "early" sympetaly (Erbar 1991; Erbar and Leins 1996). In late sympetaly, the developing petals appear free at first and are later united by a common base. In early sympetaly, a ring meristem appears first, on which the individual petals then become visible. Late sympetaly is common in Gentianales and Lamiales, whereas early sympetaly is common in Dipsacales and Asterales. An interesting problem is whether these two developmental patterns could be (functionally or organizationally) dependent on the overall shape of the floral apex at the time of organ inception (Endress 1997b; Roels 1998). Because in Gentianales and Lamiales the ovaries are predominantly superior and in Dipsacales and Asterales predominantly inferior, the early shape of the floral apex is more convex in the first group, but more concave in the latter group. The more concave shape could then give the impression of a ring meristem at petal inception, whereas this is not the case in apices that are more convex.

In some asterids, sympetaly is expressed in early floral development but is no longer evident in mature flowers (e.g., Apiaceae, Araliaceae, Pittosporaceae) (Erbar and Leins 1988, 1995a). Perhaps more than one developmental step is necessary to form distinct sympetaly: (1) confluent meristems of the initiated petals, and (2) an intercalary elongation zone below the united petal meristems. It could be that in these asterids only step 1 has been acquired but not step 2. This should also be studied in other asterids that seemingly have free petals.

In many groups with sympetalous flowers, the stamens are also fused with the corolla to a greater or lesser degree. This fusion further enhances the firmness of the flower, allowing more intimate synorganization between androecium and corolla. For instance, in polysymmetric flowers, this fusion creates five separate nectar canals (if five stamens are present), resulting in the architecture of revolver flowers, in which pollinators must move around the flower to exploit the entire nectar supply and thus most likely have extensive contact with anthers and stigmas. The most complicated revolver flowers are in Apocynaceae (Asclepiadoideae, Periplocoideae, Secamonoideae), in which five pollen units (groups of pollinia) can be attached to an insect by means of a complicated transport organ (translator with clip or glue). Other clades of Apocynaceae show much simpler flowers, and the stepwise evolutionary complication can be seen if the entire family is comparatively studied (Fallen 1986; M. Endress et al. 1996; M. Endress and Bruyns 2000; M. Endress 2001).

In some eudicots, sepals and petals are fused and form a synorganized perianth (e.g., Tropaeolaceae, Myrtaceae: especially *Eucalyptus;* Drinnan and Ladiges 1989). In monocots, the outer and inner perianth whorls may also be fused (e.g., some Asparagaceae).

Floral spurs

Floral spurs are hollow outgrowths of perianth organs. Spurs are present in corollas more often than in calyces. They do not characterize large clades. However, they have evolved in many angiosperm clades (commonly at genus level), especially in eudicots. Among noneudicots, spurs occur only among monocots in some orchids and lilies (e.g., *Tricyrtis*). Examples of genera with spurred flowers are *Aquilegia* (Ranunculaceae; Hodges 1997a, 1997b; Fulton and Hodges 1999), *Epimedium* (Berberidaceae), and *Halenia* (Gentianaceae; von Hagen and Kadereit 2002), all with four or five spurs. Most spurred flowers have only one spur, such as *Delphinium* (Ranunculaceae), some Plantaginaceae (Sutton 1988), and *Heterotoma* (Campanulaceae; Ayers 1990). These spurs have a connection with nectar production or nectar presentation, or both. In early-diverging eudicots with nectariferous petals (Ranunculaceae, Berberidaceae), nectar is produced at the end of the spur and stored in the spur, which is a part of the petals. However, in *Fumaria* and relatives (Papaveraceae), the petals are not nectariferous, and nectar is produced at the base of the stamen filaments (e.g., *Corydalis*). In some core eudicots (e.g., asterids), nectar is more often produced by a separate disk nectary and, if a spur is present, stored in the spur. An exception is *Halenia* (Gentianaceae), in which nectar is produced directly in the spur and whose closest relatives have shallow petal nectaries. Another special case is *Diascia* (Scrophulariaceae), which has oil-producing spurs from which specialized bees collect oil with their two front legs (Vogel 1974). For other examples, see Hodges (1997b).

Conspicuous adaptive radiations have often accompanied the origin of spurs. This can be seen if genera with spurs are compared with closely related genera that have flowers without spurs. Some authors suggest that the presence of spurs is a key innovation (Hodges 1997b), but others argue that this is not always the case: in *Halenia* (Gentianaceae), radiation took place long after spurs had been acquired (von Hagen and Kadereit 2002).

Androecium

The term "androecium" refers to all stamens, which are, in turn, defined as floral organs that contain microsporangia. If stamens are defined as such (i.e., as "floral organs"), they are by definition restricted to flowering plants. In a broader sense, stamens are not restricted to angiosperms (Endress 2001b). However, angiosperm stamens have a thecal organization (Endress and Stumpf 1990) not present in other plant groups. Each stamen characteristically has an anther with two lateral thecae often situated on a more or less elongate sterile part, the filament. Each theca is differentiated into a collateral pair of pollen sacs (microsporangia). At maturity each theca opens by a common longitudinal slit between the two pollen sacs. All major groups of angiosperms are characterized by this type of stamen structure (Figure 12.7A). In only rare cases do some small angiosperm clades deviate from this pattern (Endress and Hufford 1989; Hufford and Endress 1989; Endress and Stumpf 1990), as discussed below. However, the diversity of anther shapes based on variation of proportions of the basic pattern is considerable (Bernhardt 1996; Endress 1996a; Figure 12.8).

Deviation from the basic anther pattern occurs in the opening mechanism and the number of pollen sacs per stamen. In some groups, the longitudinal slit bifurcates at the ends, and then each theca opens by two valves. This type of anther is mainly known from basal angiosperms (some Magnoliales, Laurales), from some early-diverging eudicots (Trochodendraceae), and also from some Saxifragales (Hamamelidaceae; Endress and Hufford 1989; Hufford and Endress 1989). There is a strong correlation between this dehiscence pattern and the presence of a thick connective (i.e., the sterile part of the anther between the two thecae). Such anthers are prone to lose one pollen sac per theca (some Laurales,

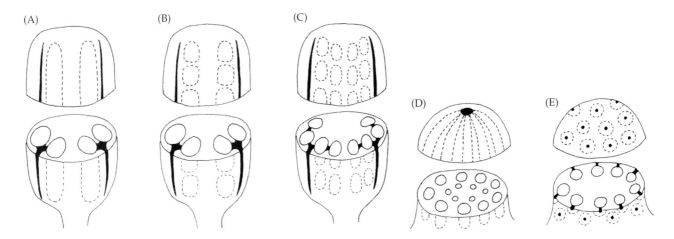

FIGURE 12.7 Androecium. Anther types in angiosperms (schematic diagrams). (A–C) with thecal organization. (D, E) without thecal organization. (A) Basic type with two disporangiate thecae. (B) Polysporangiate anther with transverse septa. (C) Polysporangiate anther with transverse and longitudinal septa. (D) Polysporangiate anther with a single pore for all sporangia in the center. (E) Polysporangiate anther with numerous pores, one for each sporangium. (From Endress and Stumpf 1990.)

some Hamamelidaceae). A loss of one pollen sac per theca also occurs in some eudicots that have otherwise typical anther structure, but there is no obvious phylogenetic pattern. Rather, a functional relationship can be seen, as this reduction occurs, for example, in some plants with cleistogamous flowers. In a hybrid between a tetra- and disporangiate species of *Microseris* (Asteraceae), one major gene and four modifier loci are responsible for the difference in microsporangium number (Gailing and Bachmann 2000). An increase in number of microsporangia also occurs in some groups but also without a distinct systematic pattern (Figure 12.7, B and C). The most extreme forms of stamens are those with a complete loss of thecal organization; an irregular number of microsporangia is present, and each opens by an individual pore or all converge to one apical pore (Figure 12.7, D and E). Such forms are known from the two parasitic (but unrelated) families Rafflesiaceae and Santalaceae, from a single genus, *Polyporandra*, of Icacinaceae (Endress and Stumpf 1990; Kårehed 2001), and perhaps *Boschia* and *Cullenia* of Malvaceae (Nyffeler and Baum 2000, and see Chapter 8).

The number of stamens in a flower is, in general, more variable than the number of the other organ categories (sepals, petals, carpels). In many groups with whorled floral phyllotaxis, stamens occur in two whorls, resulting in a total of six stamens in many monocots and 10 in many eudicots (diplostemony). A variant of diplostemony is obdiplostemony, in which the epipetalous (and not the episepalous) stamens appear to be the outer organs, with the result that the carpels—if they are isomerous with the stamens of a whorl—alternate with the epipetalous instead of the episepalous stamens. This formation is unexpected because subsequent whorls of organs normally alternate with each other. The apparent reason for this disturbance of normal alternation is that epipetalous stamens tend to be smaller, so that there is more space for the carpels between the epipetalous than between the episepalous stamens (Eckert 1965; Matthews and Endress 2002). Examples are found in many Geraniales and Oxalidales.

In many taxa, the number of stamens per flower has increased dramatically. In flowers with spiral phyllotaxis, this increase is attained by an extended period of stamen formation continuing the ontogenetic spiral. In contrast, in flowers with a whorled phyllotaxis, there are different methods of increasing the number of stamens. It might be associated with an increase in whorl number, often with double positions (e.g., in Papaveraceae; Endress 1987a, 1987b), or an increase in the number of floral sectors (e.g., in some Crassulaceae). A different mode of development is by the formation of primary primordia on which a larger number of secondary primordia are formed, each giving rise to a complete stamen. Such primary primordia may be formed at the place where single stamens would be expected; they appear as several distinct primordia, or as a single ring primordium on which the individual stamen primordia are then not formed in whorls but in a less regular fashion. Despite the irregular stamen position, in androecia with primary and secondary primordia, centripetal and centrifugal patterns can be distinguished (Figure 12.9, A and B) that are often characteristic at the family level (Corner 1946). Centripetal stamen initiation occurs in Mimosoideae–Fabaceae (Gemmeke 1982), and centrifugal initiation is found in Capparaceae (Leins and Metzenauer 1979; Ronse De Craene and Smets 1997), Dilleniaceae (Corner 1946; Endress 1997a), and Lecythidaceae (Hirmer 1918;

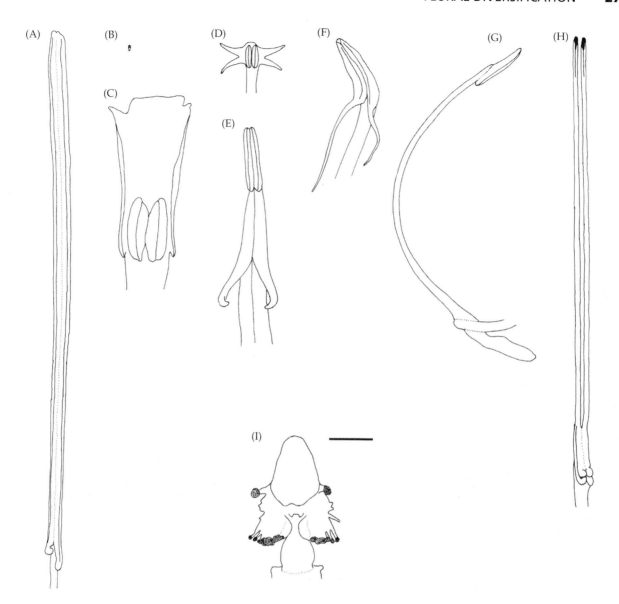

FIGURE 12.8 Androecium. Diversity in the basic type of anthers. A–B, extremes in sizes. C–I, diversity in proportions. (A) *Strelitzia reginae* (Strelitziaceae). (B) *Piper peltatum* (Piperaceae). (C) *Costus igneus* (Costaceae). (D) *Globba winitii* (Zingiberaceae). (E) *Roscoea purpurea* (Zingiberaceae). (F) *Thunbergia mysorensis* (Acanthaceae). (G) *Salvia patens* (Lamiaceae). (H) *Demosthenesia cordifolia* (Ericaceae). (I) *Psychopsis papilio* (Orchidaceae). Bar = 4 mm. (From Endress 1996a.)

Endress 1994c). More rarely, the stamens seem to be initiated almost simultaneously, as in Achariaceae (Bernhard and Endress 1999).

The centrifugal pattern of development is so unusual that it was once thought to characterize a major group of dicots, the Dilleniidae (e.g., Cronquist 1957, 1981), a group now known to be highly polyphyletic (see Chapter 2). A series of phylogenetic studies showed that centrifugal androecia evolved many times in clades of core eudicots that are well-separated phylogenetically (e.g., Caryophyllales, Saxifragales, Malvales, Brassicales, Ericales) (e.g., Chase et al. 1993; Savolainen et al. 2000a; D. Soltis et al. 2000). Thus, the derivation of the centrifugal pattern is of interest within these clades, at a much lower taxonomic level than previously suggested when this pattern of development was thought to characterize an entire subclass.

Synandry, the congenital union of stamens, is much less common than the union of organs in the other three floral organ categories. There are two types of synandry, one in which only the filament region is united, and the other (much rarer) in which the anther region is also united. Synandry occurs in many different angiosperm groups. It is more prominent in unisexual than in bisexual flowers because there is no gynoecium in the center of staminate flowers; the stamens can therefore be unit-

(A)

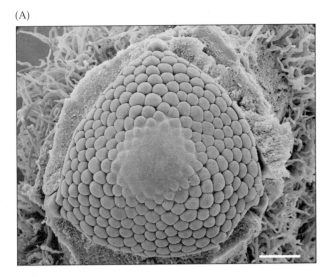

FIGURE 12.9 Androecium. Divergent development of polyandrous androecia. (A) Centripetal. *Annona squamosa*

(B)

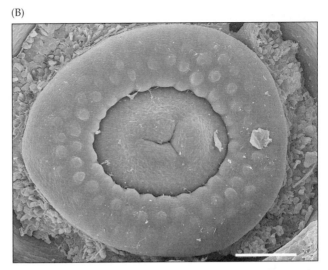

(Annonaceae). Bar = 200 μm. (B) Centrifugal. *Barringtonia* cf. *samoensis* (Lecythidaceae). Bar = 100 μm.

ed in a central structure that involves both the filaments and anthers (Figure 12.10). Such extreme cases are present in Myristicaceae, Menispermaceae, Euphorbiaceae, and Cucurbitaceae (only in a few taxa in the last three families). Cases in which only the filaments are involved in stamen union are more common in bisexual flowers (e.g., diagnostic for some Amaranthaceae, Malvaceae, and Meliaceae).

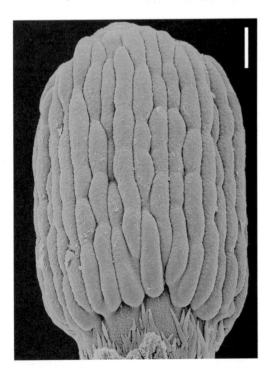

FIGURE 12.10 Androecium. Stamens completely united into a synandrium, and pollen sacs polysporangiate (portioned by transverse constrictions). *Myristica insipida* (Myristicaceae). Bar = 200 μm.

Gynoecium

Gynoecia are exclusive and omnipresent in angiosperms. Carpels, the structural elements of gynoecia, are cup-shaped or scale-shaped in early development. One to many ovules are formed near the margin of each carpel. During development, the ovule or ovules become completely enclosed in the carpel—the carpel flanks grow toward each other on the ventral side and finally become sealed, so that the inner space with the ovule(s) is secluded from the outer world (angiospermy). Closure and sealing of the carpels is attained by secretion or by postgenital fusion. Endress and Igersheim (2000) distinguished four types of angiospermy in basal angiosperms, early-diverging eudicots, and basal monocots: (1) closure by secretion, without postgenital fusion; (2) closure by postgenital fusion at the periphery, but with a continuous secretory canal without fusion; (3) closure by postgenital fusion at the entire periphery and a secretory canal that stops below the stigma; and (4) closure by complete postgenital fusion, without a secretory canal (see Chapter 3, Figure 3.16). Postgenital fusion (epidermal fusion) comes about by interdentation and sometimes also division of epidermal cells between two contiguous, adherent surfaces, with the result that the surfaces are no longer obvious (Sinha 2000). In the basalmost angiosperms, Amborellaceae, Nymphaeaceae, and Austrobaileyales, carpels are cup-shaped and sealed only by secretion (type 1); they have only one or a small number of carpels (Doyle and Endress 2000; Endress and Igersheim 2000; Endress 2001a; see also Chapter 3). In other angiosperm groups, carpel closure by partial or complete postgenital fusion is predominant (types 2 to 4) (Endress and Igersheim 1999, 2000; Igersheim et al.

2001). The process of postgenital fusion probably first evolved in the carpels of basal angiosperms. In more elaborate flowers, postgenital fusion also occurs at locations other than within individual carpels, such as among different carpels (Walker 1975; Endress et al. 1983) and between stamens and carpels (e.g., Fallen 1986; Endress 1994c).

A major functional innovation of the advent of angiospermy is pollination on the surface of the carpel (the stigma), and no longer on the surface of the ovule (the micropyle), and, concomitantly, the formation of a pollen tube transmitting tract (PTTT). The PTTT almost always differentiates along the primary morphological surface of the carpel (which, however, is hidden from the outer world because of carpel closure; Endress 1994c). Pollen grains are often deposited in groups on the stigma, and then several pollen tubes grow synchronously along the PTTT. Consequently, the PTTT acts as an area where pollen tubes (male gametophytes) are selected before fertilization (Mulcahy 1979; Mulcahy and Mulcahy 1987). Another innovation is that the stigma and PTTT act as sites where self-incompatibility reactions take place (Wheeler et al. 2001). Such prezygotic self-incompatibility is otherwise not well developed in spermatophytes in which carpels are lacking (Runions and Owens 1998).

Syncarpy

Syncarpy is the congenital union of carpels. Two major aspects of syncarpy are (1) the extent of union—whether carpels are united only at the base or all the way to the top, with many transitions between these extremes; and (2) the extent of union (confluence) of the inner morphological surface of the carpels. If the inner morphological surfaces are confluent, this facilitates the formation of a unified PTTT of all carpels, an area termed a compitum (Carr and Carr 1961; Endress 1982). Presence of a compitum is a precondition for centralized pollen tube selection, which is hypothesized to be more efficient than individual selection in each carpel (Endress 1982; Armbruster et al. 2002).

Basal angiosperms are largely apocarpous. Examples include Amborellaceae, some Nymphaeaceae, Austrobaileyaceae, and Schisandraceae. However, syncarpy does appear in some, such as Canellaceae, *Takhtajania* (Winteraceae), and *Monodora* (Annonaceae) (see Chapter 3). In contrast, syncarpy predominates in core eudicots and monocots (except for Alismatales; cf. Igersheim et al. 2001). More than 80% of all angiosperm species are syncarpous; the remainder are either apocarpous or unicarpellate. It has yet to be determined in more detailed studies how common the presence of a compitum is in syncarpous gynoecia. For instance, a compitum may not be present in at least some of those few early-diverging eudicots with a syncarpous ovary (e.g., Buxaceae, Trochodendraceae); the same may be

true for Saxifragales (Endress and Igersheim 1999). However, the presence of a compitum seems to be the normal state in other core eudicots (rosids, asterids, Caryophyllales, Santalales) and also in monocots other than some families in Alismatales (an early-diverging monocot lineage), which are partly apocarpous.

Inferior ovaries

Because gynoecia are formed as the last organs in the center of the flower, they have a unique position. Not only can the carpels unite to form a unified structure in syncarpous gynoecia without obstructing other organs, they also can evolve inferior ovaries (Figure 12.11). It has been suggested that inferior ovaries are better protected against certain pollinators with potentially destructive mouth parts, such as beetles or birds (Grant 1950). Inferior ovaries have evolved many times, as indicated in character-state reconstructions across all angiosperms (Gustafsson and Albert 1999) and are more common in more derived groups of angiosperms. However, they are also present in some basal angiosperms, including Chloranthaceae, Gomortegaceae, and Hernandiaceae. Furthermore, inferior ovaries appeared early in the fossil record of the angiosperms—they are known from the Early Cretaceous (Friis et al. 1994). In core eudicots and derived lineages of monocots, entire large clades are characterized by inferior ovaries (e.g., the eudicots Asterales, Dipsacales, Rubiaceae; the monocots Orchidaceae, Zingiberales).

Developmentally, inferior ovaries typically arise by carpel initiation on a slightly concave floral apex and by differential growth of the floral base including the gynoecium base, which results in a reinforcement of the concavity (this has been referred to as appendicular epigyny; see Chapter 6). In contrast, in most superior ovaries, the floral apex remains convex throughout development (see Chapter 6). However, ovaries that appear inferior can, in rare cases, result from a seemingly hypogynous early developmental stage, as in Cactaceae (Boke 1964, 1966).

Evolutionary reversals from epigyny to hypogyny have been revealed in several groups. Single genera with superior ovaries that occur in groups characterized by inferior ovaries suggest such reversals, and change from inferior to superior may even be seen during ontogeny, such as in *Gaertnera* (Rubiaceae; Igersheim et al. 1994) and *Tetracarpaea* (Tetracarpaeaceae; D. Soltis et al. 2003c). In Saxifragaceae there is evolutionary oscillation between ovaries that appear superior and those that are inferior, with all ranges of intermediate forms also present (Kuzoff et al. 1999; D. Soltis et al. 2001). Developmental studies in Saxifragaceae have indicated that some ovaries that appear superior are in fact inferior; these ovaries, now termed pseudosuperior, actually result from the developmental early stage (appendicular epigyny) characteristic of inferior ovaries (Chapter 6). Furthermore, differ-

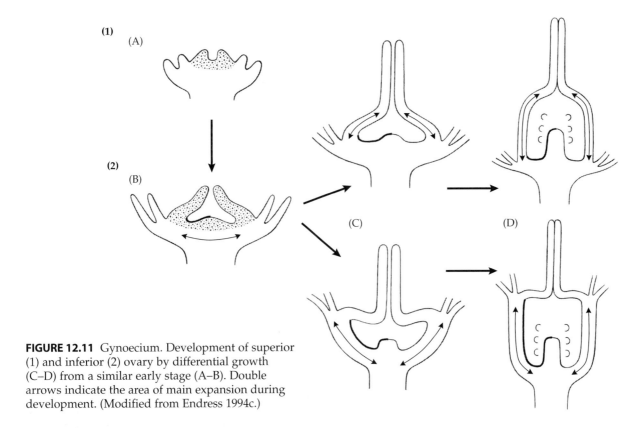

FIGURE 12.11 Gynoecium. Development of superior (1) and inferior (2) ovary by differential growth (C–D) from a similar early stage (A–B). Double arrows indicate the area of main expansion during development. (Modified from Endress 1994c.)

ences in the relative growth of the superior and inferior regions of the ovary result in the diversity of ovary positions (e.g., one-quarter inferior, half inferior, three-quarters inferior) found throughout the family.

Ovules

Ovules are constituent parts of the carpels, but they appeared before carpels in the evolutionary history of seed plants. Ovules consist of a nucellus (megasporangium) and, in angiosperms, two integuments. In the nucellus, meiosis takes place, and the embryo sac (megagametophyte) develops, in which fertilization of the egg cell occurs. One or both integuments form a micropyle, a narrow canal through which a pollen tube is chemotactically attracted (Herrero 2001). The pollen tube grows through the micropyle, reaches the nucellus apex, which it penetrates, grows into the embryo sac, and releases the male gametes. One male gamete usually fuses with the egg cell to give rise to an embryo, and the other fuses with the diploid (or haploid) nucleus of the central cell to give rise to a triploid (or diploid) endosperm (Williams and Friedman 2002).

Crassinucellate (Figure 12.12A) and tenuinucellate ovules (Figure 12.12B) can be distinguished in angiosperms. Tenuinucellate ovules are characterized by a single cell layer around the meiocyte in the nucellus, whereas crassinucellate ovules have two or more cell layers. Crassinucellate ovules are predominant in angiosperms,

but they are not uniform in structure and should not be considered a single character. However, it has long been known that some larger groups of angiosperms (asterids) are characterized by tenuinucellate ovules (Chapter 9; see also Endress et al. 2000a; Albach et al. 2001b; Endress 2002). In addition, tenuinucellate ovules commonly have only a single integument. The macrosystematic significance of this combination was emphasized by Philipson (1977) in a review of ovule forms in dicotyledons. Among monocots, the systematic distribution of these ovule types is less distinct (Rudall 1997).

Several families previously considered rosids, dilleniids, or hamamelids but recently shown to be members of the asterid clade (e.g., Balsaminaceae, Byblidaceae, Escalloniaceae, Hydrangeaceae, Lecythidaceae, Loasaceae, Marcgraviaceae, Pittosporaceae, Roridulaceae, Sarraceniaceae, Theaceae, Vahliaceae) are characterized by tenuinucellate ovules. Tenuinucellate, unitegmic ovules predominate in asterids, although they are not found in all members of the clade (Endress et al. 2000a; Albach et al. 2001b; Endress 2003b).

The ancestral nucellar state for the asterids is uncertain (Chapter 9). Most Cornales and Ericales, which are successive sisters to all other asterids (Chapter 9), have tenuinucellate ovules. However, there is not a perfect association between tenuinucellate ovules and the unitegmic condition (compare Figures 9.9 and 9.10). In Ericales, ovules are tenuinucellate, but vary between bitegmic and unitegmic (Boesewinkel and Bouman

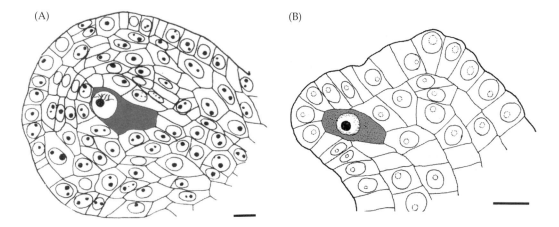

FIGURE 12.12 Ovules. Two major types of nucellus at the meiocyte stage; meiocyte highlighted in dark gray. (A) Crassinucellate. *Corylopsis willmottiae* (Hamamelidaceae). (modified from Endress 1977.) (B) Tenuinucellate.

Dermatobotrys saundersii (former Scrophulariaceae s.l.; probable member of Stilbaceae). Bars = 10 μm. (Modified from Hakki 1977.)

1991; Endress 2003b). In Garryales, which may be sister to most of the rest of euasterid I, ovules are commonly crassinucellate, but mostly unitegmic (e.g., Eckardt 1963; Kapil and Mohana Rao 1966; Satô 1976; Endress 2003b); in Cornales, they are crassinucellate or tenuinucellate and unitegmic (Satô 1976; Endress 2003b). Forms intermediate between crassinucellate and tenuinucellate may characterize families, such as Convolvulaceae (see Endress 2003b).

Tenuinucellate ovules are smaller than crassinucellate ovules, and thus the potential for flexibility in ovule number (between one and exceedingly many) per carpel is higher with tenuinucellate than with crassinucellate ovules. The largest number of ovules per ovary occurs in some Orchidaceae, which have tenuinucellate ovules—more than half a million in *Coryanthes* (see Nazarov and Gerlach 1997). In asterids with tenuinucellate ovules, some families, such as Solanaceae, also have clades with high ovule numbers. However, ovule number is a poorly explored character.

Morphological Elaborations in Flowers Pollinated by Animals

Many morphological elaborations are based on synorganization of organs. An especially successful elaboration (innovation) was the evolution of sympetaly, which allowed the easy building of different floral architectures (e.g., lip flowers, flowers with tubes of various lengths and widths, such as in Gentianales, Lamiales) and specialization for many different pollinators (e.g., bees, butterflies, sphingids, birds, bats; e.g., Vogel 1990). Another example of successful innovation is the evolution of revolver flowers—flowers that have several canals to reach the nectar, forcing pollinators to move

around the center of the flower to gather all available nectar. This increases body contact with the anthers and stigma and may enhance pollination success. Revolver flowers are common in Gentianales and Solanaceae; the number of origins of the feature has not been studied.

Another morphological elaboration is portioning of the pollen produced in a flower (or anther), resulting in staggered pollen presentation (Leins 2000). This is achieved by various means, including elaborate secondary pollen presentation, such as in Asterales (Leins and Erbar 1990; Erbar and Leins 1995b) or Fabales (Westerkamp and Weber 1999). In flowers with secondary pollen presentation, pollen is not transferred to the pollinators directly from the anthers but is first deposited on the style or in the tip of a keel before it is removed by pollinators. Secondary pollen presentation has evolved in many different clades of angiosperms (Yeo 1992), some of which are species-rich, such as Asteraceae and Fabaceae. Pollen portioning is also found in flowers with poricidal anthers, which are buzz-pollinated by bees (discussed below). Pollination by pollinia has been successful in Orchidaceae and asclepiads (Apocynaceae; Pacini and Hesse 2002). Here, the entire pollen mass of a theca is transported in a solid body, thus no pollen is lost. In some subgroups of both orchids and asclepiads, pollen portioning is also attained by the formation of massulae instead of entire pollinia.

The evolution of monosymmetric flowers allowed enhanced precision mechanisms of pollen application to the body of a pollinator and subsequently to the stigma of a flower. This innovation also led to a greater potential for diversification in pollination biology. Some large families and orders have evolved monosymmetric flowers, for example Orchidaceae, Zingiberales, Fabales, and Lamiales (Endress 1999). There have also been repeated reversals from monosymmetric to polysymmetric flowers in some clades characterized by bilateral symmetry, such as Lamiales (see Chapter 9, Figure 9.12)

Various means of pollinator attraction, in addition to pollen, have evolved in flowers, such as nectar, oil, resin, and perfume, which are used for food, nest building, or attraction of mates. Whereas pollen and nectar are used by many pollinators, oil, resin, and perfume are used by some highly specialized bees (Vogel 1974, 1988; Simpson and Neff 1981; Steiner 1991). Some flowers pollinated by certain bees that collect pollen by "buzzing"—by activity of flight muscles to stimulate the movement of pollen out of the anthers—have attained a "solanoid" overall shape, that is, they have a flat, expanded corolla, a cone-shaped androecium with large, showy, poricidal anthers and short filaments, and a punctiform stigma (Vogel 1974; Vaknin et al. 2001). Such flowers are known from many families of eudicots and monocots. In flowers that offer only pollen as a reward to pollinators, there is also an evolutionary trend to diversify stamen structure and function into showy "feeding stamens" and cryptic "pollinating stamens" (heteranthery) and, in some cases, the pollen becomes different in the two stamen morphs (Vogel 1978; Endress 1997a). Arrangement, size, shape, color, and scent of flowers are means of differential attraction of pollinators and a source for reproductive isolation, speciation, and thus diversification (Waser 1998; Chittka et al. 1999; Lunau 2000; Chittka and Thomson 2001).

A considerable repertoire of mating systems, apart from specialized relationships with different pollinators, also plays a role in the diversification of flowers in angiosperms. Examples include various kinds of gender distribution, self-incompatibility, self-compatibility, dichogamy, and herkogamy (see, e.g., Bertin and Newman 1993; Barrett 1995, 1998, 2002; Barrett et al. 1996, 2000; Holsinger 1996; Richards 1997). These specializations involve various ways of enhancing outbreeding or keeping a balance between outbreeding and inbreeding (Lloyd and Schoen 1992; Freeman et al. 1997; Schoen et al. 1997). Monoecy and dioecy are different ways of separating genders on one or on separate individuals (e.g., Renner and Won 2001). Dioecy seems to evolve commonly from monoecy (Renner and Ricklefs 1995). In dichogamy, both genders are produced by the same individual but not at the same time; both genders can be present in the same flowers (bisexual flowers) or in separate flowers (unisexual flowers); protogyny (with the female function working first) and protandry (with the male function working first) can be distinguished (Lloyd and Webb 1986; Renner 2001). In herkogamy both genders are produced by the same individual and at the same time, but they are spatially separated (Webb and Lloyd 1986). A special case of herkogamy is heterostyly, in which different floral morphs with long and short stamens or styles are pro-

duced on different individuals (Barrett 1992; see Chapter 9, Figure 9.5). Commonly, these mating systems are flexible at relatively low levels of the phylogenetic hierarchy (Korpelainen 1998). An exception is the practically universal presence of protogyny in basal angiosperms with bisexual flowers (Endress 1990b, 2001a).

Morphological Reduction in Flowers that are not Animal-Pollinated

In some angiosperm groups, wind pollination is predominant, and flowers are concomitantly very different in structure and behavior from their animal-pollinated ancestors. The main trend is floral reduction of various kinds, but with increased pollen production (Wagenitz 1975), which may lead to drastic changes in flower appearance. Interestingly, a phylogenetic analysis focused on the nitrogen-fixing clade suggests that wind pollination is more likely to evolve in those groups that have small, simple flowers and dry pollen (Culley et al. 2002; see Chapter 8). Major wind-pollinated groups are Fagales and many Rosales (the urticoid families) among eudicots, and several Poales among monocots, grasses being the most prominent (Linder 1998). Wind pollination has also evolved in many smaller groups that are phylogenetically nested in larger animal-pollinated groups, such as single genera (e.g., *Dodonaea* in Sapindaceae, *Ambrosia* in Asteraceae) or single species in species-rich genera (e.g., *Acer negundo*).

In *Acer negundo* (Figure 12.13B), in contrast to the insect-pollinated *Acer pseudoplatanus* (Figure 12.13A), the flowers are unisexual, petals and nectary are lacking, and the stigma is much larger. Thus, a dramatic change in the superficial appearance of the flower is associated with a shift to wind pollination. Fagales, which are predominantly wind-pollinated, have a suite of traits—not only in floral morphology at anthesis but also in floral development before and after anthesis—that is apparently functionally related to wind pollination. These traits include flowering early in the growing season, in the leafless state, and immaturity of the ovary and ovules at anthesis, with the effect that pollination may precede fertilization by months. In extreme cases (e.g., *Corylus*), ovules are not even present at the time of pollination (Endress 1977; Thompson 1979). Whether wind pollination evolved within Fagales or their ancestors and whether there were reversals to insect pollination in Fagales is not clear (Manos et al. 2001).

Water pollination is much less prominent in angiosperms, but it has evolved in several water plant groups (Philbrick and Les 1996). Water pollination is also accompanied by floral reduction, extreme in some groups (some genera of Hydrocharitaceae and some other small families of Alismatales, *Callitriche* of Plantaginaceae; Cook 1982).

(A)

(B)

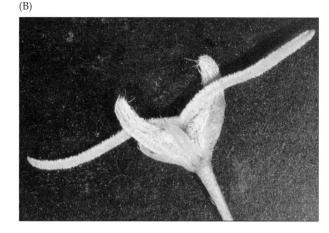

FIGURE 12.13 Drastically different appearance of flowers with two different pollination syndromes in two species of the same genus. (A) *Acer pseudoplatanus* (Sapindaceae), insect-pollinated, bisexual flower. (B) *Acer negundo* (Sapindaceae), wind-pollinated, carpellate flower.

Evolutionary Diversification in Flowers of the Major Angiosperm Groups

In the basalmost clades of extant angiosperms, consisting of Amborellaceae, Nymphaeaceae, and Austrobaileyales (Chapter 3), flowers are generally small (except for some more specialized taxa: Austrobaileyaceae, some Nymphaeaceae), floral phyllotaxis is spiral (Amborellaceae, Austrobaileyales) or whorled (Nymphaeaceae), and stamens open with longitudinal slits (not with valves; except for *Nuphar*, Nymphaeaceae; Hufford 1996a). The carpels are not postgenitally fused but closed by secretion (except for some more specialized taxa: *Illicium* and Nymphaeaceae; Doyle and Endress 2000; Endress and Igersheim 2000; Endress 2001c).

In Magnoliales and Laurales, floral phyllotaxis is diverse and flexible, and the spiral pattern is not suppressed because synorganizations are largely lacking (Endress 1987a; Chapter 3). Floral merism is highly variable not only in basal angiosperms but also in early-diverging eudicots (Chapter 5). In basal angiosperms, the perianth often lacks a clear differentiation into sepals and petals, and in some instances, a perianth is lacking altogether (see also Chapter 3). Stamens are often massive with short filaments, thick connectives, and valvate dehiscence. Uniaperturate pollen predominates, although diverse forms are present (Sampson 2000; see Chapter 3). The gynoecium is often apocarpous. Ovules are crassinucellate, bitegmic, and commonly anatropous.

In monocots, flowers are almost always trimerous (Chapters 3 and 4). There is a tendency for intimate connection of organs of different kinds within the same floral radius (Endress 1995a). The perianth is not always clearly differentiated into sepals and petals. Pollen is also uniaperturate. Nectaries are present in the septa of carpels (septal nectaries) or on the tepal surface (tepal nectaries). Nectaries are lacking in some large groups (e.g., Poales). Often, floral bracts or inflorescence bracts are elaborated and play important roles as protective organs for flowers or entire inflorescences (e.g., Poaceae and Arecaceae) or as attractive organs for pollinators (e.g., Araceae, Arecaceae, Bromeliaceae, Pandanaceae, Zingiberales).

In eudicots, floral phyllotaxis is commonly whorled, except for some groups of early-diverging eudicots (some Ranunculales, Nelumbonaceae); flowers are often pentamerous, except for the basalmost clades. The perianths of eudicots have sepals and petals, except for early-diverging members. However, in Ranunculales, petals with nectaries may have evolved separately. Stamens commonly have well-developed filaments and anthers with thin connectives, each theca opening by a longitudinal slit.

Early-diverging eudicots are labile for floral construction (see also Chapter 5). Spiral and whorled floral phyllotaxis occur in close relatives, sometimes in the same genus, but more often among genera and families. Spiral flowers occur in many Ranunculaceae (Schöffel 1932; Hiepko 1965a), some Menispermaceae (Endress 1995b), perhaps Lardizabalaceae (*Sargentodoxa*), *Nelumbo* (perianth; Hayes et al. 2000), and sometimes Trochodendraceae (*Trochodendron* is spiral and whorled; Endress 1990a). In taxa with spiral floral phyllotaxis, variation in floral organ number is relatively high; in taxa with whorled flowers, di- and trimerous flowers predominate (Ranunculales, Proteaceae, Buxaceae, *Tetracentron*; Drinnan et al. 1994). In some groups, thecae open by valves and not by longitudinal slits (a few Ranunculaceae, Eupteleaceae, Platanaceae, Trochodendraceae). Pollen is often tricolpate, in some tricolporate (e.g., some Menispermaceae; Thanikaimoni 1984). Disk nectaries are lacking except for Buxaceae and Sabiaceae (von Balthazar and Endress 2002b); nectaries are present on petals or staminodes, or are lacking. Ovules are mostly crassinucellate and bitegmic (but pseudocrassinucellate in Papaveraceae and Ranunculaceae; cf. Endress and Igersheim 1999).

In core eudicots, flowers are predominantly pentamerous or pentamerous-derived (see Chapters 5 and 6). Increase in stamen number, based on secondary stamen primordia superimposed on the primary primordia, has occurred in several families. Pollen is tricolporate or tricolporate-derived. Disk nectaries are common.

In rosids, flowers are almost always choripetalous. Ovules are mostly crassinucellate and bitegmic. In many groups, a secondary increase in stamen number is common. Many details of rosid flowers are poorly known. Comparative studies of larger clades are just beginning to be conducted (e.g., Ronse De Craene and Smets 1999; Matthews et al. 2001; Ronse De Craene et al. 2001; Schönenberger et al. 2001; Matthews and Endress 2002, 2004, 2005).

In asterids, flowers are almost always sympetalous, and, in many groups, stamens are also fused with the sympetalous corolla. At least in the euasterids (Gentianales, Lamiales of the lamiid clade; Asterales, Dipsacales of the campanulid clade), stamens are in a single whorl. Carpels are commonly reduced to two (or three). Ovules are tenuinucellate and unitegmic. The great diversity in floral forms among asterids is mainly based on the plasticity of the sympetalous corolla and the often monosymmetric flowers (Chapter 9).

Future Directions

Molecular tools in the study of plant evolution have opened new directions in the study of flower structure and evolution. Molecular systematics has provided a new framework for studying evolutionary changes in floral features. However, for most taxa, more fine-grained molecular systematic studies are obviously nec-

essary before more detailed floral evolutionary studies become possible. It is also evident that floral structure in many families is only rudimentarily known. With new analyses, the evolutionary directions of floral features may turn out to be much more intricate, and we may find that some features have changed much more than we thought, particularly if research is focused at the population level (Endress 1994a). As a rare example of a detailed study, Kuzoff et al. (2001) showed how superior and inferior ovary position can vary within smaller groups such as Saxifragaceae or even within a single genus such as *Lithophragma*. Similar variability can also be expected in other features in other clades. Detailed studies of floral structure and development in single families (e.g., Hydrangeaceae; Hufford 2001), as well as large-scale comparative studies (e.g., gynoecium in basal angiosperms, Endress and Igersheim 2000b), are badly needed. Many groups of rosids and basal asterids are still poorly known, and comparative floral studies on new assemblages of families as they appear in APG (1998) and APG II (2003), are, of course, lacking. Such large-scale morphological comparisons are much more time-consuming than comparative molecular studies and will therefore proceed at a slower pace (e.g., Matthews and Endress 2002, 2004, 2005). However, eventually, with more refinement of character scoring, improved morphological and combined molecular and morphological analyses should become possible (for examples of characters to be analyzed, see Endress 2003b). In addition, fossil flowers have been and should increasingly be included in such surveys (e.g., Crane et al. 1995; Magallón et al. 1999; Sun et al. 2002). Studies in molecular developmental genetics should also be expanded and combined with comparative studies. Such efforts will be more successful the more multifaceted they are (Baum et al. 2002; Cronk et al. 2002; D. Soltis et al. 2002; Endress 2003a, 2003b; Friedman et al. 2004).

13

Evolution of Genome Size and Base Chromosome Number

Introduction

Both chromosome number and genome size vary tremendously across the flowering plants (Fedorov 1969; see the Plant DNA C-Values Database at http://www.rbgkew.org.uk/cval/homepage.html). This variation has stimulated a great deal of speculation about the original genome size and base chromosome number of the angiosperms, as well as about the patterns of genome and chromosomal evolution. In earlier reports, some authors proposed that the original base chromosome number for angiosperms was low, between $x = 6$ and 9 (e.g., Ehrendorfer et al. 1968; Stebbins 1971; Walker 1972; Raven 1975; Grant 1981), and that the original genome size of angiosperms was small (Leitch et al. 1998; D. Soltis et al. 2003b), which would suggest a close correspondence between the evolution of genome size and chromosome number. However, analysis of the vast range of chromosome numbers and genome sizes encountered in angiosperms shows that genome size can vary independently of chromosome number.

Recent studies have revealed the complexities involved in reconstructing ancestral base chromosome numbers and genome sizes for angiosperms. A large component of this complexity is that polyploidy is rampant throughout angiosperm evolutionary history. The question of what proportion of angiosperms is of polyploid origin is almost certainly moot, given that several recent studies have demonstrated that one or more rounds of ancient genome duplication characterize most, if not all, angiosperms formerly considered to be diploid. Complete sequencing of the genome of *Arabidopsis thaliana,* which has a very small genome (157 Mb; Bennett et al. 2003), revealed many duplicate genes and suggested two or three rounds of genome duplication (Vision et al. 2000; Bowers et al. 2003). Similarly, diploid members of *Brassica* may be ancient tetraploids or hexaploids (Lagercrantz 1998). A genomics investigation of rice indicates that it too is an ancient polyploid; a polyploidy event apparently occurred after the divergence of Poales (see Chapter 4), but before the divergence of the major cereals from each other (Paterson et al. 2004). Perhaps a more appropriate question is: How many rounds of genome duplication have occurred in various lineages of angiosperms?

In this chapter, we discuss the evolution of both genome size and base chromosome number in a general sense across the angiosperms, using the phylogenetic framework now available (e.g., Hoot et al. 1999; P. Soltis et al. 1999b; Savolainen et al. 2000a, 2000b; D. Soltis et al. 2000), as updated by more recent analyses of basal angiosperms (Zanis et al. 2002) and eudicots (D. Soltis et al. 2003a).

Genome Size

The amount of DNA in an unreplicated gametic nuclear genome is referred to as the 1C-value (or simply C-value). The C-value is often loosely referred to as genome size, but strictly speaking, genome size is the amount of DNA in an unreplicated, gametic chromosome set. Genome size equals the 2C nuclear DNA amount divided by ploidal level, and this equals the C-value in diploids (Bennett et al. 1998). This formula gives an accurate estimate for individuals with constituent genomes of equal size (e.g., diploids and autopolyploids) but provides only an estimate of the mean for individuals with constituent genomes of different sizes (e.g., some diploid hybrids and allopolyploids). Note that for polyploids, genome sizes estimated in this way are always smaller than C-values. For example, in the diploid *Triticum monococcum*, 2C = 12.45 pg, so 1C = 12.45 ÷ 2 = 6.23 pg, which also equals the genome size. In the tetraploid *T. dicoccum*, in contrast, 2C = 24.05 pg, so 1C = 24.05 ÷ 2, which equals 12.03 pg, but the genome size is 24.05 ÷ 4, or 6.01 pg.

C-values have been estimated for more than 3,500 species of angiosperms (Bennett et al. 1997; Bennett and Leitch 2001, 2003; Hanson et al. 2001a, 2001b; Leitch and Hanson 2002), representing more than 1% of the approximately 250,000 to 300,000 species of flowering plants and approximately 50% of all angiosperm families (*sensu* APG 1998; APG II 2003). The Plant DNA C-Values Database (http://www.rbgkew.org.uk/cval/homepage.html) represents the largest collection of nuclear DNA amounts for any group of organism (reviewed in Leitch et al. 1998). C-values in angiosperms span a huge range. The smallest reported values are for *Cardamine amara* (Brassicaceae; 1C = 0.05 pg, Bennett and Smith 1991) and *Fragaria* (Rosaceae; 1C = 0.10 pg, Antonius and Ahokas 1996). *Arabidopsis thaliana,* a well-known model organism with a very small genome, has 1C = 0.16 pg (Bennett al. 2003); the largest value is for *Fritillaria assyriaca* (Liliaceae; 1C = 127.4 pg, Bennett and Smith 1976).

Despite this enormous range in DNA amount, the basic complement of genes required for normal growth and development appears to be essentially the same, leading to what is termed the "C-value paradox" (Thomas 1971). That is, the large variation in genome size detected in both plants and animals is not correlated with organismal complexity. There is now general agreement that genomes are not simply linear collections of genes; differences in amount of DNA can largely be attributed to changes in the proportion of noncoding, repetitive DNA.

Several mechanisms have been proposed for this large variation in genome size in the angiosperms (reviewed in Kellogg and Bennetzen 2004). Repeated cycles of polyploidy may increase genome size (e.g., Leitch and Bennett 1997; D. Soltis and Soltis 1999; Otto and Whitton 2000; P. Soltis and Soltis 2000; Wendel 2000). In addition, transposable elements also appear to contribute to increases in genome size throughout eukaryotes (e.g., Flavell 1988; Bennetzen 2000, 2002; Sankoff 2001; Kidwell 2002). Gregory (2001) has suggested that the phrase "C-value paradox" be replaced by "C-value enigma" to indicate that the current challenge is to understand the mechanisms and forces that determine the amounts of repetitive DNA in a genome.

The hypothesis of repeated cycles of polyploidy has received particular attention in flowering plants and is supported by isozyme evidence. Several basal lineages (e.g., Magnoliaceae, Lauraceae, Calycanthaceae), as well as some clades of eudicots with uniformly high chromosome numbers (e.g., Hippocastanaceae [now part of Sapindaceae], Trochodendraceae, *Salix* + *Populus* of Salicaceae), have many duplicated loci, in agreement with ancient polyploidy (D. Soltis and Soltis 1990), as suggested by Stebbins (1950, 1971). As noted, even *Arabidopsis* with its very small genome and low chromosome number appears to be an ancient polyploid. Multiple genome duplication events in angiosperm diversification are also supported by recent analyses of genomic evidence (Bowers et al. 2003). By comparing gene number and gene order for a gymnosperm (Pinaceae), a monocot, *Oryza* (Poaceae), and several eudicot lineages (Solanaceae, asterid; Fabaceae, eurosid I; Malvaceae, eurosid II; and Brassicaceae, eurosid II), Bowers et al. (2003) estimated the number and timing of genome-

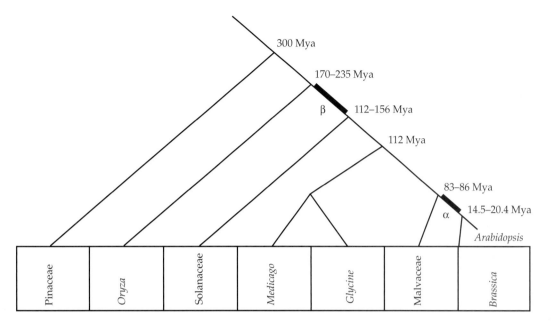

FIGURE 13.1 Multiple genome duplication events in the evolutionary history of *Arabidopsis* (modified from Bowers et al. 2003). The symbols α and β indicate two genome-wide duplication events inferred and the approximate timing of each (see text). Ages are given in million years before present (Mya).

wide duplication events (Figure 13.1; see also Kellogg 2003). Within eurosid II, a genome-wide duplication event seems to have occurred after the split between Malvaceae and *Arabidopsis* (Brassicaceae). This genome duplication event is estimated to be older than 14.5 to 20.4 Mya, an event that may characterize all Brassicaceae and perhaps their close relatives. An earlier genome duplication event is suggested after the split between rice (monocots) and eudicots. This earlier duplication event was estimated to have occured at least 112 to 156 Mya by Bowers et al. (2003); the younger end of this time range seems reasonable based on the probable time of origin of the monocots estimated from both fossil evidence (Gandolfo et al. 1998, 2002) and molecular divergence (e.g., K. Bremer 2000; see Chapters 3 and 4).

Leitch et al. (1998) calculated mean C-values for many angiosperm species and considered the evolution of genome size in light of angiosperm phylogeny. Updated calculations (in D. Soltis et al. 2003b) showed (Figure 13.2) that despite the enormous range in values, most angiosperms have small C-values, between 0.1 and 3.5 pg. In fact, the modal C-value for angiosperms is actually quite low, 0.7 pg (~675 Mbp). Large genomes are present in a few large clades: monocots, Ranunculales, Santalales, Caryophyllales, asterids, and rosids. However, even in these clades, most members have small or intermediate-sized genomes; only two groups contain members with very large genomes (≥35 pg, 50 times the modal value), Santalales and monocots. Considering just the six groups in which large to very large

genome sizes have been reported, relatively few species in each group have large genomes. Furthermore, those species with large genomes tend to be restricted to the more derived families within each of these groups (Leitch et al. 1998). The most parsimonious explanation for these observations is that the ancestral angiosperms had small genomes and that the possession of large genomes is derived.

Within extant seed plants, a small genome is unique to the angiosperms (Leitch et al. 1998, 2001). Extant gymnosperms are all characterized by larger C-values than angiosperms. The modal C-value for 152 gymnosperms is 15.8 pg compared with a modal value of 0.6 for angiosperms (Leitch et al. 2001; this modal value for angiosperms is slightly lower than that reported by Leitch et al. 1998, 0.7 pg). The modal C-value for 50 leptosporangiate ferns is 7.95 pg (I.J. Leitch, pers. comm.), which is also higher than that for angiosperms.

Since the analysis by Leitch et al. (1998), new developments have indicated that a reevaluation of the diversification of 1C-values in angiosperms is timely. Recent topologies (e.g., Qiu et al. 1999; Chase et al. 2000a; D. Soltis et al. 2000a; Zanis et al. 2002) provide greater resolution and internal support for relationships throughout the angiosperms than earlier single-gene analyses (e.g., Chase et al. 1993; see Chapter 2). Furthermore, C-values for some crucial taxa (e.g., *Amborella, Austrobaileya, Illicium, Ceratophyllum*) have been estimated (e.g., Leitch and Hanson 2002; Hanson et al. 2001a, 2001b). *Amborella* and two species of *Nymphaea*

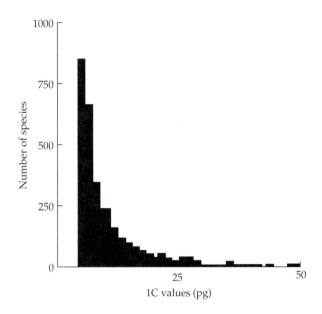

FIGURE 13.2 Distribution of 1C-values for 3,543 angiosperm species (mean 1C = 6.25 pg; mode 1C = 0.6 pg; see Leitch et al. 1998). (From D. Soltis et al. 2003b.)

were determined to have very small genomes (1C-values of 0.89, and 0.60 and 1.10, respectively; Leitch and Hanson 2002).

Using new C-values and the most recent topologies, D. Soltis et al. (2003b) reexamined genome size evolution in the angiosperms, as well as in seed plants in general. They used genome size estimates for diploids, avoided using means for large clades, and implemented the same categories of C-values established by Leitch et al. (1998): C-values less than or equal to 1.4 pg and 3.5 pg (2 and 5 times the modal C-value of 0.7 pg for angiosperms) were considered "very small" and "small," respectively. C-values of 3.51 to 13.099 pg were considered "intermediate"; C-values greater than or equal to 14.0 pg and 35 pg (20 and 50 times the modal C-value) were considered "large" and "very large" genomes, respectively. We will use these genome size categories throughout the remainder of this section of the chapter.

Average genome sizes for families can be misleading if values reported in the literature represent both diploids and polyploids, because often the highest values of genome size in a family are for polyploids. In

Magnolia, the reported C-values are 0.90 pg, 5.98 pg, and 7.1 pg. *Magnolia kobus* has a C-value of 0.90 pg and is a diploid member of the genus with $2n = 38$, which is the lowest number for *Magnolia. Liriodendron,* the sister of *Magnolia,* also with $2n = 38$, has a C-value of 0.80, which is comparable to *M. kobus.* The highest C-values for *Magnolia* are from two separate reports for *M. × soulangiana,* with $2n = 76$. The C-values for this species, 5.98 and 7.1, would therefore be attributed to polyploidy, although genome doubling alone cannot be responsible for this huge increase in C-value. These results indicate that computing a simple mean genome size for Magnoliaceae would be inappropriate. *Magnolia* and *Liriodendron* should both be represented in analyses of genome evolution, and *Magnolia* should be represented by the value for the diploid *M. kobus.* With this approach, the ancestral state for Magnoliaceae is reconstructed as a very small C-value (less than 1.5 pg). When a mean value is used for *Magnolia,* the family has a small C-value (1.5–3.6). Mean values reported for large clades at higher taxonomic levels can also be misleading. The mean for Santalales is reported as 1C = 12.7 (Leitch et al. 1998), but this value is strongly influenced by extremely high values for one genus, *Viscum* (discussed below).

Using a gymnosperm clade as outgroup and the same categories for C-values designated by Leitch et al. (1998), the ancestral genome size of the angiosperms is reconstructed as very small (D. Soltis et al. 2003b; Figure 13.3); this result is obtained regardless of the optimization method used ("all most parsimonious states," ACCTRAN, DELTRAN). Thus, parsimony reconstruction of a revised and updated matrix reinforces the earlier conclusion (Leitch et al. 1998) that the ancestral genome size of angiosperms was "very small."

Reconstruction of genome size diversification across all embryophytes (land plants) similarly revealed that the ancestral genome size of angiosperms was very small (Leitch et al., 2005; Figure 13.4). Reports from the literature indicate that all three bryophyte lineages (hornworts, liverworts, mosses) are characterized by very small genomes (1C < 1.4 pg). Reconstructions that include C-values for all three bryophyte groups indicate the original genome size of land plants was very small. However, the accuracy of the reports for hornworts and liverworts has been questioned (Leitch et al., 2004). Reconstructions that use C-values for mosses,

FIGURE 13.3 Parsimony reconstruction of genome size diversification in the angiosperms obtained using the "all most parsimonious states" option of MacClade. The general topology is from D. Soltis et al. (2000), with modifications following Chase et al. (2000), Fishbein et al. (2001), Zanis et al. (2002), and D. Soltis et al. (2003a; see text). 1C-values are from Bennett and Leitch (2001), Hanson et al.

(2001a, 2001b), Leitch and Hanson (2002); see D. Soltis et al. ▶ (2003b). Ranges of values for Asparagales, commelinids, Dioscoreales, Liliales, and Pandanales are from Leitch et al. (1998); more detail for monocots is provided in Figure 13.5. Can = Canellales; Laur = Laurales; Pro = Proteales; Sant = Santalales. The rosid clade is summarized with Vitaceae as sister to the remaining eurosids (see Chapter 8).

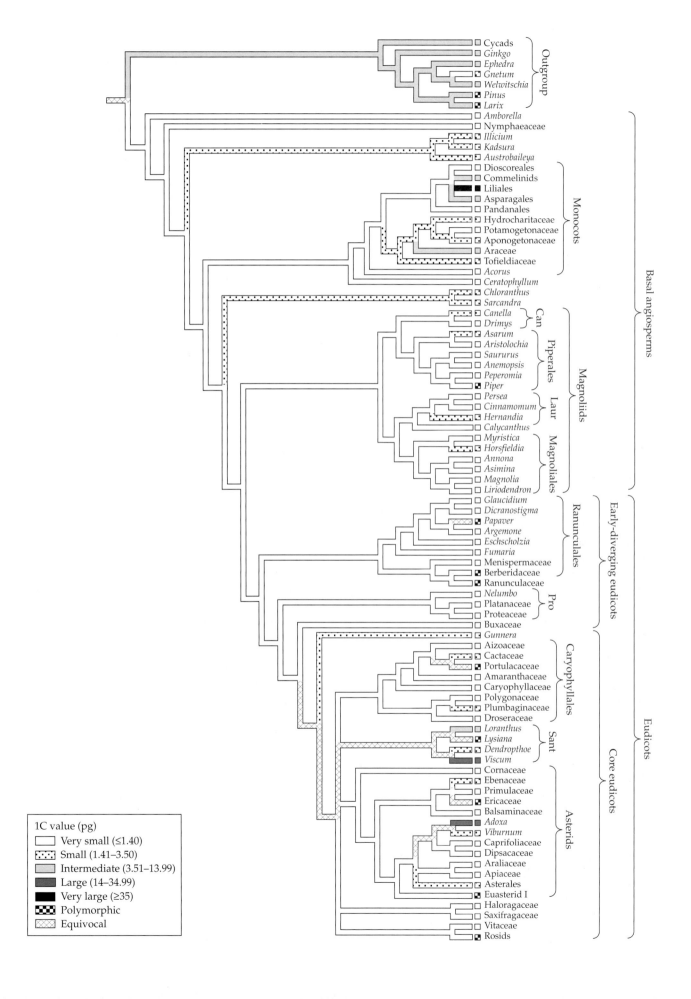

1C value (pg)	
☐	Very small (≤1.40)
⬚	Small (1.41–3.50)
▨	Intermediate (3.51–13.99)
▰	Large (14–34.99)
■	Very large (≥35)
▦	Polymorphic
▧	Equivocal

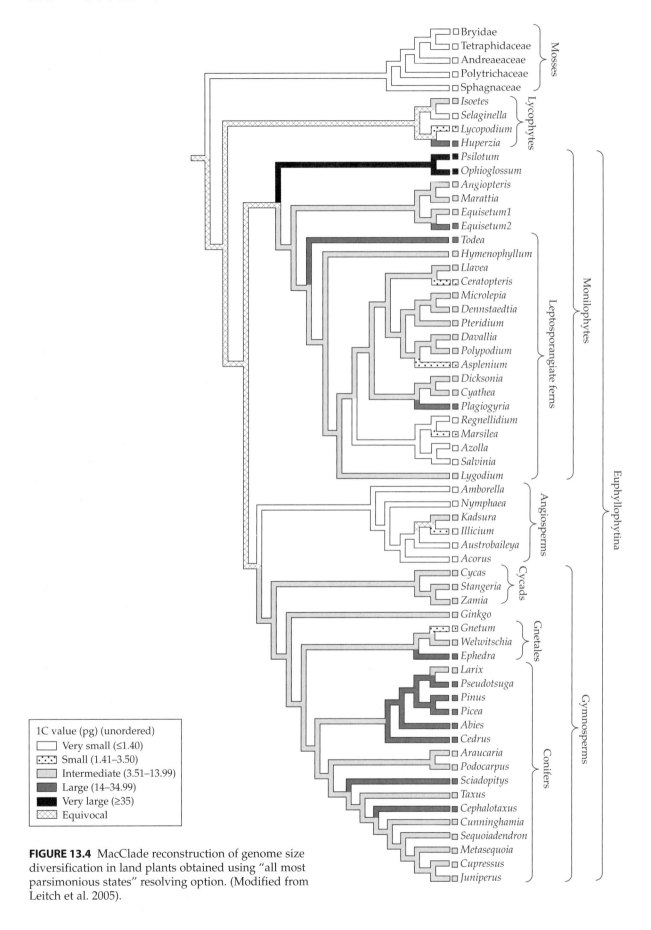

FIGURE 13.4 MacClade reconstruction of genome size diversification in land plants obtained using "all most parsimonious states" resolving option. (Modified from Leitch et al. 2005).

but with liverworts and hornworts excluded (Figure 13.4), suggest that the original genome size of all land plants was equivocal. In contrast to the very small genomes of bryophytes, diverse C-values are evident in different members of the lycophyte clade. Within the lycophyte clade, *Isoetes* has an intermediate C-value, and *Selaginella* has a very small C-value; the C-values of *Lycopodium* and *Huperzia* are small and large, respectively, based on the estimates now available. The ancestor of the lycophyte clade is reconstructed as equivocal with the "all most parsimonious states" and DELTRAN options (Figure 13.4); with ACCTRAN the ancestral state is an intermediate genome with an increase in genome size occurring in *Lycopodium* and decreases in other taxa (*Selaginella* and other species of *Lycopodium*).

Following the lycophyte clade, the ancestral state of all remaining vascular plants (the Euphyllophytina of Kenrick and Crane 1997) again depends on the "trace" option used in MacClade (Maddison and Maddison 1992). With the "all most parsimonious states" and DELTRAN options, the ancestral condition of this clade—as well as the ancestral state of the angiosperms—is ambiguous. If the accelerated transformation option (ACCTRAN) is used, then the ancestor of the Euphyllophytina had an intermediate-sized genome, and the origin of the angiosperms involved a decrease in genome size (Figure 13.4). However, in analyses with all three bryophyte lineages included and with DELTRAN optimization, the ancestral state of Euphyllophytina is a very small genome, and angiosperms simply retained the plesiomorphic condition (reconstruction not shown).

Regardless of the "trace" option used, reconstructions indicate that several independent decreases and increases in genome size have taken place across land plants. The evolution of Marsileaceae (represented by *Regnellidium* and *Pilularia*) + Salviniaceae (represented by *Salvinia*) within the ferns clearly involved a major decrease in genome size (Figure 13.4). There are other examples of apparent genome size decrease in the leptosporangiate ferns. In all reconstructions, ferns have an ancestral genome size that is intermediate; reductions to small genomes apparently occurred independently in *Ceratopteris* and *Aspleniuim* (Figure 13.4).

Additional examples of genome downsizing are found in the gymnosperms. Considering groups within the gymnosperm clade, the mean C-values are: cycads, 14.71 pg; *Ginkgo*, 9.95 pg; Gnetales, 7.23 pg; Pinaceae, 22.02 pg; and other conifers, 11.89 pg. The lowest C-values reported for gymnosperms are for *Gnetum* (1C = 3.38), which appears to represent a decrease in genome size within the clade of extant gymnosperms (Figure 13.4; D. Soltis et al. 2003b; Leitch et al. 2005). *Larix* may also represent an example of genome downsizing just within the conifers (Figure 13.4).

There is also evidence for several independent increases in genome size across land plants. One such increase may have occurred in *Huperzia* of the lycophytes; the largest increase may have occurred in the clade that contains Ophioglossaceae (*Ophioglossum*) + Psilotaceae (*Psilotum*; Figure 13.4). However, both of these examples are dependent on the sampling of bryophytes and the trace option used in the reconstructions. If all three bryophyte lineages are included, with DELTRAN optimization both *Huperzia* and Ophioglossaceae + Psilotaceae represent increases in genome size (not shown). In other reconstructions (those using only mosses and with "all most parsimonious states" and DELTRAN options) the ancestral states for these nodes are equivocal, and hence it is unclear whether they represent true increases of genome size. With all trace options, however, reconstructions indicate that gymnosperms had an ancestral genome size that was intermediate with three separate increases in genome size occurring in various conifer lineages. Multiple increases in genome size are also evident in the leptosporangiate ferns (i.e., *Todea*, *Plagiogyria*; Figure 13.4). In summary, genome size evolution across land plants appears to have been highly dynamic, with both increases and decreases (Leitch et al. 2005).

Returning to angiosperms, reconstructions indicate that a very small genome size was ancestral throughout the basal angiosperms, the early-diverging eudicots, and also some of the major clades of core eudicots (e.g., Caryophyllales, Saxifragales, asterids; Figure 13.3). Among basal angiosperms, the ancestors of the monocot and magnoliid clades also appear to have had very small genomes. However, the Austrobaileyales clade is characterized by C-values in the small and intermediate ranges, rather than the very small range (Figure 13.3; Leitch and Hanson 2002). *Austrobaileya* (Austrobaileyaceae) has a C-value of 9.52 pg. The 1C-value of *Illicium anisatum* (Schisandraceae *sensu* APG II 2003) has been reported to be 3.40; genome size estimates reported for three species of *Kadsura* (Schisandraceae) range from 7.4 to 8.9, which are in the small, and intermediate ranges (Figure 13.3), respectively. Therefore, not all members of early-branching angiosperm lineages have very small genomes. Austrobaileyales may represent an evolutionary lineage that long ago experienced an increase in genome size (Figure 13.3; D. Soltis et al. 2003b).

Reconstructions also suggest that the original genome size for monocots was very small (Figure 13.5). *Ceratophyllum*, a possible sister group to all monocots (Zanis et al. 2002), has a very small genome (C-value of 1.0 pg), as does *Acorus*, the sister to all other monocots (mean C-value of 0.52; range, 0.40 to 0.65). Following *Acorus*, Alismatales are sister to all remaining monocots (see Chapter 4). However, reconstructions at the base of the monocots are hampered by a lack of C-values for

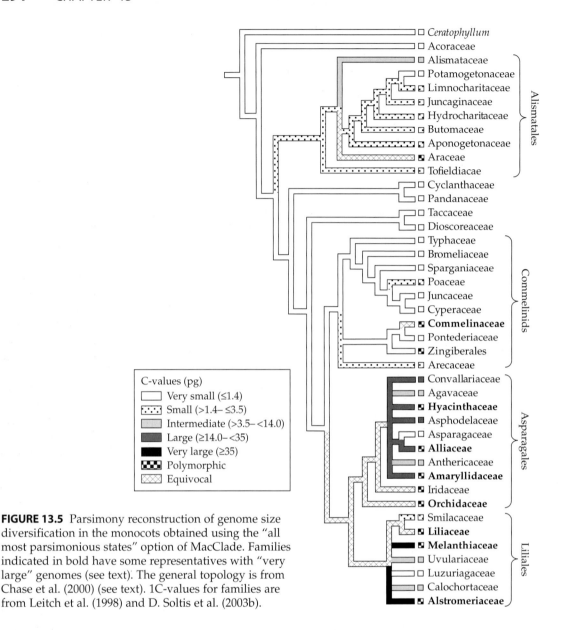

FIGURE 13.5 Parsimony reconstruction of genome size diversification in the monocots obtained using the "all most parsimonious states" option of MacClade. Families indicated in bold have some representatives with "very large" genomes (see text). The general topology is from Chase et al. (2000) (see text). 1C-values for families are from Leitch et al. (1998) and D. Soltis et al. (2003b).

critical taxa; additional estimates are needed for early-branching monocots. Nonetheless, small or very small C-values are characteristic of other monocot families such as Typhaceae, Pandanaceae, Dioscoreaceae, and Bromeliaceae (Figure 13.5).

A very large genome may have evolved at least three times independently in the monocots (Figure 13.5): (1) in some Commelinaceae of the commelinid clade; (2) in some members of Liliaceae, Melanthiaceae, and Alstroemeriaceae of Liliales; and (3) in some Alliaceae and Asparagaceae of Asparagales. However, the number of origins of very large genomes in the latter two orders is uncertain. Although some members of Asparagales and Liliales have very large genomes, most have large or smaller genomes. It remains unclear, however, if the common ancestor of these two large clades (Asparagales and Liliales) had a very large genome, or if a very

large genome originated independently in the two clades. Orchidaceae and Iridaceae (successive sisters to other Asparagales) contain an array of 1C-values, from very small to large. Clarifying ancestral genome size and subsequent diversification in Asparagales and Liliales will require additional genome size estimates as well as better resolution of relationships among the constituent members of these two clades.

Some estimates of angiosperm phylogeny place Chloranthaceae as sister to magnoliids and eudicots, but support for this placement is low (see Chapter 3). Genome sizes for Chloranthaceae are in the small or intermediate range. Estimates for two species of *Chloranthus* are 1C = 2.90 and 3.59, respectively; *Sarcandra* has a C-value of 4.30 pg. Hence, Chloranthaceae may represent another ancient lineage that, like Austrobaileyales, experienced an early increase in genome size (Figure 13.3).

Character-state reconstructions indicate that a very small genome size is ancestral for the entire magnoliid clade, as well as for the Laurales + Magnoliales and Piperales + Canellales subclades (Figure 13.3). Within Canellales, *Drimys* (Winteraceae) has 1C = 1.13 pg. Within Piperales, Saururaceae have very small genomes, with 1C estimates for *Anemopsis, Houttuynia,* and *Saururus* of 0.8 pg, 1.3 pg, and 0.5 pg, respectively. Piperaceae also have small or very small genomes. The mean value of 10 estimates for the genus *Piper* is 1C = 1.25. In Laurales, genome size estimates for Lauraceae and Calycanthaceae are very small (e.g., *Persea americana,* 1C = 1.2 pg and *Calycanthus,* 1C = 0.98). Magnoliales, including Myristicaceae and Magnoliaceae, also have very small genomes.

The reconstruction of a very small genome for the common ancestor of extant Magnoliaceae is made more noteworthy because these plants are presumed ancient polyploids (the diploid parents of which are now extinct; Stebbins 1971), a conclusion supported by isozyme data (D. Soltis and Soltis 1990). Similarly, Myristicaceae, Lauraceae, and Winteraceae are examples of other basal lineages with high base chromosome numbers (thought to be ancient polyploids) that also have very small genomes. Thus, the original genome sizes of the now-extinct "paleodiploid" Magnoliaceae and other basal lineages such as Myristicaceae, Calycanthaceae, Lauraceae, and Winteraceae, were likely even smaller than those estimated for modern taxa. Likewise, the eudicot *Aesculus hippocastanum* has a very small genome (1C = 0.1 pg), but it, as well as the entire genus *Aesculus* (Hippocastanaceae, now part of Sapindaceae), are ancient polyploids with 2*n* = 40 (D. Soltis and Soltis 1990). It is also apparent, therefore, that presumed ancient polyploids do not necessarily have large genomes.

The ancestral genome for eudicots appears to have been very small (Figure 13.3) based on reconstructions for key clades, such as Ranunculales and Proteales, as well as data for Buxaceae. New data for *Platanus* (Platanaceae; 1C = 1.30 pg) and *Nelumbo* (Nelumbonaceae; 1C = 0.24 pg), combined with an earlier report for *Grevillea* (Proteaceae; 1C = 0.83 pg), further suggest that the ancestral genome for Proteales was very small (D. Soltis et al. 2003b).

Ranunculales occupy an important position as sister to all other eudicots and may also have had a very small ancestral genome (D. Soltis et al. 2003b). Berberidaceae have a very small average genome size (1C = 1.35 pg for 11 diploids), as do Menispermaceae with 1C = 0.70 pg. Papaveraceae also appear to have a very small ancestral genome, but this is not evident from the mean value for the entire family. The average genome size for Papaveraceae is 1C = 3.05 pg, but this estimate is heavily influenced by numerous estimates for species of the large genus *Papaver,* which range from 1C = 2.33

to 8.90 pg. Polyploidy contributes to the two highest values in *Papaver* (*P. orientale,* 1C = 8.90 pg, 2*n* = 42; *P. setigerum,* 1C = 6.70 pg, 2*n* = 44). The mean 1C-value for known diploids in *Papaver* is 2.2 pg. Within Papaveraceae, many early-branching members (based on the topology of Hoot et al. 1997) have very small genome sizes. *Fumaria* (1C = 0.55 pg), *Glaucium* (1C = 0.60 pg), *Eschscholzia* (1C = 1.10 pg), and *Argemone* (1C = 0.60 pg) all have very small genomes. Genome size estimates for diploids and recent phylogenetic trees indicate that the ancestral genome size for Papaveraceae was very small (1C <1.4 pg).

The ancestral genome size for core eudicots is reconstructed as equivocal. The single estimate for Gunnerales, the sister to all other core eudicots, is 1C = 7.44 pg (using a species of *Gunnera*). Despite this relatively large C-value for *Gunnera,* a very small genome size may be ancestral for most core eudicots. However, reconstructions are hampered by the uncertainty surrounding the relationships of core eudicot lineages (see Chapters 2 and 5). A very small genome size is reconstructed as ancestral for Saxifragales, Caryophyllales, and asterids. Although only three C-values are available for Saxifragales, all are less than 1.4 pg (Figure 13.3).

Within asterids, Cornales, followed by Ericales, represent the successive sisters to the euasterids (Albach et al. 2001a; B. Bremer et al. 2002). Diploid Cornales and Ericales have 1C-values less than 1.4 pg; hence, the ancestral condition for the asterid clade is a very small genome. Within the asterids, a large genome is found only in *Adoxa* (Adoxaceae; 1C = 14.30 pg), which occupies a derived position within the asterid clade (Figure 13.3) and is a polyploid with 2*n* = 36. Adoxaceae are part of asterid II, and other diploids of this clade have either small or very small genomes.

The ancestral genome size of the rosids is unclear. Vitaceae appear to represent the sister to all other rosids, and *Vitis* has a very small genome (1C = 0.4 to 0.6 pg for 21 diploids). However, relationships within the rosids remain poorly understood (see Chapter 8), precluding the reconstruction of an ancestral genome size for this large clade at this time.

Although Santalales are often considered to have a large genome (see Leitch et al. 1998), very large genomes are confined to *Viscum* (Santalaceae) with C-values of 76.0 and 79.3 pg. However, *Viscum* is clearly derived within Santalales (Nickrent and Malécot 2001; see Chapter 6). C-values for other Santalaceae (*sensu* APG II 2003) are much lower. *Loranthus* has a C-value of 15.20 pg, and *Lysiana* has a mean C-value of 12.50 pg (based on values for five diploid species that range from 11.03 to 15.28). *Dendrophthoe* (Santalaceae) has a mean C-value of only 4.3 pg (range, 2.7 to 6.2 pg), calculated from C-values for five diploids. The ancestral state for the family is therefore reconstructed as equivocal. Unfortunately, C-values are not available for early-branching lineages of Santalales; additional C-values would be useful for inferring patterns of genome size evolution.

Genetic obesity hypothesis

Bennetzen and Kellogg (1997) proposed that genome size evolution in plants would be largely unidirectional, with an overall pattern of increase due to the combined influence of polyploidy and the accumulation of retroelements. These authors suggested that plants have a "one way ticket to genetic obesity." However, the hypothesis of unidirectional increase in genome size has rarely been critically evaluated (Bennetzen and Kellogg 1997; Cox et al. 1998; Wendel et al. 2002). Broad surveys of angiosperms appear to be in general agreement with the genetic obesity hypothesis, with very large genomes confined to taxa that occupy derived positions within larger clades (Leitch et al. 1998; D. Soltis et al. 2003b). How-

ever, these approaches are coarse-grained; careful evaluation of the genetic obesity hypothesis within individual clades (e.g., within families and genera) is required.

The evolution of genome size at lower taxonomic levels has rarely been examined in a phylogenetic context. In Pooideae (Poaceae), the ancestral grass genome size was inferred to be 2C = 3.5 pg (Kellogg 1998). Kellogg also provided evidence for a steady increase in genome size in Pooideae, ultimately leading to the much larger genome sizes (2C = 10.7 pg or more) in the Triticeae. However, genome size decreased in some genera, including *Oryza, Chloris,* and *Sorghum.* In contrast to the model of unidirectional change in genome size, in a clade of *Gossypium* (cotton) and allies (the cotton tribe, Gossypieae, Malvaceae), the frequency of genome size decreases exceeded the examples of increases (Figure 13.6; Wendel et al. 2002). In Gossypieae, the shortest tree recovered

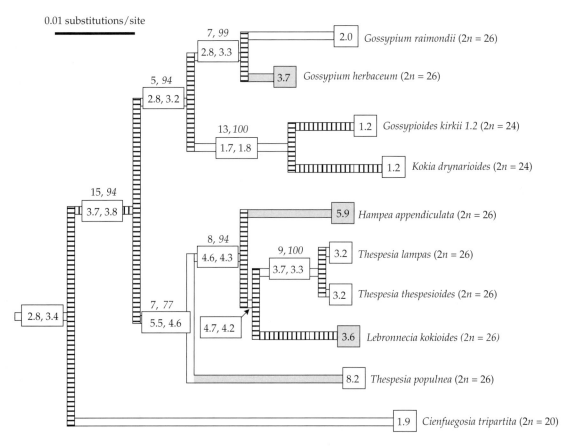

FIGURE 13.6 Reconstruction of the evolution of genome size in the cotton tribe (Gossypieae). The shortest tree recovered from parsimony analysis, inferred from CesA1 sequence data, is identical in topology to the maximum likelihood phylogeny shown here. The number of character-state changes and jackknife support (%; in italics) from maximum parsimony analysis are shown above each internal branch. The tree is rooted with the outgroup *Malva sylvestris* (ingroup–outgroup branch length = 0.12). Genome sizes (in pg) are shown at branch tips before species names, which are followed by somatic chromosome numbers. Ancestral genome sizes were estimated using sum-of-squared-changes parsimony analysis (Maddison 1991) and a generalized least squared method (Martins and Hansen 1997). A linear (=Wagner) parsimony (Swofford and Maddison 1987) reconstruction was also conducted, but these estimates are not shown. The former estimates are shown in boxes on the internal branches. Inferred genome size increases are shown by shaded branches, decreases are indicated by unfilled branches, and ambiguities or stasis are denoted by hatched branches. (Modified from Wendel et al. 2002.)

from parsimony analysis of sequence data indicated that independent decreases in genome size occurred in *Cienfuegosia, Gossypium raimondii*, the ancestor of *Gossypioides kirkii* + *Kokia drynarioides*, and the ancestor of *Thespesia*. Three separate increases are indicated: *Gossypium herbaceum, Hampea appendiculata*, and *Thespesia populnea* (Figure 13.6). In the slipper orchids (Cypripedioideae; Orchidaceae), genome size varies 5.7-fold. Cox et al. (1998) did not reconstruct genome size evolution, but visual comparison of genome size and phylogenetic position indicated that both increases and decreases in genome size have occurred in the slipper orchids.

Recent analyses of C-values in several groups having diploids and polyploids revealed additional complexities in genome size diversification (Leitch and Bennett 2004). Polyploids sometimes have genome sizes that are substantially smaller than those of diploid congeners. Using C-values for 3,021 species, Leitch and Bennett found that mean 1C-value does not increase in direct proportion with ploidy. This observation held true in comparisons conducted in several distinct clades, including asterids, rosids, and monocots. Leitch and Bennett's data provide clear evidence for genomic downsizing in many polyploids (Figure 13.7). Thus, several lines of evidence clearly indicate that genome size evolution in the angiosperms, as well as in embryophytes in general, is dynamic, with both increases and decreases having occurred (Rabinowicz 2000; Wendel et al. 2002; Leitch et al. 2005).

Mechanisms of change in genome size

There are several avenues by which changes in genome size can occur. Repeated cycles of polyploidy may substantially increase genome size in a single step. Transposable elements also appear to contribute to increases in genome size throughout eukaryotes (e.g., Bennetzen 2000; Kidwell 2002), perhaps through the large-scale accumulation of retroelements, as in Poaceae (Bennetzen 1996, 2000; SanMiguel et al. 1996) and some nonangiosperm lineages, such as *Pinus* (Elsik and Williams 2000).

Mechanisms of genome contraction continue to be poorly understood, but understanding of how decreases in genome size can take place is improving (e.g., Vicient et al. 1999; Kirik et al. 2000; Bennetzen 2002; Frank et al. 2002; Hancock 2002; Petrov 2001, 2002). Unequal recombination can slow the increase in genome size, and illegitimate recombination and other deletion processes may be the major mechanisms for decreases in genome size (Bennetzen 2002; Devos et al. 2002; Kellogg and Bennetzen 2004). Studies of microbial genomes suggest that downsizing of some genomes may be the result of homologous recombination at repeated genes, leading to the loss of large blocks of DNA as well as repeated sequences (Frank et al. 2002). Differences in double-stranded break repair may be responsible for some

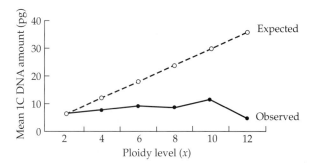

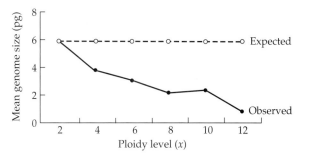

FIGURE 13.7 Relationship between genome size and ploidy for 3,021 angiosperm species. Open circles and dashed lines are expected values for 1C DNA content based on genome doubling with no loss of DNA; black dots and black lines are observed values, revealing loss of DNA as ploidy increases. (From Leitch and Bennett 2004.)

variation in genome size (Kirik et al. 2000). Differences in the processing of chromosomal ends have been reported in *Arabidopsis* and *Nicotiana* (tobacco) (Orel and Puchta 2003); the former has a genome size 20 times smaller than the latter. Free DNA ends are much more stable in tobacco, perhaps due to better protection of DNA break ends. Exonucleolytic degradation of DNA ends might be a driving force in the evolution of genome size (Orel and Puchta 2003). In some animals, insertion/deletion biases may lead to significant changes in genome size. A high rate of deletion has apparently occurred in *Drosophila* (Petrov 2002). Hancock (2002) has reconsidered the relationship between the level of repetitiveness in genomic sequences and genome size. A previously reported correlation between genome size and repetitiveness was generally confirmed, but with some deviations. Some of the variance in repetitive DNA observed may be the result of variation in the effectiveness of mechanisms for regulating slippage errors during replication.

The driving forces behind changes in genome size remain unclear. It is tempting to invoke selection, and several hypotheses have been proposed for reductions in genome size (Leitch and Bennett 2004): (1) to reduce the nucleotypic effects of increased DNA amounts; (2) to reduce the biochemical costs associated with additional DNA amounts; and (3) to enhance polyploid stability.

Base Chromosome Number

Some authors proposed that the original base chromosome number for angiosperms was low, between $x = 6$ and 9 (Ehrendorfer et al. 1968; Stebbins 1971; Walker 1972; Raven 1975; Grant 1981). For example, Raven (1975) suggested that the ancestral base chromosome number for angiosperms was $x = 7$. This estimate was obtained by examining base chromosome numbers throughout the subclasses of angiosperms, with particular attention to subclass Magnoliidae, the subclass containing those taxa considered to be the most "primitive" of all extant angiosperms (e.g., Cronquist 1968; Takhtajan 1969). Raven concluded that "The original basic chromosome number in the angiosperms seems clearly to have been $x = 7$, characteristic of all major groups of both dicots and monocots except Caryophyllidae with $x = 9$."

Paradoxically, however, many basal angiosperm families (former Magnoliidae) have high haploid chromosome numbers, including Calycanthaceae ($n = 11$, 12), Lauraceae ($n = 12$), Magnoliaceae ($n = 19$), and Myristicaceae ($n = 19$). Some early-diverging eudicots also have high haploid chromosome numbers, such as Trochodendraceae ($n = 19$); Cercidiphyllaceae, an early-branching member of the core eudicot clade Saxifragales, also have a high number ($n = 19$). Two hypotheses were proposed to explain the paradox of high chromosome numbers in angiosperms then considered "primitive" (Stebbins 1971; Grant 1981): (1) the original base chromosome number for angiosperms was in the modal range of $x = 12$ to 14 and lower base numbers were derived via descending aneuploidy; and (2) the original base chromosome numbers for angiosperms were $x = 6$ and 7, and high base numbers were derived via ancient polyploidy. Most authors favored the second hypothesis, although no genetic data were available at that time. Raven (1975), for example, concluded that most Magnoliidae were based on $x = 7$.

Using genetic evidence from isozymes, D. Soltis and Soltis (1990) determined that species from families of basal angiosperms, including Magnoliaceae ($n = 19$), Myristicaceae ($n = 19, 21$), Lauraceae ($n = 12$), and Calycanthaceae ($n = 11, 12$), exhibited extensive isozyme increases compared to the number typical of diploid seed plants. The most parsimonious explanation for these genetic results is ancient polyploidy. In contrast, *Eupomatia* ($n = 10$) and Annonaceae ($n = 7, 8, 9$) did not exhibit isozyme increases, suggesting that these lineages might be diploid rather than anciently polyploid. These genetic results are consistent with an original base number for angiosperms of $x = 10$ or lower.

Reconstructing base chromosome numbers for large clades of angiosperms is difficult. For example, *Arabidopsis* has $n = 5$, but chromosome painting using BAC contigs indicates that the *Arabidopsis* karyotype arose through perhaps six translocations and three inversions from an ancestor with $n = 8$ (Lysak et al. 2001, 2003). Chromosomes of diploid species of *Brassica* also have experienced considerable rearrangement (Lagercrantz 1998). Coupled with these complexities is the additional problem that these "diploid" numbers are themselves of ancient polyploid ancestry (see above). Complicating the issue further is the sheer size of the angiosperm clade, the diverse array of chromosome numbers reported for some families and genera, the extensive variation in chromosome number present within some species and genera, and the lack of data for other taxa. Thus, for many clades with diverse chromosome numbers, base chromosome number cannot be readily inferred.

With these caveats in mind, any efforts to reconstruct ancestral base chromosome numbers must be considered approximations at best. We focused primarily on basal angiosperms and early-diverging eudicots because these lineages will provide insight into early chromosome evolution and diversification in the flowering plants. Furthermore, an understanding of the chromosomal evolution in core eudicots is compromised by the lack of resolution of relationships among the lineages of core eudicots.

We modified the three-gene angiosperm topology (D. Soltis et al. 2000) to reflect relationships among basal angiosperms as inferred in more recent analyses (Zanis et al. 2002). Relationships among early-diverging eudicots were revised using data from recent three-gene (Hoot et al. 1999) and four-gene (D. Soltis et al. 2003a; Kim et al. 2004b) topologies. In the three-gene tree, Papaveraceae appeared as sister to other Ranunculales; in the four-gene trees, Eupteleaceae appeared as sister to all other Ranunculales (see Chapter 5). We explored the effects of both alternatives in parsimony reconstructions.

We used chromosome numbers provided in Fedorov (1969) and mapped the lowest base number for each family. For example, Magnoliaceae have $2n = 38, 76$; we used $x = 19$ for the family. For some families, such as Rosaceae ($x = 7, 8, 9$), Saxifragaceae ($x = 7, 11$), and Crassulaceae ($x = 8$), we used ancestral numbers reconstructed from phylogenetic analyses of these lineages (e.g., Rosaceae—Morgan et al. 1993; Crassulaceae—Mort et al. 2001; Saxifragaceae—D. Soltis et al. 2001b; nitrogen-fixing clade—Sytsma et al. 2002). However, for these and other core eudicot families, the ancestral base number was not critical to this analysis, which was aimed primarily at basal lineages.

When base chromosome numbers are taken directly from the literature without accounting for possible ancient polyploidy, reconstructions for the basalmost angiosperms are equivocal because many basal lineages have high chromosome numbers. For example, Amborellaceae have $2n = 26$. The ancestral state of the magnoliid clade of Magnoliales + Laurales and Canellales

+ Piperales is unambiguously reconstructed as $x = 13$ or higher by all resolving options ("all most parsimonious states," ACCTRAN, DELTRAN; not shown). When Papaveraceae are sister to other Ranunculales, reconstruction using the "all most parsimonious states" option in MacClade indicated that the original base chromosome number for eudicots is equivocal. However, with other trace options, the ancestral base chromosome number for the eudicot clade is also $x = 13$ or higher. When Eupteleaceae are considered sister to all other Ranunculales, the ancestral base number for all members of the magnoliid clade, as well as Ranunculales and all other eudicots, is again $x = 13$ or higher, regardless of the trace option used.

Basal eudicot lineages are characterized by a mix of low and high base chromosome numbers. Buxaceae have $x = 10$, 14; Sabiaceae have $x = 14$. Within Ranunculales, Berberidaceae have $x = 6, 7, 8$; Menispermaceae have $x = 11$; Ranunculaceae have $x = 6, 7, 8, 9$; Papaveraceae have $x = 6, 7, 8$; other Ranunculales (e.g., Eupteleaceae, Lardizabalaceae) have $x = 14$. Many of the high chromosome numbers for early-diverging eudicots, as well as basal angiosperms, likely represent ancient polyploidy. We therefore conducted a second optimization in which we substituted hypothetical "original" base chromosome numbers for those families for which genetic data implied ancient polyploidy (see Table 13.1 for summary). Thus, for Magnoliaceae ($n = 19$) we substituted $x = 9, 10$; for Calycanthaceae ($n = 11, 12$) we substituted $x = 5, 6$; for Lauraceae ($n = 12$), we substituted $x = 6$; for Myristicaceae ($n = 19, 21$), we used $x = 10$.

In these reconstructions, $x = 8$ was commonly reconstructed as an ancestral state in the early diversification of the angiosperms (Figure 13.8). Thus, the simple exercise of substituting lower base numbers for a few lineages revealed the plausibility of much lower ancestral base numbers for many angiosperms. When we substituted hypothetical base chromosome numbers for members of the magnoliid clade (Laurales + Magnoliales and Canellales + Piperales), the original base number of the entire clade was also reconstructed as $x = 7$. However, by varying the base numbers substituted (Table 13.1), a base number of $x = 8$ can also be reconstructed for magnoliids (Figure 13.8).

Continuing this exercise beyond the families noted above (Table 13.1) is complicated by the fact that genetic data are not available for many basal families. For example, no genetic data are available for Amborellaceae ($2n = 26$), Nymphaeaceae ($2n = 24$ and higher), Schisandraceae (including *Illicium*; $2n = 22$), or Austrobaileyaceae ($2n = 44$). Such genetic data are important because these families occupy pivotal phylogenetic positions. Isozyme evidence provides support for ancient polyploidy for all families examined having $x = 11$ or above, but not for those families having $x = 7, 8, 9, 10$ (D. Soltis and Soltis 1990). Using this information, we assumed that those basal lineages with numbers above $x = 11$ are ancient polyploids and substituted a hypothetical original "diploid" number (e.g., for Amborellaceae, $n = 6$, 7; Nymphaeaceae, $n = 6$; Schisandraceae and Austrobaileyaceae, $n = 5, 6$; see Table 13.1). Using this approach, we reconstructed the original base chromosome number for angiosperms as $x = 6$ (Figure 13.8).

Depending on the reconstruction and numbers used, a base number of $x = 7$ or $x = 8$ was reconstructed as ancestral for Ranunculales and eudicots (Figure 13.8). Our analyses provided little resolution regarding ancestral base chromosome numbers for major clades such as asterids, rosids, Saxifragales, and Caryophyllales.

Our conclusion that the original base chromosome number for angiosperms was low, between $x = 6$ and 9, is therefore in line with earlier proposals that involved examining chromosome numbers (Ehrendorfer et al. 1968; Stebbins 1971; Walker 1972; Raven 1975; Grant 1981). Because of the complexities of chromosome evolution and the caveats of these character-state reconstructions, perhaps the best that can be said is that early base chromosome numbers were probably low, no higher than the range of $x = 6$ to 9 often proposed.

TABLE 13.1   Taxa for which hypothetical base chromosome numbers were inferred

Taxon	Family	Haploid Number (n)	Haploid Number Substituted
Austrobaileya	Austrobaileyaceae	11	5, 6
Calycanthus	Calycanthaceae	11	5, 6
Illicium	Schisandraceae	11	5, 6
Schisandra	Schisandraceae	11	5, 6
Nymphaea	Nymphaeaceae	12	6
Sabia	Sabiaceae	12	6
Sassafras	Lauraceae	12	6
Amborella	Amborellaceae	13	6, 7
Canella	Canellaceae	13, 14	6, 7
Decaisnea	Lardizabalaceae	14	7
Akebia	Lardizabalaceae	14	7
Euptelea	Eupteleaceae	14	7
Myristica	Myristicaceae	19, 21	10
Magnolia	Magnoliaceae	19	9, 10
Tetracentron	Trochodendraceae	20	10
Trochodendron	Trochodendraceae	20	10

FIGURE 13.8 MacClade reconstruction of base chromosome number diversification in angiosperms obtained using the "all most parsimonious states" option. This reconstruction uses the actual numbers reported in the literature for many of the genera and families indicated. However, there is evidence for ancient polyploidy in many basal lineages. For those taxa thought to be ancient polyploids, a hypothetical original base number has been substituted (see text). Most gymnosperms (outgroups) have *n* = 11 or 12. This figure represents a simplified version of a larger reconstruction involving 172 taxa. The large rosid, asterid, Caryophyllales, Santalales, and Saxifragales clades have been reduced to a single terminal. The ancestral state shown for each of these clades is that reconstructed using the larger matrix.

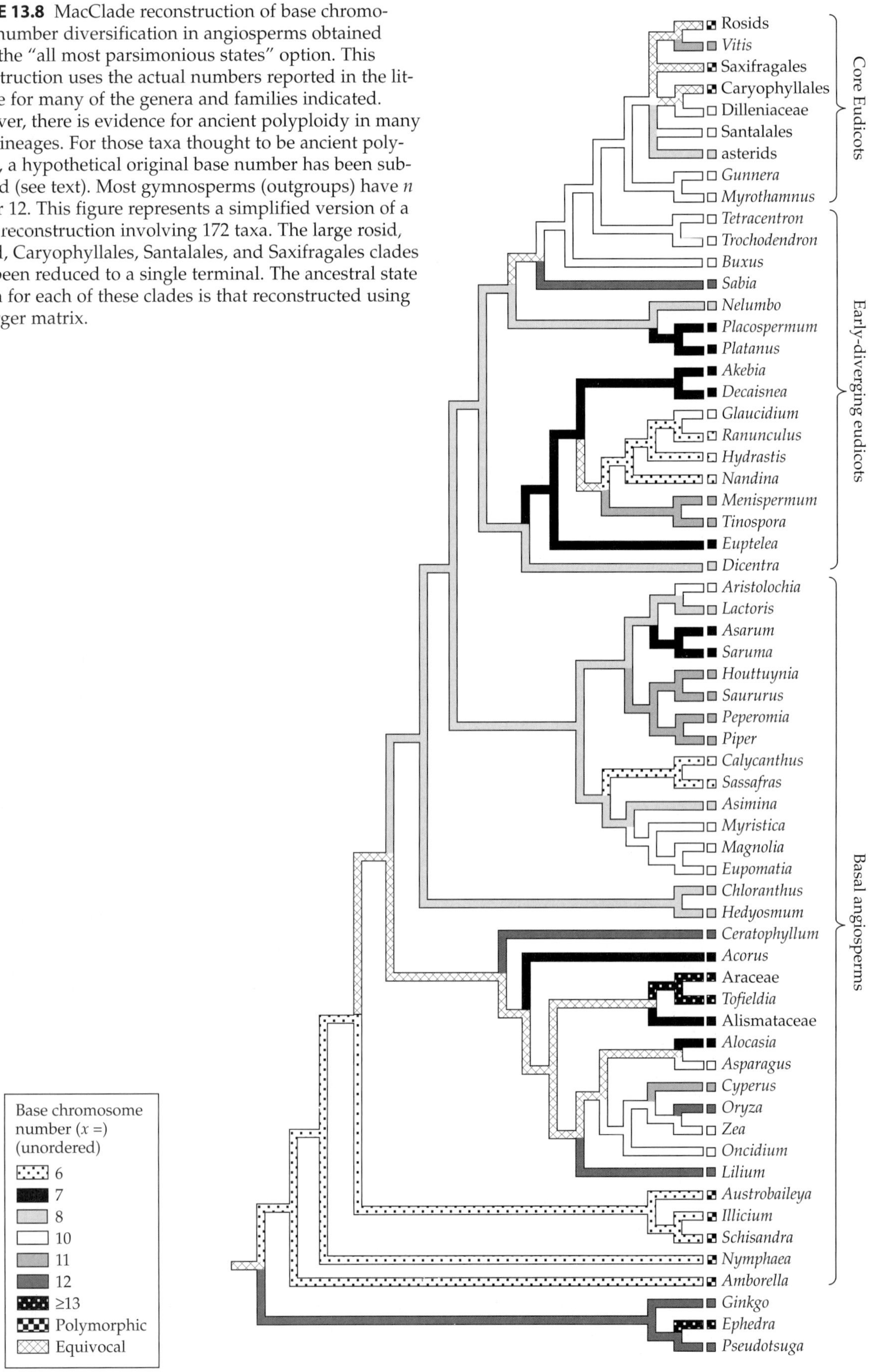

Base chromosome number (*x* =) (unordered)

- 6
- 7
- 8
- 10
- 11
- 12
- ≥13
- Polymorphic
- Equivocal

Core Eudicots:
Rosids, *Vitis*, Saxifragales, Caryophyllales, Dilleniaceae, Santalales, asterids, *Gunnera*, *Myrothamnus*

Early-diverging eudicots:
Tetracentron, *Trochodendron*, *Buxus*, *Sabia*, *Nelumbo*, *Placospermum*, *Platanus*, *Akebia*, *Decaisnea*, *Glaucidium*, *Ranunculus*, *Hydrastis*, *Nandina*, *Menispermum*, *Tinospora*, *Euptelea*, *Dicentra*

Basal angiosperms:
Aristolochia, *Lactoris*, *Asarum*, *Saruma*, *Houttuynia*, *Saururus*, *Peperomia*, *Piper*, *Calycanthus*, *Sassafras*, *Asimina*, *Myristica*, *Magnolia*, *Eupomatia*, *Chloranthus*, *Hedyosmum*, *Ceratophyllum*, *Acorus*, Araceae, *Tofieldia*, Alismataceae, *Alocasia*, *Asparagus*, *Cyperus*, *Oryza*, *Zea*, *Oncidium*, *Lilium*, *Austrobaileya*, *Illicium*, *Schisandra*, *Nymphaea*, *Amborella*

Ginkgo, *Ephedra*, *Pseudotsuga*

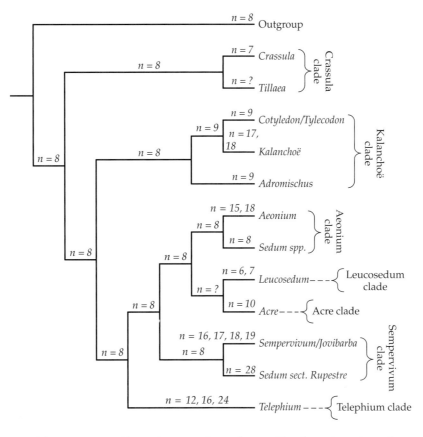

FIGURE 13.9 Reconstruction of the evolution of base chromosome number in Crassulaceae. Simplified summary topology of relationships among the major clades in Crassulaceae. Numbers above branches are the base chromosome numbers reconstructed using MacClade. (Redrawn from Mort et al. 2001.)

When the results of phylogenetic analyses are compared with other sources of data (e.g., morphology, chemistry, chromosome number), it becomes apparent that chromosome number is often an effective indicator of relationship. We provide here a sample of clades in which variation in chromosome number corresponds closely to phylogeny.

1. Rosaceae have traditionally been divided into subfamilies by fruit type (e.g., achene, pome, drupe). However, molecular phylogenies indicate that some fruit types are homoplasious in the family. In contrast, base chromosome number is in close agreement with the clades recovered (see Chapter 8; Morgan et al. 1994).

2. In Crassulaceae, a haploid number of $n = 8$ is common, and phylogenetic analyses indicate that this number is ancestral (Figure 13.9). A base number of $n = 9$ unites members of the *Kalanchoë* clade; $n = 6, 7$ unites the *Leucosedum* clade; $n = 10$ unites the *Acre* clade. Higher numbers, representing episodes of both polyploidy and aneuploidy, unite other clades: $n = 16, 17, 18, 19$ characterize the *Sempervivum–*

Jovibarba clade; $n = 12, 16, 24$ typify the *Telephium* clade (Figure 13.9; Mort et al. 2001).

3. In Saxifragaceae, molecular analyses identified clades of genera that differ from the traditional taxonomy for the family. *Chrysosplenium* and *Peltoboykinia* differ significantly in morphology (placed in different tribes or subtribes), but form a well-supported clade. They are also unusual in having $x = 11$ (D. Soltis et al. 2001a, 2001b); most genera of Saxifragaceae have $x = 7$. Another group of six genera (now called the *Darmera* group), although not previously considered closely related, form a well-supported clade; these genera are distinctive in the family in having $2n = 36$ ($x = 18$) or higher (D. Soltis et al. 2001b).

4. In subfamily Pooideae (Poaceae), the core pooids (Poeae, Aveneae, Triticeae, and Bromeae) all have $x = 7$, whereas other Pooideae exhibit a diverse array of numbers ranging from $x = 5$ to 13 (Kellogg 1998). This situation is particularly exciting because the large cytogenetic, genomic, and genetic database for Pooideae, coupled with phy-

logenetic information, will ultimately make it possible to infer the mechanisms and directionality of chromosomal change throughout the group.

5. In the nitrogen-fixing clade (Rosales, Fagales, Fabales, and Cucurbitales of eurosid I), base chromosome number agrees in large part with relationships found using DNA sequence data. Molecular phylogenetic studies indicated that Cannabaceae are derived from within Celtidaceae (Sytsma et al. 2002). APG II (2003) therefore recognized an expanded Cannabaceae that also includes members of the former Celtidaceae. An expanded Cannabaceae is also supported by other lines of evidence, including chromosome number. Celtidaceae and Cannabaceae are the only members of the nitrogen-fixing clade to possess $n = 10$ (Sytsma et al. 2002).

Other examples of the phylogenetic utility of base chromosome numbers include Onagraceae (Sytsma and Smith 1992; Levin et al. 2003) and Iridaceae (Goldblatt et al. 2002).

Future Studies

Considerable progress has been made in the past few years in achieving a better understanding of genome diversification in the angiosperms. Whereas new C-values and a phylogenetic approach permitted an improved understanding of genome size evolution in the angiosperms, to make further progress, additional C-values are needed. For example, data are needed in many clades, including basal Santalales and basal monocots. Genome size diversification is much more dynamic than the unidirectional trend toward "genomic obesity" proposed by Bennetzen and Kellogg (1997).

Frequent genome contractions—as well as expansions—have occurred in the course of angiosperm (and land plant) diversification. However, many more case-by-case investigations are needed to evaluate the dynamics of genome size evolution. We still know surprisingly little about the mechanisms and causes of changes in genome size, particularly genomic downsizing. Thus, future studies should focus more on the mechanics of alterations in genome size.

The reconstructions of ancestral base chromosome numbers that we conducted for this book are preliminary. Analyses that more rigorously reconstruct diversification of chromosome numbers within families and genera would be useful. However, because chromosomal evolution is so complex, additional large-scale reconstructions of chromosomal evolution may be of little value.

Isozyme data provide support for ancient polyploidy in some basal lineages, but no genetic data are available for many basal lineages. Genetic data are needed to test the hypothesis that many basal lineages having only high chromosome numbers are ancient polyploids. For example, is *Amborella* with $2n = 26$ an ancient polyploid, as we suspect, and if so, how many rounds of genome duplication has it experienced? With the rapidly increasing application of genomic tools to angiosperms, it should soon be possible to address these questions. In addition, genomic approaches will eventually permit the evaluation of evolutionary trends in genome size, as well as genome organization and structure. Such studies have already been conducted in some groups, such as the grasses and Brassicaceae. Genomic data for other angiosperms and seed plants will ultimately provide additional critical information about genome evolution in flowering plants.

Appendix:
Angiosperm Supertree

TABLE A	Taxonomic description of composite terminal taxa in the angiosperm supertree
Clade	**Composition of clade**
Caryophyllales A	Portulacaceae
	Cactaceae
	Didiereaceae
	Basellaceae
	Halophytaceae
Caryophyllales B	Aizoaceae
	Gisekiaceae
	Phytolaccaceae
	Agdestidaceae
	Nyctaginaceae
	Sarcobataceae
	Barbeuiaceae
	Corbichonia
	Limeum

TABLE B Taxa included within the broadly circumscribed families represented in the supertree, following APG

Terminal taxa	Included taxa
Asparagaceae	Hesperocallidaceae
Bixaceae	Cochlospermaceae Diegodendraceae
Buxaceae	Didymelaceae
Campanulaceae	Lobeliaceae
Caprifoliaceae	Diervillaceae Linnaeaceae Morinaceae Valerianaceae
Chrysobalanaceae	Dichapetalaceae Euphroniaceae Trigoniaceae
Circaeasteraceae	Kingdoniaceae
Columelliaceae	Desfontainiaceae
Garryaceae	Aucubaceae
Geraniaceae	Hypseocharitaceae
Gunneraceae	Myrothamnaceae
Haloragaceae	Penthoraceae Tetracarpaeaceae
Iteaceae	Pterostemonaceae
Juglandaceae	Rhoipteleaceae
Melastomataceae	Memecylaceae
Melianthaceae	Francoaceae
Nitrariaceae	Tetradiclidaceae
Nymphaeaceae	Cabombaceae
Ochnaceae	Medusagynaceae Quiinaceae
Papaveraceae	Fumariaceae Pteridophyllaceae
Parnassiaceae	Lepuropetalaceae
Passifloraceae	Malesherbiaceae Turneraceae
Proteaceae	Platanaceae
Rhizophoraceae	Erythroxylaceae
Schisandraceae	Illiciaceae
Tetrameristaceae	Pellicieraceae
Trochodendraceae	Tetracentraceae
Zygophyllaceae	Krameriaceae

Source: Davies et al. 2004.

ANGIOSPERM SUPERTREE One of 10,000 most parsimonious supertrees; nodes collapsing in the strict consensus are indicated with an arrowhead. Numbers preceding the names of the terminal taxa indicate the number of species within those taxa (from Davies et al. 2004). † = terminals follow the broader circumscription sensu APG (1998; Table B), ‡ = terminals represent several amalgamated families (Table A). Alternative placement of non-monophyletic families not amalgamated are indicated by a dashed line. (Modified from Davies et al. 2004.) The Overview tree (page 305) summarizes the angiosperm supertree, with two large clades, the monocots and eudicots, indicated in bold. The subsequent pages (306–310) present details of the supertree for monocots (page 306) and eudicots (pages 307–310), respectively.

OVERVIEW

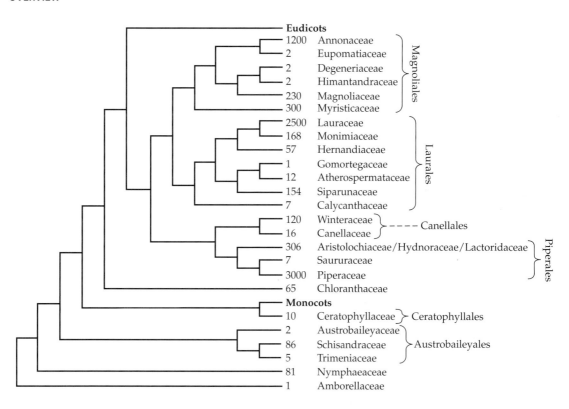

MONOCOTS

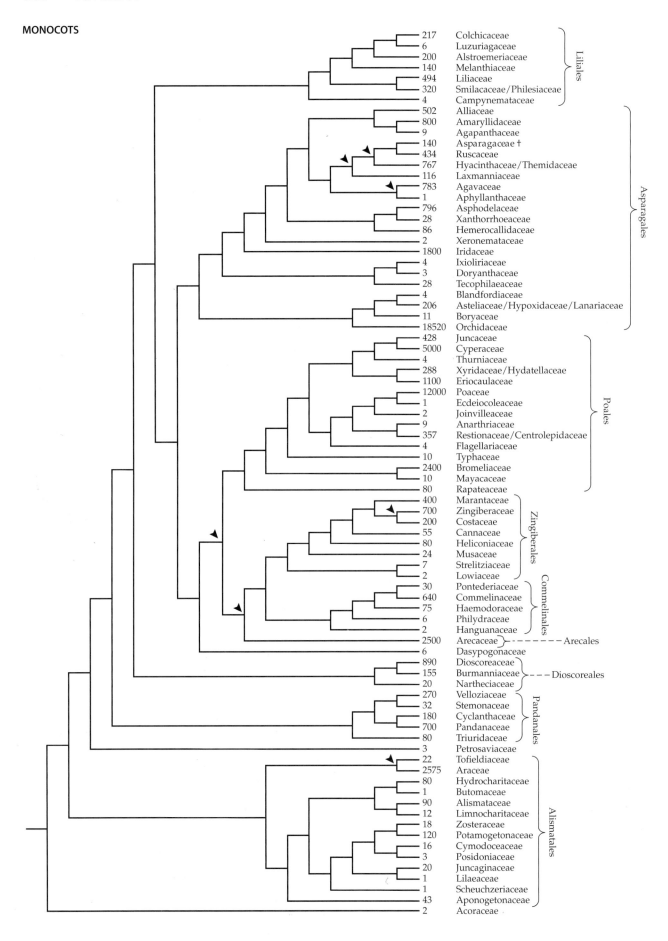

EUDICOTS

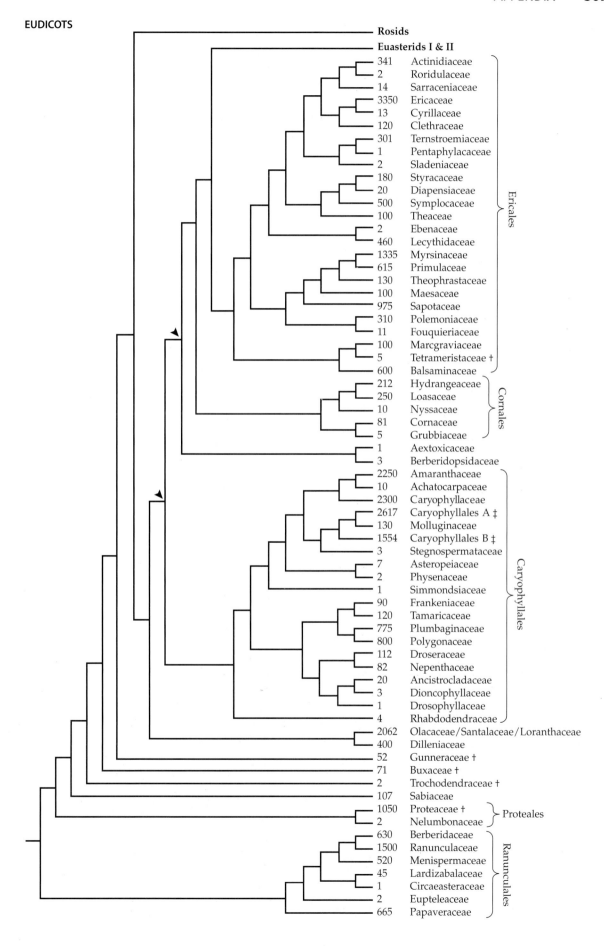

Rosids

Euasterids I & II

341	Actinidiaceae	
2	Roridulaceae	
14	Sarraceniaceae	
3350	Ericaceae	
13	Cyrillaceae	
120	Clethraceae	
301	Ternstroemiaceae	
1	Pentaphylacaceae	
2	Sladeniaceae	
180	Styracaceae	
20	Diapensiaceae	
500	Symplocaceae	
100	Theaceae	Ericales
2	Ebenaceae	
460	Lecythidaceae	
1335	Myrsinaceae	
615	Primulaceae	
130	Theophrastaceae	
100	Maesaceae	
975	Sapotaceae	
310	Polemoniaceae	
11	Fouquieriaceae	
100	Marcgraviaceae	
5	Tetrameristaceae †	
600	Balsaminaceae	

212	Hydrangeaceae	
250	Loasaceae	
10	Nyssaceae	Cornales
81	Cornaceae	
5	Grubbiaceae	

1   Aextoxicaceae
3   Berberidopsidaceae

2250	Amaranthaceae	
10	Achatocarpaceae	
2300	Caryophyllaceae	
2617	Caryophyllales A ‡	
130	Molluginaceae	
1554	Caryophyllales B ‡	
3	Stegnospermataceae	
7	Asteropeiaceae	
2	Physenaceae	
1	Simmondsiaceae	
90	Frankeniaceae	Caryophyllales
120	Tamaricaceae	
775	Plumbaginaceae	
800	Polygonaceae	
112	Droseraceae	
82	Nepenthaceae	
20	Ancistrocladaceae	
3	Dioncophyllaceae	
1	Drosophyllaceae	
4	Rhabdodendraceae	

2062   Olacaceae/Santalaceae/Loranthaceae
400   Dilleniaceae
52   Gunneraceae †
71   Buxaceae †
2   Trochodendraceae †
107   Sabiaceae

1050	Proteaceae †	Proteales
2	Nelumbonaceae	

630	Berberidaceae	
1500	Ranunculaceae	
520	Menispermaceae	
45	Lardizabalaceae	Ranunculales
1	Circaeasteraceae	
2	Eupteleaceae	
665	Papaveraceae	

EUASTERIDS I & II

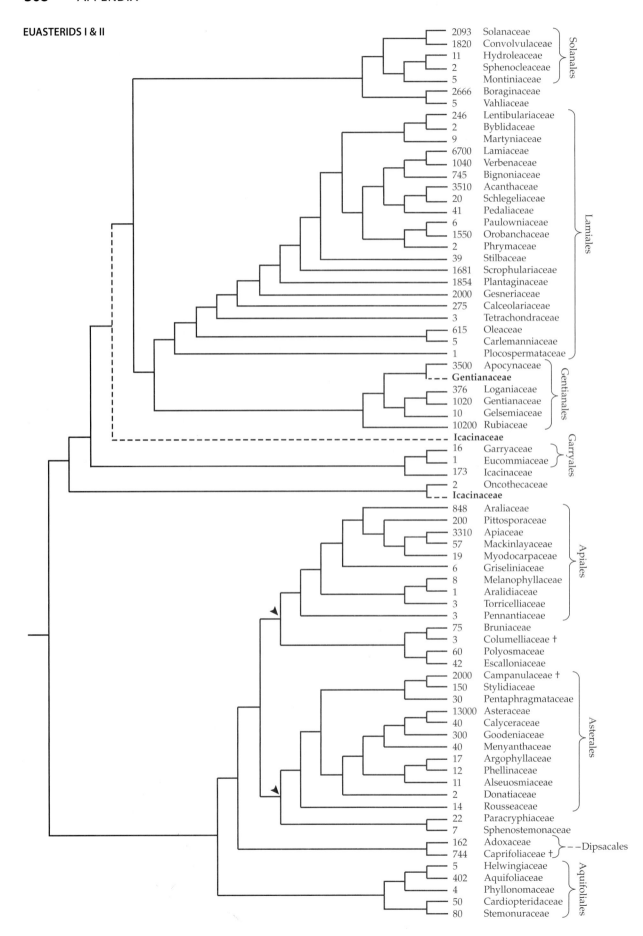

2093	Solanaceae
1820	Convolvulaceae
11	Hydroleaceae
2	Sphenocleaceae
5	Montiniaceae
2666	Boraginaceae
5	Vahliaceae
246	Lentibulariaceae
2	Byblidaceae
9	Martyniaceae
6700	Lamiaceae
1040	Verbenaceae
745	Bignoniaceae
3510	Acanthaceae
20	Schlegeliaceae
41	Pedaliaceae
6	Paulowniaceae
1550	Orobanchaceae
2	Phrymaceae
39	Stilbaceae
1681	Scrophulariaceae
1854	Plantaginaceae
2000	Gesneriaceae
275	Calceolariaceae
3	Tetrachondraceae
615	Oleaceae
5	Carlemanniaceae
1	Plocospermataceae
3500	Apocynaceae
	Gentianaceae
376	Loganiaceae
1020	Gentianaceae
10	Gelsemiaceae
10200	Rubiaceae
	Icacinaceae
16	Garryaceae
1	Eucommiaceae
173	Icacinaceae
2	Oncothecaceae
	Icacinaceae
848	Araliaceae
200	Pittosporaceae
3310	Apiaceae
57	Mackinlayaceae
19	Myodocarpaceae
6	Griseliniaceae
8	Melanophyllaceae
1	Aralidiaceae
3	Torricelliaceae
3	Pennantiaceae
75	Bruniaceae
3	Columelliaceae †
60	Polyosmaceae
42	Escalloniaceae
2000	Campanulaceae †
150	Stylidiaceae
30	Pentaphragmataceae
13000	Asteraceae
40	Calyceraceae
300	Goodeniaceae
40	Menyanthaceae
17	Argophyllaceae
12	Phellinaceae
11	Alseuosmiaceae
2	Donatiaceae
14	Rousseaceae
22	Paracryphiaceae
7	Sphenostemonaceae
162	Adoxaceae
744	Caprifoliaceae †
5	Helwingiaceae
402	Aquifoliaceae
4	Phyllonomaceae
50	Cardiopteridaceae
80	Stemonuraceae

Solanales
Lamiales
Gentianales
Garryales
Apiales
Asterales
Dipsacales
Aquifoliales

ROSIDS

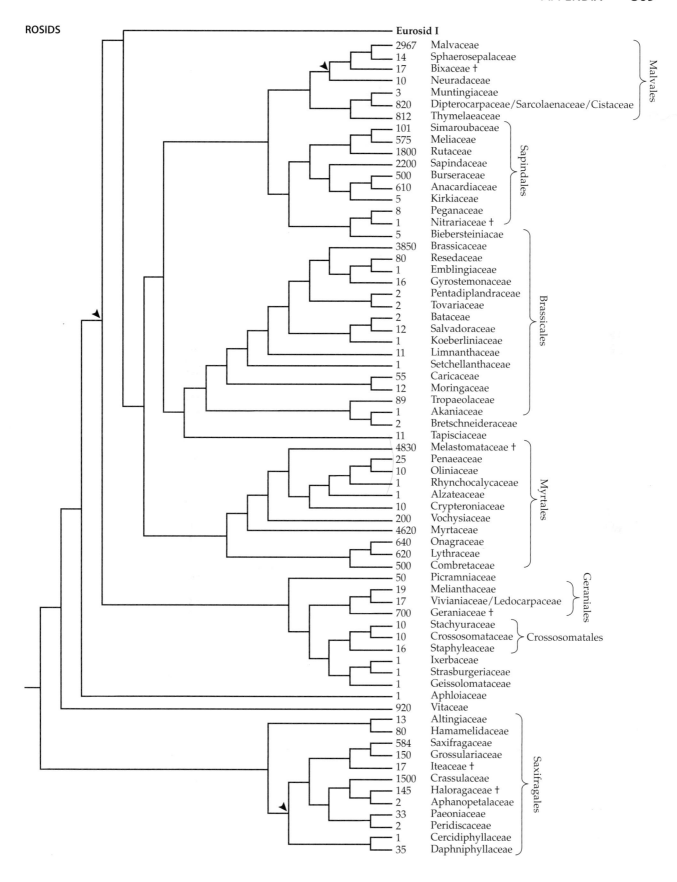

Eurosid I

2967 Malvaceae
14 Sphaerosepalaceae
17 Bixaceae †
10 Neuradaceae
3 Muntingiaceae
820 Dipterocarpaceae/Sarcolaenaceae/Cistaceae
812 Thymelaeaceae
} Malvales

101 Simaroubaceae
575 Meliaceae
1800 Rutaceae
2200 Sapindaceae
500 Burseraceae
610 Anacardiaceae
5 Kirkiaceae
8 Peganaceae
1 Nitrariaceae †
} Sapindales

5 Biebersteiniacae
3850 Brassicaceae
80 Resedaceae
1 Emblingiaceae
16 Gyrostemonaceae
2 Pentadiplandraceae
2 Tovariaceae
2 Bataceae
12 Salvadoraceae
1 Koeberliniaceae
11 Limnanthaceae
1 Setchellanthaceae
55 Caricaceae
12 Moringaceae
89 Tropaeolaceae
1 Akaniaceae
2 Bretschneideraceae
} Brassicales

11 Tapisciaceae

4830 Melastomataceae †
25 Penaeaceae
10 Oliniaceae
1 Rhynchocalycaceae
1 Alzateaceae
10 Crypteroniaceae
200 Vochysiaceae
4620 Myrtaceae
640 Onagraceae
620 Lythraceae
500 Combretaceae
} Myrtales

50 Picramniaceae
19 Melianthaceae
17 Vivianiaceae/Ledocarpaceae
700 Geraniaceae †
} Geraniales

10 Stachyuraceae
10 Crossosomataceae
16 Staphyleaceae
} Crossosomatales

1 Ixerbaceae
1 Strasburgeriaceae
1 Geissolomataceae
1 Aphloiaceae

920 Vitaceae

13 Altingiaceae
80 Hamamelidaceae
584 Saxifragaceae
150 Grossulariaceae
17 Iteaceae †
1500 Crassulaceae
145 Haloragaceae †
2 Aphanopetalaceae
33 Paeoniaceae
2 Peridiscaceae
1 Cercidiphyllaceae
35 Daphniphyllaceae
} Saxifragales

EUROSIDS I

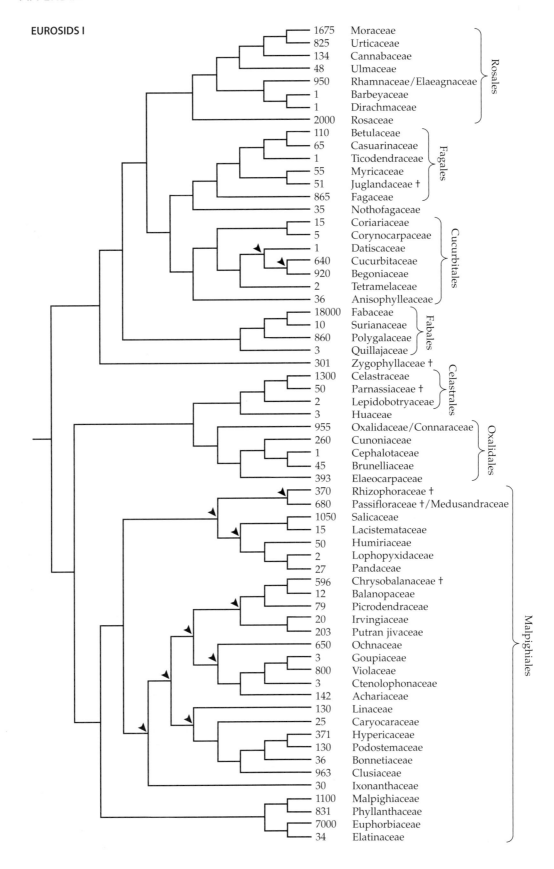

1675	Moraceae
825	Urticaceae
134	Cannabaceae
48	Ulmaceae
950	Rhamnaceae/Elaeagnaceae
1	Barbeyaceae
1	Dirachmaceae
2000	Rosaceae
110	Betulaceae
65	Casuarinaceae
1	Ticodendraceae
55	Myricaceae
51	Juglandaceae †
865	Fagaceae
35	Nothofagaceae
15	Coriariaceae
5	Corynocarpaceae
1	Datiscaceae
640	Cucurbitaceae
920	Begoniaceae
2	Tetramelaceae
36	Anisophylleaceae
18000	Fabaceae
10	Surianaceae
860	Polygalaceae
3	Quillajaceae
301	Zygophyllaceae †
1300	Celastraceae
50	Parnassiaceae †
2	Lepidobotryaceae
3	Huaceae
955	Oxalidaceae/Connaraceae
260	Cunoniaceae
1	Cephalotaceae
45	Brunelliaceae
393	Elaeocarpaceae
370	Rhizophoraceae †
680	Passifloraceae †/Medusandraceae
1050	Salicaceae
15	Lacistemataceae
50	Humiriaceae
2	Lophopyxidaceae
27	Pandaceae
596	Chrysobalanaceae †
12	Balanopaceae
79	Picrodendraceae
20	Irvingiaceae
203	Putran jivaceae
650	Ochnaceae
3	Goupiaceae
800	Violaceae
3	Ctenolophonaceae
142	Achariaceae
130	Linaceae
25	Caryocaraceae
371	Hypericaceae
130	Podostemaceae
36	Bonnetiaceae
963	Clusiaceae
30	Ixonanthaceae
1100	Malpighiaceae
831	Phyllanthaceae
7000	Euphorbiaceae
34	Elatinaceae

Rosales

Fagales

Cucurbitales

Fabales

Celastrales

Oxalidales

Malpighiales

References

Aboy, H. E. 1936. A Study of the Anatomy and Morphology of *Ceratophyllum demersum*. MSc thesis, Cornell University, Ithaca, NY.

Ackerman, J. D. 1983. On the evidence for a primitively epiphytic habit in orchids. *Systematic Botany* 8:474–476.

———. 2000. Abiotic pollen and pollination: Ecological, functional, and evolutionary perspectives. *Plant Systematics and Evolution* 222:167–185.

Adams, K. C., Y.-L. Qiu, M. Stoutemyer, and J. D. Palmer. 2002. Punctuated evolution of mitochondrial gene content: high and variable rates of mitochondrial gene loss and transfer to the nucleus during angiosperm evolution. *Proceedings of the National Academy of Sciences USA* 99:9905–9912.

Adams, S. P., T. P. V. Hartman, K. Y. Lim, M. W. Chase, M. D. Bennett, I. J. Leitch, and A. R. Leitch. 2001. Loss and recovery of *Arabidopsis*-type telomere repeat sequences 5′-(TTTAGGG)n-3′ in the evolution of a major radiation of flowering plants. *Proceedings of the Royal Society of London, B,* 268:1541–1546.

(AGI) *Arabidopsis* Genome Initiative. 2000. Analysis of the genome sequence of the flowering plant *Arabidopsis thaliana*. *Nature* 408:796–815.

Airy Shaw, H. K. 1964. Plagiopteraceae. *Kew Bulletin* 18:266.

———. 1965. On a new species of the genus *Silvianthus* Hook. f., and on the family Carlemanniaceae. *Kew Bulletin* 19:507–512.

Airy Shaw, H. K., and J. C. Willis. 1966. A Dictionary of the Flowering Plants and Ferns. 7th ed., revised H. K. Airy Shaw. Cambridge University Press, Cambridge.

Akkermans, A. D. L., and C. van Dijk. 1981. Non-leguminous root-nodule symbioses with actinomycetes and *Rhizobium*. *In* W. Broughton [ed.], Nitrogen Fixation, 1. Oxford University Press, London.

Albach D. C. 1998. Phylogeny of the Asteridae s. l. Master's thesis, Washington State University, Pullman, WA.

Albach, D. C., D. E. Soltis, M. W. Chase, and P. S. Soltis. 2001c. Phylogenetic placement of the enigmatic *Hydrostachys*. *Taxon* 50:781–805.

Albach, D. C., P. S. Soltis, and D. E. Soltis. 2001b. Patterns of embryological and biochemical evolution in the asterids. *Systematic Botany* 26:242–262.

Albach, D. C., P. S. Soltis, D. E. Soltis, and R. G. Olmstead. 2001a. Phylogenetic analysis of the Asteridae s. l. based on sequences of four genes. *Annals of the Missouri Botanical Garden* 88:163–212.

Albert, V. A., A. Backlund, K. Bremer, M. W. Chase, J. R. Manhart, B. D. Mishler, and K. C. Nixon. 1994. Functional constraints and *rbcL* evidence for land plant phylogeny. *Annals of the Missouri Botanical Garden* 81:534–567.

Albert, V. A., M. H. G. Gustafsson, and L. Di Laurenzio. 1998. Ontogenetic systematics, molecular developmental genetics, and the angiosperm petal. *In* D. E. Soltis, P. S. Soltis, and J. J. Doyle [eds.], Molecular Systematics of Plants, II. Kluwer, Boston.

Albert, V. A., D. Oppenheimer, and C. Lindqvist. 2002. Pleiotropy, redundancy and the evolution of flowers. *Trends in Plant Science* 7:297–301.

Albert, V. A., S. E. Williams, and M. W. Chase. 1992. Carnivorous plants: phylogeny and structural evolution. *Science* 257:1491–1495.

Albertson, R. C., J. A. Markert, P. D. Danley, and T. D. Kocher. 1999. Phylogeny of a rapidly evolving clade: The cichlid fishes of Lake Malawi, East Africa. *Proceedings of the National Academy of Sciences USA* 96:5107–5110.

Alexandersson, R., and S. D. Johnson. 2002. Pollinator-mediated selection on flower-tube length in a hawk-moth-pollinated *Gladiolus* (Iridaceae). *Proceedings of the Royal Society of London B* 269:631–636.

Allen, O. N., and E. K. Allen. 1976. Symbiotic Nitrogen Fixation in Plants. Cambridge University Press, Cambridge.

Al-Shammary, K. I. 1991. Systematic studies of the Saxifragaceae, chiefly from the Southern Hemisphere. PhD thesis, University of Leicester.

Al-Shammary, K. I. A., and R. J. Gornall. 1994. Trichome anatomy of the Saxifragaceae s. l. from the Southern Hemisphere. *Botanical Journal of the Linnean Society* 114:99–131.

Al-Shehbaz, I. A. 1984. The tribes of Cruciferae (Brassicaceae) in the southeastern United States. *Journal of the Arnold Arboretum* 65:343–373.

Al-Shehbaz, I. A., and B. G. Schubert. 1989. The Dioscoreaceae of the southeastern United States. *Journal of the Arnold Arboretum* 70:57–95.

Alverson, W. S., K. G. Karol, D. A. Baum, M. W. Chase, S. M. Swensen, R. McCourt, and K. J. Sytsma. 1998. Circumscription of the Malvales and relationships to other Rosidae: evidence from *rbcL* sequence data. *American Journal of Botany* 85:876–887.

Alverson, W. S., B. A. Whitlock, R. Nyffeler, C. Bayer, and D. A. Baum. 1999. Phylogeny of the core Malvales: evidence from *ndhF* sequence data. *American Journal of Botany* 86:1474–1486.

Ambrose, B. A., D. R. Lerner, P. Ciceri, C. M. Padilla, M. F. Yanofsky, and R. J. Schmidt. 1999. Genes specifying grass floral organ identity (abstract), XVI International Botanical Congress, St. Louis.

Ambrose, B. A., D. R. Lerner, P. Ciceri, C. M. Padilla, M. F. Yanofsky, and R. J. Schmidt. 2000. Molecular and genetic analyses of the *Silky1* gene reveal conservation in floral organ specification between eudicots and monocots. *Molecular Cell* 5:569–579.

Anderberg, A. A. 1992. The circumscription of the Ericales, and their cladistic relationships to other families of "higher" dicotyledons. *Systematic Botany* 17:660–675.

———. 1993. Cladistic relationships among major clades of the Ericales. *Plant Systematics and Evolution* 184:207–231.

Anderberg, A. A., and B. Stahl. 1995. Phylogenetic interrelationships in the order Primulales, with special emphasis on the family circumscriptions. *Canadian Journal of Botany* 73:1699–1730.

Anderberg, A. A., C. Rydin, and M. Källersjö. 2002. Phylogenetic relationships in the order Ericales s. l. : analyses of molecular data from five genes from the plastid and mitochondrial genomes. *American Journal of Botany* 89: 677–687.

Anderberg, A. A., B. Stähl, and M. Källersjö. 1998. Phylogenetic relationships in the Primulales inferred from *rbcL* sequence data. *Plant Systematics and Evolution* 211:93–102.

———. 2000. Maesaceae, a new primuloid family in the order Ericales s. l. *Taxon* 49:183–187.

Andersson, L., and M. W. Chase. 2001. Phylogeny, relationships and classification of Marantaceae. *Botanical Journal of the Linnean Society* 135:275–287.

Andersson, L., and J. H. E. Rova. 1999 The *rps16* intron and the phylogeny of the Rubioideae. *Plant Systematics and Evolution* 214:161–186.

Andreasen, K., and B. Bremer. 2000. Combined phylogenetic analysis in the Rubiaceae-Ixoroideae: morphology, nuclear and chloroplast DNA data. *American Journal of Botany* 87:1731–1748.

Antonius, K., and H. Ahokas. 1996. Flow cytometric determination of polyploidy level in spontaneous clones of strawberries. *Hereditas* 124:285.

APG (Angiosperm Phylogeny Group). 1998. An ordinal classification for the families of flowering plants. *Annals of the Missouri Botanical Garden* 85:531–553.

APG II (Angiosperm Phylogeny Group). 2003. An update of the Angiosperm Phylogeny Group classification for the orders and families of flowering plants. *Botanical Journal of the Linnean Society* 141:399–436.

Applequist, W. L., and R. S. Wallace. 2001. Phylogeny of the portulacaceous cohort based on *ndhF* sequence data. *Systematic Botany* 26:406–419.

Arber, A. 1920. Water Plants: A Study of Aquatic Angiosperms. Cambridge University Press, Cambridge.

———. 1921. The leaf structure of Iridaceae, considered in relation to the phyllode theory. *Annals of Botany (London)* 35:301–336.

———. 1925. Monocotyledons: A Morphological Study. Cambridge University Press, Cambridge.

Arber, E. A. N., and J. Parkin. 1907. On the origin of angiosperms. *Journal of the Linnean Society, Botany* 38:29–80.

———. 1908. Studies on the evolution of the angiosperms: the relationship of the angiosperms to the Gnetales. *Annals of Botany* 22:489–515.

Argus, G. W. 1997. Infrageneric classification of *Salix* (Salicaceae) in the New World. *Systematic Botany Monographs* 52:1–121.

Armbruster, W. S. 1992. Phylogeny and the evolution of plant-animal interactions. *BioScience* 42:12–20.

———. 1994. Evolution of plant pollination systems: hypotheses and tests with the Neotropical vine *Dalechampia*. *Evolution* 47:1480–1505.

———. 1996. Evolution of floral morphology and function: an integrative approach to adaptation, constraint, and compromise in *Dalechampia* (Euphorbiaceae). *In* D. G. Lloyd and S. C. H. Barrett [eds.], Floral Biology. Studies on Floral Evolution in Animal-Pollinated Plants, 241–272. Chapman and Hall, New York.

Armbruster, W. S., and B. G. Baldwin. 1998. Switch from specialized to generalized pollination. *Nature* 394:632.

Armbruster, W. S., E. M. Debevec, and M. F. Willson. 2002. Evolution of syncarpy in angiosperms: theoretical and phylogenetic analyses of the effects of carpel fusion on offspring quantity and quality. *Journal of Evolutionary Biology* 15:657–672.

Asmussen, C. B., and M. W. Chase. 2001. Coding and noncoding plastid DNA in palm systematics. *American Journal of Botany* 88:1103–1117.

Asmussen, C. B., W. J. Baker, and J. Dransfield. 2000. Phylogeny of the palm family (Arecaceae) based on *rps16* intron and *trnL-trnF* plastid DNA sequences. *In* K. L. Wilson and D. A. Morrison [eds.], Monocots: Systematics and Evolution, 525–537. CSIRO, Melbourne.

Avise, J. C. 1994. Molecular Markers, Natural History, and Evolution. Chapman and Hall, New York.

Ax, P. 1987. The Phylogenetic System: The Systematization of Organisms on the Basis of their Phylogenesis. John Wiley and Sons, Chichester.

Axelrod, D. I. 1952. A theory of angiosperm evolution. *Evolution* 6:29–60.

———. 1970. Mesozoic paleo-geography and early angiosperm history. *Botanical Review* 36:277–319.

———. 1972. Edaphic aridity as a factor in angiosperm evolution. *American Naturalist* 106:311–320.

Axsmith, B. J., E. L. Taylor, and T. N. Taylor. 1998. The limitations of molecular systematics: a palaeobotanical perspective. *Taxon* 47:105–108.

Ayers, T. J. 1990. Systematics of *Heterotoma* (Campanulaceae) and the evolution of nectar spurs in the New World Lobelioideae. *Systematic Botany* 15:296–327.

Azuma, T., T. Kajita, J. Yokoyama, and H. Ohashi. 2000. Phylogenetic relationships of *Salix* (Salicaceae) based on *rbcL* sequence data. *American Journal of Botany* 87:67–75.

Baas, P. 1984. Vegetative anatomy of *Berberidopsis* and *Streptothamnus* (Flacourtiaceae). *Blumea* 30:39–44.

Baas, P., R. Geesink, W. A. Van Heel, and H. J. Muller. 1979. The affinities of *Plagiopteron suaveolens* Griff. (Plagiopteraceae). *Grana* 18:69–89.

Baas, P., E. Wheeler, and M. Chase. 2000. Dicotyledonous wood anatomy and the APG system of angiosperm classification. *Botanical Journal of the Linnean Society* 134:3–17.

Bachmann, K. 1983. Evolutionary genetics and the genetic control of morphogenesis in flowering plants. *Evolutionary Biology* 15:157–208.

Backlund, A., and B. Bremer. 1997. Phylogeny of the Asteridae s. str. based on *rbcL* sequences, with particular reference to the Dipsacales. *Plant Systematics and Evolution* 207:225–254.

———. 1998. To be or not to be: principles of classification and monotypic plant families. *Taxon* 47:391–401.

Backlund, A., and N. Pyck. 1998. Diervillaceae and Linnaeaceae, two new families of caprifolioids. *Taxon* 47:657–661.

Backlund, M., B. Oxelman, and B. Bremer. 2000. Phylogenetic relationships within the Gentianales based on *ndhF* and *rbcL* sequences with particular reference to the Loganiaceae. *American Journal of Botany* 87:1029–1043.

Bailey, I. W. 1944a. The development of vessels in angiosperms and its significance in morphological research. *American Journal of Botany* 31:421–428.

———. 1944b. The comparative morphology of the Winteraceae. *Journal of the Arnold Arboretum* 25:97–103.

———. 1953. Evolution of the tracheary tissue of land plants. *American Journal of Botany* 40:4–8.

———. 1956. Nodal anatomy in retrospect. *Journal of the Arnold Arboretum* 37:269–287.

Bailey, I. W., and B. G. L. Swamy. 1948. *Amborella trichopoda* Baill., a new morphological type of vesselless dicotyledon. *Journal of the Arnold Arboretum* 29:245–254.

———. 1951. The conduplicate carpel and its initial trends of specialization. *American Journal of Botany* 38:373–379.

Baillon, H. 1866–1895. Histoire des Plantes. Hachette, Paris.

Baker, S. C., K. Robinson-Beers, J. M. Villanueva, J. C. Gaiser, and C. S. Gasser. 1997. Interactions among genes regulating ovule development in *Arabidopsis thaliana*. *Genetics* 145:1109–1124.

Baker, W. J., C. B. Asmussen, S. C. Barrow, J. Dransfield, and T. A. Hedderson. 1999. A phylogenetic study of the palm family (Palmae) based on chloroplast DNA sequences from the *trnL-trnF* region. *Plant Systematics and Evolution* 219:111–126.

Bakker, F. T., D. D. Vassiliades, C. M. Morton, and V. Savolainen. 1988. Phylogenetic relationships of *Biebersteinia* Stephen (Geraniaceae) inferred from *rbcL* and *atpB* sequence comparisons. *Botanical Journal of the Linnean Society* 127:149–158.

Balogh, P. 1982. Generic redefinition in subtribe Spiranthinae (Orchidaceae). *American Journal of Botany* 69:1119–1132.

Barkman, T. J., G. Chenery, J. R. McNeal, J. Lyons-Weiler, and C. W. dePamphilis. 2000. Independent and combined analysis of sequences from all three genomic compartments converge to the root of flowering plant phylogeny. *Proceedings of the National Academy of Sciences USA* 97:13166–13171.

Barkman, T. J., S.-H. Lim, K. M. Salleh, N. D . J. Nais. 2004. Mitochondrial DNA sequences reveal the photosynthetic relatives of *Rafflesia*, the world's largest flower. *Proceedings of the National Academy of Sciences USA* 101:787–792.

Barkman, T. J., J. McNeal, S.-H. Lim, G. Coat, H. B. Croom, N. Young, and C. W. dePamphilis. Mitochondrial DNA sequences suggest 12 origins of parasitism in angiosperms and implicate parasitic plants as vectors of horizontal gene transfer. In prep.

Barrett, S. C. H. 1992. Evolution and Function of Heterostyly. Springer, Berlin.

———. 1995. Mating-system evolution in flowering plants: micro- and macroevolutionary approaches. *Acta Botanica Neerlandica* 44:385–402.

———. 1998. The evolution of mating strategies in flowering plants. *Trends in Plant Science* 3:335–341.

———. 2002. The evolution of plant sexual diversity. *Nature Reviews Genetics* 3:274–284.

Barrett, S. C. H., and S. W. Graham. 1997. Adaptive radiation in the aquatic plant family Pontederiaceae. Insights from phylogenetic analysis. *In* T. J. Givnish and K. J. Sytsma [eds.], Molecular Evolution and Adaptive Radiation, 225–258. Cambridge University Press, Cambridge.

Barrett, S. C. H., L. D. Harder, and A. C. Worley. 1996. The comparative biology of pollination and mating in flowering plants. *Philosophical Transactions of the Royal Society of London B* 351:1271–1280.

Barrett, S. C. H., L. K. Jesson, and A. M. Baker. 2000. The evolution and function of stylar polymorphisms in flowering plants. *Annals of Botany (Supplement A)* 85:253–265.

Barthlott, W., and I. Theissen. 1995. Epicuticular wax ultrastructure and classification of Ranunculiflorae. *Plant Systematics and Evolution*, Supplement 9:39–45.

Barthlott, W., S. Porembski, E. Fischer, and B. Gemmel. 1998. First protozoa-trapping plant found. *Nature* 392:447.

Basinger, J. F., and D. L. Dilcher. 1984. Ancient bisexual flowers. *Science* 224:511–513.

Bateman, R. M., and W. A. DiMichele. 1994. Saltational evolution of form in plants: a neoGoldschmidtian synthesis. *In* D. S. Ingram and A. Hudson [eds.], Shape and Form in Plants and Fungi, 63–100. Academic Press, London.

Bate-Smith, E. C. 1965. Recent progress in the chemical taxonomy of some phenolic constituents of plants. *Bulletin de la Société Botanique de France, Mémoire* 1965:16–28.

Bate-Smith, E. C., I. K. Ferguson, K. Hutson, S. R. Jensen, B. J. Nielsen, and T. Swain. 1975. Phytochemical interrelationships in the Cornaceae. *Biochemical Systematics and Ecology* 3:79–89.

Baum, B. R. 1992. Combining trees as a way of combining data sets for phylogenetic inference, and the desirability of combining gene trees. *Taxon* 41:3–10.

Baum, D. A. 1998. The evolution of plant development. *Current Opinions in Plant Biology* 1:79–86.

Baum, D. A., and M. J. Donoghue. 2002. Transference of function, heterotopy and the evolution of plant development. *In* Q. C. B. Cronk [ed.], Developmental Genetics and Plant Evolution, 52–69. Taylor and Francis, London.

Baum, D. A., and B. A. Whitlock. 1999. Plant development: genetic clues to petal evolution. *Current Biology* 9:R525–R527.

Baum, D. A., J. Doebley, V. F. Irish, and E. M. Kramer. 2002. Response: missing links: The genetic architecture of flower and floral diversification. *Trends in Plant Science* 7:31–34.

Baum, D. A., R. L. Small, and J. F. Wendel. 1998. Biogeography and floral evolution of baobabs (*Adansonia*, Bombaceae) as inferred from multiple data sets. *Systematic Biology* 47:181–207.

Bawa, S. B. 1970. Haloragaceae. *Bulletin of the Indian National Science Academy* 41:226–229.

Bayer, C. 1999. The bicolor unit: homology and transformation of an inflorescence structure unique to core Malvales. *Plant Systematics and Evolution* 214:187–198.

Bayer, C., and K. Kubitzki. 1996. Inflorescence morphology of some Australian Lasiopetaleae (Sterculiaceae). *Telopea* 6:721–728.

Bayer, C., M. W. Chase, and M. F. Fay. 1998. Muntingiaceae, a new family of dicotyledons with malvalean affinities. *Taxon* 47:37–42.

Bayer, C., M. F. Fay, A. de Bruijn, V. Savolainen, C. M. Morton, K. Kubitzki, and M. W. Chase. 1999. Support for an expanded family concept of Malvaceae within a recircumscribed order Malvales: a combined analysis of plastid *atpB* and *rbcL* sequences. *Botanical Journal of the Linnean Society* 129:267–303.

Beardsley, P. M., and R. G. Olmstead. 2002. Redefining Phrymaceae: the placement of *Mimulus*, tribe Mimuleae, and *Phryma*. *American Journal of Botany* 89:1093–1102.

Beardsley, P. M., S. Schoenig, and R. G. Olmstead. 2001. Radiation of *Mimulus* (Phrymaceae) in western North America: Evolution of polyploidy, woodiness, and pollination syndromes. *In* Botany 2001: Plants and People, Abstracts, p. 100. Albuquerque, NM.

Bedell, H. G. 1980. A taxonomic and morphological re-evaluation of the Stegnospermaceae (Caryophyllales). *Systematic Botany* 5:419–431.

Behnke, H.-D. 1967a. A tabulated survey of some characters of systematic importance in centrospermous families. *Plant Systematics and Evolution* 125:95–98.

———. 1967b. Ultrastructure of sieve-element plastids in Caryophyllales (Centrospermae): Evidence for the delimitation and classification of the order. *Plant Systematics and Evolution* 126:31–54.

———. 1969. Die Siebröhren-Plastiden bei Monocotylen. *Naturwissenschaften* 55:140–141.

———. 1971. Zum Feinbau der Siebröhrenplastiden von *Aristolochia* und *Asarum* (Aristolochiaceae). *Planta* 97:62–69.

———. 1976a. A tabulated survey of some characters of systematic importance in centrospermous families. *Plant Systematics and Evolution* 125:95–98.

———. 1976b. Ultrastructure of sieve-element plastids in Caryophyllales (Centrospermae): evidence for the delimitation and classification of the order. *Plant Systematics and Evolution* 126:31–54.

———. 1977. Zur Skulptur der Pollen-Exine bei drei Centrospermen (*Gisekia, Limeum, Hectorella*), bei Gyrostemonaceen und Rhabdodendraceen. *Plant Systematics and Evolution* 128:227–235.

———. 1981. Sieve-element characters. *Nordic Journal of Botany* 1:381–400.

———. 1982. *Geocarpon minimum*: Sieve-element plastids as additional evidence for its inclusion in Caryophyllaceae. *Taxon* 31:45–47.

———. 1984. Ultrastructure of sieve-element plastids of Myrtales and allied groups. *Annals of the Missouri Botanical Garden* 71:824–831.

———. 1988a. Sieve-element plastids, phloem protein, and evolution of the flowering plants: III. Magnoliidae. *Taxon* 37:699–732.

———. 1988b. Sieve-element plastids and systematic relationships of Rhizophoraceae, Anisophylleaceae and allied groups. *Annals of the Missouri Botanical Garden* 75:1387–1409.

———. 1991. Sieve-element characters of *Ticodendron*. *Annals of the Missouri Botanical Garden* 78:131–134.

———. 1993. Further studies of the sieve-element plastids of the Caryophyllales including *Barbeuia, Corrigiola, Lyallia, Microtea, Sarcobatus,* and *Telephium*. *Plant Systematics and Evolution* 186:231–243.

———. 1995. Sieve-element plastids, phloem proteins, and evolution in the Ranunculanae. *Plant Systematics and Evolution, Supplement* 9:25–37.

———. 1997. Sarcobataceae: a new family of Caryophyllales. *Taxon* 46:495–507.

———. 1999. P-type sieve-element plastid present in members of the tribes Triplareae and Coccolobeae (Polygonaceae) renew the links between the Polygonales and the Caryophyllales. *Plant Systematics and Evolution* 214:15–27.

———. 2000. Forms and sizes of sieve-element plastids and evolution of the monocotyledons. *In* K. L. Wilson and D. A. Morrison [eds.], Monocots: Systematics and Evolution, pp. 163–188. CSIRO, Melbourne.

Behnke, H. D., and W. Barthlott. 1983. New evidence from ultrastructural and micro-morphological fields in angiosperm classification. *Nordic Journal of Botany* 3:343–66.

Behnke, H. D., J. Treutlein, M. Wink, K. Kramer, C. Schneider, and P. C. Kao. 2000. Systematics and evolution of Velloziaceae, with special reference to sieve-element plastids and *rbcL* sequence data. *Botanical Journal of the Linnean Society* 134:93–129.

Bell, C. D., D. E. Soltis, and P. S. Soltis. 2005. The age of the angiosperms: A molecular time-scale without a clock. *Evolution,* in press.

Bennett, M. D. 1972. Nuclear DNA content and minimum generation time in herbaceous plants. *Proceedings of the Royal Society of London B* 181:109–135.

———. 1987. Variation in genomic form in plants and its ecological implications. *New Phytologist* 106:177–200.

———. 1998. Plant genome values: How much do we know? *Proceedings of the National Academy of Sciences USA* 95: 2011–2016.

Bennett, M. D., and I. J. Leitch. 1995. Nuclear DNA amounts in angiosperms. *Annals of Botany* 80:169–196.

———. 2001. Angiosperm DNA C-values database (release 3. 1 Sept. 2001). Available at http://www. rbgkew. org. uk/cval /hompage. html.

Bennett, M. D., and J. B. Smith. 1976. Nuclear DNA amounts in angiosperms. *Philosophical Transactions of the Royal Society of London B* 274:227–274.

———. 1991. Nuclear DNA amounts in angiosperms. *Philosophical Transactions of the Royal Society of London B* 334: 309–345.

Bennett, M. D., A. Cox, and I. J. Leitch. 1997. Angiosperms DNA C-values database. Available at http://www. rbgkew. org. uk/cval/database1. html.

Bennett, M. D., I. J. Leitch, and L. Hanson. 1998. DNA amounts in two samples of angiosperm weeds. *Annals of Botany* 82:121–134.

Bennett, M. D., J. B. Smith, and J. S. Heslop-Harrison. 1982. Nuclear DNA amounts in angiosperms. *Proceedings of the Royal Society of London B* 216:179–199.

Bennett, M. D., I. J. Leitch, H. J. Price, and J. S. Johnston. 2003. Comparions with *Caenorhabditis* (~100 Mb) and *Drosophila* (~175 Mb) using flow cytometry show genome size in *Arabidopsis* to be ~157 MB and this is 25% larger than the *Arabidopsis* genome initiative of ~125 Mb. *Annals of Botany* 91:547–557.

Bennetzen, J. L. 1996. The contributions of retroelements to plant genome organization, function, and evolution. *Trends in Microbiology* 4:347–353.

———. 2000. Transposable element contributions to plant gene and genome evolution. *Plant Molecular Biology* 42:251–269.

———. 2002. Mechanisms and rates of genome expansion and contraction in flowering plants. *Genetica* 115:29–36.

Bennetzen, J. L., and E. A. Kellogg. 1997. Do plants have a one-way ticket to genomic obesity? *The Plant Cell* 9:1509–1514.

Bensel, C. R., and B. F. Palser. 1975. Floral anatomy in the Saxifragaceae sensu lato. II. Saxifragoideae and Iteoideae. *American Journal of Botany* 62:661–675.

Bentham, G. 1880. Notes on Euphorbiaceae. *Botanical Journal of the Linnean Society* 17:185–267.

Bentham, G., and J. D. Hooker. 1862–1883. Genera Plantarum. Reeve, London.

Benzing, D. H. 1967. Developmental patterns in stem primary xylem of woody Ranales. I and II. *American Journal of Botany* 54: 805–813, 813–820.

Benzing, D. H., and J. T. Atwood. 1984. Orchidaceae: ancestral habitats and current status in forest canopies. *Systematic Botany* 9:155–165.

Benzing, D. H., T. Givnish, and D. Bermudes. 1985. Absorptive trichomes in *Brocchinia reducta* (Bromeliaceae) and their evolutionary and systematic significance. *Systematic Botany* 10:81–91.

Berger, A. 1930. Crassulaceae. *In* A. Engler and K. Prantl. [eds.], Die Natürlichen Pflanzenfamilien, ed. 2, 18a, 352–483. Engelmann, Leipzig.

Bernays, E., and C. De Luca. 1981. Insect antifeedant properties of an iridoid glycoside: ipolamiide. *Experientia* 37:1289–1290.

Bernhard, A. 1999. Flower structure, development, and systematics in Passifloraceae and in *Abatia* (Flacourtiaceae). *International Journal of Plant Sciences* 160:135–150.

Bernhard, A., and P. K. Endress. 1999. Androecial development and systematics in Flacourtiaceae s. l. *Plant Systematics and Evolution* 215:141–155.

Bernhardt, P. 1996. Anther adaptations in animal pollination. *In* W. G. D'Arcy and R. C. Keating [eds.], The Anther: Form,

Function and Phylogeny, 192–220. Cambridge University Press, Cambridge.

———. 2000. Convergent evolution and adaptive radiation of beetle-pollinated angiosperms. *Plant Systematics and Evolution* 222:293–320.

Berry, P. E., V. Savolainen, K. Sytsma, J. C. Hall, and M. W. Chase. 2001. *Lissocarpa* is sister to *Diospyros* (Ebenaceae). *Kew Bulletin* 56:725–729.

Bertin, R., and C. Newman. 1993. Dichogamy in angiosperms. *Botanical Review* 59:112–152.

Bessey, C. E. 1897. The phylogeny and taxonomy of angiosperms. *Botanical Gazette* 24:145–178.

———. 1915. The phylogenetic taxonomy of flowering plants. *Annals of the Missouri Botanical Garden* 2:109–164.

Beusekom, C. F. V. 1971. Revision of *Meliosma* (Sabiaceae), section *Lorenzanea* excepted, living and fossil, geography and phylogeny. *Blumea* 19:355–529.

Bharathan, G. 1996. Reproductive development and nuclear DNA content in angiosperms. *American Journal of Botany* 83:440–451.

Bharathan, G., and E. A. Zimmer. 1995. Early branching events in monocotyledons–partial 18S ribosomal DNA sequence analysis. *In* P. J. Rudall, P. J. Cribb, D. F. Cutler, and C. J. Humphries [eds.], Monocotyledons: Systematics and Evolution, 750. Royal Botanic Gardens, Kew.

Bhatnagar, A. K., and M. Garg. 1977. Affinities of *Daphniphyllum*: a palynological approach. *Phytomorphology* 27:92–97.

Blanc, P. 1986. Edification d'arbres par croissance d'etablissement de type monocotylédonien: l'exemple de Chloranthaceae. Colloque international sur l'arbre, 101–123. Naturalia Monspeliensia, numéro hors série, Montpellier.

Blarer, A., D. L. Nickrent, and P. K. Endress. 2004. Comparative floral structure and systematics in Apodanthaceae (Rafflesiales). *Plant Systematics and Evolution* 245:119–142.

Boesewinkel, F. D., and F. Bouman. 1991. The development of bi- and unitegmic ovules and seeds in *Impatiens* (Balsaminaceae). *Botanische Jahrbücher für Systematik* 113:87–104.

Bogle, A. L. 1969. The genera of Portulacaceae and Basellaceae in the southeastern United States. *Journal of the Arnold Arboretum* 50:566–598.

———. 1970. Floral morphology and vascular anatomy of the Hamamelidaceae: the apetalous genera of Hamamelidoideae. *Journal of the Arnold Arboretum* 51:310–366.

———. 1986. The floral morphology and vascular anatomy of the Hamamelidaceae subfamily Liquidambaroideae. *Annals of the Missouri Botanical Garden* 73:325–347.

———. 1989. The floral morphology, vascular anatomy, and ontogeny of the Rhodoleioideae (Hamamelidaceae) and their significance in relation to the 'lower' hamamelids. *In* P. R. Crane and S. Blackmore [eds.], Evolution, Systematics and Fossil History of the Hamamelidae, 1, 201–226. Clarendon Press, Oxford.

Bohs, L., and R. G. Olmstead. 1997. Phylogenetic relationships in *Solanum* (Solanaceae) based on *ndhF* sequences. *Systematic Botany* 22:5–17.

Boke, N. H. 1963. Anatomy and development of the flower and fruit of *Pereskia pititache*. *American Journal of Botany* 50:843–858.

———. 1964. The cactus gynoecium: a new interpretation. *American Journal of Botany* 51:598–610.

———. 1966. Ontogeny and structure of the flower and fruit of *Pereskia aculeata*. *American Journal of Botany* 53:534–542.

Boros, C. A., and F. R. Stermitz. 1990. Iridoids. An updated review, I. *Journal of Natural Products* 53:1055–1147.

———. 1991. Iridoids. An updated review, II. *Journal of Natural Products* 54:1172–1246.

Borsch, T. 1998. Pollen types in the Amaranthaceae: morphology and evolutionary significance. *Grana* 37:129–142.

Borsch, T., K. W. Hilu, D. Quandt, V. Wilde, C. Neinhuis, and W. Barthlott. 2003. Non-coding plastid *trnT-trnF* sequences

reveal a well resolved phylogeny of basal angiosperms. *Journal of Evolutionary Biology* 16:558–576.

Bortenschlager, S., A. Auinger, J. Blaha, and P. Simonsburger. 1972. Pollen morphology of Achatocarpaceae (Centrospermae). *Berichte des Naturwissenschaftlich Medizinischen Vereins Innsbruck* 59:7–13.

Bouman, F. 1984. The ovule. *In* B. M. Johri [ed.], Embryology of Angiosperms, 123–157. Springer, Berlin.

Bouman, F., and S. Schier. 1979. Ovule ontogeny and seed coat development in *Gentiana* with a discussion on the evolutionary origin of the single integument. *Acta Botanica Neerlandica* 28:467–478.

Bowe, L. G. Coat, and C. dePamphilis. 2000. Phylogeny of seed plants based on all three genomic compartments: extant gymnosperms are monophyletic and Gnetales' closest relatives are conifers. *Proceedings of the National Academy of Sciences USA* 97:4092–4097.

Bowers, J. E., B. A. Chapman, J. Rong, and A. H. Paterson. 2003. Unravelling angiosperm genome evolution by phylogenetic analysis of chromosomal duplication events. *Nature* 422:433–438.

Bowers, M. D., and G. M. Puttick. 1988. Response of generalist and specialist insects to qualitative allelochemical variation. *Journal of Chemical Ecology* 14:319–334.

Bowman, J. L. 1997. Evolutionary conservation of angiosperm flower development at the molecular and genetic levels. *Journal of Bioscience* 22:515–527.

———. 2000. Axial patterning in leaves and other lateral organs. *Current Opinions in Genetics and Development* 10:399–404.

Bowman, J. L., H. Brüggemann, J.-Y. Lee, and K. Mummenhoff. 1999. Evolutionary changes in floral structure within *Lepidium* L. (Brassicaceae). *International Journal of Plant Sciences* 160:917–929.

Bowman, J. L., D. R. Smyth, and E. M. Meyerowitz. 1989. Genes directing flower development in *Arabidopsis*. *Plant Cell* 1:37–52.

———. 1991. Genetic interactions among floral homeotic genes of *Arabidopsis*. *Development* 112:1–20.

Bradley, D., R. Carpenter, H. Sommer, N. Hartley, and E. Coen. 1993. Complementary floral homeotic phenotypes result from opposite orientations of a transposon at the *PLENA* locus of *Antirrhinum*. *Cell* 72:85–95.

Braem, G. 1993. Studies in the Oncidiinae. Discussion of some taxonomic problems with description of *Gudrania* Braem, gen. nov., and reinstatement of the genus *Lophiaris* Raffinesque. *Schlechteriana* 4:8–21.

Brandl, R., W. Mann, and M. Sprintzl. 1992. Estimation of the monocot-dicot age through tRNA sequences from the chloroplast. *Proceedings of the Royal Society of London B* 249:13–17.

Bremer, B. 1987. The sister group of the paleo-tropical tribe Argostemmateae: a redefined neotropical tribe Hamelieae (Rubiaceae, Rubioideae). *Cladistics* 3:35–51.

———. 1996. Phylogenetic studies within Rubiaceae and relationships to other families based on molecular data. *Opera Botanica Belgica* 7:33–50.

Bremer, B., and L. Struwe. 1992. Phylogeny of the Rubiaceae and Loganiaceae: Congruence or conflict between morphological and molecular data. *American Journal of Botany* 79:1171–1194.

Bremer, B., K. Andreason, and D. Olsson. 1995. Subfamilial and tribal relationships in the Rubiaceae based on *rbcL* sequence data. *Annals of the Missouri Botanical Garden* 82:383–397.

Bremer, B., K. Bremer, N. Heirdari, P. Erixon, R. G. Olmstead, M. Källersjö, A. A. Anderberg, and E. Barkhordarian. 2002. Phylogenetics of asterids based on 3 coding and 3 non-coding chloroplast DNA markers and the utility of non-coding DNA at higher taxonomic levels. *Molecular Phylogenetics and Evolution* 24:274–301.

Bremer, B., R. K. Jansen, B. Oxelman, M. Backlund, H. Lantz, and K.-J. Kim. 1999. More characters or more taxa for a robust phylogeny–case study from the coffee family (Rubiaceae). *Systematic Biology* 48:413–435.

Bremer, B., R. G. Olmstead, L. Struwe, and J. A. Sweere. 1994. *rbcL* sequences support exclusion of *Retzia*, *Desfontainia*, and *Nicodemia* from the Gentianales. *Plant Systematics and Evolution* 190:213–230.

Bremer, K. 1987. Tribal interrelationships of the Asteraceae. *Cladistics* 3:210–253.

———. 1988. The limits of amino acid sequence data in angiosperm phylogenetic reconstruction. *Evolution* 42:795–803.

———. 1994. *Asteraceae: Cladistics and classification*. Timber Press, Portland.

———. 1996. Major clades and grades of the Asteraceae. *In* D. J. N. Hind and H. J. Beentje [eds.], Compositae: Systematics, 1–7. Royal Botanic Gardens, Kew.

———. 2000. Early Cretaceous lineages of monocot flowering plants. *Proceedings of the National Academy of Sciences USA* 97:4707–4711.

———. 2002. Gondwanan evolution of the grass alliance of families (Poales). *Evolution* 56:1374–1387.

Bremer, K., and M. H. Gustafsson. 1997. East Gondwana ancestry of the sunflower alliance of families. *Proceedings of the National Academy of Sciences USA* 94:9188–9190.

Bremer, K., and R. K. Jansen. 1992. A new subfamily of the Asteraceae. *Annals of the Missouri Botanical Garden* 79:414–415.

Bremer, K., A. Backlund, B. Sennblad, U. Swenson, K. Andreasen, M. Hjertson, J. Lundberg, M. Backlund, and B. Bremer. 2001. A phylogenetic analysis of 100+ genera and 50+ families of euasterids based on morphological and molecular data with notes on possible higher level morphological synapomorphies. *Plant Systematics and Evolution* 229:137–169.

Brenner, G. J. 1996. Evidence for the earliest stage of angiosperm pollen evolution: a paleoequatorial section from Israel. *In* D. W. Taylor and L. J. Hickey [eds.], Flowering Plant Origin, Evolution, and Phylogeny, 91–115. Chapman and Hall, New York.

Breuer, B., T. Stuhlfauth, H. Fock, and H. Huber. 1987. Fatty acids of some Cornaceae, Hydrangeaceae, Aquifoliaceae, Hamamelidaceae and Styracaceae. *Phytochemistry* 26:1441–1445.

Brewbaker, J. L. 1957. Pollen cytology and self-incompatibility systems in plants. *Journal of Heredity* 48:271–277.

———. 1967. The distribution and phylogenetic significance of binucleate and trinucleate pollen grains in the angiosperms. *American Journal of Botany* 54:1069–1083.

Briggs, B., and L. Johnson. 1979. Evolution of Myrtaceae—evidence from inflorescence structure. *Proceedings of the Linnean Society New South Wales* 102:157–256.

Briggs, B. G., A. D. Marchant, S. Gilmore, and C. L. Porter. 2000. A molecular phylogeny of Restionaceae and allies. *In* K. L. Wilson and D. Morrison, A, [eds.], Monocots: Systematics and Evolution, 661–671. CSIRO, Melbourne.

Britton, T., B. Oxelman, A. Vinnersten, and K. Bremer. 2002. Phylogenetic dating with confidence intervals using mean pathlengths. *Molecular Phylogenetics and Evolution* 24:58–65.

Brizicky, G. K. 1962. The genera of Rutaceae in the southeastern United States. *Journal of the Arnold Arboretum* 43:1–22.

———. 1963. The genera of the Sapindales in the southeastern United States. *Journal of the Arnold Arboretum* 44:462–501.

Brückner, C. 2000. Clarification of the carpel number in Papaverales, Capparales, and Berberidaceae. *Botanical Review* 66:155–307.

Brummitt, R. K. 1997. Taxonomy versus cladonomy, a fundamental controversy in biological systematics. *Taxon* 46:723–734.

Brummitt, R. K., H. Banks, M. A. T. Johnson, K. A. Doherty, K. Jones, M. W. Chase, and P. J. Rudall. 1998. Taxonomy of Cyanastroideae (Tecophilaeaceae): a multidisciplinary approach. *Kew Bulletin* 53:769–803.

Bruneau, A., F. Forest, P. S. Herendeen, B. B. Klitgaard, and G. P. Lewis. 2001. Phylogenetic relationships in the Caesalpinioideae (Leguminosae) as inferred from chloroplast *trnL* intron sequences. *Systematic Botany* 26:487–514.

Bryant, H. N. 1994. Comments on the phylogenetic definition of taxon names and conventions regarding the naming of crown clades. *Systematic Biology* 43:124–130.

———. 1996. Explicitness, stability, and universality in the phylogenetic definition and usage of taxon names: a case study of the phylogenetic taxonomy of the Carnivora (Mammalia). *Systematic Biology* 45:174–189.

Burger, W. 1977. The Piperales and monocots—alternative hypotheses for the origin of the monocotyledonous flowers. *Botanical Review* 43:345–393.

Burleigh J. G., and S. Mathews. 2004. Phylogenetic signal in nucleotide data from seed plants: implications for resolving the seed plant tree of life. *American Journal of Botany*, 91:1599–1613.

Burns-Balogh, P., and V. Funk. 1986. A phylogenetic analysis of the Orchidaceae. *Smithsonian Contributions to Botany* 61:1–79.

Burtt, B. L. 1974. Patterns of structural change in the flowering plants. *Transactions of the Botanical Society of Edinburgh* 42:133–142.

———. 1994. A commentary on some recurrent forms and changes of form in angiosperms. *In* D. S. Ingram and A. Hudson [eds.], Shape and Form in Plants and Fungi, 143–152. Academic Press, London.

Buzgo, M. 2001. Flower structure and development of Araceae compared with alismatids and Acoraceae. *Botanical Journal of the Linnean Society* 136:393–425.

Buzgo, M., and P. K. Endress. 2000. Floral structure and development of Acoraceae and its systematic relationships with basal angiosperms. *International Journal of Plant Sciences* 161:23–41.

Buzgo, M., P. S. Soltis, and D. E. Soltis. 2004. Floral developmental morphology of *Amborella trichopoda* (Amborellaceae). *International Journal of Plant Sciences* 165: 925-947.

Caddick, L. R., P. J. Rudall, P. Wilkin, T. A. J. Hedderson, and M. W. Chase. 2002a. Phylogenetics of Dioscoreales based on combined analyses of morphological and molecular data. *Botanical Journal of the Linnean Society* 138:123–144.

Caddick, L. R., P. Wilkin, P. J. Rudall, T. A. J. Hedderson, and M. W. Chase. 2002b. Yams reclassified: a recircumscription of Dioscoreaceae and Dioscoreales. *Taxon* 51:103–114.

Cameron, K. M., and W. C. Dickison. 1998. Foliar architecture of vanilloid orchids: insights into the evolution of reticulate leaf venation in monocotyledons. *Botanical Journal of the Linnean Society* 128:45–70.

Cameron, K. M., M. W. Chase, W. R. Anderson, and H. G. Hills. 2001. Molecular systematics of Malpighiaceae: evidence from plastid *rbcL* and *matK* sequences. *American Journal of Botany* 88:1847–1862.

Cameron, K. M., M. W. Chase, and P. J. Rudall. 2003. Recognition and recircumscription of Petrosaviales to include *Petrosavia* and *Japonolirion* (Petrosaviaceae s. l.). *Brittonia* 55:214–225.

Cameron, K. M., M. W. Chase, W. M. Whitten, P. J. Kores, D. C. Jarrell, V. A. Albert, T. Yukawa, H. G. Hills, and D. H. Goldman. 1999. A phylogenetic analysis of the Orchidaceae: evidence from *rbcL* nucleotide sequences. *American Journal of Botany* 86:208–224.

Cameron, K. M., K. J. Wurdack, and R. W. Jobson. 2002. Molecular evidence for the common origin of snap-traps among carnivorous plants. *American Journal of Botany* 89:1503–1509.

Campbell, C. S. 1985. The subfamilies and tribes of Gramineae (Poaceae) in the southeastern United States. *Journal of the Arnold Arboretum* 66:123–199.

Canright, J. E. 1952. The comparative morphology and relationships of the Magnoliaceae. I. Trends of specialization in the stamens. *American Journal of Botany* 39:484–497.

———. 1955. The comparative morphology and relationships of the Magnoliaceae. IV. Wood and nodal anatomy. *Journal of the Arnold Arboretum* 36:119–140.

Cantino, P. D. 1992a. Toward a phylogenetic classification of the Labiatae. *In* R. M. Harley and T. Reynolds [eds.], Advances in Labiate Science, 27–32. Royal Botanic Garden, Kew, London.

———. 1992b. Evidence for a polyphyletic origin of the Labiatae. *Annals of the Missouri Botanical Garden* 79:361–379.

———. 1998. Binomials, hyphenated uninomials, and phylogenetic nomenclature. *Taxon* 47:425–429.

———. 2000. Phylogenetic nomenclature: addressing some concerns. *Taxon* 49:85–93.

Cantino, P. D., R. G. Olmstead, and S. J. Wagstaff. 1997. A comparison of phylogenetic nomenclature with the current system: a botanical case study. *Systematic Biology* 46:313–331.

Capasso, A., R. Urrnaga, L. Garofala, L. Sorrentino, and R. Aquino. 1996. Phytochemical and pharmacological studies on medicinal herb *Acicarpha tribuloides*. *International Journal of Pharmacognosy.* 34:255–261.

Carlquist, S. 1975. Ecological Strategies of Xylem Evolution. University of California Press, Berkeley.

———. 1978. Vegetative anatomy and systematics of Grubbiaceae. *Botaniska Notiser* 131:117–126.

———. 1981. Wood anatomy of Nepenthaceae. *Bulletin of the Torrey Botanical Club* 108:324–330.

———. 1982. *Exospermum stipitatum* (Winteraceae): observations on wood, leaves, flowers, pollen, and fruit. *Aliso* 10:257–277.

———. 1983. Wood anatomy of *Bubbia* (Winteraceae), with comments on origin of vessels in dicotyledons. *American Journal of Botany* 70:578–590.

———. 1988. Comparative Wood Anatomy. Springer, Berlin.

———. 1990. Wood anatomy of *Ascarina* (Chloranthaceae) and the tracheid-vessel element transition. *Aliso* 13:447–462.

———. 1991. Wood and bark anatomy of *Ticodendron*: comments on relationships. *Annals of the Missouri Botanical Garden* 78:97–104.

———. 1992. Wood anatomy and stem of *Chloranthus*: summary of wood anatomy of Chloranthaceae, with comments on relationships, vessellessness, and the origin of monocotyledons. *IAWA Bulletin II* 13:3–16.

———. 1996. Wood anatomy of primitive angiosperms: new perspectives and syntheses. *In* D. W. Taylor and L. J. Hickey [eds.], Flowering Plant Origin, Evolution, and Phylogeny, 68–90. Chapman and Hall, New York.

———. 1999. Wood, stem, and root anatomy of Basellaceae with relation to habit, systematics, and cambial variants. *Flora* 194:1–12.

———. 1999a. Wood and stem anatomy of *Stegnosperma* (Caryophyllales): phylogenetic relationships, nature of lateral meristems and successive cambial activity. *IAWA Journal* 20:149–163.

———. 1999b. Wood anatomy, stem anatomy, and cambial activity of *Barbeuia* (Caryophyllales). *IAWA Journal* 20:431–440.

———. 2000. Wood and bark anatomy of *Takhtajania* (Winteraceae); phylogenetic and ecological implications. *Annals of the Missouri Botanical Garden* 87:317–322.

———. 2001a. Comparative Wood Anatomy: Systematic, Ecological, and Evolutionary Aspects of Dicotyledon Wood. 2nd ed. Springer, Berlin.

———. 2001b. Wood and stem anatomy of Rhabdodendraceae is consistent with placement in Caryophyllales sensu lato. *IAWA Journal* 22:171–181.

———. 2003. Wood anatomy of Aextoxicaceae and Berberidopsidaceae is compatible with their inclusion in Berberidopsidales. *Systematic Botany* 28:317–325.

Carlquist, S., and C. J. Boggs. 1996. Wood anatomy of Plumbaginaceae. *Bulletin of the Torrey Botanical Club* 123:135–147.

Carlquist, S., and E. L. Schneider. 2001. Vegetative anatomy of the New Caledonia endemic *Amborella trichopoda*: relationships with the Illiciales and implications for vessel origin. *Pacific Science* 55:305–312.

———. 2002. The tracheid-vessel element transition in angiosperms involves multiple independent features: cladistic consequences. *American Journal of Botany* 89:185–195.

Carlquist, S., and E. J. Wilson. 1995. Wood anatomy of *Drosophyllum* (Droseraceae): ecological and phylogenetic considerations. *Bulletin of the Torrey Botanical Club* 122:185–189.

Carpenter, R., and E. Coen. 1990. Floral and homeotic mutations produced by transposon mutagenesis in *Antirrhinum majus*. *Genes and Development* 4:1483–1493.

Carr, S. G. M., and D. J. Carr. 1961. The functional significance of syncarpy. *Phytomorphology* 11:249–256.

Cavalier-Smith, T. 1985a. The Evolution of Genome Size. John Wiley, Chichester.

———. 1985b. Eukaryotic gene numbers, non-coding DNA and genome size. *In* T. Cavalier-Smith [ed.], The Evolution of Genome Size, 69–103. John Wiley, Chichester.

Chadwell, T. B., S. J. Wagstaff, and P. D. Cantino. 1992. Pollen morphology of *Phryma* and some putative relatives. *Systematic Botany* 17:210–219.

Chalk, L., and M. M. Chattaway. 1937. Identification of woods with included phloem. *Tropical Woods* 50:1–31.

Chanderbali, A. S., H. van der Werff, and S. S. Renner. 2001. Phylogeny and historical biogeography of Lauraceae: evidence from the chloroplast and nuclear genomes. *Annals of the Missouri Botanical Garden* 88:104–134.

Chandler, G. T., and R. J. Bayer. 2000. Phylogenetic placement of the enigmatic western Australian genus *Emblingia* based on *rbcL* sequences. *Plant Species Biology* 15:67–72.

Chandler, G. T., and G. M. Plunkett. 2004. Evolution in Apiales: nuclear and chloroplast markers together in (almost) perfect harmony. *Botanical Journal of the Linnean Society* 144:123–147.

Charlesworth, B., P. Sniegowski, and W. Stephan. 1995. The evolutionary dynamics of repetitive DNA in eukaryotes. *Nature* 371:215–220.

Chase, M. W. 1986. A reappraisal of the oncidioid orchids. *Systematic Botany* 11:477–491.

———. 1987. Systematic implications of pollinarium morphology in *Oncidium* Sw., *Odontoglossum* Kunth, and allied genera (Orchidaceae). *Lindleyana* 2:8–28.

———. 1988. Obligate twig epiphytes: a distinct subset of Neotropical orchidaceous epiphytes. *Selbyana*: 24–30.

———. 2001. The origin and biogeography of Orchidaceae. *In* A. M. Pridgeon, P. J. Cribb, M. W. Chase, and F. Rasmussen [eds.], Genera Orchidacearum, Vol. II: Orchidoideae (part I), 1–5. Oxford University Press, Oxford.

———. 2004. Monocot relationships: an overview. *American Journal of Botany* 91:1645–1655.

Chase, M. W., and V. A. Albert. 1998. A perspective on the contribution of plastid *rbcL* DNA sequences to angiosperm phylogenetics. *In* D. E. Soltis, P. S. Soltis, and J. J. Doyle [eds.], Molecular Systematics of Plants II: DNA Sequencing, 488–507. Kluwer, Boston.

Chase, M. W., and A. V. Cox. 1998. Gene sequences, collaboration, and analysis of large data sets. *Australian Systematic Botany* 11:215–229.

Chase, M. W., and H. G. Hills. 1992. Orchid phylogeny, flower sexuality, and fragrance seeking. *BioScience* 42:43–49.

Chase, M. W., and J. D. Palmer. 1988. Chloroplast DNA variation, geographical distribution and morphological parallelism in subtribe Oncidiinae (Orchidaceae). *American Journal of Botany* 75:163–164.

———. 1989. Chloroplast DNA systematics of the lilioid monocots: feasibility, resources, and an example from the Orchidaceae. *American Journal of Botany* 76:1720–1730.

————. 1992. Floral morphology and chromosome number in subtribe Oncidiinae (Orchidaceae): evolutionary insights from a phylogenetic analysis of chloroplast DNA restriction site variation. *In* P. S. Soltis, D. E. Soltis, and J. J. Doyle [eds.], Molecular Systematics of Plants 324–339. Chapman and Hall, New York.

————. 1997. Leapfrog radiation in floral and vegetative traits among twig epiphytes in the orchid subtribe Oncidiinae. *In* T. J. Givnish and K. J. Sytsma [eds.], Molecular Evolution and Adaptive Radiation, 331–352. Cambridge University Press, Cambridge.

Chase, M. W., and J. S. Pippen. 1988. Seed morphology in the subtribe Oncidiinae (Orchidaceae). *Systematic Botany* 13:313–323.

Chase, M. W., K. M. Cameron, R. L. Barrett, and J. V. Freudenstein. 2003. A phylogenetic classification of Orchidaceae. *In* K. M. Dixon, S. P. Kell, R. L. Barrett, and P. J. Cribb [eds.], Orchid Conservation, 69–89. Natural History Publications, Kota Kinabalu, Sabah, Malaysia.

Chase, M. W., K. M. Cameron, H. G. Hills, and D. Jarrell. 1994. Molecular systematics of the Orchidaceae and other lilioid monocots. *In* A. Pridgeon [ed.], Proceedings of the 14th World Orchid Conference, 61–73. HMSO, London.

Chase, M. W., A. de Bruijn, G. Reeves, A. V. Cox, P. J. Rudall, M. A. T. Johnson, and L. E. Eguiarte. 2000c. Phylogenetics of Asphodelaceae (Asparagales): an analysis of plastid *rbcL* and *trnL-F* DNA sequences. *Annals of Botany* 86:935–951.

Chase, M. W., M. R. Duvall, H. G. Hills, J. G. Conran, A. V. Cox, L. E. Eguiarte, J. Hartwell, M. F. Fay, L. R. Caddick, K. M. Cameron, and S. Hoot. 1995a. Molecular phylogenetics of Lilianae. *In* P. J. Rudall, P. J. Cribb, D. F. Cutler, and C. J. Humphries [ed.], Monocotyledons: Systematics and Evolution, 109–137. Royal Botanic Gardens, Kew.

Chase, M. W., M. F. Fay, D. S. Devey, O. Maurin, N. Rønsted, J. Davies, Y. Pillon, G. Petersen, O. Seberg, M. N. Tamura, C. B. Asmussen, K. Hilu, T. Borsch, J. I. Davis, D. W. Stevenson, J. C. Pires, T. J. Givnish, K. J. Sytsma, and S. W. Graham. 2005. Multi-gene analyses of monocot relationships: a summary. *In* J. T. Columbus, E. A. Friar, C. W. Hamilton, J. M. Porter, L. M. Prince, and M. G. Simpson [eds.], Monocots: Comparative Biology and Evolution. 2 vols. Rancho Santa Ana Botanic Garden, Claremont, CA. in press.

Chase, M. W., M. F. Fay, and V. Savolainen. 2000b. Higher-level classification in the angiosperms: new insights from the perspective of DNA sequence data. *Taxon* 49:685–704.

Chase, M., C. M. Morton, and J. A. Kallunki. 1999. Phylogenetic relationships of Rutaceae: a cladistic analysis of the subfamilies using evidence from *rbcL* and *atpB* sequence variation. *American Journal of Botany* 86:1191–1199.

Chase, M. W., P. J. Rudall, and J. G. Conran. 1996. New circumscriptions and a new family of asparagoid lilies: genera formerly included in Anthericaceae. *Kew Bulletin* 51:667–680.

Chase, M. W., D. E. Soltis, R. G. Olmstead, D. Morgan, D. H. Les, B. D. Mishler, M. R. Duvall, R. A. Price, H. G. Hills, Y.-L. Qiu, K. A. Kron, J. H. Rettig, E. Conti, J. D. Palmer, J. R. Manhart, K. J. Sytsma, H. J. Michaels, W. J. Kress, K. G. Karol, W. D. Clark, M. Hedrén, B. S. Gaut, R. K. Jansen, K.-J. Kim, C. F. Wimpee, J. F. Smith, G. R. Furnier, S. H. Strauss, Q.-Y. Xiang, G. M. Plunkett, P. S. Soltis, S. M. Swensen, S. E. Williams, P. A. Gadek, C. J. Quinn, L. E. Eguiarte, E. Golenberg, G. H. Learn Jr., S. W. Graham, S. C. H. Barrett, S. Dayanandan, and V. A. Albert. 1993. Phylogenetics of seed plants: an analysis of nucleotide sequences from the plastid gene *rbcL*. *Annals of the Missouri Botanical Garden* 80:526–580.

Chase, M. W., D. E. Soltis, P. S. Soltis, P. J. Rudall, M. F. Fay, W. H. Hahn, S. Sullivan, J. Joseph, T. Givnish, K. J. Sytsma, and J. C. Pires. 2000a. Higher-level systematics of the monocotyledons: An assessment of current knowledge and a new classification. *In* K. L. Wilson and D. A. Morrison [eds.], Monocots: Systematics and Evolution, 3–16. CSIRO, Melbourne.

Chase, M. W., D. W. Stevenson, P. Wilkin, and P. J. Rudall. 1995b. Monocot systematics: a combined analysis. *In* P. J. Rudall, P. J. Cribb, D. F. Cutler, and C. J. Humphries [eds.], Monocotyledons: Systematics and Evolution, 685–730. Royal Botanic Gardens, Kew.

Chase, M. W., S. Zmarzty, M. D. Lledo, K. J. Wurdack, S. M. Swensen, and M. F. Fay. 2002. When in doubt, put it in Flacourtiaceae: a molecular phylogenetic analysis based on plastid *rbcL* DNA sequences. *Kew Bulletin* 57:141–181.

Chaw, S. M., C. L. Parkinson, Y. Cheng, T. M. Vincent, and J. D. Palmer. 2000. Seed plant phylogeny inferred from all three plant genomes: monophyly of extant gymnosperms and origin of Gnetales from conifers. *Proceedings of the National Academy of Sciences USA* 97:4086–4091.

Chaw, S. M., A. Zharkikh, H.-M. Sung, T.-C. Lau, and W.-H. Li. 1997. Molecular phylogeny of extant gymnosperms and seed plant evolution: analysis of nuclear 18S rRNA sequences. *Molecular Biology and Evolution* 14:56–58.

Chen, Z. D., X. Q. Wang, H. Y. Sun, Y. Han, Z. X. Zhang, Y. P. Zou, and A.M. Lu. 1998. Systematic position of Rhoipteleaceae: evidence from nucleotide sequences of *rbcL* gene. *Acta Phytotaxonomica Sinica* 36:1–7.

Chittka, L., and J. D. Thomson. 2001. Cognitive Ecology of Pollination. Animal Behavior and Evolution. Cambridge University Press, Cambridge.

Chittka, L., J. D. Thomson, and N. M. Waser. 1999. Flower constancy, insect psychology, and plant evolution. *Naturwissenschaften* 86:361–377.

Chorinsky, F. 1931. Vergleichende morphologishe Untersuchungen der Haargebilde bei Portulacaceae und Cactaceae. *Österreichische Botanische Zeitschrift* 80:308–327.

Civeyrel, L., A. Le Thomas, K. Ferguson, and M. W. Chase. 1998. Critical reexamination of palynological characters used to delimit Asclepiadaceae in comparison to the molecular phylogeny obtained from plastid *matK* sequences. *Molecular Phylogenetics and Evolution* 9:517–527.

Clausing, G., and S. S. Renner. 2001. Molecular phylogenetics of Melastomataceae and Memecylaceae: implications for character evolution. *American Journal of Botany* 88:486–498.

Clausing, G., K. Meyer, and S. S. Renner. 2000. Correlations among fruit traits and evolution of different fruits within Melastomataceae. *Botanical Journal of the Linnean Society* 133: 303–326.

Clegg, M. 1990. Dating the monocot-dicot divergence. *Trends in Ecology and Evolution* 5:1–2.

Clement, J. S., and T. J. Mabry. 1996. Pigment evolution in the Caryophyllales: a systematic overview. *Botanica Acta* 109:360–367.

Coen, E. S., and E. M. Meyerowitz. 1991. The war of the whorls: genetic interactions controlling flower development. *Nature* 353:31–37.

Conran, J. G., M. W. Chase, and P. J. Rudall. 1997. Two new monocotyledon families Anemarrhenaceae and Behniaceae (Lilianae: Asparagales). *Kew Bulletin* 52:995–999.

Conti, E., A. Fischbach, and K. J. Sytsma. 1993. Tribal relationships in Onagraceae: implications from *rbcL* sequence data. *Annals of the Missouri Botanical Garden* 80:672–685.

Conti, E., A. Litt, and K. J. Sytsma. 1996. Circumscription of Myrtales and their relationships to other rosids: evidence from *rbcL* sequence data. *American Journal of Botany* 83:221–233.

Contreras, V. R., R. Scogin, and C. T. Philbrick. 1993. A phytochemical study of selected Podostemaceae: systematic implications. *Aliso* 13:513–520.

Cook, C. D. K. 1982. Pollination mechanisms in the Hydrocharitaceae. *In* J. J. Symoens, S. S. Hooper, and P. Compère [eds.], Studies on Aquatic Vascular Plants, 1–15. Royal Botanical Society of Belgium, Brussels.

Cordell, G. A. 1974. The biosynthesis of indole alkaloids. *Lloydia* 37:219–298.

Corner, E. J. H. 1946. Centrifugal stamens. *Journal of the Arnold Arboretum* 27:423–437.

———. 1976. The Seeds of Dicotyledons. Cambridge University Press, New York.

Cornet, B., 1989. Late Triassic angiosperm-like pollen from the Richmond Rift Basin of Virginia, U. S. A. *Palaeontographica, Abt. B.* 213:37–87.

Cosner, M. E., R. K. Jansen, and T. G. Lammers. 1994. Phylogenetic relationships in the Campanulales based on *rbcL* sequences. *Plant Systematics and Evolution* 190:79–95.

Coulter, J. M., and C. J. Chamberlain. 1903. Morphology of Angiosperms. Appleton, New York.

Cox, A. V., G. J. Abdelnour, M. D. Bennett, and I. J. Leitch. 1998. Genome size and karyotype evolution in the slipper orchids (Cypripedioideae: Orchidaceae). *American Journal of Botany* 85:681–687.

Cox, A. V., A. M. Pridgeon, V. A. Albert, and M. W. Chase. 1997. Phylogenetics of the slipper orchids (Cypripedioideae: Orchidaceae): nuclear rDNA ITS sequences. *Plant Systematics and Evolution* 208:197–223.

Cox, P. A. 1988. Hydrophilous pollination. *Annual Review of Ecology and Systematics* 19:261–280.

Cox, P. A., and C. J. Humphries. 1993. Hydrophilous pollination and breeding system evolution in seagrasses: a phylogenetic approach to the evolutionary ecology of the Cymodoceaceae. *Botanical Journal of the Linnean Society* 113:217–226.

Crane, P. R. 1985. Phylogenetic analysis of seed plants and the origin of angiosperms. *Annals of the Missouri Botanical Garden* 72:716–793.

———. 1989. Patterns of evolution and extinction in vascular plants. *In* K. C. Allen and D. E. G. Briggs [eds.], Evolution and the Fossil Record. Belhaven Press, London.

Crane, P. R., and S. Blackmore. 1989. Evolution, Systematics and Fossil History of the Hamamelidae. 2 vols. Clarendon Press, Oxford.

Crane, P. R., and P. Kenrick. 1997b. Diverted development of reproductive organs: a source of morphological innovation in land plants. *Plant Systematics and Evolution* 206:161–174.

———. 1997a. Problems in cladistic classification: Higher-level relationships in land plants. *Aliso* 15:87–104.

Crane, P. R., E. M. Friis, and K. R. Pedersen. 1989. Reproductive structure and function in Cretaceous Chloranthaceae. *Plant Systematics and Evolution* 165:211–226.

———. 1995. The origin and early diversification of angiosperms. *Nature* 374:27–33.

Crane, P. R., P. Herendeen, and E. M. Friis. 2004. Fossils and plant phylogeny. *American Journal of Botany*, 91:1683–1699.

Crane, P. R., K. R. Pedersen, E. M. Friis, and A. N. Drinnan. 1993. Early Cretaceous (Early to Middle Albian) platanoid inflorescences associated with *Sapindopsis* leaves from the Potomac Group of eastern North America. *Systematic Botany* 18:328–344.

Crepet, W. L. 1996. Timing in the evolution of derived floral characters: upper Cretaceous (Turonian) taxa with tricolpate and tricolpate-derived pollen. *Review of Palaeobotany and Palynology* 90:339–359.

———. 2000. Progress in understanding angiosperm history, success and relationships: Darwin's abominably "perplexing phenomenon". *Proceedings of the National Academy of Sciences USA* 97:12939–12941.

Crepet, W. L., and K. C. Nixon. 1998. Fossil Clusiaceae from the Late Cretaceous (Turonian) of New Jersey and implications regarding the history of bee pollination. *American Journal of Botany* 85:1122–1133.

Crepet, W. L., K. C. Nixon, and M. A. Gandolfo. 2004. Fossil evidence and phylogeny: the age of major angiosperm clades based on mesofossil and macrofossil evidence from Cretaceous deposits. *American Journal of Botany* 91:1666–1682.

Croizat, L. 1941. On the systematic position of *Daphniphyllum* and its allies. *Lingnan Science Journal* 20:79–103.

Cronk, Q. C. B, R. M. Bateman, and J. A. Hawkins [eds.]. 2002. Developmental Genetics and Plant Evolution. Taylor and Francis, London.

Cronquist, A. 1957. Outline of a new system of families and orders of dicotyledons. *Bulletin du Jardin Botanique National de Belgique* 27:13–40.

———. 1968. The Evolution and Classification of Flowering Plants. Houghton Mifflin, Boston.

———. 1981. An Integrated System of Classification of Flowering Plants. Columbia University Press, New York.

———. 1983. Some realignments in the dicotyledons. *Nordic Journal of Botany* 3:75–83.

———. 1984. A commentary on the definition of the order Myrtales. *Annals of the Missouri Botanical Garden* 71:780–782.

———. 1988. The Evolution and Classification of Flowering Plants. New York Botanical Garden, Bronx.

Cronquist A., and R. F. Thorne. 1994. Nomenclatural and taxonomic history. *In* H.-D. Behnke and T. J. Mabry [eds.], Caryophyllales: Evolution and Systematics, 87–121. Springer, Berlin.

Cuénoud, P., M. A. D. Martinez, P.-A. Loizeau, R. E. Spichiger, and J.-F. Manen. 2000. Molecular phylogeny and biogeography of the genus *Ilex* L. (Aquifoliaceae). *Annals of Botany* 85:111–122.

Cuénoud, P., V. Savolainen, L. W. Chatrou, M. Powell, R. J. Grayer, and M. W. Chase. 2002. Molecular phylogenetics of Caryophyllales based on nuclear 18S rDNA and plastid *rbcL, atpB,* and *matK* DNA sequences. *American Journal of Botany* 89:132–144.

Cuerrier, A., L. Brouillet, and D. Barabé. 1997. Numerical analyses of the modern classifications of the flowering plants. *American Journal of Botany* 84:185 [abstract].

Culley, T. M., S. G. Weller, and A. K. Sakai. 2002. The evolution of wind pollination in angiosperms. *Trends in Ecology and Evolution* 17:361–369.

Dafni, A. 1983. Pollination of *Orchis caspia*–a nectarless plant which deceives the pollinators of nectariferous species from other plant families. *Journal of Ecology* 71:467–474.

———. 1987. Pollination in *Orchis* and related genera: evolution from reward to deception. *In* J. Arditti [ed.], Orchid Biology, Reviews and Perspectives. Cornell University Press, Ithaca.

Dafni, A., and D. M. Calder. 1987. Pollination by deceit and floral mimesis in *Thelymitra antennifera* (Orchidaceae). *Plant Systematics and Evolution* 158:11–22.

Dafni, A., and Y. Ivri. 1981. Floral mimicry between *Orchis israelitica* Baumann and Dafni (Orchidaceae) and *Bellevalia flexuosa* Boiss. (Liliaceae). *Oecologia* 49:229–232.

Dahlgren, G. 1989. The last Dahlgrenogram: system of classification of the dicotyledons. *In* K. Tan, R. R. Mill, and T. S. Elias [eds.], Plant Taxonomy, Phylogeography and Related Subjects, 249–260. Edinburgh University Press, Edinburgh.

———. 1991. Steps towards a natural system of the dicotyledons: embryological characters. *Aliso* 13:107–165.

Dahlgren, R. M. T. 1975. The distribution of characters within an angiosperm system. I. Some embryological characters. *Botaniska Notiser* 128:181–197.

———. 1977. A commentary on a diagrammatic presentation of the angiosperms in relation to the distribution of character states. *Plant Systematics and Evolution Supplement* 1:253–283.

———. 1980. A revised system of classification of the angiosperms. *Botanical Journal of the Linnean Society* 80:91–124.

———. 1983. General aspects of angiosperm evolution and macrosystematics. *Nordic Journal of Botany* 3:119–149.

———. 1988. Rhizophoraceae and Anisophylleaceae: Summary statement, relationships. *Annals of the Missouri Botanical Garden* 75:1259–1277.

Dahlgren, R. M. T., and F. Rasmussen. 1983. Monocotyledon evolution: Characters and phylogenetic estimation. *Evolutionary Biology* 16:255–395.

Dahlgren, R. M. T., and R. Thorne. 1984. The order Myrtales: Circumscription, variation, and relationships. *Annals of the Missouri Botanical Garden* 71:633–699.

Dahlgren, R. M. T., H. T. Clifford, and P. F. Yeo. 1985. The Families of the Monocotyledons: Structure, Evolution and Taxonomy. Springer, Berlin.

Dahlgren, R. M. T., S. Rosendal-Jensen, and B. J. Nielsen. 1981. A revised classification of the angiosperms with comments on correlation between chemical and other characters. *In* D. A. Young and D. S. Seigler [eds.], Phytochemistry and Angiosperm Phylogeny, 149–204. Praeger Publishers, New York.

Damtoft, S., S. R. Jensen, and B. J. Nielsen. 1992. Biosynthesis of iridoid glucosides in *Lamium album*. *Phytochemistry* 31:135–137.

Dandy, J. E. 1927. The genera of Saxifragaceae. *Kew Bulletin of Miscellaneous Information* 1927:100–118.

Danilova, M. E. 1996. Anatomia seminum comparativa. Tomus 5. Rosidae I. [In Russian.]. NAUKA, Leningrad.

Darwin, C. 1859. The Origin of Species. John Murray, London.

———. 1875. Insectivorous Plants. John Murray, London.

Davies, T. J., T. G. Barraclough, M. W. Chase, P. S. Soltis, D. E. Soltis, and V. Savolainen. 2004. Darwin's abominable mystery: insights from a supertree of the angiosperms. *Proceedings of the National Acadamy of Sciences USA* 101:1904–1909.

Davioud, E., F. Bailleul, P. Delaveau, and H. Jacquemin. 1985. Iridoids of Guyanese species of *Stigmaphyllon*. *Planta Medica* 51:78–79.

Davis, C. C., and K. J. Wurdack. 2005. Host-to-parasite gene transfer in flowering plants: Phylogenetic evidence from Malpighiales. *Science* 305:676-677.

Davis, C. C., and M. W. Chase. 2004. Elatinaceae are sister to Malpigiaceae; Peridiscaceae belong to Saxifragales. *American Journal of Botany* 91: 262–273.

Davis, C. C., W. R. Anderson, and M. J. Donoghue. 2001. Phylogeny of Malpighiaceae: evidence from chloroplast *ndhF* and *trnL-F* nucleotide sequences. *American Journal of Botany* 88:1830–1846.

Davis, J. I., and R. J. Soreng. 1993. Phylogenetic structure in the grass family (Poaceae), as determined from chloroplast DNA restriction site variation. *American Journal of Botany* 80:1444–1454.

Davis, J. I., M. P. Simmons, D. W. Stevenson, and J. F. Wendel. 1998. Data decisiveness, data quality, and incongruence in phylogenetic analysis: an example from the monocotyledons using mitochondrial *atpA* sequences. *Systematic Biology* 47:282–310.

Davis, J. I., D. W. Stevenson, G. Petersen, O. Seberg, L. M. Campbell, J. V. Freudenstein, D. H. Goldman, C. R. Hardy, F. A. Michelangeli, M. P. Simmons, C. D. Specht, F. Vergara-Silva, and M. A. Gandolfo. 2004. A phylogeny of the monocots, as inferred from *rbcL* and *atpA* sequence variation and a comparison of methods for calculating jackknife and bootstrap values. *Systematic Biology,* 29:467–510.

Dayanandan, S., P. S. Ashton, and R. B. Primack. 1999. Phylogeny of the tropical tree family Dipterocarpaceae based on nucleotide sequences of the chloroplast *rbcL* gene. *American Journal of Botany* 86:1182–1190.

Debry, R., and R. G. Olmstead. 2000. A simulation study of reduced tree-search effort in bootstrap resampling analysis. *Systematic Biology* 49:171–179.

de Candolle, A. P. 1813. Théorie Elémentaire de Botanique. Deterville, Paris.

———. 1824–1873. Prodromus Systematis Naturalis Regni Vegetabilis. Treuttel et Würtz, Paris.

Degut, A. V., and N. S. Fursa. 1980. Phenol compounds of *Digitalis ferruginea*. *Khimiya Prirodnykh Soedinenii*: 417–418.

de Jussieu, A. L. 1789. Genera Plantarum Secundum Ordines Naturales Disposita. Herissant and Barrios, Paris.

Dellaporta, S. L., and A. Calderon-Urrea. 1993. Sex determination in flowering plants. *Plant Cell* 5:1241–1251.

———. 1994. The sex determination process in maize. *Science* 266:1501–1505.

dePamphilis, C. W., and J. D. Palmer. 1990. Loss of photosynthetic and chlororespiratory genes from the plastid genome of a parasitic flowering plant. *Nature* 348:337–339.

deQueiroz, K. 1997. The Linnaean hierarchy and the evolutionization of taxonomy, with emphasis on the problem of nomenclature. *Aliso* 15:125–144.

deQueiroz, K., and J. Gauthier. 1990. Phylogeny as central principle in taxonomy: Phylogenetic definitions of taxon names. *Systematic Zoology* 39:307–322.

———. 1992. Phylogenetic taxonomy. *Annual Review of Ecology and Systematics* 23:449–480.

———. 1994. Toward a phylogenetic system of biological nomenclature. *Trends in Ecology and Evolution* 9:27–31.

Deroin, T. 2000. Notes on the vascular anatomy of the fruit of *Takhtajania* (Winteraceae) and its interpretation. *Annals of the Missouri Botanical Garden* 87:398–406.

Devos, K. M., J. K. M. Brown, and J. L. Bennetzen. 2002. Genome size reduction through illegitimate recombination counteracts genome expansion in *Arabidopsis*. *Genome Research* 12:1075–1079.

De Wilde, W. 1971. The systematic position of tribe Paropsieae, in particular the genus *Ancistrothyrsus*, and a key to the genera of Passifloraceae. *Blumea* 19:99–104.

Dickinson, T. A., and R. Sattler. 1974. Development of the epiphyllous inflorescence of *Phyllonoma integerrima* (Turez.) Loes.: implications for comparative morphology. *Botanical Journal of the Linnean Society* 69:1–13.

Dickison, W. 1981. Evolutionary relationships of the Leguminosae. In R. M. M. Polhill and P. H. Raven [eds.], Advances in Legume Systematics,1, 35–54. Royal Botanical Gardens, Kew.

Dickison, W. C., and P. Baas. 1977. The morphology and relationships of *Paracryphia* (Paracryphiaceae). *Blumea* 23:417–438.

Dickison, W. C., and P. K. Endress. 1983. Ontogeny of the stem-node-leaf vascular continuum of *Austrobaileya*. *American Journal of Botany* 70:906–911.

Dickison, W. C., M. H. Hils, T. W. Lucansky, and W. L. Stern. 1994. Comparative anatomy and systematics of woody Saxifragaceae Endl. *Botanical Journal of the Linnean Society* 114:167–182.

Diels, D. 1930. Iridaceae. *In* A. Engler and K. Prantl [eds.], Die Natürlichen Pflanzenfamilien, 2nd ed., Part 15a, 469–505. Engelmann, Leipzig.

Dilcher, D. L. 1989. The occurrence of fruits with affinities to Ceratophyllaceae in lower and mid-Cretaceous sediments. *American Journal of Botany* 76:162.

———. 2000. Toward a new synthesis: Major evolutionary trends in the angiosperm fossil record. *Proceedings of the National Academy of Sciences USA* 97:7030–7036.

Dilcher, D. L., and P. R. Crane. 1984. *Archaeanthus*: an early angiosperm from the Cenomanian of the western interior of North America. *Annals of the Missouri Botanical Garden* 71:351–383.

Dobzhansky, T. 1973. Nothing in biology makes sense except in the light of evolution. *American Biology Teacher* 35:125–129.

Dodd, M. E., J. Silvertown, and M. W. Chase. 1999. Phylogenetic analysis of trait evolution and species diversity variation among angiosperm families. *Evolution* 53:732–744.

Donoghue, M. J. 1983. The phylogenetic relationships of *Viburnum*. *In* N. Platnick and V. A. Funk [eds.], Advances in Cladistics, 143–166. Columbia University Press, New York.

———. 1989. Phylogenies and the analysis of evolutionary sequences, with examples from seed plants. *Evolution* 43:1137–1156.

Donoghue, M. J., and J. A. Doyle. 1989b. Phylogenetic analysis of angiosperms and the relationships of Hamamelidae. *In* P. R. Crane and S. Blackmore [eds.], Evolution, Systematics, and Fossil History of Hamamelidae, 1, 17–45. Clarendon Press, Oxford.

———. 1989a. Phylogenetic studies of seed plants and angiosperms based on morphological characters. *In* K. Bremer and H. Jörnvall [eds.], The Hierarchy of Life: Molecules and Morphology in Phylogenetic Studies, 181–193. Elsevier Science Publishers, Amsterdam.

———. 2000. Seed plant phylogeny: demise of the anthophyte hypothesis? *Current Biology* 10:R106–R109.

Donoghue, M. J., T. Eriksson, P. A. Reeves, and R. G. Olmstead. 2001. Phylogeny and phylogenetic taxonomy of Dipsacales, with special reference to *Sinadoxa* and *Tetradoxa* (Adoxaceae). *Harvard Papers in Botany* 6:459–479.

Donoghue, M. J., R. G. Olmstead, J. F. Smith, and J. D. Palmer. 1992. Phylogenetic relationships of Dipsacales based on *rbcL* sequences. *Annals of the Missouri Botanical Garden* 79:333–345.

Donoghue, M. J., R. H. Ree, and D. A. Baum. 1998. Phylogeny and evolution of flower symmetry in the Asteridae. *Trends in Plant Science* 3:311–317.

Douglas, A. W., and S. C. Tucker. 1996b. Comparative floral ontogenies among Persoonioideae including *Bellendena* (Proteaceae). *American Journal of Botany* 83:1528–1555.

———. 1996a. The developmental basis of diverse carpel orientations in Grevilleoideae (Proteaceae). *International Journal of Plant Sciences* 157:373–397.

Douglas, G. E. 1957. The inferior ovary. II. *Botanical Review* 23:1–46.

Doust, A. N. 2000. Comparative floral ontogeny in Winteraceae. *Annals of the Missouri Botanical Garden* 87:366–379.

———. 2001. The developmental basis of floral variation in *Drimys winteri* (Winteraceae). *International Journal of Plant Sciences* 162:697–717.

Douzery, E. J. P., A. M. Pridgeon, P. J. Kores, H. P. Linder, H. Kurzweil, and M. W. Chase. 1999. Molecular phylogenetics of Diseae (Orchidaceae): a contribution from nuclear ribosomal ITS sequences. *American Journal of Botany* 86:887–899.

Doweld, A. B. 1998. Carpology, seed anatomy and taxonomic relationships of *Tetracentron* (Tetracentraceae) and *Trochodendron* (Trochodendraceae). *Annals of Botany* 82:413–443.

Downie, S. R., and J. D. Palmer. 1992. Restriction site mapping of the chloroplast DNA inverted repeat: a molecular phylogeny of Asteridae. *Annals of the Missouri Botanical Garden* 79:266–238.

———. 1994. A chloroplast DNA phylogeny of the Caryophyllales based on structural and inverted repeat restriction site variation. *Systematic Botany* 19:236–252.

Downie, S. R., D. S. Katz-Downie, and K. Cho. 1997. Relationships in the Caryophyllales as suggested by phylogenetic analyses of partial chloroplast DNA ORF2280 homolog sequences. *American Journal of Botany* 84:253–273.

Downie, S. R., D. S. Katz-Downie, and K. Spalik. 2000a. A phylogeny of Apiaceae tribe Scandiceae: evidence from nuclear ribosomal DNA internal transcribed spacer sequences. *American Journal of Botany* 87:76–95.

Downie, S. R., D. S. Katz-Downie, and M. F. Watson. 2000b. A phylogeny of the flowering plant family Apiaceae based on chloroplast DNA *rpl16* and *rpoC1* intron sequences: towards a suprageneric classification of subfamily Apioideae. *American Journal of Botany* 87:273–292.

Downie, S. R., S. Ramanath, D. S. Katz-Downie, and E. Llanas. 1998. Molecular systematics of Apiaceae subfamily Apioideae: phylogenetic analyses of nuclear ribosomal DNA internal transcribed spacer and plastid *rpoC1* intron sequences. *American Journal of Botany* 85:563–591.

Doyle, J. A. 1978. Origin of angiosperms. *Annual Review of Ecology and Systematics* 9:365–392.

———. 1994. Origin of the angiosperm flower: a phylogenetic perspective. *Plant Systematics and Evolution, Supplement* 8:7–29.

———. 1996. Seed plant phylogeny and the relationships of Gnetales. *International Journal of Plant Sciences* 157:S3–S39.

———. 1998a. Phylogeny of vascular plants. *Annual Review of Ecology and Systematics* 29:567–599.

———. 1998b. Molecules, morphology, fossils, and the relationships of angiosperms and Gnetales. *Molecular Phylogenetics and Evolution* 9:448–462.

———. 2000. Paleobotany, relationships, and geographic history of Winteraceae. *Annals of the Missouri Botanical Garden* 87:303–316.

———. 2001. Significance of molecular phylogenetic analyses for paleobotanical investigations on the origin of angiosperms. *Palaeobotanist* 50:167–188.

Doyle, J. A., and M. J. Donoghue. 1986. Seed plant phylogeny and the origin of the angiosperms: an experimental cladistic approach. *Botanical Review* 52:321–431.

———. 1987. The importance of fossils in elucidating seed plant phylogeny and macroevolution. *Review of Palaeobotany and Palynology* 50:63–95.

———. 1992. Fossils and seed plant phylogeny reanalyzed. *Brittonia* 44:89–106.

———. 1993. Phylogenies and angiosperm diversification. *Paleobiology* 19:141–167.

Doyle, J. A., and P. K. Endress. 2000. Morphological phylogenetic analysis of basal angiosperms: comparison and combination with molecular data. *International Journal of Plant Sciences* 161:S121–S153.

Doyle, J. A., and L. J. Hickey. 1976. Pollen and leaves from the mid-Cretaceous Potomac Group and their bearing on early angiosperm evolution. *In* C. B. Beck [ed.], Origin and Early Evolution of Angiosperms, 139–206. Columbia University Press, New York.

Doyle, J. A., and C. L. Hotton. 1991. Diversification of early angiosperm pollen in a cladistic context. *In* S. Blackmore and S. H. Barnes [eds.], Pollen and Spores: Patterns of Diversification, 169–195. Clarendon, Oxford.

Doyle, J. A., and A. Le Thomas. 1997. Phylogeny and geographic history of Annonaceae. *Géographie Physique et Quaternaire* 51:353–361.

Doyle, J. A., M. J. Donoghue, and E. Zimmer. 1994. Integration of morphological and ribosomal RNA data on the origin of angiosperms. *Annals of the Missouri Botanical Garden* 81:419–450.

Doyle, J. A., H. Eklund, and P. S. Herendeen. 2003. Floral evolution in Chloranthaceae: implications of a morphological phylogenetic analysis. *International Journal of Plant Sciences* 164:S365–S382.

Doyle, J. A., P. Bygrave, and A. Le Thomas. 2000. Implications of molecular data for pollen analysis in Annonaceae. *In* M. M. Harley, C. M. Morton, and S. Blackmore [eds.], Pollen and Spores: Morphology and Biology, 259–284. Royal Botanic Gardens, Kew.

Doyle, J. J. 1994. Phylogeny of the legume family: An approach to understanding the origin of nodulation. *Annual Review of Ecology and Systematics* 25:325–349.

Doyle, J. J., and M. S. Luckow. 2003. The rest of the iceberg: legume diversity and evolution in a phylogenetic context. *Plant Physiology* 131:900–910.

Doyle, J. J., J. L. Doyle, J. A. Ballenger, E. E. Dickson, T. Kajita, and H. Ohashi. 1997. A phylogeny of the chloroplast gene *rbcL* in the Leguminosae: Taxonomic correlations and insights into the evolution of nodulation. *American Journal of Botany* 84:541–554.

Dressler, R. L. 1981. The Orchids: Natural History and Classification. Harvard University Press, Cambridge, MA.

———. 1983. Classification of the Orchidaceae and their probable origin. *Telopea* 2:413–424.

———. 1993. Phylogeny and Classification of the Orchid Family. Cambridge University Press, Cambridge.

Dressler, R. L., and C. H. Dodson. 1960. Classification and phylogeny in the Orchidaceae. *Annals of the Missouri Botanical Garden* 47:25–68.

Drewes, S. E., L. Kayonga, T. E. Clark, T. D. Brackenbury, and C. C. Appleton. 1996. Iridoid molluscicidal compounds from *Apodytes dimidiata*. *Journal of Natural Products* 59:1169–1170.

Drinnan, A. N., and P. Ladiges. 1989. Operculum development in *Eucalyptus cloeziana* and *Eucalyptus* informal subg. *Monocalyptus* (Myrtaceae). *Plant Systematics and Evolution* 166:183–196.

Drinnan, A. N., P. R. Crane, and S. B. Hoot. 1994. Patterns of floral evolution in the early diversification of non-magnoliid dicotyledons (eudicots). *Plant Systematics and Evolution, Supplement* 8:93–122.

Du, L., J. Lykkesfeldt, C. Olsen, and B. Halkier. 1995. Involvement of cytochrome P450 in oxime production in glucosinolate biosynthesis as demonstrated by an *in vitro* microsomal enzyme system isolated from jasmonic acid-induced seedlings of *Sinapis alba* L. *Proceedings of the National Academy of Sciences USA* 92:12505–12509.

Duvall, M. R. 2000. Seeking the dicot sister group of the monocots. *In* K. L. Wilson and D. A. Morrison [eds.], Monocots: Systematics and Evolution, 25–32. CSIRO, Melbourne.

Duvall, M. R., M. T. Clegg, M. W. Chase, W. D. Lark, W. J. Kress, H. G. Hills, L. E. Eguiarte, J. F. Smith, B. S. Gaut, E. A. Zimmer, and G. H. Learn, Jr. 1993a. Phylogenetic hypotheses for the monocotyledons constructed form *rbcL* sequences. *Annals of the Missouri Botanical Garden* 80:607–619.

Duvall, M. R., G. H. Learn, L. E. Eguiarte, and M. T. Clegg. 1993b. Phylogenetic analysis of *rbcL* sequences identifies *Acorus calamus* as the primal extant monocotyledon. *Proceedings of the National Academy of Sciences USA* 90:4611–4644.

Duvall, M. R., S. Mathews, N. Mohammad, and T. Russell. 2004. Placing the monocots; conflicting signal from trigenomic analyses. *In* J. T. Columbus, E. A. Friar, C. W. Hamilton, J. M. Porter, L. M. Prince, and M. G. Simpson [eds.]. Monocots: Comparative Biology and Evolution. 2 vols. Rancho Santa Ana Botanic Garden, Claremont, CA. in press.

Eames, A. J. 1931. The vascular anatomy of the flower with refutation of the theory of carpel polymorphism. *American Journal of Botany* 47:147–188.

———. 1952. Relationships of the Ephedrales. *Phytomorphology* 2:79–100.

———. 1961. Morphology of the Angiosperms. McGraw-Hill, New York.

Eber, E. 1934. Karpellbau und Plazentationsverhältnisse in der Reihe der Helobiae. *Flora* 127:273–330.

Eckardt, T. 1963. Some observations on the morphology and embryology of *Eucommia ulmoides* Oliv. *Journal of the Indian Botanical Society* 42A:27–34.

———. 1964. Das Homologieproblem und Fälle strittiger Homologien. *Phytomorphology* 14:79–92.

———. 1976. Classical morphological features of centrospermous families. *Plant Systematics and Evolution* 126:5–25.

Eckenwalder, J. E. 1996. *Salix*. Biology of *Populus* and its implications for management and conservation. *In* R. F. Stettler [ed.], National Research Council of Canada, Ottawa.

Eckert, G. 1965. Entwicklungsgeschichtliche und blütenanatomische Untersuchungen zum Problem der Obdiplostemonie. *Botanische Jahrbücher für Systematik* 85:523–604.

Edwards, G. E., and D. A. Walker. 1983. C_3 and C_4: Mechanism, and Cellular and Environmental Regulation, of Photosynthesis. Blackwell Scientific Publications, Oxford.

Ehrendorfer, F., F. Krendl, E. Habeler, and W. Sauer. 1968. Chromosome numbers and evolution in primitive angiosperms. *Taxon* 17:337–468.

Eichler, A. 1875–1878. Blüthendiagramme, I/II. Engelmann, Leipzig.

Ekabo, O. A., N. R. Farnsworth, T. Santisuk, and V. Reutrakul. 1993. Phenolic, iridoid and ionyl glycosides from *Homalium ceylanicum*. *Phytochemistry* 32:747–754.

Eklund, H. 1999. Big survivors with small flowers: fossil history and evolution of Laurales and Chloranthaceae. PhD dissertation, Uppsala University.

Eklund, H., J. A. Doyle, and P. S. Herendeen. 2004. Morphological phylogenetic analysis of living and fossil Chloranthaceae. *International Journal of Plant Sciences* 165:107–151.

Eklund, H., E. M. Friis, and K. R. Pedersen. 1997. Chloranthaceous floral structures from the Late Cretaceous of Sweden. *Plant Systematics and Evolution* 207:13–42.

Eldredge, N., and J. Cracraft. 1980. Phylogenetic Patterns and the Evolutionary Process. Columbia University Press, New York.

Eldredge, N., and S. J. Gould. 1972. Punctuated equilibria: an alternative to phyletic gradualism. *In* T. J. M. Schopf [ed.], Models in Paleobiology, 82–115. Freeman, San Francisco.

Elias, T. S. 1970. The genera of Ulmaceae in the southeastern United States. *Journal of the Arnold Arboretum* 51:18–40.

———. 1971. The genera of Fagaceae in the southeastern United States. *Journal of the Arnold Arboretum* 52:159–195.

———. 1972. The genera of Juglandaceae in the southeastern United States. *Journal of the Arnold Arboretum* 53:26–51.

Eliasson, U. H. 1988. Floral morphology and taxonomic relations among the genera of Amaranthaceae in the New World and the Hawaiian Islands. *Botanical Journal of the Linnean Society* 96:235–283.

El-Naggar, L. J., and J. L. Beal. 1980. Iridoids. A review. *Journal of Natural Products* 43:649–707.

Elsik, C. G., and C. G. Williams. 2000. Retroelements contribute to the excess low-copy-number DNA in pine. *Molecular General Genetics* 264:47–55.

Elvander, P. E. 1984. The taxonomy of *Saxifraga* (Saxifragaceae) section *Boraphila* subsection *Integrifoliae* in western North America. *Systematic Botany Monographs* 3:1–44.

Endress, M. E. 2001. Apocynaceae and Asclepiadaceae: united they stand. *Haseltonia* 8: 2–9.

Endress, M. E., and P. Bruyns. 2000. A revised classification of the Apocynaceae s. l. *Botanical Review* 66:1–56.

Endress, M. E., and W. D. Stevens. 2001. The renaissance of the Apocynaceae s. l. : recent advances in systematics, phylogeny and evolution: introduction. *Annals of the Missouri Botanical Garden* 88:517–522.

Endress, M. E., B. Sennblad, S. Nilsson, L. Civeyrel, M. W. Chase, S. Huysmans, E. Grafström, and B. Bremer. 1996. A phylogenetic analysis of Apocynaceae s. str. and some related taxa in Gentianales: a multidisciplinary approach. *Opera Botanica Belgica* 7:59–102.

Endress, P. K. 1967. Systematische Studie über die verwandtschaftlichen Beziehungen zwischen den Hamamelidaceen und Betulaceen. *Botanische Jahrbücher für Systematik* 87:431–525.

———. 1976. Die Androeciumanlage bei polyandrischen Hamamelidaceen und ihre systematische Bedeutung. *Botanische Jahrbücher für Systematik* 97:436–457.

———. 1977. Evolutionary trends in the Hamamelidales-Fagales-group. *Plant Systematics and Evolution, Supplement* 1:321–347.

———. 1978. Blütenontogenese, Blütenabgrenzung und systematische Stellung der perianthlosen Hamamelidoideae. *Botanische Jahrbücher für Systematik* 100:249–317.

————. 1980. Ontogeny, function and evolution of extreme floral construction in Monimiaceae. *Plant Systematics and Evolution* 134:79–120.

————. 1982. Syncarpy and alternative modes of escaping disadvantages of apocarpy in primitive angiosperms. *Taxon* 31:48–52.

————. 1984. The flowering process in the Eupomatiaceae (Magnoliales). *Botanische Jahrbücher für Systematik* 104:297–319.

————. 1986a. Reproductive structures and phylogenetic significance of extant primitive angiosperms. *Plant Systematics and Evolution* 152:1–28.

————. 1986b. Floral structure, systematics and phylogeny in Trochodendrales. *Annals of the Missouri Botanical Garden* 73:297–324.

————. 1987a. Floral phyllotaxis and floral evolution. *Botanische Jahrbücher für Systematik* 108:417–438.

————. 1987b. The early evolution of the angiosperm flower. *Trends in Ecology and Evolution* 2:300–304.

————. 1987c. The Chloranthaceae: reproductive structures and phylogenetic position. *Botanische Jahrbücher für Systematik* 109:153–226.

————. 1989a. Aspects of evolutionary differentiation of the Hamamelidaceae and lower Hamamelididae. *Plant Systematics and Evolution* 162:193–211.

————. 1989b. Chaotic floral phyllotaxis and reduced perianth in *Achlys* (Berberidaceae). *Botanica Acta* 102:159–163.

————. 1990a. Patterns of floral construction in ontogeny and phylogeny. *Biological Journal of the Linnean Society* 39:153–175.

————. 1990b. Evolution of reproductive structures and functions in primitive angiosperms (Magnoliidae). *Memoirs of the New York Botanical Garden* 55:5–34.

————. 1992a. Evolution and floral diversity: the phylogenetic surroundings of *Arabidopsis* and *Antirrhinum*. *International Journal of Plant Sciences* 153:S106–S122.

————. 1992b. Primitive Blüten: sind Magnolien noch zeitgemäss? *Stapfia* 28:1–10.

————. 1993. Cercidiphyllaceae. *In* K. Kubitzki, J. G. Rohwer, and V. Bittrich. [eds.], The Families and Genera of Vascular Plants, II, 250–252. Springer, Berlin.

————. 1994a. Floral structure and evolution of primitive angiosperms: recent advances. *Plant Systematics and Evolution* 192:79–97.

————. 1994b. Evolutionary aspects of the floral structure in *Ceratophyllum*. *Plant Systematics and Evolution, Supplement* 8:175–183.

————. 1994c. Diversity and Evolutionary Biology of Tropical Flowers. Cambridge University Press, Cambridge.

————. 1995a. Major evolutionary trends of monocot flowers. *In* P. J. Rudall, P. J. Cribb, D. F. Cutler, and C. J. Humphries [eds.], Monocotyledons: Systematics and Evolution, 43–79. Royal Botanic Gardens, Kew.

————. 1995b. Floral structure and evolution in Ranunculanae. *Plant Systematics and Evolution, Supplement* 9:47–61.

————. 1996a. Diversity and evolutionary trends in angiosperm anthers. *In* W. G. D'Arcy and R. C. Keating [eds.], The Anther: Form, Function and Phylogeny, 92–110. Cambridge University Press, Cambridge.

————. 1996b. Homoplasy in angiosperm flowers. *In* M. J. Sanderson and L. Hufford [eds.], Homoplasy and the Evolutionary Process, 301–323. Academic Press, San Diego.

————. 1997a. Relationships between floral organization, architecture, and pollination mode in *Dillenia* (Dilleniaceae). *Plant Systematics and Evolution* 206:99–118.

————. 1997b. Evolutionary biology of flowers: prospects for the next century. *In* K. Iwatsuki and P. H. Raven [eds.], Evolution and Diversification of Land Plants, 99–119. Springer, Tokyo.

————. 1998. *Antirrhinum* and Asteridae: evolutionary changes of floral symmetry. *Symposium Series Society of Experimental Biology* 53:133–140.

————. 1999. Symmetry in flowers: diversity and evolution. *International Journal of Plant Sciences* 160:S3–S23.

————. 2001a. The flowers in extant basal angiosperms and inferences on ancestral flowers. *International Journal of Plant Sciences* 162:1111–1140.

————. 2001b. Origins of flower morphology. *Journal of Experimental Zoology (Molecular Development and Evolution)* 291:105–115.

————. 2003a. Morphology and angiosperm systematics in the molecular era. *Botanical Review* 68:545–570.

————. 2003b. What should a "complete" morphological phylogenetic analysis entail? Problems and promises. *In* T. F. Stuessy, E. Hörandl, and V. Mayer [eds.], Deep Morphology, 131–164. Gantner, Ruggell.

————. 2003c. Early floral development and the nature of calyptra in Eupomatiaceae. *International Journal of Plant Sciences* 164:489–503.

Endress, P. K., and L. Hufford. 1989. The diversity of stamen structures and dehiscence patterns among Magnoliidae. *Botanical Journal of the Linnean Society* 100:45–85.

Endress, P. K., and A. Igersheim. 1997. Gynoecium diversity and systematics of the Laurales. *Botanical Journal of the Linnean Society* 125:93–168.

————. 1999. Gynoecium diversity and systematics of the basal eudicots. *Botanical Journal of the Linnean Society* 130:305–393.

————. 2000a. The reproductive structures of the basal angiosperm *Amborella trichopoda* (Amborellaceae). *International Journal of Plant Sciences* 161:S237–S248.

————. 2000b. Gynoecium structure and evolution in basal angiosperms. *International Journal of Plant Sciences* 161:S211–S223.

Endress, P. K., and F. B. Sampson. 1983. Floral structure and relationships of the Trimeniaceae (Laurales). *Journal of the Arnold Arboretum* 64:447–473.

Endress, P. K., and S. Stumpf. 1990. Non-tetrasporangiate stamens in the angiosperms: structure, systematic distribution and evolutionary aspects. *Botanische Jahrbücher für Systematik* 112:193–240.

Endress, P. K., P. Baas, and M. Gregory. 2000a. Systematic plant morphology and anatomy—50 years of progress. *Taxon* 49: 401–434.

Endress, P. K., A. Igersheim, F. B. Sampson, and G. E. Schatz. 2000b. Floral structure of *Takhtajania* and its systematic position in Winteraceae. *Annals of the Missouri Botanical Garden* 87:347–365.

Endress, P. K., M. Jenny, and M. E. Fallen. 1983. Convergent elaboration of apocarpous gynoecia in higher advanced dicotyledons (Sapindales, Malvales, Gentianales). *Nordic Journal of Botany* 3:293–300.

Engler, A. 1930. Saxifragaceae. *In* A. Engler and K. Prantl [eds.], Die Natürlichenn Pflanzenfamilien, 2nd ed., 18a, 74–226. Engelmann, Leipzig.

Engler, A., and K. Prantl [eds.]. 1887–1915. Die Natürlichenn Pflanzenfamilien, various volumes. Engelmann, Leipzig.

Erbar, C. 1991. Sympetaly: a systematic character. *Botanische Jahrbücher für Systematik* 112:417–451.

Erbar, C., and P. Leins. 1981. Zur Spirale in Magnolienblüten. *Beiträge zur Biologie der Pflanzen* 56:225–241.

————. 1983. Zur Sequenz von Blütenorganen bei einigen Magnoliiden. *Botanische Jahrbücher für Systematik* 103: 433–449.

————. 1985. Studien zur Organsequenz in Apiaceen-Blüten. *Botanische Jahrbücher für Systematik* 105:379–400.

————. 1988. Blütenentwicklungsgeschichtliche Studien an *Aralia* und *Hedera* (Araliaceae). *Flora* 180:391–406.

————. 1995a. An analysis of the early floral development of *Pittosporum tobira* (Thunb.) Aiton and some remarks on the systematic position of the family Pittosporaceae. *Feddes Repertorium* 106:463–473.

————. 1995b. Portioned pollen release and the syndromes of secondary pollen presentation in the Campanulales-Asterales-complex. *Flora* 190:323–338.

————. 1996. Distribution of the character states "early" and "late" sympetaly within the "Sympetalae Tetracyclicae" and presumably related groups. *Botanica Acta* 109:427–440.

————. 1997. Different patterns of floral development in whorled flowers, exemplified by Apiaceae and Brassicaceae. *International Journal of Plant Sciences* 158:S49–S64.

Erbar, C., S. Kusma, and P. Leins. 1999. Development and interpretation of nectary organs in Ranunculaceae. *Flora* 194:317–332.

Erdtman, G. 1952. Pollen Morphology and Plant Taxonomy. Almqvist and Wiksell, Stockholm.

————. 1966. Pollen Morphology and Plant Taxonomy. Hafner, New York.

Erdtman, G., P. Leins, R. Melville, and C. Metcalfe. 1969. On the relationships of *Emblingia*. *Botanical Journal of the Linnean Society* 62:169–186.

Ereshefsky, M. 1994. Some problems with the Linnean hierarchy. *Philosophy of Science* 61:186–205.

Eriksson, O., E. M. Friis, K. R. Pederson, and P. R. Crane. 2000b. Seed size and dispersal systems of Early Cretaceous angiosperms from Famalicão, Portugal. *International Journal of Plant Sciences* 161:319–329.

Ernst, W. R. 1963. The genera of Hamamelidaceae and Platanaceae in the southeastern United States. *Journal of the Arnold Arboretum* 44:193–210.

————. 1964. The genera of Berberidaceae, Lardizabalaceae, and Menispermaceae in the southeastern United States. *Journal of the Arnold Arboretum* 45:1–35.

Ettlinger, M., and A. Kjaer. 1968. Sulfur compounds in plants. *Recent Advances in Phytochemistry* 1:59–144.

Evans, R. C., and T. A. Dickinson. 1999a. Floral ontogeny and morphology in subfamily Amygdaloideae T. and G. (Rosaceae). *International Journal of Plant Sciences* 160:955–979.

————. 1999b. Floral ontogeny and morphology in subfamily Spiraeoideae Endl. (Rosaceae). *International Journal of Plant Sciences* 160:981–1012.

Evans, R. C., C. Evans, and C. S. Campbell. 2002. The origin of the apple subfamily (Maloideae; Rosaceae) is clarified by DNA sequence data from duplicated GBSSI genes. *American Journal of Botany* 89:1478–1484.

Eyde, R. H. 1964. Inferior ovary and generic affinities of *Garrya*. *American Journal of Botany* 51:1083–1092.

————. 1988. Comprehending *Cornus*: puzzles and progress in the systematics of the dogwoods. *Botanical Review* 54:233–351.

Eyde, R. H., and C. C. Tseng. 1971. What is the primitive floral structure of Araliaceae? *Journal of the Arnold Arboretum* 52:205–239.

Eyde, R. H., and Q.-X. Xiang. 1990. Fossil mastixioid (Cornaceae) alive in eastern Asia. *American Journal of Botany* 52:205–239.

Fallen, E. M. 1986. Floral structure in the Apocynaceae: morphological, functional, and evolutionary aspects. *Botanische Jahrbücher für Systematik* 106:245–286.

Fan, C., and Q.-X. Xiang. 2003. Phylogenetic analyses of Cornales based on 26S rRNA and combined 26S rDNA-*matK-rbcL* sequence data. *American Journal of Botany* 90:1357–1372.

Farris, J. S. 1976. Phylogenetic classification of fossils with recent species. *Systematic Zoology* 25:271–282.

————. 1988. Computer program and documentation Hennig 86. Port Jefferson, New York.

————. 1996. *Jac*. Swedish Museum of Natural History, Stockholm.

————. 2002. RASA attributes highly significant structure to randomized data. *Cladistics* 18:334–353.

Farris, J. S., V. A. Albert, M. Källersjö, D. Lipscomb, and A. G. Kluge. 1996. Parsimony jackknifing outperforms neighbor-joining. *Cladistics* 12:99–124.

Fay, M. F., K. M. Cameron, G. T. Prance, M. D. Lledo, and M. W. Chase. 1997a. Familial relationships of *Rhabdodendron* (Rhabdodendraceae): plastid *rbcL* sequences indicate a caryophyllid placement. *Kew Bulletin* 52:923–932.

Fay, M. F., S. M. Swensen, and M. W. Chase. 1997b. Taxonomic affinities of *Medusagyne oppositifolia* (Medusagynaceae). *Kew Bulletin* 52:111–120.

Fay, M. F., C. Bayer, W. Alverson, A. de Bruijn, S. Swensen, and M. Chase. 1998a. Plastid *rbcL* sequences indicate a close affinity between *Diegodendron* and *Bixa*. *Taxon* 47:43–50.

Fay, M. F., R. G. Olmstead, J. E. Richardson, E. Santiago, G. T. Prance, and M. W. Chase. 1998b. Molecular data support the inclusion of *Duckeodendron celastroides* in Solanaceae. *Kew Bulletin* 53:203–212.

Fay, M. F., P. J. Rudall, S. Sullivan, K. L. Stobart, A. Y. de Bruijn, G. Reeves, F. Qamaruz-Zaman, W.-P. Hong, J. Joseph, W. J. Hahn, J. G. Conran, and M. W. Chase. 2000a. Phylogenetic studies of Asparagales based on four plastid DNA loci. *In* K. L. Wilson and D. A. Morrison [eds.], Monocots: Systematics and Evolution, 360–371. CSIRO, Melbourne.

Fay, M. F., B. Bremer, G. T. Prance, M. Van Der Bank, D. Bridson, and M. W. Chase. 2000b. Plastid *rbcL* sequence data show *Dialypetalanthus* to be a member of Rubiaceae. *Kew Bulletin* 55:853–864.

Fedorov, A. 1969. Chromosome Numbers of Flowering Plants. Academy of Sciences, Leningrad.

Feild, T. S., N. C. Arens, and T. E. Dawson. 2003b. The ancestral ecology of angiosperms: emerging perspectives from extant basal lineages. *International Journal of Plant Sciences* 164:S129–S142.

Feild, T. S., N. C. Arens, J. A. Doyle, T. E. Dawson, and M. J. Donoghue. 2004. Dark and disturbed: a new image of early angiosperm ecology. *Paleobiology* 30:82–107.

Feild, T. S., P. J. Franks, and T. L. Sage. 2003a. Ecophysiological shade adaptation in the basal angiosperm, *Austrobaileya scandens* (Austrobaileyaceae). *International Journal of Plant Sciences* 164:313–324.

Feild, T. S., M. A. Zwieniecki, T. Brodribb, T. Jaffre, M. J. Donoghue, and N. M. Holbrook. 2000a. Structure and function of tracheary elements in *Amborella trichopoda*. *International Journal of Plant Sciences* 161:705–712.

Feild, T. S., M. A. Zwieniecki, and N. M. Holbrook. 2000b. Winteraceae evolution: an ecophysiological perspective. *Annals of the Missouri Botanical Garden* 87:323–334.

Felsenstein, J. 1978a. The number of evolutionary trees. *Systematic Zoology* 27:27–33.

————. 1978b. Cases in which parsimony or compatibility methods will be positively misleading. *Systematic Zoology* 27:401–410.

————. 1985. Confidence limits on phylogenies: an approach using the bootstrap. *Evolution* 39:783–791.

————. 1988. Phylogenies from molecular sequences: inferences and reliability. *Annual Review of Genetics* 22:521–565.

Ferguson, D. K. 1989. A survey of the Liquidambaroideae (Hamamelidaceae) with a view to elucidating its fossil record. *In* P. R. Crane and S. Blackmore. [eds.], Evolution, Systematics and Fossil History of the Hamamelidae, 1, 249–272. Clarendon Press, Oxford.

Ferguson, I. K. 1966. The Cornaceae in the southeastern United States. *Journal of the Arnold Arboretum* 47:106–116.

————. 1977. Cornaceae. *World Pollen and Spore Flora* 6:1–34.

Feuer, S. 1991. Pollen morphology and the systematic relationships of *Ticodendron incognitum*. *Annals of the Missouri Botanical Garden* 78:143–151.

Fishbein, M. 2001. Evolutionary innovation and diversification in the flowers of Asclepiadaceae. *Annals of the Missouri Botanical Garden* 88:603–623.

Fishbein, M., and D. E. Soltis. 2004. Further resolution of the rapid radiation of Saxifragales (angiosperms, eudicots) supported by mixed-model Bayesian analysis. *Systematic Botany* 29:883–991.

Fishbein, M., C. Hibsch-Jetter, D. E. Soltis, and L. Hufford. 2001. Phylogeny of Saxifragales (angiosperms, eudicots): Analysis of a rapid, ancient radiation. *Systematic Biology* 50:817–847.

Flavell, R. 1988. Repetitive DNA and chromosome evolution in plants. *Philosophical Transactions of the Royal Society of London B* 312:227–242.

Floyd, S. K., and W. E. Friedman. 2000. Evolution of endosperm developmental patterns among basal flowering plants. *International Journal of Plant Sciences* 161:S57–S81.

Floyd, S. K., V. T. Lerner, and W. E. Friedman. 1999. A developmental and evolutionary analysis of embryology in *Platanus* (Platanaceae), a basal eudicot. *American Journal of Botany* 86:1523–1537.

Frank, A. C., H. Amiri, and S. G. E. Andersson. 2002. Genome deterioration: loss of repeated sequences and accumulation of junk DNA. *Genetica* 115:1–12.

Franz, E. 1908. Beiträge zur Kenntnis der Portulacaceen und Basellaceen. *Botanische Jahrbücher für Systematik* 97:1–46.

Freeman, D., J. Doust, A. El-Keblawy, K. Migla, and E. McArthur. 1997. Sexual specialization and inbreeding avoidance in the evolution of dioecy. *Botanical Review* 63:65–92.

Freshwater, D. W., S. Fredericq, B. S. Butler, M. H. Hommersand, and M. W. Chase. 1994. A gene phylogeny of the red algae (Rhodophyta) based on plastic *rbcL*. *Proceedings of the National Academy of Sciences USA* 91:7281–7285.

Freudenstein, J. V., D. M. Senyo, and M. W. Chase. 2000. Mitochondrial DNA and relationships in the Orchidaceae. *In* K. L. Wilson and D. A. Morrison [eds.], Monocots: Systematics and Evolution, 421–429. CSIRO, Melbourne.

Friedman, W. E. 1990. Sexual reproduction in *Ephedra nevadensis* (Ephedraceae): further evidence of double fertilization in a nonflowering seed plant. *American Journal of Botany* 77:1582–1598.

———. 1992. Evidence of a pre-angiosperm origin of endosperm: implications for the evolution of flowering plants. *Science* 255:336–339.

———. 1994. The evolution of embryogeny in seed plants and the developmental origin and early history of endosperm. *American Journal of Botany* 81:1468–1486.

———. 1996. Biology and evolution of the Gnetales. *International Journal of Plant Sciences* 157:S1–S125.

Friedman, W. E., and S. K. Floyd. 2001. Perspective: the origin of flowering plants and their reproductive biology: a tale of two phylogenies. *Evolution* 55:217–231.

Friedman, W. E., C. Moore, and M. D. Purugganan. 2004. The evolution of plant development. *American Journal of Botany* 91:1726–1741.

Friis, E. M., and P. R. Crane. 1989. Reproductive structures of Cretaceous Hamamelidae. *In* P. R. Crane and S. Blackmore [eds.], Evolution, Systematics and Fossil History of the Hamamelidae, 1, 155–174. Clarendon Press, Oxford.

Friis, E. M., and P. K. Endress. 1990. Origin and evolution of angiosperm flowers. *Advances in Botanical Research* 17:99–162.

Friis, E. M., P. R. Crane, and K. R. Pedersen. 1986. Floral evidence for Cretaceous chloranthoid angiosperms. *Nature* 320:163–164.

———. 1997. Fossil history of magnoliid angiosperms. *In* K. Iwatsuki and P. H. Raven [eds.], Evolution and Diversification of Land Plants, 121–156. Springer, New York.

———. 1999. Early angiosperm diversification: the diversity of pollen associated with angiosperm reproductive structures in early Cretaceous floras from Portugal. *Annals of the Missouri Botanical Garden* 86:259–296.

Friis, E. M., P. R. Crane, and K. R. Pedersen. 1988. Reproductive structures of Cretaceous Platanaceae. *Biologiske Meddelelser Kongelige Danske Videnskabernes Selskab* 31:1–56.

Friis, E. M., J. A. Doyle, P. K. Endress, and Q. Leng. 2003. *Archaefructus*–angiosperm precursor or specialized early angiosperm? *Trends in Plant Science* 8:369–373.

Friis, E. M., K. R. Pedersen, and P. R. Crane. 1994. Angiosperm floral structures from the Early Cretaceous of Portugal. *Plant Systematics and Evolution, Supplement* 8:31–49.

———. 2000. Reproductive structure and organization of basal angiosperms from the early Cretaceous (Barremian or Aptian) of western Portugal. *International Journal of Plant Sciences* 161:S169–S182.

———. 2001. Fossil evidence of water lilies (Nymphaeales) in the Early Cretaceous. *Nature* 410:357–360.

Fritsch, P. W. 2001. Phylogeny and biogeography of the flowering plant genus *Styrax* (Styracaceae) based on chloroplast DNA restriction sites and DNA sequences of the internal transcribed spacer region. *Molecular Phylogenetics and Evolution* 19:387–408.

Frohlich, M. W. 1999. MADS about Gnetales. *Proceedings of the National Academy of Sciences USA* 96:8811–8813.

Frohlich, M. W., and E. M. Meyerowitz. 1997. The search for flower homeotic gene homologs in basal angiosperms and Gnetales: a potential new source of data on the evolutionary origin of flowers. *International Journal of Plant Sciences* 158:S131–S142.

Frohlich, M. W., and D. S. Parker. 2000. The mostly male theory of flower evolutionary origins: from genes to fossils. *Systematic Botany* 25:155–170.

Fulton, M., and S. A. Hodges. 1999. Floral isolation between *Aquilegia formosa* and *Aquilegia pubescens*. *Proceedings of the Royal Society of London B* 266:2247–2252.

Furness C. A., and P. J. Rudall. 2004. Pollen aperture evolution—a crucial factor for eudicot success? *Trends in Plant Science* 9:1360–1385.

Furness, C. A., P. Rudall, and F. B. Sampson. 2002. Evolution of microsporogenesis in angiosperms. *International Journal of Plant Sciences* 163:235–260.

Fuse, S., and M. N. Tamura. 2000. A phylogenetic analysis of the plastid *matK* gene with an emphasis on Melanthiaceae sensu lato. *Plant Biology* 2:415–427.

Gadek, P., and C. J. Quinn. 1993. An analysis of relationships within the Cupressaceae sensu stricto based on *rbcL* sequences. *Annals of the Missouri Botanical Garden* 80:581–586.

Gadek, P. A., E. S. Fernando, C. J. Quinn, S. B. Hoot, T. Terrazas, M. C. Sheahan, and M. W. Chase. 1996. Sapindales: molecular delimitation and infraordinal groups. *American Journal of Botany* 83:802–811.

Gadek, P. A., C. J. Quinn, J. E. Rodman, K. G. Karol, E. Conti, R. A. Price, and E. S. Fernando. 1992. Affinities of the Australian endemic Akaniaceae: new evidence from *rbcL* sequences. *Australian Systematic Botany* 5:717–734.

Gaffney, E. S., and P. A. Meylan. 1988. A phylogeny of turtles. *In* M. J. Benton [ed.], The Phylogeny and Classification of Tetrapods, I: Amphibians, Reptiles, and Birds, 157–219. Clarendon Press, Oxford.

Gailing, O., and K. Bachmann. 2000. The evolutionary reduction of microsporangia in *Microseris* (Asteraceae): transition genotypes and phenotypes. *Plant Biology* 2:455–461.

Gaiser, J. C., K. Robinson-Beers, and C. S. Gasser. 1995. The *Arabidopsis SUPERMAN* gene mediates asymmetric growth of the outer integument of ovules. *Plant Cell* 7:333–345.

Galen, C. 1999. Why do flowers vary? *BioScience* 49:631–640.

Gandolfo, M. A., K. C. Nixon, and W. L. Crepet. 2002. Triuridaceae fossil flowers from the Upper Cretaceous of New Jersey. *American Journal of Botany* 89:1940–1957.

————. 2004. The oldest complete fossil flowers of Nymphaeaceae and implications for the complex insect entrapment pollination mechanisms in early Angiosperms. *Proceedings of the National Academy of Sciences USA* 101: 8056–8060.

Gandolfo, M. A., K. C. Nixon, W. L. Crepet, D. W. Stevenson, and E. M. Friis. 1998. Oldest known fossils of monocotyledons. *Nature* 394:532–533.

Gastony, G. J., and D. E. Soltis. 1977. Chromosome studies of *Parnassia* and *Lepuropetalon* from the eastern United States. A new base number for *Parnassia. Rhodora* 79:573–578.

Gaussen, H. 1946. Les Gymnospermes, actuelles et fossiles. *Travaux du Laboratoire Forestier Toulouse* 1:1–26.

Gauthier R. 1950 The nature of the inferior ovary in the genus *Begonia. Contributions de l'Institute de Botanique, Université de Montréal* 66:1–93.

Gemmeke, V. 1982. Entwicklungsgeschichtliche Untersuchungen an Mimosaceen-Blüten. *Botanische Jahrbücher für Systematik* 103:185–210.

Giannasi, D. E., G. Zurawski, G. Learn, and M. T. Clegg. 1992. Evolutionary relationships of the Caryophyllidae based on comparative *rbcL* sequences. *Systematic Botany* 17:1–15.

Gibbs, R. D. 1957. The Mäule reaction, lignin, and the relationships between woody plants, K. V. Thimann [ed.], The Physiology of Forest Trees, 269–312. Ronald, New York.

Gibson, A. C. 1979. Anatomy of *Koeberlinia* and *Canotia* revisited. *Madroño* 26:1–12.

Gibson, A. C., and P. S. Nobel. 1986. The Cactus Primer. Harvard University Press, Cambridge, MA.

Gilg, E. 1925. Flacourtiaceae. *In* A. Engler and K. Prantl [eds.], Die Natürlichen Pflanzenfamilien, 2nd ed., 21, 377–457. Engelmann, Leipzig.

Gilg, E., and R. Pilger. 1905. Rutaceae. *In* R. Pilger [ed.], Beiträge zur Flora der Hylea nach den Sammlungen von E. Ule. *Verhandlungen des Botanischen Vereins der Provinz Brandenburg* 47:152–154.

Giussani, L. M., J. H. Cota-Sanchez, F. O. Zuloaga, and E. A. Kellogg. 2001. A molecular phylogeny of the grass subfamily Panicoideae (Poaceae) shows multiple origins of C_4 photosynthesis. *American Journal of Botany* 88:1993–2012.

Givnish, T. J. 1997. Adaptive radiation and molecular systematics: issues and approaches. *In* T. J. Givnish and K. J. Sytsma [eds.], Molecular Evolution and Adaptive Radiation, 1–54. Cambridge University Press, Cambridge.

Givnish, T. J., and K. J. Sytsma. 1997. Molecular Evolution and Adaptive Radiation. Cambridge University Press, Cambridge.

Givnish, T. J., E. L. Burkhardt, R. Happel, and J. Weintraub. 1984. Carnivory in the bromeliad *Brocchinia reducta,* with a cost/benefit model for the general restriction of carnivorous plants to sunny, moist, nutrient-poor habitats. *American Naturalist* 124:479–497.

Givnish, T. J., T. M. Evans, J. C. Pires, and K. J. Sytsma. 1999. Polyphyly and convergent morphological evolution in Commelinales and Commelinidae: evidence from *rbcL* sequence data. *Molecular Phylogenetics and Evolution* 12:360–385.

Givnish, T. J., J. C. Pires, S. W. Graham, M. A. McPherson, L. M. Prince, T. B. Patterson, H. S. Rai, E. H. Roalson, T. M. Evans, W. J. Hahn, K. C. Millam, A. W. Meerow, M. Molvray, P. J. Kores, H. E. O'Brien, L. C. Hall, W. J. Kress, and K. J. Sytsma. 2005. Phylogenetic relationships of monocots based on the highly informative plastid gene *ndhF*: evidence for widespread concerted convergence. *In* J. T. Columbus, E. A. Friar, C. W. Hamilton, J. M. Porter, L. M. Prince, and M. G. Simpson [eds.]. Monocots: Comparative Biology and Evolution. 2 vols. Rancho Santa Ana Botanic Garden, Claremont, CA, in press.

Givnish, T. J., K. J. Sytsma, J. F. Smith, W. J. Hahn, D. H. Benzing, and E. M. Burkhardt. 1997. Molecular evolution and adaptive radiation in *Brocchinia* (Bromeliaceae: Pitcairnioideae) atop tepuis of the Guyana Shield. *In* T. J. Givnish and K. J. Sytsma [eds.], Molecular Evolution and Adaptive Radiation, 259–312. Cambridge University Press, Cambridge.

Glisic, L. M. 1928. Development of the female gametophyte and endosperm in *Haberlea rhodopensis* (Friv.). *Bulletin de l'Institute et Jardin Botanique, Université Belgrade* 1:1–13.

Goldblatt, P. 1976. Cytotaxonomic studies in the tribe Quillajeae (Rosaceae). *Annals of the Missouri Botanical Garden* 63:200–206.

————. 1990. Phylogeny and classification of the Iridaceae. *Annals of the Missouri Botanical Garden* 77:607–627.

————. 1991. An overview of the systematics, phylogeny and biology of the southern African Iridaceae. *Contributions from the Bolus Herbarium* 13:1–74.

Goldblatt, P., J. E. Henrich, and P. J. Rudall. 1984. Occurrence of crystals in Iridaceae and allied families and their phylogenetic significance. *Annals of the Missouri Botanical Garden* 71:1013–1020.

Goldblatt, P., J. C. Manning, and P. Bernhardt. 1998. Adaptive radiation of bee-pollinated *Gladiolus* species (Iridaceae) in Southern Africa. *Annals of the Missouri Botanical Garden* 85:492–517.

————. 2001. Radiation of pollination systems in *Gladiolus* (Iridaceae: Crocoideae) in Southern Africa. *Annals of the Missouri Botanical Garden* 88:713–734.

Goldblatt, P., V. Savolainen, O. Porteous, I. Sostaric, M. Powell, G. Reeves, J. C. Manning, and T. G. Barraclough. 2002. Radiation in the Cape flora and the phylogeny of peacock irises *Moraea* (Iridaceae) based on four plastid DNA regions. *Molecular Phylogenetics and Evolution* 25:341–360.

Goloboff, P. 1993. NONA. Computer programs and documentation. Tucumán, Argentina.

Gómez-Laurito, J., and P. Gómez. 1989. *Ticodendron*: A new tree from Central America. *Annals of the Missouri Botanical Garden* 76:1148–1151.

González, F. A., and P. Rudall. 2001. The questionable affinities of *Lactoris*: evidence from branching pattern, inflorescence morphology, and stipule development. *American Journal of Botany* 88:2143–2150.

————. 2003. Structure and development of the ovule and seed in Aristolochiaceae, with particular reference to *Saruma. Plant Systematics and Evolution* 241:223–244.

González, F. A., and D. W. Stevenson. 2002. A phylogenetic analysis of the subfamily Aristolochioideae (Aristolochiaceae). *Botanical Review* 26:25–58.

Goremykin, V., V. Bobrova, J. Pahnke, A. Troitsky, A. Antonov, and W. Martin. 1996. Noncoding sequences from the slowly evolving chloroplast inverted repeat in addition to *rbcL* data do not support gnetalean affinities of angiosperms. *Molecular Biology and Evolution* 13:383–396.

Goremykin, V., S. Hansmann, and W. Martin. 1997. Evolutionary analysis of 58 proteins encoded in six completely sequenced chloroplast genomes: revised molecular estimates of two seed plant divergence times. *Plant Systematics and Evolution* 206:337–351.

Goremykin, V., K. I. Hirsch-Ernst, S. Wölfl, and F. H. Hellwig. 2003. Analysis of the *Amborella trichopoda* chloroplast genome sequence suggests that *Amborella* is not a basal angiosperm. *Molecular Biology and Evolution* 20:1499–1505.

————. 2004. The chloroplast genome of *Nymphaea alba*, whole genome analysis and the problem of identifying the most basal angiosperm. *Molecular Biology and Evolution* 21:1445–1454.

Gottsberger, G. 1977. Some aspects of beetle pollination in the evolution of flowering plants. *Plant Systematics and Evolution,* Supplement 1:211–226.

————. 1988. The reproductive biology of primitive angiosperms. *Taxon* 37:630–643.

Gottwald, H., and N. Parameswaran. 1968. Das sekundäre Xylem und die systematische Stellung der Ancistrocladaceae und Dioncophyllaceae. *Botanische Jahrbücher für Systematik* 88:49–69.

GPWG (Grass Phylogeny Working Group). 2000. A phylogeny of the grass family (Poaceae), as inferred from eight character sets. *In* S. W. L. Jacobs and J. Everett [eds.], Grasses: Systematics and Evolution, 3–7. CSIRO, Melbourne.

Graham, L. E., C. F. Delwiche, and B. D. Mishler. 1991. Phylogenetic connections between the "green algae" and the "bryophytes". *Advances in Bryology* 4:213–214.

Graham, S. A. 1964. The genera of Rhizophoraceae and Combretaceae in the southeastern United States. *Journal of the Arnold Arboretum* 45:285–301.

———. 1984. Alzateaceae, a new family of Myrtales in the American tropics. *Annals of the Missouri Botanical Garden* 71: 757–779.

Graham, S. A., J. V. Crisci, and P. C. Hoch. 1993b. Cladistic analysis of the Lythraceae sensu lato based on morphological characters. *Botanical Journal of the Linnean Society* 113:1–33.

Graham, S. W., and R. G. Olmstead. 2000. Utility of 17 chloroplast genes for inferring the phylogeny of the basal angiosperms. *American Journal of Botany* 87:1712–1730.

Graham, S. W., P. A. Reeves, A. C. E. Burns, and R. G. Olmstead. 2000. Microstructural changes in noncoding chloroplast DNA: interpretation, evolution, and utility of indels and inversions in basal angiosperm phylogenetic inference. *International Journal of Plant Sciences* 161:S83–S96.

Grant, D. P., P. Cregan, and R. C. Shoemaker. 2000. Genome organization in dicots: genome duplication in *Arabidopsis* and synteny between soybean and *Arabidopsis*. *Proceedings of the National Academy of Sciences USA* 97:4168–4173.

Grant, V. 1950. The protection of ovules in flowering plants. *Evolution* 4:179–201.

———. 1981. Plant Speciation. Columbia University Press, New York.

Grant, V., and K. A. Grant. 1965. Flower Pollination in the *Phlox* Family. Columbia University Press, New York.

Graur, D., L. Duret, and M. Gouy. 1996. Phylogenetic position of the order Lagomorpha (rabbits, hares and allies). *Nature* 379:333–335.

Gray, A. 1875. A conspectus of the North American Hydrophyllaceae. *Proceedings of the American Academy* 10:312–332.

Graybeal, A. 1998. Is it better to add taxa or characters to a difficult phylogenetic problem? *Systematic Biology* 47:9–17.

Grayum, M. H. 1987. A summary of evidence and arguments supporting the removal of *Acorus* from the Araceae. *Taxon* 36:723–729.

Gregory, T. R. 2001. Coincidence, coevolution, or causation? DNA content, cell size, and the C-value enigma. *Biological Reviews* 76:65–101.

Griffiths, G. C. D. 1973. Some fundamental problems in biological classification. *Systematic Zoology* 22:338–343.

———. 1974. On the foundations of biological systematics. *Acta Biotheoretica* 23:85–131.

———. 1976. The future of Linnean nomenclature. *Systematic Zoology* 25:168–173.

Grimaldi, D. 1999. The co-radiations of pollinating insects and angiosperms in the Cretaceous. *Annals of the Missouri Botanical Garden* 86:373–406.

Grime, J., and M. Mowforth. 1982. Variation in genome size—an ecological interpretation. *Nature* 299:151–153.

Grudzinskaja, I. A. 1967. Ulmaceae and reasons for distinguishing Celtidoideae as a separate family Celtidaceae Link. *Botanicheskii Zhurnal* 52:1723–1749.

Gustafsson, M. H. G. 1995. Petal venation in the Asterales and related orders. *Botanical Journal of the Linnean Society* 118:1–18.

Gustafsson, M. H. G., and V. A. Albert. 1999. Inferior ovaries and angiosperm diversification. *In* P. M. Hollingsworth, R. M. Bateman, and R. J. Gornall [eds.], Molecular Systematics and Plant Evolution, 403–431. Taylor and Francis, London.

Gustafsson, M. H. G., and K. Bremer. 1995. Morphology and phylogenetic interrelationships of the Asteraceae, Calyceraceae, Campanulaceae, Goodeniaceae and related families (Asterales). *American Journal of Botany* 82:250–265.

———. 1997. The circumscription and systematic position of Carpodetaceae. *Australian Systematic Botany* 10:855–872.

Gustafsson, M. H. G., A. Backlund, and B. Bremer. 1996. Phylogeny of the Asterales sensu lato based on *rbcL* sequences with particular reference to the Goodeniaceae. *Plant Systematics and Evolution* 199:217–242.

Gustafsson, M. H. G., V. Bittrich, and P. F. Stevens. 2002. Phylogeny of Clusiaceae based on *rbcL* sequences. *International Journal of Plant Sciences* 163:1045–1054.

Haber, J. M. 1966. The comparative anatomy and morphology of the flowers and inflorescences of the Proteaceae. III. Some African taxa. *Phytomorphology* 16:490–527.

Hahn, W. J. 2002. A molecular phylogenetic study of the Palmae (Arecaceae) based on *atpB, rbcL* and 18S nrDNA sequences. *Systematic Biology* 51:91–112.

Hakki, M. I. 1977. Über die Embryologie, Morphologie und systematische Zugehörigkeit von *Dermatobotrys saundersii* Bolus. *Botanische Jahrbücher für Systematik* 98:93–119.

Hall, A. E., A. Fiebig, and D. Preuss. 2002. Beyond the *Arabidopsis* genome: opportunities for comparative genomics. *Plant Physiology* 129:1439–1447.

Hall, J. C., K. J. Sytsma, and H. H. Iltis. 2002. Phylogeny of Capparaceae and Brassicaceae based on chloroplast sequence data. *American Journal of Botany* 89:1826–1842.

Hallier, H. 1905. Provisional scheme for the natural (phylogenetic) system of the flowering plants. *New Phytologist* 4:151–162.

Hamby, R. K., and E. A. Zimmer. 1988. Ribosomal RNA sequences for inferring phylogeny within the grass family (Poaceae). *Plant Systematics and Evolution* 160:29–37.

———. 1992. Ribosomal RNA as a phylogenetic tool in plant systematics. *In* P. S. Soltis, D. E. Soltis, and J. J. Doyle [eds.], Molecular Systematics of Plants, 50–91. Chapman and Hall, New York.

Hancock, J. M. 2002. Genome size and the accumulation of simple sequence repeats: implications of new data from genome sequencing projects. *Genetica* 115:93–103.

Hansen, A., S. Hansmann, T. Samigullin, A. Antonov, and W. Martin. 1999. *Gnetum* and the angiosperms: molecular evidence that their shared morphological characters are convergent, rather than homologous. *Molecular Biology and Evolution* 16:1006–1009.

Hansen, C. H., L. Du, P. Naur, C. E. Olsen, K. B. Axelsen, A. J. Hick, J. A. Pickett, and B. A. Halkier. 2001a. CYP83B1 is the oxime-metabolizing enzyme in the glucosinolate pathway. *Journal of Biological Chemistry* 276:24790–24796.

Hansen, C. H., U. Wittstock, C. E. Olsen, A. J. Hick, J. A. Pickett, and B. A. Halkier. 2001b. Cytochrome P450 CYP79F1 from *Arabidopsis* catalyzes the conversion of dihomomethionine and trihomomethionine to the corresponding aldoximes in the biosynthesis of aliphatic glucosinolates *Journal of Biological Chemistry* 276:11078–11085.

Hansen, H. V. 1992. Studies in the Calyceraceae with a discussion of its relationships to Compositae. *Nordic Journal of Botany* 12:63–75.

Hansen, T. F., W. S. Armbruster, and L. Antonsen. 2000. Comparative analysis of character displacement and spatial adaptations as illustrated by the evolution of *Dalechampia* blossoms. *American Naturalist* 156 (Supplement):S17–S34.

Hanson, L., A. McMahon, M. A. T. Johnson, and M. D. Bennett. 2001a. First nuclear DNA C-values for 25 angiosperm families. *Annals of Botany* 87:251–258.

———. 2001b. First nuclear DNA C-values for another 25 angiosperm families. Annals of Botany 88:851-858.

Hao, G., R. M. K. Saunders, and M.-L. Chye. 2000. A phylogenetic analysis of the Illiciaceae based on sequences of internal transcribed spacers (ITS) of nuclear ribosomal DNA. *Plant Systematics and Evolution* 223:81–90.

Hapeman, J. R., and K. Inoue. 1997. Plant-pollinator interactions and floral radiation in *Platanthera* (Orchidaceae). *In* T. J. Givnish and K. J. Sytsma [eds.], Molecular Evolution and Adaptive Radiation, 433–454. Cambridge University Press, Cambridge.

Harborne, J. 1982. Introduction to Ecological Chemistry 2nd ed. Academic Press, London.

Harland, W. B., R. L. Armstrong, A. V. Cox, L. E. Craig, A. G. Smith, and D. G. Smith. 1989. A Geologic Timescale. Cambridge University Press, Cambridge.

Harms, H. 1930. Hamamelidaceae. *In* A. Engler and K. Prantl [eds.], Die Natürlichen Pflanzenfamilien, 2nd ed., 18a, 330–343. Engelmann, Leipzig.

———. 1934. Reihe Centrospermae. *In* A. Engler and K. Prantl [eds.], Die Natürlichen Pflanzenfamilien, 2nd ed., 16c, 1–6. Engelmann, Leipzig.

Harris, S. A., J. P. Robinson, and B. E. Juniper. 2002. Genetic clues to the origin of the apple. *Trends in Genetics* 18:426–430.

Harris. T. M. 1941. *Caytonanthus*, the microsporophyll of *Caytonia*. Annals of Botany 5:47–58.

Harris, T. M. 1964. The Yorkshire Jurassic Flora. II. Caytoniales, Cycadales, and Pteridosperms. British Museum (Natural History), London.

Hasebe, M., R. Kofuji, M. Ito, M. Kato, K. Iwatsuki, and K. Ueda. 1992. Phylogeny of gymnosperms inferred from *rbcL* gene sequences. *Botanical Magazine, Tokyo* 105:673–679.

Hasegawa, M., H. Kishino, and T. Yano. 1985. Dating of the human-ape split by a molecular clock of mitochondrial DNA. *Journal of Molecular Evolution* 21:160–174.

Hatch, M. D., T. Kagawa, and S. Craig. 1975. Subdivision of C_4-pathway species based on differing C_4 acid decarboxylating systems and ultrastructural features. *Australian Journal of Plant Physiology* 2:111–118.

Hayes, V., E. L. Schneider, and S. Carlquist. 2000. Floral development of *Nelumbo nucifera* (Nelumbonaceae). *International Journal of Plant Sciences* 161:S183–S191.

Heard, J., and K. Dunn. 1995. Symbiotic induction of a MADS-box gene during development of alfalfa root nodules. *Proceedings of the National Academy of Sciences USA* 92:5273–5277.

Heckman, D. S., D. M. Geiser, B. R. Eidell, R. L. Stauffer, N. L. Kardos, and S. B. Hedges. 2001. Molecular evidence for the early colonization of land by fungi and plants. *Science* 293:1129–1133.

Hegnauer, R. 1962–1994. Chemotaxonomie der Pflanzen, 1–11a. Birkhäuser, Basel.

Heimsch, C. Jr. 1942. Comparative anatomy of the secondary xylem in the Gruinales and Terebinthales of Wettstein with reference to taxonomic grouping. *Lilloa* 8:83–198.

Hempel, A. L., P. A. Reeves, R. G. Olmstead, and R. K. Jansen. 1995. Implications of *rbcL* sequence data for higher order relationships of the Loasaceae and the anomalous aquatic plant *Hydrostachys* (Hydrostachyaceae). *Plant Systematics and Evolution* 194:25–37.

Hennig, W. 1950. Grundzüge einer Theorie der Phylogenetischen Systematik. Deutscher Zentralverlag, Berlin.

———. 1965. Phylogenetic systematics. *Annual Review of Entomology* 10:97–116.

———. 1966. Phylogenetic Systematics. University of Illinois Press, Urbana.

———. 1969. Die Stammesgeschichte der Insekten. Kramer, Frankfurt.

———. 1981. Insect Phylogeny. John Wiley and Sons, Chichester.

———. 1983. Stammesgeschichte der Chordaten. *Fortschritte der Zoologischen Systematik und Evolutionsforschung* 2:1–208.

Henrickson, J. 1967. Pollen morphology of the Fouquieriaceae. *Aliso* 6:137–160.

Henslow, G. 1893. A theoretical origin of the endogens from the exogens through self-adaptation to an aquatic habitat. *Journal of the Linnean Society, Botany,* 29:485–528.

Henwood, M. J., and J. M. Hart. 2001. Towards an understanding of the phylogenetic relationships of Australian Hydrocotyloideae (Apiaceae). *Edinburgh Journal of Botany* 58:269–289.

Herendeen, P. S., and P. R. Crane. 1995. The fossil history of the monocotyledons. *In* P. J. Rudall, P. J. Cribb, D. F. Cutler, and C. J. Humphries [eds.], Monocotyledons: Systematics and Evolution, 1–21. Royal Botanic Gardens, Kew.

Herendeen, P. S., P. R. Crane, and A. Drinnan. 1995. Fagaceous flowers, fruits, and cupules from the Campanian (Late Cretaceous) of central Georgia, USA. *International Journal of Plant Sciences* 156:93–116.

Herendeen, P. S., W. L. Crepet, and K. C. Nixon. 1993. *Chloranthus*-like stamens from the Upper Cretaceous of New Jersey. *American Journal of Botany* 80:865–871.

Herrero, M. 2001. Ovary signals for directional pollen tube growth. *Sexual Plant Reproduction* 14:3–17.

Hershkovitz, M. A. 1993. Revised circumscriptions and subgeneric taxonomies of *Calandrinia* and *Montiopsis* (Portulacaceae) with notes on phylogeny of the portulacaceous alliance. *Annals of the Missouri Botanical Garden* 80:333–365.

Hershkovitz, M. A., and E. A. Zimmer. 1997. On the evolutionary origins of the cacti. *Taxon* 46:217–232.

Heywood, V. A. 1993. Flowering Plants of the World. Batsford, London.

———. 1998. Flowering Plants of the World, 2nd ed. Batsford, London.

Hibbett, D. S., and M. J. Donoghue. 1998. Integrating phylogenetic analysis and classification in fungi. *Mycologia* 90:347–356.

Hibsch-Jetter, C., D. E. Soltis, and T. D. Macfarlane. 1997. Phylogenetic analysis of *Eremosyne pectinata* (Saxifragaceae s. l.) based on *rbcL* sequence data. *Plant Systematics and Evolution* 204:225–232.

Hickey, L. J., and J. A. Doyle. 1977. Early Cretaceous fossil evidence for angiosperm evolution. *Botanical Review* 43:3–104.

Hickey, L. J., and D. W. Taylor. 1991. The leaf architecture of *Ticodendron* and the application of foliar characters in discerning its relationships. *Annals of the Missouri Botanical Garden* 78:105–130.

———. 1996. Origin of the angiosperm flower. *In* D. W. Taylor and L. J. Hickey [eds.], Flowering Plant Origin, Evolution, and Phylogeny, 176–231. Chapman and Hall, New York.

Hickey, L. J., and A. D. Wolfe. 1975. The bases of angiosperm phylogeny: vegetative morphology. *Annals of the Missouri Botanical Garden* 62:538–589.

Hiepko, P. 1965b. Das zentrifugale Androeceum der Paeoniaceae. *Berichte der Deutschen Botanischen Gesellschaft* 77:427–435.

———. 1965a. Vergleichend-morphologische und entwicklungsgeschichtliche Untersuchungen über das Perianth bei den Polycarpicae. *Botanische Jahrbücher für Systematik* 84:359–508.

Hill, C. R., and P. R. Crane. 1982. Evolutionary cladistics and the origin of angiosperms. *In* K. A. Joysey and A. E. Friday [eds.], Problems of Phylogenetic Reconstruction. Academic Press, New York.

Hillis, D. M. 1995. Approaches for assessing phylogenetic accuracy. *Systematic Biology* 44:3–16.

———. 1996. Inferring complex phylogenies. *Nature* 383:130.

Hillis, D. M., J. Huelsenbeck, and D. Swofford. 1994. Hobgoblin of phylogenetics? *Nature* 369:363–364.

Hillis, D. M., D. D. Pollock, J. A. McGuire, and D. J. Zwickl. 2003. Is sparse taxon sampling a problem for phylogenetic inference? *Systematic Biology* 52:124–126.

Hilu, K. W., T. Borsch, K. Muller, D. E. Soltis, P. S. Soltis, V. Savolainen, M. Chase, M. Powell, L. Alice, R. Evans, H. Sauquet, C. Neinhuis, T. Slotta, J. Rohwer, and L. Chatrou. 2003. Inference of angiosperm phylogeny based on *matK* sequence information. *American Journal of Botany* 90:1758–1776.

Hirmer, M. 1918. Beiträge zur Morphologie der polyandrischen Blüten. *Flora* 110:140–192.

Hitchcock, C. L., A. Cronquist, M. Ownbey, and J. W. Thompson. 1961. Vascular Plants of the Pacific Northwest. University of Washington Press, Seattle, WA.

Hodges, S. A. 1997a. Floral nectar spurs and diversification. *International Journal of Plant Sciences* 158:S81–S88.

———. 1997b. Rapid radiation due to a key innovation in columbines (Ranunculaceae: *Aquilegia*). *In* T. J. Givnish and K. J. Sytsma [eds.], Molecular Evolution and Adaptive Radiation, 391–405. Cambridge University Press, Cambridge, UK.

Hodges, S. A., J. B. Whittall, M. Fulton, and J. Y. Yang. 2002. Genetics of floral traits influencing reproductive isolation between *Aquilegia formosa* and *Aquilegia pubescens*. *American Naturalist* 159:S51–S60.

Hofmann, U. 1977. Die Stellung von *Stegnosperma* innerhalb der Centrospermen. *Berichte der Deutschen Botanischen Gesellschaft* 90:39–52.

Holsinger, K. E. 1996. Pollination biology and the evolution of mating systems in flowering plants. *Evolutionary Biology* 29:107–149.

Honma, T., and K. Goto. 2001. Complexes of MADS-box proteins are sufficient to convert leaves into floral organs. *Nature* 409:525–529.

Hooker, J. D. 1862–67. Ampelideae. *In* G. Bentham and J. D. Hooker [eds.], Genera Plantarum I, 386–388. Reeve & Co., London.

Hoot, S. B. 1991. Phylogeny of the Ranunculaceae based on epidermal microcharacters and micromorphology. *Systematic Botany* 16:741–755.

———. 1995. Phylogeny of the Ranunculaceae based on *atpB*, *rbcL*, and 18S nuclear ribosomal DNA sequence data. *Plant Systematics and Evolution, Supplement* 9:241–251.

Hoot, S. B., and P. R. Crane. 1995. Interfamilial relationships in the Ranunculidae based on molecular systematics. *Plant Systematics and Evolution, Supplement* 9:119–131.

Hoot, S. B., and A. W. Douglas. 1998. Phylogeny of the Proteaceae based on *atpB* and *atpB-rbcL* intergenic spacer region sequences. *Australian Systematic Botany* 11:301–320.

Hoot, S. B., A. Culham, and P. R. Crane. 1995a. The utility of *atpB* gene sequences in resolving phylogenetic relationships: Comparison with *rbcL* and 18S ribosomal DNA sequences in the Lardizabalaceae. *Annals of the Missouri Botanical Garden* 82:194–207.

———. 1995b. Phylogenetic relationships of the Lardizabalaceae and Sargentodoxaceae: Chloroplast and nuclear DNA sequence evidence. *Plant Systematics and Evolution, Supplement* 9:195–199.

Hoot, S. B., J. W. Kadereit, F. R. Blattner, K. B. Jork, A. E. Schwarzbach, and P. R. Crane. 1997. The phylogeny of the Papaveraceae s. l. based on four data sets: *atpB* and *rbcL* sequences, *trnK* restriction sites and morphological characters. *Systematic Botany* 22:575–590.

Hoot, S. B., S. Magallón, and P. R. Crane. 1999. Phylogeny of basal eudicots based on three molecular data sets: *atpB*, *rbcL*, and 18S nuclear ribosomal DNA sequences. *Annals of the Missouri Botanical Garden* 86:1–32.

Hoot, S. B., A. A. Reznicek, and J. D. Palmer. 1994. Phylogenetic relationships in *Anemone* (Ranunculaceae) based on morphology and chloroplast DNA. *Systematic Botany* 19:169–200.

Horak, K. E. 1981. Anomalous secondary thickening in *Stegnosperma* (Phytolaccaceae). *Bulletin of the Torrey Botanical Club* 108:189–197.

Hu, J.-M., M. Lavin, M. F. Wojciechowski, and M. J. Sanderson. 2000. Phylogenetic systematics of the tribe Millettieae (Leguminosae) based on chloroplast *trnK/matK* sequences and its implications for evolutionary patterns in Papilionoideae. *American Journal of Botany* 87:418–430.

———. 2002. Phylogenetic analysis of nuclear ribosomal ITS/5.8S sequences in the tribe Millettieae (Fabaceae): *Poecilanthe-Cyclolobium*, the core Millettieae, and the *Callerya* group. *Systematic Botany* 27:722–733.

Huber, H. 1969. Die Samenmerkmale und Verwandtschaftsverhältnisse der Liliifloren. *Mitteilungen der Botanischen Staatssammlung München* 8:219–538.

———. 1977. The treatment of monocotyledons in an evolutionary system of classification. *Plant Systematics and Evolution, Supplement* 1:285–298.

———. 1993. *Neurada*, eine Gattung der Malvales. *Sendtnera* 1:7–10.

Huber, K. A. 1980. Morphologische und entwicklungsgeschichtliche Untersuchungen an Blüten und Blütenständen von Solanaceen und von *Nolana paradoxa* Lindl. (Nolanaceae). *Dissertationes Botanicae* 55. Cramer, Vaduz.

Huelsenbeck, J. P., B. Larget, R. E. Miller, and F. Ronquist. 2002. Potential applications and pitfalls of Bayesian inference of phylogeny. *Systematic Biology* 51:673–688.

Huelsenbeck, J. P., F. Ronquist, R. Nielsen, and J. P. Bollback. 2001. Bayesian inference of phylogeny and its impact on evolutionary biology. *Science* 294:2310–2314.

Huether, C. A. Jr. 1968. Exposure of natural genetic variability underlying the pentamerous corolla constancy in *Linanthus androsaceus* ssp. *androsaceus*. *Genetics* 60:123–146.

———. 1969. Constancy of the pentamerous corolla phenotype in natural populations of *Linanthus*. *Evolution* 23:572–588.

Hufford, L. 1992. Rosidae and their relationships to other non-magnoliid dicotyledons: A phylogenetic analysis using morphological and chemical data. *Annals of the Missouri Botanical Garden* 79:218–248.

———. 1995. Patterns of ontogenetic evolution in perianth diversification of *Besseya* (Scrophulariaceae). *American Journal of Botany* 82:655–680.

———. 1996a. The origin and early evolution of angiosperm stamens. *In* W. G. D'Arcy and R. C. Keating [eds.], The Anther: Form, Function, and Phylogeny, 58–91. Cambridge University Press, Cambridge.

———. 1996b. Ontogenetic evolution, clade diversification, and homoplasy. *In* M. J. Sanderson and L. Hufford [eds.], Homoplasy: The Recurrence of Similarity in Evolution, 271–301. Academic Press, San Diego.

———. 2001. Ontogeny and morphology of the fertile flowers of Hydrangeaceae and allied genera of tribe Hydrangeae (Hydrangeaceae). *Botanical Journal of the Linnean Society* 137:139–187.

Hufford, L., and P. R. Crane. 1989. A preliminary phylogenetic analysis of the "lower" Hamamelidae. *In* P. R. Crane and S. Blackmore [eds.], Evolution, Systematics and Fossil History of the Hamamelidae, 1, 175–192. Clarendon Press, Oxford.

Hufford, L., and W. C. Dickison. 1992. A phylogenetic analysis of Cunoniaceae. *Systematic Botany* 17:181–200.

Hufford, L., and P. K. Endress. 1989. The diversity of anther structures and dehiscence patterns among Hamamelididae. *Botanical Journal of the Linnean Society* 99:301–346.

Hufford, L., M. L. Moody, and D. E. Soltis. 2001. A phylogenetic analysis of Hydrangeaceae based on sequences of the plastid gene *matK* and their combination with *rbcL* and morphological data. *International Journal of Plant Sciences* 162:835–846.

Hughes, N. F. 1994. The Enigma of Angiosperm Origins. Cambridge University Press, Cambridge.

Hürlimann, H., and H. U. Stauffer. 1957. *Daenikera*, eine neue Santalaceen-Gattung. *Vierteljahrsschrift der Naturforschenden Gesellschaft Zürich* 102:332–336.

Hutchinson, J. 1934. The Families of Flowering Plants. Oxford University Press, Oxford.

———. 1959. The Families of Flowering Plants, 2nd ed. Oxford University Press, Oxford.

———. 1967. The Genera of Flowering Plants. Clarendon Press, Oxford.

———. 1973. The Families of Flowering Plants, 3rd ed. Clarendon Press, Oxford.

Igersheim, A., and P. K. Endress. 1997. Gynoecium diversity and systematics of the Magnoliales and winteroids. *Botanical Journal of the Linnean Society* 124:213–271.

Igersheim, A., M. Buzgo, and P. K. Endress. 2001. Gynoecium diversity and systematics in basal monocots. *Botanical Journal of the Linnean Society* 136:1–65.

Igersheim, A., C. Puff, P. Leins, and C. Erbar. 1994. Gynoecial development of *Gaertnera* Lam. and of presumably allied taxa of the Psychotrieae (Rubiaceae): secondarily "superior" vs. inferior ovaries. *Botanische Jahrbücher für Systematik* 116:401–414.

Ingrouille, M. J., M. W. Chase, M. F. Fay, D. Bowman, M. Van Der Bank, and A. de Bruijn. 2002. Systematics of Vitaceae from the viewpoint of plastid *rbcL* DNA sequence data. *Botanical Journal of the Linnean Society* 138:421–432.

Inouye, H., S. Ueda, M. Hirabayashi, and N. Shimoka-Wa. 1966. Studies on the monoterpene glucosides of *Daphniphyllum macropodum*. *Yakugaku Zasshi* 86:943–947.

Irish, V. F., and E. M. Kramer. 1998. Genetic and molecular analysis of angiosperm flower development. *Advances in Botanical Research* 28:199–230.

Jack, T. 2001. Plant development going MADS. *Plant Molecular Biology* 46:515–520.

Jäger-Zürn, I. 1966. Infloreszenz- und blütenmorphologische, sowie embryologische Untersuchungen an *Myrothamnus* Welw. *Beiträge zur Biologie der Pflanzen* 42:241–271.

———. 1997. Embryological and floral studies in *Weddellina squamulosa* Tul. (Podostemaceae, Tristichoideae). *Aquatic Botany* 57:151–182.

Jaing, Z., and R. Zhou. 1992. Distribution of the iridoid compounds in the Hamamelidae. *Zhongguo Yaoke Daxue Xuebao* 23:140–143.

Jansen, R. K., and J. D. Palmer. 1988. Phylogenetic implications of chloroplast DNA restriction site variation in the Mutisieae (Asteraceae). *American Journal of Botany* 75:751–764.

Jansen, R. K., and K.-J. Kim. 1996. Implications of chloroplast DNA data for the classification and phylogeny of the Asteraceae. *In* D. Hind and H. Beentje [eds.], Compositae: Systematics, 317–339. Proceedings of the International Compositae Conference, Royal Botanic Gardens, Kew.

Jansen, R. K., and J. D. Palmer. 1987. A chloroplast DNA inversion marks an ancient evolutionary split in the sunflower family, Asteraceae. *Proceedings of the National Academy of Sciences USA* 84:5818–5822.

Jansen, R. K., H. J. Michaels, and J. D. Palmer. 1991. Phylogeny and character evolution in the Asteraceae based on chloroplast DNA restriction site mapping. *Systematic Botany* 16:98–115.

Jansen, R. K., H. J. Michaels, R. S. Wallace, K. J. Kim, S. C. Keeley, L. E. Watson, and J. D. Palmer. 1992. Chloroplast DNA Variation in the Asteraceae: Phylogenetic and Evolutionary Implications. *In* P. S. Soltis, D. E. Soltis, and J. J. Doyle [eds.], *Molecular Systematics of Plants* 252–279. Chapman and Hall, New York.

Jensen, S. R. 1991. Plant iridoids, their biosynthesis and distribution in angiosperms. *In* J. B. Harborne and F. A. Tomas-Barberan [eds.], Ecological Chemistry and Biochemistry of Plant Terpenoids. Clarendon Press, Oxford.

———. 1992. Systematic implications of the distribution of iridoids and other chemical compounds in the Loganiaceae and other families of the Asteridae. *Annals of the Missouri Botanical Garden* 79:284–302.

Jensen, S. R., S. E. Lyse-Peterson, and B. J. Nielsen. 1979. Novel bis-iridoid glucosides from *Dipsacus sylvestris*. *Phytochemistry* 18:273–277.

Jensen, S. R., B. J. Nielsen, and R. Dahlgren. 1975. Iridoid compounds, their occurrence and systematic importance in the angiosperms. *Botaniska Notiser* 128:148–180.

Jensen, U. 1995. Secondary compounds of the Ranunculiflorae. *Plant Systematics and Evolution, Supplement* 9:85–97.

Jensen, U., S. B. Hoot, J. T. Johansson, and K. Kosuge. 1995. Systematics and phylogeny of the Ranunculaceae—a revised family concept on the basis of molecular data. *Plant Systematics and Evolution, Supplement* 9:273–280.

Jiang, Z., and R. Zhou. 1992. Distribution of the iridoid compounds in the Hamamelidae. *Zhongguo Yaoke Daxue Xuebao* 23:140–143.

Jobson, R. W., and V. A. Albert. 2002. Molecular rates parallel diversification contrasts between carnivorous plant sister lineages. *Cladistics* 18:127–136.

Jobson, R. W., J. Playford, K. M. Cameron, and V. A. Albert. 2003. Molecular phylogenetics of Lentibulariaceae inferred from plastid *rps*16 intron and *trnL-F* DNA sequences: implications for character evolution and biogeography. *Systematic Botany* 28:157–171.

Johansen, B., L. B. Pedersen, M. Skipper, and S. Frederiksen. 2002. MADS-box gene evolution-structure and transcription patterns. *Molecular Phylogenetics and Evolution* 23:458–480.

Johansson, J. T. 1995. A revised chloroplast DNA phylogeny of the Ranunculaceae. *Plant Systematics and Evolution, Supplement* 9:253–261.

Johansson, J. T., and R. K. Jansen. 1993. Chloroplast DNA variation and phylogeny of the Ranunculiflorae. *Plant Systematics and Evolution* 187:29–49.

Johnson, C. A. S., and B. G. Briggs. 1984. Myrtales and Myrtaceae—a phylogenetic analysis. *Annals of the Missouri Botanical Garden* 71:700–756.

Johnson, L. A., and D. E. Soltis. 1995. Phylogenetic inference using *matK* sequences. *Annals of the Missouri Botanical Garden* 82:149–175.

Johnson, L. A., J. L. Schultz, D. E. Soltis, and P. S. Soltis. 1996. Monophyly and generic relationships of Polemoniaceae based on *matK* sequences. *American Journal of Botany* 83:1207–1224.

Johnson, L. A., D. E. Soltis, and P. S. Soltis. 1999. Phylogenetic relationships of Polemoniaceae inferred from 18S ribosomal DNA sequences. *Plant Systematics and Evolution* 214:65–89.

Johnson, S., and K. Steiner. 2000. Generalization versus specialization in plant pollination systems. *Trends in Ecology and Evolution* 15:140–143.

Johnson, S. D., H. P. Linder, and K. E. Steiner. 1998. Phylogeny and radiation of pollination systems in *Disa* (Orchidaceae). *American Journal of Botany* 85:402–411.

Johri, B. M., K. B. Ambegaokar, and S. Srivastava. 1992. Comparative Embryology of Angiosperms, 2 vols. Springer, Berlin.

Jones, E., T. Hodkinson, J. Parnell, and M. W. Chase. 2005. The Juncaceae-Cyperaceae interface: a combined plastid gene analysis. *In* J. T. Columbus, E. A. Friar, C. W. Hamilton, J. M. Porter, L. M. Prince, and M. G. Simpson [eds.]. Monocots: Comparative Biology and Evolution, 2 vols. Rancho Santa Ana Botanic Garden, Claremont, CA, in press.

Jones, J. H. 1986. Evolution of the Fagaceae: the implications of foliar features. *Annals of the Missouri Botanical Garden* 73:228–275.

Judd, W. S. 1998. The Smilacaceae in the southeastern United States. *Harvard Papers in Botany* 3:147–169.

Judd, W. S., and S. R. Manchester. 1998. Circumscription of Malvaceae (Malvales) as determined by preliminary cladistic analysis employing morphological, palynological, and chemical characters. *Brittonia* 49:384–405.

Judd, W. S., and R. G. Olmstead. 2004. A survey of tricolpate (eudicot) phylogeny. *American Journal of Botany* 91: 1627–1644.

Judd, W. S., and K. A. Kron. 1993. Circumscription of Ericaceae (Ericales) as determined by preliminary cladistic analyses based on morphological, anatomical and embryological features. *Brittonia* 45:99–114.

Judd, W. S., and S. R. Manchester. 1997. Circumscription of Malvaceae (Malvales) as determined by a preliminary cladistic analysis of morphological, anatomical, palynological, and chemical characters. *Brittonia* 49:384–405.

Judd, W. S., C. S. Campbell, E. A. Kellogg, and P. F. Stevens. 1999. Plant Systematics: A Phylogenetic Approach. Sinauer, Sunderland, MA.

Judd, W. S., C. S. Campbell, E. A. Kellogg, P. F. Stevens, and M. J. Donoghue. 2002. Plant Systematics: a Phylogenetic Approach. Sinauer, Sunderland, MA.

Judd, W. S., R. W. Sanders, and M. J. Donoghue. 1994. Angiosperm family pairs: preliminary cladistic analyses. *Harvard Papers in Botany* 5:1–51.

Juniper, B. E., R. J. Robins, and D. M. Joel. 1989. The Carnivorous Plants. Academic Press, London.

Kadereit, J. W. 1993. Papaveraceae. *In* K. Kubitzki, J. Rohwer, and V. Bittrich [eds.], The Families and Genera of Vascular Plants, II, 494–506. Springer, Berlin.

Kadereit, J. W., and K. J. Sytsma. 1992. Disassembling *Papaver*: A restriction site analysis of chloroplast DNA. *Nordic Journal of Botany* 12:205–217.

Kadereit, J. W., F. R. Blattner, K. B. Jork, and A. E. Schwarzbach. 1994. Phylogenetic analysis of the Papaveraceae s. 1. (incl. Fumariaceae, Hypecoaceae and *Pteridophyllum*) based on morphological characters. *Botanische Jahrbücher für Systematik* 116:361–390.

———. 1995. The phylogeny of the Papaveraceae sensu lato: Morphological, geographical and ecological implications. *Plant Systematics and Evolution, Supplement* 9:133–145.

Kajita, T., H. Ohashi, Y. Tateishi, C. D. Bailey, and J. J. Doyle. 2001. *RbcL* and legume phylogeny, with particular reference to Phaseoleae, Millettieae, and allies. *Systematic Botany* 26:515–536.

Källersjö, M., V. A. Albert, and J. S. Farris. 1999. Homoplasy increases phylogenetic structure. *Cladistics* 15:91–93.

Källersjö, M., G. Bergqvist, and A. A. Anderberg. 2000. Generic realignment in primuloid families of the Ericales s. 1. : a phylogenetic analysis based on DNA sequences from three chloroplast genes and morphology. *American Journal of Botany* 87:1325–1341.

Källersjö, M., J. S. Farris, M. Chase, B. Bremer, M. F. Fay, C. J. Humphries, G. Petersen, O. Seberg, and K. Bremer. 1998. Simultaneous parsimony jackknife analysis of 2538 *rbcL* DNA sequences reveals support for major clades of green plants, land plants, seed plants, and flowering plants. *Plant Systematics and Evolution* 213:259–287.

Källersjö, M., J. S. Farris, A. G. Kluge, and C. Bult. 1992. Skewness and permutation. *Cladistics* 8:275–287.

Kamelina, O. P. 1984. On the embryology of the genus *Escallonia* (Escalloniaceae). *Botanicheskii Zhurnal* 69:1304–1316.

Kanno, A., H. Saeki, T. Kameya, H. Saedler, and G. Theissen. 2003. Heterotopic expression of class B floral homeotic genes supports a modified ABC model for tulip (*Tulipa gesneriana*). *Plant Molecular Biology* 52:831–841.

Kapil, R. N., and P. R. Mohana Rao. 1966. Studies on the Garryaceae. II. Embryology and systematic position of *Garrya* Douglas ex Lindley. *Phytomorphology* 16:564–578.

Kaplan, D. R. 1967. Floral morphology, organogenesis and interpretation of the inferior ovary in *Downingia bacigalupii*. *American Journal of Botany* 54:1274–1290.

Kaplan, M. A. C., and O. R. Gottlieb. 1982. Iridoids as systematic markers in dicotyledons. *Biochemical Systematics and Ecology* 10:239–347.

Kaplan, M. A. C., J. Ribeiro, and O. R. Gottlieb. 1991. Chemogeographical evolution of terpenoids in Icacinaceae. *Phytochemistry* 30:2671–2676.

Kårehed, J. 2001. Multiple origins of the tropical forest tree family Icacinaceae. *American Journal of Botany* 88:2259–2274.

———. 2002. Evolutionary Studies in Asterids Emphasizing Euasterids. PhD dissertation, Uppsala University.

Kårehed, J., J. Lundberg, B. Bremer, and K. Bremer. 1999. Evolution of the Australasian families Alseuosmiaceae, Argophyllaceae and Phellinaceae. *Systematic Botany* 24:660–682.

Karol, K. G., Y. Suh, G. E. Schatz, and E. Zimmer. 2000. Molecular evidence for the phylogenetic position of *Takhtajania* in the Winteraceae: Inference from nuclear ribosomal and chloroplast gene spacer sequences. *Annals of the Missouri Botanical Garden* 87:414–432.

Karoly, K., and J. K. Conner. 2000. Heritable variation in a family-diagnostic trait. *Evolution* 54:1433–1438.

Kato, M. 1990. Ophioglossaceae: a hypothetical archetype for the angiosperm carpel. *Botanical Journal of the Linnean Society* 102:303–311.

Keating, R. 1973. Pollen morphology and relationships of Flacourtiaceae. *Annals of the Missouri Botanical Garden* 60:273–305.

———. 2000. Anatomy of the young vegetative shoot of *Takhtajania perrieri* (Winteraceae). *Annals of the Missouri Botanical Garden* 87:335–346.

Keefe, J. M., and J. M. F. Moseley. 1978. Wood anatomy and phylogeny of *Paeonia* section *Moutan*. *Journal of the Arnold Arboretum* 59:274–297.

Keller, J. A., P. S. Herendeen, and P. R. Crane. 1996. Fossil flowers of the Actinidiaceae from the Campanian (Late Cretaceous) of Georgia. *American Journal of Botany* 83:528–541.

Kellogg, E. A. 1998. Relationships of cereal crops and other grasses. *Proceedings of the National Academy of Sciences USA* 95:2005–2010.

———. 1999. Phylogenetic aspects of the evolution of C_4 photosynthesis. *In* R. F. Sage and R. K. Monson [eds.], C_4 Plant Biology. Academic Press, New York.

———. 2000. The grasses: a case study in macroevolution. *Annual Review of Ecology and Systematics* 31:217–238.

———. 2001. Evolutionary history of the grasses. *Plant Physiology* 125:1198–1205.

———. 2002. Are macroevolution and microevolution qualitatively different? *In* Q. C. B. Cronk, R. M. Bateman, and J. A. Hawkins [eds.], Developmental Genetics and Plant Evolution, 70–84. Taylor and Francis, London.

———. 2003. It's all relative. *Nature* 422:383–384.

Kellogg, E. A., and J. L. Bennetzen. 2004. The evolution of nuclear genome structure in plants. *American Journal of Botany* 91:1709–1725.

Kellogg, E. A., and N. D. Juliano. 1997. The structure and function of RUBISCO and their implications for systematic studies. *American Journal of Botany* 84:413–428.

Kellogg, E. A., and H. P. Linder. 1995. Phylogeny of Poales. *In* P. J. Rudall, P. J. Cribb, D. F. Cutler, and C. J. Humphries [eds.], Monocotyledons: Systematics and Evolution, 750. Royal Botanic Gardens, Kew.

Kenrick, P., and P. R. Crane. 1997. The Origin and Early Diversification of Land Plants. Smithsonian Institution Press, Washington.

Kidwell, M. G. 2002. Transposable elements and the evolution of genome size in eukaryotes. *Genetica* 115:49–63.

Kim, K.-J., and R. K. Jansen. 1995. *ndhF* sequence evolution and the major clades in the sunflower family. *Proceedings of the National Academy of Sciences USA* 92:10379–10383.

———. 1998. Chloroplast DNA restriction site variation and phylogeny of the Berberidaceae. *American Journal of Botany* 85:1766–1778.

Kim, K.-J., R. K. Jansen, and R. G. Olmstead. 1994. Multiple origins of sympetaly in dicots. *American Journal of Botany* 81(6, Supplement):165.

Kim, K.-J., R. K. Jansen, R. S. Wallace, H. J. Michaels, and J. D. Palmer. 1992. Phylogenetic implications of *rbcL* sequence variation in the Asteraceae. *Annals of the Missouri Botanical Garden* 79:428–445.

Kim, S., M.-J. Yoo, V. A. Albert, J. S. Farris, P. S. Soltis, and D. E. Soltis. 2004a. Phylogeny and diversification of B-function MADS-box genes in angiosperms: evolutionary and functional implications of a 260-million-year-old duplication. *American Journal of Botany* 91: 2102–2118

Kim, S., D. E. Soltis, P. S. Soltis, M. J. Zanis, and Y. Suh. 2004b. Phylogenetic relationships among early-diverging eudicots based on four genes: were the eudicots ancestrally woody? *Molecular Phylogenetics and Evolution* 31:16–30.

Kim, S., C.-W. Park, Y.-D. Kim, and Y. Suh. 2001. Phylogenetic relationships in family Magnoliaceae inferred from *ndhF* sequences. *American Journal of Botany* 88:717–728.

Kim, Y.-D., and R. K. Jansen. 1995. Phylogenetic implications of chloroplast DNA variation in the Berberidaceae. *Plant Systematics and Evolution, Supplement* 9:341–349.

Kirik, A., S. Salomon, and H. Puchta. 2000. Species-specific double-strand break repair and genome evolution in plants. *EMBO Journal* 19:5562–5566.

Kishino, H., and M. Hasegawa. 1989. Evaluation of the maximum likelihood estimate of the evolutionary tree topologies from DNA sequence data, and the branching order in Hominoidea. *Journal of Molecular Evolution* 29:170–179.

Kishino, H., J. L. Thorne, and W. J. Bruno. 2001. Performance of a divergence time estimation method under a probabilistic model of rate evolution. *Molecular Biology and Evolution* 18:352–361.

Kjaer, A. 1973. The natural distribution of glucosinolates: a uniform group of sulfur containing glucosides. *In* G. Bendz and J. Santesson [ed.], Chemistry in Botanical Classification, 229–234. Academic Press, New York.

Klavins, S. D., T. N. Taylor, and E. L. Taylor. 2002. Anatomy of *Umkomasia* (Corystospermales) from the Triassic of Antarctica. *American Journal of Botany* 89:664–676.

Klopfer, K. 1973. Florale Morphogenese und Taxonomie der Saxifragaceae sensu lato. *Feddes Repertorium* 84:475–516.

Koch, M. 2003. Molecular phylogenetics, evolution and population biology in the Brassicaceae. *In* V. K. Sharma and A. Sharma [eds.], Plant Genome: Biodiversity and Evolution. Science Publishers, Inc., Enfield, NH.

Koch, M., J. Bishop, and T. Mitchell-Olds. 1999. Molecular systematics and evolution of *Arabidopsis* and *Arabis*. *Plant Biology* 1:529–537.

Koch, M., B. Haubold, and T. Mitchell-Olds. 2001. Molecular systematics of the Brassicaceae: evidence from coding plastid *matK* and nuclear *Chs* sequences. *American Journal of Botany* 88:534–544.

Koch, M., K. Mummenhoff, and I. A. Al-Shehbaz. 2003. Molecular systematics, evolution, and population biology in the mustard family: a review of a decade of studies. *Annals of the Missouri Botanical Garden* 90:151–171.

Kolpalova, M. V., and D. M. Popov. 1994. Study of the amounts of iridoids in *Paeonia anomala* L. (Paeoniaceae). *Khimiko-Farmatsevticheskii Zhurnal* 28:24–26.

Kong, H.-Z., and Z. Chen. 2000. Phylogeny of *Chloranthus* (Chloranthaceae) inferred from sequence analysis of nrDNA ITS region. *Acta Botanica Sinica* 42:762–764.

Kong, H.-Z., Z. Chen, and A.-M. Lu. 2002a. Phylogeny of *Chloranthus* (Chloranthaceae) based on nuclear ribosomal ITS and plastid *trnL-F* sequence data. *American Journal of Botany* 89:940–946.

Kong, H.-Z., A.-M. Lu, and P. K. Endress. 2002b. Floral organogenesis of *Chloranthus sessilifolius*, with special emphasis on the morphological nature of the androecium of *Chloranthus* (Chloranthaceae). *Plant Systematics and Evolution* 232:181–188.

Koontz, J. A., and D. E. Soltis. 1999. DNA sequence data reveal polyphyly of Brexioideae (Brexiaceae; Saxifragaceae sensu lato). *Plant Systematics and Evolution* 219:199–208.

Kopriva, S., C.-C. Chu, and H. Bauwe. 1996. Molecular phylogeny of *Flaveria* as deduced from the analysis of nucleotide sequences encoding the H-protein of the glycine cleavage system. *Plant Cell and Environment* 19:1028–1036.

Kores, P. J., P. H. Weston, M. Molvray, and M. W. Chase. 2000. Phylogenetic relationships within the Diurideae (Orchidaceae): inferences from plastid *matK* DNA sequences. *In* K. L. Wilson and D. A. Morrison [eds.], Monocots: Systematics and Evolution, 449–456. CSIRO, Melbourne.

Korpelainen, H. 1998. Labile sex expression in plants. *Biological Review* 73:157–180.

Kostecka-Madalska, O., and A. Rymkiewick. 1971. Further research for plants containing aucubin. *Farmacia Polonica* 27:899–903.

Kosuge, K. 1994. Petal evolution in Ranunculaceae. *Plant Systematics and Evolution, Supplement* 8:185–191.

Kosuge, K., and M. Tamura. 1989. Ontogenetic studies on petals of the Ranunculaceae. *Journal of Japanese Botany* 64:65–67.

Kosuge, K., K. Mistunaga, K. Loike, and T. Ohmoto. 1994. Studies on the constituents of *Ailanthus integrifolia*. *Chemical and Pharmaceutical Bulletin* 42:1669–1671.

Kosuge, K., F.-D. Pu, and M. Tamura. 1989. Floral morphology and relationships of *Kingdonia*. *Acta Phytotaxonomica Geobotanica* 40:61–67.

Kotilainen, M., P. Elomaa, A. Uimari, V. Albert, D. Yu, and T. H. Teeri. 2000. *GRCD1*, an *AGL2*-like MADS box gene, participates in the C function during stamen development in *Gerbera hybrida*. *Plant Cell* 12:1893–1902.

Kral, R. B. 1983. The Xyridaceae in the southeastern United States. *Journal of the Arnold Arboretum* 64:421–429.

Kramer, E. M., and V. F. Irish. 1999. Evolution of genetic mechanisms controlling petal development. *Nature* 399:144–148.

———. 2000. Evolution of the petal and stamen developmental programs: evidence from comparative studies of the lower eudicots and basal angiosperms. *International Journal of Plant Sciences* 161:S29–S40.

Kramer, E. M., V. S. Di Stilio, and P. M. Schlüter. 2003. Complex patterns of gene duplication in the *APETALA3* and *PISTILLATA* lineages of the Ranunculaceae. *International Journal of Plant Sciences* 164:1–11.

Kramer, E. M., R. L. Dorit, and V. F. Irish. 1999. Molecular evolution of genes controlling petal and stamen development: Duplication and divergence within the *APETALA3* and *PISTILLATA* MADS-box gene lineages. *Genetics* 149:765–783.

Kramer, E. M., M. A. Jaramillo, and V. Di Stilio. 2004. Patterns of gene duplication and functional evolution during the diversification of the *AGAMOUS* subfamily of MADS box genes in angiosperms. *Genetics* 166:1011–1023.

Kraus R., P. Trimborn, and H. Ziegler. 1995. *Tristerix aphyllus*, a holoparasitic Loranthaceae. *Naturwissenschaften* 82:150–151l.

Kress, W. J. 1990. The phylogeny and classification of the Zingiberales. *Annals of the Missouri Botanical Garden* 77:698–721.

Kress, W. J., L. M. Prince, W. J. Hahn, and E. A. Zimmer. 2001. Unraveling the evolutionary radiation of the families of the Zingiberales using morphological and molecular evidence. *Systematic Biology* 50:926–944.

Kron, K. A. 1996. Phylogenetic relationships of Empetraceae, Epacridaceae, Ericaceae, Monotropaceae, and Pyrolaceae: Evidence from nucleotide ribosomal 18S sequence data. *Annals of Botany* 77:293–303.

———. 1997. Exploring alternative systems of classification. *Aliso* 15:105–112.

Kron, K. A., and M. W. Chase. 1993. Systematics of the Ericaceae, Empetraceae, Epacridaceae and related taxa based on *rbcL* sequence data. *Annals of the Missouri Botanical Garden* 80:735–741.

Kron, K. A., and W. S. Judd. 1997. Systematics of the *Lyonia* group (Andromedeae, Ericacae) and the use of species as terminals in higher-level cladistic analyses. *Systematic Botany* 22:479–492.

Kron, K. A., W. S. Judd, and D. M. Crayn. 1999. Phylogenetic relationships of Andromedeae (Ericaceae subfam. Vaccinioideae). *American Journal of Botany* 86:1290–1300.

Kron, K. A., W. S. Judd, P. F. Stevens, D. M. Crayn, A. A. Anderberg, P. A. Gadek, C. J. Quinn, and J. L. Luteyn. 2002. Phylogenetic classification of Ericaceae: molecular and morphological evidence. *Botanical Review* 68:335–423.

Kubitzki, K. 1987. Origin and significance of trimerous flowers. *Taxon* 36:21–28.

———. 1993. Myrothamnaceae. *In* K. Kubitzki, O. Rohwer, and V. Bittrich [eds.], The Families and Genera of Vascular Plants, II, 468–469. Springer, Berlin.

Kubitzki, K., J. Rohwer, and V. Bittrich [eds.]. 1993. The Families and Genera of Vascular Plants, II. Springer, Berlin.

Kubitzki, K., P. Von Sengbusch, and H.-H. Poppendiek. 1991. Parallelism, its evolutionary origin and systematic significance. *Aliso* 13:191–206.

Kuijt, J. 1969. The Biology of Parasitic Flowering Plants. University of California Press, Berkeley, California.

———. 1982. The Viscaceae in the southeastern United States. *Journal of the Arnold Arboretum* 63:401–410.

Kuzoff, R. K., and C. S. Gasser. 2000. Recent progress in reconstructing angiosperm phylogeny. *Trends in Plant Science* 5:330–336.

Kuzoff, R. K., L. Hufford, and D. E. Soltis. 2001. Structural homology and developmental transformations associated with ovary diversification in *Lithophragma* (Saxifragaceae). *American Journal of Botany* 88:196–205.

Kuzoff, R. K., D. E. Soltis, L. Hufford, and P. S. Soltis. 1999. Phylogenetic relationships within *Lithophragma* (Saxifragaceae): hybridization, allopolyploidy, and ovary diversification. *Systematic Botany* 24:598–615.

Kuzoff, R. K., J. Sweere, D. Soltis, P. Soltis, and E. Zimmer. 1998. The phylogenetic potential of entire 26S rDNA sequences in plants. *Molecular Biology and Evolution* 15:251–263.

Lacroix, C., and R. Sattler. 1988. Phyllotaxis theories and tepal-stamen superposition in *Basella rubra*. *American Journal of Botany* 75:906–917.

Lagercrantz, U. 1998. Comparative mapping between *Arabidopsis thaliana* and *Brassica nigra* indicates that *Brassica* genomes have evolved through extensive genome replication accompanied by chromosome fusions and frequent rearrangements. *Genetics* 150:1217–1228.

Lagercrantz, U., and D. J. Lydiate. 1996. Comparative genome mapping in *Brassica*. *Genetics* 144:1903–1910.

Lammel, G., and H. Rimpler. 1981. Iridoids in *Clerodendrum thomsonae*, Verbenaceae. *Zeitschrift für Naturforschung C: Bioscience* 36:708–713.

Lammers, T. G. 1992. Circumscription and phylogeny of the Campanulales. *Annals of the Missouri Botanical Garden* 81:388–413.

Lammers, T. T., T. F. Stuessy, and M. Silva. 1986. Systematic relationships of the Lactoridaceae, an endemic family of the Juan Fernandez Islands, Chile. *American Journal of Botany* 152:243–266.

Landsmann, J., E. S. Dennis, T. J. V. Higgins, C. A. Appleby, A. A. Kortt, and W. J. Peacock. 1986. Common evolutionary origin of legume and non-legume plant haemoglobins. *Nature* 324:166–168.

Langley, C. H., and W. Fitch. 1974. An estimation of the constancy of the rate of molecular evolution. *Journal of Molecular Evolution* 3:161–177.

Laroche, J., P. Li, L. Maggia, and J. Bousquet. 1997. Molecular evolution of angiosperm mitochondrial introns and exons. *Proceedings of the National Academy of Sciences USA* 94:5722–5727.

Lee, I., D. S. Wolfe, S. Nilsson, and D. Weigel. 1997. A *LEAFY* co-regulator encoded by *UNUSUAL FLORAL ORGANS*. *Current Biology* 7:95–104.

Lee, M. S. Y. 1996. The phylogenetic approach to biological taxonomy: Practical aspects. *Zoologica Scripta* 25:187–190.

Lee, S. L., and J. Wen. 2001. A phylogenetic analysis of *Prunus* and the Amygdaloideae (Rosaceae) using ITS sequences of nuclear ribosomal DNA. *American Journal of Botany* 88:150–160.

Lehman, N. L., and R. Sattler. 1993. Homeosis in floral development of *Sanguinaria canadensis* and *S. canadensis* "multiplex" (Papaveraceae). *American Journal of Botany* 80:1323–1335.

Leinfellner, W. 1969. Über die Karpelle verschiedener Magnoliales. VIII. Überblick über alle Familien der Ordnung. *Österreichische Botanische Zeitschrift* 117:107–127.

Leins, P. 1972. Das Karpell im ober- und unterständigen Gynoeceum. *Berichte der Deutschen Botanischen Gesellschaft* 85:291–294.

———. 2000. Blüte und Frucht. Schweizerbart, Stuttgart.

Leins, P., and C. Erbar. 1985. Ein Beitrag zur Blütenentwicklung der Aristolochiaceen, einer Vermittlergruppe zu den Monokotylen. *Botanische Jahrbücher für Systematik* 107: 343–368.

———. 1990. On the mechanisms of secondary pollen presentation in the Campanulales-Asterales-complex. *Botanica Acta* 103:87–92.

———. 1991. Fascicled androecia in Dilleniidae and some remarks on the *Garcinia* androecium. *Botanica Acta* 104:336–344.

———. 1994. Flowers in Magnoliidae and the origin of flowers in other subclasses of the angiosperms. II. The relationships between the flowers of Magnoliidae, Dilleniidae, and Caryophyllidae. *Plant Systematics and Evolution, Supplement* 8:209–218.

Leins, P., and G. Metzenauer. 1979. Entwicklungsgeschichtliche Untersuchungen an *Capparis*-Blüten. *Botanische Jahrbücher für Systematik* 100:542–554.

Leitch, I. J., and M. D. Bennett. 1997. Polyploidy in angiosperms. *Trends in Plant Science* 2:470–476.

———. 2004. Genomic downsizing in polyploid plants. *Botanical Journal of the Linnean Society* 82:651–663.

Leitch, I. J., and L. Hanson. 2002. DNA C-values in seven families fill phylogenetic gaps in the basal angiosperms. *Botanical Journal of the Linnean Society* 140:175–179.

Leitch, I. J., M. W. Chase, and M. D. Bennett. 1998. Phylogenetic analysis of DNA C-values provides evidence for a small ancestral genome size in flowering plants. *Annals of Botany* 82:85–94.

Leitch, I. J., L. Hanson, M. Winfield, J. Parker, and M. D. Bennett. 2001. Nuclear DNA C-values complete familial representation in gymnosperms. *Annals of Botany* 88:843–849.

Leitch I. J., D. E. Soltis, P. S. Soltis and M. D. Bennett. 2005. Evolution of DNA amounts across land plants (Embryophyta). *Annals of Botany* 95:207–217.

Lemke, D. 1988. A synopsis of Flacourtiaceae. *Aliso* 12:29–43.

Leon-Kloosterziel, K. M., C. L. Keijzer, and M. Koorneef. 1994. A seed shape mutant of *Arabidopsis* that is affected in integument development. *The Plant Cell* 6:385–392.

Les, D. H. 1988. The origin and affinities of the Ceratophyllaceae. *Taxon* 37:326–435.

Les, D. 1993. Ceratophyllaceae. *In* K. Kubitzki, J. G. Rohwer, and V. Bittrich [eds.], The Families and Genera of Vascular Plants, II, 246–250. Springer, Berlin.

Les, D. H., and R. R. Haynes. 1995. Systematics of the subclass Alismatidae: a synthesis of approaches. *In* P. J. Rudall, P. J. Cribb, D. F. Cutler, and C. J. Humphries [eds.], Monocotyledons: Systematics and Evolution, 353–377. Royal Botanic Gardens, Kew.

Les, D. H., and E. L. Schneider. 1995. The Nymphaeales, Alismatidae, and the theory of an aquatic monocotyledon origin. *In* P. J. Rudall, P. J. Cribb, D. F. Cutler, and C. J. Humphries [eds.], Monocotyledons: Systematics and Evolution, 23–42. Royal Botanic Gardens, Kew.

Les, D. H., M. A. Cleland, and M. Waycott. 1997a. Phylogenetic studies in Alismatidae, II: Evolution of marine angiosperms (seagrasses) and hydrophily. *Systematic Botany* 22:443–463.

Les, D. H., E. Landolt, and D. J. Crawford. 1997b. Systematics of the Lemnaceae (duckweeds): Inferences from micromolecular and morphological data. *Plant Systematics and Evolution* 204:161–177.

Les, D. H., C. T. Philbrick, and R. A. Novelo. 1997c. The phylogenetic position of riverweeds (Podostemaceae): insights from *rbcL* sequence data. *Aquatic Botany* 57:5–27.

Les, D. H., D. K. Garvin, and C. F. Wimpee. 1991. Molecular evolutionary history of ancient aquatic angiosperms. *Proceedings of the National Academy of Sciences USA* 88:10119–10123.

Les, D. H., E. L. Schneider, D. J. Padgett, P. S. Soltis, D. E. Soltis, and M. Zanis. 1999. Phylogeny, classification and floral evolution of water lilies (Nymphaeaceae; Nymphaeales): a synthesis of non-molecular, *rbcL*, *matK*, and 18S rDNA data. *Systematic Botany* 24:28–46.

Les, D. H., M. L. Moody, S. W. L. Jacobs, and R. J. Bayer. 2002. Systematics of seagrasses (Zosteraceae) in Australia and New Zealand. *Systematic Botany* 27:468–484.

Levin, R. A., W. L. Wagner, P. C. Hoch, M. Nepokroeff, J. C. Pires, E. A. Zimmer, and K. J. Sytsma. 2003. Family-level relationships of Onagraceae based on chloroplast *rbcL* and *ndhF* data. *American Journal of Botany* 90:107–115.

Lewis, G. P. 1998. *Caesalpinia*, a revision of the *Poincianella-Erythrostemon* group. Royal Botanic Gardens, Kew.

Lewis, G. P., B. B. Simpson, and J. L. Neff. 2000. Progress in understanding the reproductive biology of the Caesalpinioideae (Leguminosae). *In* P. S. Herendeen and A. Bruneau [eds.], Advances in Legume Systematics, part 9, 65–78. Royal Botanic Gardens, Kew.

Li, J., A. L. Bogle, and A. S. Klein. 1999a. Phylogenetic relationships in the Hamamelidaceae: evidence from the nucleotide sequences of the plastid gene *matK*. *Plant Systematics and Evolution* 218:205–219.

———. 1999b. Phylogenetic relationships of the Hamamelidaceae inferred from sequences of internal transcribed spacers (ITS) of nuclear ribosomal DNA. *American Journal of Botany* 86:1027–1037.

Li, J., H. Huang, and T. Sang. 2002. Molecular phylogeny and infrageneric classification of *Actinidia* (Actinidiaceae). *Systematic Botany* 27:408–415.

Li, R.-Q., Z.-D. Chen, A.-M. Lu, D. E. Soltis, P. S. Soltis, and P. S. Manos. 2004. Phylogenetic relationships in Fagales based on DNA sequences from three genomes. *International Journal of Plant Sciences* 165:311–324.

Liden, M., and B. Oxelman. 1996. Do we need "phylogenetic taxonomy"? *Zoologica Scripta* 25:183–185.

Linder, H. P. 1998. Morphology and the evolution of wind pollination. *In* S. J. Owens and P. J. Rudall [eds.], Reproductive Biology in Systematics, Conservation and Economic Botany, 123–135. Royal Botanic Gardens, Kew.

———. 2000. Vicariance, climate change, anatomy and phylogeny of Restionaceae. *Botanical Journal of the Linnean Society* 134:159–177.

Linder, H. P., B. G. Briggs, and L. A. S. Johnson. 2000. Restionaceae: a morphological phylogeny. *In* K. L. Wilson and D. A. Morrison [eds.], Monocots: Systematics and Evolution, 653–660. CSIRO, Melbourne.

Lippok, B., A. A. Gardine, P. S. Williamson, and S. S. Renner. 2000. Pollination by flies, bees, and beetles of *Nuphar ozarkana* and *N. advena* (Nymphaeaceae). *American Journal of Botany* 87:898–902.

Litt, A. J. 1999. Floral Morphology and Phylogeny of Vochysiaceae. PhD dissertation, City University of New York.

Litt, A., and V. F. Irish. 2003. Duplication and diversification in the *APETALA1/FRUITFULL* floral homeotic gene lineage: implications for the evolution of floral development. *Genetics* 165:821–833.

Litt, A., and D. W. Stevenson. 2003. Floral development and morphology of Vochysiaceae. I. The structure of the gynoecium. *American Journal of Botany* 90:1533–1547.

Liu, F.-G., R. Liu, M. Miyamoto, N. Freire, P. Ong, M. Tennant, T. Young, and K. Gugel. 2001. Molecular and morphological supertrees for eutherian (placental) mammals. *Science* 291: 1786–1789.

Lledó, M. D., M. B. Crespo, K. M. Cameron, M. F. Fay, and M. W. Chase. 1998. Systematics of Plumbaginaceae based upon cladistic analysis of *rbcL* sequence data. *Systematic Botany* 23:21–29.

Lledó, M. D., P. O. Karis, M. B. Crespo, M. F. Fay, and M. W. Chase. 2001. Phylogenetic position and taxonomic status of the genus *Aegialitis* and subfamilies Staticoideae and Plumbaginoideae (Plumbaginaceae): evidence from plastid DNA sequences and morphology. *Plant Systematics and Evolution* 229:107–124.

Lloyd, D. G. 1982. Selection of combined versus separate sexes in seed plants. *American Naturalist* 120:571–585.

Lloyd, D. G., and S. C. H. Barrett. 1996. Floral Biology. Studies on Floral Evolution in Animal-Pollinated Plants. Chapman and Hall, New York.

Lloyd, D. G., and D. Schoen. 1992. Self- and cross-fertilization in plants. I. Functional dimensions. *International Journal of Plant Sciences* 153:358–369.

Lloyd, D. G., and C. J. Webb. 1986. The avoidance of interference between the presentation of pollen and stigmas in angiosperms. I. Dichogamy. *New Zealand Journal of Botany* 24:135–162.

Loconte, H. 1996. Comparison of alternative hypotheses for the origin of the angiosperms. *In* D. W. Taylor and L. J. Hickey [eds.], Flowering Plant Origin, Evolution, and Phylogeny. Chapman and Hall, New York.

Loconte, H., and J. R. Estes. 1989. Phylogenetic systematics of Berberidaceae and Ranunculales (Magnoliidae). *Systematic Botany* 14:565–579.

Loconte, H., and D. W. Stevenson. 1990. Cladistics of the Spermatophyta. *Brittonia* 42:197–211.

———. 1991. Cladistics of the Magnoliidae. *Cladistics* 7:267–296.

Loconte, H., M. Campbell, and D. W. Stevenson. 1995. Ordinal and familial relationships of ranunculid genera. *Plant Systematics and Evolution, Supplement* 9:99–118.

Long, A. G. 1977. Lower carboniferous pteridosperm cupules and the origin of angiosperms. *Royal Society of Edinburgh Transactions* 70:37–61.

Long, R. W. 1970. The genera of Acanthaceae in the southeastern United States. *Journal of the Arnold Arboretum* 51:257–309.

Løvtrup, S. 1977. The Phylogeny of Vertebrata. Wiley, London.

Lowry, P. P., G. M. Plunkett, and A. A. Oskolski. 2001. Early lineages in *Apiales*: insights from morphology, wood anatomy and molecular data. *Edinburgh Journal of Botany* 58:207–220.

Lunau, K. 2000. The ecology and evolution of visual pollen signals. *Plant Systematics and Evolution* 222:89–111.

Lundberg, J. 2001. The asteralean affinity of the Mauritian *Roussea* (Rousseaceae). *Botanical Journal of the Linnean Society* 137:267–276.

Lundberg, J., and K. Bremer. 2002. A phylogenetic study of the order Asterales using one large morphological and three molecular data sets. *International Journal of Plant Sciences* 164:553–578.

Lüthy, B., and P. Matile. 1984. The mustard oil bomb: rectified analysis of the subcellular organisation of the myrosinase system. *Biochemie und Physiologie der Pflanzen* 179:5–12.

Lyons-Weiler, J., G. A. Hoelzer, and R. J. Tausch. 1996. Relative apparent synapomorphy analysis (RASA). I. The statistical measure of phylogenetic signal. *Molecular Biology and Evolution* 13:749–757.

Lysak, M. A., P. F. Franz, H. B. M. Ali, and I. Schubert. 2001. Chromosome painting in *Arabidopsis thaliana*. *The Plant Journal* 28:689–697.

Lysak, M. A., A. Pecinka, and I. Schubert. 2003. Recent progress in chromosome painting of *Arabidopsis* and related species. *Chromosome Research* 11:195–204.

Ma, H. 1998. To be, or not to be, a flower-control of floral meristem identity. *Trends in Genetics* 14:26–32.

Ma, H., and C. W. dePamphilis. 2000. The ABCs of floral evolution. *Cell* 101:5–8.

Mabberley, D. J. 1993. The Plant Book: A Portable Dictionary of the Vascular Plants, 2nd ed. Cambridge University Press, Cambridge.

Maddison, D. R. 1991. The discovery and importance of multiple islands of most-parsimonious trees. *Systematic Zoology* 40:315–328.

Maddison, W. P. 1990. A method for testing the correlated evolution of two binary characters: are gains or losses concentrated on certain branches of a phylogenetic tree? *Evolution* 44:539–557.

Maddison, W. P., and D. R. Maddison. 1992. MacClade: Analysis of phylogeny and character evolution. Sinauer, Sunderland, MA.

Magallón, S., P. R. Crane, and P. S. Herendeen. 1999. Phylogenetic pattern, diversity, and diversification of eudicots. *Annals of the Missouri Botanical Garden* 86:297–372.

Magallón, S., and M. J. Sanderson. 2002. Relationships among seed plants inferred from highly conserved genes: sorting conflicting phylogenetic signals among ancient lineages. *American Journal of Botany* 89:1991–2006.

Magallón-Puebla, S., P. S. Herendeen, and P. R. Crane. 1997. *Quadriplatanus georgianus* gen. et sp. nov. : Staminate and pistillate platanaceous flowers from the late Cretaceous (Coniacian-Santonian) of Georgia, U. S. A. *International Journal of Plant Sciences* 158:373–394.

Magallón-Puebla, S., P. S. Herendeen, and P. K. Endress. 1996. *Allonia decandra*: floral remains of the tribe Hamamelideae (Hamamelidaceae) from Campanian strata of Southeastern U. S. A. *Plant Systematics and Evolution* 202:177–198.

Magin, N. 1977. Das Gynoeceum der Apiaceae: Modell und Ontogenie. *Berichte der Deutschen Botanischen Gesellschaft* 90:S53–S66.

Makboul, A. M. 1986. Chemical constituents of *Verbena officinalis*. *Fitoterapia* 57:50–51.

Malek, O., K. Lattig, R. Hiesel, A. Brennicke, and V. Knoop. 1996. RNA editing in bryophytes and a molecular phylogeny of land plants. *EMBO Journal* 14:1403–1411.

Manchester, S. R. 1986. Vegetative and reproductive morphology of an extinct plane tree (Platanaceae) from the Eocene of western North America. *Botanical Gazette* 147:200–226.

Manchester S. R., and D. L. Dilcher. 1997. Reproductive and vegetative morphology of *Polyptera* (Juglandaceae) from the Paleocene of Wyoming and Montana. *American Journal of Botany* 84:649–663.

Manchester, S. R., and R. B. Miller. 1978. Tile cells and their occurrence in malvalean fossil woods. *IAWA Bulletin II* 2–3:23–28.

Manen, J. F., A. Natali, and F. Ehrendorfer. 1994. Phylogeny of Rubiaceae-Rubieae inferred from the sequence of a cp-DNA intergene region. *Plant Systematics and Evolution* 190:195–211.

Manhart, J. R., and J. H. Rettig. 1994. Gene sequence data. *In* H.-D. Behnke and T. J. Mabry [eds.], Caryophyllales: Evolution and Systematics, 235–246. Springer, Berlin.

Manning, W. E. 1938. The morphology of the flowers of the Juglandaceae. I. The inflorescence. *American Journal of Botany* 27:839–852.

———. 1940. The morphology of the flowers of the Juglandaceae. II. The pistillate flowers and fruit. *American Journal of Botany* 27:839–852.

Manos, P. S. 1997. Systematics of *Nothofagus* (Nothofagaceae) based on rDNA spacer sequences (ITS): Taxonomic congruence with morphology and plastid sequences. *American Journal of Botany* 84:1137–1155.

Manos, P. S., and A. M. Stanford. 2001. The biogeography of Fagaceae: tracking the Tertiary history of temperate and subtropical forests of the Northern Hemisphere. *International Journal of Plant Sciences* 162:S77–S93.

Manos, P. S., and K. P. Steele. 1997. Phylogenetic analyses of "higher" Hamamelidae based on plastid sequence data. *American Journal of Botany* 81:1407–1419.

Manos, P. S., and D. E. Stone. 2001. Evolution, phylogeny, and systematics of the Juglandaceae. *Annals of the Missouri Botanical Garden* 88:231–269.

Manos, P. S., Z.-K. Zhou, and C. H. Cannon. 2001. Systematics of Fagaceae: phylogenetic tests of reproductive trait evolution. *International Journal of Plant Sciences* 162:1361–1379.

Markgraf, F. 1955. Über Laubblatt-Homologien und verwandtschaftliche Zusammenhänge bei Sarraceniales. *Planta* 46: 414–446.

Márquez-Guzmán, J., M. Engleman, A. Martínez-Mena, E. Martínez, and C. Ramos. 1989. Anatomia reproductiva de *Lacandonia schismatica* (Lacandoniaceae). *Annals of the Missouri Botanical Garden* 76:124–127.

Marsden, M. P. F., and I. W. Bailey. 1955. A fourth type of nodal anatomy in dicotyledons, illustrated by *Clerodendron trichotomum* Thunb. *Journal of the Arnold Arboretum* 36:1–51.

Martin, W., A. Gierl, and H. Saedler. 1989. Molecular evidence for pre-Cretaceous angiosperm origins. *Nature* 339:46–48.

Martin, W., D. J. Lydiate, H. Brinkmann, G. Forkmann, H. Saedler, and R. Cerff. 1993. Molecular phylogenies in angiosperm evolution. *Molecular Biology and Evolution* 10:140–162.

Martins, E. P., and T. F. Hansen. 1997. Phylogenies and the comparative method: a general approach to incorporating phylogenetic information into the analysis of interspecific data. *American Naturalist* 149:646–667.

Mast, A. R., S. Kelso, A. J. Richards, D. Lang, D. M. S. Feller, and E. Conti. 2001. Phylogenetic relationships in *Primula* L. and related genera (Primulaceae) based on noncoding chloroplast DNA. *International Journal of Plant Sciences* 162:1381–1400.

Masterson, J. 1994. Stomatal size in fossil plants: evidence for polyploidy in majority of angiosperms. *Science* 264:421–424.

Mathews, S., and M. J. Donoghue. 1999. The root of angiosperm phylogeny inferred from duplicate phytochrome genes. *Science* 286:947–950.

———. 2000. Basal angiosperm phylogeny inferred from duplicate phytochromes A and C. *International Journal of Plant Sciences* 161:S41–S55.

Matthews, M. L., and P. K. Endress. 2002. Comparative floral structure and systematics in Oxalidales (Cunoniaceae, Cephalotaceae, Connaraceae, Elaeocarpaceae, Tremandraceae, Oxalidaceae). *Botanical Journal of the Linnean Society* 140:321–381.

———. 2004. Comparative floral structure and systematics in Cucurbitales (Corynocarpaceae, Coriariaceae, Tetramelaceae, Datiscaceae, Begoniaceae, Cucurbitaceae, Anisophylleaceae). *Botanical Journal of the Linnean Society* 145:129–185.

———. 2005. Comparative floral structure and systematics in Crossosomatales (Crossosomataceae, Stachyuraceae, Staphyleaceae, Aphloiaceae, Geissolomataceae, Ixerbaceae, Strasburgeriaceae). *Botanical Journal of the Linnean Society* 147:1–46.

Matthews, M. L., P. K. Endress, J. Schönenberger, and E. M. Friis. 2001. A comparison of floral structures of Anisophylleaceae and Cunoniaceae and the problem of their systematic position. *Annals of Botany* 88:439–455.

Mayer, V., M. Moller, M. Perret, and A. Weber. 2003. Phylogenetic position and generic differentiation of Epithemateae (Gesneriaceae) inferred from plastid DNA sequence data. *American Journal of Botany* 90:321–329.

McDade, L. 1992. Pollinator relationships, biogeography, and phylogenetics. *BioScience* 42:21–26.

McDade, L., S. E. Masta, M. L. Moody, and E. Waters. 2000. Phylogenetic relationships among Acanthaceae: Evidence from two genomes. *Systematic Botany* 25:106–121.

McKenna, M. C. 1975. Toward a phylogenetic classification of the Mammalia. *In* W. P. Luckett and F. S. Szalay [eds.], Phylogeny of the Primates: A Multidisciplinary Approach, 21–46. Plenum, New York.

McLean, R., and M. Evans. 1934. The Mäule Reaction and the systematic position of the Gnetales. *Nature* 134:936–937.

Meerow, A. W., M. F. Fay, C. L. Guy, Q.-B. Li, F. Q. Qamaruz-Zaman, and M. W. Chase. 1999. Systematics of Amaryllidaceae based on cladistic analyses of plastid *rbcL* and *trnL-F* sequence data. *American Journal of Botany* 86:1325–1345.

Meeuse, A. D. J. 1975. Floral evolution as a key to angiosperms descent. *Acta Botanica Indica* 3:1–18.

Meimberg, H., P. Dittrich, G. Bringmann, J. Schlauer, and G. Heubl. 1999. Molecular phylogeny of Caryophyllidae s. l. based on *matK* sequences with special emphasis on carnivorous taxa. *Plant Biology* 2:218–228.

Meimberg, H., A. Wistuba, P. Dittrich, and G. Heubl. 2001. Molecular phylogeny of Nepenthaceae based on cladistic analysis of plastid *trnK* intron sequence data. *Plant Biology* 3:164–175.

Meinhardt, H. 1982. Models of Biological Pattern Formation. Academic Press, London.

Melchior, H. [ed.]. 1964. A. Engler's Syllabus der Pflanzenfamilien, 12th ed. Gebrüder Borntraeger, Berlin-Nikolassee.

Melville, R. 1962. A new theory of the angiosperm flower. I. The gynoecium. *Kew Bulletin* 17:1–50.

———. 1963. A new theory of the angiosperm flower. II. The androecium. *Kew Bulletin* 17:51–63.

———. 1969. Leaf venation patterns and the origin of angiosperms. *Nature* 224:121–125.

———. 1971. Red data book. 5. Angiospermae. International Union for the Conservation of Nature and Natural Resources, Survival Service Commission. Arts Graphiques, Heliographia, Lausanne.

Mennega, A. M. W. 1980. Anatomy of the secondary xylem. *In* A. J. M. Leeuwenberg [ed.], Die Natürlichen Pflanzenfamilien: Fam. Loganiaceae, 112–161. Duncker and Humblot, Berlin.

Merino Sutter, D., and P. K. Endress. 2003. Female flower and cupule structure in Balanopaceae, an enigmatic rosid family. *Annals of Botany* 92:459–469.

Metcalfe, C. R., and L. Chalk. 1950. Anatomy of the Dicotyledons. Leaves, Stem, and Wood in Relation to Taxonomy with Notes on Economic Uses. Clarendon Press, Oxford.

———. 1988/1989. Anatomy of the Dicotyledons, 2nd ed. Oxford University Press, Oxford.

Meyerowitz, E. M. 1997. The search for homeotic gene homologs in basal angiosperms and Gnetales: a potential new source of data on the evolutionary origin of flowers. *International Journal of Plant Sciences* 158:S131–S142.

Michelangeli, F. A., D. S. Penneys, J. Giza, D. E. Soltis, M. H. Hils, and J. D. Skean, Jr. 2004. A preliminary phylogeny of the tribe Miconieae (Melastomataceae) based on *nrITS* sequence data and its implication on inflorescence position. *Taxon* 53: 279–290.

Miller, N. G. 1971. The genera of Polygalaceae in the southeastern United States. *Journal of the Arnold Arboretum* 52:267–284.

Miller, R. B. 1975. Systematic anatomy of the xylem and comments on the relationships of the Flacourtiaceae. *Journal of the Arnold Arboretum* 56:20–102.

Miller, R. E., T. R. Buckley, and P. S. Manos. 2002. An examination of the monophyly of morning glory taxa using Bayesian phylogenetic inference. *Systematic Biology* 51:740–753.

Misa Ward, N., and R. Price. 2002. Phylogenetic Relationships of Marcgraviaceae: insights from three chloroplast genes. *Systematic Botany* 27:149–160.

Mishler, B. D. 1994. Cladistic analysis of molecular and morphological data. *American Journal of Physical Anthropology* 94:143–156.

———. 1999. Getting rid of species? *In* R. A. Wilson [ed.], Species. New Interdisciplinary Essays, 307–315. The MIT Press, Cambridge.

Mishler, B. D., M. J. Donoghue, and V. A. Albert. 1991. The decay index as a measure of relative robustness within a cladogram (abstract). *Hennig X (Annual meeting of the Willi Hennig Society)*.

Mitchell-Olds, T., and M. J. Clauss. 2002. Plant evolutionary genomics. *Current Opinions in Plant Biology* 5:74–79.

Mølgaard, P. 1985. Caffeic acid as a taxonomic marker in dicotyledons. *Nordic Journal of Botany* 5:203–213.

Mols, J. B., B. Gravendeel, L. W. Chatrou, M. D. Pirie, P. C. Bygrave, M. W. Chase, and P. A. Kessler. 2004. Identifying clades in Asian Annonaceae: monophyletic genera in the polyphyletic Miliusieae. *American Journal of Botany* 91:590–600.

Money, L. L., I. W. Bailey, and B. G. L. Swamy. 1950. The morphology and relationships of the Monimiaceae. *Journal of the Arnold Arboretum* 31:372–404.

Moody, M., L. Hufford, D. E. Soltis, and P. S. Soltis. 2001. Phylogenetic relationships of Loasaceae subfamily Gronovioideae inferred from *matK* and ITS sequence data. *American Journal of Botany* 88:236–336.

Morgan, D. R., and D. E. Soltis. 1993. Phylogenetic relationships among Saxifragaceae sensu lato based on *rbcL* sequence data. *Annals of the Missouri Botanical Garden* 80:631–660.

Morgan, D. R., D. Soltis, and K. Robertson. 1994. Systematic and evolutionary implications of *rbcL* sequence variation in Rosaceae. *American Journal of Botany* 81:890–903.

Morgan, M. 2000. Evolution of interactions between plants and their pollinators. *Plant Species Biology* 15:249–259.

Mort, M. E., and D. E. Soltis. 1999. Phylogenetic relationships and the evolution of ovary position in *Saxifraga* section *Micranthes*. *Systematic Botany* 24:139–147.

Mort, M. E., D. E. Soltis, P. S. Soltis, J. Francisco-Ortega, and A. Santos-Guerra. 2001. Phylogenetic relationships and evolution of Crassulaceae inferred from *matK* sequence data. *American Journal of Botany* 88:76–91.

Mort, M. E., P. S. Soltis, D. E. Soltis, and M. L. Mabry. 2000. Comparison of three methods for estimating internal support on phylogenetic trees. *Systematic Biology* 49:160–171.

Morton, C. M., M. W. Chase, and K. G. Karol. 1997. Phylogenetic relationships of two anomalous dicot genera, *Physena* and *Asteropeia*: evidence from *rbcL* plastid DNA sequences. *Botanical Review* 63:231–239.

Morton, C. M., M. W. Chase, K. A. Kron, and S. M. Swensen. 1996. A molecular evaluation of the monophyly of the order Ebenales based upon *rbcL* sequence data. *Systematic Botany* 21:567–586.

Mouradov, A. T., T. Glassick, L. Murphy, B. Fowler, S. Majla, and R. D. Teasdale. 1998. *NEEDLY*, a *Pinus radiata* ortholog of *FLORICAULA/LEAFY* genes, expressed in both reproductive and vegetative meristems. *Proceedings of the National Academy of Sciences USA* 95:6537–6542.

Muasya, A. M., D. A. Simpson, M. W. Chase, and A. Culham. 2001. A phylogenetic analysis of *Isolepis* and allied genera (Cyperaceae) based on plastid *rbcL* and *trnL-F* DNA sequence data. *Systematic Botany* 26:342–353.

Muasya, A., D. A. Simpson, A. Culham, and M. W. Chase. 1998. An assessment of suprageneric phylogeny in Cyperaceae using *rbcL* DNA sequences. *Plant Systematics and Evolution* 211:257–271.

Muellner, A. N., R. Samuel, S. A. Johnson, M. Cheek, T. D. Pennington, and M. W. Chase. 2003. Molecular phylogenetics of Meliaceae (Sapindales) based on nuclear and plastid DNA sequences. *American Journal of Botany* 90:471–480.

Mulcahy, D. 1979. The rise of the angiosperms: a genecological factor. *Science* 206:20–23.

Mulcahy, D., and G. Mulcahy. 1987. The effects of pollen competition. *American Science* 75:44–50.

Müller, G. B., and G. P. Wagner. 1991. Novelty in evolution. *Annual Reviews of Ecology and Systematics* 22:229–256.

Mummenhoff, K., H. Brüggemann, and J. L. Bowman. 2001. Chloroplast DNA phylogeny and biogeography of *Lepidium* (Brassicaceae). *American Journal of Botany* 88:2051–2063.

Munro, S. L., and H. P. Linder. 1997. The embryology and systematic relationships of *Prionium serratum* (Juncaceae: Juncales). *American Journal of Botany* 84:850–860.

Murata, J., T. Ohi, S. G. Wu, D. Darnaedi, T. Sugawara, T. Nakanishi, and H. Murata. 2001. Molecular phylogenetics of *Aristolochia* (Aristolochiaceae) inferred from *matK* sequences. *Acta Phytotaxonomica Geobotanica* 52:75–83.

Murbeck, S. 1912. Untersuchungen über den Blütenbau der Papaveraceen. *Kungl Svenska Vetenskapsakademiens Handlingar* 50:1–168.

Nagasawa, N., M. Miyoshi, Y. Sano, H. Satoh, H. Hirano, H. Sakai, and Y. Nagato. 2003. *SUPERWOMAN1* and *DROOPING LEAF* genes control floral organ identity in rice. *Development* 130:705–718.

Nakai, T. 1942. Notulae ad plantas Asiae orientalis. XVIII. *Journal of Japanese Botany* 18:91–120.

Nandi, O. I. 1998a. Ovule and seed anatomy of Cistaceae and related Malvanae. *Plant Systematics and Evolution* 209:239–264.

———. 1998b. Floral development and systematics of Cistaceae. *Plant Systematics and Evolution* 212:107–134.

Nandi, O. I., M. W. Chase, and P. K. Endress. 1998. A combined cladistic analysis of angiosperms using *rbcL* and nonmolecular data sets. *Annals of the Missouri Botanical Garden* 85:137–212.

Nash, G. 1903. A revision of the family Fouquieriaceae. *Bulletin of the Torrey Botanical Club* 30:449–459.

Naylor, G. J. P., and W. M. Brown. 1997. Structural biology and phylogenetic estimation. *Nature* 388:527–528.

Nazarov, V. V., and G. Gerlach. 1997. The potential seed productivity of orchid flowers and peculiarities of their pollination systems. *Lindleyana* 12:188–204.

Nei, M., S. Kumar, and K. Takahashi. 1998. The optimization principle in phylogenetic analysis tends to give incorrect topologies when the number of nucleotides or amino acids used is small. *Proceedings of the National Academy of Sciences USA* 95:12390–12397.

Nelson, G. 1972. Phylogenetic relationship and classification. *Systematic Zoology* 21:227–231.

———. 1973. Classification as an expression of phylogenetic relationships. *Systematic Zoology* 22:344–359.

Nickerson, J., and G. Drouin. 2004. The sequence of the largest subunit of RNA polymerase II is a useful marker for inferring seed plant phylogeny. *Molecular Phylogenetics and Evolution* 31:404–415.

Nickrent, D. L. 2002. Orígenes filogenéticos de las plantas parásites. *In* J. A. López-Sáez, P. Catalán, and L. Sáez [eds]. Plantas parásites de la Peninsula Ibérica e Islas Baleares, 29–56. Mundi-Prensa Libros, S. A., Madrid.

Nickrent, D. L., and R. J. Duff. 1996. Molecular studies of parasitic plants using ribosomal RNA. *In* M. T. Moreno, J. I. Cubero, D. Berner, D. Joel, L. J. Musselman, and C. Parker [eds.], Advances in Parasitic Plant Research, 28–52. Junta de Andalucia, Direccion General de Investigacion Agraria, Cordoba, Spain.

Nickrent, D. L., and V. Malécot. 2001. A molecular phylogeny of Santalales. *In* A. Fer, P. Thalouarn, D. Joel, L. J. Musselman, C. Parker, and J. A. C. Verklejj [eds.], Proceedings of the 7th International Parasitic Weed Symposium, 69–74. Faculté des Sciences, Université de Nantes, Nantes, France.

Nickrent, D. L., and D. E. Soltis. 1995. A comparison of angiosperm phylogenies from nuclear 18S rDNA and *rbcL* sequences. *Annals of the Missouri Botanical Garden* 82:208–234.

Nickrent, D. L., and E. M. Starr. 1994. High rates of nucleotide substitution in nuclear small-subunit (18S) rDNA from holoparasitic flowering plants. *Journal of Molecular Evolution* 39:62–70.

Nickrent, D. L., A. Blarer, Y.-L. Qiu, D. E. Soltis, P. S. Soltis, and M. Zanis. 2002. Molecular data and the relationships of the enigmatic angiosperm Hydnoraceae. *American Journal of Botany* 89:1809–1817.

Nickrent, D. L., R. J. Duff, A. F. Colwell, A. D. Wolfe, N. D. Young, K. E. Steiner, and C. W. dePamphilis. 1998. Molecular phylogenetics and evolutionary studies of parasitic plants. *In* P. S. Soltis, D. E. Soltis, and J. J. Doyle [eds.], Molecular Systematics of Plants II: DNA Sequencing, 211–241. Kluwer, Boston.

Nicoletti, M., A. Di Fabio, M. Seralini, J. A. Garbarino, and M. C. Chamy. 1991. Iridoids from *Loasa tricolor*. *Biochemical Systematics and Evolution* 19:167–170.

Nishino, E. 1978. Corolla tube formation in four species of Solanaceae. *Botanical Magazine, Tokyo* 91:263–277.

———. 1983. Corolla tube formation in the Primulaceae and Ericales. *Botanical Magazine, Tokyo* 96:319–342.

Nixon, K. C. 1999. The parsimony ratchet: a rapid means for analyzing large data sets. *Cladistics* 15:407–414.

———. 2000. Winclada. Program and documentation. K. C. Nixon, Ithaca, NY.

Nixon, K. C., W. L. Crepet, D. W. Stevenson, and E. M. Friis. 1994. A reevaluation of seed plant phylogeny. *Annals of the Missouri Botanical Garden* 81:484–533.

Nowicke, J. W. 1968. Palynotaxonomic study of the Phytolaccaceae. *Annals of the Missouri Botanical Garden* 55:294–364.

———. 1975. Pollen morphology in the order Centrospermae. *Grana* 15:51–77.

———. 1994. Pollen morphology and exine ultrastructure. *In* H.-D. Behnke and T. J. Mabry [eds.], Caryophyllales: Evolution and Systematics, 167–222. Springer, Berlin.

———. 1996. Pollen morphology, exine structure and the relationships of Basellaceae and Didiereaceae to Portulacaceae. *Systematic Botany* 21:187–208.

Nowicke, J. W., and J. J. Skvarla. 1977. Pollen morphology and the relationship of the Plumbaginaceae, Polygonaceae and Primulaceae to the order Centrospermae. *Smithsonian Contributions to Botany* 37:1–64.

———. 1984. Pollen morphology and the relationships of *Simmondsia chinensis* to the order Euphorbiales. *American Journal of Botany* 71:210–215.

Nyffeler, R., and D. L. Baum. 2000. Phylogenetic relationships of the durians (Bombacaceae-Durioneae or /Malvaceae/Helicteroideae/Durioneae) based on chloroplast and nuclear ribosomal DNA sequences. *Plant Systematics and Evolution* 224:55–82.

Oginuma, K., Z. Giu, and Z.-S. Yue. 1995. Karyomorphology of *Rhioptelea* (Rhiopteleaceae). *Acta Phytotaxonomica et Geobotanica* 46:147–151.

Ohashi, K., T. Tanikawa, Y. Okumura, K. Kawazoe, N. Tatara, M. Minato, H. Shibuya, I. Kitagawa, A. Shimoyama, M. Yamadaki, Y. Nakazawa, S. Yahara, and T. Nohara. 1993. Indonesian medicinal plants: X. Chemical structures of four new triterpene-glycosides, gongganosides D, E, F, and G, and two secoiridoid glucosides from the bark of *Bhesa paniculata* (Celastraceae). *Shoyakugaku Zasshi* 47:56–59.

O'Kane, S. L., and I. A. Al-Shehbaz. 2003. Phylogenetic position and generic limits of *Arabidopsis* (Brassicaccae) based on sequences of nuclear ribosomal DNA. *Annals of the Missouri Botanical Garden* 90:603–612.

Olmstead, R. G., and P. A. Reeves. 1995. Evidence for the polyphyly of the Scrophulariaceae based on chloroplast *rbcL* and *ndhF* sequences. *Annals of the Missouri Botanical Garden* 82:176–193.

Olmstead, R. G., and J. A. Sweere. 1994. Combining data in phylogenetic systematics: an empirical approach using three molecular data sets in the Solanaceae. *Systematic Biology* 43:467–481.

Olmstead, R. G., B. Bremer, K. M. Scott, and J. D. Palmer. 1993. A parsimony analysis of the Asteridae sensu lato based on *rbcL* sequences. *Annals of the Missouri Botanical Garden* 80:700–722.

Olmstead, R. G., C. W. dePamphilis, A. D. Wolfe, N. D. Young, W. J. Elisons, and A. Reeves. 2001. Disintegration of the Scrophulariaceae. *American Journal of Botany* 88:348–361.

Olmstead, R. G., K.-J. Kim, R. K. Jansen, and S. J. Wagstaff. 2000. The phylogeny of the Asteridae sensu lato based on chloroplast *ndhF* gene sequence. *Molecular Phylogenetics and Evolution* 16:96–112.

Olmstead, R. G., H. J. Michaels, K. M. Scott, and J. D. Palmer. 1992. Monophyly of the Asteridae and identification of their major lineages inferred from DNA sequences of *rbcL*. *Annals of the Missouri Botanical Garden* 79:249–265.

Olmstead, R. G., P. A. Reeves, and A. C. Yen. 1998. Patterns of sequence evolution and implications for parsimony analysis of chloroplast DNA. *In* D. E. Soltis, P. S. Soltis, and J. J. Doyle [eds.], Molecular Systematics of Plants II: DNA Sequencing, 164–187. Kluwer, Boston.

Olmstead, R. G., J. A. Sweere, R. E. Spangler, L. Bohs, and J. D. Palmer. 1999. Phylogeny and provisional classification of the Solanaceae based on chloroplast DNA. *In* M. Nee, D. E. Symon, R. N. Lester, and J. P. Jessop [eds.], Solanaceae IV, 111–137. Royal Botanic Gardens, Kew.

Olson, M. E. 2002a. Combining data from DNA sequences and morphology for a phylogeny of Moringaceae (Brassicales). *Systematic Botany* 27:55–73.

———. 2002b. Intergeneric relationships within the Caricaceae-Moringaceae clade (Brassicales) and potential morphological synapomorphies of the clade and its families. *International Journal of Plant Sciences* 163:51–65.

Orel, N., and H. Puchta. 2003. Differences in the processing of DNA ends in *Arabidopsis thaliana* and tobacco: possible implications for genome evolution. *Plant Molecular Biology* 51:523–531.

Orgel, F. 1980. Selfish DNA: the ultimate parasite. *Nature* 284:604–607.

Osborn, J. M. 2000. Pollen morphology and ultrastructure of gymnospermous anthophytes. *In* M. M. Harley, C. M. Morton, and S. Blackmore [eds.], Pollen and Spores: Morphology and Biology, 163–185. Royal Botanic Gardens, Kew.

Otto, S. P., and J. Whitton. 2000. Polyploidy incidence and evolution. *Annual Review of Genetics* 34:401–437.

Oxelman, B., M. Backlund, and B. Bremer. 1999. Relationships of the Buddlejaceae s. l. investigated using parsimony jack-knife and branch support analysis of chloroplast *ndhF* and *rbcL* sequence data. *Systematic Botany* 24:164–182.

Oxelman, B., M. Liden, and D. Berglund. 1997. Chloroplast *rps16* intron phylogeny of the tribe Sileneae (Caryophyllaceae). *Plant Systematics and Evolution* 206:393–410.

Pacini, E., and M. Hesse. 2002. Types of pollen dispersal units in orchids, and their consequences for germination and fertilization. *Annals of Botany* 89:653–664.

Page, M., and R. Johnstone. 1992. Variation across species in the size of the nuclear genome supports the junk-DNA explanation for the C-value paradox. *Proceedings of the Royal Society of London B* 249:119–124.

Pagel, M. D. 1998. Inferring evolutionary processes from phylogenies. *Zoologica Scripta* 26:331–348.

———. 1999. The maximum likelihood approach to reconstructing ancestral character states of discrete characters on phylogenies. *Systematic Biology* 48:612–622.

Palmer J. D., D. E. Soltis, and M. W. Chase. 2004. The plant tree of life: an overview and some points of view. *American Journal of Botany* 91:1437–1445.

Pant, D. D. 1977. The plant of *Glossopteris*. *Journal of the Indian Botanical Society* 56:1–23.

Pardo, F., F. Perich, R. Torres, and F. D. Monache. 1998. Phytotoxic iridoid glucosides form the roots of *Verbascum thapsus*. *Journal of Chemical Ecology* 24:645–653.

Parenti, L. R. 1980. A phylogenetic analysis of the land plants. *Biological Journal of the Linnean Society* 13:225–242.

Parkinson, C. L., K. L. Adams, and J. D. Palmer. 1999. Multigene analyses identify the three earliest lineages of extant flowering plants. *Current Biology* 9:1485–1488.

Patel, V. C., J. J. Skvarla, and P. H. Raven. 1984. Pollen characters in relation to the delimitation of Myrtales. *Annals of the Missouri Botanical Garden* 71:858–969.

Paterson, A. H., J. E. Bowers, and B. A. Chapman. 2004. Ancient polyploidization predating divergence of the cereals, and its consequences for comparative genomics. *Proceedings of the National Academy of Sciences USA* 101:9903–9908.

Patterson, C. D., D. M. Williams, and C. J. Humphries. 1993. Congruence between molecular and morphological phylogenies. *Annual Review of Ecology and Systematics* 24:153–188.

Pedersen, K. R., E. M. Friis, P. R. Crane, and A. N. Drinnan. 1994. Reproductive structures of an extinct platanoid from the Early Cretaceous (latest Albian) of eastern North America. *Review of Palaeobotany and Palynology* 80:291–303.

Pelaz, S., G. S. Ditta, E. Baumann, E. Wisman, and M. F. Yanofsky. 2000. B and C floral organ identity functions require *SEPALLATA* MADS-box genes. *Nature* 405:200–203.

Pennington, R. T., M. Lavin, H. Ireland, B. B. Klitgaard, J. Preston, and J.-M. Hu. 2001. Phylogenetic relationships of basal papilionoid legumes based upon sequences of the chloroplast intron *trnL*. *Systematic Botany* 26:537–556.

Perkins, J. 1925. Übersicht über die Gattungen der Monimiaceae. Engelmann, Leipzig.

Perrier de la Bâthie, H. 1933. Les Brexiées de Madagascar. *Bulletin de la Société Botanique de France* 80:198–204.

———. 1942. Au sujet des affinités des *Brexia*, des Celastracées, et de deux *Brexia* nouveaux de Madagascar. *Bulletin de la Société Botanique de France* 89:219–221.

Persson, C. 2001. Phylogenetic relationships in Polygalaceae based on plastid DNA sequences from the *trnL-F* region. *Taxon* 50:763–779.

Petrov, D. A. 2002. DNA loss and evolution of genome size in *Drosophila*. *Genetica* 115:81–91.

Philbrick, C. T., and D. H. Les. 1996. Evolution of aquatic angiosperm reproductive systems. *BioScience* 46:813–826.

Philipson, W. R. 1970. Constant and variable features of the Araliaceae. *In* N. K. B. Robson, D. F. Cutler, and M. Gregory [eds.], New Research in Plant Anatomy, 87–100, Supplement 1. Botanical Journal of the Linnean Society, Academic Press, London.

———. 1974. Ovular morphology and the major classification of the dicotyledons. *Botanical Journal of the Linnean Society* 68:89–108.

———. 1977. Ovular morphology and the classification of dicotyledons. *Plant Systematics and Evolution, Supplement* 1:123–140.

Pires, J. C., and K. Sytsma. 2002. A phylogenetic evaluation of a biosystematic framework: *Brodiaea* and related petaloid monocots (Themidaceae). *American Journal of Botany* 89:1342–1359.

Pires, J. C., M. F. Fay, W. S. Davis, L. Hufford, J. Rova, M. W. Chase, and K. J. Sytsma. 2001. Molecular and morphological phylogenetic analyses of Themidaceae (Asparagales). *Kew Bulletin* 56:691–626.

Planchon, J. E. 1887. Ampelideae. Vitaceae. In A. de Candolle and C. de Candolle [eds.], Vitaceae. *Monographiae Phanerogamarum* 5(2):305–654. Masson, Paris.

Plouvier, V. 1964. Recherche de l'Arbutoside et de l'Asperuloside chez quelques Rubiacées. Présence du Monotropeoside chez le *Liquidambar* (Hamamelidacées). *Comptes Rendus de l'Académie des Sciences Paris* 258:735–737.

———. 1992. Chimiotaxonomie des Caprifoliaceae et relations avec quelques familles voisines. *Bulletin du Muséum National d'Histoire Naturelle, sect. B. Adansonia* 14:461–472.

Plouvier, V., and J. Favre-Bonvin. 1971. Les iridoides et sécoiridoides: repartition, structure, propriétés, biosynthèse. *Phytochemistry* 10:1697–1722.

Plumstead, E. P. 1956. Bisexual fructifications borne on *Glossopteris* leaves from South Africa. *Palaeontographica* B100:1–25.

Plunkett, G. M. 2001. Relationship of the order Apiales to subclass Asteridae: a re-evaluation of morphological characters based on insights from molecular data. *Edinburgh Journal of Botany* 58:183–200.

Plunkett, G. M., and S. R. Downie. 1999. Major lineages within Apiaceae subfamily Apioideae: a comparison of chloroplast restriction site and DNA sequence data. *American Journal of Botany* 86:1014–1026.

Plunkett, G. M., and P. P. Lowry, Jr. 2001. Relationships among "ancient araliads" and their significance for the systematics of Apiales. *Molecular Phylogenetics and Evolution* 19:259–276.

Plunkett, G. M., G. T. Chandler, P. P. Lowry II, S. M. Pinney, and T. S. Sprenkle. 2004. Recent advances in understanding Apiales and a revised classification. *South African Journal of Botany* 70:371–381.

Plunkett, G. M., D. E. Soltis, and P. S. Soltis. 1995. Phylogenetic relationships between Juncaceae and Cyperaceae: insights from *rbcL* sequence data. *American Journal of Botany* 82:520–525.

———. 1996a. Evolutionary patterns in Apiaceae: inferences based on *matK* sequence data. *Systematic Botany* 21:477–495.

———. 1996b. Higher level relationships of Apiales (Apiaceae and Araliaceae) based on phylogenetic analysis of *rbcL* sequences. *American Journal of Botany* 83:499–515.

———. 1997a. Clarification of the relationship between Apiaceae and Araliaceae based on *matK* and *rbcL* sequence data. *American Journal of Botany* 84:567–580.

———. 1997b. Evolutionary patterns in Apiaceae: inferences based on *matK* sequence data. *Systematic Botany* 21:477–495.

Pollock, D. D., D. J. Zwickl, J. A. McGuire, and D. M. Hillis. 2002. Increased taxon sampling is advantageous for phylogenetic inference. *Systematic Biology* 51:664–671.

Poser, G. V., M. E. Toffoli, M. Sobral, and A. Heniques. 1997. Iridoid glucosides substitution patterns in Verbenaceae and their taxonomic implication. *Plant Systematics and Evolution* 205:265–287.

Posluszny, U. B., and P. B. Tomlinson. 2003. Aspects of inflorescence and floral development in the putative basal angiosperm *Amborella trichopoda* (Amborellaceae). *Canadian Journal of Botany* 81:28–39.

Potgieter, K., and V. A. Albert. 2001. Phylogenetic relationships within Apocynaceae s. l. based on *trnL* intron and *trnL-F* spacer sequences and propagule characters. *Annals of the Missouri Botanical Garden* 88:523–549.

Prance, G. T. 1968. The systematic position of *Rhabdodendron* Gilg and Pilg. *Bulletin du Jardin Botanique National de Belgique* 38:127–146.

Price, H. J. 1988. Nuclear DNA content variation within angiosperm species. *Evolutionary Trends in Plants* 2:53–60.

Price, R., and J. Palmer. 1993. Phylogenetic relationships of the Geraniaceae and Geraniales from *rbcL* sequence comparisons. *Annals of the Missouri Botanical Garden* 80:661–671.

Pryer, K. M., H. Schneider, A. R. Smith, R. Cranfill, P. Wolf, J. S. Hunt, and S. D. Sipes. 2001. Horsetails and ferns are a monophyletic group and the closest living relatives to seed plants. *Nature* 409:618–622.

Puff, C., and A. Weber. 1976. Contributions to the morphology, anatomy, and karyology of *Rhabdodendron* and a reconsideration of the systematic position of Rhabdodendraceae. *Plant Systematics and Evolution* 125:195–222.

Purvis, A. 1995. A modification to Baum and Ragan's method for combining phylogenetic trees. *Systematic Biology* 44:251–255.

Pyankov, V. I., E. G. Artyusheva, G. E. Edwards, C. C. J. Black, and P. S. Soltis. 2001. Phylogenetic analysis of tribe Salsoleae (Chenopodiaceae) based on ribosomal ITS sequences: implications for the evolution of photosynthesis types. *American Journal of Botany* 88:1189–1198.

Qiu, Y.-L., M. W. Chase, S. B. Hoot, E. Conti, P. R. Crane, K. J. Sytsma, and C. R. Parks. 1998. Phylogenetics of the Hamamelidae and their allies: parsimony analyses of nucleotide sequences of the plastid gene *rbcL*. *International Journal of Plant Sciences* 159:891–905.

Qiu, Y.-L., M. W. Chase, D. H. Les, and C. R. Parks. 1993. Molecular phylogenetics of the Magnoliidae: Cladistic analyses of nucleotide sequences of the plastid gene *rbcL*. *Annals of the Missouri Botanical Garden* 80:587–606.

Qiu, Y.-L., J.-Y. Lee, F. Bernasconi-Quadroni, D. E. Soltis, P. S. Soltis, M. Zanis, Z. Chen, V. Savolainen, and M. W. Chase. 1999. The earliest angiosperms: evidence from mitochondrial, plastid and nuclear genomes. *Nature* 402:404–407.

Qiu, Y.-L., J.-Y. Lee, F. Bernasconi-Quadroni, D. E. Soltis, P. S. Soltis, M. Zanis, E. Zimmer, Z. Chen, V. Savolainen, and M. Chase. 2000. Phylogeny of basal angiosperms: analyses of five genes from three genomes. *International Journal of Plant Sciences* 161:S3–S27.

Qiu, Y.-L., J. Lee, B. A. Whitlock, F. Bernasconi-Quadroni, and O. Dombrovska. 2001. Was the ANITA rooting of the angiosperm phylogeny affected by long branch attraction? *Molecular Biology and Evolution* 18:1745–1753.

Quibell, C. F. 1972. Comparative and Systematic Anatomy of Carpenterieae (Philadelphiaceae). PhD dissertation, University of California, Berkeley.

Rabinowicz, P. D. 2000. Are obese plant genomes on a diet? *Genome Research* 10:893–894.

Ragan, M. A. 1992. Phylogenetic inference based on matrix representation of trees. *Molecular Phylogenetics and Evolution* 1:53–58.

Rahn, K. 1996. A phylogenetic study of the Plantaginaceae. *Botanical Journal of the Linnean Society* 120:145–198.

Ramshaw, J. A. M., D. L. Richardson, B. T. Meatyard, R. Brown, H. M. Richardson, E. W. Thompson, and D. Boulter. 1972. The time of origin of the flowering plants determined by using amino acid sequence data of cytochrome C. *New Phytologist* 71:773–779.

Rao, P. R. M. 1972. Embryology of *Nyssa sylvatica*, and systematic consideration of the family Nyssaceae. *Phytomorphology* 22:8–21.

Raubeson, L. A., and R. K. Jansen. 1992. A rare chloroplast-DNA structural mutation is shared by all conifers. *Biochemical Systematics and Ecology* 20:17–24.

Raven, P. H. 1975. The bases of angiosperm phylogeny: cytology. *Annals of the Missouri Botanical Garden* 62:724–764.

———. 1979. Onagraceae as a model of plant evolution. *In* L. Gottlieb and S. Jain [eds.], Plant Evolutionary Biology, 85–107. Chapman and Hall, London.

Ray, J. 1703. Methodus Plantarum, Emendata et Aucta. Smith and Walford, London.

Record, S. J. 1933. The woods of *Rhabdodendron* and *Duckeodendron*. *Tropical Woods* 33:6–10.

Ree, R. H., and M. J. Donoghue. 2000. Inferring rates of change in flower symmetry in asterid angiosperms. *Systematic Biology* 48:633–641.

Reeves, G., M. W. Chase, P. Goldblatt, P. J. Rudall, M. F. Fay, A. V. Cox, B. Lejeune, and T. Souza-Chies. 2001. Molecular systematics of Iridaceae: evidence from four plastid DNA regions. *American Journal of Botany* 88:2074–2087.

Reeves, P. A., and R. G. Olmstead. 1998. Evolution of novel morphological, ecological, and reproductive traits in a clade containing *Antirrhinum*. *American Journal of Botany* 86:1301–1315.

Reichenbach, H. G. L. 1827–1829. *In* D. J. Christ [ed.], J. C. Moessler's Handbuch der Gewächskunde, 2nd ed., 3 vols. Hammerich, Atona.

Remane, A. 1956. Die Grundlagen des natürlichen Systems, der vergleichenden Anatomie und der Phylogenetik. Akademische Verlagsbuchhandlung Geest & Portig, Leipzig.

Renner, S. S. 1999. Circumscription and phylogeny of the Laurales: evidence from molecular and morphological data. *American Journal of Botany* 86:1301–1315.

———. 2001. How common is heterodichogamy? *Trends in Ecology and Evolution* 16:595–597.

Renner, S. S., and A. S. Chanderbali. 2000. What is the relationship among Hernandiaceae, Lauraceae, and Monimiaceae, and why is this question so difficult to answer? *International Journal of Plant Sciences* 161:S109–S119.

Renner, S. S., and K. Meyer. 2001. Melastomataceae come full circle: biogeographic reconstruction and molecular clock dating. *Evolution* 55:1315–1324.

Renner, S. S., and R. E. Ricklefs. 1995. Dioecy and its correlates in the flowering plants. *American Journal of Botany* 82:596–606.

Renner, S. S., and H. S. Won. 2001. Repeated evolution of dioecy from monoecy in Siparunaceae (Laurales). *Systematic Biology* 50:700–712.

Renner, S. S., G. Clausing, and K. Meyer. 2001. Historical biogeography of Melastomataceae: the roles of Tertiary migration and long-distance dispersal. *American Journal of Botany* 88:1290–1300.

Retallack, G., and D. L. Dilcher. 1981. Arguments for a glossopterid ancestry of angiosperms. *Palaeobiology* 7:54–67.

Rettig, J. H., H. D. Wilson, and J. M. Manhart. 1992. Phylogeny of the Caryophyllales—Gene sequence data. *Taxon* 41:201–209.

Reymanówna, M. 1968. On seeds containing *Eucommiidites troedssonii* pollen from the Jurassic of Grojec, Poland. *Botanical Journal of the Linnean Society* 61:147–152.

———. 1973. The Jurassic flora from Grojec near Krakow in Poland. Part II. Caytoniales and anatomy of *Caytonia*. *Acta Palaeobotanica* 14:45–87.

Rice, K. A., M. J. Donoghue, and R. G. Olmstead. 1997. Analyzing large data sets: *rbcL* 500 revisited. *Systematic Biology* 46:157–178.

Richards, A. J. 1997. Plant Breeding Systems. Chapman and Hall, London.

Richardson, J. E., M. F. Fay, Q. C. B. Cronk, D. Bowman, and M. W. Chase. 2000. A molecular phylogenetic analysis of Rhamnaceae using *rbcL* and *trnL-F* plastid DNA sequences. *American Journal of Botany* 87:1309–1324.

Riechmann, J. L., and E. M. Meyerowitz. 1997. MADS domain proteins in plant development. *Biological Chemistry* 378:1079–1109.

Ritland, K., and M. T. Clegg. 1987. Evolutionary analysis of plant DNA sequences. *The American Naturalist* 130:S74–S100.

Robertson, K. R. 1972a. The genera of Geraniaceae in the southeastern United States. *Journal of the Arnold Arboretum* 53:182–201.

———. 1972b. The Malpighiaceae in the southeastern United States. *Journal of the Arnold Arboretum* 53:101–112.

———. 1974. The genera of Rosaceae in the southeastern United States. *Journal of the Arnold Arboretum* 55:303–332, 344–401, 600–662.

———. 1975. The genera of Oxalidaceae in the southeastern United States. *Journal of the Arnold Arboretum* 57:205–216.

Robinson, H. 1985. Observations on fusion and evolutionary variability in the angiosperm flower. *Systematic Botany* 10:105–109.

Robinson, H., and P. Burns-Balogh. 1982. Evidence for a primitively epiphytic habit in Orchidaceae. *Systematic Botany* 7:353–358.

Robinson-Beers, K., R. E. Pruitt, and C. S. Gasser. 1992. Ovule development in wild-type *Arabidopsis* and two female-sterile mutants. *The Plant Cell* 4:1237–1249.

Rodenburg, W. F. 1971. A revision of the genus *Trimenia* (Trimeniaceae). *Blumea* 19:3–15.

Rodman, J. E. 1990. Centrospermae revisited, part 1. *Taxon* 39:383–393.

———. 1991a. A taxonomic analysis of glucosinolate-producing plants. I. Phenetics. *Systematic Botany* 16:598–618.

———. 1991b. A taxonomic analysis of glucosinolate-producing plants. II. Cladistics. *Systematic Botany* 16:619–699.

———. 1994. Cladistic and phenetic studies. *In* H.-D. Behnke and T. J. Mabry [eds.], Caryophyllales: Evolution and Systematics, 279–301. Springer, Berlin.

Rodman, J. E., K. G. Karol, R. A. Price, E. Conti, and K. J. Sytsma. 1994. Nucleotide sequences of *rbcL* confirm the capparalean affinity of the Australian endemic Gyrostemonaceae. *Australian Systematic Botany* 7:57–69.

Rodman, J. E., K. G. Karol, R. A. Price, and K. J. Sytsma. 1996. Molecules, morphology, and Dahlgren's expanded order Capparales. *Systematic Botany* 21:289–307.

Rodman, J. E., M. K. Oliver, R. R. Nakamura, J. U. J. McClammer, and A. H. Bledsoe. 1984. A taxonomic analysis and revised classification of Centrospermae. *Systematic Botany* 9:297–323.

Rodman, J. E., R. A. Price, K. Karol, E. Conti, K. J. Sytsma, and J. D. Palmer. 1993. Nucleotide sequences of the *rbcL* gene indicate monophyly of mustard oil plants. *Annals of the Missouri Botanical Garden* 80:686–699.

Rodman, J. E., P. S. Soltis, D. E. Soltis, K. J. Sytsma, and K. G. Karol. 1998. Parallel evolution of glucosinolate biosynthesis inferred from congruent nuclear and plastid gene phylogenies. *American Journal of Botany* 85:997–1006.

Roehls, P. 1998. Phylogenetic Position and Delimitation of the Order Dipsacales. A Multidisciplinary Approach. Doctoral dissertation, University of Leuven.

Roehls, P., and E. Smets. 1996. A floral ontogenetic study in Dipsacales. *International Journal of Plant Sciences* 157:203–218.

Roehls, P., L. P. Ronse De Craene, and E. F. Smets. 1997. A floral ontogenetic investigation of the Hydrangeceae. *Nordic Journal of Botany* 17:235–254.

Rogers, G. K. 1983. The genera of Alismataceae in the southeastern United States. *Journal of the Arnold Arboretum* 64:383–420.

———. 1984. The Zingiberales (Cannaceae, Marantaceae, and Zingiberaceae) in the southeastern United States. *Journal of the Arnold Arboretum* 65:5–55.

———. 1985. The genera of Phytolaccaceae in the southeastern United States. *Journal of the Arnold Arboretum* 66:1–37.

Rohweder, O. 1967. Centrospermen-Studien 3. Blütenentwicklung und Blütenbau bei Silenoideen (Caryophyllaceen). *Botanische Jahrbücher für Systematik* 86:130–185.

Rohweder, O., and P. K. Endress. 1983. Samenpflanzen. Thieme, Stuttgart.

Rohwer, J. G. 1993. Lauraceae. *In* K. Kubitzki, J. Rohwer, and V. Bittrich [eds.], The Families and Genera of Vascular Plants, II, 366–391. Springer, Berlin.

Rokas, A., B. L. Williams, N. King, and S. B. Carroll. 2003. Genome-scale approaches to resolving incongruence in molecular phylogenies. *Nature* 425:798–804.

Romeike, A. 1978. Tropane alkaloids: Occurrence and systematic importance in angiosperms. *Botanische Notisier* 131:85–96.

Rönblom, K., and A. A. Anderberg. 2002. Phylogeny of Diapensiaceae based on molecular data and morphology. *Systematic Botany* 27:383–395.

Ronquist, F. 1996. Matrix representation of trees, redundancy, and weighting. *Systematic Biology* 45:247–253.

Ronse De Craene, L. P. 1992. The androecium of the Magnoliophytina: characterization and systematic importance. Doctoral dissertation, University of Leuven.

Ronse De Craene, L. P., and E. F. Smets. 1992. Complex polyandry in the Magnoliatae: definition, distribution and systematic value. *Nordic Journal of Botany* 12:621–649.

———. 1993. The distribution and systematic relevance of the androecial character polymery. *Botanical Journal of the Linnean Society* 113:285–350.

———. 1994. Merosity in flowers: definition, origin, and taxonomic significance. *Plant Systematics and Evolution* 191:83–104.

———. 1995. The distribution and systematic relevance of the androecial character oligomery. *Botanical Journal of the Linnean Society* 118:193–247.

———. 1997. A floral ontogenetic study of some species of *Capparis* and *Boscia*, with special emphasis on the androecium. *Botanische Jahrbücher für Systematik* 119:231–255.

———. 1998a. Meristic changes in gynoecium morphology, exemplified by floral ontogeny and anatomy. *In* S. J. Owens and P. J. Rudall [eds.], Reproductive Biology in Systematics, Conservation and Economic Botany, 85–112. Royal Botanic Gardens, Kew.

———. 1998b. Notes on the evolution of androecial organisation in the Magnoliophytina (angiosperms). *Botanica Acta* 111:77–86.

———. 1999. Similarities in floral ontogeny and anatomy between the genera *Francoa* (Francoaceae) and *Greyia* (Greyiaceae). *International Journal of Plant Sciences* 160:377–393.

Ronse De Craene, L. P., H. P. Linder, T. Dlamini, and E. F. Smets. 2001. Evolution and development of floral diversity of Melianthaceae, an enigmatic Southern African family. *International Journal of Plant Sciences* 162:59–82.

Ronse De Craene, L. P., P. S. Soltis, and D. E. Soltis. 2003. Evolution of floral structure in basal angiosperms. *International Journal of Plant Sciences* 164:S329–S363.

Ronse De Craene, L. P., E. F. Smets, and P. Vanvinckenroye. 1998. Pseudodiplostemony, and its implications for the evolution of the androecium in the Caryophyllaceae. *Journal of Plant Research* 111:25–43.

Ronse De Craene, L. P., P. Vanvinckenroye, and E. F. Smets. 1997. A study of floral morphological diversity in *Phytolacca* (Phytolaccaceae) based on early floral ontogeny. *International Journal of Plant Sciences* 158:57–72.

Rosatti, T. J. 1984. The Plantaginaceae in the southeastern United States. *Journal of the Arnold Arboretum* 65:533–562.

———. 1986. The genera of Sphenocleaceae and Campanulaceae in the southeastern United States. *Journal of the Arnold Arboretum* 67:1–64.

———. 1987. The genera of Pontederiaceae in the southeastern United States. *Journal of the Arnold Arboretum* 68:35–71.

———. 1989. The genera of suborder Apocynineae (Apocynaceae and Asclepiadaceae) in the southeastern United States. *Journal of the Arnold Arboretum* 70:307–401, 443–514.

Rosenberg, M. S., and S. Kumar. 2001. Incomplete taxon sampling is not a problem for phylogenetic inference. *Proceedings of the National Academy of Sciences USA* 98:10751–10756.

Rothwell, G. W., and R. Serbet. 1994. Lignophyte phylogeny and the evolution of spermatophytes: a numerical cladistic analysis. *Systematic Botany* 19:443–482.

Rothwell, G. W., and R. A. Stockey. 2002. Anatomically preserved *Cycadeoidea* (Cycadeoidaceae), with a reevaluation of systematic characters for the seed cones of Bennettitales. *American Journal of Botany* 89:1447–1458.

Rothwell G. W., M. R. Van Atta, H. E. Ballard, Jr., and R. A. Stockey. 2003. Molecular phylogenetic relationships among Lemnaceae and Araceae using the chloroplast *trnL-trnF* intergenic spacer. *Molecular Phylogenetics and Evolution* 30:378–385.

Rudall, P. J. 1994. Anatomy and systematics of Iridaceae. *Botanical Journal of the Linnean Society* 114:1–21.

———. 1997. The nucellus and chalaza in monocotyledons: structure and systematics. *Botanical Review* 63:140–181.

———. 2000. 'Cryptic' characters in monocotyledons: homology and coding. *In* R. Scotland and R. T. Pennington [eds.], Homology and Systematics: Coding Characters for Phylogenetic Analysis 114–123. Taylor and Francis, London.

———. 2003. Monocot pseudanthia revisited: floral structure of the mycoheterotrophic family Triuridaceae. *International Journal of Plant Sciences* 164:S307–320.

Rudall, P. J., and R. M. Bateman. 2002. Roles of synorganisation, zygomorphy and heterotopy in floral evolution: the gynostemium and labellum of orchids and other lilioid monocots. *Biological Reviews* 77:403–441.

Rudall, P. J., and M. W. Chase. 1996. Systematics of Xanthorrhoeaceae sensu lato: evidence for polyphyly. *Telopea* 6:629–647.

Rudall, P. J., P. J. Cribb, D. F. Cutler, and C. J. Humphries [eds.]. 1995. Monocotyledons: Systematics and Evolution. Royal Botanical Gardens, Kew.

Rudall, P. J., C. A. Furness, M. W. Chase, and M. F. Fay. 1997. Microsporogenesis and pollen sulcus type in Asparagales (Lilianae). *Canadian Journal of Botany* 75:408–430.

Rudall, P. J., D. W. Stevenson, and H. P. Linder. 1999. Structure and systematics of *Hanguana*, a monocotyledon of uncertain affinity. *Australian Systematic Botany* 12:311–330.

Rudall, P., K. L. Stobart, W.-P. Hong, J. G. Conran, C. A. Furness, G. C. Kite, and M. W. Chase. 2000. Consider the lilies: systematics of Liliales. *In* K. L. Wilson and D. A. Morrison [eds.], Monocots: Systematics and Evolution, 347–359. CSIRO, Melbourne.

Runions, C. J., and J. N. Owens. 1998. Evidence of pre-zygotic self-incompatibility in a conifer. *In* S. J. Owens and P. J. Rudall [eds.], Reproductive Biology in Systematics, Conservation and Economic Botany, 255–264. Royal Botanic Gardens, Kew.

Rydin, C., M. Källersjö, and E. M. Friis. 2002. Seed plant relationships and the systematic position of Gnetales based on nuclear and chloroplast DNA: conflicting data, rooting problems and the monophyly of conifers. *International Journal of Plant Sciences* 163:197–214.

Sage, R. F., M. Li, and R. K. Monson. 1999. The taxonomic distribution of C_4 photosynthesis. *In* R. F. Sage and R. K. Monson [eds.], The Biology of C_4 Photosynthesis. Academic Press, San Diego.

Salamin, N., M. W. Chase, T. R. Hodkinson, and V. Savolanien. 2003. Assessing internal support with large phylogenetic DNA matrices. *Molecular Phylogenetics and Evolution* 27:528–539.

Salamin, N., T. R. Hodkinson, and V. Savolainen. 2002. Building supertrees: an empirical assessment using the grass family (Poaceae). *Systematic Biology* 51:136–150.

Salazar, G. A., M. W. Chase, M. A. Soto Arenas, and M. J. Ingrouille. 2003. Phylogenetics of Cranichideae with an emphasis on Spiranthinae (Orchidaceae: Orchidoideae): evidence from plastid and nuclear DNA sequences. *American Journal of Botany* 90:777–795.

Salisbury, E. J. 1919. Variation in *Eranthis hyemalis, Ficaria verna,* and other members of the Ranunculaceae, with special reference to trimery and the origin of the perianth. *Annals of Botany* 33:47–79.

Sampson, F. B. 2000. Pollen diversity in some modern magnoliids. *International Journal of Plant Sciences* 161:S193–S210.

Samuel, R., H. Kathriarachi, P. Hoffmann, M. Barfuss, K. J. Wurdack, and M. W. Chase. 2005. Molecular phylogenetics of Phyllanthaceae: evidence from plastid *matK* and nuclear PHYC sequences. *American Journal of Botany* 92:132–141.

Sanderson, M. J. 1991. In search of homoplastic tendencies: statistical inference of topological patterns in homoplasy. *Evolution* 45:351–358.

———. 1993. Reversibility in evolution: a maximum likelihood approach to character gain/loss bias in phylogenies. *Evolution* 47:236–252.

———. 1997. A nonparametric approach to estimating divergence times in the absence of rate constancy. *Molecular Biology and Evolution* 14:1218–1231.

———. 1998. Estimating rate and time in molecular phylogenies: beyond the molecular clock? *In* D. E. Soltis, P. S. Soltis, and J. J. Doyle [eds.], Molecular Systematics of Plants II: DNA sequencing, 242–264. Kluwer, Boston.

———. 2002. Estimating absolute rates of molecular evolution and divergence times: A penalized likelihood approach. *Molecular Biology and Evolution* 19:101–109.

Sanderson, M. J., and M. J. Donoghue. 1992. The suitability of molecular and morphological evidence in reconstructing plant phylogeny. *In* P. S. Soltis, D. E. Soltis, and J. J. Doyle [eds.], Molecular Systematics of Plants, 340–368. Chapman and Hall, New York.

———. 1994. Shifts in diversification rate with the origin of angiosperms. *Science* 264:1590–1593.

Sanderson, M. J., and J. A. Doyle. 2001. Sources of error and confidence intervals in estimating the age of angiosperms from *rbcL* and 18S rDNA data. *American Journal of Botany* 88:1499–1516.

Sanderson, M. J., A. Purvis, and C. Henze. 1998. Phylogenetic supertrees: assembling the tree of life. *Trends in Ecology and Evolution* 13:105–109.

Sanderson, M. J., J. L. Thorne, N. Wikström, and K. Bremer. 2004. Molecular evidence on plant divergence times. *American Journal of Botany*, 91:1656–1665.

Sanderson, M. J., M. F. Wojciechowski, J.-M. Hu, T. Sher Khan, and S. G. Brady. 2000. Error, bias, and long-branch attraction in data for two chloroplast photosystem genes in seed plants. *Molecular Biology and Evolution* 17:782–797.

Sang, T., D. J. Crawford, and T. F. Stuessy. 1995. Documentation of reticulate evolution in peonies (*Paeonia*) using internal transcribed spacer sequences of nuclear ribosomal DNA: implications for biogeography and concerted evolution. *Proceedings of the National Academy of Sciences USA* 92:6813–6817.

———. 1997. Chloroplast DNA phylogeny, reticulate evolution, and biogeography of *Paeonia* (Paeoniaceae). *American Journal of Botany* 84:1120–1136.

Sankoff, D. 2001. Gene and genome duplication. *Current Opinions in Genetics and Development* 11:681–684.

SanMiguel, P. A., A. Tikhonov, Y. K. Jin, N. Motochoulskaia, D. Zakharov, A. Melake-Berhan, P. S. Springer, K. J. Edwards, M. Lee, Z. Avramova, and J. Bennetzen. 1996. Nested retrotransposons in the intergenic regions of the maize genome. *Science* 274:765–768.

Sarich, V., and A. C. Wilson. 1967. Rates of albumin evolution in primates. *Proceedings of the National Academy of Sciences USA* 58:142–148.

Satô, Y. 1976. Embryological studies of some cornaceous plants. *Science Reports of the Tôhoku University, Series IV* 37:117–130.

Sattler, R. 1973. Organogenesis of Flowers: A Photographic Text-Atlas. University of Toronto Press, Toronto.

Sauquet, H., J. A. Doyle, T. Scharaschkin, T. Borsch, K. W. Hilu, L. W. Chatrou, and A. Le Thomas. 2003. Phylogenetic analysis of Magnoliales and Myristicaceae based on multiple data sets: implications for character evolution. *Botanical Journal of the Linnean Society* 142:125–186.

Savolainen, V., M. Chase, C. M. Morton, D. E. Soltis, C. Bayer, M. F. Fay, A. de Bruijn, S. Sullivan, and Y.-L. Qiu. 2000a. Phylogenetics of flowering plants based upon a combined analysis of plastid *atpB* and *rbcL* gene sequences. *Systematic Biology* 49:306–362.

Savolainen, V., M. W. Chase, N. Salamin, D. E. Soltis, P. S. Soltis, A. Lopez, O. Fedrigo, and G. J. P. Naylor. 2002. Phylogeny reconstruction and functional constraints in organellar genomes: plastid versus animal mitochondrion. *Systematic Biology* 51:638–647.

Savolainen, V., M. F. Fay, D. C. Albach, A. Backlund, M. van der Bank, K. M. Cameron, S. A. Johnson, M. D. Lledo, J.-C. Pintaud, M. Powell, M. C. Sheahan, D. E. Soltis, P. S. Soltis, P. Weston, W. M. Whitten, J. Wurdack, and M. W. Chase. 2000b. Phylogeny of the eudicots: a nearly complete familial analysis based on *rbcL* gene sequences. *Kew Bulletin* 55:257–309.

Savolainen, V., J. F. Manen, E. Douzery, and R. Spichiger. 1994. Molecular phylogeny of families related to Celastrales based on *rbcL* 5' flanking regions. *Molecular Phylogenetics and Evolution* 3:27–37.

Savolainen, V., R. Spichiger, and J. F. Manen. 1997. Polyphyletism of Celastrales deduced from a non-coding chloroplast DNA region. *Molecular Phylogenetics and Evolution* 7:145–157.

Sawada, M. 1971. Floral vascularization of *Paeonia japonica* with some consideration on systematic position of the Paeoniaceae. *Botanical Magazine, Tokyo* 84:51–60.

Schaal, B. A., and W. J. Leverich. 2001. Plant population biology and systematics. *Taxon* 50:679–696.

Schatz, G. E. 2000. The rediscovery of a Malagasy endemic: *Takhtajania perrieri* (Winteraceae). *Annals of the Missouri Botanical Garden* 87:297–302.

Schick, B. 1980. Untersuchungen über Biotechnik der Apocynaceenblüte. I. Morphologie und Funktion des Narbenkopfes. *Flora, Morphology, Geobotany, Oekophysiology* 170:394–432.

———. 1982. Untersuchungen über die Biotechnik der Apocynaceenblüte. II. Bau und Funktion des Bestäubungsapparates. *Flora, Morphology, Geobotany, Oekophysiology* 172:347–371.

Schinz, H. 1893. Amaranthaceae. *In* A. Engler and K. Prantl [eds.], Die Natürlichen Pflanzenfamilien, 1st ed., Vol. III, 1a, 91–118. Engelmann, Leipzig.

Schmid, R. 1964. Die systematische Stellung der Dioncophyllaceae. *Botanische Jahrbücher für Systematik* 83:1–56.

———. 1978. Actinidiaceae, Davidiaceae, and Paracryphiaceae: systematic considerations. *Botanische Jahrbücher für Systematik* 100:196–204.

Schneider, E. L. 1979. Pollination biology of the Nymphaeaceae. *In* D. M. Caron [ed.], Proceedings of the Fourth International Symposium on Pollination, 419–430. Md Agricultural Experimental Station Special Miscellaneous Publication 1.

Schneider, E. L., and S. Carlquist. 1996. Vessels in *Brasenia* (Cabombaceae): new perspectives on vessel origin in primary xylem of angiosperms. *American Journal of Botany* 83:1236–1240.

Schneider, E. L., S. Carlquist, and A. Kohn. 1995. Vessels in Nymphaeaceae: *Nuphar, Nymphaea,* and *Ondinea. International Journal of Plant Sciences* 156:857–862.

Schneider, E. L., S. C. Tucker, and P. S. Williamson. 2003. Floral development in the Nymphaeales. *International Journal of Plant Sciences* 164:S279–S292.

Schodde, R. 1970. Two new suprageneric taxa in the Monimiacaeae alliance (Laurales). *Taxon* 19:324–332.

Schoen, D. J., M. T. Morgan, and T. Bataillon. 1997. How does self-pollination evolve? Inferences from floral ecology and molecular genetic variation. *In* J. Silvertown, M. Franco, and J. L. Harper [eds.], Plant Life Histories, 77–101. Cambridge University Press, Cambridge.

Schöffel, K. 1932. Untersuchungen über den Blütenbau der Ranunculaceen. *Planta* 17:315–371.

Scholz, H. 1964. Geraniales, Rutales, Sapindales, Celastrales. *In* H. Melchior [ed.], A. Engler's Syllabus der Pflanzenfamilien, 12th ed., 267–268, 246–262. Gebrüder Borntraeger, Berlin-Nikolassee.

Schönenberger, J., and E. Conti. 2003. Molecular phylogeny and floral evolution of Penaeaceae, Oliniaceae, Rhynchocalycaceae, and Alzateaceae (Myrtales). *American Journal of Botany* 90:293–309.

Schönenberger, J., E. M. Friis, M. L. Matthews, and P. K. Endress. 2001. Cunoniaceae in the Cretaceous of Europe: evidence from fossil flowers. *Annals of Botany* 88:423–437.

Schopf, J. M. 1976. Morphologic interpretations of fertile structures in glossopterid gymnosperms. *Review of Palaeobotany and Palynology* 21:25–64.

Schuettpelz, E., S. B. Hoot, R. Samuel, and F. Ehrendorfer. 2002. Multiple origins of Southern Hemisphere *Anemone* (Ranunculaceae) based on plastid and nuclear sequence data. *Plant Systematics and Evolution* 231:143–151.

Schürhoff, P. N. 1926. Die Zytologie der Blütenpflanzen. Ferdinand Enke Verlag, Stuttgart.

Schwarz-Sommer, Z., P. Huijser, W. Nacken, H. Saedler, and H. Sommer. 1990. Genetic control of flower development in *Antirrhinum majus. Science* 250:931–936.

Schwarzbach, A. E., and L. A. McDade. 2002. Phylogenetic relationships of the mangrove family Avicenniaceae based on chloroplast and nuclear ribosomal DNA sequences. *Systematic Botany* 27:84–98.

Schwarzbach, A. E., and R. E. Ricklefs. 2000. Systematic affinities of Rhizophoraceae and Anisophylleaceae, and intergeneric relationships within Rhizophoraceae, based on chloroplast DNA, nuclear ribosomal DNA, and morphology. *American Journal of Botany* 87:547–564.

Scogin, R. 1977. Anthocyanins of the Fouquieriaceae. *Biochemical Systematics and Ecology* 6:297–298.

———. 1978. Leaf phenolics of the Fouquieriaceae. *Biochemical Systematics and Ecology* 6:297–298.

Scotland, R. W., J. A. Sweere, P. A. Reeves, and R. G. Olmstead. 1995. Higher level systematics of Acanthaceae determined by chloroplast DNA sequences. *American Journal of Botany* 82:266–275.

Seelanan, T., A. Schnabel, and J. Wendel. 1997. Congruence and consensus in the cotton tribe (Malvaceae). *Systematic Botany* 22:259–290.

Segraves, K. A., and J. N. Thompson. 1999. Plant polyploidy and pollination: floral traits and insect visits to diploid and tetraploid *Heuchera grossulariifolia. Evolution* 53:1114–1121.

Sennblad, B., and B. Bremer. 1996. The familial and subfamilial relationships of Apocynaceae and Asclepiadaceae evaluated with *rbcL* data. *Plant Systematics and Evolution* 202:153–175.

———. 2002. Classification of Apocynaceae s. l. according to a new approach combining Linnean and phylogenetic taxonomy. *Systematic Biology* 51:389–409.

Senters, A. E., and D. E. Soltis. 2003. Phylogenetic relationships in *Ribes* (Grossulariaceae) inferred from ITS sequence data. *Taxon* 52:51–66.

Sharma, V. K. 1968. Floral morphology, anatomy, and embryology of *Coriaria nepalensis* Wall. with a discussion of the interrelationships of the family Coriariaceae. *Phytomorphology* 18:143–153.

Sheahan, M. C., and M. W. Chase. 1996. A phylogenetic analysis of Zygophyllaceae R. Br. based on morphological, anatomical, and *rbcL* DNA sequence data. *Botanical Journal of the Linnean Society* 122:279–300.

———. 2000. Phylogenetic relationships within Zygophyllaceae based on DNA sequences of three plastid regions, with special emphasis on Zygophylloideae. *Systematic Botany* 25:371–384.

Shindo, S., K. Sakakibara, R. Sano, K. Ueda, and M. Hasebe. 2001. Characterization of a *FLORICAULA/LEAFY* homologue of *Gnetum parviflorum* and its implications for the evolution of reproductive organs in seed plants. *International Journal of Plant Sciences* 162:1199–1209.

Shore, J. S., K. L. McQueen, and S. L. Little. 1994. Inheritance of plastid DNA in the *Turnera ulmifolia* complex. *American Journal of Botany* 81:1636–1639.

Simmons, M. P., V. Savolainen, C. C. Clevinger, R. H. Archer, and J. I. Davis. 2001. Phylogeny of the Celastraceae inferred from 26S nuclear ribosomal DNA, phytochrome B, *rbcL, atpB,* and morphology. *Molecular Phylogenetics and Evolution* 19:353–366.

Simpson, B. B., and J. L. Neff. 1981. Floral rewards: alternatives to pollen and nectar. *Annals of the Missouri Botanical Garden* 68:301–322.

Simpson, D. 1995. Relationships within Cyperales. *In* P. J. Rudall, P. J. Cribb, D. F. Cutler, and C. J. Humphries [eds.], Monocotyledons: Systematics and Evolution, 750. Royal Botanic Gardens, Kew.

Sims, H. J., P. S. Herendeen, and P. R. Crane. 1998. New genus of fossil Fagaceae from the Santonian (Late Cretaceous) of central Georgia, USA. *International Journal of Plant Sciences* 159:391–404.

Sims, H. J., P. S Herendeen, R. Lupia, R. A. Christopher, and P. R. Crane. 1999. Fossil flowers with Normapolles pollen from the Upper Cretaceous of southeastern North America. *Review of Palaeobotany and Palynology* 106:131–151.

Sinha, N. 2000. The response of epidermal cells to contact. *Trends in Plant Science* 5:233–234.

Sinha, N., and E. Kellogg. 1996. Parallelism and diversity in multiple origins of C_4 photosynthesis in the grass family. *American Journal of Botany* 83:1458–1470.

Skipper, M. 2002. Genes from the *APETALA3* and *PISTILLATA* lineages are expressed in developing vascular bundles of the tuberous rhizome, flowering stem and flower primordia of *Eranthis hyemalis. Annals of Botany* 89:83–88.

Skvarla, J. J., and J. W. Nowicke. 1976. Ultrastructure of pollen exine in centrospermous families. *Plant Systematics and Evolution* 126:55–78.

———. 1982. Pollen fine structure and relationships of *Achatocarpus triana* and *Phaulothamnus* A. Gray. *Taxon* 31:244–249.

Skvarla, J. J., B. L. Turner, V. C. Patel, and A. S. Tomb. 1977. Pollen morphology in the Compositae and in morphologically related families. *In* V. H. Heywood, J. B. Harborne, and B. L. Turner [eds.], The Biology and Chemistry of the Compositae, 141–201. Academic Press, London.

Sleumer, H. 1954. Flacourtiaceae. *In* C. G. G. J. van Steenis [ed.], Flora Malesiana, series 1, 2–106, pp. 2–106. Noordhoff-Kolff N. V., Djakarta.

———. 1980. Flacourtiaceae. *Flora Neotropica* 22:1–499.

Smissen, R. D., J. C. Clement, P. J. Garnock-Jones, and G. K. Chambers. 2002. Subfamilial relationships within Caryophyllaceae as inferred from 5' *ndhF* sequences. *American Journal of Botany* 89:1336–1341.

Smith, A. R., K. M. Pryer, P. G. Wolf, R. Cranfill, and H. Schneider. An ordinal and familial classification for extant lycophytes and moniliophytes. *Taxon* (submitted).

Smith, D., T. J. Barkman, and C. W. dePamphilis. 2001. Hemiparasitism, Encyclopedia of Biodiversity, vol. 3, 317–328. Academic Press, New York.

Smith, G. H. 1928. Vascular anatomy of Ranalian flowers. II. Ranunculaceae (continued), Menispermaceae, Calycanthaceae, Annonaceae. *Botanical Gazette* 85:152–177.

Smith, J. F. 1996. Tribal relationships within the Gesneriaceae: A cladistic analysis of morphological data. *Systematic Botany* 21:497–513.

———. 2000a. A phylogenetic analysis of tribes Beslerieae and Napeantheae (Gesneriaceae) and evolution of fruit types: parsimony and maximum likelihood analyses of *ndhF* sequences. *Systematic Botany* 25:72–81.

———. 2000b. Phylogenetic resolution within the tribe Episcieae (Gesneriaceae): congruence of ITS and *ndhF* sequences from parsimony and maximum-likelihood analyses. *American Journal of Botany* 87:883–897.

Smith, J. F., and S. Atkinson. 1998. Phylogenetic analysis of the tribes Gloxinieae and Gesnerieae (Gesneriaceae): Data from *ndhF* sequences. *Selbyana* 19:122–131.

Smith, J. F., and C. L. Carroll. 1997. Phylogenetic relationships of the Episcieae (Gesneriaceae) based on *ndhF* sequences. *Systematic Botany* 22:713–724.

Smith, J. F., W. J. Kress, and E. A. Zimmer. 1993. Phylogenetic analysis of the Zingiberales based on *rbcL* sequences. *Annals of the Missouri Botanical Garden* 80:620–630.

Smith, L. B., and C. E. Wood. 1975. The genera of Bromeliaceae in the southeastern United States. *Journal of the Arnold Arboretum* 56:375–397.

Soltis, D. E., and L. Hufford. 2002. Ovary development and diversification in Saxifragaceae. *International Journal of Plant Sciences* 163:277–293.

Soltis, D. E., and P. S. Soltis. 1990. Isozyme evidence for ancient polyploidy in primitive angiosperms. *Systematic Botany* 15:328–337.

———. 1997. Phylogenetic relationships in Saxifragaceae sensu lato: a comparison of topologies based on 18S rDNA and *rbcL* sequences. *American Journal of Botany* 84:504–522.

———. 1998. Choosing an approach and an appropriate gene for phylogenetic analysis. In D. E. Soltis, P. S. Soltis, and J. J. Doyle [eds.], Molecular Systematics of Plants II: DNA Sequencing, 1–42. Kluwer, Boston.

———. 1999. Phylogenetic analyses of large data sets: Approaches using the angiosperms. In M. Kato [ed.], The Biology of Diversity, 91–103. Springer, Tokyo.

———. 2000. Contributions of molecular systematics to studies of molecular evolution. *Plant Molecular Biology* 42:45–75.

———. 2003. The role of phylogenetics in comparative genetics. *Plant Physiology* 132:1790–1800.

Soltis, D. E., V. A. Albert, S. Kim, M.-J. Yoo, P. S. Soltis, M. W. Frohlich, J. Leebens-Mack, H. Kong, K. Wall, C. dePamphilis, and H. Ma. 2005. Evolution of the flower. In R. Henry [ed.], Diversity and Evolution of Plants 165–200. CABI Publishing, Wallingford, UK.

Soltis, D. E., V. A. Albert, V. Savolainen, K. Hilu, Y-L. Qiu, M. W. Chase, J. S. Farris, S. Stefanovic, D. W. Rice, J. D. Palmer, and P. S. Soltis. 2004. Genome-scale data, angiosperm relationships, and "ending incongruence": a cautionary tale in phylogenetics. *Trends in Plant Science* 9:477–483.

Soltis, D. E., M. Fishbein, and R. K. Kuzoff. 2003c. Reevaluating the evolution of epigyny: Data from phylogenetics and floral ontogeny. *International Journal of Plant Sciences* 164:S251–S264.

Soltis, D. E., A. Grable, D. Morgan, P. S. Soltis, and R. Kuzoff. 1993. Molecular systematics of Saxifragaceae sensu stricto. *American Journal of Botany* 80:1056–1081.

Soltis, D. E., C. Hibsch-Jetter, P. S. Soltis, M. Chase, and J. S. Farris. 1997b. Molecular phylogenetic relationships among angiosperms: an overview based on *rbcL* and 18S rDNA sequences. In K. Iwatsuki and P. H. Raven [eds.], Evolution and Diversification of Land Plants, 157–178. Springer, Tokyo.

Soltis, D. E., M. E. Mort, R. Kuzoff, M. Zanis, M. Fishbein, and L. Hufford. 2001b. Elucidating deep-level phylogenetic relationships in Saxifragaceae using sequences for six chloroplastic and nuclear DNA regions. *Annals of the Missouri Botanical Garden* 88:669–693.

Soltis, D. E., M. E. Mort, P. S. Soltis, C. Hibsch-Jetter, E. A. Zimmer, and D. Morgan. 1999. Phylogenetic relationships of the enigmatic angiosperm family Podostemaceae inferred from 18S rDNA and *rbcL* sequence data. *Molecular Phylogenetics and Evolution* 11:261–272.

Soltis, D. E., A. E. Senters, M. Zanis, S. Kim, J. D. Thompson, P. S. Soltis, L. P. Ronse De Craene, P. K. Endress, and J. S. Farris. 2003a. Gunnerales are sister to other core eudicots: implications for the evolution of pentamery. *American Journal of Botany* 90:461–470.

Soltis, D. E., P. S. Soltis, V. Albert, C. dePamphilis, M. Frohlich, H. Ma, D. Oppenheimer, and G. Theissen. 2002b. Missing links: the genetic architecture of the flower and floral diversification. *Trends in Plant Science* 7:22–30.

Soltis, D. E., P. S. Soltis, M. D. Bennett, and I. J. Leitch. 2003b. Evolution of genome size in the angiosperms. *American Journal of Botany* 90:1596–1603.

Soltis, D. E., P. S. Soltis, M. W. Chase, M. Mort, D. Albach, M. Zanis, V. Savolainen, W. Hahn, S. Hoot, M. Fay, M. Axtell, S. Swensen, K. Nixon, and J. Farris. 2000. Angiosperm phylogeny inferred from a combined data set of 18S rDNA, *rbcL* and *atpB* sequences. *Botanical Journal of the Linnean Society* 133:381–461.

Soltis, D. E., P. S. Soltis, D. R. Morgan, S. M. Swensen, B. C. Mullin, J. M. Dowd, and P. G. Martin. 1995. Chloroplast gene sequence data suggest a single origin of the predisposition for symbiotic nitrogen fixation in angiosperms. *Proceedings of the National Academy of Sciences USA* 92:2647–2651.

Soltis, D. E., P. S. Soltis, M. E. Mort, M. W. Chase, V. Savolainen, S. B. Hoot, and C. M. Morton. 1998. Inferring complex phylogenies using parsimony: an empirical approach using three large DNA data sets for angiosperms. *Systematic Biology* 47:32–42.

Soltis, D. E., P. S. Soltis, D. L. Nickrent, L. A. Johnson, W. J. Hahn, S. B. Hoot, J. A. Sweere, R. K. Kuzoff, K. A. Kron, M. Chase, S. M. Swensen, E. Zimmer, S. M. Chaw, L. J. Gillespie, W. J. Kress, and K. Sytsma. 1997a. Angiosperm phylogeny inferred from 18S ribosomal DNA sequences. *Annals of the Missouri Botanical Garden* 84:1–49.

Soltis, D. E., P. S. Soltis, and M. Zanis. 2002a. Phylogeny of seed plants based on evidence from eight genes. *American Journal of Botany* 89:1670–1681.

Soltis, D. E., M. Tago-Nakazawa, Q.-X. Xiang, S. Kawano, J. Murata, M. Wakabayashi, and C. Hibsch-Jetter. 2001a. Phylogenetic relationships and evolution in *Chrysosplenium* (Saxifragaceae) based on *matK* sequence data. *American Journal of Botany* 88:883–893.

Soltis, P. S., and D. E. Soltis. 1998. Molecular evolution of 18S rDNA in angiosperms: implications for character weighting in phylogenetic analysis. In D. E. Soltis, P. S. Soltis, and J. J. Doyle [eds.], Molecular Systematics of Plants II: DNA Sequencing, 188–210. Kluwer, Boston.

———. 2000. The role of genetic and genomic attributes in the success of polyploids. *Proceedings of the National Academy of Sciences USA* 97:7051–7057.

———. 2001. Molecular systematics: assembling and using the tree of life. *Taxon* 50:663–678.

———. 2004. The origin and diversification of angiosperms. *American Journal of Botany* 91:1614–1626.

Soltis, P. S., D. E. Soltis, P. Wolf, D. Nickrent, S.-M. Chaw, and R. L. Chapman. 1999a. The phylogeny of land plants inferred from 18S rDNA sequences: Pushing the limits of rDNA signal? *Molecular Biology and Evolution* 16:1774–1784.

Soltis, P. S., D. E. Soltis, and M. W. Chase. 1999b. Angiosperm phylogeny inferred from multiple genes as a research tool for comparative biology. *Nature* 402:402–404.

Soltis, P. S., D. E. Soltis, M. W. Chase, P. K. Endress, and P. R. Crane. 2004. The diversification of flowering plants. *In* M. Donoghue and J. Cracraft [eds.], Assembling the Tree of Life, 154–167. Oxford University Press, Oxford.

Soltis, P. S., D. E. Soltis, V. Savolainen, P. R. Crane, and T. G. Barraclough. 2002. Rate heterogeneity among lineages of tracheophytes: integration of molecular and fossil data and evidence for molecular living fossils. *Proceedings of the National Academy of Sciences USA* 99:4430–4435.

Soltis, P. S., D. E. Soltis, M. J. Zanis, and S. Kim. 2000. Basal lineages of angiosperms: Relationships and implications for floral evolution. *International Journal of Plant Sciences* 161:S97–S107.

Soreng, R. J., and J. I. Davis. 1998. Phylogenetics and character evolution in the grass family (Poaceae): simultaneous analysis of morphological and chloroplast DNA restriction site character sets. *Botanical Review* 64:1–85.

Soros, C. L., and D. H. Les. 2002. Phylogenetic relationships in the Alismataceae. *Botany 2002: botany in the curriculum, Abstracts* [Madison, Wisconsin], 152.

Sosa, V., and M. W. Chase. 2003. Phylogenetics of Crossosomataceae based on *rbcL* sequence data. *Systematic Botany* 28: 96–105.

Spencer, K. C., and D. S. Seigler. 1985a. Cyanogenic glycosides of *Malesherbia*. *Biochemical Systematics and Ecology* 13:421–431.

———. 1985b. Cyanogenic glycosides and systematics of the Flacourtiaceae. *Biochemical Systematics and Ecology* 13:433–435.

Sperling, C. R., and V. Bittrich. 1993. Basellaceae. *In* K. Kubitzki, J. G. Rohwer, and V. Bittrich [eds.], The Families and Genera of Vascular Plants, II, 143–146. Springer, Berlin.

Sperry, J. S. 1983. Observations on the structure and function of hydathodes in *Blechnum lehmanii*. *American Fern Journal* 73:65–72.

Sperry, J. S., K. L. Nicols, J. E. M. Sullivan, and S. E. Eastlack. 1994. Xylem embolism in ring-porous, diffuse-porous, and coniferous trees of northern Utah and interior Alaska. *Ecology* 75:1736–1752.

Spongberg, S. A. 1971. The Staphyleaceae in the southeastern United States. *Journal of the Arnold Arboretum* 52:196–203.

———. 1972. The genera of Saxifragaceae in the southeastern United States. *Journal of the Arnold Arboretum* 53:409–498.

———. 1978. The genera of Crassulaceae in the southeastern United States. *Journal of the Arnold Arboretum* 59:197–248.

Spooner, D. M., G. J. Anderson, and R. K. Jansen. 1993. Chloroplast DNA evidence for the interrelationships of tomatoes, potatoes, and pepinos (Solanaceae). *American Journal of Botany* 80:676–688.

Sporne, K. R. 1954. A note on nuclear endosperm as a primitive character among dicotyledons. *Phytomorphology* 4:275–278.

Sprague, E. F. 1925. The classification of dicotyledons. II. Evolutionary progressions. *Journal of Botany* 63:105–113.

Sprent, J. I. 1994. Evolution and diversity in the legume-rhizobium symbiosis: Chaos theory? *Plant and Soil* 161:1–10.

Springer, M. S., E. C. Teeling, O. Madsen, M. J. Stanhope, and W. W. De Jong. 2001. Integrated fossil and molecular data reconstruct bat echolocation. *Proceedings of the National Academy of Sciences USA* 98:6241–6246.

Stafford, H. A. 1994. Anthocyanins and betalains: evolution of the mutually exclusive pathways. *Plant Science* 101:91–98.

Stanley, S. M. 1975. A theory of evolution above the species level. *Proceedings of the National Academy of Sciences USA* 72:646–650.

———. 1979. Macroevolution: Pattern and Process. Freeman, San Francisco.

Stebbins, G. L. 1967. Adaptive radiation and trends of evolution in higher plants. *Evolutionary Biology* 1:101–142.

———. 1970. Adaptive radiation in angiosperms. I. Pollination mechanisms. *Annual Review of Ecology and Systematics* 1:307–326.

———. 1971. Chromosomal Variation in Higher Plants. Arnold, London.

———. 1974. Flowering Plants: Evolution Above the Species Level. Belknap Press, Cambridge.

———. 1976a. Chromosome, DNA and plant evolution. *Evolutionary Biology* 9:1–34.

———. 1976b. Chromosomal Evolution in Higher Plants. Addison-Wesley, London.

Stefanovic, S., M. Jager, J. Deutsch, J. Broutin, and M. Masselot. 1998. Phylogenetic relationships of conifers inferred from partial 28S rRNA gene sequences. *American Journal of Botany* 85:688–697.

Stefanovic, S., L. Krueger, and R. G. Olmstead. 2002. Monophyly of the Convolvulaceae and circumscription of their major lineages based on DNA sequences of multiple chloroplast loci. *American Journal of Botany* 89:1510–1522.

Steiner, K. E. 1991. Oil flowers and oil bees: further evidence for pollinator adaptation. *Evolution* 45:1493–1501.

Stellari, G. M., A. Jaramillo, and E. M. Kramer. 2004. Evolution of the *APETALA3* and *PISTILLATA* lineages of MADS-box containing genes in the basal angiosperms. *Molecular Biology and Evolution* 21:506–519.

Stern, W. L. 1974. Comparative anatomy and systematics of woody Saxifragaceae: *Escallonia*. *Botanical Journal of the Linnean Society* 68:1–20.

Stern, W. L., G. K. Brizicky, and R. H. Eyde. 1969. Comparative anatomy and relationships of Columelliaceae. *Journal of the Arnold Arboretum* 50:36–75.

Stevens, P. F. 1984. Metaphors and topology in the development of botanical systematics 1690–1960, or the art of putting new wine in old bottles. *Taxon* 33:169–211.

———. 2004. Angiosperm Phylogeny Website. Version 5, May 2004. Available at http://www. mobot.org/MOBOT/ research/APweb/.

———. Clusiaceae. *In* K. Kubitzki [ed.], The Families and Genera of Vascular Plants. III. Springer, Berlin, in press.

Stevenson, D. W., and H. Loconte. 1995. Cladistic analysis of monocot families. *In* P. J. Rudall, P. J. Cribb, D. F. Cutler, and C. J. Humphries [eds.], Monocotyledons: Systematics and Evolution, 543–578. Royal Botanic Gardens, Kew.

Stewart, W. N., and G. W. Rothwell. 1993. Paleobotany and the Evolution of Plants, 2nd ed. Cambridge University Press, Cambridge.

Struck, M. 1997. Floral divergence and convergence in the genus *Pelargonium* (Geraniaceae) in southern Africa: ecological and evolutionary considerations. *Plant Systematics and Evolution* 208:71–97.

Struwe, L., V. A. Albert, and B. Bremer. 1994. Cladistics and family level classification of the Gentianales. *Cladistics* 10:175–206.

Stuessy, T. F. 2000. Taxon names are *not* defined. *Taxon* 49:231–233.

Stuhlfauth, T., H. Fock, H. Huber, and K. Klug. 1985. The distribution of fatty acids including petroselinic and tariric acids in the fruit and seed oils of the Pittosporaceae, Araliaceae, Umbelliferae, Simarubaceae and Rutaceae. *Phytochemistry* 13:447–453.

Suh, Y., L. B. Thien, H. E. Reeve, and E. A. Zimmer. 1993. Molecular evolution and phylogenetic implications of internal transcribed spacer sequences of ribosomal DNA in Winteraceae. *American Journal of Botany* 80:1042–1055.

Sun, G., D. L. Dilcher, S. Zheng, and Z. Zhou. 1998. In search of the first flower: A Jurassic angiosperm, *Archaefructus*, from northeast China. I. *Science* 282:1692–1695.

Sun, G., Q. Li, D. L. Dilcher, S. Zheng, K. C. Nixon, and X. Wang. 2002. Archaefructaceae, a new basal angiosperm family. *Science* 296:899–904.

Sunberg, P., and G. Pleijel. 1994. Phylogenetic classification and the definition of taxon names. *Zoologica Scripta* 23:19–25.

Sundstrom, J., and P. Engström. 2002. Conifer reproductive development involves B-type MADS-box genes with distinct and different activities in male organ primordia. *Plant Journal* 31:161–169.

Surange, K. R., and S. Chandra. 1975. Morphology of the gymnospermous fructifications of the *Glossopteris* flora and their relationships. *Palaeontographica, Abt. B, Palaeophytology* 149:153–180.

Sutton, D. A. 1988. A Revision of the Tribe Antirrhineae. British Museum (Natural History), London/Oxford University Press, Oxford.

———. 1989. The Daphniphyllales: a systematic review. *In* P. R. Crane and S. Blackmore [eds.], Evolution, Systematics and Fossil History of the Hamamelidae, 1, 285–291. Clarendon Press, Oxford.

Suzuki, Y., G. V. Glazko, and M. Nei. 2002. Overcredibility of molecular phylogenies obtained by Bayesian phylogenetics. *Proceedings of the National Academy of Sciences USA* 99:16138–16143.

Swamy, B. G. L., and I. W. Bailey. 1949. The morphology and relationships of *Cercidiphyllum. Journal of the Arnold Arboretum* 30:187–210.

Swamy, B. G. L., and P. M. Ganapathy. 1957. On endosperm in dicotyledons. *Botanical Gazette* 119:47–50.

Swensen, S. M. 1996. The evolution of actinorhizal symbioses: evidence for multiple origins of the symbiotic association. *American Journal of Botany* 83:1503–1512.

Swensen, S. M., and B. C. Mullin. 1997. The impact of molecular systematics on hypotheses for the evolution of root nodule symbioses and implications for expanding symbioses to new host plant genera. *Plant and Soil* 194:185–192.

Swensen, S. M., J. N. Luthi, and L. H. Rieseberg. 1998. Datiscaceae revisited: monophyly and the sequence of breeding system evolution. *Systematic Botany* 23:157–169.

Swensen, S. M., B. C. Mullin, and M. W. Chase. 1994. Phylogenetic affinities of the Datiscaceae based on an analysis of nucleotide sequences from a plastid *rbcL* gene. *Systematic Botany* 19:157–168.

Swisher, C. C., Y.-Q. Wang, X.-L. Wang, X. X. Wang, and Y. Wang. 1999. Cretaceous age for the feathered dinosaurs of Liaoning, China. *Nature* 400:58–61.

Swofford, D. L. 1991. Phylogenetic Analysis Using Parsimony, v 3.0s. Illinois Natural History Survey, Champaign.

———. 1993. Phylogenetic Analysis Using Parsimony. v. 3.1.1. Illinois Natural History Survey, Champaign.

———. 1999. PAUP* Phylogenetic Analysis Using Parsimony, v. 4.0. Sinauer, Sunderland, MA.

Swofford, D. L., and W. P. Maddison. 1987. Reconstructing ancestral character states under Wagner parsimony. *Mathematical Biosciences* 87:199–229.

Swofford, D. L., G. J. Olsen, P. J. Waddell, and D. M. Hillis. 1996. Phylogenetic inference. *In* D. M. Hillis, C. Moritz, and B. K. Mable [eds.], Molecular Systematics, 407–514. Sinauer, Sunderland, MA.

Sykorova, E., K. Y. Lim, Z. Kunicka, M. W. Chase, M. D. Bennett, J. Fajkus, and A. R. Leitch. 2003. Telomere variability in the monocotyledonous plant order Asparagales. *Proceedings of the Royal Society of London B* 270:1893–1904.

Sytsma, K. J., and J. C. Pires. 2001. Plant systematics in the next 50 years- re-mapping the new frontier. *Taxon* 50:713–732.

Sytsma, K. J., and J. F. Smith. 1992. Molecular systematics of Onagraceae: examples from *Clarkia* and *Fuchsia. In* P. S.

Soltis, D. E. Soltis, and J. J. Doyle [eds.], Molecular Systematics of Plants. Chapman and Hall, New York.

Sytsma, K. J., J. Morawetz, J. C. Pires, M. Nepokroeff, E. Conti, M. Zjhra, J. C. Hall, and M. W. Chase. 2002. Urticalean rosids: circumscription, rosid ancestry, and phylogenetics based on *rbcL*, *trnL-F*, and *ndhF* sequences. *American Journal of Botany* 89:1531–1546.

Sytsma, K. J., J. Smith, and L. Gottlieb. 1990. Phylogenetics in *Clarkia* (Onagraceae): restriction site mapping of chloroplast DNA. *Systematic Botany* 15:280–295.

Takahashi, K., and M. Nei. 2000. Efficiencies of fast algorithms of phylogenetic inference under the criteria of maximum parsimony, minimum evolution and maximum likelihood when a large number of sequences are used. *Molecular Biology and Evolution* 17:1251–1258.

Takhtajan, A. 1969. Flowering Plants: Origin and Dispersal. Smithsonian Institution Press, Washington.

———. 1980. Outline of the classification of flowering plants (Magnoliophyta). *Botanical Review* 46:225–359.

———. [ed.]. 1985. Anatomia seminum comparativa. Tomus 1. Liliopsida seu Monocotyledones. NAUKA, Leningrad. [In Russian.]

———. 1987. Systema Magnoliophytorum. NAUKA, Leningrad. [In Russian.]

———. 1991. Evolutionary Trends in Flowering Plants. Columbia University Press, New York.

———. 1997. Diversity and Classification of Flowering Plants. Columbia University Press, New York.

Tamura, M. 1965. Morphology, ecology, and phylogeny of the Ranunculaceae. IV. Reproductive organs. *Science Reporter Osaka University* 14:53–71.

———. 1993. Ranunculaceae. *In* K. Kubitzki, J. Rohwer, and V. Bittrich [eds.], Families and Genera of Vascular Plants, II, 563–583. Springer, Berlin.

Tamura, M. N. 1998. Calochortaceae, Liliaceae, Melanthiaceae, Nartheciaceae, and Trilliaceae. *In* K. Kubitzki, [ed.], Families and Genera of Vascular Plants, III, 164–172, 343–353, 369–391, 444–451. Springer, Berlin.

Tamura, M. N., J. Yamashita, S. Fuse, and M. Haraguchi. 2004. Molecular phylogeny of monocotyledons inferred from combined analysis of *matK* and *rbcL* gene sequences. *Journal of Plant Research* 117:109–120.

Tang, Y. 1994. Embryology of *Plagiopteron suaveolens* Griffith (Plagiopteraceae) and its systematic implications. *Botanical Journal of the Linnean Society* 116:145–157.

Taylor, D. W. 1990. Paleobiogeographic relationships of angiosperms from the Cretaceous and early Tertiary of the North American area. *Botanical Review* 56:249–417.

Taylor, D. W., and L. J. Hickey. 1992. Phylogenetic evidence for the herbaceous origin of angiosperms. *Plant Systematics and Evolution* 180:137–156.

———. 1996a. Evidence for and implications of an herbaceous origin for angiosperms. *In* D. W. Taylor and L. J. Hickey [eds.], Flowering Plant Origin, Evolution and Phylogeny, 232–266. Chapman and Hall, New York.

———. 1996b. Flowering Plant Origin, Evolution and Phylogeny. Chapman and Hall, New York.

Taylor, E. L., and T. N. Taylor. 1992. Reproductive biology of the Permian Glossopteridales and their suggested relationship to flowering plants. *Proceedings of the National Academy of Sciences USA* 89:11495–11497.

Taylor, R. L. 1965. The genus *Lithophragma* (Saxifragaceae). University of California Publications in Botany, Vol. 37.

Taylor, T. N., and S. Archangelsky. 1985. The Cretaceous pteridosperms *Ruflorinia* and *Ktalenia* and implications of cupule and carpel evolution. *American Journal of Botany* 72:1842–1853.

Taylor, T. N., G. M. Del Fueyo, and E. L. Taylor. 1994. Permineralized seed fern cupules from the Triassic of Antarctica: Implications for cupule and carpel evolution. *American Journal of Botany* 81:666–677.

Taylor, W. C., and R. J. Hickey. 1992. Habitat, evolution, and speciation in *Isoetes*. *Annals of the Missouri Botanical Garden* 79:613–622.

Templeton, A. R. 1983. Phylogenetic inference from restriction endonuclease cleavage site maps with particular reference to the evolution of human and apes. *Evolution* 37:221–244.

Ter Welle, B. J. H. 1976. Silica grains in woody plants of the neotropics, especially Surinam. *In* P. Baas, A. J. Bolton, and M. Catling [eds.], Wood Structure in Biological and Technological Research, 107–142. Leiden Botanical Series Number 3. Leiden University Press, Leiden.

Thanikaimoni, G. 1984. Ménispermacées: palynologie et systématique. *Travaux de l'Institut Français de Pondichéry, Section Scientifique et Technique* 18:1–135.

Theissen, G., A. Becker, A. Di Rosa, A. Kanno, J. T. Kim, T. Munster, K. U. Winter, and H. Saedler. 2000. A short history of MADS-box genes in plants. *Plant Molecular Biology* 42:115–149.

Theissen, G., and H. Saedler. 1999. The golden decade of molecular floral development (1990–1999): a cheerful obituary. *Developmental Genetics* 25:181–193.

———. 2001. Floral quartets. *Science* 409:469–471.

Thieret, J. W., and J. O. Luken. 1996. The Typhaceae in the southeastern United States. *Harvard Papers in Botany* 8:27–56.

Thomas, C. A. 1971. The genetic organization of chromosomes. *Annual Review of Genetics* 5:237–256.

Thomas, H. H. 1934. The nature and origin of the stigma. *New Phytologist* 33:173–198.

———. 1936. Paleobotany and the origin of the angiosperms. *Botanical Review* 2:397–418.

———. 1955. Mesozoic pteridosperms. *Phytomorphology* 5:177–185.

———. 1958. *Lidgettonia*, a new type of fertile *Glossopteris*. *Bulletin of the British Museum Natural History, Geology* 3:179–189.

Thomasson, J. R. 1987. Fossil grasses: 1820–1986 and beyond. *In* T. R. Soderstrom, K. W. Hilu, C. S. Campbell, and M. E. Barkworth, [eds.], Grass Systematics and Evolution, 159–171. Smithsonian Institution Press, Washington.

Thompson, J. D., T. J. Gibson, F. Plewniak, F. Jeanmougin, and D. G. Higgins. 1997. The ClustalX windows interface: flexible strategies for multiple sequence alignment aided by quality analysis tools. *Nucleic Acids Research* 24:4876–4882.

Thompson J. N. 1994. The Co-evolutionary Process. Chicago University Press, Chicago.

Thompson, J. N., and O. Pellmyr. 1992. Mutualism with pollinating seed parasites amid co-pollinators: constraints on specialization. *Ecology* 73:1780–1791.

Thompson, M. M. 1979. Growth and development of the pistillate flower and nut in "Barcelona" filbert. *Journal of the American Horticultural Society* 104:427–432.

Thorne, J. L., and H. Kishino. 2002. Divergence time estimation and rate evolution with multilocus data sets. *Systematic Biology* 51:689–702.

Thorne, R. F. 1968. Synopsis of a putatively phylogenetic classification of the flowering plants. *Aliso* 6:57–56.

———. 1974. A phylogenetic classification of the Annoniflorae. *Aliso* 8:147–209.

———. 1976. A phylogenetic classification of the Angiospermae. *Evolutionary Biology* 9:35–106.

———. 1983. Proposed new realignments in the angiosperms. *Nordic Journal of Botany* 3:85–117.

———. 1992a. Classification and geography of the flowering plants. *Botanical Review* 58:225–348.

———. 1992b. An updated phylogenetic classification of the flowering plants. *Aliso* 13:365–389.

———. 2000. The classification and geography of the monocotyledon subclasses Alismatidae, Liliidae, and Commelinidae. *In* B. Nordenstam, G. El Ghazaly, and M. Kassas [eds.], Plant Systematics for the 21st Century, 75–124. Portland, OR.

———. 2001. The classification and geography of the flowering plants: dicotyledons of the class Angiospermae (subclasses Magnoliidae, Ranunculidae, Caryophyllidae, Dilleniidae, Rosidae, Asteridae, and Lamiidae). *Botanical Review* 66:441–647.

Thulin, M., B. Bremer, J. Richardson, J. Niklasson, M. F. Fay, and M. W. Chase. 1998. Family relationships of the enigmatic rosid genera *Barbeya* and *Dirachma* from the Horn of Africa Region. *Plant Systematics and Evolution* 213:103–119.

Tillich, H.-J. 1985. Keimlingsbau und verwandtschaftliche Beziehungen der Araceae. *Gleditschia* 13:63–73.

———. 1995. Seedling and systematics in monocotyledons. *In* P. J. Rudall, P. J. Cribb, D. F. Cutler, and C. J. Humphries [eds.], Monocotyledons: Systematics and Evolution. Royal Botanic Gardens, Kew.

Tobe, H. 1991. Reproduction morphology, anatomy, and relationship of *Ticodendron*. *Annals of the Missouri Botanical Garden* 78:135–142.

Tobe, H., and N. R. Morin. 1996. Embryology and circumscription of Campanulaceae and Campanulales: a review of literature. *Journal of Plant Research* 109:425–435.

Tobe, H., and P. H. Raven. 1983. An embryological analysis of the Myrtales: its definition and characteristics. *Annals of the Missouri Botanical Garden* 70:71–94.

———. 1984. The embryology and relationships of *Rhynchocalyx* Olv. (Rhynchocalycaceae). *Annals of the Missouri Botanical Garden* 71:836–843.

———. 1988. Seed morphology and anatomy of Rhizophoraceae, inter- and infrafamilial relationships. *Annals of the Missouri Botanical Garden* 75:1319–1342.

———. 1989. The embryology and systematic position of *Rhabdodendron* (Rhabdodendraceae). *In* K. Tan, R. R. Mill, and T. S. Elias [eds.], Plant Taxonomy, Phytogeography, and Related Subjects, 233–248. Edinburgh University Press, Edinburgh.

Tobe, H., and B. Sampson. 2000. Embryology of *Takhtajania* (Winteraceae) and a summary statement of embryological features for the family. *Annals of the Missouri Botanical Garden* 87:389–397.

Tobe, H., T. Jaffré, and P. H. Raven. 2000. Embryology of *Amborella* (Amborellaceae): descriptions and polarity of character states. *Journal of Plant Research* 113:271–280.

Tokuoka, T., and H. Tobe. 2001. Ovules and seeds in subfamily Phyllanthoideae (Euphorbiaceae): structure and systematic implications. *Journal of Plant Research* 114:75–92.

Tomlinson, P. B. 1962. Phylogeny of the Scitamineae–morphological and anatomical considerations. *Evolution* 10:192–213.

———. 1969. Commelinales-Zingiberales. *In* C. R. Metcalfe [ed.], Anatomy of Monocotyledons. Clarendon Press, Oxford.

———. 1974. Development of the stomatal complex as a taxonomic character in the monocotyledons. *Taxon* 23:109–128.

———. 1995. Non-homology of vascular organisation in monocotyledons and dicotyledons. *In* P. J. Rudall, P. J. Cribb, D. F. Cutler, and C. J. Humphries [eds.], Monocotyledons: Systematics and Evolution, 589–622. Royal Botanic Gardens, Kew.

Torrey, J. G., and R. H. Berg. 1988. Some morphological features for generic characterization among the Casuarinaceae. *American Journal of Botany* 75:684–874.

Townrow, J. A. 1962. On *Pteruchus* a microsporophyll of the Corystospermaceae. *Bulletin of the British Museum (Natural History), Geology* 6:287–320.

Trevisan, L. 1988. Angiosperm pollen (monosulcate-trichotomosulcate phase) from the very early Lower Cretaceous of Southern Tuscany (Italy): some aspects. *Seventh International Palynological Congress (Brisbane) Abstracts* 165.

Trift, I., M. Källersjö, and A. A. Anderberg. 2002. The monophyly of *Primula* (Primulaceae) evaluated by analysis of sequences from the chloroplast gene *rbcL*. *Systematic Botany* 27:396–407.

Tsou, C.-H. 1995. Embryology of Theaceae — Anther and ovule development of *Adinandra*, *Cleyera*, and *Eurya*. *Journal of Plant Research* 108:77–86.

Tucker, G. C. 1987. The genera of Cyperaceae in the southeastern United States. *Journal of the Arnold Arboretum* 68:361–445.

———. 1989. The genera of Commelinaceae in the southeastern United States. *Journal of the Arnold Arboretum* 70:97–130.

Tucker, S. C. 1960. Ontogeny of the floral apex of *Michelia fuscata*. *American Journal of Botany* 47:266–277.

———. 1988. Loss versus suppression of floral organs. *In* P. Leins, S. C. Tucker, and P. K. Endress [eds.], Aspects of Floral Development, 69–82. Cramer, Berlin.

———. 1989. Overlapping organ initiation and common primordia in flowers of *Pisum sativum* (Leguminosae: Papilionoideae). *American Journal of Botany* 76:714–729.

———. 1991. Helical floral organogenesis in *Gleditsia*, a primitive caesalpinioid legume. *American Journal of Botany* 78:1130–1149.

———. 1997. Floral evolution, development, and convergence: the hierarchical-significance hypothesis. *International Journal of Plant Sciences* 158:S143–S161.

———. 1999. Evolutionary lability of symmetry in early floral development. *International Journal of Plant Sciences* 160:S25–S39.

———. 2000. Organ loss in detarioid and other leguminous flowers, and the possibility of saltatory evolution. *In* P. S. Herendeen and A. Bruneau [eds.], Advances in Legume Systematics, 9, 107–120. Royal Botanic Gardens, Kew.

Tucker, S. C., and A. W. Douglas. 1996. Floral structure, development, and relationships of paleoherbs: *Saruma*, *Cabomba*, *Lactoris*, and selected Piperales. *In* D. W. Taylor and L. J. Hickey [eds.], Flowering Plant Origin, Evolution, and Phylogeny. Chapman and Hall, New York.

Tucker, S. C., O. L. Stein, and K. S. Derstine. 1985. Floral development in *Caesalpinia* (Leguminosae). *American Journal of Botany* 72:1424–1434.

Tyree, M. T., and F. Ewers. 1991. The hydraulic architecture of trees and other woody plants. *New Phytologist* 119:345–360.

———. 1996. Hydraulic architecture of wood tropical plants. *In* S. S. Mulkey, R. L. Chazdon, and A. P. Smith [eds.], Tropical Forest Ecophysiology, 217–243. Chapman and Hall, New York.

Tzeng, T. Y., and C. H. Yang. 2001. A MADS box gene from lily (*Lilium longiflorum*) is sufficient to generate dominant negative mutation by interacting with *PISTILLATA (PA)* in *Arabidopsis thaliana*. *Plant and Cell Physiology* 42:1156–1168.

Ueda, K., A. Hanyun, T. Shiuchi, A. Seo, H. Okubo, and M. Hotta. 1997. Origin of Podostemaceae, a marvelous aquatic flowering plant family. *Plant Species Biology* 110:87–92.

Uhl, N. W., J. Dransfield, J. I. Davis, M. A. Luckow, K. S. Hansen, and J. J. Doyle. 1995. Phylogenetic relationships among palms: cladistic analyses of morphological and chloroplast DNA restriction site variation. *In* P. J. Rudall, P. J. Cribb, D. F. Cutler, and C. J. Humphries [eds.], Monocotyledons: Systematics and Evolution, 623–661. Royal Botanic Gardens, Kew.

Upchurch, G. R. 1984. Cuticle evolution in Early Cretaceous angiosperms from the Potomac Group of Virginia and Maryland. *Annals of the Missouri Botanical Garden* 71:522–550.

Vaknin, Y., S. Gan-Mor, A. Bechar, B. Ronen, and D. Eisikowitch. 2001. Are flowers morphologically adapted to take advantage of electrostatic forces in pollination? *New Phytologist* 152:301–306.

van der Bank, M., M. F. Fay, and M. W. Chase. 2002. Molecular phylogenetics of Thymelaeaceae with particular reference to African and Australian genera. *Taxon* 51:329–339.

———. 2002. Molecular phylogenetics of Thymelaeaceae with particular reference to African and Australian genera. *Taxon* 51:329–339.

van der Pijl, L., and C. H. Dodson. 1966. Orchid Flowers, Their Pollination and Evolution. University of Miami Press, Coral Gables, FL.

van Heel, W. A. 1981. A SEM-investigation on the development of free carpels. *Blumea* 27:499–522.

———. 1983. The ascidiform early development of free carpels, a S. E. M.-investigation. *Blumea* 28:231–270.

———. 1984. Flowers and fruits in Flacourtiaceae. V. The seed anatomy and pollen morphology of *Berberidopsis* and *Streptothamnus*. *Blumea* 30:31–37.

Vanvinckenroye, P., and E. Smets. 1996. Floral ontogeny of five species of *Talinum* and of related taxa (Portulacaceae). *Journal of Plant Research* 109:387–402.

van Vliet, G. J. C. M., and P. Baas. 1984. Wood anatomy and classification of the Myrtales. *Annals of the Missouri Botanical Garden* 71:783–800.

Vergara-Silva, F., S. Espinosa-Matias, B. A. Ambrose, S. Vazquez-Santana, A. Martinez-Mena, J. Marquez-Guzman, E. Martinez, E. M. Meyerowitz, and E. R. Alvarez-Buylla. 2003. Inside-out flowers characteristic of *Lacandonia schismatica* evolved at least before its divergence from a closely related taxon, *Triuris brevistylis*. *International Journal of Plant Sciences* 164:345–357.

Vicient, C. M., A. Suoniemi, K. Anamtamat-Jonsson, J. Tanskanen, A. Beharav, E. Nevo, and A. H. Schulman. 1999. Retrotransposon BARE-1 and its role in genome evolution in the genus *Hordeum*. *Plant Cell* 11:1769–1784.

Vink, W. 1988. Taxonomy in Winteraceae. *Taxon* 37:691–698.

———. 1993. Winteraceae. *In* K. Kubitzki, J. Rohwer, and V. Bittrich [eds.], The Genera and Families of Vascular Plants, II, 630–638. Springer, Berlin.

Vinnersten, A., and K. Bremer. 2001. Age and biogeography of major clades in Liliales. *American Journal of Botany* 88:1695–1703.

Vishnu-Mittre. 1953. A male flower of the Pentoxyleae with remarks on the structure of the female cones of the group. *Palaeobotanist* 2:75–84.

———. 1963. Pollen morphology of Indian Amaranthaceae. *Journal of the Indian Botanical Society* 42:86–101.

Vision, T. J., D. G. Brown, and S. D. Tanksley. 2000. The origins of genomic duplications in *Arabidopsis*. *Science* 290:2114–2117.

Vöchting, H. 1886. Über Zygomorphie und deren Ursachen. *Jahrbuch der Wissenschaftlichen Botanik* 17:297–346.

Vogel, S. 1954. Blütenbiologische Typen als Elemente der Sippengliederung, dargestellt anhand der Flora Südafrikas. *Botanische Studien* 1:1–338.

———. 1959. Organographie der Blüten kapländischer Ophrydeen mit Bemerkungen zum Koaptations-Problem I/II. *Abhandlungen der Akademie der Wissenschaften und der Literatur, Mainz, Jahrbuch* 1960:81–94.

———. 1969. Über synorganisierte Blütensporne bei einigen Orchideen. *Österreichische Botanische Zeitschrift* 116:244–262.

———. 1974. Ölblumen und ölsammelnde Bienen. *Tropische und Subtropische Pflanzenwelt* 7:1–267.

———. 1978. Evolutionary shifts from reward to deception in pollen flowers. *In* A. J. Richards [ed.], The Pollination of Flowers by Insects, 89–96. Academic Press, London.

———. 1988. Die Ölblumensymbiosen: Parallelismus und andere Aspekte ihrer Entwicklung in Raum und Zeit. *Zeitschrift fur Zoologische Systematik und Evolutions-Forschung* 26:341–362.

———. 1990. Radiación adaptiva del sindrome floral en las familias neotropicales. *Boletín de la Academia Nacional de Ciencias (Córdoba, Argentina)* 59:5–30.

———. 2000. The floral nectaries of Malvaceae s. l. : a conspectus. *Kurtziana* 28:155–171.

von Balthazar, M., and P. K. Endress. 2002a. Development of inflorescences and flowers in Buxaceae and the problem of perianth interpretation. *International Journal of Plant Sciences* 163:847–876.

———. 2002b. Reproductive structures and systematics of Buxaceae. *Botanical Journal of the Linnean Society* 140:193–228.

von Balthazar, M., P. K. Endress, and Y.-L. Qiu. 2000. Phylogenetic relationships in Buxaceae based on nuclear internal transcribed spacers and plastid *ndhF* sequences. *International Journal of Plant Sciences* 161:785–792.

von Balthazar, M., G. E. Schatz, and P. K. Endress. 2003. Female flowers and inflorescences of Didymelaceae. *Plant Systematics and Evolution* 237:199–208.

von Hagen, B., and J. W. Kadereit. 2002. Phylogeny and flower evolution of the Swertiinae (Gentianaceae-Gentianeae): homoplasy and the principle of variable proportions. *Systematic Botany* 27:548–572.

von Poser, G. L., M. E. Toffoli, M. Sobral, and A. T. Henriques. 1997. Iridoid glucosides substitution patterns in Verbenaceae and their taxonomic implication. *Plant Systematics and Evolution* 205:265–287.

Voznesenskaya, E. V., V. R. Franceschi, O. Kiirats, H. Freitag, and G. E. Edwards. 2001. Kranz anatomy is not essential for terrestrial C_4 plant photosynthesis. *Nature* 414:543–546.

Wagenitz, G. 1959. Die systematische Stellung der Rubiaceae. *Botanische Jahrbücher für Systematik* 79:17–35.

———. 1975. Blütenreduktion als ein zentrales Problem der Angiospermen-Systematik. *Botanische Jahrbücher für Systemtik* 96:448–470.

Wagenitz, G., and B. Laing. 1984. Die Nektarien der Dipsacales und ihre systematische Bedeutung. *Botanische Jahrbücher für Systematik* 104:483–507.

Wagner, G. P. 1996. Homologues, natural kinds and the evolution of modularity. *American Zoology* 36:36–43.

———. 2001. The Character Concept in Evolutionary Biology. Academic Press, San Diego.

Wagstaff, S. J., L. Hickerson, R. Spangler, P. A. Reeves, and R. G. Olmstead. 1998. Phylogeny of Labiatae s. l., inferred from cpDNA sequences. *Plant Systematics and Evolution* 209:265–274.

Wagstaff, S. J., and R. G. Olmstead. 1997. Phylogeny of Labiatae and Verbenaceae inferred from *rbcL* sequences. *Systematic Botany* 22:165–179.

Walbot, V. 2000. A green chapter in the book of life. *Nature* 408:794–795.

Walker, D. B. 1975. Postgenital carpel fusion in *Catharanthus roseus* (Apocynaceae). I. Light and scanning electron microscope study of gynoecial ontogeny. *American Journal of Botany* 62:457–467.

Walker, J. W. 1972. Chromosome numbers, phylogeny, phytogeography of the Annonaceae and their bearing on the (original) basic chromosome number of angiosperms. *Taxon* 21:57–65.

Walker, J. W., and J. A. Doyle. 1975. The bases of angiosperm phylogeny: palynology. *Annals of the Missouri Botanical Garden* 62:664–723.

Walker, J. W., and A. G. Walker. 1984. Ultrastructure of Lower Cretaceous angiosperm pollen and the origin and early evolution of flowering plants. *Annals of the Missouri Botanical Garden* 71:464–521.

Walker-Larsen, J., and L. D. Harder. 2000. The evolution of staminodes in angiosperms: patterns of stamen reduction, loss,

and functional re-invention. *American Journal of Botany* 87:1367–1384.

———. 2001. Vestigial organs as opportunities for functional innovation: the example of the *Penstemon* staminode. *Evolution* 55:477–487.

Wallander, E., and V. A. Albert. 2000. Phylogeny and classification of Oleaceae based on *rps16* and *trnL-F* sequence data. *American Journal of Botany* 87:1827–1841.

Wang, C., and D. Yu. 1997. Diterpenoid, sesquiterpenoid and secoiridoid glucosides from *Aster auriculatus*. *Phytochemistry* 45:1483–1487.

Wang, Z.-Q. 2004. A new Permian gnetalean cone as fossil evidence supporting current molecular phylogeny. *Annals of Botany* 94:281–288.

Wannan, B. S., J. T. Waterhouse, P. A. Gadek, and C. J. Quinn. 1985. Biflavonyls and the affinities of *Blepharocarya*. *Biochemical Systematics and Ecology* 13:105–108.

Wanntorp, L., H.-E. Wanntorp, and M. Källersjö. 2002. Phylogenetic relationships of *Gunnera* based on nuclear ribosomal DNA ITS region, *rbcL* and *rps16* intron sequences. *Systematic Botany* 27:512–521.

Ward, N. M., and R. A. Price. 2002. Phylogenetic relationships of Marcgraviaceae: Insights from three chloroplast genes. *Systematic Botany* 27:149–160.

Warming, E. 1879. Haandbog i den systematiske botanik. P. G. Philipsens Forlag, Copenhagen.

Waser, N. M. 1998. Pollination, angiosperm speciation, and the nature of species boundaries. *Oikos* 81:198–201.

Waser, N. M., L. Chittka, M. V. Price, N. M. Williams, and J. Ollerton. 1996. Generalization in pollination systems, and why it matters. *Ecology* 77:1043–1060.

Webb, C. J., and D. G. Lloyd. 1986. The avoidance of interference between the presentation of pollen and stigmas in angiosperms. II. Herkogamy. *New Zealand Journal of Botany* 24:163–178.

Weber, A. 1980. Die Homologie des Perigons der Zingiberaceen. Ein Beitrag zur Morphologie und Phylogenie des Monokotylen-Perigons. *Plant Systematics and Evolution* 133:149–179.

Weberling, F. 1970. Weitere Untersuchungen zur Morphologie des Unterblattes bei den Dikotylen: V. Piperales. *Beiträge zur Biologie der Pflanzen* 46:403–434.

Webster, G. L. 1967. The genera of Euphorbiaceae in the southeastern United States. *Journal of the Arnold Arboretum* 48:303–430.

Weigend, M., O. Mohr, and T. J. Motley. 2002. Phylogeny and classification of the genus *Ribes* (Grossulariaceae) based on 5S-NTS sequences and morphological and anatomical data. *Botanische Jahrbücher für Systematik* 124:163–182.

Weiland, G. R. 1918. The origin of dicotyls. *Science* 48:18–21.

Weller, S., A. Sakai, A. Rankin, A. Golonka, B. Kutcher, and K. Ashby. 1998. Dioecy and the evolution of pollination systems in *Schiedea* and *Alsinidendron* (Caryophyllaceae: Alsinoideae) in the Hawaiian Islands. *American Journal of Botany* 85:1377–1388.

Wendel, J. F. 2000. Genome evolution in polyploids. *Plant Molecular Biology* 42:225–249.

Wendel, J. F., R. C. Cronn, J. S. Johnston, and H. J. Price. 2002. Feast and famine in plant genomes. *Genetica* 115:37–47.

Westerkamp, C., and A. Weber. 1999. Keel flowers of the Polygalaceae and Fabaceae: a functional comparison. *Botanical Journal of the Linnean Society* 129:207–221.

Wettstein, R. V. 1907. Handbuch der systematischen Botanik, II. Franz Deuticke, Vienna.

Wheeler, M. J., V. E. Franklin-Tong, and F. C. H. Franklin. 2001. The molecular and genetic basis of pollen-pistil interactions. *New Phytology* 151:565–584.

Whitehead, D. R. 1969. Wind pollination in the angiosperms: evolutionary and environmental considerations. *Evolution* 23:28–35.

———. 1983. Wind pollination: some ecological and evolutionary perspectives. *In* L. Real [ed.], Pollination Biology, 97–108. Academic Press, Orlando.

Whitten, W. M., N. H. Williams, and M. W. Chase. 2000. Subtribal and generic relationships of Maxillarieae (Orchidaceae) with emphasis on Stanhopeinae: combined molecular evidence. *American Journal of Botany* 87:1842–1856.

Wiegrefe, S. J., K. J. Sytsma, and R. P. Guries. 1998. The Ulmaceae, one family or two? Evidence from chloroplast DNA restriction site mapping. *Plant Systematics and Evolution* 210:249–270.

Wiens, D., and B. A. Barlow. 1971. The cytogeography and relationships of the viscaceous and eremolepidaceous mistletoes. *Taxon* 20:313–332.

Wiens, J. J. 1998. Combining data sets with different phylogenetic histories. *Systematic Biology* 47:568–581.

Wikström, N., and P. Kenrick. 2001. Evolution of Lycopodiaceae (Lycopsida): Estimating divergence times from *rbcL* gene sequences by use of nonparametric rate smoothing. *Molecular Phylogenetics and Evolution* 19:177–186.

Wikström, N., V. Savolainen, and M. W. Chase. 2001. Evolution of the angiosperms: calibrating the family tree. *Proceedings of the Royal Society of London B* 268:2211–2220.

———. 2003. Angiosperm divergence times: congruence and incongruence between fossils and sequence divergence estimates. *In* P. C. J. Donoghue and M. P. Smith [eds.]. Telling the Evolutionary Time: Molecular Clocks and the Fossil Record, 142–165. Taylor and Francis, London.

Wilcox, T. P., D. J. Zwickl, T. A. Heath, and D. M. Hillis. 2002. Phylogenetic relationships of the dwarf boas and a comparison of Bayesian and bootstrap measures of phylogenetic support. *Molecular Phylogenetics and Evolution* 25:361–371.

Wiley, E. O. 1979. An annotated Linnean hierarchy, with comments on natural taxa and competing systems. *Systematic Zoology* 28:308–337.

———. 1981. Phylogenetics: The Theory and Practice of Phylogenetic Systematics. John Wiley and Sons, New York.

Williams, J. H., and W. E. Friedman. 2002. Identification of diploid endosperm in an early angiosperm lineage. *Nature* 415:522–526.

Williams, N. H., M. W. Chase, T. Fulcher, and W. M. Whitten. 2001. Molecular systematics of the Oncidiinae based on evidence from four DNA regions: expanded circumscriptions of *Cyrtochilum*, *Erycina*, *Otoglossum* and *Trichocentrum* and a new genus (Orchidaceae). *Lindleyana* 16:113–139.

Williams, S. E. 1976. Comparative sensory physiology of the Droseraceae: evolution of a plant sensory system. *Proceedings of the American Philosophical Society* 120:187–204.

Williams, S. E., V. A. Albert, and M. W. Chase. 1994. Relationships of Droseraceae: a cladistic analysis of *rbcL* sequence and morphological data. *American Journal of Botany* 81:1027–1037.

Williamson, P. S., and E. L. Schneider. 1993a. Cabombaceae. *In* K. Kubitzki, J. G. Rohwer, J. Bittrich [eds.], The Families and Genera of Vascular Plants, II, 157–161. Springer, Berlin.

———. 1993b. Nelumbonaceae. *In* K. Kubitzki, J. Rohwer, and V. Bittrich [eds.], The Families and Genera of Vascular Plants, II, 470–473. Springer, Berlin.

Wilson, K. A. 1960a. The genera of Arales in the southeastern United States. *Journal of the Arnold Arboretum* 41:47–72.

Wilson, K. A. 1960b. The genera of Hydrophyllaceae and Polemoniaceae in the southeastern United States. *Journal of the Arnold Arboretum* 41:197–212.

Wilson, K. L., and D. A. Morrison [eds.]. 2000. Monocot Systematics and Evolution. CSIRO, Melbourne.

Wilson, M. A., B. Gaut, and M. T. Clegg. 1990. Chloroplast DNA evolves slowly in the palm family (Arecaceae). *Molecular Biology and Evolution* 7:303–314.

Wilson, P., and J. D. Thomson. 1996. How do flowers diverge? *In* D. G. Lloyd and S. C. H. Barrett [eds.], Floral Biology. Studies on Floral Evolution in Animal-Pollinated Plants, 88–111. Chapman and Hall, New York.

Wilson, P. G., M. M. O'Brien, P. A. Gadek, and C. J. Quinn. 2001. Myrtaceae revisited: a reassessment of infrafamilial groups. *American Journal of Botany* 88: 2013–2025.

Wilson, T. K. 1966. The comparative anatomy of the Canellaceae: IV. Floral morphology and conclusions. *American Journal of Botany* 53:336–343.

Winter, K. U., A. Becker, T. Munster, J. T. Kim, H. Saedler, and G. Theissen. 1999. MADS-box genes reveal that gnetophytes are more closely related to conifers than to flowering plants. *Proceedings of the National Academy of Sciences USA* 96:7342–7347.

Withner, C. L. 1941. Stem anatomy and phylogeny of the Rhoipteleaceae. *American Journal of Botany* 28:872–878.

Wolfe, K. H., M. Gouy, Y.-W. Yang, P. M. Sharp, and W.-H. Li. 1989. Date of the monocot-dicot divergence estimated from chloroplast DNA sequence data. *Proceedings of the National Academy of Sciences USA* 86:6201–6205.

Wolfe, K. H., C. W. Morden, and J. D. Palmer. 1992. Function and evolution of a minimal plastid genome from a nonphotosynthetic parasitic plant. *Proceedings of the National Academy of Sciences USA* 89:10648–10652.

Wolter-Filho, W., A. I. Da Rocha, M. Yoshida, and O. R. Gottlieb. 1985. Ellagic-acid derivatives from *Rhabdodendron macrophyllum*. *Phytochemistry* 24:1991–1993.

———. 1989. Chemosystematics of *Rhabdodendron*. *Phytochemistry* 28:2355–2357.

Wood, C. E. 1961. The genera of Ericaceae in the southeastern United States. *Journal of the Arnold Arboretum* 42:10–80.

———. 1974. A Student's Atlas of Flowering Plants: Some Dicotyledons of Eastern North America, Generic Flora of the Southeastern U. S. Project. Harper and Row, New York.

Wood, C. E., and R. B. Channell. 1960. The genera of Ebenales in the southeastern United States. *Journal of the Arnold Arboretum* 41:1–35.

Worsdell, W. C. 1908. The affinities of *Paeonia*. *Journal of Botany* 46:114–116.

Wu, C. I., and W. H. Li. 1985. Evidence for higher rates of nucleotide substitution in rodents than in man. *Proceedings of the National Academy of Sciences USA* 82:1741–1745.

Wunderlich, R. 1959. Zur Frage der Phylogenie der Endospermtypen bei den Angiospermen. *Österreichische Botanische Zeitschrift* 106:203–483.

Wurdack, J. J., and R. B. Kral. 1982. The genera of Melastomataceae in the southeastern United States. *Journal of the Arnold Arboretum* 63:429–439.

Wurdack, K. J., and J. W. Horn. 2001. A reevaluation of the affinities of the Tepuianthaceae: molecular and morphological evidence for placement in the Malvales. *In* Botany 2001: Plants and People, Abstracts p. 151. Albuquerque, NM.

Wurdack, K. J., P. Hoffmann, R. Samuel, A. de Bruijn, M. Van Der Bank, and M. W. Chase. 2004. Molecular phylogenetic analysis of Phyllanthaceae (Phyllanthoideae pro parte, Euphorbiaceae sensu lato) using plastid *rbcL* DNA sequences. *American Journal of Botany* 91:1882–1900.

Wyss, A. R., and J. Meng. 1996. Application of phylogenetic taxonomy to poorly resolved crown clades: a stem-modified node-based definition of Rodentia. *Systematic Biology* 45:559–568.

Xiang, Q.-X., M. L. Moody, D. E. Soltis, C.-Z. Fan, and P. S. Soltis. 2002. Relationships within Cornales and circumscription of Cornaceae—*matK* and *rbcL* sequence data and effects of outgroups and long branches. *Molecular Phylogenetics and Evolution* 24:35–57.

Xiang, Z. Y., D. E. Soltis, D. R. Morgan, and P. S. Soltis. 1993. Phylogenetic relationships of *Cornus* L. sensu lato and putative relatives inferred from *rbcL* sequence data. *Annals of the Missouri Botanical Garden* 80:723–734.

Yakovlev, M. S., and M. D. Yoffe. 1957. On some peculiar features in the embryogeny of *Paeonia* L. *Phytomorphology* 7:74–82.

Yamada, T., R. Imaichi, and M. Kato. 2001b. Developmental morphology of ovules and seeds of Nymphaeales. *American Journal of Botany* 88:963–974.

Yamada, T., H. Tobe, R. Imaichi, and M. Kato. 2001a. Developmental morphology of the ovules of *Amborella trichopoda* (Amborellaceae) and *Chloranthus serratus* (Chloranthaceae). *Botanical Journal of the Linnean Society* 137:277–290.

Yamada, T., K. Uehara, M. Ito, and M. Kato. 2003. Expression pattern of an *INNER NO OUTER* homologue in genus *Nymphaea* (Nymphaeaceae) and its implication for the evolution of outer integument. *Development, Genes and Evolution* 213:510–513.

Yamashita, J., and M. N. Tamura. 2000. Molecular phylogeny of the Convallariaceae (Asparagales). *In* K. L. Wilson and D. A. Morrison [eds.], Monocots: Systematics and Evolution, 387–400. CSIRO, Melbourne.

Yamazaki, T. 1974. A system of Gamopetalae based on the embryology. *Journal of the Faculty of Science, University of Tokyo, III,* 11:263–281.

Yang, Z. S. 1997. PAML: a program for package for phylogenetic analysis by maximum likelihood. *CABIOS* 15:555–556.

Yang, Z. S., S. Kumar, and M. Nei. 1995. A new method of inference of ancestral nucleotide and amino acid sequences. *Genetics* 141:1641–1650.

Yeo, P. F. 1992. Secondary pollen presentation: form, function and evolution. *Plant Systematics and Evolution, Supplement* 6:1–268.

Young, D. A. 1981. Are the angiosperms primitively vesselless? *Systematic Botany* 6:313–330.

Young, N. D., K. E. Steiner, and C. W. dePamphilis. 1999. The evolution of parasitism in the Scrophulariaceae/Orobanchaceae: plastid gene sequences refute an evolutionary transition series. *Annals of the Missouri Botanical Garden* 86:876–893.

Yu, D., M. Kotilainen, E. Pollanen, M. Mehto, P. Elomaa, Y. Helariutta, V. A. Albert, and T. H. Teeri. 1999. Organ identity genes and modified patterns of flower development in *Gerbera hybrida. Plant Journal* 17:51–62.

Zanis, M. J., P. S. Soltis, Y.-L. Qiu, E. Zimmer, and D. E. Soltis. 2003. Phylogenetic analyses and perianth evolution in basal angiosperms. *Annals of the Missouri Botanical Garden* 90:129–150.

Zanis, M., D. E. Soltis, P. S. Soltis, S. Mathews, and M. J. Donoghue. 2002. The root of the angiosperms revisited. *Proceedings of the National Academy of Sciences USA* 99:6848–6853.

Zavada, M. S., and D. L. Dilcher. 1986. Comparative pollen morphology and its relationship to phylogeny of pollen in the Hamamelidae. *Annals of the Missouri Botanical Garden* 73:348–381.

Zhang, L. B., and S. Renner. 2003. The deepest splits in Chloranthaceae as resolved by chloroplast sequences. *International Journal of Plant Sciences* 164:S383–S392.

Zhao, D., Q. Yu, C. Chen, and H. Ma. 2001. Genetic control of reproductive meristems. *In* M. T. McManus and B. Veit. [eds.], Meristematic Tissues in Plant Growth and Development, 89–142. Sheffield Academic Press, Sheffield.

Zimmer, E. A., E. H. Roalson, L. E. Skog, J. K. Boggan, and A. Idnurm. 2002. Phylogenetic relationships in the Gesnerioideae (Gesneriaceae) based on nrDNA ITS and cpDNA *trnL-F* and *trnE-T* spacer region sequences. *American Journal of Botany* 89:296–311.

Zomlefer, W. B., N. H. Williams, W. M. Whitten, and W. S. Judd. 2001. Generic circumscription and relationships in the tribe Melanthieae (Liliales, Melanthiaceae), with emphasis on *Zigadenus*: evidence from ITS and *trnLF* sequence data. *American Journal of Botany* 88:1657–1669.

Zona, S. 1997. The genera of Palmae (Arecaceae) in the southeastern United States. *Harvard Papers in Botany* 11:71–107.

Zuckerkandl, E. 2002. Why so many noncoding nucleotides? The eukaryote genome as an epigenetic machine. *Genetica* 115:105–129.

Zuckerkandl, E., and L. Pauling. 1962. Molecular disease, evolution, and genetic heterogeneity. *In* M. Kasha and B. Pullman [eds.], Horizons in Biochemistry, 198–225. Academic Press, New York.

Zwickl, D. J., and D. M. Hillis. 2002. Increased taxon sampling greatly reduces phylogenetic error. *Systematic Biology* 51:588–598.

Taxonomic
Index

Subject
Index

Page numbers in *italics* refer to material found in a figure or a table.